Thermodynamics of the Earth and Planets

This textbook provides an intuitive yet mathematically rigorous introduction to thermodynamics and thermal physics focused on the rich variety of planetary processes. It demonstrates how the workings of planetary bodies can be understood in depth by reducing them to fundamental physics and chemistry.

The book is based on two courses that the author has taught for many years at the University of Georgia. It offers a strong "first-principles" theoretical foundation in classical thermodynamics, yet it also provides many examples of numerical calculations, including a large number of *Maple* procedures that the reader can use and modify, limited only by their interests and imagination. The book assumes that the reader is proficient in calculus and introductory college physics, chemistry and Earth science. All required material beyond this is introduced in the book. The book also includes a large number of bibliographic references and suggestions for further study. There are many "Worked Example" boxes interspersed in the text, and end-of-chapter exercises, which in many cases expand upon the topics covered in the worked examples. Worked solutions to the problems are provided online. The book also includes Software Boxes, which show the reader how to implement numerical solutions to many of the problems, using *Maple*.

As well as being an ideal textbook on planetary thermodynamics for advanced students in the Earth and planetary sciences, it also provides an innovative and quantitative complement to more traditional courses in geological thermodynamics, igneous and metamorphic petrology, chemical oceanography, mineral deposits, planetary geology and planetary atmospheres. In addition to its use as a textbook, it is also of great interest to researchers looking for a "one-stop" source of concepts and techniques that they can apply to their research problems.

- Ties together the physics and chemistry of planetary systems into a single textbook.
- Emphasizes first principles and foundations.
- Contains rigorous mathematical derivation of ALL results.
- Applications and examples are drawn from many different branches of Earth and planetary sciences.
- Contains many worked numerical examples and end-of-chapter problems.
- Includes many downloadable *Maple* routines, explained in Software Boxes.

ALBERTO PATIÑO DOUCE is a Professor in the Department of Geology at the University of Georgia, where he has been for over 20 years. He has very wide research interests, which include the origin of Earth's continents, the nature of the early terrestrial atmosphere, the volatile contents of the Martian mantle and of meteorite parent bodies, the origin of life, foundational issues in classical and non-equilibrium thermodynamics, applications of statistical physics to human societies, and the reasons why human cultures repeatedly ignore the finiteness of natural resources, overreach and self-destroy. His teaching in geology, planetary sciences and thermodynamics at all levels is recognized as being unusually intense, quantitative, energetic, demanding and uncompromising, but despite this he is a frequent recipient of his Department's "Professor of the Year" award, chosen by student vote. His former Ph.D. students are pursuing a wide range of careers in academia, at NASA, and in national security. He is an Associate Editor for the journal *Lithos* and formerly for *Mineralogy & Petrology*.

Thermodynamics of the Earth and Planets

ALBERTO PATIÑO DOUCE

University of Georgia

CAMBRIDGE UNIVERSITY PRESS
Cambridge, New York, Melbourne, Madrid, Cape Town,
Singapore, São Paulo, Delhi, Tokyo, Mexico City

Cambridge University Press
The Edinburgh Building, Cambridge CB2 8RU, UK

Published in the United States of America by Cambridge University Press, New York

www.cambridge.org
Information on this title: www.cambridge.org/9780521896214

First published 2011

Printed in the United Kingdom at the University Press, Cambridge

A catalog record for this publication is available from the British Library

Library of Congress Cataloging in Publication data
Patiño Douce, Alberto.
Thermodynamics of the Earth and planets / Alberto Patiño Douce.
p.   cm.
Includes bibliographical references and index.
ISBN 978-0-521-89621-4 (hardback)
1. Laws of thermodynamics – Textbooks.   2. Planetary theory – Textbooks.   I. Title.
QC311.P236 2011
551.5′22–dc22   2011009935

ISBN 978-0-521-89621-4 Hardback

Additional resources for this publication at www.cambridge.org/patino_douce

Contents

Preface

My first words when meeting a new class at the beginning of every semester, whether an introductory physical geology course or a graduate seminar, are always more or less the same: "*Geology does not exist* !". Some students start frantically going over their schedules, wondering whether they are in the right room, but most of them just stare at me, wondering whether I am a lunatic. While they do this I explain that what I meant was that the Earth and other planets are complex systems in which every process can, and must, be dismantled until we can understand it in terms of the simplest possible physics and chemistry. This does little to put them at ease, but over the course of the first few weeks of class many of them come to understand what I mean, and even to agree with it.

This brings me to the several reasons why I decided to write this book. First, although a few good textbooks on thermodynamics applied to Earth systems are available, I find that none of them goes into the fundamentals of thermodynamics with the depth that I am convinced is necessary. Rather, they tend to discuss the foundational principles of thermodynamics on a "need to know" basis. My approach is exactly the opposite: build a solid understanding of the foundations of thermodynamics first, and explain everything else in terms of this understanding. Second, many students in Earth and Planetary Sciences have a tendency to think of our science as standing in splendid isolation of the fundamental sciences, and of the laws of nature that they have painstakingly codified. Yet ultimately everything is physics, and must be understood as such. Third, there is too much of a terrestrial emphasis in all current books on thermodynamics for Earth and Planetary scientists. The diversity of bodies and environments in the Solar System provides a wealth of opportunities to demonstrate the unifying explanatory power of thermodynamics. To name just a few, why not look at cryolavas in Triton and brines on the Martian surface as examples of eutectic melting? At methane–ethane fractionation in Titan's atmosphere and Fe–Ni fractionation during crystallization of planetary cores as examples of binary T–X loops? At the conditions in the Martian mantle to discuss equations of state for solids and the concept of thermal pressure? Or at the evolution of atmospheric composition while making atmospheric mass and gravitational acceleration, and therefore pressure, adjustable parameters? You will find these and other unusual examples in this book. Finally, I think that in order to be complete and useful, a textbook must not only cover the fundamental principles but also show the *full* details of how those fundamental principles and mathematical relationships are converted to numerical results. In this book I show the derivation of every single equation that is used to obtain numerical estimates, and in those cases in which the equations cannot be solved by hand I include open *Maple* procedures that you are free to use, modify and expand as far as your interests and abilities will take you.

Throughout the book I try to emphasize the importance of physical intuition and mathematical rigor, and of striking a balance between the two. The book is organized in 14 chapters, some of them more orthodox than others. The table of contents is fairly self-explanatory, but I wish to highlight some points. The First Law is introduced in Chapter 1, but only after a lengthy discussion of conservation laws in general, and other associated concepts. At the end of Chapter 1 there is a discussion of the nature of planetary materials that I believe can be found nowhere else. The Second and Third Laws must wait until Chapter 4, as Chapter 2 focuses on energy conservation, and the discussions of energy dissipation and thermodynamic cycles in Chapter 3 lead more or less naturally to the need for the Second Law. Thermodynamic potentials, including Gibbs free energy, are introduced in Chapter 4 in a mathematically formal way, i.e. I first demonstrate (rigorously) the properties of the Legendre transform, and only then apply it to derive the various thermodynamic potentials. There is no other way that makes sense. Chapter 2 is a thorough quantitative presentation of all sources of planetary internal energy. I know of no comparable treatment anywhere else. Chapter 3 covers heat diffusion and advection in planetary bodies; it is not intended to substitute the various excellent textbooks that are available on this topic (cited in Chapter 3), but rather as a complement that may add some new twists to this fundamental field. The way I define activity in Chapter 5 is unusual, and owes not a small amount to Guggenheim's unique insight, but it is in my view the most mathematically appealing way of doing it (you may feel Guggenheim's presence in this statement too). Discussion of non-ideal activity is focused chiefly on the concept itself, and on the computational difficulties associated with its mathematical representation. There is simply no space to delve in detail into the huge fields of solution theory in crystals (Chapter 5), fluids (Chapter 9), melts (Chapter 10) and electrolyte solutions (Chapter 11), but I hope that I provide enough background for students to be able to jump directly into the relevant research literature. The equations needed to calculate phase boundaries are developed in full and implemented in a series of progressively more complete *Maple* procedures (Chapters 1, 5, 6, 7, 8 and 9). Feel free to use "black box software" if you wish, but first make sure that you understand where those results come from. I have tried to emphasize the concept of universality of critical behavior by highlighting the similarities among critical mixing phenomena (Chapter 7), lambda phase transitions (Chapter 7) and the critical point of fluids (Chapter 9), without using the word "universality" nor introducing the concept of critical exponents. The interested students should be able to pick it up from here. I emphasize the use of non-dimensional variables as much as possible, including in unusual contexts such as discussion of critical mixing and solvi (Chapter 7), high pressure and temperature behavior of solids (Chapter 8), critical phenomena in fluids (Chapter 9), concentration of non-equilibrium atmospheric species (Chapter 12) and energy dissipation by planetary differentiation (Chapter 2). I have done my best to lay out in simple yet rigorous mathematical terms the sometimes confusing topics of equations of state for solids and thermal pressure (Chapter 8); again, I know of no comparable treatment elsewhere. There is a full discussion of the calculation of species distribution in fluids, both by chemical equilibrium and by Gibbs free energy minimization, including the derivation of all necessary equations and the accompanying implementation in *Maple* (Chapter 9) – I hope that this will demystify what is no more than relatively simple algebra. Chapter 10 is a somewhat unusual take on igneous petrology. My debt to the work of M. Hirschmann, P. Asimow and E. Stolper should be evident here, but I hope to have added some new insights especially with regards to volatile-fluxed melting. I present "toy models" of ozone depletion (Chapter 12) and greenhouse warming (Chapter 13) as applications of chemical kinetics and radiative heat transfer, respectively. Chapter 12 also contains an introduction to non-equilibrium thermodynamics, that I hope to be able to expand upon in the not too distant future. Finally, there is a simplistic but, I believe, fundamental discussion of the origin of life in Chapter 14.

All chapters contain Worked Examples. These are also a fundamental part of the text. Do not skip them, or you will miss explanations and discussions that are not repeated elsewhere. There are also Boxes that contain accessory material that may be skipped without loss of continuity, but to which you should return at some point to clarify the contents of the main text. Finally, there are Software Boxes that contain generally succinct documentation for the *Maple* procedures that can be downloaded from the book's website. I plan on updating and adding to these procedures periodically, and any additional documentation or (inevitable) corrections will also be posted on the website. Please feel free to contact me with suggested changes or additions to the *Maple* library. If you use any of these procedures in published work please cite this book as reference. All figures in this book are my original artwork, and were designed to match specifically the associated content in the text – in fact, in not a few occasions the text was written around the figure. I think that this makes the content much more understandable.

Mike Roden read some of the text and made perceptive comments that I did my best to incorporate. Matt Lloyd at Cambridge was instrumental in getting this project started. I wish to express my gratitude to him and all his staff for their support, encouragement, patience and understanding in the face of missed deadlines. In particular, I wish to thank Assistant Editor Laura Clark, Production Editor, Emma Walker and Copy-editor Beverley Lawrence. My wife Marta has been more patient throughout this project than I have any right to expect, and our son Javier has contributed to it more than he knows – all my love to both of you! Writing this book took about three years and would have been impossible without the company and encouragement offered by our cats, Kali and Watson (to the memory of both of whom I dedicate this book), Ajax, Marx and Engels, and Leonidas and Dottie.

Energy in planetary processes and the First Law of Thermodynamics

1

This book is about the physical chemistry of planetary processes. Although in detail each planetary body in the Solar System looks very different, all of the planets and moons have reached their current states as a result of the same fundamental laws of nature, which are codified into the sciences that we know as physics and chemistry. A real understanding of the nature and evolution of the bodies that make up the Solar System requires that we immerse ourselves in physics and chemistry, and that we come to think of planetary processes as specific applications of these sciences. These applications can be more complex, in the sense of the number of variables involved, than those that physicists and chemists deal with when working under controlled laboratory conditions. Perhaps for this reason students of geological and planetary sciences tend to view these sciences as separate or "stand alone". This is not so, however. Using an analogy that most of us are likely to be familiar with (and that, admittedly, may be a bit stretched), the sciences that we know as geology and planetary science (and their "sub-fields" such as petrology, mineralogy, or oceanography, to name just a few) are the "user interface", the set of graphics and icons and mnemonics that we see on our computer screens. This user interface is supported and made possible by a rich and complex operating system (e.g. Linux, Windows, Mac, according to our tastes). The "operating system" of geology and planetary sciences consists of physics and chemistry. By immersing ourselves in the "operating system" we will be able to see connections among planetary processes that we might not have suspected, and we will be able to better understand what makes each planetary body unique.

We have talked about planetary processes – but exactly what is a planetary process? Simply stated, anything in a planet, physical or chemical, that changes with time ("geological", and also "biological", are specific instances of "physical or chemical"). Planetary processes must be driven by some source of energy, otherwise they would stop and cease to be processes. We can furthermore make the general statement that, the higher the rate of energy supply, the more active a planetary process, and the system in which it occurs, are. Our first task, then, is to formalize our understanding of energy and to lay out the mathematical framework that makes it possible to track energy conversion processes.

In this chapter we will examine the ways in which energy is manifested in planetary processes. Along the way, we shall introduce a number of fundamental physical concepts and tools that must become part of our physical intuition and operating practice as we attack problems in planetary sciences. The chapter culminates with a formal development of the First Law of Thermodynamics and a discussion of some of the relationships between macroscopic phenomena and their microscopic underpinnings.

1.1 Some necessary definitions

It is customary to start books on thermodynamics by devoting several sections to defining much of the terminology that will be used throughout the work. It is my experience that if one defines all terms at the beginning but some of those terms are not used until much later, the meaning of many of these "deferred" terms does not become clear until they are applied in a specific example. Some backtracking necessarily ensues, with some loss of efficiency (energy dissipation, if we were defining that term here!) and, worse, with a tendency to forget the rigorous meaning of the terms. I have thus chosen to define terms as their need is first encountered, so that definitions will always be given in specific contexts that will make their meaning clear and easy to assimilate. There is, however, a minimum set of definitions that we must consider at this point.

Thermodynamics is the subset of physics that studies energy transformation processes, and in particular processes in which thermal energy is involved. As we shall see, some of the manifestations of thermal energy may not entail changes in temperature or exchanges of heat, so the meaning of thermodynamics is more subtle than what this definition may suggest.

In order to make problems mathematically tractable, it is customary when studying thermodynamics to subdivide the universe into systems. We will then call that part of the universe that we are studying the *system*, and the surrounding parts of the universe with which our system may be able to interact becomes the *environment*. We are usually free to define our system in any way we please, so that we can tailor it to the problem that we are trying to understand. For example, our system may be a mineral assemblage within a thin section, a parcel of volcanic gas expanding during an eruption, a magma chamber, a planetary atmosphere or a planetary core. Systems may be *open* or *closed*, depending on whether or not matter can move across their boundaries. Of course, most systems in nature are open to some extent. The basic thermodynamic relationships, however, are most easily introduced and developed for closed systems. Throughout this book, unless we explicitly state that we are considering an open system, it must be understood that we are dealing with closed systems. The terms "open" and "closed" refer **only** to mass transfer and **not** to energy transfer. Energy transfer can take place across the boundaries of both closed and open systems.

Classical thermodynamics concerns itself with systems that are at *equilibrium*. A rigorous definition of thermodynamic equilibrium is necessary, but cannot be implemented until we have developed much of the mathematical formalism of thermodynamics. For now, we will state that, in order for a closed system in which gravity can be ignored to be at equilibrium, its temperature and pressure must be uniform and if, in addition, there is no energy transfer across its boundaries, then the amounts and compositions of physically distinguishable subsets of the system (that we will call *phases*) must not change with time. The effect of gravity cannot be ignored over planetary length-scales, however, and is discussed in Chapters 2, 3 and 13. This definition is *not* complete but it is consistent with the formal definition of chemical equilibrium and will allow us to "feel our way" around thermodynamics until we can construct that formal definition (in Chapter 4). An example may clarify our intuitive definition. Suppose that our system consists of a certain amount of liquid water and a certain amount of ice, each of which is a different phase, held in a container that is a perfect thermal insulator. This is an example of a *heterogeneous* system, because it has more than one distinguishable part or phase. If the pressure and temperature everywhere in the system are 1 bar and $0\,°C$, then the system is at equilibrium because the relative amounts of ice and water will not change with time. If the temperature and the pressure are any other combination of values, then the system is *not* at equilibrium, because as time goes by the amount of one of the phases will increase at the expense of the other. If the temperatures of the phases are different (e.g. we open the container and dump ice at $-20\,°C$ into water at $20\,°C$, then close the container), then the system is also not at equilibrium because one of the phases will grow at the expense of the other. Let us assume that the relative amounts of ice and water are such that all of the water in the container freezes. We now have a *homogeneous system*, i.e., one that consists of a single phase. This system will be at equilibrium only once its temperature is uniform and heat flow within the system ceases. In general, a homogeneous system is at equilibrium if there are no gradients in temperature, pressure nor composition (Chapters 4 and 12), although in the presence of an external field, such as a gravitational field, this requirement must be relaxed (Chapter 13).

We will often make reference to the *state* of a system. The implication when we do so, unless we say otherwise, is that we refer to the state of a system at equilibrium. The state of a system at equilibrium can be fully characterized by the values of a small number of variables, of which pressure and temperature are the most familiar ones. If we have a homogeneous system, for example a given amount of liquid water, then the state of the system is fully characterized once we specify its pressure, P, and temperature, T. For every combination of pressure and temperature liquid water has a single and well defined set of values for its physical properties, such as density, ρ (or its inverse, molar volume, V), refractive index or dielectric constant. What this says is that we only need to specify P and T to specify the state of the system. In principle, we can specify the state of the system equally well by choosing another pair of variables, such as molar volume and refractive index. Thermodynamics allows us to do this (the reasons will become clear in Chapter 6), even if it may not be the most sensible choice. For more complex systems we may need additional variables, but whether this is the case, and how many more variables we need, is not intuitively obvious (again, we will develop this formally in Chapter 6). For now, we note that if we go back to the system consisting of ice and water in an insulated container, we need just two variables (P and T) to specify the state of that system. The proportion of the two phases does not affect the thermodynamic state of the system as long as it is at equilibrium, even though it may be important in other contexts.

One final subject that must be covered in this introductory section is that of the thermodynamic temperature scale. Temperature is a fundamental physical quantity, in the sense that it is irreducible to a combination of simpler quantities. Other fundamental quantities are length, mass, time, and electric charge. The units in which these quantities are measured are defined in terms of conventional values such as the meter, kilogram and second. The absolute nature of these units is immediately evident because it is easy to grasp, at least in principle, what we mean by zero length, zero mass or zero time, and because it is also self evident that these three fundamental dimensions cannot take negative values. Temperature is different, because the temperature scales that are used in everyday life, and in many engineering applications, are based on arbitrary references which do not establish absolute values and, in particular, give no special meaning to the value of zero. In the Celsius or centigrade scale, zero is the temperature at which ice and water are at equilibrium at 1 bar pressure, and in this scale negative temperatures are obviously possible. An absolute temperature scale exists but is not easy to define until we have studied thermodynamics in some depth. We shall not worry here about how absolute thermodynamic temperatures are defined, but good discussions can be found, for example, in the classic textbooks by Lewis and Randall (1961), Glasstone (1946) and Guggenheim (1967). What we will do is emphasize that *all thermodynamic calculations must be carried out in this absolute temperature scale*, in which temperatures are measured in kelvins (symbolized K). Conversion is accomplished by adding 273.15 to the temperature in ∘C in order to obtain the temperature in K (note no "°" in K!).

1.2 Conservation of energy and different manifestations of energy

Conservation of energy (or, more accurately, mass-energy) is an example of a law of nature. By a "law of nature" we mean a statement that summarizes a large number of empirical observations (large enough that we are confident that we will not come across a contradictory observation) and that cannot be derived from simpler concepts, principles or laws. It is just a statement of a specific way in which our universe works. Whether or not it may be possible to understand why our universe works as it does is the subject of considerable argument among physicist working at the frontiers of knowledge, but it is a topic that is beyond the scope of this book. One possible statement of the law of conservation of energy is that any change in the total energy content of a system must equal the amount of energy received by the system minus the amount of energy extracted from the system. In order to be useful, however, a law of nature must be expressed as a mathematical statement. Why? Because mathematics is the only unambiguous and universally intelligible language. The language of mathematics has the additional advantages of being economical (i.e. concepts are expressed with the minimum possible number of symbols) and of being accompanied by a well-defined set of operation and manipulation rules. The First Law of Thermodynamics, which we will introduce in Section 1.10, is the mathematical statement of the law of conservation of energy and the starting point for our thermodynamic exploration of planetary processes.

Implicit in the law of conservation of energy is the concept that there are different kinds of energy that are equivalent to each other. Equivalence, however, does not mean unrestricted convertibility, leading to the concept that there are two fundamentally different manifestations of energy. There are those kinds of energy that can be fully and freely converted to other kinds. This category comprises all types of energy but one. Examples include: mechanical energy (in its various manifestations), electric energy, electromagnetic energy, and relativistic rest mass. Thermal energy is the one type of energy that belongs in a different category because it cannot be freely nor fully converted to other types of energy *although unrestricted conversion of other types of energy to thermal energy is always possible*. This distinction between thermal and other types of energy is at the root of another law of nature, called the Second Law of Thermodynamics. In Chapter 4 we will examine this law, which is perhaps one of the most fundamental, and mysterious, laws of nature. In the meantime, we need to formalize the definitions and mathematical formulations of the various types of energy that are important in planetary processes.

1.3 Mechanical energy. An introduction to dissipative and non-dissipative transformations

Although we all have an intuitive knowledge of what energy is, it is a concept that is notoriously difficult to define. For example, in elementary classical physics we learn that "energy is the capability to do work". You will have to make up your own mind on whether or not this definition is useful (or, indeed, whether it is a definition at all!). The concept of work, in contrast, has a unique and simple mathematical definition which, in differential form, is:

$$dW = \bar{f} \cdot d\bar{x},\tag{1.1}$$

where W is work, $\bar{f}$ is force and $d\bar{x}$ is the distance over which the point of application of the force moves. The use of lowercase bold symbols with an overstrike for $\bar{f}$ and $d\bar{x}$ means that these two quantities are vectors, and the dot between the two vectors represents a mathematical operation called "inner product" or "dot product" (Box 1.1). The inner product of two vectors yields a scalar quantity (W in this case), according to equation (1.1.1).

Box 1.1	Vectors, fields and the inner product

Physical magnitudes that have orientation, such as force and distance, are represented by geometric objects that are loosely called vectors. There are, in fact, two distinct types of vectors. Distance is an example of what is called a contravariant vector, and force is an example of a covariant vector. In modern mathematical language, distance is a *vector* and force is a *one-form*. Although both are oriented geometrical objects, one difference between the two is that one-forms are gradients of scalar fields, and vectors are not. A field is a mathematical function that gives the value of a variable as a function of space and time. If the variable is a scalar, such as temperature or density, the field is called a scalar field and the function returns a single number at each point in space and time. Suppose now that we keep time fixed. We can then determine the rate of change of the field intensity (e.g. temperature) relative to each of the three orthogonal spatial directions (see also Box 1.3). The set of the three derivatives $(\partial T/\partial x, \partial T/\partial y, \partial T/\partial z)$ is the gradient of the temperature field. This geometrical object is a one-form. Just as a vector, it has orientation (which is the direction of maximum rate of change) but it differs from a vector, among other things, in that its magnitude is given not by a length but by the separation between contour lines. The more closely spaced the contour lines are, the greater the magnitude of the one-form is. Force is the gradient of a potential energy field. In particular, the gravitational force is the gradient of the gravitational potential energy field. The more steeply gravitational potential energy varies, the stronger the gravitational force is. Excellent and comprehensive introductions to these concepts are given in the first chapters of the massive text *Gravitation* by Misner, Thorne and Wheeler (1973), and in Burke (1985). A classical and very accessible exposition (using the terminology of contravariant and covariant vectors, rather than vectors and one-forms) is given by Kreyszig (1991).

The magnitude of a vector or a one-form is a scalar that measures its "intensity" and is symbolized by $|x|$, where $\bar{x}$ is the vector or one-form. The magnitude of a vector corresponds to the intuitive concept of length, but, as I mentioned above, the magnitude of a one-form is better thought of as the spacing between contour lines – the more closely spaced the contour lines are, the greater the magnitude of the one-form (i.e. the steeper the gradient). The inner product is an operation that combines a vector and a one-form and returns a scalar. Geometrically, the inner product of $\bar{x}$ and $\bar{y}$ is the scalar A defined by:

$$A = \bar{x} \cdot \bar{y} \equiv |x||y|\cos\theta,\tag{1.1.1}$$

where θ is the angle between the two objects. Thus, if $\bar{x}$ is a vector oriented perpendicular ($\theta = \pi/2$) to the gradient of a scalar field ($=$ the one-form $\bar{y}$), the inner product vanishes. The inner product attains its maximum absolute value if the vector points in the direction of the field gradient ($\theta = 0$) or exactly opposite ($\theta = \pi$). It is positive if $0 \leq \theta < \pi/2$ and negative for $\pi/2 < \theta \leq \pi$.

We can use the concept of work to turn the definition of energy on its head and in the process make it clearer. Whenever work is performed, there is a force interaction between a system and its environment, or between different parts of a system. Work is responsible for energy transfer between the objects that interact, so some of these objects give up energy, and the same amount of energy, *measured by the magnitude of the work performed*, is stored in others. This statement may not be a definition of energy, but at least it tells us how to calculate changes in energy content. From this statement we also rescue the fact that energy has the same dimension as work (Box 1.2).

Box 1.2	Dimensional analysis

The dimension of a physical quantity is a fundamental and immutable property that defines what the quantity is. Thus, distance has dimension of length, inertia has dimension of mass, and duration has dimension of time. Units are arbitrary scales that are used to measure the magnitude of a physical quantity, for instance, distance can be measured in meters, kilometers, parsecs, etc. These are different units that measure the dimension length. Some physical quantities are fundamental in the sense that their dimension cannot be reduced to combinations of other dimensions. Examples are length, mass, electric charge, time and temperature. The key idea of dimensional analysis is that in any equation relating physical quantities the identity applies to dimension as well. The fundamental dimensions length, mass, electric charge, time and temperature are labeled $[L]$, $[M]$, $[Q]$, $[T]$ and $[\Theta]$, respectively. Dimensions of derived physical quantities are reduced to combinations of these fundamental dimensions. For instance, acceleration has dimension $[L] \times [T]^{-2}$, and force has dimension $[M] \times [L] \times [T]^{-2}$.

In the notation of dimensional analysis, enclosing the name or symbol of a quantity in square brackets means that we are referring to the dimension of the quantity. From equation (1.1) we have:

$$[work] = [force] \times [distance] \tag{1.2.1}$$

or:

$$[work] = [M] \times [L]^2 \times [T]^{-2}. \tag{1.2.2}$$

The units of the fundamental dimensions in the SI system are meter (m, length), kilogram (kg, mass), coulomb (C, electric charge), second (s, time) and kelvin (K, temperature). The SI unit of force is the newton (N), and from dimensional analysis we see that $1\,N = 1\,kg\,m\,s^{-2}$. Similarly, the SI unit of work, or energy, is the joule (J) and $1\,J = 1\,N\,m = 1\,kg\,m^2\,s^{-2}$. Note the subtle but important difference between the concepts of dimension and units. The dimension of a quantity is unique and universal, for instance, $[M] \times [L]^2 \times [T]^{-2}$ for work or energy. The units can be anything we agree upon, as long as they conform to the required dimensional equation, such as (1.2.2). In general I will use SI units throughout this book, with one important exception, which is that I use bars and its multiples and submultiples (kbar, Mbar, mbar, etc.) as the unit of pressure, instead of pascal (Pa), which is the SI unit. The conversion factor is $1\,bar = 10^5\,Pa$.

Just as it is difficult to define energy, it is also somewhat unsatisfactory to try to pigeonhole different types of energy into strict categories. In this and subsequent chapters we will come across examples that will highlight this difficulty. As a matter of convenience and tradition, however, we will include in our discussion of mechanical energy only the energy associated with motion (kinetic energy) and that one associated with position in a gravitational field (gravitational potential energy). Both of these types of energy play crucial roles in planetary processes. We must keep in mind, however, that many other types of energy, such as the energy associated with a change in shape or size of an object, the energy associated with a magnetic or electrostatic field, or the energy contained in a crystalline lattice, can ultimately be reduced to specific manifestations of mechanical energy. The one type of energy that is distinct is heat, and in this section we will begin our journey towards our understanding of the reasons for, and consequences of, this difference – this is, indeed, what much of thermodynamics is all about.

1.3.1 Gravitational potential energy

Gravitational potential energy is perhaps the most familiar kind of mechanical energy – it is responsible, for instance, for the fact that it is harder to hike up a mountain than down. Gravitational potential energy arises from the existence of a gravitational attractive force that acts between objects with mass. The magnitude of the gravitational force $|\bar{f}_g|$ between two bodies with masses m_1 and m_2 separated by a distance x is given by Newton's law of universal gravitation:

$$|\bar{f}_g| = \frac{Gm_1 m_2}{x^2}, \tag{1.2}$$

where G is the universal gravitational constant (physical constants and other important numerical data are given in Appendix 1). Force is a vector (more precisely, a covariant vector or one-form, Box 1.1) and this expression yields only its magnitude. The gravitational attraction caused by a body is described by a *vector field*, which is a function that assigns a vector to each point of space. The magnitude of the vector, called the field intensity, is the gravitational force per unit mass, i.e. the gravitational acceleration. The orientation of the vector depends on the distribution of mass in the body that generates the field. For a point-like mass it is oriented towards the mass.

Let us consider an object of mass m, say a rock, in the gravitational field of another object, for instance a planet, of mass M. If the rock experiences a displacement $d\bar{x}$, the gravitational force $\bar{f}_g$ performs an amount of work given by equation (1.1). Gravitational potential energy, U_g, depends only on the position of an object in a gravitational field. The work performed by the gravitational force represents energy that is extracted from the object's gravitational potential energy. Because of the law of conservation of energy the changes in the two variables must balance out, which in differential form we write as follows:

$$dU_g + \bar{f}_g \cdot d\bar{x} = 0. \tag{1.3}$$

We consider a displacement that is directed radially outwards from the planet with mass M (Fig. 1.1). If we define the direction away from the planet as being the positive orientation,

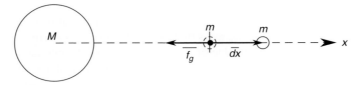

Fig. 1.1 Work performed by the gravitational force $\bar{f}_g$ between two bodies with masses M and m, when body m moves a distance $d\bar{x}$ towards infinity.

then $\bar{f}_g$ always has a negative orientation – this is the mathematical expression of the fact that gravity is always an attractive force. We can then write the inner product in equation (1.3) as:

$$\bar{f}_g \cdot d\bar{x} = -\frac{GMm}{x^2}dx, \tag{1.4}$$

where the negative sign arises from the fact that we are calculating the inner product between a vector and a one-form that subtend an angle of $180°$ (Box 1.1). This is the work *performed by the gravitational force* (the general three-dimensional case requires more complex notation, but the physics are well summarized in (1.4)). The change in gravitational potential energy that corresponds to a finite displacement between two positions, x_a and x_b can be obtained by substituting (1.4) into (1.3) and integrating:

$$U_{g,b} - U_{g,a} = -\int_a^b \bar{f}_g \cdot d\bar{x} = GMm\int_a^b \frac{dx}{x^2} = -GMm\left(\frac{1}{x_b} - \frac{1}{x_a}\right). \tag{1.5}$$

By convention, we define gravitational potential energy as being 0 when the objects are separated by an infinite distance, i.e. we make $U_{g,a} = 0$ as $x_a \to \infty$. With this convention we then define the gravitational potential energy of an object with mass m in the gravitational field of a planet with mass M as:

$$U_g = -\frac{GMm}{r}, \tag{1.6}$$

where r is the distance between the centers of mass of the two bodies. For any finite value of r, U_g is negative, and U_g approaches a maximum value of 0 as the separation between the two bodies approaches infinity. When writing equation (1.4) I justified the negative sign on purely mathematical grounds, as arising from the inner product of two oppositely pointing vectors, but we can now see the physical meaning of this negative sign. Suppose mass m is moved away from M ($\Delta r > 0$). In order for this to occur an external agent must perform work. By conservation of energy this work becomes stored as potential energy in the gravitational field. Therefore it must be $\Delta U_g > 0$, which is what results from equation (1.6) with $\Delta r > 0$. Conversely, if $\Delta r < 0$ the gravitational field gives up energy ($\Delta U_g < 0$) which is transferred to mass m and appears, for example, as kinetic energy.

I will introduce here two other equations that we use in the analysis of gravitationally driven planetary heating in Chapter 2. First, we can see from equation (1.2) that the gravitational acceleration, g, due to a body of mass M at a distance r from its center of mass is given by:

$$g = -\frac{GM}{r^2}, \tag{1.7}$$

where the negative sign expresses the fact that gravitational acceleration is always attractive (directed towards the mass that causes it, where r is positive away from the mass). Recall that the numerical value of g is the intensity of the gravitational field. We also define the gravitational potential, Φ_g, as the gravitational potential energy per unit mass:

$$\Phi_g = -\frac{GM}{r}. \tag{1.8}$$

Gravitational potential is a scalar field (Box 1.1). Gravitational acceleration is a one-form (or covariant vector) that is the gradient of the potential field. The magnitude of the one-form is given by:

$$g = -\frac{d\Phi_g}{dr}.$$

(1.9)

Force is another one-form, that is the product of a scalar (mass) times acceleration (Newton's second law of motion). Mass, or inertia, is the scaling factor between force and acceleration; this is the origin of the term *scalar*.

We are generally concerned with *differences* in gravitational potential energy between different states of a system. For example, when tectonic processes elevate a mountain range, or when lava flows build a volcano, gravitational potential energy is stored in the rocks. How much potential energy is stored in a mountain range? This depends on the distance that the rocks can move towards the center of the planet before they get to a level below which they can move no further. How do we define such a level? Sea level may be a good first approximation, but we can give a more general answer, that will also hold for planets without oceans. We begin by looking at two additional questions. First, where did this gravitational potential energy associated with topography come from? Conservation of energy requires that we identify an energy source, which in this case entails conversion of some of the planet's internal heat to gravitational potential energy (more on this in Chapter 3). Second, what happens to this potential energy as the mountain loses elevation? The short answer is that this gravitational potential energy ultimately becomes heat and is dissipated to space, but the pathway may entail some intermediate steps, depending on how the mountain loses elevation. In general, elevation is lost by a combination of three processes: erosion, isostatic adjustment and tectonic collapse. During erosion, potential energy becomes heat as a result of friction during sediment transport and also when particles come to their final resting place in a sedimentary basin. Isostatic adjustment may return gravitational potential energy to the mantle, either as heat or as mechanical energy. Tectonic collapse results in gravitational potential energy either being dissipated directly as heat during ductile deformation, or being stored as elastic energy (another type of mechanical energy that we will discuss) to be eventually dissipated, ultimately as heat too, during earthquakes. All of these processes drive towards converting the surface of the planet to one over which there are no differences in gravitational potential energy. Such a surface is called an *equipotential surface*. A well-defined reference level for potential energy on Earth is thus the *geoid*, which is defined as the equipotential surface that is as close as possible (e.g. in a least square sense) to mean sea level (see Worked Example 1.1). In planets without oceans, we may choose as our reference the equipotential surface that is as close as possible to the mean planetary radius (and, if we were to follow the same convention used for Earth, we should call such surfaces: areoid, aphrodoid, hermoid, selenoid, etc.). We will generally be concerned with differences between the value of U_g at the geoid and its value at any other level that we may be interested in.

> **Worked Example 1.1   Gravitational potential energy and topography**

(a) Consider a mass m of rock initially located on Earth's geoid. Let R be the geoid's mean radius. The rock is then moved to an elevation h above the geoid, such that $h \ll R$, i.e. we stay close to the planet's surface. The initial distance between the two centers of mass (the rock's and the Earth's) is R, and the final distance, after the rock is raised, is $R + h$. We first use equation (1.6) to show that the gravitational potential energy of the rock in its final state, relative to the geoid, is given by mgh, the equation that you probably remember from introductory physics.

Calling the gravitational potential energies at the geoid and at an elevation h above the geoid $U_{g,geoid}$ and $U_{g,h}$, we have:

$$U_{g,h} - U_{g,geoid} = -\frac{GMm}{R+h} - \left(-\frac{GMm}{R}\right) = \frac{GMmh}{R(R+h)}. \tag{1.10}$$

If we stay close to the planet's surface, then $R(R+h) \approx R^2$. We can also consider the planet's gravitational acceleration to be constant over the interval R to $R + h$. Using equation (1.7) to calculate gravitational acceleration at the geoid and substituting in (1.10):

$$U_{g,h} - U_{g,geoid} \approx \frac{GMmh}{R^2} = -mgh = m|g|h. \tag{1.11}$$

With our sign convention g is always a negative quantity. The "g" in mgh is thus the magnitude of g, as shown in the last term of equation (1.11).

(b) The Sierra Nevada of California is the largest uninterrupted mountain range in the United States. It is roughly 600 km long, 100 km wide and has a mean elevation of 1.5 km (averaged over this horizontal extent). Assuming that this average elevation represents the center of mass of the mountain range, and that the rocks making up the Sierra Nevada have an average density of 2800 kg m^{-3}, what is the potential energy stored in the Sierra Nevada relative to the geoid? At the geoid, $g \approx 9.8$ m s^{-2}. Plugging in these values into equation (1.11) we find that the Sierra Nevada stores approximately 3.7×10^{21} J of gravitational potential energy, approximately equivalent to an explosive yield of one million megatons (see Section 1.12.2).

(c) Uplift of the Sierra Nevada has occurred over the past 5 million years. Assuming a constant rate of uplift over that time interval, the energy flux (energy per unit area per unit time) that went into elevating the Sierra Nevada is approximately 3.9×10^{-4} W m^{-2} = 0.39 mW m^{-2}, where 1 W(Watt) = 1 J s^{-1}. Typical terrestrial heat fluxes are of the order of 50–100 mW m^{-2}, i.e. two orders of magnitude greater. There is plenty of energy in the Earth to elevate mountain ranges.

1.3.2 Kinetic energy

Bodies in motion have kinetic energy, U_k, that arises from their speed and is given by:

$$U_k = \frac{1}{2}mv^2, \tag{1.12}$$

where v is the magnitude of the body's velocity, i.e. its speed. Kinetic energy is a function of the *relative* speed between a body and an observer. For example, if we observe, from a location at rest on the Earth, an asteroid of mass m moving towards Earth with speed v, the kinetic energy of the asteroid in our reference frame is given by equation (1.12) and the Earth has no kinetic energy in our reference frame. Measured from a reference frame attached to the asteroid, on the other hand, the asteroid has no kinetic energy, and the Earth has kinetic energy given by equation (1.12), but with m in this case being the mass of the Earth. Although the two values of kinetic energy are very different, if the two bodies collide the result is unique: conversion to heat of an amount of energy equivalent to most of the kinetic energy of the asteroid, as measured from the Earth's reference frame (Worked Example 1.2). In order to see why this is the case we need to introduce the law of conservation of momentum, which is a law of nature that, in classical physics, is distinct from energy conservation. In reality, energy and momentum conservation are different manifestations of a single conservation law that arises from the geometrical properties of spacetime, but this becomes an issue only for objects moving at relativistic speeds.

Momentum, $\bar{p}$, is a vector quantity, and is given by:

$$\bar{p} = m\bar{v},   \tag{1.13}$$

where $\bar{v}$ is the velocity vector. The law of conservation of momentum states that the total momentum of a system is conserved. In contrast to energy, that has many different manifestations, momentum is unique and is always conserved; it cannot be converted to other "types" of momentum. In a perfectly elastic collision both momentum and *kinetic* energy are conserved, i.e., there is no conversion of kinetic energy into other types of energy. The concept of elastic collisions will enter into our discussion of a thermodynamic variable known as internal energy, later in this chapter. In an inelastic collision, in contrast, momentum is conserved (as it must always be) but *kinetic* energy is not. Of course, the total energy of the system must be conserved, so that during an inelastic collision some kinetic energy is converted to other types of energy (ultimately heat). Conversion of kinetic energy to heat during inelastic collisions takes place, for example, when celestial bodies collide. This process was responsible for accretionary heating during the Solar System's formative period (Chapter 2). On a very different scale, inelastic collisions of subatomic particles with atoms in a crystal are the cause of radioactive heating, one of the key sources of energy in terrestrial planets (Chapter 2).

Worked Example 1.2   Dissipation of kinetic energy during collisions of celestial bodies

(a) An asteroid of mass m_a moves directly towards Earth (mass M) with velocity $\bar{v}_a$, as measured from Earth. Consider a reference frame relative to which the Earth is initially at rest (i.e. the Earth's initial velocity is zero), but that will remain fixed even if the state of motion of the Earth were to change. Figure 1.2a shows the initial situation as seen from this reference frame, choosing a leftward-directed velocity as positive. We consider the collision of the two bodies as being perfectly inelastic, meaning that momentum, $\bar{p}$, is conserved, but kinetic energy is not. After the collision, the Earth and the asteroid merge into a single body of mass $(m_a + M)$, moving with velocity $\bar{v}_f$ relative to the same reference frame as before, which has not been affected by the collision. We seek the magnitude and the direction (leftwards or rightwards) of $\bar{v}_f$, the change in the velocity of the Earth relative to the external reference frame, and the change in kinetic energy of the Earth + asteroid system resulting from the collision.

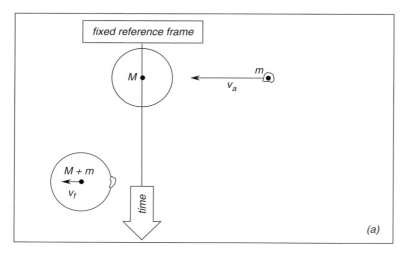

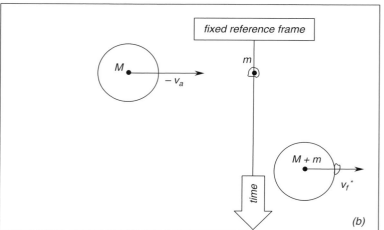

Fig. 1.2 Asteroid (mass m) colliding with Earth (mass M), as seen from Earth (a) and from the asteroid (b).

The final velocity is calculated from momentum conservation:

$$\bar{p}_{initial} = m\bar{v}_a = \bar{p}_{final} = (m+M)\bar{v}_f$$

$$\bar{v}_f = \frac{m}{m+M}\bar{v}_a. \tag{1.14}$$

The change in the Earth's velocity is simply v_f, as its initial velocity in the chosen frame of reference is zero. The amount of kinetic energy that is converted to other forms of energy (chiefly heat) is given by:

$$U_{k,final} - U_{k,initial} = \frac{1}{2}\left[(m+M)|\bar{v}_f|^2 - m|\bar{v}_a|^2\right] = -\frac{1}{2}m|\bar{v}_a|^2\left(\frac{M}{m+M}\right). \tag{1.15}$$

If the asteroid is small compared to Earth, then the expression in parentheses ≈ 1, and the heat dissipated is essentially equal to the kinetic energy of the asteroid before the collision.

(b) We repeat the calculation from the point of view of a reference frame that is initially at rest relative to the asteroid, and that, as before, is not affected by the collision (Fig. 1.2b). In this case the Earth's initial velocity is $-\bar{v}_a$ and the asteroid's initial velocity is zero.

Conservation of momentum is in this instance:

$$\bar{p}_{initial} = M(-\bar{v}_a) = \bar{p}_{final} = (m + M)\bar{v}_f^*$$

$$\bar{v}_f^* = \frac{M}{m + M}(-\bar{v}_a), \tag{1.16}$$

where the asterisk is used to remind ourselves that the final velocity in this case is measured relative to a different reference frame, initially fixed to the asteroid. The change in the Earth's velocity is then given by:

$$\Delta\bar{v}_{Earth} = -\frac{M}{m + M}(\bar{v}_a) - (-\bar{v}_a) = \frac{m}{m + M}\bar{v}_a, \tag{1.17}$$

which is the same as the change calculated in part (a), compare equation (1.14). Note carefully the meaning of the signs: equation (1.16) shows that after the collision the Earth is still moving towards the right, but at a lower speed than before the collision. Equation (1.17) shows that the velocity change is positive, i.e. leftwards. As seen from the asteroid's reference frame, the Earth has slowed down by the same amount as the Earth has speeded up (from rest) when seen from the Earth's reference frame. You should verify that the loss of kinetic energy as calculated from the asteroid reference frame also agrees with the loss calculated from the Earth's reference frame, equation (1.15).

1.3.3 Energy dissipation

Energy dissipation is the conversion of any kind of non-thermal energy to heat. A process that accomplishes this conversion is called a *dissipative transformation*. In the example of an asteroid colliding with Earth kinetic energy is dissipated, i.e. it is converted to heat. The full import of this definition will become clear in Chapter 4. For now, it is important to realize that, although energy is conserved during a dissipative process (this is the First Law of Thermodynamics), a full reversal of a dissipative transformation is impossible. In other words, it may be possible to reconvert some of the heat back to mechanical energy, but there is always a fraction of the heat generated by dissipation that can never be converted back to non-thermal energy (this is the Second Law of Thermodynamics). Transformations between different types of non-thermal energy are called *non-dissipative*, meaning that the inverse transformation is, in principle, possible.

1.4 Expansion work. Introduction to equations of state

1.4.1 The concept of expansion work

Change in volume of any substance entails an energy transfer, in other words, a system that undergoes a volume change either absorbs or releases energy. Energy transfer mediated by a change in volume is called expansion work. If a substance expands, it is performing mechanical work on its environment and, by the law of conservation of energy, this energy must come from somewhere. Expansion of volcanic gas during a pyroclastic eruption is a process in which a large amount of energy is transferred from a magmatic system to the atmosphere by performing expansion work on the latter. Other examples are perhaps less eye-catching but not less important: discontinuous phase transitions, such as vaporization and freezing of water, are accompanied by expansion work, as are many mineral reactions. For instance, consider the difference in density between the Al_2SiO_5 polymorphs kyanite and sillimanite. When kyanite transforms to sillimanite the volume of a fixed amount of Al_2SiO_5 increases, because its density decreases. This mineral transformation performs expansion work on its environment. The converse is also true, when a system is compressed it absorbs mechanical energy from its environment and this energy is stored in the substance. Transformation of sillimanite to kyanite absorbs mechanical energy from the environment, and this energy is stored as chemical energy in the crystal of kyanite (the expression "chemical energy" is for the time being quite vague, it will take on a more precise meaning in subsequent sections and chapters).

Dissipative compressions are an important source of heat in a wide range of planetary processes. For example, air masses that descend rapidly in the atmosphere and protostellar clouds of gas (such as the one that gave rise to our Solar System over 4.55 Ga) that contract under their own gravitational pull, heat up as a result of compression, i.e. compression transforms mechanical energy into thermal energy. In contrast, elastic materials, such as rocks at relatively low temperatures, are able to undergo approximately non-dissipative changes in volume, storing compression work (= negative expansion work) as elastic energy.

Expansion work, then, plays an important role in many planetary processes, but how do we measure it? Figure 1.3 shows a system of arbitrary shape expanding into an environment that exerts a uniform pressure P. If the system's surface area is A, then as the system expands it must overcome a total force PA over its entire surface. Although PA is an oriented object we are not using vector notation to represent it. This is careless notation, that we can justify informally by noting that during uniform expansion pressure and force have the same orientation as displacement, so that the inner product of the two objects equals the scalar product of their magnitudes (Box 1.1).

Consider an infinitesimal expansion that causes the surface of the system to move outwards by an amount dx (Figure 1.3). Then the total amount of work performed by the system during this infinitesimal transformation is:

$$dW = PAdx. \qquad (1.18)$$

The product Adx is the change in volume of the system, dV (because this is an infinitesimal expansion, the total surface area can be considered to remain constant). The differential expression for expansion work is then:

$$dW = PdV. \qquad (1.19)$$

The SI unit of pressure is the pascal (Pa), defined as a pressure of $1\ N\ m^{-2}$, so that a volume change of $1\ m^3$ against a pressure of 1 Pa performs 1 J of work. Another unit of pressure that I find much preferable is the bar, where $1\ bar = 10^5\ Pa$. One good reason to prefer the bar over the pascal is that atmospheric pressure on Earth at sea level is approximately 1 bar, so it is easy to develop an intuitive feeling for the magnitude of a pressure expressed in bar,

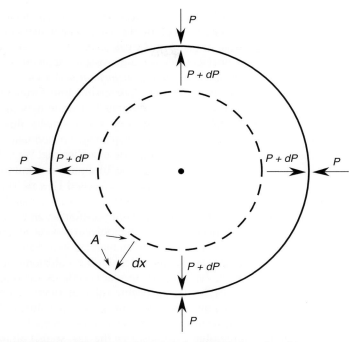

Fig. 1.3   Expansion work. The shell defined by the two spheres is the increase in volume $dv = A\,dx$.

kbar or mbar. A convenient unit of volume is, then, J bar^{-1}, where 1 J bar$^{-1} = 10^{-5}$ m^3 $= 10$ cm^3. Expressing volume in J bar^{-1} and pressure in bar, W comes out in J.

1.4.2 Quasi-static, reversible and irreversible processes

In the derivation of the expansion work formula I have been rather sloppy with the sign convention, and this sloppiness must be corrected before we can advance any further. In order to do so, we need to address another issue which is often the source of much confusion and consternation when encountering thermodynamics for the first time. I stated that the system expands "into an environment that exerts a uniform pressure P" – but what is the pressure *inside* the system? Classical thermodynamics is concerned only with states of equilibrium. Because of this constraint, the only situation that can be rigorously addressed with classical thermodynamics is one in which the pressure inside the system equals the external pressure. But if the two pressures are the same, then why is the system expanding at all? Here lies the problem, if the two pressures are the same then the system is static, and there is no work being performed. Classical thermodynamics deals with this riddle by inventing an idealized type of transformation called a *quasi-static transformation*. This is a process that takes a system from an initial equilibrium state to a final equilibrium state by means of an infinite number of intermediate equilibrium states that are infinitesimally close to one another. Such a transformation is, of course, physically impossible. In our example of an expanding system, we could visualize a situation in which the pressure inside the system is always greater than the external pressure by an infinitesimal amount, dP. By this we mean that the internal pressure is barely higher than the external pressure, so that the system does expand, but not so much higher as to make any difference in the amount of expansion work given by equation (1.19), i.e. we would get the same numerical result whether we used the internal or the external pressure in our calculations.

A concept that is distinct to that of a quasi-static transformation, but is sometimes confused with it, is that of a *reversible transformation*. By definition this is a process during which there is no energy dissipation. A reversible transformation must be quasi-static, *but the converse is not true:* a quasi-static transformation may dissipate energy, and thus be *irreversible*. An example of this would be infinitesimally slow expansion of a gas (i.e. a quasi-static process) accomplished by displacing a piston that slides inside a cylinder with friction. The irreversibility arises from frictional heating. We will come across other examples that will gradually make the concepts of dissipation and irreversibility clear, and will define these concepts rigorously using the Second Law of Thermodynamics (Chapters 4 and 12).

In contrast, the concept of a quasi-static transformation can be clarified by means of an example. A gas is a collection of moving molecules. Consider what happens at the microscopic level when a gas is subjected to an increase in pressure. The gas is at equilibrium if, as a result of interactions among its molecules (i.e. collisions), the distribution of molecular kinetic energies (or, equivalently, molecular speeds) is smooth, with a unique peak value (Fig. 1.4) that specifies the most probable value of kinetic energy. The temperature of the gas is a measure of this most probable kinetic energy value. Equilibrium in the gas is possible only if there has been sufficient time to allow its molecules to exchange information throughout the entire volume occupied by the gas, so that all molecules "know" what the most probable kinetic energy is. Information is transported by molecular collisions, so that it travels at a rate comparable to molecular speeds. If the gas is now compressed at a rate that is very slow relative to most molecular speeds then the gas will remain at equilibrium throughout the compression, because information about perturbations in one part of the system reach the entire system before the magnitude of the perturbations can change significantly. This is an example of a quasi-static transformation. In contrast, if the gas is compressed at a rate that is much faster than the speed with which most molecules move, then the gas cannot be at equilibrium. In this case a perturbation in one part of the system can grow significantly before the rest of the system "knows" about it. Molecules in different parts of the system will have different speed distributions, so that the speed distribution for the entire volume of gas will not display a unique maximum value (dashed curve in Fig. 1.4). The fast compression is not quasi-static and, therefore, it is an irreversible transformation. During such an irreversible transformation it is not possible to define the temperature nor the pressure of the gas, as there is no unique most probable molecular kinetic energy (Fig. 1.4). Sonic booms, which occur when air is compressed at a rate that is faster than characteristic molecular speeds, are examples of irreversible transformations of this type.

1.4.3 Measuring expansion work. Material properties and equations of state

We are now able to come to a rigorous definition of the sign of the expansion work. If the pressure exerted by the system, P (we assume a quasi-static expansion), and the change in volume, $dV = A\,dx$, have the same orientation, then $P\,dV$ is a positive quantity. The work performed by the system on the environment during expansion, dW, is a positive quantity. This work represents energy transferred from the system to its environment, so that the energy content of the environment changes by an amount dW, and the energy content of the system that is expanding changes by an amount $-dW$. It is important to understand

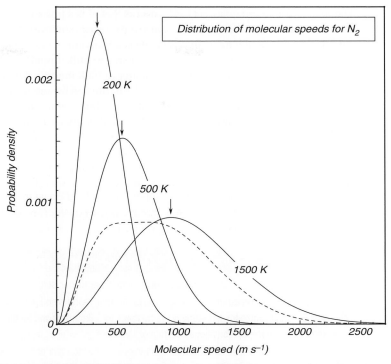

Fig. 1.4
Maxwell–Boltzmann probability density distribution for molecular speeds in N_2 (assumed to be an ideal gas) in the absence of a gravitational field. The areas under all the curves are unity. As temperature increases the probability distribution becomes wider but there is always a unique most probable speed, that depends on temperature only (arrows). The dashed line shows a hypothetical non-equilibrium distribution of molecular speeds – the most probable speed, and hence the temperature, cannot be defined in this case.

exactly what equation (1.19) is measuring and how to interpret the sign of the expansion work that one calculates upon integrating this equation. Because P is always a positive quantity, a positive value of W means that the system has expanded and performed work on its environment; the system's energy content has therefore decreased. Compression, i.e. a decrease in the system's volume, results in a negative value of W which *implies that the system has gained energy*. This energy was extracted from its environment. This sign convention is not universally followed, but it leads to equations that represent physical processes in an intuitively satisfying way.

The expansion work associated with a finite change in volume is calculated by integrating equation (1.19). Except for the special case of isobaric processes, volume and pressure are not independent variables. In order to integrate equation (1.19), then, we need a function that relates volume with pressure. One such function is the bulk modulus, K, which is a *material property*, i.e. a parameter that has a well-defined value for every substance and is independent of system size:

$$K = -V \frac{\partial P}{\partial V}.$$

(1.20)

The negative sign in equation (1.20) insures that K is a positive quantity, as volume always decreases under an increase in the applied pressure. The symbol ∂ indicates partial differentiation, and is used whenever a variable is a function of more than one other variable (see Box 1.3). Equation (1.20) defines the dimension of bulk modulus as being the same as that of pressure. A substance has an infinite quantity of bulk moduli, depending on how one specifies the volume change to take place, but the definition becomes precise if we specify which variables are to be held constant during the transformation. For example, the bulk modulus measured at constant temperature, called the isothermal bulk modulus, is defined as:

$$K_T = -V \left(\frac{\partial P}{\partial V} \right)_T,$$ (1.21)

where the subscript T denotes that temperature is held constant. We shall see that isothermal processes are not the only ones that we must consider, so that other bulk moduli will eventually be introduced.

| Box 1.3 | Functions of several variables and partial derivatives |

Manipulation of functions of several variables, and in particular some basic proficiency in multivariate calculus, are essential requirements for working in thermodynamics and many other branches of physics and chemistry. A good source for self-study is the textbook by Sokolnikoff and Redheffer (1966). Here I will cover only some basic concepts. Consider a function of i variables, $y = y(x_1, x_2, ..., x_i)$. The function y has i partial derivatives, each of which is taken with respect to one of the independent variables, while holding all of the other independent variables constant. A simple analogy is to think of the two-dimensional surface of the Earth, and to consider the slope of the terrain along the N–S direction at constant longitude, and then along the E–W direction at constant latitude. Partial derivatives are denoted with the symbol ∂, and the partial derivative of y relative to x_n is written as follows:

$$\frac{\partial y}{\partial x_n}.$$ (1.3.1)

From the operational point of view finding the value of (1.3.1) poses no problems: we just consider all of the other variables, $x_{i \neq n}$ to be constants, and calculate the derivative following the usual rules. The notation of (1.3.1) is mathematically complete, but in thermodynamics it is customary to specify which are the variable or variables that are held constant when performing a calculation. Thus, in thermodynamics (1.3.1) is usually written as follows:

$$\left(\frac{\partial y}{\partial x_n} \right)_{x_1, x_2, ..., x_{i \neq n}},$$ (1.3.2)

where the list of subscripts after the parenthesis specifies the variable or variables that are held constant during differentiation. Whether or not this notation is strictly necessary is the subject of some contention, to which I return in Box 1.4.

The total differential of a function of several variables, dy, is the sum of the changes in the value of y arising from the change in each of the independent variables. It is defined as follows:

$$dy = \left(\frac{\partial y}{\partial x_1} \right)_{x_2, x_3, ..., x_i} dx_1 + \left(\frac{\partial y}{\partial x_2} \right)_{x_1, x_3, ..., x_i} dx_2 + \cdots + \left(\frac{\partial y}{\partial x_i} \right)_{x_1, x_2, ..., x_{i-1}} dx_i.$$ (1.3.3)

Box 1.3   Continued

Recalling the example of the surface of the Earth, we can calculate the total change in elevation between two points by first moving N–S at constant longitude, and then E–W at constant latitude. Or consider the volume of a fixed amount of substance, V. This is generally a function of P and T, so V is a function of two variables, $V = V(P,T)$. If we change both pressure and temperature, the total change in volume is given by:

$$dV = \left(\frac{\partial V}{\partial T}\right)_P dT + \left(\frac{\partial V}{\partial P}\right)_T dP. \tag{1.3.4}$$

There are a number of identities among partial derivatives that are useful in thermodynamic derivations. Consider four independent variables, x, y, z and w, such that the following functions exist and are continuous and differentiable:

$$w = w(x,z); \quad z = z(x,y); \quad \text{thus,} \quad w = w(x,y), \tag{1.3.5}$$

we can write two different equations for the total differential of w, dw:

$$dw = \left(\frac{\partial w}{\partial x}\right)_z dx + \left(\frac{\partial w}{\partial z}\right)_x dz \tag{1.3.6}$$

and:

$$dw = \left(\frac{\partial w}{\partial x}\right)_y dx + \left(\frac{\partial w}{\partial y}\right)_x dy. \tag{1.3.7}$$

The total differential of z is given by:

$$dz = \left(\frac{\partial z}{\partial x}\right)_y dx + \left(\frac{\partial z}{\partial y}\right)_x dy. \tag{1.3.8}$$

Substituting into the first equation for dw (1.3.6):

$$dw = \left(\frac{\partial w}{\partial x}\right)_z dx + \left(\frac{\partial w}{\partial z}\right)_x \left[\left(\frac{\partial z}{\partial x}\right)_y dx + \left(\frac{\partial z}{\partial y}\right)_x dy\right], \tag{1.3.9}$$

which, collecting terms in dx, dy, becomes:

$$dw = \left[\left(\frac{\partial w}{\partial x}\right)_z + \left(\frac{\partial w}{\partial z}\right)_x \left(\frac{\partial z}{\partial x}\right)_y\right] dx + \left(\frac{\partial w}{\partial z}\right)_x \left(\frac{\partial z}{\partial y}\right)_x dy. \tag{1.3.10}$$

Now, because x and y are independent variables, the only way in which equations (1.3.7) and (1.3.10) can be simultaneously true is if the coefficients of dx and dy are equal in each of the equations. In other words:

$$\left(\frac{\partial w}{\partial y}\right)_x = \left(\frac{\partial w}{\partial z}\right)_x \left(\frac{\partial z}{\partial y}\right)_x \tag{1.3.11}$$

and:

$$\left(\frac{\partial w}{\partial x}\right)_y = \left(\frac{\partial w}{\partial x}\right)_z + \left(\frac{\partial w}{\partial z}\right)_x \left(\frac{\partial z}{\partial x}\right)_y. \tag{1.3.12}$$

Box 1.3	Continued

The first of these equations, (1.3.11), is known as the *chain rule of partial differentiation* and is easy to remember. The second one has no name, and is harder to memorize – it is best to look it up every time that one may want to use it.   When dealing with equations of state that relate three variables (P, V, T) one is often faced with the following problem. Here we have three variables, x, y, z, and by virtue of the equation of state we can write:

$$x = x(y,z); z = z(x,y),$$
(1.3.13)

we then have the following total differentials:

$$dx = \left(\frac{\partial x}{\partial y}\right)_z dy + \left(\frac{\partial x}{\partial z}\right)_y dz$$
(1.3.14)

$$dz = \left(\frac{\partial z}{\partial x}\right)_y dx + \left(\frac{\partial z}{\partial y}\right)_x dy.$$
(1.3.15)

Substituting the second one into the first one:

$$dx = \left(\frac{\partial x}{\partial y}\right)_z dy + \left(\frac{\partial x}{\partial z}\right)_y \left[\left(\frac{\partial z}{\partial y}\right)_x dy + \left(\frac{\partial z}{\partial x}\right)_y dx\right]$$
(1.3.16)

collecting terms and rearranging:

$$dx = \left[\left(\frac{\partial x}{\partial y}\right)_z + \left(\frac{\partial x}{\partial z}\right)_y \left(\frac{\partial z}{\partial y}\right)_x\right] dy + \left(\frac{\partial x}{\partial z}\right)_y \left(\frac{\partial z}{\partial x}\right)_y dx.$$
(1.3.17)

Because x and y are independent variables, the coefficient of dx in the right-hand side of this equation must be equal to 1, and the coefficient of dy must be 0, so:

$$\left(\frac{\partial x}{\partial z}\right)_y = \frac{1}{\left(\frac{\partial z}{\partial x}\right)_y}$$
(1.3.18)

and:

$$\left(\frac{\partial x}{\partial y}\right)_z = -\left(\frac{\partial x}{\partial z}\right)_y \left(\frac{\partial z}{\partial y}\right)_x.$$
(1.3.19)

An equation of state (EOS) is another type of function that relates volume with pressure and temperature. EOS are of fundamental importance in many branches of planetary sciences, and we will discuss different types of EOS in subsequent chapters. Every substance has an EOS. This is just another way of saying that a fixed amount of any substance at equilibrium at a given pressure and temperature has a unique and well defined volume. Some EOS have physical fundament and at least their general form can be derived from first principles. In other cases, EOS are more or less arbitrary best-fit equations to experimentally measured P–V–T relationships. In some cases equations of state can be mathematically very complex, or their explicit mathematical form may be unknown, so that experimental

P–V–T data are summarized in tables. There is, however, one EOS that is mathematically very simple, that can be derived from first principles, and that has important applications in classical thermodynamics and some areas of planetary sciences. This is the ideal gas EOS:

$$PV = RT, \tag{1.22}$$

where V is molar volume, i.e. the volume of one mol of gas, T is absolute temperature, and R is the universal gas constant (see Appendix 1). The behavior of real gases at low pressure (≤ 1 bar) and temperatures well above their condensation temperatures is reasonably well approximated by this EOS. Up to this point we have been writing volume in bold type, $\boldsymbol{V}$, but we have dropped the bold attribute in the ideal gas equation of state, (1.22). This is so because we need to distinguish between situations in which we are referring to the total volume of a system ($\boldsymbol{V}$) from those in which we are referring to a specific amount of matter, for instance, 1 mol (V). For variables whose value depends on the amount of substance considered we will use non-bold uppercase symbols to denote the value normalized to one mol (e.g. V for molar volume) and bold uppercase symbols to denote the value of the variable for the entire system (e.g. $\boldsymbol{V}$ for total volume). V is a material property, independent of system size, but $\boldsymbol{V}$ is not. The scalar variables pressure and temperature, that do not depend on the size of a system, will also be written with non-bold symbols, whereas other scalar quantities such as various types of mechanical energy (e.g. $\boldsymbol{U}_g$ and $\boldsymbol{U}_k$, equations (1.6) and (1.12)) that depend on system size are written in bold type.

In principle, it is always possible to derive the isothermal bulk modulus from the EOS. This is particularly simple for the ideal gas EOS, as shown in the following Worked Example, which compares the magnitude of expansion work associated with pressure changes in gases, liquids and solids.

Worked Example 1.3 Expansion work of gases, liquids and solids

What is the isothermal bulk modulus of an ideal gas? From the ideal gas EOS:

$$P = \frac{RT}{V} \tag{1.23}$$

so:

$$\left(\frac{\partial P}{\partial V} \right)_T = -\frac{RT}{V^2}. \tag{1.24}$$

Inserting this expression into equation (1.21) and using the ideal gas EOS to simplify the result yields $K_T = P$. In contrast to gases, liquids and solids have bulk moduli that are nearly constant over restricted pressure ranges. For example, for pressures of the order of 1 bar, K_T for liquid water is approximately 10^4 bar and K_T for quartz $\approx 4 \times 10^5$ bar. Compared to K_T for an ideal gas at 1 bar, these values mean that liquids and solids are several orders of magnitude less compressible than gases, and quartz is somewhat less compressible than water (greater pressure changes are needed to cause a given volume change). The energy consequences of these differences are important. We seek to compare the expansion work performed by an ideal gas, a liquid (water) and a solid (quartz) undergoing the same pressure change.

From the ideal gas EOS:

$$dV = -\frac{RT}{P^2} dP \qquad (1.25)$$

substituting in equation (1.19):

$$dW = P dV = -RT \frac{dP}{P} \qquad (1.26)$$

and integrating:

$$W = RT \ln \left(\frac{P_0}{P} \right), \qquad (1.27)$$

where P_0 is the initial pressure, and P the final pressure. For an expansion process, $P_0 > P$ and hence $W > 0$, in agreement with our sign convention for expansion work (a positive value of W means that the system is performing work on its environment, and hence that the system's energy content decreases). Equation (1.27) yields the amount of work for expansion of 1 mol of ideal gas. For any other amount of gas, it must be multiplied by the number of mols.

Calculating the expansion work of an ideal gas is straightforward because of the simplicity of the ideal gas EOS. The general expression for expansion work as a function of bulk modulus, or, equivalently, for substances with an EOS more complicated than that for ideal gases, is not that simple (Chapters 8 and 9). If the bulk modulus is large enough and we are only interested in a rough approximation then the following simplifying assumption becomes possible. We integrate equation (1.21) assuming constant bulk modulus:

$$\frac{V}{V_0} = \exp \left(\frac{P_0 - P}{K_T} \right). \qquad (1.28)$$

This equation shows that, for a pressure drop of, say, two orders of magnitude, the volume of water increases by a factor of ~ 1.01, and the volume of quartz by a factor of ~ 1.00025. For the same pressure change, an ideal gas increases in volume by a factor of 100. As a first approximation, we may neglect the volume change of liquids and solids when integrating equation (1.19), but the attentive reader will immediately notice a seeming contradiction here: if volume is constant, then there should be no expansion work. The answer to this riddle is that volume is not constant, but assuming that it is constant leads to a much simpler mathematical expression that, under certain circumstances, yields a numerical answer that is an acceptable approximation to the exact answer. The limits of applicability of the simplified treatment that follows is explored in Exercise 1.9. We re-write equation (1.21) as follows:

$$dV = -\frac{V}{K} dP \qquad (1.29)$$

and substitute into equation (1.19):

$$dW = P dV = -\frac{V}{K} P dP. \qquad (1.30)$$

Integrating this expression under the assumption of constant volume yields:

$$W \approx \frac{V}{2K}\left(P_0{}^2 - P^2\right),\qquad (1.31)$$

where expansion implies $P_0 > P$, so that $W > 0$.

Consider expansion of 1 mol of each substance from $P_0 = 10$ bar to $P = 1$ bar, at a constant temperature of $25\,°C = 298$ K. For water and quartz, we need their molar volumes, which are 1.8 J bar^{-1} and 2.27 J bar^{-1}, respectively. The values of expansion work are $W_{water} \approx 8.9 \times 10^{-3}$ J mol^{-1} and $W_{quartz} \approx 2.8 \times 10^{-4}$ J mol^{-1}. For an ideal gas, equation 1.27 yields $W \approx 5.7 \times 10^{3}$ J mol^{-1}. The difference between the amount of work performed by an expanding gas and the work performed by an equivalent amount of expanding liquid or solid is huge – six or seven orders of magnitude in the above example. This is a simple mathematical statement of the greater destructive power of a pyroclastic eruption compared to a lava flow. Condensed phases (solids and liquids) and non-condensed phases (gases) differ in many aspects of their physico-chemical behaviors (Section 1.15). At the root of many of these differences is the wide gap that exists between their respective abilities to store or dispense energy in response to changes in pressure, which is expressed numerically in their vastly different compressibilities.

1.5 Isothermal and adiabatic processes. Dissipative vs. non-dissipative transformations redux

An isothermal volume change is a dissipative transformation. To see why, consider what must happen in order for the temperature of a gas to remain constant as it is compressed. If heat were not allowed to escape the system, then the temperature of the gas would increase during compression. In order for the temperature to remain constant heat must be allowed to leave the system, and energy dissipation takes place.

Let us repeat our thought experiment more carefully. First, we imagine a system that is perfectly insulated to heat transfer, such that all of the mechanical energy that we transfer into the system by compressing it is stored in the system, where it goes to increase its *internal energy*. Internal energy is a thermodynamic variable of a kind called *state function*, which we will define more precisely later in this chapter. I will symbolize it by E and, as a first approximation, we may think of temperature as a measure of the internal energy content of a system. The increase in temperature that we observe in the first part of our thought experiment measures the increase in its internal energy, which, if the system's thermal insulation is perfect and the process is quasi-static, must exactly match the amount of work that was performed on the system. *As long as no heat leaves the system, no energy dissipation takes place if the compression is quasi-static.* Energy dissipation occurs only when heat is exchanged between the system and its environment. A system that is perfectly impervious to heat transfer is called an *adiabatic* system and a process that takes place without exchange of heat is an adiabatic process. Although a perfectly adiabatic process is an abstraction, many natural processes entail heat exchanges that are small enough that they can be considered to be adiabatic to a good approximation. Examples include convection (e.g. in a planet's core, mantle, oceans and atmosphere) and mechanical wave propagation (seismic waves and sound). What is common to all of these processes is that the rate of heat diffusion is negligible compared to the rate at which mechanical energy is exchanged – this topic will be explored further in Chapter 3.

Returning to our thought experiment, we now repeat it with a system that allows heat to move freely across its boundaries, and we compress it quasi-statically once again. In order for the compression to be isothermal, the mechanical energy that we perform on the system must be allowed to leave the system at the same rate as it is being added to it. This energy leaves the system as heat, so that during an isothermal and quasi-static compression *all of the mechanical energy is dissipated as heat*. As we shall see, this last statement is in fact true only for ideal gases, for which internal energy is not a function of volume. Internal energy varies with volume in real substances, but a slightly modified version of that statement is true in general: isothermal processes are not just dissipative processes, they are the processes that result in the maximum possible heat dissipation.

1.6 Elastic energy

The work associated with an adiabatic volume change is calculated by using the adiabatic bulk modulus, K_S, to integrate equation (1.19), where K_S is defined as:

$$K_S = -V \left(\frac{\partial P}{\partial V} \right)_S.$$

(1.32)

The subscript S denotes a special type of adiabatic transformation during which a thermodynamic variable known as entropy remains constant. The meaning of this, and the need to distinguish between different types of adiabatic processes, will become clear in Chapter 4.

The adiabatic bulk modulus is one of a set of parameters that describe the elastic behavior of materials. A substance is said to deform elastically if: (i) the magnitude of the deformation, or *strain*, is a linear function of the applied *stress* (force per unit area) and (ii) the deformation is reversible, in the sense that the body recovers its initial shape and size once the stress is removed. A perfectly elastic deformation is also reversible in the sense of being non-dissipative: when stress is released the substance returns the same amount of mechanical energy that was used to cause the deformation. If energy is dissipated then by the law of energy conservation the amount of mechanical energy returned during the process of *elastic rebound* will be less than the amount of mechanical energy used to cause the deformation.

The elastic behavior of a material in general is described by a parameter called the elastic modulus, λ, defined as:

$$\lambda = \frac{stress}{strain}.$$

(1.33)

Stress is force per unit area and strain is the ratio of the change in the volume, length or shape of a body (caused by stress), to the value of the corresponding variable in the original, pre-strained state. Stress has dimension of pressure, and in fact pressure is one particular type of stress (also called uniform or isotropic stress). Because strain is defined in such a way as to be dimensionless, the dimension of λ is also that of pressure. Materials can have three distinct elastic moduli, that measure different properties of a substance. All substances (solids, liquids and gases) have an adiabatic bulk modulus, K_S, that relates adiabatic pressure changes to the corresponding relative changes in volume. The ratio dV/V is the volumetric strain, and dP is the stress responsible for this strain, so that the definition of adiabatic bulk modulus given by equations (1.32) is consistent with the general definition of elastic modulus given in equation (1.33).

Fluids can undergo elastic volume changes, but they cannot stretch nor shear elastically over time- and length-scales characteristic of planetary processes. The adiabatic bulk modulus is thus the only elastic modulus that can be defined for fluids. Solids, in contrast, are also able to deform elastically by stretching and by shearing, in addition to undergoing elastic volume changes. This fundamental difference between solids and fluids can be understood on the basis of their microscopic structures. The atoms in a solid are arranged in a crystalline lattice and are held in well-defined positions in the lattice by a balance of electrostatic forces acting among neighboring atoms – these forces are what we call chemical bonds. When a solid is stretched or sheared, the bonds are distorted but, up to a point, they can hold and will bring back the atoms to their equilibrium positions if the stress is removed. Elastic strain is the macroscopic expression of the microscopic behavior of the atomic bonds in a crystal. For solids, then, one can define two other elastic moduli in addition to the bulk modulus. *Young's modulus, E*, measures relative elongation as a function of applied tensile stress (i.e. stretching). *Shear modulus, μ*, relates shear stress to shear strain (defined as a change of shape at constant volume). Fluids, in contrast, do not have crystalline structure and interatomic attractive forces are weaker than in a crystal (and altogether nonexistent in an ideal gas). When a fluid is sheared its atoms change position without preserving any memory of where they were before the stress was applied, so that there is no restorative force that would cause elastic rebound. The macroscopic expression of this behavior is that fluids flow under shear, rather than deforming elastically. There are geological environments, however, in which the distinction between solids and fluids becomes blurred. For example, mantle convection and glacier flow are two situations in which crystalline solids flow under shear stress, rather than undergoing elastic strain.

Exact calculation of the energy stored during a generalized elastic deformation is mathematically complex and computationally intensive (see, for example, the classic text by Malvern, 1969), but an order of magnitude estimate is straightforward and useful in many planetary sciences applications. Consider an elastic body of characteristic linear dimension x. Application of a force F causes this dimension to change by an amount $\delta x \ll x$. The characteristic stress on the body is given by $\sigma \approx F/x^2$, and the strain by $\varepsilon \approx \delta x/x$, so that:

$$\lambda = \frac{\sigma}{\varepsilon} \approx \frac{F}{x \delta x}. \tag{1.34}$$

For a perfectly elastic material, i.e. one that can undergo non-dissipative deformation, the stored elastic energy, U_c, must equal the work performed in accomplishing the deformation. Thus:

$$d\boldsymbol{U}_c = F\,d(\delta x) = \lambda\, x\, \delta x\, d(\delta x). \tag{1.35}$$

Because $\delta x \ll x$ we consider x to be constant and integrate with respect to δx, as follows:

$$\boldsymbol{U}_c = \lambda x \int \delta x\, d(\delta x) = \frac{1}{2}\lambda x\,(\delta x)^2 = \frac{1}{2}\lambda x^3 \varepsilon^2. \tag{1.36}$$

The elastic energy stored per unit volume, U_c, is then given by:

$$U_c \approx \frac{1}{2}\lambda\varepsilon^2. \tag{1.37}$$

For stretching of a thin elastic body, such as a wire or spring, this formula is exact, with $\lambda = E$ (Young's modulus). In the case of shear strain of elastic materials the formula yields only an order of magnitude approximation with $\lambda = \mu$, the shear modulus.

Planetary lithospheres store vast amounts of elastic energy as they bend and buckle under shear stress. A fraction of this energy is released as seismic waves during earthquakes, ultimately to be dissipated as heat: seismic waves are adiabatic as a first approximation, but, as all natural processes, they are not perfectly adiabatic and energy dissipation takes place. Changes in ground elevation accompanying an earthquake represent the transformation of another fraction of the stored elastic energy to gravitational potential energy. The source of this energy is not the lithosphere itself, however, but the planet's internal heat. Planetary lithospheres act as energy storage and transfer mechanisms, much as enormous clock springs. Elastic energy is also stored when a planetary body is deformed by tidal forces, and its dissipation is in this case a potentially important source of planetary internal energy, which we will discuss in Chapter 2.

1.7 Two complementary descriptions of nature: macroscopic and microscopic

The definition of mechanical work, equation (1.1), makes no requirements as to the nature of the force that is responsible for transfer of mechanical energy. In the present-day universe there are four distinct forces: gravitational forces, electromagnetic forces, and strong and weak nuclear forces. By "distinct" we mean that each of these forces arises from a different property of matter, and each of them is described by its own physical law. The first two are the most familiar ones. Gravity arises from an object's mass and is described in classical physics by equation (1.2); electromagnetic forces arise from an object's electric charge and also follow an inverse square law. There are good reasons to believe that, at very high energy levels (think of this as extremely high temperatures, such as those that prevailed in the very early Universe), the four forces may become indistinguishable, but this is something that need not concern us as planetary scientists.

So far I have made explicit mention of the importance of gravitational forces in planetary processes (e.g. Section 1.3), and I have hinted at the relevance of electrostatic forces as the explanation for the elastic behavior of solids. There is a fundamental difference between these two examples, and we must now place this discussion on firmer ground. We begin by stating explicitly what is the difference between the ways in which we addressed gravitational potential energy on the one hand, and elastic energy in a solid on the other. In the first case, we used the actual gravitational force to define and calculate potential energy. In the other case, we defined macroscopic properties of the material (the elastic moduli) and then explained these properties in terms of electrostatic forces acting at the atomic level. The description of elasticity in terms of atoms, crystalline structure and electrostatic forces may be helpful, but is not required in order to have a full quantitative description of elastic strain.

The elastic modulus is an example of a macroscopic property. Macroscopic properties provide a phenomenological description of the behavior of a system. By this we mean that we measure and understand the behavior of the system from macroscopic observations (the phenomena that we can observe), without regard to the ultimate physical mechanism that explains the observed behavior. In this example, the physical mechanism can only be understood from a microscopic description of the system at the atomic or molecular level. Many natural processes besides elasticity can be understood from these two distinct and complementary points of view, the macroscopic or phenomenological and the microscopic or atomic. Both descriptions are correct and useful, and combining both views usually allows us to gain a deeper understanding of a process. The two descriptions must, of course, be mutually consistent. Thermodynamics is the phenomenological description of energy conversion processes and of many of its consequences, including chemical equilibrium. The complementary microscopic description is provided by statistical mechanics. Although we will not encounter formal statistical mechanic derivations in this book, I will introduce some of the key concepts of statistical mechanics and use them to understand the physical foundation of some thermodynamic concepts and relationships that may otherwise appear obscure or arbitrary.

In contrast to our discussion of elasticity, where complementary macroscopic and microscopic descriptions are possible, our description of gravitational processes is purely phenomenological and we do not seek a microscopic description of, for example, gravitational potential energy. Why this difference? The answer rests partly on the fact that gravitational and electromagnetic forces have vastly different magnitudes. Physicists describe this difference by stating that gravity is exceedingly weak (by about 40 orders of magnitude) compared to electrostatic force. This is not to say that physicists are not interested in a microscopic description of gravity – they are, but this is still one of the major unresolved problems of physics. As planetary scientists, we may recast the distinction between gravitational and electrostatic forces by saying that there is commonly little overlap between the respective scales of length and mass over which each of these two forces is the dominant one. Gravity is negligible compared to electrostatic forces for distances and masses such as those characteristic of atomic and molecular structure (e.g. the structure of a crystal or the behavior of a fluid at the molecular level), in other words, what we call the microscopic description of nature. Conversely, gravity is the dominant force when we consider masses and length scales such as those that are typical of our macroscopic view of nature. This is certainly true of masses of the order of planetary bodies, but also of much smaller systems, down to, for example, grains of sand in a clastic sediment – for what is the force that drives sedimentation? There are macroscopic planetary environments in which electrostatic forces are important, however. A simple example is that of flocculation of clay particles.

At length scales much smaller than those of individual atoms, such as those characteristic of the atomic nucleus, both gravitational and electrostatic forces become negligible compared to the strong nuclear force. Otherwise, how could a nucleus made up of a large number of equally charged protons be stable? The strong nuclear force, which is responsible for binding protons and neutrons in atomic nuclei, is involved in the liberation of energy by nuclear processes. One example of this is the conversion of mass to electromagnetic energy by nuclear fusion in stellar cores. Another example is spontaneous radioactive decay, in which mass is converted to either kinetic energy of subatomic particles (during alpha and beta decays) or electromagnetic energy (gamma radiation). Radioactive decay is of critical importance in the energy budget of many planetary bodies.

1.8 Energy associated with electric and magnetic fields

1.8.1 Electrostatic forces

The magnitude of the electrostatic force, $|\bar{f}_e|$, between two electric charges, q_1 and q_2, separated by a distance x in vacuum is given by Coulomb's law:

$$|\bar{f}_e| = \frac{1}{4\pi\epsilon_0} \frac{|q_1||q_2|}{x^2},$$

$$(1.38)$$

where ϵ_0 is a constant called the permittivity of free space (Appendix 1), and the SI unit of charge is the coulomb, symbolized C. The intensity of the electric field, E, is the force exerted per unit of charge (e.g. in units of $N\,C^{-1}$). For a point charge of magnitude q in vacuum it is given by:

$$E = \frac{q}{4\pi\epsilon_0 x^2}. \tag{1.39}$$

Coulomb's law is formally identical to Newton's law of gravitation. For instance, compare the equations for the gravitational and electric field intensities, g and E (1.7) and (1.39). Both field intensities decrease with the square of distance, and both are proportional to the magnitudes of the property that generates the field (mass or electrical charge). A crucial difference between the two physical laws is that gravity is always attractive, whereas electrostatic force can be attractive or repulsive, depending on whether the charges are different or alike, respectively. This is the reason why equation (1.38) uses the absolute value of the charges, $|q|$, whereas no such specification is needed in Newton's law of gravitation. Another important difference is that all materials behave identically relative to gravity, so that the universal gravitation constant, G, is unique. In contrast, different materials behave differently in the presence of an electric charge, so that if the charges are separated by any medium other than vacuum the constant ϵ_0 must be replaced by ϵ, the permittivity of the material in question. For all materials, $\epsilon > \epsilon_0$, so that the force between electric charges is maximum in vacuum.

Electrostatic and gravitational forces are also of vastly different magnitudes, as we mentioned in the previous section. We quantify this statement in the following Worked Example.

Worked Example 1.4   Relative magnitudes of gravitational and electrostatic forces

Let us compare the relative strengths of gravitational and electrostatic forces at atomic length scales and planetary length scales. From Newton's law, equation (1.2), and Coulomb's law, equation (1.38), we derive the ratio of the gravitational force $|\bar{f}_g|$ to the electrostatic force $|\bar{f}_e|$ between two bodies with masses m_1 and m_2, carrying electric charges q_1 and q_2 and immersed in a medium of permittivity ϵ:

$$\frac{|\bar{f}_g|}{|\bar{f}_e|} = 4\pi\epsilon G \frac{m_1 m_2}{|q_1||q_2|}. \tag{1.40}$$

Because gravitational and electrostatic forces are both described by inverse square laws this ratio is independent of the separation between the objects. It depends only on their mass to charge ratio and on the permittivity of the surrounding medium. Given that electrons are the smallest stable charged particles (they carry the unit of electric charge) we can use the mass and charge of the electron to get an idea of the intrinsic difference between gravitational and electric forces. Substituting the appropriate values for the electron (Appendix 1), and taking $\epsilon = \epsilon_0$ (i.e. electrons in vacuum), we find:

$$\left(\frac{|\bar{f}_g|}{|\bar{f}_e|}\right)_{electron} \approx 2.4 \times 10^{-43}. \tag{1.41}$$

In other words, gravity is some 42 orders of magnitude weaker than the electrostatic force! A major unsolved problem of physics is why this difference is so vast (see, for example, Randall, 2007). As planetary scientists, we can use this result to reassure ourselves that when we study planetary systems and materials at the microscopic level we can neglect gravity.

But if gravity is so much weaker than electrostatic forces, then under what conditions does gravity become the dominant force? Let us consider two bodies of identical mass, m, carrying identical charges, q, and immersed in a medium of permittivity ϵ. We make $|\bar{f}_g| = |\bar{f}_e|$ and we obtain from equation (1.40):

$$q = (4\pi\epsilon G)^{1/2} m. \tag{1.42}$$

We seek an estimate of the characteristic size, λ, of a particle for which this equality is likely to be satisfied. For particle diameters greater than λ gravity will prevail over electrostatic forces (although of course the size has to be considerably greater than λ for the difference to be significant). Let each of the bodies be a sphere of radius λ and density ρ. The mass of each sphere is given by:

$$m = \frac{4}{3}\pi\lambda^3\rho. \tag{1.43}$$

The *dielectric strength* of a material is defined as the intensity of the electric field, E, under which the material breaks down and starts conducting electricity. For example, lightning occurs when the dielectric strength of air is exceeded. In our example, the dielectric strength of the medium separating the spheres is what controls the maximum amount of charge that can be stored in them. We next state without demonstration the following relationship between the electric charge q in a sphere of radius λ and the intensity of the electric field, E, at the surface of the sphere

$$E = \frac{q}{4\pi\epsilon\lambda^2}. \tag{1.44}$$

This result follows from a theorem of vector calculus known as Gauss's theorem (see, for example, Sokolnikoff & Redheffer, 1966, p. 397). It says that the charge behaves as if it were concentrated in the center of the sphere (compare equation (1.39)). The electric field is maximum at the surface of the sphere and falls off away from it following the inverse square law. If the sphere is immersed in a medium of dielectric strength k_m, then it can sustain a maximum electric field $E = k_m$, and can thus store a maximum q of:

$$q = 4\pi\epsilon\lambda^2 k_m. \tag{1.45}$$

Substituting equations (1.43) and (1.45) in equation (1.42) and solving for λ:

$$\lambda = \frac{3\epsilon}{\rho(4\pi\epsilon G)^{1/2}} k_m. \tag{1.46}$$

Let us assume that the medium separating the spheres is a gas at low pressure ≤ 1 bar. The permittivity of gases at low pressure does not differ significantly from the permittivity of vacuum, so we make $\epsilon = \epsilon_0$. Air at 1 bar has a dielectric strength $k_m \approx 3 \times 10^6$ N C^{-1}. This value decreases with decreasing pressure, attains a minimum at pressures of about 0.01 bar, and then increases again approaching a value of $\sim 10^9\,\mathrm{N\,C^{-1}}$ for very rarefied gases, such as those in interplanetary space (see, for example, https://commons.lbl.gov/display/ALSBL6/Dielectric+strength+of+air). Taking a characteristic value of ρ for rocks of $3000\,\mathrm{kg\,m^{-3}}$, we get values for λ ranging from 300 m for air at 1 bar to 100 km for the interplanetary medium. In between, there is a range of gas pressures (≈ 0.01 bar) for which λ becomes very small, perhaps of the order of millimeters.

These numbers mean that for planetary bodies larger than about 100 km electrostatic forces can never be greater than gravitational forces, because the charges that would be required to produce such forces would be dissipated in electrical discharges. For rock bodies in an atmosphere such as Earth's, the calculated size limit is about 300 m. In practice the crossover lengths must be much smaller than these, perhaps by several orders of magnitude, because even if the dielectric strength of the medium is not approached there are few processes in planetary environments that can generate electrical charges of the high magnitudes that would be required to compete with gravitational forces.

1.8.2 Atomic bonding

Electrostatic forces are responsible for atomic bonding. Charged particles in an electrostatic field have electrostatic potential energy, U_e, which accounts for interaction between two oppositely charge particles separated by a distance r in vacuum that we write as:

$$U_e = \frac{1}{4\pi\epsilon_0}\frac{q_1 q_2}{r}. \tag{1.47}$$

This equation is analogous to (1.6), for gravitational potential energy. As in that case we set $U_e = 0$ at infinity, but the sign convention is opposite to that of gravitational potential. This is convenient because electrostatic forces can be both attractive and repulsive (e.g., Griffiths, 1999, p. 90–96), but it makes no difference for the present discussion.

In an ionic compound, such as the mineral halite, the crystalline structure is held together by electrostatic attraction between oppositely charged ions. The *lattice energy* of an ionic crystal is defined as the energy released when the free ions in a gas phase come together to form the solid (e.g., Holbrook *et al.*, 1990). By "free ions" we mean an ideal situation in which the ions are infinitely distant in vacuum. If only one ion each of sodium and chloride were involved, the lattice energy of sodium chloride would be the electrostatic potential energy U_e (given by equation (1.47)) for a Cl^- anion and a Na^+ cation separated by the interatomic distance in halite, r. Note that because the ions have opposite charges $U_e < 0$. This is the amount of electrostatic potential energy that was "lost" when the ions moved from infinity ($U_e = 0$) to their equilibrium position. Of course, the energy is not lost but converted to some other type of energy. In this case, the electrostatic potential energy of ions, which is a microscopic property of the system, is dissipated as heat when the crystal forms. This heat is what we call the lattice energy of the crystal, which is a macroscopic property.

In practice, a calculation based on equation (1.47) does not yield the correct value of lattice energy because one must consider not just the force between individual ions (or, in the language of physics, *point charges*) but rather interactions among electrostatic fields arising from a distribution of point charges. Each anion and cation in the structure of halite interacts electrostatically not just with one nearest neighbor of the opposite charge, but with many ions of both equal and opposite charges, located at different distances. If the crystalline

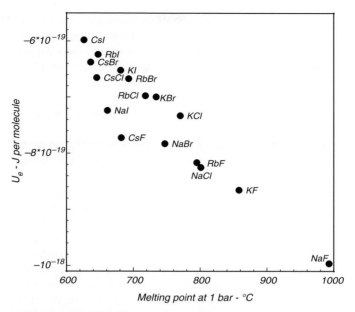

Fig. 1.5   Correlation of single-pair interatomic electrostatic potential energies (U_e) with melting points of alkali halides (melting points from *CRC Handbook of Chemistry and Physics*).

structure of an ionic compound is well known, then its lattice energy can be calculated on the basis of equation (1.47) and a set of geometric parameters known as Madelung constants, that account for interactions among distant ions (see Moody & Thomas, 1965). A simpler approach is Kapustinskii's equation (Kapustinskii, 1956; Moody & Thomas, 1965), which is based on equation (1.47) and a single constant that depends on the number of atoms per formula unit and that essentially sums up the contributions of all significant Madelung constants. Both of these approaches entail the calculation of a macroscopic property, the lattice energy, from considerations of the microscopic structure of the system.

Even if equation (1.47) does not yield exact lattice energy values, the general validity of this discussion can be gauged by comparing values of U_e calculated with equation (1.47) to the melting points of simple ionic compounds (Fig. 1.5). As anions and cations get larger, the value of r increases and the absolute value of U_e decreases. For example, given that ionic radii of both halogens and alkali metals increase with increasing atomic number, the electrostatic potential energy calculated with equation (1.47) should become a negative number of smaller absolute magnitude for alkali halides of progressively higher atomic number combinations. NaF has the shortest atomic bond among all the halides shown in Fig. 1.5, so that its formation releases the greater amount of electrostatic potential energy. NaF should thus have the strongest atomic bond, or, in other words, the highest melting point. This is indeed the case (Fig 1.5). Melting points are negatively correlated with U_e, as we should expect from the fact that formation of shorter bonds releases a greater amount of electrostatic potential energy, so that they require higher energy to be broken.

Pure ionic compounds are rare among planetary materials. In particular, silicates contain a complex and variable combination of covalent, ionic and van der Waal's bonds. Definition of lattice energy in such compounds is less straightforward (e.g. Glasser & Brooke Jenkins, 2000; Yoder & Flora, 2005), but chemical bonding in them is nonetheless a consequence of the fact that, in order to break apart their crystalline structure, it is necessary to perform work against electrostatic forces. Electrostatic forces are also responsible for the divergence of the behavior of real gases from that predicted by the ideal gas EOS. For instance, real gases are able to condense as liquids but ideal gases are not. Electrostatic forces are responsible for this difference.

1.8.3 Magnetic forces

Coulomb's law describes the force between static electric charges. If electric charges are in motion, which is what we call an electric current, an additional force arises between moving charges. This is what we observe as a magnetic force. The magnetic field generated by an electric current is described by a mathematical equation known as Biot–Savart law, which is more complicated than the equations that describe the gravitational and electrostatic fields. I will not present this equation explicitly (see, for example, Griffiths, 1999), but I will state two important properties of the magnetic field, symbolized by **B**. First, the intensity of the field generated by a current flowing in an electrical conductor is directly proportional to the intensity of the current (amount of charge moving per unit of time). Second, the orientation of the magnetic field is perpendicular to the current direction.

Magnetic fields act only on moving electric charges. Stationary electric charges, or particles with no electric charge, do not interact with a magnetic field and are subject to no magnetic force. A moving charge is subject to a magnetic force that has a magnitude proportional to: the intensity of the magnetic field, the magnitude of the electric charge, and the component of the charge's velocity *perpendicular to the field*. The magnetic force, described mathematically by the Lorentz force law, is oriented perpendicular to the direction of motion of the charge, i.e. the current direction (Fig. 1.6). If a charged particle moves in a direction parallel to that of the magnetic field then there is no magnetic force.

Because of the way in which magnetic forces act on moving charges, a stationary magnetic field (one that does not change, in intensity nor orientation with time) performs no work. This is so because the magnetic force is always perpendicular to the displacement of the particles that feel the force (Box 1.1). In principle, then, no energy would appear to be required in order to maintain a magnetic field, but there are some hidden liabilities here. In the first place, the magnetic field exists as a result of an electric current and all electrical conductors

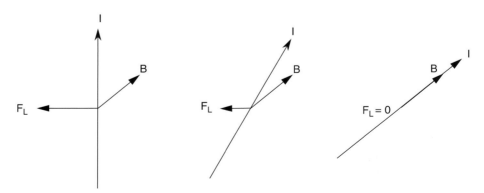

Fig. 1.6 Relationship between the magnetic field, B, the electric current, I, and the Lorentz force, F_L. When B and I are parallel $F_L = 0$.

have a finite resistance that dissipates some of the electrical energy in the current (kinetic energy of moving charges). This particular type of energy conversion (electrical energy to heat) is called *ohmic heating* (because Ohm's law relates current intensity and electrical resistance to energy dissipation). Sustaining a magnetic field requires a constant supply of electrical energy that is dissipated as heat. Secondly, the Lorentz force is perpendicular to the direction of motion of charges (a microscopic concept) but not necessarily to the direction of motion of macroscopic parcels of the conductor that carries the charges. If the conductor moves in response to the Lorentz force then work is certainly being performed. The energy must come from the electric current that generates the magnetic field, i.e. the intensity of the current must increase in order to balance the work performed by the Lorentz force. The magnetic field acts as a transfer medium for this energy.

All planets in the Solar System with the exception of Venus and Mars have magnetic fields (Pluto is not a planet). Even if a planetary magnetic field performed no work, and as we shall see this is not the case, its existence implies that there must be a source of energy that generates the electric current responsible for the magnetic field, and that at least some of this energy is dissipated by ohmic heating. There are good reasons to hypothesize that the origin of planetary magnetic fields is circulation of electrical currents in the deep interiors of the planets, and that the currents are generated by a process that is described as a self-excited planetary dynamo (Bullard & Gellman, 1954). The details of this process are fiendishly complex and a full discussion is beyond the scope of this book – see for example Buffett and Bloxham (2002), Jones *et al.* (1995), Kuang and Bloxham (1997), Olson *et al.* (1999).

The existence of planetary dynamos is based on the fact that there is an inverse to Biot–Savart law: if a material that contains free electric charges (e.g. electrons in an electrical conductor) moves in a magnetic field, then an electric current will flow in the conductor, with an intensity proportional to the intensity of the magnetic field. The energy that appears as electrical energy does not come from the magnetic field, but from whatever is the source of the force that moves the conductor in the magnetic field. The explanation for planetary magnetic fields is that electrical conductors move in the planet's magnetic field, inducing electric currents which in turn generate the magnetic field – hence the term "self-excited". The nature of the electrical conductor varies among different planets. It is likely to be molten metal in the Earth's core, pressure-ionized hydrogen in Jupiter and Saturn and electrolyte-rich aqueous solutions in Uranus and Neptune.

Planetary dynamos would come to a stop, and planetary magnetic fields would collapse, in the absence of an energy source that keeps the electrical conductor moving. That energy source is heat, so we must look for processes that can convert thermal energy to mechanical energy. One such process, and the one that is thought to be responsible for planetary magnetic fields, is thermal convection. We discuss convection in Chapters 3 and 4. For now we notice that one of the outcomes of convection is to transform thermal energy to kinetic energy. In an electrically conductive layer, such as the Earth's core, this kinetic energy is dissipated by a combination of processes (Fig. 1.7). Some of it is dissipated as heat by friction in the convecting medium – this process is called viscous heating and occurs in any convecting material, whether or not it is electrically conductive. The rest of the kinetic energy is converted to electric current (i.e. electric energy) and some of it is dissipated by ohmic heating. The intensity of the current, and hence the rate of conversion of kinetic energy that is required to sustain it, is a function of the magnitude of the work performed by Lorentz forces. Some of this work is performed by the planet's magnetic field outside of the electrically conductive layer in which the magnetic field is generated, and dissipates

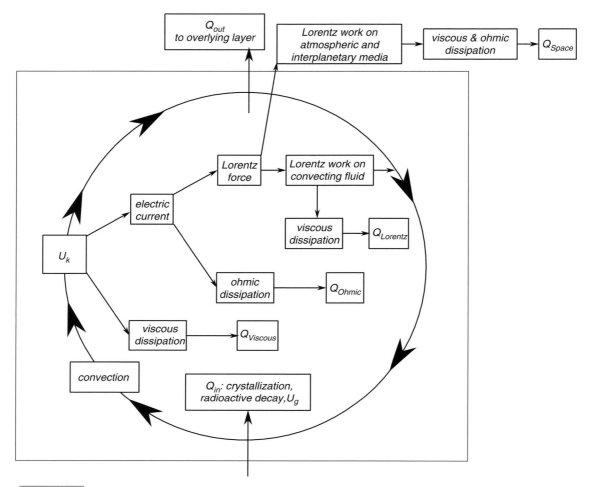

Fig. 1.7 The energetics of planetary magnetic fields. The box is the conductive layer in which the field is generated. The layer absorbs thermal energy (Q_{in}) from sources such as crystallization and radioactive decay, and expels thermal energy (Q_{out}) to the overlying cooler environment. U_k is kinetic energy of the moving conductor. This energy is dissipated by viscous and ohmic heating both within the convective layer and outside of it.

energy, for instance, by heating of conductive plasma in the planet's atmosphere or in the interplanetary environment. The fraction of energy that is dissipated in this way is probably negligible. Most of the work done by Lorentz forces translates to motion of the conductive material that generates the magnetic field and is ultimately dissipated by viscous heating.

    Suppose that no energy is dissipated externally to the layer in which the magnetic field originates, i.e. we somehow manage to eliminate viscous and ohmic heating in the atmosphere and space environment (see Fig. 1.7). Then a combination of viscous and ohmic dissipation would return all the kinetic energy of convection as heat to the same thermal reservoir from which heat was derived to drive convection in the first place. This mechanism would appear to be capable of operating indefinitely without influx of energy. There is nothing in the law of conservation of energy that would prevent it. And yet such a perpetual motion machine is impossible. The Second Law of Thermodynamics requires a thermal gradient in order for conversion of heat to mechanical energy to take place, which means that the convective layer must lose heat to its environment, or convection will stop.

1.9 Thermal energy and heat capacity

Heat and thermal energy are not the same. The relationship between the two concepts parallels that between mechanical energy and work. Heat is a quantity that measures the exchange of thermal energy between systems, or parts of a system, that are at different temperatures. I have taken for granted that thermal energy, and heat, are equivalent to other types of energy, and that they are measured in the same units (Joules in the SI system). The validity of this statement is not evident *a priori*, and is based on experimental observations, beginning with those of James P. Joule in the mid nineteenth century.

Given that heat flows down a temperature gradient, and that heat flow is a transfer of thermal energy, it follows that temperature is an indicator of the thermal energy content of a system. It is important to emphasize: temperature is not energy, but it must vary directly with thermal energy content. A somewhat imperfect analogy is the position of a body in a gravitational field, e.g. elevation above the surface of a planet. Elevation is not energy, but it is related in a direct way to potential energy (equation (1.6)). Left alone, a body will fall to a lower elevation and transfer potential energy, via work of the gravitational force, to kinetic energy. A body at high temperature relative to another one will transfer thermal energy to the latter, via heat flow.

Quantifying the relationship between temperature and thermal energy begins with the definition of *heat capacity*, symbolized by C. The heat capacity of a system is the ratio of the heat absorbed by the system to its temperature increase. Because heat capacity is generally not a constant, the definition is cast in differential form, i.e.:

$$C = \frac{dQ}{dT},$$ (1.48)

where dQ symbolizes heat absorbed by the system. The sign convention is important and must be emphasized: $dQ > 0$ means that the system absorbs heat. Heat capacity is therefore always a positive quantity. Definition (1.48) is incomplete, because, unless one precisely specifies under what conditions heat transfer takes place, the value of C is not well defined. Consider a system made up of a fixed amount of liquid water. If the system absorbs the same amount of heat at constant volume (e.g. the water is held in a perfectly rigid container) and at constant pressure (the container is perfectly flexible and transmits atmospheric pressure), the temperature increase will be greater in the former case than in the latter. This is so because, if the volume is not allowed to change, then all the heat becomes thermal energy. On the other hand, volume increases at constant pressure, so that some of the heat becomes expansion work. In this case the increase in thermal energy, and hence in temperature, will be less than in the constant volume case. Consider now the same system in the flexible container but at the liquid–gas phase transition temperature (i.e. the boiling point of water). As long as both a liquid and a gas phase are present, the heat capacity is infinite, for in this case $dT = 0$ for finite values of dQ.

We must specify the exact conditions under which the heat transfer process takes place. This applies to all thermodynamic processes, and is known as *imposing constraints*. A possible constraint is that no work be performed during heat transfer, so that all heat becomes thermal energy. We symbolize the no work constraint with the subindex $\{W\}$, so that:

$$C_{\{W\}} = \left(\frac{d\mathbf{Q}}{dT}\right)_{\{W\}} \tag{1.49}$$

means heat capacity measured under conditions such that no work is performed. If the only kind of work that we are concerned with is expansion work then the constraint $\{\mathbf{W}\}$ becomes simply a constant volume condition (equation (1.19)). We define the constant volume heat capacity for a homogeneous system as follows:

$$C_V = \left(\frac{d\mathbf{Q}}{dT}\right)_V. \tag{1.50}$$

If the system is able to perform other types of work, such as gravitational or electrostatic work, elastic shear work or work by a Lorentz force in a magnetic field, then the $\{\mathbf{W}\}$ constraint must be defined differently.

If a system absorbs heat under conditions such that no work is performed then its energy content must increase by an amount equal to the heat absorbed. This energy content is measured by the thermodynamic state variable (or state function) *internal energy,* which we will symbolize by $\mathbf{E}$ (this symbol is not universally used; some authors use U for internal energy but I reserve U for various types of mechanical energy). The meaning of *state variable* is that, as long as a system is at equilibrium, the value of the variable depends only on the state of the system (as defined, for example, by its pressure and temperature) and *not* on the path that the system followed to reach that state. In other words, given P and T, the value of the internal energy of a system at equilibrium is unique. State variables preserve no memory of the system's history.

Internal energy, as all other thermodynamic state variables, is a macroscopic property. Its physical interpretation requires that we discuss the system from a microscopic point of view (Section 1.14), but its macroscopic definition is simple and follows from equation (1.50). For systems for which the only possible type of work is expansion work, and in which no phase transitions nor chemical reactions take place, the heat exchanged at constant volume must be equal to the change in internal energy, so we can rewrite equation (1.50) as follows:

$$C_V = \left(\frac{\partial \mathbf{E}}{\partial T}\right)_V. \tag{1.51}$$

This equation can be taken as the definition of either internal energy or constant volume heat capacity. Which of the two we choose is not important. What matters is that it allows us to calculate things – this is the essence of what an *operational definition* is. Note that this is the definition of internal energy *only* if expansion work is the only possible type of work for the system of interest. If other types of work are possible, then the general definition of internal energy is the following:

$$C_{\{W\}} = \left(\frac{\partial \mathbf{E}}{\partial T}\right)_{\{W\}}. \tag{1.52}$$

The derivative symbols in equation (1.50) have become partial derivative symbols in equations (1.51) and (1.52). This is so because $\mathbf{E}$ is a state function whose value is fixed by the values of the independent variables that define the state of the system. In other words, $\mathbf{E}$ is a function of several variables (see Box 1.3) and the notation in equation (1.51) is an unambiguous indication: (i) that the two independent variables that we are using in this particular case to define the state of the system are temperature and volume, and (ii) that we want to measure the rate of change in internal energy with respect to temperature while holding the system's volume constant. Volume is the *only* variable that is constrained to be constant. Other variables, such as pressure, vary during constant volume heating, but the energetic consequences of changes in these other variables are implicitly built into equation (1.51) (see also Box 1.4).

Box 1.4   **Comments on the notation of thermodynamics**

It has been pointed out that the notation:

$$\left(\frac{\partial y}{\partial x}\right)_z \tag{1.4.1}$$

is unique to thermodynamics and is at odds with standard mathematical usage. Some authors (for example, Truesdell, 1984) have called for the abolition of this supposed abomination, and a recasting of thermodynamics in standard mathematical notation. This would be desirable from a formal point of view, but in my opinion it ignores the physical essence of thermodynamics, or at least of its practical applications. Although the state of a system is defined by the values of only a few intensive variables (at least two) there are other quantities that do not vary independently.

Say that $y = y(x,z)$ is a state function. Geometrically this is a two-dimensional surface that exists in a space of *at least* three dimensions. In thermodynamics the number of dimensions of the embedding space is always more than three, meaning that y is a function not only of x and z, but also of other variables, u, v, w, etc. But once we specify that we are interested in the behavior of y as a function of x and z we lose the freedom to choose the values of these other variables – they are determined by the intersection of the y surface with the corresponding coordinates. The surface $y = y(x,z)$ is only one of many possible surfaces that determine the value of y. We could also have chosen $y = y(v,w)$, and then the values of x and z would be fixed by the intersections of this other surface. The expression:

$$dy = \left(\frac{\partial y}{\partial x}\right)_z dx + \left(\frac{\partial y}{\partial z}\right)_x dz \tag{1.4.2}$$

tells us unequivocally which is the particular y surface that we are considering. The advantages of thermodynamics's peculiar notation will become clear in subsequent chapters.

The partial derivative notation was not used in equation (1.50) because Q is not necessarily a function of any other variable. In fact, an equation such as (1.50) is mathematically sloppy, though useful from a physical point of view. All this equation is saying is that we want to track how absorption of an infinitesimal amount of heat changes the temperature of a system, even if in general there is no function $Q = Q(T,V)$ that we can differentiate to obtain C_V. This is not true of state variables: the function $E = E(T,V)$ exists and is differentiable. This distinction is commonly formulated as one between exact and inexact differentials. The derivatives of state variables are exact differentials, which means that the values of state variables are given by differentiable functions. In contrast, dQ and dW are inexact differentials: they represent infinitesimal amounts of heat transferred or work performed, but there may not exist a function of other thermodynamic variables that may allow us to calculate derivatives of Q and W. A corollary of this statement is that Q and W cannot be state variables – they may not even be functions! This distinction is certainly correct and has further mathematical implications (see, for example, Lewis & Randall, 1961, or Glasstone, 1946) but these will not be pursued here, as they are not essential to our goals.

We now have a precise definition of heat capacity at constant volume. Another heat capacity that is well defined is the heat capacity at constant pressure, C_P, which is defined in terms of heat transferred as:

$$C_P = \left(\frac{d\boldsymbol{Q}}{dT} \right)_P .$$
(1.53)

As with C_V, we wish to recast this expression in terms of a state variable in order for it to be of use in thermodynamic calculations. Internal energy is not a convenient choice, because when heat is absorbed at constant pressure there is a change in volume and hence expansion work takes place. Some of the heat does not become internal energy, but it is not *a priori* obvious how much.

In order to place this discussion on firm ground it is necessary to give a precise mathematical definition of the First Law of Thermodynamics. Before we do that in the next section, there is one additional issue that needs to be clarified. Our definitions of heat capacity and internal energy make no mention of the size of the system. Clearly, the amount of heat that is required to change the temperature of a system by a certain amount depends both on some intrinsic material property of the system and on the size of the system. In thermodynamic calculations we commonly wish to do so independently of system size. One way to do that is to work with material properties and other thermodynamic variables on a molar (per mol) basis. Variables such as molar heat capacity, molar volume and bulk modulus are intrinsic characteristics of a substance or system: they are material properties that are independent of the system's size. A thermodynamic variable or property that is independent of system size is called an *intensive variable*. Molar heat capacity, molar volume, molar internal energy and bulk modulus are examples of intensive variables, and are thus material properties. The total heat capacity of a system, its total internal energy, or its total volume, are examples of *extensive variables*, and are the product of an intensive variable times the size of the system. Pressure and temperature are intensive variables too, but they are not material properties. Intensive variables will be represented with uppercase non-bold symbols (e.g. C_V, E, V) and their extensive counterparts with uppercase bold symbols (e.g. $\boldsymbol{C_V}$, $\boldsymbol{E}$, $\boldsymbol{V}$). The intensive variables P and T, of course, have no extensive counterparts. In some applications we will find it convenient to consider thermodynamic properties per unit mass. These are called *specific* properties, and will be represented with lowercase non-bold symbols, e.g. c_P for the *specific heat capacity* or e for specific internal energy. Molar and specific properties are simply related to one another by the molecular weight of the substance. For example, for constant pressure heat capacities:

$$c_P = \frac{C_P}{w} = \frac{C_P}{\rho V},$$
(1.54)

where w is molecular weight, V is molar volume and ρ is density. I note in passing that Guggenheim (1967) has pointed out the sloppiness of the term "molecular weight", given, among other things, that what this quantity actually refers to is mass. He suggested the more precise term *proper mass*. Unfortunately, it does not appear that this suggestion has gained much acceptance.

1.10 The First Law of Thermodynamics

The First Law of Thermodynamics is the mathematical statement of the law of conservation of energy. If we restrict ourselves to systems and processes in which work and heat are the only pathways for energy transfer, then the statement of the First Law is simply:

$$dE = dQ - dW, \tag{1.55}$$

where dE is an (infinitesimal) change in the internal energy of the system, dQ is the *heat absorbed* by the system and dW is the *work performed* by the system. This sign convention must be clearly understood and adhered-to rigorously. It is not universally used, but it is the one that is most intuitively appealing. It is consistent with the various sign conventions that we have established in previous sections. For example, if a system performs expansion work on its environment in the absence of any heat exchange ($dQ = 0$), then $dW > 0$ and its internal energy decreases. If the system loses heat without performing work ($dW = 0$), then $dQ < 0$ and its internal energy also decreases.

The First Law as expressed by equation (1.55) is not specific about the type of work represented by dW, but it excludes some types of energy. Processes in which there are mass-energy conversions, such as radioactive decay and nuclear reactions, are not included in (1.55). The First Law of Thermodynamics applies to such processes, but the equation has to be modified in order to account for the mass-energy term. In systems that contain non-negligible amounts of (macroscopic) kinetic energy this energy must also be added to equation (1.51) as an extra term.

When solving an energy balance problem one starts by writing down the First Law of Thermodynamics in a form that is appropriate to the nature of the problem at hand. For example, if we restrict ourselves to chemical systems in which only expansion work is possible, then equation (1.55) can be re-written as follows:

$$dE = dQ - PdV. \tag{1.56}$$

Once we have chosen the appropriate expression for the First Law, the next step is to consider whether the constraints of the problem allow any simplification. For example, for a constant volume (*isochoric*) process $dV = 0$, so the First Law of Thermodynamics becomes:

$$dE_V = dQ_V, \tag{1.57}$$

where the subscript V reminds us of the constant volume constraint. One could then be tempted to make the substitution $dE_V = C_V dT$, but before doing so the nature of the system under study must be carefully considered. For a system that does not undergo a phase transition (e.g. melting or vaporization), nor a chemical reaction, this substitution is indeed valid, because in this case all of the heat that is exchanged is manifested as a temperature change. This is called *sensible* heat. Heat associated with a phase transition or with a chemical transformation is not reflected in a temperature change, and is sometimes called *latent* heat. If latent heat is involved in the transformation then the substitution $dE_V = C_V dT$ is not valid.

The First Law applied to a constant pressure (isobaric) process:

$$dE_P = dQ_P - PdV \tag{1.58}$$

leads to the definition of a new state variable. Using internal energy as a thermodynamic state variable for isobaric processes leads to cumbersome equations, because isobaric heat transfer is accompanied by a change in volume and internal energy is generally a function of volume. To see why this must be so, consider a system from the microscopic point of view. In general, there are electrostatic attractions and repulsions among the system's microscopic components (e.g. molecules in a gas, ions in a crystal) and any work performed against those forces during a change in volume becomes electrostatic potential energy, which must be part of the system's internal energy. Internal energy would be independent of volume only if there were no interatomic forces in the system. This is true *only* in ideal gases (Section 1.14). For all real substances E is a function of V. Given the simplicity of equation (1.57), however, we would like to obtain a similarly compact equation that states the First Law of Thermodynamics for isobaric processes. In order to do this, we design a new state variable with the desired properties. This new state variable, called *enthalpy*, is universally symbolized by $\boldsymbol{H}$, and is defined as follows:

$$\boldsymbol{H} \equiv \boldsymbol{E} + P\boldsymbol{V}. \tag{1.59}$$

This trick – inventing new functions to make calculations easier – is a common one in thermodynamics. It may not be *a priori* obvious why we would choose to define enthalpy in this way. You can think of it initially as a useful mathematical device, but it will become apparent that enthalpy has a clear physical meaning. Enthalpy has two important properties. First, given that internal energy is a state variable, enthalpy is a state variable too, as pressure and volume depend only on the equilibrium state of a system. Second, from the definition of enthalpy it follows immediately that it has dimension of energy.

Applying the product rule for differentiation to (1.59) we see that an infinitesimal change in enthalpy is given by:

$$d\boldsymbol{H} = d\boldsymbol{E} + Pd\boldsymbol{V} + \boldsymbol{V}dP, \tag{1.60}$$

which, for an isobaric process ($dP = 0$), simplifies to:

$$d\boldsymbol{H}_P = d\boldsymbol{E}_P + Pd\boldsymbol{V}. \tag{1.61}$$

Substituting equation (1.58) in (1.61):

$$d\boldsymbol{H}_P = d\boldsymbol{Q}_P \tag{1.62}$$

shows that the physical meaning of enthalpy is heat exchanged at constant pressure. The enthalpy change includes a contribution from a change in internal energy and a contribution from expansion work (equation (1.61)). Recalling the definition of heat capacity at constant pressure (equation (1.53)), we see that, for a process in which only sensible heat is involved:

$$C_p = \left(\frac{\partial \boldsymbol{H}}{\partial T}\right)_P. \tag{1.63}$$

1.11 Independent variables and material properties

The state of a thermodynamic system is fully determined once we fix the values of a small number of variables that we can think of as the independent variables, or control variables.

In Chapter 6 we will prove that the state of a homogeneous system of fixed composition at equilibrium is fully defined once we fix the values of two intensive variables. Thermodynamics allows us remarkable freedom for choosing which intensive variables to use as the independent variables. We can use this freedom to our advantage by choosing variable combinations that are best suited to each specific problem. In chemical thermodynamics, and especially in its applications to planetary sciences, pressure and temperature are the two most "fundamental" independent variables, largely because they are the easiest ones to measure, but also because understanding how a system changes as a function of these variables, or how one of these variables changes as a function of the other (e.g. with depth in a planetary body), tend to be some of our chief preoccupations. The molar volume of a homogeneous systems is related to P and T through an equation of state (Section 1.4.3). Given an EOS, we can always express one of the three variables, $P - V - T$, as a function of the other two and substitute as needed in thermodynamic equations.

Equations of state are well known for some substances but not for many others. A material, however, always has an equation of state, even if we don't know what this equation is. This is equivalent to saying that a given substance at a given pressure and temperature has a single and well-defined molar volume, or, equivalently, that molar volume is a state variable. But because EOS for many substances are either unknown or mathematically unwieldy, it is also possible, and often advantageous, to express $P - V - T$ relations as a set of material properties. In practice, there are three material properties that are relatively straightforward to measure and that show up repeatedly in thermodynamic calculations. In Chapters 4 and 8 we will see that this last attribute is not a coincidence. Two of these properties describe the mechanical properties of the material. They are the bulk modulus K, or its inverse, the compressibility, β, and the coefficient of thermal expansion at constant pressure, α. Recall that K (and β) can be defined at either isothermal or adiabatic conditions (Section 1.4.3). In order to complete the thermodynamic characterization of a material we need to describe its thermal properties as well, and this additional information is contained in the heat capacity, either C_V or C_P. We have already given the mathematical definition of K_T (equation (1.21)). We repeat it here along with the definitions of molar heat capacity, C_P, coefficient of thermal expansion, α, and isothermal compressibility, β_T:

$$C_P = \left(\frac{\partial H}{\partial T} \right)_P \tag{1.64}$$

$$K_T = -V \left(\frac{\partial P}{\partial V} \right)_T ; \quad \beta_T = -\frac{1}{V} \left(\frac{\partial V}{\partial P} \right)_T \Rightarrow K_T = \frac{1}{\beta_T} \tag{1.65}$$

$$\alpha = \frac{1}{V} \left(\frac{\partial V}{\partial T} \right)_P . \tag{1.66}$$

Material properties can be derived from the EOS (see Exercises 1.11 and 1.12) and they are not independent of one another (Exercise 1.13).

1.12 Some applications of the First Law of Thermodynamics

1.12.1 Discontinuous phase transitions and latent heat

A phase transition is a change in the physical nature of a system that occurs in response to a change in the values of some of the controlling intensive variables (e.g. temperature or pressure) but without a change in chemical composition. We will see in Chapter 7 that there are different types of phase transitions. The most familiar ones are discontinuous phase transitions. As their name implies, these are phase transitions at which there is an observable discontinuity in physical properties: melting, boiling and sublimation are familiar examples. Latent heat is always absorbed or liberated during discontinuous phase transitions – in fact, the existence of non-zero latent heat is the formal thermodynamic definition of a discontinuous phase transition (Chapter 7).

Suppose that we want to know how much heat is required to convert, isobarically, 1 mol of liquid H_2O at a temperature T_1 to 1 mol of H_2O gas at a temperature $T_2 > T_{vap} > T_1$, where T_{vap} is the temperature at which the liquid to gas phase transition for H_2O takes place for the pressure of interest. Because this is an isobaric transformation, the heat required to effect this transformation equals the change in enthalpy between liquid H_2O at T_1 and H_2O gas at T_2 (equation (1.62)). Calling this total enthalpy change ΔH we can write it as follows:

$$\Delta H = \Delta H_{liquid, T_1 \Rightarrow T_{vap}} + \Delta H_{liquid\ to\ gas} + \Delta H_{gas, T_{vap} \Rightarrow T_2}. \tag{1.67}$$

The first and last terms on the right-hand side of the equation are, respectively, the enthalpy changes of the liquid, from the initial temperature to the vaporization temperature, and of the gas, from the vaporization temperature to the final temperature. They involve only sensible heat, so we make $dH = C_P dT$ (equation (1.63)) and integrate:

$$\Delta H_{liquid, T_1 \Rightarrow T_{vap}} = \int_{T_1}^{T_{vap}} C_{P, liquid} dT \tag{1.68}$$

and:

$$\Delta H_{gas, T_{vap} \Rightarrow T_2} = \int_{T_{vap}}^{T_2} C_{P, gas} dT. \tag{1.69}$$

Note that if pressure were not constant, then the substitution $dH = C_P dT$ would **not** be valid. Heat capacities are not constant. They vary with temperature and, much less strongly, with pressure. In order to integrate equations (1.68) and (1.69) it is necessary to express C_P as a function of temperature, and then integrate this function. This is discussed in Software Box 1.1, and also in Chapter 5. The middle term on the right-hand side of equation (1.67), $\Delta H_{liquid\ to\ gas}$, is the energy associated with the phase transition, which for a liquid to gas phase transition is called the enthalpy of vaporization, symbolized by ΔH_{vap}.

Software Box 1.1   An introduction to *Maple*: calculation of heat capacity integrals and enthalpy of reaction as a function of temperature

Thermodynamic calculations are not difficult, but they can be tedious. In my opinion, there is no point in doing any calculations or routine algebraic manipulations by hand if they can be accomplished much faster and with much less possibility of errors creeping in by relying on symbolic algebra software. Of the several products available, I use *Maple*. Throughout this book I rely on a number of *Maple* procedures that I have written myself. All of the code is available in *Windows* format from www.cambridge.org/patino_douce, from where the files can be downloaded and run.

All you need to do is install *Maple* on your computer. The files can of course be opened and edited. I am a self-taught programmer, however, so it is likely that many readers will find better and more elegant ways of accomplishing the same tasks. If you prefer different software, such as *Mathematica* or *Matlab*, and you are proficient in it, then it will probably be very easy to translate the *Maple* code. If you have never used symbolic algebra software before you may need to spend a few hours learning the basics of how *Maple* works before you attempt to understand and run the procedures that accompany this book. *Maple* can do many things. It can perform numeric calculations. It can perform algebraic manipulations. It can differentiate and integrate functions. It can solve equations and systems of equations, linear and non-linear, algebraic and differential. It can plot functions. It can read and write files. And it is a powerful programming language, so that all of the things that *Maple* does can be part of a program, which is called a *Maple procedure*.

A good introduction to the use of Maple in thermodynamic calculations is to apply it to solve the integral of the heat capacity function, which is needed to calculate the enthalpy change of a chemical reaction at any arbitrary temperature (equation (1.100)). The heat capacity of all substances varies with temperature but there is generally no strong physical basis to predict the form of the function $C_P = C_P(T)$. The approach that is universally used is to measure heat capacity over a range of temperatures (see, for example, Anderson, 2005) and fit the data empirically with a polynomial function. Different functions are in use. The following, which is sometimes called the Shomate equation (see Shomate, 1954; Shomate & Cohen, 1955), appears to work well for minerals and fluids of geological interest:

$$C_P = a_0 + a_1 T + a_2 T^{-2} + a_3 T^{-1/2} + a_4 T^2, \qquad (S1.1.1)$$

where T is temperature in Kelvin and the a_is are empirical best-fit coefficients. Two geologically oriented data bases that use this equation are those of Robie and Hemingway (1995) and Holland and Powell (1998). Many, *but not all*, species listed in the NIST Chemistry WebBook also use this equation. Throughout much of this book I use Holland and Powell's data base, so equation (S1.1.1) is the one that we will initially implement in *Maple*. Holland and Powell truncate the heat capacity function after the fourth term, but there is no harm in programming the full equation and setting $a_4 = 0$ in Holland and Powell's data.

The problem that we wish to solve is to find the value of the following definite integral:

$$\int_{298}^{T} C_P \, dT = \int_{298}^{T} \left[a_0 + a_1 T + a_2 T^{-2} + a_3 T^{-1/2} + a_4 T^2 \right] dT, \qquad (S1.1.2)$$

where T is the temperature of interest, and 298 stands for 298.15 K, the temperature at which thermodynamic data are tabulated. The beauty of *Maple* is that we do not have to deal with any cumbersome algebra and arithmetic. All we need to do is tell *Maple* what is the function that we want to integrate and what the integration limits are, and then ask it to integrate. Out pops a number.

The *Maple* procedure that performs the heat capacity integral is placed in a *package*, which is a file that contains a collection of *Maple* procedures that can be called from other procedures. The name of this file is **th_shomate.mw**. The package contains several procedures. The first one, named cp_sh, calculates the value of the heat

Software Box 1.1   Continued

capacity, i.e., of function (S1.1.1). The second procedure is named `intcp_sh` and calculates the definite integral of the heat capacity equation, i.e., function (S1.1.2). The fourth procedure, named `HT_sh`, calculates the enthalpy at the temperature of interest by adding the heat capacity integral to the enthalpy at 298.15 K (e.g. equation (1.100)). Each of these procedures receives from the calling procedure an array that contains the values of the thermodynamic properties, and a separate variable that contains the temperature in K. The remaining procedures work in the same way but calculate other thermodynamic functions. They will be introduced in later chapters. The procedures in **th_shomate.mw** are placed in a table (one of the many *Maple* data structures – we won't get into that here), and this table is saved into a package, which is an executable file that other *Maple* procedures can call. The last line in **th_shomate.mw** takes care of this, and here is a very important point: packages must be saved in a directory, or library, that must be known by the calling procedure. The name of this directory can be anything you want. I have chosen **c:/thcalc**, so you should create a directory (or "folder") with this name before attempting to run any of the *Maple* procedures from the website. When you download **th_shomate.mw** you can put it in this or any directory you wish. Then open the file and execute it (you can do this in the "Edit" pull down menu). Execution of the file places the **th_shomate** executable package in **c:/thcalc**, and the package is ready to be used by other procedures.

The file **th_template_1.mw** contains a number of commands and procedures that are used by many of the thermodynamic calculation worksheets that we will discuss in this and subsequent chapters. It is important to understand what each of these commands does. I describe them in the order in which they appear in the *Maple* worksheet.

`libname := ...` tells Maple the location of the library where the packages are stored. It must match the name used when the package was created (see above).

`with (th_shomate)` tells Maple to load the th_shomate package.

`with (spread)` tells Maple to load a standard package that enables spreadsheet functions.

`RefStateData := CreateSpreadsheet()` creates a spreadsheet named `RefStateData`, that is used to store standard state thermodynamic properties at the reference conditions, 298.15 K and 1 bar. Each row in the spreadsheet contains data for one chemical species. For now we will only use the first nine columns, but more data columns will be occupied in later chapters. The first column contains correlative numbers that will serve to identify the species (more on this below). The second column stores the name of the species. Each additional column after the second one corresponds to a thermodynamic property, generally at 298K and 1 bar. Successive columns store the values of $\Delta_f H^0_{1,298}$, $S^0_{1,298}$ $\Delta_f G^0_{1,298}$ (the functions S and G will be defined in later chapters) and the a_i coefficients of the C_P function. Enthalpy and Gibbs free energy must be entered in kJoules, all other quantities in Joules. The data can be entered directly in *Maple*, but I find it easier to store the data (e.g. from Holland & Powell, 1998, or some other data base) in a regular spreadsheet such as *QuattroPro* and then copy whatever is needed to *Maple*. Once you have all of the data that you will need for the calculations in the Maple spreadsheet it is best to save it in *Tab Delimited* format by right-clicking anywhere in the spreadsheet and then *Export Data*. It can then be imported directly into *Maple* by performing the inverse operation. The data for this example are stored in tab-delimited format in a file named `spgrt`.

> The next block of statements is a procedure called `load` that loads a one-dimensional array with the thermodynamic data for a species, identified by the number in the first column of the spreadsheet. This is used by other procedures, for example the following.
>
> `deltareax`, that calculates $\Delta_r H^0_{1,298}$, $\Delta_r S^0_{1,298}$, $\Delta_r G^0_{1,298}$ and the $\Delta_r a_i$ coefficients that are used to integrate the $\Delta_r C_P$ equation (see equation (1.100)). These are the differences in thermodynamic properties at the 298.15 K reference temperature (entropy and Gibbs free energy will be covered in later chapters).
>
> Finally, procedure `delH` calculates the enthalpy of reaction at the temperature of interest and 1 bar, $\Delta_r H_{1,T}$ (see equation (1.100)) simply by calling `HT_sh` in the **th_shomate** package with the thermodynamic properties specific to the reaction of interest.
>
> All that remains now is to tell *Maple* what is the reaction that we want to calculate, and where to find the data. We do this by creating a table with two columns. The first column contains the stoichiometric coefficient of each chemical species in the reaction, positive if it is a product, negative for a reactant. The second column contains the row number that identifies the species in the spreadsheet that contains the data (i.e. the number in the first column of the spreadsheet). The example given in **th_template_1.mw** is for the reaction:
>
> $$MgAl_2O_4 + 2Mg_2Si_2O_6 \Rightarrow Mg_2SiO_4 + Mg_3Al_2Si_3O_{12}. \qquad (1.70)$$
>
> Suppose we enter the properties for spinel in row 1, enstatite in row 2, forsterite in row 3 and pyrope in row 4. We name the table that identifies this reaction `spgrt`, and the four entries in the table are: `[1,3], [1,4], [-1,1], [-2,1]`. All we need to do now to calculate the enthalpy change for this reaction at any temperature we wish is to run the procedure `delH`, providing it with the name of the reaction and the temperature (see **th_template_1.mw**).

An isobaric discontinuous phase transition is associated with expansion work, as density changes across discontinuous phase transition. Because pressure is constant, the magnitude of this expansion work, $W_{transition}$, is given by:

$$W_{transition} = P\Delta V_{transition}. \qquad (1.70)$$

The molar volume of liquid H_2O at 1 bar and 373 K is 1.88 J bar^{-1} mol^{-1} (convert this number to density in kg m^{-3}, so as to get a feeling for the relationship between density and molar volume). The molar volume of H_2O vapor at the same pressure and temperature is 3.05×10^3 J bar^{-1} mol^{-1}. The expansion work associated with the phase transition is thus ~ 3 kJ mol^{-1}. From equation (1.61) we have:

$$\Delta E_{vap} = \Delta H_{vap} - W_{transition}. \qquad (1.71)$$

The enthalpy of vaporization of H_2O is 40.7 kJ mol^{-1}. Thus, when H_2O boils its internal energy increases by $\sim$37.7 kJ mol^{-1}. This is about one order of magnitude greater than the expansion work. The increase in internal energy reflects the fact that molecules in the gas state carry significantly more translational kinetic energy than in the liquid state (Section 1.14). The increase in internal energy (i.e. molecular kinetic energy) is a microscopic contribution to the enthalpy of vaporization, that is distinct from the macroscopic expansion work that is performed against the pressure exerted by the environment.

Let us define a variable $\Delta T_{pot} = \Delta H_{vap}/C_P$ with dimension of temperature. We can call this variable a *potential temperature*, because it represents the temperature difference that would be caused by full conversion of latent heat to sensible heat. The heat capacities of liquid water and water vapor at temperatures in the neighborhood of the boiling point are $C_{P,liquid} \approx 75.9$ J K^{-1} mol^{-1} and $C_{P,gas} \approx 37.4$ J K^{-1} mol^{-1}. Thus, for vaporization of water ΔT_{pot} is of order 100–1000 K. This means that the first and last terms on the right-hand side of equation (1.67) are negligible compared to the thermal effect of the phase transition. Water vapor in the terrestrial atmosphere stores a large amount of thermal energy, and is a major factor in driving the planet's weather patterns (more on this in Chapter 4). Enthalpies of crystallization are generally of smaller magnitude than enthalpies of condensation, with ΔT_{pot} of order 10–100 K (Chapter 10). These values are nevertheless large enough to have important effects on the energetics of planetary systems undergoing melting or crystallization. For instance, enthalpy of crystallization of metallic Fe is almost certainly the immediate source of heat that drives convection in the Earth's core, and may also drive convection in large silicate magma chambers (as an aside, I see no reason why a partially molten metallic planetary core cannot be called a magma chamber, except perhaps for tradition). Conversely, the large energy requirement of melting plays an important role in the origin of planetary magmas (Chapter 10).

1.12.2 Adiabatic expansion of gases

The destructive power of pyroclastic eruptions, or of any conventional (i.e., chemical) explosion, results from fast expansion of a gas driven by conversion of internal energy of the gas to expansion work. Although in reality the very fast rate of expansion means that the process is not an equilibrium one, we can at least get an idea of the magnitude of the energy involved by treating the problem as a reversible adiabatic expansion. The First Law of Thermodynamics for an adiabatic process ($dQ = 0$) is:

$$dE = -PdV. \tag{1.72}$$

We equate the energy liberated by a pyroclastic eruption, $W_{pyroclastic}$, to the expansion work of the volcanic gas. Because volume is not constant during this process we cannot make the substitution: $dE = C_V \, dT$, as in general internal energy is a function of volume as well as temperature. We need to consider the total change in internal energy as a function of the partial derivatives of E relative to the variables that we choose as independent variables (Box 1.3). Choosing temperature and volume as the independent variables:

$$dE = \left(\frac{\partial E}{\partial T}\right)_V dT + \left(\frac{\partial E}{\partial V}\right)_T dV = C_V dT + \left(\frac{\partial E}{\partial V}\right)_T dV. \tag{1.73}$$

The energy liberated by a pyroclastic eruption is then given by:

$$W_{pyroclastic} = PdV = -\left[\int_{T_e}^{T_a} C_V dT + \int_{V_e}^{V_a} \left(\frac{\partial E}{\partial V}\right)_T dV\right], \tag{1.74}$$

where T_e, V_e are the eruption temperature and the molar volume of the gas at that temperature, and T_a, V_a are atmospheric temperature and the molar volume of the gas at that temperature. In order to solve the integrals we need explicit expressions for $C_V = C_V(T)$ and for $E = E(V)$. The problem is simplified enormously if we assume that volcanic gases behave as ideal gases, because for an ideal gas C_V is constant and internal energy is a function of temperature only, i.e. $(\partial E/\partial V)_T = 0$. We derive these properties of ideal gases in Section 1.14.1. For an ideal gas, then, **but only for an ideal gas:**

$$W_{pyroclastic} = -C_V(T_a - T_e). \tag{1.75}$$

Because eruption temperature is greater than atmospheric temperature, $W_{pyroclastic} > 0$. According to our sign convention a positive value means that the system (expanding volcanic gas) performs work on its environment (hapless bystanders). Typical eruption temperatures for silicic magmas are $\sim$850 °C, and we can take a typical atmospheric temperature to be 15 °C. The major component of volcanic gases is H_2O, which consists of polyatomic molecules. In Section 1.14 we shall see that for such gases $C_V = 3R$. Substituting these numerical values we get an energy release per mol of erupted volcanic gas of:

$$W_{pyroclastic} \approx 2.08 \times 10^4 \text{J mol}^{-1}. \tag{1.76}$$

As an example, the 1980 eruption of Mt. St. Helen's extruded ~ 1 km^3 of tephra. Assuming a pre-eruptive H_2O content of 3 wt% and a magma density of 2700 kg m^{-3} we see that some 4.5×10^{12} mols of H_2O were erupted. The estimated energy yield of the Mt. St. Helen's eruption is $\sim 9.36 \times 10^{16}$ J.

The yield of nuclear weapons is measured in kilotons (kt), where 1 kt = 4.184×10^{12} J. The 1980 Mt. St. Helen's eruption was thus equivalent (less the high energy electromagnetic radiation and radioactive fallout) to a 22.4 megaton nuclear device – or about *fifteen hundred times larger* than the bomb that destroyed Hiroshima. This yield is also larger than that of the largest thermonuclear weapon ever detonated by the USA (the $\sim$18 megaton Bravo test), but less than half the size of the largest man made thermonuclear explosion: the 50 megaton Soviet "Tsar Bomba" (I don't cease to be amazed by the fact that the pilots who dropped this bomb, and the crews flying observer aircraft nearby, managed to survive).

You may have noticed that we calculated the magnitude of the mechanical energy liberated by a pyroclastic eruption without actually integrating the expansion work term. Two conditions made this possible: (i) our assumption that the expansion is adiabatic, and (ii) our knowledge (or reasonable assumption) of the initial and final temperatures of the process. Lacking this last piece of information, we would have had to derive the magnitude of the expansion work by some other means, independent of temperature.

1.12.3 Frictional heating in faults and shear zones

When displacement occurs along a brittle fault or a ductile shear zone the frictional force that opposes motion performs work. Frictional forces are dissipative, which means that the work that they perform is converted to heat. What is the maximum temperature increase at a fault as a result of this energy dissipation? Let us assume that frictional heat is distributed uniformly throughout a finite volume of width z, bracketing the fault or shear zone (Fig. 1.8). The actual magnitude of z depends on the rate of dissipation of mechanical energy (i.e. the strain rate) – more on this in Chapter 3. We consider the volume of rock enclosed within width z to be an adiabatic system. The change in internal energy of the heated rock volume is then given by (compare with equation (1.72)):

$$d\boldsymbol{E} = -Pd\boldsymbol{V} - (-d\boldsymbol{W}). \tag{1.77}$$

In this equation we distinguish between expansion work arising from the change in temperature of the rocks, and work performed by the frictional force. The work performed by

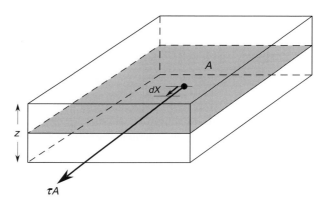

Fig. 1.8 Frictional heating of a volume Az by dissipation of frictional work on a fault plane of area A (grey).

the frictional force on the system is $d\mathbf{W}$. The work performed by the system against the frictional force is thus $-d\mathbf{W}$. A fault or shear zone moves when the force generated by the shear stress component, τ, equals the frictional force. If the total surface area of the fault plane is A then the frictional force is τA. Calling the displacement along the fault dx, $d\mathbf{W}$ is given by:

$$dW = \tau A dx, \tag{1.78}$$

where, because frictional forces always act in the direction of displacement, I have been sloppy and neglected the vector notation (see Box 1.1). Faulting can be treated as a constant pressure process, because pressure inside a planet is determined by depth and the magnitude of the instantaneous motion along a fault or shear zone is generally negligible compared to depth. We thus write dE as follows:

$$dE = dH - PdV = C_P dT - PdV \tag{1.79}$$

so:

$$C_P dT - PdV = -PdV - (-dW). \tag{1.80}$$

Substituting in (1.78):

$$C_P dT = \tau A dx. \tag{1.81}$$

Note that C_P in this equation is the extensive variable, i.e. the total heat capacity of the mass of heated rock. Let $V = Az$ be the total volume of heated rock (Fig. 1.8), and N be the number of mols of substance contained in this volume. If we symbolize the molar volume of the substance with V and its molar heat capacity with C_P, then:

$$N = \frac{V}{V} = \frac{Az}{V} = \frac{C_P}{C_P}. \tag{1.82}$$

Substituting in (1.81) and rearranging we arrive at an equation in terms of intensive variables:

$$dT = \frac{\tau V}{zC_P}dx, \tag{1.83}$$

from which we can also derive the rate of temperature increase, dT/dt (where $t =$ time):

$$\frac{dT}{dt} = \frac{\tau V}{z C_P} \frac{dx}{dt}.$$ (1.84)

The value of τ is determined either by the brittle shear strength for rocks (for a brittle fault) or by the yield strength for a given ductile strain rate for a ductile shear zone; typical values in both cases are of the order of 100 to a few 100 bar. Given that all three parameters: τ, V and C_P vary within fairly restricted ranges for planetary materials, the rate of temperature increase is controlled primarily by the width of the heated rock volume, z. Because the length scale z is controlled by heat diffusion (Chapter 3) we can expect that it varies inversely with the strain rate, which we can represent by dx/dt. We can thus expect a very strong dependency of dT/dt on dx/dt, because as strain rate increases, z decreases and both effects combine to make dT/dt larger. The ductile behavior of rocks is strongly temperature-dependent, with viscosity decreasing exponentially with increasing temperature (see for instance Turcotte & Schubert, 2002, Chapter 7). A positive feedback mechanism results, that focuses ductile deformation in relatively narrow shear zones, in which most of the temperature increase, and consequent lowering of the viscosity, take place.

As a numerical example, consider "slickensides", that probably form as a result of instantaneous heating at a fault during brittle failure (i.e. earthquakes). For a slickenside width $z = 5$ mm, and fault motion during an earthquake $x = 1$ m we obtain a maximum temperature increase (for τ of a few hundred bars) $\Delta T \approx 2000$ K. Fault motions during earthquakes may last for only a few seconds, so that the heating rate may be of the order of hundreds of degrees per second.

1.13 Enthalpy associated with chemical reactions

1.13.1 Enthalpy of reaction and enthalpy of formation

Phenomena such as condensation of solids in the solar nebula, generation and crystallization of magmas, metamorphism, weathering, diagenesis, precipitation of carbonate minerals from seawater and crystallization of evaporites, are chemical reactions. The essence of a chemical reaction is that atoms of certain elements are transferred among phases, or rearrange themselves into different molecules inside a homogeneous phase (e.g. chemical reactions between gas species in a homogeneous atmosphere). Atomic bonds are broken and form during chemical reactions, as a result of which there is a net surplus or deficit of energy that is exchanged with the environment. The macroscopic manifestation of these microscopic energy transactions is the enthalpy change associated with a chemical reaction. This enthalpy is called the *enthalpy of reaction* and is denoted by $\Delta_r H$. Enthalpy of reaction is the amount of heat liberated or absorbed during a chemical reaction at constant pressure and constant temperature. In Chapters 4 and 5 we will arrive at a precise and rigorous definition of chemical equilibrium and we will see that enthalpy of reaction is an important part of that definition and of the algebraic procedure by which we calculate the location of chemical equilibria. Enthalpy of reaction behaves as latent heat: it is heat exchanged without a change in temperature. You may think that this statement cannot possibly be correct, as you are probably familiar with endothermic and exothermic reactions, during which the temperature of the system decreases and increases, respectively. The answer to this puzzle is that these temperature changes are caused by the fact that the rate of heat transfer is much slower than the rate of chemical reaction. For example, during an exothermic reaction enthalpy is liberated at a rate faster than the rate at which heat can be carried away from the site where the reaction takes place. As a result, latent heat (enthalpy of reaction) is converted to sensible heat (temperature increase).

We can write the algebraic definition of enthalpy of reaction by considering a generic chemical reaction between reactant A and product B:

$$A \rightarrow B. \tag{1.85}$$

The enthalpy of reaction at a given T and P is the difference between the enthalpy of B and the enthalpy of A at those conditions:

$$\Delta_r H = H_B - H_A. \tag{1.86}$$

But what *are* the enthalpies of A and B? The First Law of Thermodynamics codifies the law of energy conservation: it states that the total amount of energy is conserved, but says nothing about the absolute magnitude of energy. A more formal statement of this fact is that the First Law of Thermodynamics (equation (1.55)) is a differential equation and, as such, it admits an infinite number of solutions that differ by an additive constant (the integration constant). This is not a problem, because all we care about is the *difference* in enthalpy between different states of a system. This is true in general: even though we did not state it explicitly, in all examples that we have discussed thus far we have calculated differences in enthalpy, internal energy or other types of energy, and not their absolute values. In the case of a chemical reaction, we are not interested in the absolute magnitudes of H_A or H_B, all we are interested in is $\Delta_r H$. Because enthalpy is a state variable, the First Law of Thermodynamics assures us that, as long as we define individual enthalpies relative to the same reference level, the value of $\Delta_r H$ at any given T and P is unique and well defined.

We need to define some arbitrary reference level relative to which we will measure enthalpies – the integration constant, if you wish. We did just this when we defined potential energy $= 0$ at infinity (Section 1.3.1), or when we specified that we were measuring kinetic energy relative to a reference frame fixed to the Earth (Section 1.3.2). The universal convention for chemical systems is to set the enthalpy of all pure chemical elements in their stable configuration at 298.15 K and 1 bar equal to zero. The "stable configuration" requirement is important. We cannot define rigorously what this means until after we have defined chemical equilibrium (Chapters 4 and 5), but we can study examples that make the meaning clear. The enthalpy of diatomic oxygen at 298.15 K and 1 bar is zero, because O_2 is the stable oxygen species at those conditions. In contrast, the enthalpies of atomic oxygen (O) and of ozone (O_3) at 298.15 K and 1 bar are **not** zero. For carbon, graphite has zero enthalpy at 298.15 K and 1 bar. The enthalpy of pure elements in their stable configurations at any other P–T combination is not zero either.

We define the *enthalpy of formation* of a substance (compound or element) at the *reference pressure and temperature* as the value of $\Delta_r H$ for the reaction that forms the substance from the elements (in their stable configurations) at 298.15K and 1 bar. In the thermodynamics literature there are, unfortunately, different symbols in use for this quantity. In this book we will symbolize the enthalpy of formation at the reference pressure (1 bar) and temperature (298.15 K) by $\Delta_f H^0_{1,298}$. This notation is not in widespread use. Enthalpy of formation at 298.15 K and 1 bar is one of the values that are listed in tables of thermodynamic properties of substances, and is commonly symbolized $\Delta_f H^0$. The convenience of adding the actual values of the reference pressure and temperature, as I do here, will become clear in due course, as will the meaning of the ^{0} superscript which appears in both notations.

We will not discuss in this book how enthalpies of formation are measured, as excellent discussions are available in classical textbooks on chemical and geochemical thermodynamics (see, for example, Anderson, 2005, Chapter 5), but we will look at some examples to make the concept clear. The enthalpy of formation, $\Delta_f H^0_{1,298}$, of liquid H_2O is equal to $\Delta_r H$ for the reaction:

$$H_2 + \frac{1}{2}O_2 \Rightarrow H_2O_{liquid}. \tag{1.87}$$

at 298.15 K and 1 bar. Because this is a constant pressure process, $\Delta_f H^0_{1,298}$ is simply the heat liberated when 1 mol of liquid water forms from a stoichiometric mixture of hydrogen and oxygen gas at 298.15 K and 1 bar (see equation (1.62)). Similarly, $\Delta_f H^0_{1,298}$ of diamond is the heat exchanged by the following polymorphic transformation at 298.15 K and 1 bar, per mol of carbon:

$$C_{graphite} \Rightarrow C_{diamond}. \tag{1.88}$$

The enthalpy of formation of graphite is zero. Whether or not the formation reaction can actually take place at 298.15 K and 1 bar, or whether or not the substance in question is stable at those conditions, is not important. For example, we can define $\Delta_f H^0_{1,298}$ for H_2O gas even though H_2O gas is not stable at 298.15 K and 1 bar – although of course, the enthalpy of formation of H_2O gas at the reference conditions 298.15 K and 1 bar is different from that of liquid H_2O at the same conditions. Defining and evaluating $\Delta_f H^0_{1,298}$ for a substance that is not stable at 298.15 K and 1 bar is no more a problem than the fact that it is not possible to locate a body at infinity, yet infinity provides a convenient reference level for potential energy.

The quantity $\Delta_f H^0_{1,298}$ is sometimes called the standard state enthalpy of formation. Unfortunately, the phrase "standard state" applied to thermodynamic variables (including enthalpy) can also have another, quite different, meaning. We will come across this alternate meaning in Chapter 5 and we will see that the intended meaning is commonly (but not always) evident from the context. I will make the meaning explicit whenever there is a possibility of confusion (for an in-depth look at this and other terminology issues, see Anderson, 2005). More trivially, why choose the reference conditions at a strange value such as 298.15 K? Because this temperature corresponds to 25 °C, which historically has been considered "standard" room temperature (arbitrarily so, and uncomfortably warm in the view of this writer). I usually abbreviate 298.15 as 298.

1.13.2 Enthalpy of reaction as a function of temperature

I will use this example to describe the procedure that is used to calculate $\Delta_r H$ at any arbitrary temperature and 1 bar. These calculation procedures will be used in later chapters to calculate chemical equilibrium.

Consider the problem of condensation of solids in the solar nebula. During the formative period of planetary systems the chemical elements are progressively extracted from a gas phase by formation of solid phases. Which phases and elements condense depends on temperature. Refractory phases such as perovskite ($CaTiO_3$) and corundum (Al_2O_3) condense at temperatures of the order of 1600 K, whereas very volatile phases (generically called "ices") in which the chief constituents are C, O, N and H condense at much lower temperatures, 300 K or less. Water ice is one of the most abundant of planetary "ices". If condensation takes place at pressures lower than the pressure of the triple point of H_2O (~ 0.006 bar), then the solid phase forms directly by reaction of chemical species in the gas phase, because the liquid phase is not stable at those conditions (Chapter 6). Suppose that we want to know the enthalpy change (heat released) during condensation of H_2O ice from gaseous H_2 and O_2 at 273 K and 10^{-4} bar. The strategy to calculate the enthalpy change for this reaction is to break up the problem into two parts. First, we calculate the enthalpy of reaction at the temperature of interest (273 K in this case) and the reference pressure of 1 bar. Let us call this enthalpy of reaction $\Delta_r H_{1,T}$. Then, we calculate how enthalpy of reaction varies as a function of pressure at a constant temperature of 273 K, integrate this function from 1 bar to the pressure of interest (in this case, 10^{-4} bar) and add the result to $\Delta_r H$ at 273 K and 1 bar. Symbolically:

$$\Delta_r H_{P,T} = \Delta_r H_{1,T} + \int_1^P \left(\frac{\partial (\Delta_r H)}{\partial P} \right)_T dP, \tag{1.89}$$

where, for the example that we are considering, $T = 273$ K and $P = 10^{-4}$ bar. Here we will look only at the first part of the problem, i.e. calculation of the enthalpy of reaction at 273 K and 1 bar: $\Delta_r H_{1,273}$. The pressure integral (second term on the right-hand side of equation (1.100)) will be discussed in Chapters 5, 8 and 9, because it relies on concepts that we have not discussed yet. Here we will focus on the fact that, even though equilibrium condensation of the solid from the gas does not occur at 1 bar (because liquid is stable at that pressure), calculation of the enthalpy change at 1 bar is always possible.

Relative to $\Delta_f H^0_{1,298}$ of liquid H_2O (equation (1.87)), there are two differences in $\Delta_r H$ for the reaction that forms H_2O ice from H_2 and O_2 gas at 273 K: the temperature is different and the H_2O phase is different. In order to visualize how to proceed it is best to draw a diagram (Fig. 1.9). Because enthalpy is a state variable its value is independent of

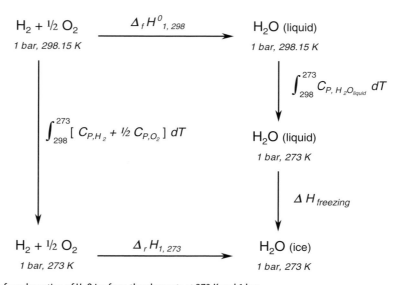

Fig. 1.9 Enthalpy of condensation of H_2O ice from the elements at 273 K and 1 bar.

the path that we use to calculate it, it only depends on the state of the system that we are considering. In our case the state of interest is H_2O ice at 273 K and 1 bar. The diagram (Fig. 1.9) shows that, starting from an initial state consisting of hydrogen and oxygen gas at 298 K and 1 bar (the reference conditions at which the enthalpies of the elements are zero), there are two paths by which we can arrive at the desired final state. One way is to form liquid H_2O at 298 K and 1 bar, then cool the liquid to 273 K, then freeze the liquid at 273 K to ice at 273 K. The total enthalpy change along this path is:

$$\Delta H_1 = \Delta_f H^0_{1,298} + \int_{298}^{273} C_{P,liquid H_2O}\, dT + \Delta H_{freezing},\qquad(1.90)$$

where $\Delta_f H^0_{1,298}$ is the enthalpy of formation of liquid H_2O (equation (1.87)), $\Delta H_{freezing} = -\Delta H_{melting}$ is the heat given off when water freezes to ice at 273 K and 1 bar, and the change of enthalpy associated with an isobaric change in temperature, i.e. the middle term in the right-hand side of equation (1.90), is given in general by:

$$H(T_1) = H(T_0) + \int_{T_0}^{T_1} \left(\frac{\partial H}{\partial T}\right)_P dT = H(T_0) + \int_{T_0}^{T_1} C_P dT.\qquad(1.91)$$

Another way to arrive at ice at 273 K and 1 bar is to cool the gas mixture to 273 K and then form ice directly by reaction between the gases at this temperature. This is the process that takes place in the solar nebula, and the one that we want to calculate the enthalpy of reaction for: $\Delta_r H_{1,273}$. The enthalpy change along this path (Fig. 1.9) is:

$$\Delta H_2 = \int_{298}^{273} C_{P,H_2}\, dT + \frac{1}{2}\int_{298}^{273} C_{P,O_2}\, dT + \Delta_r H_{1,273}.\qquad(1.92)$$

The heat capacity integral for O_2 is preceded by a factor of $\frac{1}{2}$, which is the stoichiometric coefficient for O_2 in the balanced chemical reaction – recall that C_p is the *molar* heat capacity. Stoichiometric coefficients are always present, but they are 1 for H_2 and H_2O. Now, because enthalpy is a state variable, it must be $\Delta H_1 = \Delta H_2$, and we see immediately that:

$$\Delta_r H_{1,273} = \Delta_f H^0_{1,298} + \Delta H_{freezing} + \int_{298}^{273}\left[C_{P,liquid H_2O} - C_{P,H_2} - \frac{1}{2}C_{P,O_2}\right] dT.$$
$$(1.93)$$

We are also interested in $\Delta_r H$ for reactions in which no pure elements take part. For example, the following reaction among spinel ($MgAl_2O_4$), enstatite ($Mg_2Si_2O_6$), forsterite (Mg_2SiO_4) and pyrope ($Mg_3Al_2Si_3O_{12}$) is an end-member model for the transition between spinel lherzolites and garnet lherzolites in the mantles of the Earth and other terrestrial planets:

$$MgAl_2O_4 + 2Mg_2Si_2O_6 \Rightarrow Mg_2SiO_4 + Mg_3Al_2Si_3O_{12}.\qquad(1.94)$$

Starting from the elements, we can reach the assemblage forsterite + pyrope at an arbitrary temperature T (and $P = 1$ bar) by either (1) forming forsterite + pyrope at 298 K and then heating this assemblage to T, or (2) forming spinel + 2 enstatite at 298 K, heating this assemblage to T, and then reacting it to form forsterite + pyrope (Fig. 1.10). The enthalpy

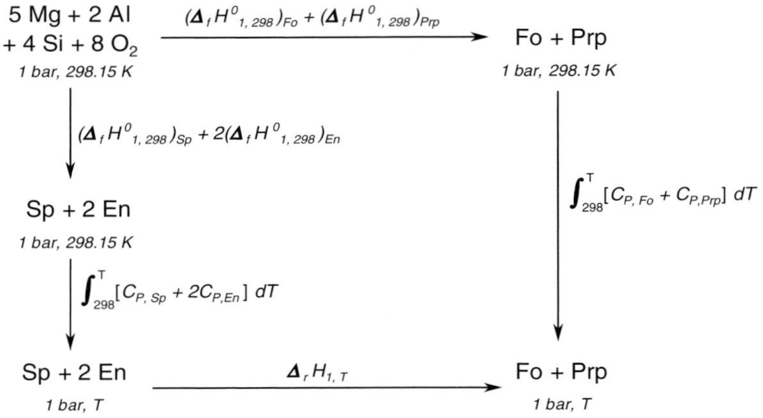

Fig. 1.10   Enthalpy of the reaction spinel + 4 estatite $\rightarrow$ forsterite + pyrope, at T and 1 bar.

change along the first path is:

$$\Delta H_1 = \left(\Delta_f H^0_{1,298}\right)_{prp} + \left(\Delta_f H^0_{1,298}\right)_{fo} + \int_{298}^{T} C_{P,prp}\, dT + \int_{298}^{T} C_{P,fo}\, dT \qquad (1.95)$$

and along the second path (note the explicit stoichiometric coefficient of enstatite):

$$\Delta H_2 = \left(\Delta_f H^0_{1,298}\right)_{sp} + 2\left(\Delta_f H^0_{1,298}\right)_{en} + \int_{298}^{T} C_{P,sp}\, dT + 2\int_{298}^{T} C_{P,en}\, dT + \Delta_r H_{1,T}. \qquad (1.96)$$

As enthalpy is a state variable, $\Delta H_1 = \Delta H_2$, so:

$$\Delta_r H = \left(\Delta_f H^0_{1,298}\right)_{prp} + \left(\Delta_f H^0_{1,298}\right)_{fo} + \int_{298}^{T} C_{P,prp}\, dT + \int_{298}^{T} C_{P,fo}\, dT$$
$$- \left[\left(\Delta_f H^0_{1,298}\right)_{sp} + 2\left(\Delta_f H^0_{1,298}\right)_{en} + \int_{298}^{T} C_{P,sp}\, dT + 2\int_{298}^{T} C_{P,en}\, dT\right]. \qquad (1.97)$$

Let us now introduce some additional notation that will simplify this equation. The sum of the four enthalpies of formation is simply the enthalpy of reaction at 298 K and 1 bar. Let us call this sum $\Delta_r H^0_{1,298}$:

$$\Delta_r H^0_{1,298} = \left(\Delta_f H^0_{1,298}\right)_{prp} + \left(\Delta_f H^0_{1,298}\right)_{fo} - \left[\left(\Delta_f H^0_{1,298}\right)_{sp} + 2\left(\Delta_f H^0_{1,298}\right)_{en}\right]. \qquad (1.98)$$

Because all the integrals are evaluated over the same temperature interval we can also collect all the heat capacity functions (assuming that they are all described by the same function) in a single function, which we call $\Delta_r C_P$:

$$\Delta_r C_P = C_{P,prp} + C_{P,fo} - \left(C_{P,sp} + 2C_{P,en}\right). \qquad (1.99)$$

Finally, we call the enthalpy of reaction at the temperature and pressure of interest $\Delta_r H_{P,T}$. In this case $P = 1$ bar, so our equation for $\Delta_r H_{1,T}$ becomes:

$$\Delta_r H_{1,T} = \Delta_r H^0_{1,298} + \int_{298}^{T} \Delta_r C_P dT. \qquad (1.100)$$

This equation is completely general. It applies to any chemical reaction, as long as there are no phase transitions along any of the paths. If phase transitions occur then their enthalpies are simply added separately, as was done with $\Delta H_{freezing}$ in equation (1.93), which is otherwise identical to (1.100). These calculations will very quickly become second nature and you will be able to do away with the diagrams. When in doubt, however, sketching diagrams such as Figs. 1.9 and 1.10 will always point you to the correct result. Simple *Maple* procedures to carry out the numerical calculations are described in Software Box 1.1.

Calculating the effect of pressure on $\Delta_r H$ is less straightforward than calculating the effect of temperature. This is so partly because the equation for $(\partial H/\partial P)_T$ is not a simple function of other thermodynamic variables or material properties (compare to $(\partial H/\partial T)_P = C_P$). The partial derivative $(\partial H/\partial P)_T$ is a function of molar volume, so that integating $H = H(P)$ requires an EOS. These are different for gases and condensed phases, and we defer their discussion to later chapters.

1.14 Internal energy and the relationship between macroscopic thermodynamics and the microscopic world

The internal energy of a system is the sum of energy contributions from translation, rotation and vibration of molecules, their electronic configurations, their nuclear configurations, and their electrostatic interactions (i.e. chemical bonding). If we do not consider nuclear reactions, including radioactive decay, then the nuclear contribution to E stays constant. If we also exclude chemical reactions and excitation of electronic shells by high-energy electromagnetic radiation then the electronic and electrostatic contributions to E also stay constant. With these restrictions, changes in E arise from changes in the translational, rotational and vibrational energies of the molecules, in response to changes in the macroscopic variables temperature and pressure. We seek a mathematical relationship between internal energy and temperature as a first step towards understanding some of the ways in which the macroscopic language of thermodynamics reflects processes that take place at the microscopic scale. A simple derivation which relies on classical deterministic physics, called the kinetic theory of gases, has been known since the mid nineteenth century. The kinetic theory of gases is only applicable to the simplest of systems: a gas made up of non-interacting particles, i.e. an ideal gas. For all other systems, the link between phenomenological thermodynamics and microscopic processes can only be constructed on the basis of statistical mechanics and must include considerations of quantum mechanics as well. This is beyond the scope of this book.

1.14.1 Internal energy of a monatomic ideal gas

Consider a gas made up of non-interacting point-like particles, i.e. a monatomic ideal gas. In such a substance, the only possible kind of energy that can exist at the microscopic level is the kinetic energy of motion, or translation, of the particles. The internal energy of such a gas must therefore be equal to the sum of the translational kinetic energies of all of the particles that make up the gas. In this case the link between the macroscopic state variable E and the microscopic repository of energy admits no other possible interpretation, because of the restrictions that we have imposed on the nature of the substance: non-interacting, point-like particles.

Not all particles have the same kinetic energy. As we saw in Section 1.4.2, molecular speeds follow a statistical distribution known as the Maxwell–Boltzmann distribution (Fig. 1.4). A consequence of this distribution of molecular speeds is that if a gas is at equilibrium at a given pressure and temperature, then the average kinetic energy of the particles, $\langle U_k \rangle$, is a well defined and unique value. This is the microscopic reason why E is a state variable. The molar internal energy is then given by:

$$E = N \langle U_k \rangle = N \frac{1}{2} m \langle c^2 \rangle, \tag{1.101}$$

where N is Avogadro's number, m is the particle mass and $\langle c^2 \rangle$ is the mean-square particle speed. This is not the square of the mean, but the mean of the squares. The reason for this is that we are averaging kinetic energy, which scales as the square of the speed.

The velocity of each particle can be written in terms of three independent components: c_x, c_y, c_z, along three orthogonal directions (Fig. 1.11) such that:

$$c^2 = c_x^2 + c_y^2 + c_z^2. \tag{1.102}$$

Now, by symmetry, it must be:

$$\langle c_x^2 \rangle = \langle c_y^2 \rangle = \langle c_z^2 \rangle = \langle u^2 \rangle \tag{1.103}$$

so that:

$$\langle c^2 \rangle = 3 \langle u^2 \rangle. \tag{1.104}$$

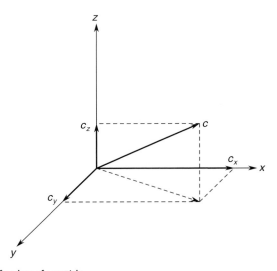

Fig. 1.11   Translational degrees of freedom of a particle.

This result also arises from Maxwell's *principle of equipartition of energy*. What the principle states is that, in a collection of particles, energy is distributed evenly among all of the *degrees of freedom* of the particles. By degrees of freedom we mean, in general, variables that can vary independently of one another. We will come across different ways in which this terminology is used, and the actual meaning will in every case be explained and then become obvious from the context. When studying the behavior of particles in a microscopic system, by degrees of freedom we mean independent ways in which particles can carry energy. If particles only have translational kinetic energy, then they have three degrees of freedom, corresponding to the three perpendicular directions in which they can move. The principle of equipartition of energy then says that the average kinetic energy along each of the three directions must be the same, for if particles are moving more slowly in one of the three directions then collisions will eventually increase their speed in that direction at the expense of the speed (= kinetic energy) in the other directions. This intuitive explanation makes sense, but it does not prove the validity of the principle. Equipartition of energy remains a principle, i.e. a statement whose truth is assumed *a priori*, and that is accepted because it leads to results that agree with observations. The same is true, of course, of the laws of thermodynamics, the principle of conservation of momentum, and all other physical laws and principles.

Because of the principle of equipartition of energy, we can use equation (1.104) in order to re-write equation (1.101) as follows:

$$E = \frac{3}{2} N m \langle u^2 \rangle.   \tag{1.105}$$

This equation says that the average kinetic energy of the particles for each degree of freedom is $\frac{1}{2} m \langle u^2 \rangle$, and the internal energy, which equals the total kinetic energy of the particles, equals this energy times three (one for each degree of freedom) times the total number of particles.

We know that internal energy is a function of temperature and we want to find out what this function is on the basis of the microscopic description of the system. It is a remarkable fact that this is in principle *always* possible, no matter how complicated the system may be. The formal method for doing this relies on finding a fundamental function that is defined in statistical mechanics and is called the *partition function* (see, for example, the textbooks by Hill, 1986, or Glazer & Wark, 2001).

All thermodynamic state variables can be derived from a substance's partition function. The problem is that it is seldom easy to find what the partition function for a given substance is. For the simplest of cases, ideal gases, it is possible to derive the value of the state functions both from the partition function and from the much simpler formalism of kinetic theory, as we do here. This is not true in general, though.

We begin by working out an expression for the pressure of an ideal gas, which must arise from collisions of the particles against the walls of the container. Pressure is force per unit area, and we recall from elementary physics (Newton's second law of motion) that force equals the rate of change of momentum. We assume that the particles that make up the ideal gas are perfectly elastic. Therefore, when a particle collides with a surface perpendicular to any one of the three independent directions of translation it reverses direction but does not lose speed, so that its momentum changes by:

$$mu - (-mu) = 2mu.   \tag{1.106}$$

Choosing a surface perpendicular to one of the three axes that define the degrees of freedom does not cause any loss of generality, because we can choose the orientation of the axes (Fig. 1.11) any way we want, as long as they are mutually perpendicular. In other words, pressure is isotropic.

If the molar volume of the gas is V then opposite walls are separated by a distance of order $V^{1/3}$ and the time between consecutive collisions is $(2V^{1/3})/u$, because the particle must move to the opposite wall and back. The rate of change of momentum of a particle when it collides with a wall, i.e. the force exerted by a particle on a wall, is then given by:

$$\frac{2mu}{2V^{1/3}u^{-1}} = \frac{mu^2}{V^{1/3}}. \tag{1.107}$$

The total pressure must equal the average force per particle, times the number of particles, divided by the surface area of the wall. In order to account for the average force per particle we must use the mean-square speed, $\langle u^2 \rangle$, because we are averaging rates of change of momentum, which scale as the square of speed (equation (1.107)). The surface area of the wall is of order $V^{2/3}$, so that the pressure is given by:

$$P = N \frac{m \langle u^2 \rangle}{V^{1/3}} \frac{1}{V^{2/3}} = \frac{Nm \langle u^2 \rangle}{V} \tag{1.108}$$

hence:

$$PV = Nm \langle u^2 \rangle. \tag{1.109}$$

Using the ideal gas EOS, $PV = RT$, we obtain:

$$T = \frac{Nm \langle u^2 \rangle}{R} = \frac{m \langle u^2 \rangle}{k_B}, \tag{1.110}$$

where $k_B = R/N$ is the gas constant per molecule, known as Boltzmann's constant. We can also write equation (1.110) as:

$$\frac{1}{2} k_B T = \frac{1}{2} m \langle u^2 \rangle. \tag{1.111}$$

This equation relates the temperature of the gas to the average kinetic energy of the particles along each degree of freedom. It allows us to calculate a characteristic value for molecular speeds, the root-mean-square speed (RMS) of the molecules, $\langle c^2 \rangle^{\frac{1}{2}} = (3 \langle u^2 \rangle)^{\frac{1}{2}}$. Molecular speeds were first calculated in this fashion by James Joule in the mid nineteenth century. The root-mean-square speed is not the same as the most probable molecular speed mentioned in Section 1.4.2. The relationship between the two quantities is shown in Fig. 1.12. The RMS molecular speed is always greater than the most probable speed, reflecting the long tail on the high speed end of the Maxwell–Boltzmann distribution. This long tail, which implies that there always are a significant number of molecules that are moving with speeds that are much faster (by a factor of 2–3) than the most probable value, plays an important role in the escape of light gas species from planetary atmospheres. This is especially the case for escape of hydrogen from primordial planetary atmospheres, and will be discussed in Chapter 13.

Equations (1.105) and (1.111) each relates a microscopic variable (E or T) to the kinetic energy of the particles. We can substitute one equation into the other to arrive at an expression

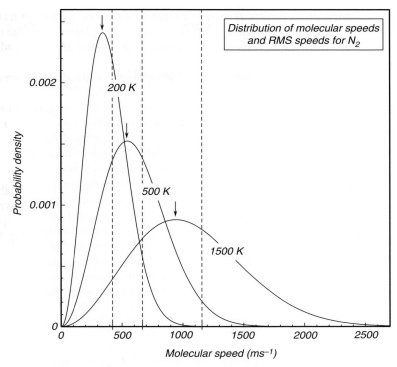

Fig. 1.12   Most probable speed (arrows) and RMS speed (dashed lines) for N_2 at 200, 500 and 1500 K.

for the molar internal energy of a monatomic ideal gas as a function of absolute temperature:

$$E_{monatomic} = \frac{3}{2}RT. \qquad (1.112)$$

This equation shows that the internal energy of an ideal gas is a function of temperature only, so that for any ideal gas $(\partial E/\partial P)_T = (\partial E/\partial V)_T = 0$. For real substances these derivatives do not vanish. They vanish for ideal gases only because of our assumptions: non-interacting, perfectly elastic particles with no volume. Also, since $C_V = (\partial E/\partial T)_V$, we get $C_V = 3/2\,R$ for a monatomic ideal gas.

Note what we have done here: from a simple hypothesis about the microscopic nature of an ideal gas and the principle of equipartition of energy we have derived a mathematical expression for the relationship between two macroscopic variables. The validity of our assumptions can then be tested by comparing predicted internal energies of ideal gases with measurements for real gases at conditions such that their behavior approaches that of ideal gases. We will do this in the next section, using C_V rather than E, because C_V can be obtained from direct experimental measurements whereas E cannot.

We could have derived these same results from statistical mechanics, with one crucial difference. In our derivation using only kinetic theory we had to use the ideal gas EOS (in equation (1.110)), which arises from macroscopic observations. Statistical mechanics allows us to do away with this last piece of macroscopic physics – we can derive both the expression for internal energy (equation (1.112)) *and the ideal gas EOS* solely from the (microscopic) partition function. The importance of this is that the same principles and formalism of statistical mechanics that would allow us to do that for an ideal gas can be used to derive equations of state for materials of planetary interest at very high pressures and temperatures, that may be experimentally inaccessible.

1.14.2 Degrees of freedom and heat capacities of polyatomic gases

The internal energy function given by equation (1.112) is appropriate for a monatomic ideal gas, but it must be modified for diatomic and polyatomic gases. This is so because, if the molecule contains more than one atom, then additional degrees of freedom are possible, and by the principle of equipartition of energy each of these additional degrees of freedom carries an additional amount of energy. For example (Fig. 1.13), a diatomic molecule has non-negligible moment of inertia relative to two orthogonal axes perpendicular to the atomic bond, and a polyatomic molecule (unless the atoms are arranged in a straight line) has non-negligible moment of inertia relative to three orthogonal axes. This means that diatomic and polyatomic molecules have kinetic energy of rotation in addition to that of translation, and each of the rotation axes adds a degree of freedom. As molecules collide with one another kinetic energy is exchanged between translational and rotational degrees of freedom, and by the principle of equipartition of energy all degrees of freedom carry the same amount of energy: $\frac{1}{2}k_B T$. The corresponding internal energies are, thus:

$$E_{diatomic} = \frac{3}{2}RT\,(translational) + \frac{2}{2}RT\,(rotational) = \frac{5}{2}RT$$
$$E_{polyatomic} = \frac{3}{2}RT\,(translational) + \frac{3}{2}RT\,(rotational) = \frac{6}{2}RT.$$

(1.113)

The constant volume heat capacities of ideal gases are, therefore: $3/2R$, $5/2R$ and $3R$, for monatomic, diatomic and non-linear polyatomic ideal gases, respectively. Heat capacities can be determined directly by experiment and are a test of the predictions of the microscopic description of ideal gases. Figure 1.14 shows measured constant volumes heat capacities at low pressure as a function of temperature, for a monatomic gas (neon), a diatomic gas (nitrogen) and a polyatomic gas with a three-dimensional molecule (ammonia). The low pressure constraint is what makes the behavior of the gases approach that of ideal gases, by keeping them far from their condensation conditions. The agreement of the measured

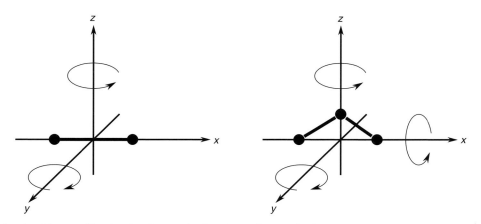

Fig. 1.13 Rotational degrees of freedom for a linear molecule and a non-linear molecule.

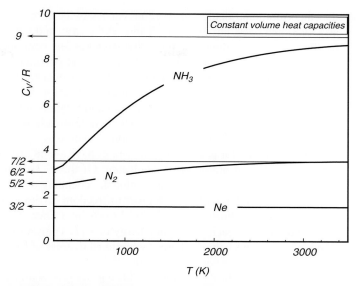

Fig. 1.14 Measured constant volume heat capacities for a monatomic gas (Ne), a diatomic gas (N$_2$) and a polyatomic gas with non-linear molecules (NH$_3$). Data from *NIST Chemistry WebBook*. The values expected from kinetic theory are 3/2 R, 7/2 R and 9 R, respectively. Heat capacity is plotted as the non-dimensional variable C_V/R.

values with those predicted by kinetic theory is essentially perfect for neon. For nitrogen the agreement is good at temperatures up to approximately room temperature ($\sim$300 K) but then C_V increases smoothly and reaches a plateau of approximately $7/2R$. In the case of ammonia there is good agreement at low temperature (200–300 K), and then, as for nitrogen, C_V increase and reaches a plateau, in this case close to $9R$.

Do these discrepancies mean that the premises that we used to calculate C_V for gases other than monatomic gas are incorrect? Well, no, what it means in this case is that they are incomplete. In other words, gases that contain more than one atom in their molecules have other degrees of freedom in addition to those corresponding to translation and rotation. These are called vibrational degrees of freedom, and they arise from the fact that atomic bonds are not rigid, but can stretch, contract and bend as an elastic material (e.g. a spring), and behave like harmonic oscillators. These oscillations carry energy and thus represent additional degrees of freedom. The principle of equipartition of energy applies to vibrational degrees of freedom too, but there are two important differences with translational and rotational degrees of freedom, which we can see by examining Fig. 1.14 in some detail.

First, vibrational degrees of freedom in gases are inactive at low temperature and begin to carry energy only as temperature rises. In other words, gas molecules must have a minimum amount of kinetic energy before collisions are able to transfer some of this energy to vibrations of the atomic bonds (if you suspect a connection with quantum mechanics here you are right). The situation is different for solids and liquids, in which vibrational modes are important at relatively low temperatures, simply because translational and rotational degrees of freedom do not exist in them (although liquids have additional complexities).

Second, in the diatomic gas there is only one atomic bond, and thus there can only be one vibrational degree of freedom, which corresponds to stretching of the atomic bond. Heat capacity for diatomic nitrogen levels off at a value of approximately $7/2R$, suggesting that the vibrational degree of freedom carries twice as much energy as the translational and rotational ones, i.e. $k_B T$. This is indeed the case, and the explanation is that harmonic oscillators (such as an atomic bond that stretches and contracts, or a pendulum, which is mathematically equivalent) carry both kinetic and potential energy, and equipartition of energy causes *each* of these energy modes to have an energy content of $\frac{1}{2}k_B T$. In the case of ammonia the difference between the measured high-temperature heat capacity ($9R$) and the heat capacity predicted from translation and rotation only ($3R$) indicates that there are six vibrational degrees of freedom, which include stretching of the N–H bonds, changes in the angles between N–H bonds, and bending relative to a plane containing all four atoms.

The general rule for gases is that molecules have a total of $3n$ degrees of freedom, where n is the number of atoms in the molecule. Of these, 3 degrees of freedom are always translational, and either 2 or 3 are rotational, depending on whether the molecule is linear or not (Fig. 1.13). Each of these degrees of freedom carries an energy equal to $\frac{1}{2}k_B T$. The remaining $3n - 5$ (linear molecule) or $3n - 6$ (non-linear molecule) degrees of freedom are vibrational and each carries an energy of $k_B T$. Translational and rotational degrees of freedom are always active, whereas vibrational degrees of freedom only become active (or "excited") above a certain temperature.

Constant volume heat capacities for ideal gases follow immediately from the generalizations given in the previous paragraph. In most planetary sciences applications, however, we are interested in C_P rather than C_V, and these are also easily derived. From the definition of enthalpy we have:

$$C_P = \left(\frac{\partial H}{\partial T}\right)_P = \left(\frac{\partial E}{\partial T}\right)_P + P\left(\frac{\partial V}{\partial T}\right)_P. \tag{1.114}$$

Using identity (1.3.12) in Box 1.3 and the fact that, for an ideal gas, $(\partial E/\partial V)_T = 0$, we see that:

$$\left(\frac{\partial E}{\partial T}\right)_P = \left(\frac{\partial E}{\partial T}\right)_V + \left(\frac{\partial E}{\partial V}\right)_T \left(\frac{\partial V}{\partial T}\right)_P = C_V \tag{1.115}$$

and from the ideal gas EOS:

$$P\left(\frac{\partial V}{\partial T}\right)_P = P\frac{R}{P} = R \tag{1.116}$$

so, for an ideal gas:

$$C_P = C_V + R. \tag{1.117}$$

The extra amount of energy, R per Kelvin per mol, is a macroscopic contribution from expansion work and does not reflect a change in the microscopic state of the system. For gases that cannot be assumed to behave ideally equation (1.117) is not valid, but the appropriate equation can always be derived from the equation of state, starting from equation (1.114) which is true in general. Thus, with some important additional considerations, the physical concepts developed in this section are the basis for understanding and quantifying the behavior of real fluids at high pressures and temperatures, something that we will do in Chapter 9.

1.14.3 Heat capacities of solids

In a crystalline solid there are only vibrational degrees of freedom. The atoms have fixed positions in the crystalline lattice. They vibrate about their equilibrium positions, but they can neither translate not rotate. Each atom can vibrate in three independent directions. We could thus expect the heat capacity of crystalline solids to be approximately $3R$ per gram atom, reflecting the three vibrational degrees of freedom of each atom, and the fact that each vibrational degree of freedom carries an energy of $k_B T$ per atom. The molar heat capacity of a substance that contains n atoms in its molecule should thus be $3nR$. This has been empirically known to be the case since the nineteenth century. It is known as Dulong and Petit's law, but it works only for temperatures comparable to or greater than room temperature.

Figure 1.15 shows constant pressure heat capacities for a few minerals (C_P and C_V for solids differ by a very small amount, because of the incompressibility of solids, see Chapter 8). They all approach Dulong and Petit's behavior at high temperature ($\sim 10^3$ K), but heat capacities fall off fairly steeply with decreasing temperature. The data confirm the general microscopic picture (atoms in a solid have vibrational degrees of freedom only) but also show that some fundamental physics is missing from the picture. The missing ingredient is quantum mechanics, and although the details are beyond the scope of this book a qualitative understanding of the behavior depicted in Fig. 1.15 will become important in subsequent discussions. Atomic vibrations in a crystal are quantized, meaning that only some definite energy levels are allowed. At low temperature the allowed energy levels are few and far apart, and degenerate to a single energy level as absolute zero is approached. The number of allowed energy levels increases with temperature, and as this happens so does the heat

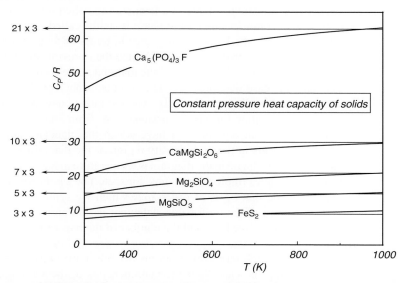

Fig. 1.15 Constant pressure heat capacities for pyrite, enstatite, forsterite, diopside and fluorapatite. Horizontal lines show Dulong and Petit's approximation, labeled next to the arrows: the number of atoms in the formula times the number of vibrational degrees of freedom per atom ($= 3$). Measured heat capacity data from Robie and Hemingway (1995). Heat capacity plotted as the non-dimensional variable C_P/R.

capacity. Above a certain temperature the quantum energy levels become so close together that the classical continuous approximation – Dulong and Petit's law – becomes valid. This temperature is characteristic for each substance and is known as the Debye temperature, θ_D. The Debye temperature for most minerals is of the order of several hundred to $\sim 10^3$ K. Heat capacity becomes a weak function of temperature above θ_D, as shown by Fig. 1.15.

Interestingly, Dulong and Petit's approximation also works for some liquids. For example, the heat capacity of water is ~ 75 J K^{-1} mol^{-1} $\approx 9R$, corresponding to $3R$ for each of the three atoms in the H_2O molecule.

1.15 An overview of the properties of matter and equations of state

Equations of state (EOS) are an essential component in the study of the physics and chemistry of planetary bodies. We will discuss different types of EOS in considerable detail in Chapters 8 and 9. Here we define some concepts about the possible states of matter in planetary environments. This will allow us to better define the $P-V-T$ ranges over which different types of EOS are applicable. Solids and liquids are *condensed phases*, whereas gases are *non-condensed phases*. Both liquids and gases are fluids. Gases are *non-condensed fluids*, which means that they expand indefinitely as pressure decreases, whereas we call liquids *condensed fluids* because, as solids, they do not behave in this manner. These contrasting macroscopic behaviors arise from differences between condensed and non-condensed fluids in the relative magnitudes of the intermolecular potentials (or forces), compared to thermal energy. A *vapor* is a gas in equilibrium with its liquid, and a *melt* is a liquid in equilibrium with its solid. For all substances it is an empirical observation that, as temperature increases, the material properties of liquid and vapor at equilibrium approach each other, until the two phases become indistinguishable at a temperature called the *critical temperature*. Above the critical temperature a single fluid phase is stable, called a *supercritical fluid*.

Here we describe the behavior of matter in terms of temperature and density (i.e. the inverse of volume). Density is a better choice than pressure for this exercise because we can relate it directly to a description of the material at the atomic scale, something which is not generally the case with pressure. The arguments and conclusions are summarized in a density–temperature diagram, Fig. 1.16, which is rather busy and will take some explaining. The physical inspiration for this discussion comes from Shalom *et al.* (2002), who present detailed mathematical arguments; I have added the planetary applications. A review of recent developments in the high density region of the diagram is given by Drake (2010). The horizontal coordinate in the diagram is temperature in Kelvin. Let the thermal energy of microscopic particles (molecules, atoms, ions, etc.) be ε_T. From our discussion in Section 1.14 we conclude that ε_T is of order $k_B T$, which we symbolize with the $\sim$ symbol, i.e. $\varepsilon_T \sim k_B T$. The justification for this statement is that particles carry an energy equal to $\frac{1}{2} k_B T$ per degree of freedom, and the number of degrees of freedom is a small number, of order 1–10.

At sufficiently high temperature atoms become ionized as a result of interatomic collisions. This happens when their thermal energy is of the same order as their ionization energy, i.e., the energy required to detach an electron from an atom. By equating the ionization energy of an element to $k_B T$ we can estimate the temperature at which thermal ionization takes place. Note, however, that because energies of individual atoms are not all the same but follow a statistical distribution (e.g., Fig. 1.12), ionization actually takes

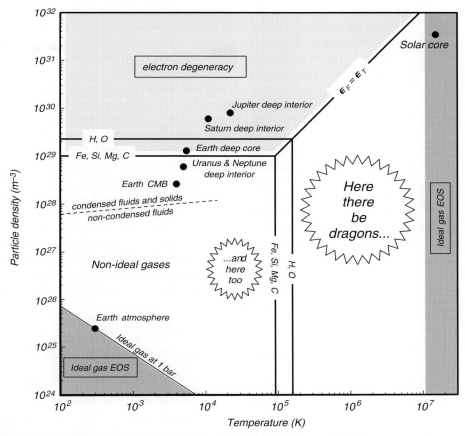

Fig. 1.16 Density–temperature diagram of matter. Dark gray regions correspond to ideal gas behavior. Pressure ionization occurs in the light gray region. The lines labeled "H, O" and "Fe, Si, Mg, C" represent approximate ionization conditions caused by temperature and pressure. CMB = core–mantle boundary.

place over a temperature range. The vertical lines labeled "H, O" and "Fe, Si, Mg, C" show characteristic ionization temperatures for these most important of planet-forming elements. Many other abundant elements plot within this range, the only important exception being helium, which has an ionization energy almost twice that of hydrogen. Thermal ionization occurs at temperatures of order 10^5 K, which are not attained in any planet in the Solar System, but may be possible inside large extrasolar planets and brown dwarfs (failed stars). Most of the solar interior is of course ionized.

Density is shown on the vertical axis as *particle density*, i.e. number of particles per unit volume. This is simply Avogadro's number divided by molar volume, and has units of m^{-3}. Using these units we can compare the degree of packing independently of particle mass. The figure shows some reference values. An ideal gas at 1 bar and 298 K (e.g., the terrestrial atmosphere) has a particle density of $\sim 2.4 \times 10^{25}$ m^{-3}. The variation of this number with temperature at a constant pressure of 1 bar, i.e. the ideal gas law, is shown in the diagram with a negatively sloping line. The ideal gas EOS reproduces P–V–T relations to a high degree of accuracy for conditions below this line ($P < 1$ bar) and perhaps over a narrow interval above the line. With increasing particle density the behavior of real gases diverges from the predictions of the ideal gas EOS, owing to the existence of interatomic potentials, i.e. attractive and repulsive forces. Up to particle densities of $\sim 10^{27}$–10^{28} m^{-3}, substances behave as non-condensed fluids (i.e., gases) even if their densities are higher than what we typically associate with a gas.

Condensation occurs at particle densities of $\sim 10^{28}$ m^{-3}, with overlapping ranges for liquids and solids. Condensed matter is orders of magnitude less compressible than gases, a fact that is expressed in Fig. 1.16 by the (apparent) proximity of a wide range of condensed planetary materials. The examples shown are: $Mg_{1.8}Fe_{0.2}SiO_4$ at the conditions of the Earth's core–mantle boundary (Chapter 8), an H_2O–CH_4–NH_3 mixture at conditions thought to represent the deep interiors of Uranus and Neptune, Fe in the Earth's inner core, and H–He mixtures in Jupiter's and Saturn's deep interiors. These examples are representative of the upper density–temperature bounds likely to be encountered in the Solar System, with the exception of the solar interior, which is also shown for comparison.

At particle densities somewhat above 10^{29} m^{-3} it is necessary to consider quantum effects. We define the quantum energy, ε_q, for a particle of mass m and momentum p as $\varepsilon_q = p^2/m$. If particles are packed with a particle density n, then each particle occupies a volume n^{-1}, which means that there is an uncertainty in the position of the particle of order $n^{-1/3}$. According to the uncertainty principle the relationship between p and n is given (approximately) by:

$$pn^{-1/3} \approx h, \tag{1.118}$$

where h is Planck's constant. We then get for the quantum energy:

$$\varepsilon_q \approx \frac{h^2 n^{2/3}}{2m}. \tag{1.119}$$

For an electron with mass m_e, accounting for the fact that each volume element can be occupied by at most two electrons with opposite spins, and allowing for spherical symmetry, the quantum energy becomes the Fermi energy, ε_F, given by:

$$\varepsilon_F = \frac{h^2 n^{2/3}}{8m_e} \left(\frac{3}{\pi} \right)^{2/3}. \tag{1.120}$$

If the Fermi energy exceeds the ionization energy and $\varepsilon_F > \varepsilon_T$ then the material undergoes *pressure ionization*. This takes place even if the thermal energy of the material is lower than its ionization energy, because bound electrons are excluded by the uncertainty principle from occupying the same volume element. The resulting condition is termed *electron degeneracy*. The horizontal lines labeled "H, O" and "Fe, Si, Mg, C" show the approximate densities at which electron degeneracy can be expected to occur in planetary materials. Much of Jupiter's and Saturn's interiors are likely to consist of pressure-ionized hydrogen, and some degree of pressure ionization may also occur in the Earth's deeper core. Most of the core, as well as the Earth's mantle, and the entire volume of the ice giants (Uranus and Neptune) are, in contrast, composed of non-degenerate matter. Given that pressures in the other terrestrial planets are lower than in the Earth's interior this conclusion also applies to them. EOS for degenerate matter are needed in order to model the interiors of the giant planets.

In contrast to "cold" dense matter, "hot" dense matter ($\varepsilon_T > \varepsilon_F$) is not degenerate. This includes the entire solar interior. At temperatures of order 10^7 K interatomic potentials are negligible compared to thermal energy. Under such conditions matter consists of a highly ionized plasma that is accurately described as a monatomic ideal gas (more on this in Chapter 2). We are left, as the figure shows, with a wide range of intermediate temperatures and densities in which the behavior of matter is poorly understood and for which equations of state are very complex, if they exist at all. Fortunately, these regions are of little interest to planetary scientists. Focusing now on the range of conditions for non-degenerate matter in planetary interiors ($T < 10^4$ K, $n < 10^{29}$ m^{-3}), which is what we will mostly be concerned with, we can identify three regions that require distinct equations of state. At very low densities the ideal gas EOS is all we need. At the other end, a class of EOS that will be introduced in Chapter 8 is generally appropriate for the entire range of densities of non-degenerate condensed matter, both solids and liquids. This leaves the intermediate density range, roughly between 10^{26} and 10^{28} m^{-3}, occupied by non-ideal gases. Equations of state for non-ideal gases are discussed in Chapter 9.

Exercises for Chapter 1

1.1 A mass m_i of ice at temperature T_i is added to a mass m_w of water at temperature T_w inside a perfectly isolated container. Let the heat capacities of ice and water be $C_{p,i}$ and $C_{p,w}$, and the enthalpy of fusion of ice be $\Delta_f H > 0$. Derive a set of criteria to decide what the final equilibrium state of the system is, i.e. pure ice, pure water or ice + water.

1.2 Calculate the gravitational potential energy stored in the Earth's largest active volcano (Mauna Loa) and in the Solar System's largest volcano (Olympus Mons in Mars). Assume that Mauna Loa is a cone with a basal radius (on the floor of the Pacific Ocean) of 50 km, and an elevation of 10 km relative to the ocean floor, and Olympus Mons a cone with a radius of 600 km and an elevation of 27 km. The gravitational accelerations of Earth and Mars are 9.8 and 3.7 m s^{-2}. Assume that Mauna Loa is 2×10^5 years old. Calculate the average power expended in building up Mauna Loa. Assuming that this rate of storage of potential energy is characteristic of large basaltic shield volcanoes, estimate how long the building of Olympus Mons may have taken. How does this compare with the building time of Olympus Mons if you assume the same mass supply rate as in Mauna Loa? Discuss the relative merits of extrapolating terrestrial mass supply rate or energy supply rate to Mars. Compare the rate of storage of potential energy in Mauna Loa with the power output of the Hawaiian hot spot, $\sim 10^{11}$ J s^{-1}. Comment on your results.

1.3 What is the potential energy stored in a rod-shaped asteroid, $100 \times 50 \times 50$ km, with density = 3000 kg m^{-3}? The stable shape of the asteroid is a sphere with the same volume.

1.4 Derive the formula for kinetic energy (equation (1.12)) by considering conservation of energy of a free-falling body in a uniform gravitational field.

1.5 Consider the asteroid–Earth collision (Worked Example 1.2). What is the relationship between m and M that results in the greatest dissipation of kinetic energy? You can do this formally, by maximizing the function in equation (1.15), or intuitively, from symmetry considerations.

1.6 What is the maximum amount of power that can in principle be harvested from wind with a velocity v, by a wind turbine with diameter d ? The actual power output of the turbine must be less than this, why?

1.7 Compare the expansion work associated with vaporization of 1 kg of water at 1 bar to that of conversion of 1 kg of kyanite to sillimanite at 3 kbar. The molar volumes in J bar^{-1} mol^{-1} are: liquid water $= 1.807$, sillimanite $= 4.986$, kyanite $= 4.415$. Assume that water vapor at 373 K and 1 bar is an ideal gas.

1.8 Starting from equation (1.18), derive the complete equation for expansion work of a sphere, and show why (1.19) is an acceptable approximation for an infinitesimal expansion.

1.9 Derive the exact expression for the energy stored during a change in volume from V_0 to V, caused by a change in pressure from P_0 to P, of an elastic solid with constant bulk modulus K_s. (Hint: this is a good opportunity to practice your *Maple*.) Compare this expression with equation (1.31), and discuss the conditions under which the constant volume assumption is no longer tenable.

1.10 Derive an equation that relates elastic energy stored per unit volume (equation (1.37)) to change in gravitational potential per unit volume (mass stays constant, what changes is elevation). The bulk modulus of silicate minerals is of order 1000 kbar, their density ~ 3000 kg m^{-3}, and $g = 9.8$ m s^{-2}. Assume that an earthquake causes uplift of 1 m. Estimate the strain that was released (the non-dimensional quantity ε). How does this value vary with bulk modulus? Comment on your results.

1.11 Derive K_T, β_T and α for an ideal gas, starting from the ideal gas EOS.

1.12 Bulk modulus and coefficient of thermal expansion are two ways of relating the three variables, P, V, T. There is a third one, that is to take the derivative of P relative to T at constant V. This gives rise to another material property known as the *Grüneisen* parameter and symbolized by γ. The Grüneisen parameter measures the increase in pressure caused by heating at constant volume. It can be defined as follows:

$$\gamma = \frac{V}{C_V}\left(\frac{\partial P}{\partial T}\right)_V.$$

Show that the Grüneisen parameter of an ideal gas is a constant, and that it is related to α and K_T by:

$$\gamma = \frac{\alpha K_T V}{C_V}.$$

1.13 Derive a general equation that relates Cp to Cv in terms of material properties. Use it to show that for an ideal gas $Cp = Cv + R$. Discuss why for solids and liquids it is $Cp \approx Cv$.

1.14 Estimate a possible range of heating rates (dT/dt) by viscous dissipation along subduction zone megashears. Can this be a significant contribution to the origin of arc magmatism?

For the following problems use standard state thermodynamic data from Holland and Powell (1998).

1.15 Calculate the enthalpy of formation of diamond at 1 bar, 298.15 K and at 1 bar, 1000 K. Do this by hand and using the *Maple* procedures described in Software Box 1.1.

1.16 Use the *Maple* procedures described in Software Box 1.1 to calculate the enthalpy change of the spinel–garnet transition (reaction (1.94)) at 1 bar, 298.15 K and at 1 bar, 1000 K. You should get ~ 23.29 kJ and 11.76 kJ, respectively. If you don't, you need to practise your *Maple*.

1.17 Calculate the enthalpy change of the oxidation of fayalite to hematite + quartz at 1 bar, 298.15 K. Is this reaction endothermic or exothermic?

1.18 Average terrestrial heat flux is 80 mW m^{-2}. At what rate must a 1 m thick slab of fayalite oxidize in order to supply energy at this rate? (Hint: you need the density fayalite but Holland and Powell supply its molar volume.) Supposing that the energy yield of fayalite oxidation is typical of silicate chemical reactions, what can you conclude about the source of the Earth's internal heat? You will need to develop some scaling arguments, and to ignore for now heat transfer complications.

Energy sources in planetary bodies

Planetary bodies can be thought of as combinations of heat reservoirs and heat engines. The heat reservoirs store internal energy, E, and the heat engines convert some of this thermal energy into various types of mechanical, electrical and chemical energies. This simple physical picture is true of all active planetary bodies, regardless of their composition (rocks, gases or ices) or size. The details, however, vary widely throughout the Solar System. In this chapter we discuss the storage of thermal energy in planetary bodies.

We begin by distinguishing internal from external heat reservoirs, and we define the latter as those that derive their energy from solar electromagnetic radiation. External heat reservoirs occur in surface and near-surface environments. Examples include the Earth's oceans and atmosphere. Internal heat reservoirs store energy at various depths, from near-surface environments to the planet's core. They are fed by dissipation of various types of non-thermal energy but there is one unifying characteristic, which is that dissipation takes place deep enough that the rate of heating exceeds the rate of heat transfer to the planet's surface (Chapter 3). The relative magnitudes of the energy fluxes from external and internal reservoirs at a planet's surface vary widely among the bodies of the Solar System. In solid planetary bodies (rocky and icy) surface energy flux is typically dominated by solar radiation, despite the fact that internal energy reservoirs in some of them are large enough to make noticeable, perhaps dominant, contributions to the planet's surface features. At the present time the surfaces of Earth, Venus, Io, Europa, Ganymede, Titan, Enceladus and Triton are being affected by processes fueled by internal energy reservoirs. In contrast, internal energy reservoirs for Mars, Mercury, the Moon, most other satellites of the giant planets and the asteroids appear to be negligible, but this was certainly not true at earlier times, as shown for instance by the huge Martian volcanoes and the Lunar maria. Internal energy reservoirs dominate surface energy fluxes in the fluid planets Jupiter, Saturn and Neptune. The relentlessly violent weather patterns of Jupiter and Saturn are driven by energy extracted from those planets' internal heat reservoirs, in stark contrast with Earth's weather, which is driven by solar energy. The contributions of internal and external reservoirs may be more or less evenly matched in Uranus.

Here we focus on the sources of planetary internal energy and on the pathways by which non-thermal energy is dissipated. Ultimately, the possible sources of thermal energy are just three: dissipation of gravitational potential energy, dissipation of nuclear binding energy, and dissipation of electrical energy. Gravitation was the chief internal energy source in the formative stages of all planets, and is still today the dominant internal energy source in some of the moons of the giant planets, and perhaps in Saturn as well. Radioactive heating is dissipation of nuclear binding energy. At present it is an important source of energy in some rocky planets (notably Earth, almost certainly Venus as well), and was even more important in the early Solar System, as a result of the greater abundances of long-lived radioactive isotopes (^{40}K, ^{232}Th, ^{235}U and ^{238}U) and of the existence of short-lived radioisotopes that are now extinct (chiefly, ^{26}Al, ^{60}Fe and ^{53}Mn). Dissipation of electrical energy by ohmic heating occurs if a planet with an electrically conductive layer is immersed in a time-varying magnetic field (Section 1.8.3). It may have been important in the early Solar System if the magnetic field of the nascent Sun was orders of magnitude stronger than today's.

2.1 Planetary heat flows

Planets are bathed by solar electromagnetic radiation with a spectrum that corresponds to black body emission at the temperature of the solar photosphere, ~6000 K (Section 12.2). This spectrum peaks in the range of visible wavelengths (~0.4–0.8 μm) or, more accurately, this is the range of visible wavelengths because we evolved on a planet in which most of the energy that reaches us from the central star is in this range of wavelengths. Some of this energy is immediately reflected back to space. The balance is *thermalized*, which means that it is absorbed by the planet's atmosphere and solid or liquid surface (if the planet has one) and eventually radiated back to space in the infrared part of the spectrum, with wavelengths of the order of 5–200 μm.

Electromagnetic radiation is described by the Stefan–Boltzmann law, which we will discuss in more detail in Chapter 13. The total amount of energy radiated per unit area and per unit time (called the radiated *energy flux,* where flux means quantity per unit of area per unit of time), also called the *irradiance,* is symbolized by F and given by:

$$F = \epsilon \sigma T^4. \tag{2.1}$$

In this equation σ is a constant known as the Stefan–Boltzmann constant (Appendix 1), T is the absolute temperature and ϵ is a parameter ($\epsilon \leq 1$), called emissivity, which describes the efficiency with which the body radiates electromagnetic energy. For a perfect black body, i.e. a body that emits radiation with equal efficiency at all wavelengths, $\epsilon = 1$. If a planet is in thermal equilibrium with solar radiation and the planet has no internal heat flow then its energy output, F_t, would derive only from thermalized sunlight and would correspond to emission from a black body at a temperature T_{eq}, given by equation (2.1) and called the planet's *equilibrium temperature* (we will discuss this in detail in Section 13.2). The values of F_t and T_{eq} are calculated on the basis of the incident solar flux and the planet's albedo, A, which is the fraction of incident sunlight that is reflected back to space (see, for example, Hubbard, 1984; de Pater & Lissauer, 2001).

The *measured* infrared emission of a planet, F_m, is not necessarily equal to the *calculated* value of the equilibrium thermalized flux, F_t. The measured value of F_m yields the planet's effective temperature, T_{ef} (equation (2.1)). If $T_{ef} > T_{eq}$ then the planet liberates internal heat. The internal heat flux, q, is given by (this equation is derived rigorously in Section 13.3.2, equation (13.35)):

$$q = F_m - F_t = \epsilon \sigma \left(T_{ef}^4 - T_{eq}^4 \right). \tag{2.2}$$

In practice, given the uncertainties in the measurements of F_m, in the albedo, A (which enters in the calculation of T_{eq}), and in the emissivity ϵ, a planet's internal heat flux can be determined in this way only if q, F_m and F_t are of the same order of magnitude. This is true for Jupiter, Saturn, Neptune and Io, and is how the average internal heat fluxes for these bodies (given in Table 2.1) have been determined. For Uranus T_{ef} appears to be only slightly greater than T_{eq}, so that Uranus's internal heat flux remains somewhat uncertain.

Table 2.1 Planetary heat flow parameters				
Planet	q(W m^{-2})	q_M(W kg^{-1})	T_{eq} (K)	T_{ef}(K)
Earth	0.08	6.40×10^{-12}	263	
Moon	0.03	1.34×10^{-11}	277	
Mars	0.04	9.00×10^{-12}	222	
Io	2.50	1.17×10^{-9}	100	109
Jupiter	5.44	1.76×10^{-10}	113	124
Saturn	2.01	1.51×10^{-10}	83	95
Uranus	0.04	3.91×10^{-12}	58	59
Neptune	0.43	3.22×10^{-11}	48	59

Values of q from Lodders and Fegley (1998), except Mars, from Carr (1999).
Values of q_M calculated from q and planetary data from Lodders and Fegley (1998).
Values of T_{eq} and T_{ef} from de Pater and Lissauer (2001), except Io from Hubbard (1984).

For all the terrestrial planets $q \ll F_m$, so that q must be measured directly (in Chapter 3 we will discuss how). Such direct measurements have only been carried in the Earth and Moon, but whereas the terrestrial value listed in Table 2.1 is an average of hundreds of thousands (if not millions) of measurements, the lunar value is an average of only two, at the Apollo 15 and 17 landing sites, and thus of very uncertain significance. The Martian heat flux listed in Table 2.1 (after Carr, 1999) is an estimate based on geochemistry (areochemistry?), which we discuss further below and in Chapter 3. No numerical estimates are available for the other terrestrial planets nor for the major moons of the giant planets (except Io), but some qualitative statements are possible. Given Venus's many active volcanoes and similarity to Earth in size and density (probably reflecting a similarity in composition) it is not unreasonable to suppose that its time-averaged heat flux may be comparable to that of Earth – the reason for the qualifier "time-averaged" will become clear in Chapter 3. The Jovian moons Europa and Ganymede, Saturn's moons Titan and Enceladus and Neptune's moon Triton all display youthful surfaces, and in some cases visibly active processes of internal origin as well, implying the existence of non-negligible internal heat flows.

The contributions of internal and external heat reservoirs in shaping a planet's surface can be gauged by comparing the equilibrium thermalized solar flux F_t, to its internal heat flux, q (Fig 2.1). With the notable exception of Io, the thermalized solar flux on the surfaces of the solid planets is 10^3–10^4 times greater than the internal heat flux. A way of interpreting this difference is that, *on average*, the rate of external processes is a few orders of magnitude faster than that of internal processes. An equivalent statement for Earth is that active mountain building and active volcanic areas affect a small fraction of the planet's surface, whereas climate, the hydrologic cycle and the biosphere cover the entire planet. In contrast, the entire surface of Io is volcanically active (see, for example, Davies, 2001). Solar energy fluxes in the giant planets are 2–3 orders of magnitude less than in the terrestrial planets. The Dutch–American planetary astronomer Gerard Kuiper noted in the 1950's that this presented a problem in view of the weather systems observed in Jupiter and Saturn, which are much more violent and constant than those of Earth. He inferred that these planets must have a large internal energy output (Kuiper, 1952), and this is indeed the case. In the giant planets (except perhaps Uranus) and Io internal heat flux is 10–100 times greater than in

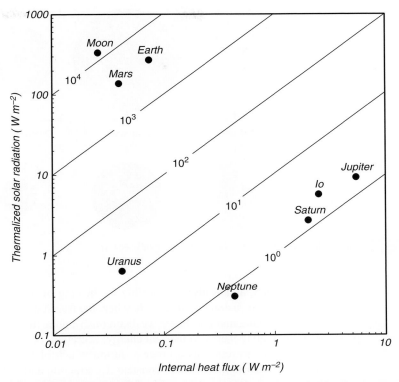

Fig. 2.1 Thermalized solar radiation compared to internal heat flux of some Solar System objects. Internal heat fluxes of Mars and Moon are highly uncertain.

the terrestrial planets, and of the same order of magnitude as the thermalized solar flux (Fig. 2.1).

A useful parameter when comparing planets and energy dissipation mechanisms is the planetary heat flow per unit mass, q_M, which is also listed in Table 2.1. Io is the largest anomaly, with a heat flow per unit mass some three orders of magnitude greater than that of Earth, and 1–2 orders of magnitude greater than those of the giant planets.

2.2 Dissipation of gravitational potential energy

Assembly of planets by coalescence of dust particles and gas molecules is accompanied by liberation of gravitational potential energy (Section 1.3.1). Planetary growth almost certainly takes place in different stages. The details of the various processes involved are not fully understood and may have differed for rocky vs. fluid planets (see, for example, Chambers, 2005). Regardless of the exact pathway, however, the gravitational potential energy of the diffuse mass of nebular material that ends up being a planet is ultimately dissipated as heat. Part of this heat is stored in the planet as internal energy, E. Dissipation of gravitational potential energy is the most important source of internal energy during

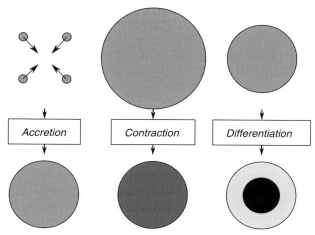

Fig. 2.2   Three pathways for dissipation of gravitational potential energy.

planetary formation, and is also responsible for heating of proto-stars up to the point at which thermonuclear fusion begins and they become true stars.

For the sake of mathematical clarity, we distinguish among three mechanisms that convert gravitational potential energy to thermal energy: accretion, contraction and differentiation (Fig. 2.2). I define accretion as a process during which planetary mass increases while density and mass distribution remain constant. Contraction and differentiation entail mass redistribution at constant total mass. The following is more specific.

Accretion corresponds to assemblage of the planet from particles that are initially considered to be gravitationally unbound to one another. Planetary mass grows with time and the gravitational potential energy of the added matter is dissipated as heat. We will make the simplifying assumption that the result of accretion is a planetary body of homogeneous composition and density. Assembly of the terrestrial planets is probably best modeled as accretion.

Contraction dissipates gravitational potential energy at constant mass as a result of a decrease in planetary radius, and consequent increase in density (equation (1.6)). This mechanism may be an appropriate model for the formation and early evolution of fluid planets, and is also the mechanism responsible for heating proto-stars.

Differentiation of an initially homogeneous planet entails separation of a dense phase that sinks towards the center of the planet relative to a less dense phase that comes to occupy its outer layer. The mass redistribution that accompanies differentiation dissipates gravitational potential energy. The resulting thermal energy is in this case deposited in the deep planetary interior, so that this mechanism may be particularly efficient in augmenting a planet's internal energy content. It is thought to be of critical importance in the thermal evolution of both solid and fluid planets.

Tidal dissipation is another source of internal energy that, ultimately, relies on gravitational potential energy. The pathway by which mechanical energy is dissipated by tides is, however, different from the other processes and will be discussed separately.

Accretion, contraction and differentiation may operate simultaneously or successively during planetary evolution. The terrestrial planets almost certainly formed by accretion, but it is less clear whether the fluid planets are chiefly the result of accretion or of contraction of a mass of diffuse nebular material, or a combination of both processes. The purpose of the following sections is to develop the mathematical description of the various aspects of gravitational heating, rather than to elucidate the mode of formation of any particular planet.

2.3 Gravitational binding energy

The total amount of gravitational potential energy liberated during formation and subsequent evolution of a planetary body to a given point in time is measured by its *gravitational binding energy*, U_B, at that time. The magnitude of a planet's gravitational binding energy provides clues and constraints about the planet's thermal history and internal energy sources. The concept of gravitational binding energy is analogous to that of lattice energy of a crystal (Section 1.8.2), except that during formation of a crystal the potential energy that is dissipated arises from work performed by electrostatic rather than gravitational forces. Calculation of gravitational binding energy is generally much simpler than calculation of lattice energies.

Consider the assembly of a self-gravitating celestial body by infinitesimal mass increments. *Self-gravitating* means that the body is held together only by gravitational forces arising from the mass of the body itself. Planets and stars are obvious examples of self-gravitating celestial bodies, and so are larger structures such as galaxies and galaxy clusters. Consider a body with spherical symmetry and let m be the mass of the sphere of radius r (Fig. 2.3). Addition of an infinitesimally thin shell, dr, results in a mass increment dm and changes the gravitational binding energy of the planet by an amount dU_B equal to the gravitational potential energy of the added mass. From equation (1.6) we have:

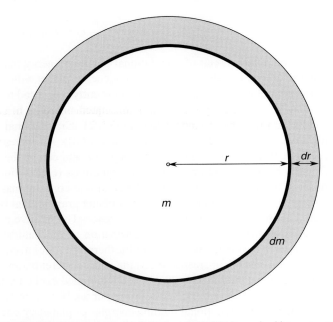

Fig. 2.3 Calculation of gravitational binding energy. Each infinitesimal shell of thickness dr adds an amount of mass dm to the planet.

$$dU_B = -\frac{Gm\,dm}{r}.$$ (2.3)

The gravitational binding energy of the fully assembled planet is obtained by integrating this expression over the mass of the planet, M:

$$U_B = -G\int_0^M \frac{m}{r}dm.$$ (2.4)

We assume initially that the planet is homogeneous and with mean density ρ. Radius is then simply related to mass by:

$$r = \left(\frac{3m}{4\pi\rho}\right)^{\frac{1}{3}}.$$ (2.5)

Substituting into equation (2.4) and integrating:

$$U_B = -\frac{3}{5}G\left(\frac{4\pi\rho}{3}\right)^{\frac{1}{3}}M^{\frac{5}{3}}.$$ (2.6)

Calling R the radius of the planet of mass M, we use equation (2.5) in order to simplify equation (2.6) to:

$$U_B = -\frac{3}{5}G\frac{M^2}{R},$$ (2.7)

which yields the gravitational binding energy U_B of a homogeneous planet of mass M and radius R. In terms of R and ρ:

$$U_B = -\frac{16}{15}G\pi^2 R^5 \rho^2.$$ (2.8)

This is the total amount of gravitational potential energy that has been converted to other types of energy in order to make a homogeneous planet with density $\rho = M(4/3\pi)^{-1}R^{-3}$. Gravitational binding energy may be transiently converted to kinetic energy of dust grains or planetesimals, but it is ultimately dissipated as heat, when these particles stick together following inelastic collisions (Section 1.3.2). Gravitational binding energy is always a negative quantity (see equations (2.7) and (2.8)). An increase in its absolute magnitude ($\Delta U_B < 0$) corresponds to energy dissipation and can be the result of an increase in mass (accretion), a decrease in radius at constant mass (contraction) or redistribution of denser material towards the center of the planet, also at constant mass (differentiation).

During formation and evolution of a planet gravitational binding energy may be either stored as internal energy or radiated to space. The relative magnitude of these two terms depends on the rate and mode of dissipation of mechanical energy and on the material properties of the planet. We will examine thermal evolution during accretion and contraction in Sections 2.4 and 2.5, respectively. Planetary differentiation leads to an additional release of gravitational binding energy that is not included explicitly in equation (2.7), and that we calculate in Section 2.6. Before refining our analysis, however, we will carry out a rough estimate of the gravitational binding energies of planetary bodies on the basis of only their sizes and mean densities, and draw some conclusions about their thermal histories and possible energy sources.

Let us compare observed planetary heat flows (Section 2.1) with the heat flows that could
be expected if all the gravitational binding energy of a planet had been stored as internal
energy during formation of the planet and then released at a constant rate over the age of
the Solar System. If the average heat flux (energy per unit of area per unit of time) of a
planet of mass M and radius R is q, the heat flow per unit mass q_M (listed in Table 2.1) is
given by:

$$q_m = \frac{4\pi R^2 q}{M}.$$ (2.9)

The binding energy per unit mass is (from equation (2.7)):

$$u_B = -\frac{3}{5} G \frac{M}{R}.$$ (2.10)

If this gravitational binding energy was dissipated at the beginning of the Solar System and
stored as internal energy, and the internal energy was then released at a constant rate since
then, we should observe an internal heat flow per unit mass of order $q_B \approx u_B/t_S$, where
$t_S = 4.55 \times 10^9$ years $= 1.44 \times 10^{17}$ s is the age of the Solar System. Values of q_M and
q_B for various planetary bodies are plotted in Fig. 2.4. Ignoring for a moment Mars and
the Moon (because heat fluxes for these two bodies are poorly constrained, see Section 2.1),

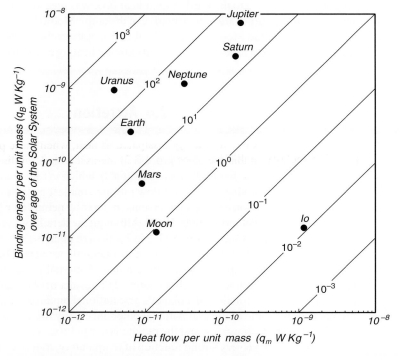

Fig. 2.4   Gravitational binding energy compared to planetary heat flow. Specific values (i.e. per unit mass) are shown, and u_B is
divided by the age of the Solar System. Only Io emits more energy than can be accounted for from its gravitational
binding energy. Heat flow values for Mars and the Moon are highly uncertain.

we see that, with one notable exception, planetary binding energies are one to two orders of magnitude *greater* than the observed energy outputs or, in other words, observed heat fluxes are *significantly less* than those that could be expected from steady release of all of the gravitational binding energy over the age of the Solar System. This discrepancy can be the result of a combination of at least three factors: (i) early in their histories the planets cooled at rates that were much faster than today's, (ii) heat transfer inside the planets is very inefficient, so that a large fraction of the dissipated binding energy is still stored inside the planets, and (iii) only a fraction of their gravitational binding energies were retained as internal energy, with the balance being radiated to space during planet formation.

The one notable exception to these conclusions is Io, for which measured heat flow is two orders of magnitude *higher* than what can be accounted for from gravitational binding energy. The conclusion in this case is that Io, which is the most volcanically active planetary body in the Solar System (see, for example, Davies, 2001) must have a major heat source that is not active to any significant extent in any of the other bodies included in Fig. 2.4.

Focusing now on the planets that appear to show an excess of gravitational binding energy, we see that there are some subtle differences. The ratio of binding energy to heat flow for Earth, Jupiter and Neptune is almost exactly the same ($q_B/q_M \approx 40$). Compared to these three planets, Saturn puts out twice as much heat relative to its binding energy ($q_B/q_M \approx 18$), whereas Uranus radiates one sixth less heat ($q_B/q_M \approx 240$). Saturn's large energy output is thought to reflect a process that is not significant at present in any other planet (see Stevenson & Salpeter, 1977; Hubbard, 1980; see also Section 2.6).

Martian heat flow has not been measured yet. The value quoted in Table 2.1 and plotted in Fig. 2.4 is an estimate based on the bulk chemical composition of Mars, which appears to be enriched in K relative to the Earth. Lunar heat flow was measured in two of the *Apollo* landing sites. If these values are representative of the entire Moon then, by comparison with other terrestrial planets the Moon's internal heat flux is greater than expected. This is rather surprising, and may be a relic of a short-lived extreme heating event in the Moon's early history.

2.4 Accretion

It is virtually certain that the chief process in the assemblage of the terrestrial planets was accretion of smaller bodies (called planetesimals), which in turn were the products of accretion of yet smaller bodies, and so on down to the smallest particles of dust and ice which formed by condensation of gaseous elements in the solar nebula (see Weidenschilling, 1974, 1976, 1980, 2000; Wetherill, 1990, 1994). Although the details of the accretionary processes that led from microscopic dust particles to Earth-size planets may not be fully understood it is possible to place some constraints on its energetic aspects. The gravitational binding energy of a planet is an upper bound to the amount of internal energy that can be derived from dissipation of gravitational potential energy. The results of Worked Example 2.1 suggest, however, that only a fraction of a planet's gravitational binding energy may be converted to internal energy. Some of the gravitational potential energy dissipated during accretion must be radiated to space, however, and the radiated energy flux, given by the Stefan–Boltzmann law (equation 2.1) is a very strong function of temperature, as it scales with the fourth-power of T.

We want to find out what fraction of a planet's gravitational binding energy is stored as internal energy during accretion. Let us assume that a planetary body grows by accretion at a rate that is fast enough that no significant heat transfer occurs from the planetary interior to its surface, so that heat is lost only by radiation from the material that makes up its surface at any given instant. We will revisit this assumption in Chapter 3 and convince ourselves that it is indeed a very good model for planetary accretion. We consider an infinitesimal increment in the planet's mass, dm, taking place in an infinitesimal time dt. The gravitational binding energy dissipated by accretion of this mass raises its temperature from the initial temperature that it had in the solar nebula, T_0, to the instantaneous temperature at the current planetary surface, T. Our assumption about negligible heat transfer rate from the planet's interior means that once this surface layer is buried by continued accretion its temperature remains constant. We describe energy balance during accretion with the general formulation of the First Law of Thermodynamics:

$$dE = dQ - dW. \tag{2.11}$$

The change in internal energy, dE, corresponds to the difference between the internal energy of the added mass at the instantaneous temperature of the planet's surface and its internal energy at the background temperature of the solar nebula. We consider accretion to be a constant pressure process for which $P = 0$, as it occurs at the surface of a planet with no atmosphere. We need an expression for $(\partial E / \partial T)_P$. The method to obtain this expression is explained in Appendix 2, where we show that:

$$\left(\frac{\partial E}{\partial T} \right)_P = C_P - P\alpha V = C_P, \tag{2.12}$$

and the second identity results from setting $P = 0$. In terms of specific heat capacity (equation (1.54)) and assuming that c_P is constant between T and T_0 we rewrite (2.12) as:

$$dE = c_P (T - T_0) dm. \tag{2.13}$$

The change in gravitational binding energy, dU_B, is balanced by mechanical work performed *on* the planet, $-dW$, i.e:

$$dU_B + (-dW) = 0 \tag{2.14}$$

so that the work performed *by* the system (i.e. the planet) is, from equation (2.3):

$$dW = dU_B = -\frac{Gmdm}{r} = -\frac{4}{3}\pi Gr^2 \rho dm, \tag{2.15}$$

where r is the instantaneous radius of the growing planet at the time when its mass is m and its surface temperature is T. A way of seeing why the signs must balance as in equation (2.14) is to realize that the system always performs work against gravity. The work performed by the system, dW, must then equal the change in gravitational potential energy. In order for the gravitational potential energy to increase the system must perform (positive) work, whereas negative work by the system corresponds to a decrease in gravitational potential energy.

The planet loses thermal energy by radiation. The infalling mass would also radiate at its nebular temperature T_0, so that the heat loss that corresponds to dissipated gravitational binding energy is the difference between the energy radiated at T and that radiated at T_0, $F - F_0$ (equation (2.1)). The net thermal energy lost over the entire surface area of the planet and over the time increment dt is thus given by:

$$dQ = -4\pi r^2 (F - F_0) dt = -4\pi r^2 \epsilon\sigma \left(T^4 - T_0^4 \right) dt. \tag{2.16}$$

Substituting equations (2.13), (2.15) and (2.16) in (2.11) we obtain:

$$c_P \left(T - T_0 \right) dm = -4\pi r^2 \epsilon \sigma \left(T^4 - T_0{}^4 \right) dt + \frac{4}{3}\pi G r^2 \rho \, dm. \qquad (2.17)$$

A slight rearrangement of this equation highlights its meaning: that the rate of dissipation of gravitational binding energy equals the rate of increase in internal energy plus the rate at which thermal energy is radiated from the planet's surface:

$$\frac{4}{3}\pi G r^2 \rho \frac{dm}{dt} = c_P \left(T - T_0 \right) \frac{dm}{dt} + 4\pi r^2 \epsilon \sigma \left(T^4 - T_0{}^4 \right). \qquad (2.18)$$

This is a fourth-degree equation in T, the temperature at the surface of a body when its radius is r and the mass accretion rate is dm/dt. We can use equation (2.18) to construct the temperature profile of a planet resulting from accretion, $T = T(r)$, by specifying an accretionary growth model that gives dm/dt as a function of r, or, equivalently, of accreted mass, m. We do this in the following Worked Example.

Worked Example 2.2   Energy dissipation, radiation and storage during planetary accretion

As an example of the application of equation (2.18), we will use it to calculate possible initial temperature profiles for the Earth, and to estimate what fraction of the Earth's gravitational binding energy might have been stored as internal energy during accretion. The problem is solved numerically, by considering short time increments, calculating the mass, radius and accretion rate at the end of each time increment, using those values to solve equation (2.18), and iterating until the total mass of the planet has been accreted. A *Maple* procedure that accomplishes this is described in Software Box 2.1.

Software Box 2.1   Calculation of temperature profile and internal energy storage during planetary accretion

The calculations discussed in Worked Example 2.2 are implemented in the *Maple* worksheet `accretion.mw`, which contains several procedures. The first procedure sets the values of physical constants and model parameters, except for the accretion time. Planetary mass and density, nebular temperature (T_0), specific heat and emissivity must be changed in this procedure, if desired. Procedure `solveT` uses *Maple*'s numerical solver to find a positive real value of T from equation (2.18). It requires the values of r and dm/dt. These are calculated by one of the following procedures, `linear_growth`, `exponential_growth` or `sine_growth`, that implement equations (2.20) through (2.22) and pass the values of m and dm/dt to `solveT` via the utility procedure `calc_out`. The three procedures iterate to the specified accretion time using a time step that can be adjusted by modifying the total number of steps desired. Output is sent to a text file which can be read and manipulated by plotting or spreadsheet programs. The name and format of the output file can be modified inside each of the procedures that describe the growth models. When execution of a model is finished *Maple* returns total gravitational binding energy, energy stored as internal energy, and the ratio between these two quantities – this information is not saved in the output files.

The first task is to specify an accretionary growth model so as to obtain numerical values for the mass accretion rate, dm/dt as a function of radius, r. We will consider three distinct models: (i) a linear model in which accretion rate is constant, (ii) a model in which accretion rate decreases exponentially with time (see Wetherill, 1990), and (iii) a model in which accretion rate increases rapidly, reaches a maximum and then decreases rapidly to zero (Hanks & Anderson, 1969). The accretion models are best described in terms of accretion rates normalized to the total mass of the assembled planet, M. The three models are described by the following equations, where $\mu = m/M$:

$$\left(\frac{d\mu}{dt}\right)_{(i)} = \frac{C}{M}$$

$$\left(\frac{d\mu}{dt}\right)_{(ii)} = \frac{C}{M}e^{-kt} \tag{2.19}$$

$$\left(\frac{d\mu}{dt}\right)_{(iii)} = \frac{C}{M}t^2 \sin kt,$$

where the dimension of $d\mu/dt$ is $[T]^{-1}$, and the constants C and k are functions of the accretion time t_a. In every case we set $\mu = 0$ at $t = 0$ and $\mu = m$ at $t = t_a$. For the linear model we easily find:

$$\left(\frac{d\mu}{dt}\right)_{(i)} = \frac{1}{t_a}$$

$$(\mu)_{(i)} = \frac{t}{t_a}. \tag{2.20}$$

For the exponential model we set $k = 1/t_a$, meaning that the accretion rate becomes exponentially slow as the accretion time is reached. Integrating between 0 and t_a we find:

$$\left(\frac{d\mu}{dt}\right)_{(ii)} = \frac{1}{t_a}\frac{e^{-t/t_a}}{\left(1 - e^{-1}\right)}$$

$$(\mu)_{(ii)} = \frac{1 - e^{-t/t_a}}{1 - e^{-1}}. \tag{2.21}$$

For the sinusoidal model we make $dm/dt = 0$ at $t = t_a$ (see also Exercise 2.1) and we get:

$$\left(\frac{d\mu}{dt}\right)_{(iii)} = \frac{1}{t_a}\frac{\pi^3}{\pi^2 - 4}\left(\frac{t}{t_a}\right)^2 \sin\left(\pi\frac{t}{t_a}\right)$$

$$(\mu)_{(iii)} = \frac{1}{\pi^2 - 4}\left[-\pi^2\left(\frac{t}{t_a}\right)^2 \cos\left(\pi\frac{t}{t_a}\right) + 2\cos\left(\pi\frac{t}{t_a}\right) + 2\pi\frac{t}{t_a}\sin\left(\pi\frac{t}{t_a}\right) - 2\right]. \tag{2.22}$$

Figure 2.5 shows $d\mu/dt$ and m/M as a function of time for the three accretion models, for an accretion time $t_a = 10$ million years. The function for planetary radius, $r = r(t)$ is obtained by substituting $m = m(t)$ in equation (2.5), with the simplifying assumption that the density of the planet remains constant during accretion – we will tackle density inhomogeneities caused by contraction and differentiation in later sections.

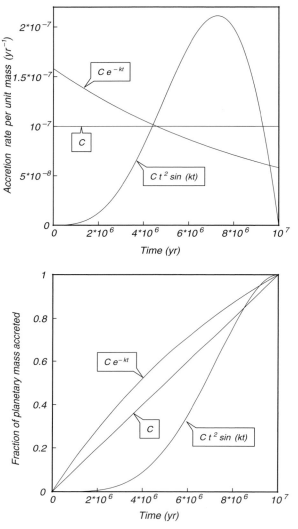

Fig. 2.5 Specific accretion rates (per unit mass) and accretionary histories for linear, exponential and sinusoidal accretion models. Accretion time is 10 million years.

Figure 2.6 shows accretion temperature profiles for the Earth for an accretion time of 10 million years and nebular temperature $T_0 = 100$ K, and assuming that the emissivity of the accreting planet was 1. This means that the figure shows minimum temperatures, as any value of $\varepsilon < 1$ will cause a larger fraction of the binding energy to be stored as internal energy (equation (2.18)). Accretion models predict that the center of the planet was initially much colder than its outer layers. This is to be expected from the fact that gravitational binding energy varies as m^2/r or, equivalently, as the fifth power of the radius (equation (2.8)), so that little heating takes place during the initial stages of growth. Thermal convection would not take place with these temperature profiles (Chapter 3), even if the planet were fully molten (which at these temperatures an iron–silicate planet would not). Linear and exponentially decaying growth predict a very steep inverted temperature gradient out to a radius of ∼1000 km, and relatively constant temperatures from that depth to the planet's surface. The differences between both models are minor.

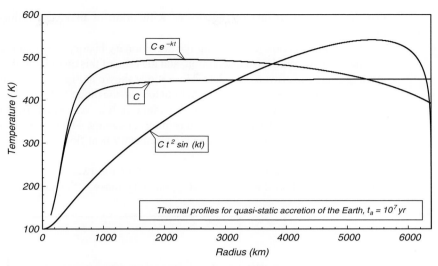

Fig. 2.6 Initial temperature profiles for quasi-static accretion of the Earth, assuming an accretion time of 10 million years and nebular temperature = 100 K. Other parameter values are: $\rho = 5500$ kg m^{-3}, $c_P = 900$ J kg^{-1} K^{-1} and $\epsilon = 1$.

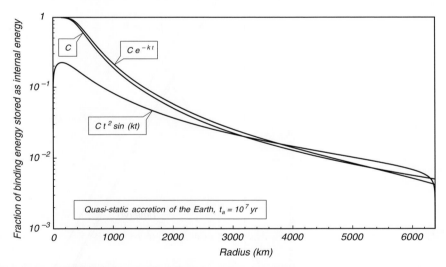

Fig. 2.7 Fraction of gravitational binding energy that is converted to internal energy during accretion of the Earth. Model parameters as in Figure 2.6.

The sinusoidal growth model results in a more gradual temperature increase outwards from the planet's center, and a marked temperature peak near the planet's surface, which reflects the late peak in mass accretion rate (compare Fig. 2.5, see also Hanks & Anderson, 1969).

If release of gravitational binding energy increases strongly with radius (equation (2.8)), then why do accretionary temperatures flatten out and eventually decrease close to the planet's surface? The answer lies in the fact that thermal radiation varies as the fourth power of temperature, so that as temperature increases a rapidly growing fraction of the gravitational energy that is dissipated is radiated rather than converted to internal energy. This can be seen in Fig. 2.7, which shows the fraction of binding energy that is converted to internal energy as a function of planetary radius. The balance is radiated.

Integrating equation (2.13) along the accretionary history we obtain the total amount of internal energy stored during accretion, and the ratio of this quantity to the total gravitational binding energy (Software Box 2.1). This ratio is shown in Fig. 2.8 as a function of accretion time, which for terrestrial planets is thought to be in the range 10^5–10^7 years (see Wetherill & Inaba, 2000). For accretion times in this range only a very small fraction of the binding energy, of the order of 10^{-2}, is stored as internal energy. The rest is radiated to space. It is thus not necessary to postulate exorbitant planetary heat flows in the early solar system to explain the discrepancies noted in Fig. 2.4.

Internal energy varies by a factor of ~3 over the two orders of magnitude range in accretion times (Fig. 2.8). As the total planetary mass is the same in all cases, this must

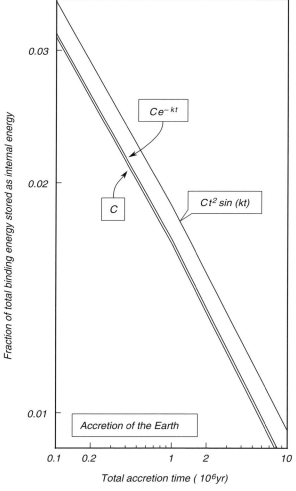

Fig. 2.8 Total fraction of binding energy converted to internal energy for the three Earth accretion models in Figure 2.7, but for variable accretion times, 10^5–10^7 years.

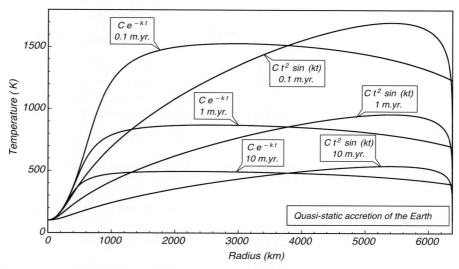

Effect of accretion time (10^5–10^7 years) on initial temperature profile of the Earth, for the exponential and sinusoidal models in Figure 2.6.

reflect a difference of comparable magnitude in planetary temperatures (equation (2.13)). Figure 2.9 shows that even for very fast accretion (10^5 years), however, maximum temperatures attained in the absence of other energy sources ($\sim$1500 K) may not be sufficient to melt an undifferentiated metal–silicate planet. Recall that these calculations assume $\varepsilon = 1$. You can explore other parameter combinations in the end-of-chapter exercises.

A conclusion that we can draw from this example is that accretion may not in general be an efficient means of converting gravitational binding energy to planetary internal energy, unless proto-planetary material has an emissivity significantly lower than 1, but there is an important exception to this statement, which we discuss in the following section.

2.4.1 The effect of large impacts

The picture of accretion as a quasi-static process, in which mass is added to the growing planet in infinitesimal increments, ignores the possibility that catastrophic planet-sized impacts may take place late in the assembly of planetary bodies (see Wetherill, 1985). Such impacts can dissipate enormous amounts of kinetic energy virtually instantaneously. Consider two celestial bodies that collide inelastically. Kinetic energy is dissipated and becomes thermal energy, which is stored as a combination of sensible heat (increase in temperature) and latent heat (melting). Assume that two bodies of masses m and M, and moving with relative speed v, collide head on. In this one-dimensional geometry v is simply the magnitude of the velocity vector $\overline{v_a}$ (Section 1.3.2). The kinetic energy dissipated is given by equation (1.15):

$$\Delta \boldsymbol{U}_k = -\frac{1}{2}mv^2\left(\frac{M}{M+m}\right). \tag{2.23}$$

Because this process transforms kinetic energy to internal energy, we must add a kinetic energy term to the First Law of Thermodynamics (equation (1.56)):

$$dE = dQ - PdV - dU_k, \tag{2.24}$$

where dU_k is the change in the system's kinetic energy. The impact is an adiabatic process, so $dQ = 0$. If we also assume that the energetic consequences of volume changes are negligible compared to the energy dissipated during a collision, we see that the change in internal energy equals the kinetic energy that is dissipated:

$$dE = -dU_k. \tag{2.25}$$

The increase in internal energy per unit total mass of the system, ΔE_m, is given by:

$$\Delta E_m = \frac{1}{2}mv^2 \left(\frac{M}{M+m}\right)\frac{1}{M+m} = \frac{v^2}{2}\left[\frac{Mm}{(M+m)^2}\right]. \tag{2.26}$$

Let the ratio between M and m be k, such that $m = kM$. Equation (2.26) can then be written as follows:

$$\Delta E_m = \frac{v^2}{2}\frac{k}{(k+1)^2}. \tag{2.27}$$

This equation shows that, for a given collision speed, the increase in internal energy per unit mass, and therefore the *average* temperature increase, depends only on the *ratio* of the masses of the colliding bodies, and not on the absolute values of M and m. ΔE_m vanishes as $k \to 0$, so that the *average* temperature increase is negligible when one of the colliding bodies is much smaller than the other one. Of course, this is a global result that ignores the local details: the K–T impact may have caused a negligible increase in the total internal energy content of the Earth, but the local increase in temperature was certainly not negligible. But this is also the point: accretion of a small body onto a much larger one does not lead to a significant increase in the latter's thermal budget. We then ask whether there is a value of k for which ΔE_m (and ΔT) is maximum. On symmetry grounds, we can argue that this must happen when the two bodies have the same mass, i.e. for $k = 1$. In order to check this we look for the extrema of the function:

$$y = \frac{k}{(k+1)^2}. \tag{2.28}$$

The first derivative of this function vanishes at $k = 1$, and the second derivative is negative at $k = 1$, showing that this extremum is a maximum. The increase in internal energy per unit mass is indeed maximum when the masses of the colliding bodies are equal. The maximum value for a given collision speed is obtained by setting $k = 1$ in equation (2.27):

$$\Delta E_{m,\text{max}} = \frac{v^2}{8}. \tag{2.29}$$

We must specify the relative speed of the collision. In general, the maximum speed for an inelastic collision is of the order of the escape velocity at the surface of a body with a mass equal to the combined masses of the impactors, $M + m$. The escape velocity of a planet is the minimum velocity that a free-falling object must have in order to be able to move away from the planet indefinitely. By "free falling" we mean that no other forces besides the gravitational attraction of the attracting planet act on the moving body. The escape velocity is therefore the speed at which the kinetic energy of the body equals its gravitational potential energy (Exercise 2.4). If two bodies collide with a relative speed that is significantly greater than the escape velocity of their combined masses then gravity is not capable of holding the mass together, and the impactors will tend to shatter rather than merge. I will give a more quantitative demonstration of this in a moment. First, note that the escape velocity at the surface of a body of mass M and radius r is given (Exercise 2.4) by $V_e = \sqrt{(2GM)/r}$, which we can write in terms of planetary radius and density as:

$$V_e = \left(\frac{8}{3} \pi G \rho r^2 \right)^{1/2}. \tag{2.30}$$

Substituting in (2.29) we get the following expression for the increase in internal energy per unit mass, for an impact at the escape velocity of the combined masses of the impactors:

$$\Delta E_{m,V_e} = \frac{1}{3} \pi G \rho r^2. \tag{2.31}$$

From equation (2.10) we write the binding energy per unit mass in terms of density and radius:

$$u_B = -\frac{4}{5} \pi G \rho r^2. \tag{2.32}$$

We see that (2.31) and (2.32) are of the same order of magnitude. If the energy dissipated were significantly greater than the binding energy then there would not be enough gravitational attraction to hold the mass together during the collision. But (2.31) is the energy dissipated by an impact at the escape velocity of the combined masses, showing that the escape velocity is the approximate limiting velocity for an inelastic collision.

Because $\Delta E_{m,Ve}$ corresponds to a collision between two bodies of equal mass (and, we will assume, equal density) we can write (2.31) in terms of the radius of each of the impactors, r_i, as follows:

$$\Delta E_{m,V_e} = \frac{2^{2/3}}{3} \pi G \rho r_i^2. \tag{2.33}$$

This function is plotted in Fig. 2.10 for bodies of two compositions: silicate–metal mixtures with $\rho = 4500$ kg m^{-3}, corresponding to chondritic impactors (undifferentiated early Solar System material), and ice bodies with $\rho = 1000$ kg m^{-3}. We can compare these energy values to the specific heat capacities and enthalpies of melting of planetary materials. The composition of chondritic meteorites is roughly one third metal and two thirds silicate. Specific heat capacities per unit mass of silicate minerals are of the order of 10^3 J K^{-1} kg^{-1} and of iron about 5×10^2 J K^{-1} kg^{-1}, so that a typical value for chondritic material may be 0.85 kJ K^{-1} kg^{-1}. The value for H$_2$O at typical nebular temperatures of 200 K is $\sim$1.5 kJ K^{-1} kg^{-1}. The vertical scales on the left of the graph show characteristic temperature increases for both types of materials, based on these heat capacities and the energy dissipation per unit mass given in the vertical axis. We see that for bodies up to 100–200 km in radius the kinetic energy dissipated by collisions is insignificant on a global scale, even though local effects can be devastating. A global temperature increase of the order of 100 K, which may be important in ice-rich bodies at nebular temperatures as it may heat them to their melting point, requires a collision between 1000-km sized bodies.

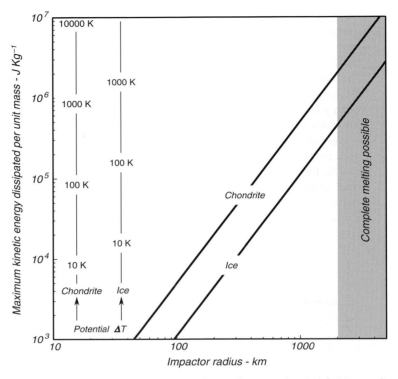

Fig. 2.10 Maximum kinetic energy dissipated per unit mass during inelastic collisions as a function of impactor radius. Values correspond to collision of two identical bodies of the given radius. The temperature scales show how these energies would convert to sensible heat in chondritic and ice bodies. Global melting requires energy dissipation of between 5×10^5 and 2×10^6 J kg^{-1}.

In order for the merged body that results from an inelastic collision to become completely molten we can specify that the temperature increase must be of the order of 100 K (for ice bodies) to a few 1000 K (for rock and metal bodies). We then need to add the specific enthalpies of melting, which are of the order of 3.4×10^5 J kg^{-1} and 5×10^5 J kg^{-1} for ice and rock–metal mixtures, respectively. The result is that a collision between chondritic impactors must dissipate $\sim 2 \times 10^6$ J kg^{-1} in order for the resulting body to become completely molten, whereas complete melting of an ice body requires some 5×10^5 J kg^{-1}. Figure 2.10 shows that for both types of materials the minimum impactor radius that produces complete melting is $\sim$2000 km.

The *global* thermal effects of collisions are generally less severe than what one is perhaps prone to imagine on the basis of the unavoidably catastrophic *local* effects. This changes as the colliding bodies approach planetary size ($\sim$2000 km), so that the very last stages of planetary accretion may have had global thermal consequences (see Wetherill, 1985). The hypothesis for the formation of the Moon that best fits geochemical, geophysical and astrophysical evidence is that it was the result of a collision between a Mars-sized body and the proto-Earth (Hartmann & Davis, 1975; Cameron & Benz, 1991). Such a collision would have dissipated more than enough energy to completely melt the Earth–Moon system (Fig. 2.10).

2.5 Contraction

We define contraction as a constant-mass process in which the initial lengthscale of the contracting body may take a finite value (it differs from accretion in both of these respects). By equation (2.7), contraction ($\Delta R < 0$) at constant mass lowers gravitational binding energy (*increases* its absolute magnitude). We seek the relationship between the change in gravitational binding energy and the change in the body's internal energy content.

Consider a spherical shell in the interior of a planet, of infinitesimal thickness dr. Take a sector of this shell of unit cross-sectional area. The difference in pressure between the bottom and the top of the shell is given by $dP = -g\, dm$, where dm is the mass of the infinitesimally thick sector of unit cross-sectional area, and the negative sign arises because, if $r > 0$ moving outwards then g points in the negative direction (see equation (1.7)). Now $dm = \rho dr$, which leads to the following relationship, known as the *condition of hydrostatic equilibrium*:

$$\frac{dP}{dr} = -g\rho = -\frac{Gm}{r^2}\rho, \tag{2.34}$$

where m is the planetary mass contained inside a sphere of radius r (see Box 2.1). The density, ρ, is in general a function of pressure and temperature, both of which vary with r.

Box 2.1	**Gravitational potential of a solid sphere**

The gravitational potential of spherical bodies turns up in many calculations of gravitational heating. In order to derive the fundamental equations it is necessary to calculate: (a) the gravitational potential of a ring, (b) the gravitational potential of a spherical shell (seen as an assemblage of infinitesimal rings – see Fig. 2.11) and, finally, (c) the gravitational potential of a sphere, which is an assemblage of infinitesimal shells (Figure 2.12).

(a) Consider a ring of radius a and a point P at a distance r along the ring's axis. The distance from P to an infinitesimal element of a ring of mass dm is x. The gravitational potential at P due to this mass element is:

$$d\Phi = -\frac{G}{x}dm. \tag{2.1.1}$$

Given the symmetry of the problem x is constant (Fig. 2.11), so the gravitational potential of the entire ring must be:

$$\Phi = -\frac{G}{x}\int dm = -\frac{GM}{x} = -\frac{GM}{\left(a^2 + r^2\right)^{\frac{1}{2}}}, \tag{2.1.2}$$

where M is the mass of the ring. As an exercise, show (i) that the gravitational acceleration vanishes at the ring's center ($r = 0$) but the gravitational potential does not and (ii) that the gravitational acceleration far away from the ring ($r \to \infty$) is equivalent to the gravitational acceleration of a point mass located at the ring's center – see equation (1.11).

(b) A spherical shell is an assemblage of infinitesimally thin rings (Fig. 2.11). Now, r is the distance between P and the center of the spherical shell, and a is the radius of the shell. The radius of the rings is, $z = a \sin \beta$, and x is the distance between a ring element and P, as before. Let $dx = ad\beta$ be the width of the ring, and M the mass of the shell. The mass of the ring, dm, is given by the ratio between its surface area and

Box 2.1 Continued

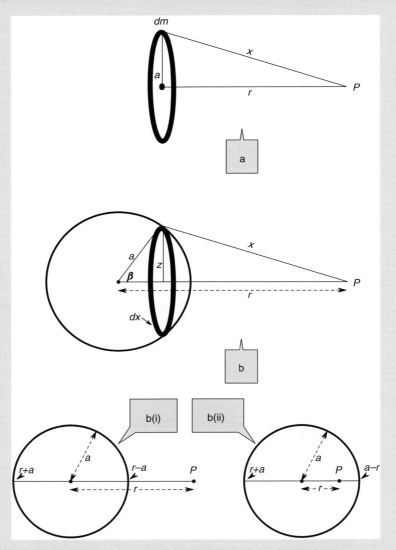

Fig. 2.11 Gravitational potential of a ring (a) and a spherical shell (b), at points outside (b(i)) and inside (b(ii)) the shell.

that of the shell:

$$dm = \frac{2\pi a \sin\beta\, dx}{4\pi a^2} M = \frac{M \sin\beta\, dx}{2a} = \frac{1}{2} M \sin\beta\, d\beta \tag{2.1.3}$$

and, from equation (2.1.2), the gravitational potential caused by the ring at P is:

$$d\Phi = -\frac{GM \sin\beta\, d\beta}{2x}. \tag{2.1.4}$$

Box 2.1	Continued

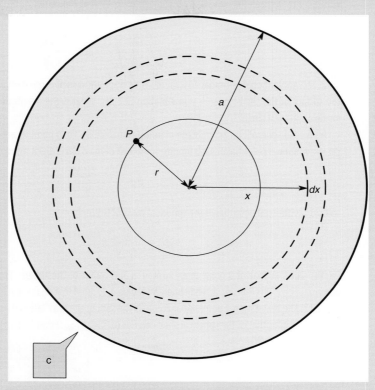

Fig. 2.12   Gravitational potential inside a solid sphere.

From the law of cosines:

$$x^2 = a^2 + r^2 - 2ar\cos\beta. \tag{2.1.5}$$

Differentiating (2.1.5) and solving for $\sin\beta\,d\beta$:

$$\sin\beta\,d\beta = \frac{x\,dx}{ar} \tag{2.1.6}$$

and substituting in (2.1.4):

$$d\Phi = -\frac{GM}{2ar}dx. \tag{2.1.7}$$

We now need to decide on the integration limits, and these are different depending on whether P is located inside or outside the spherical shell (Fig. 2.11). For P outside the shell, case b(i), the gravitational potential is given by:

$$\Phi = -\frac{GM}{2ar}\int_{r-a}^{r+a} dx = -\frac{GM}{r}, \tag{2.1.8}$$

Box 2.1   **Continued**

which is the same as that caused by a point mass located at the center of the shell. At a point P inside the shell, case b(ii):

$$\Phi = -\frac{GM}{2ar} \int_{a-r}^{r+a} dx = -\frac{GM}{a}. \tag{2.1.9}$$

Thus, the gravitational potential anywhere inside a spherical shell is constant, and equal to the potential at the surface of the shell (set $r = a$ in equation (2.1.8)). What is the gravitational acceleration everywhere inside a spherical shell?

(c) A solid sphere is an assemblage of infinitesimal shells. For a point located outside the sphere ($r \geq a$) the gravitational potential of each infinitesimal shell of mass dm is given by equation (2.1.8):

$$d\Phi = -\frac{Gdm}{r} \tag{2.1.10}$$

so the gravitational potential of the sphere of mass M is simply:

$$\Phi = -\frac{G}{r} \int dm = -\frac{GM}{r}, \tag{2.1.11}$$

i.e. the same as that of a point mass located at the center of the sphere. We now seek the gravitational potential at a point P inside a solid sphere of radius a, located at a distance r from the center of the sphere, Fig. 2.12. The gravitational potential at P equals the sum of the gravitational potential of the sphere interior to P, Φ_{inside} plus that of the spherical shell exterior to P, $\Phi_{outside}$. From (2.1.11), we have:

$$\Phi_{inside} = -\frac{GM_{inside}}{r} = -\frac{G}{r}\left(\frac{r^3}{a^3}M\right) = -\frac{GMr^2}{a^3}. \tag{2.1.12}$$

The external (finite) shell is an assemblage of infinitesimal shells, each of them of thickness dx and (variable) radius x – see Fig. 2.12. The mass of each shell is given by:

$$dm = \frac{4\pi x^2 dx}{\frac{4}{3}\pi a^3} M = \frac{3M}{a^3} x^2 dx \tag{2.1.13}$$

and its gravitational potential by equation (2.1.9):

$$d\Phi = -\frac{G}{x}\left(\frac{3M}{a^3}x^2 dx\right) = -\frac{3GM}{a^3} x dx. \tag{2.1.14}$$

The gravitational potential of the shell external to P is, then:

$$\Phi_{external} = -\frac{3GM}{a^3} \int_r^a x dx = -\frac{3GM}{2a^3}\left(a^2 - r^2\right) \tag{2.1.15}$$

so that the gravitational potential at P, $\Phi = \Phi_{inside} + \Phi_{outside}$ is:

$$\Phi = -\frac{GM}{2a^3}\left(3a^2 - r^2\right). \tag{2.1.16}$$

As a final exercise, prove that the gravitational acceleration at P is derived only from the mass internal to P, i.e. from the sphere of radius r in Fig. 2.12. Newton was the first one to prove this (oh, by the way, he had to come up with calculus in the process).

Planetary radius changes during contraction so it is more convenient to choose mass as the independent variable and write our equations in terms of infinitesimal mass increments, dm. In this way the integrals that describe the planetary interior are always taken between zero, at the planet's center, and the total (constant) planetary mass M at the planet's surface. From equation (2.5) we obtain:

$$\frac{dr}{dm} = \frac{1}{4\pi r^2 \rho} \tag{2.35}$$

and, by the chain rule:

$$\frac{dP}{dm} = \frac{dP}{dr}\frac{dr}{dm} = -\frac{Gm}{4\pi r^4}. \tag{2.36}$$

Mass increments dm are positive outwards (i.e. towards the planet's surface) and this equation states that pressure decreases in the same direction, as expected. Equation (2.36) was derived exclusively from the condition of hydrostatic equilibrium, but its right-hand side contains the factor Gm/r, which is the integrand in the equation for gravitational binding energy (equation (2.4)). Multiplying the right-hand side of equation (2.36) by $4\pi r^3$ and integrating over the planetary mass we obtain:

$$-\int_0^M \frac{Gm}{4\pi r^4} 4\pi r^3 dm = -\int_0^M \frac{Gm}{r} dm = U_B. \tag{2.37}$$

Applying the same operations to the left-hand side of equation (2.36) yields the expression:

$$\int_0^M 4\pi r^3 \frac{dP}{dm} dm, \tag{2.38}$$

which can be integrated by parts. By the chain rule:

$$\frac{d}{dm}\left(4\pi r^3\right) = 12\pi r^2 \frac{dr}{dm} \tag{2.39}$$

so that:

$$\int_0^M 4\pi r^3 \frac{dP}{dm} dm = \left[4\pi r^3 P\right]_0^M - \int_0^M 12\pi r^2 P \frac{dr}{dm} dm. \tag{2.40}$$

The first term on the right-hand side of this equation vanishes, because at the planet's center ($m = 0$) it is $r = 0$, and at the planet's surface ($m = M$) it is $P = 0$. Using equation (2.35) to eliminate dr/dm from the integrand in the second term we get:

$$\int_0^M 4\pi r^3 \frac{dP}{dm} dm = -\int_0^M 12\pi r^2 P \frac{1}{4\pi r^2 \rho} dm = -3\int_0^M \frac{P}{\rho} dm \tag{2.41}$$

which, by using (2.36) and (2.37), yields:

$$U_B = -3\int_0^M \frac{P}{\rho} dm. \tag{2.42}$$

This is a fundamental result. It states that, in a self-gravitating body at hydrostatic equilibrium, the gravitational binding energy and the body's thermodynamic state are not independent variables. Both pressure and density are thermodynamic state variables that are related to temperature, and hence to internal energy, by an equation of state. It is thus always possible, at least in principle, to relate the body's gravitational binding energy, U_B, to its internal energy, E. A special case of this result, known as the virial theorem (Worked Example 2.3) was first derived by Clausius in the 1850s. Equation (2.42) is applicable to the formation of celestial bodies, including planets, stars, gravitationally bound star clusters and even galaxies, provided that they evolve by contraction at a rate such that hydrostatic equilibrium is maintained throughout the entire process. *If this is the case then equation (2.42) determines how much of the gravitational binding energy dissipated during contraction becomes internal energy and how much becomes thermal radiation that is lost to space.* This result does not apply to accretion, which is a non-equilibrium process.

Worked Example 2.3   Thermal evolution of a contracting self-gravitating sphere of ideal gas

In order to calculate the relationship between gravitational binding energy and internal energy it is necessary to use an equation of state to express the ratio P/ρ as a function of E. This is very straightforward for an ideal gas, but is the ideal gas equation of state applicable to any type of celestial bodies? It turns out that P–T conditions in many low mass stars and proto-stellar clouds are such that the ideal gas equation of state is indeed a very good description of their material properties, because the characteristic energy of particle interactions is negligible compared to their thermal energy (Section 1.15). The fully ionized matter inside a star with mass comparable to that of the Sun behaves likes a monatomic ideal gas.

The density of the gas is given by $\rho = w/V$, where w is its molecular weight (proper mass!) and V the molar volume. We then have:

$$\frac{P}{\rho} = \frac{PV}{w} = \frac{RT}{w}. \tag{2.43}$$

From Section 1.14.2 we see that the molar internal energy of an ideal gas is given by:

$$E = \frac{f}{2} RT \tag{2.44}$$

and its molar heat capacities by:

$$C_V = \frac{f}{2} R; \quad C_P = \frac{f+2}{2} R, \tag{2.45}$$

where f is the number of translational + rotational degrees of freedom of the gas molecules. The ratio of the heat capacities is a variable that occurs often in thermodynamics so it is given its own symbol, γ, which for ideal gases is:

$$\gamma = \frac{C_P}{C_V} = 1 + \frac{2}{f} \tag{2.46}$$

from which it follows that:

$$RT = (\gamma - 1) E. \tag{2.47}$$

Substituting in (2.43) and then in (2.42) we arrive at:

$$U_B = -3 \int_0^M \frac{P}{\rho} dm = -3(\gamma - 1) \int_0^M e \, dm, \tag{2.48}$$

where $e = E/w$ is the internal energy per unit mass. The integral is taken over the entire mass of the body, so that even though e may vary for each mass increment, the value of the integral is simply E, the total internal energy content of the self-gravitating body of ideal gas. We thus arrive at the following simple yet very powerful result, which is known as the virial theorem (see the classic textbook by Chandrasekhar, 1958, for a beautifully elegant alternative derivation):

$$U_B = -3(\gamma - 1)E. \tag{2.49}$$

Any self-gravitating body of ideal gas at hydrostatic and thermodynamic equilibrium must follow this relationship between its gravitational binding energy and its total internal energy.

The change in gravitational binding energy during contraction ($\Delta U_B < 0$) equals the mechanical work performed by the planet (Exercise 2.5) so that the First Law of Thermodynamics applied to a contracting planet can be written as follows:

$$dE = dQ - dU_B, \tag{2.50}$$

which, rearranging and substituting equation (2.49), becomes:

$$dQ = dE - 3(\gamma - 1)dE = (4 - 3\gamma)dE. \tag{2.51}$$

There is a strict relationship between the heat exchanged by the body, dQ, the change in its internal energy, dE, and the change in its gravitational binding energy, dU_B. For a monatomic ideal gas we have $\gamma = 5/3$, so that in this case:

$$dQ = -dE = \frac{1}{2}dU_B. \tag{2.52}$$

This is a remarkable result. If a self-gravitating body of monatomic ideal gas radiates energy to space ($dQ < 0$), then it must contract. The gravitational potential energy dissipated by this contraction ($dU_B < 0$) is split into two equal parts. One half balances the radiated energy and the other half becomes internal energy ($dE > 0$). As the surface of the body radiates heat to space the temperature in its interior increases. This process is called Kelvin–Helmholtz cooling and is responsible for heating proto-stars to the temperatures needed to ignite thermonuclear fusion reactions in their cores.

It is important to realize that this specific result (equation (2.52)) rests on the assumption that the self-gravitating body is composed of monatomic ideal gas, which is an excellent approximation to the behavior of hot proto-stars (see Section 1.15) but not, for example, of cold fluid planets. Consider for the sake of argument a cold body made up of polyatomic molecules with translational + rotational six degrees of freedom, for which $\gamma = 4/3$. In this case:

$$dQ = 0, \quad -dE = dU_B. \tag{2.53}$$

Such a body would not radiate any of the gravitational potential energy that would be dissipated during contraction. Of course, if the surface temperature of the body is above 2.7 K (Chapter 13) then it will radiate heat to space and contract, but in this case all of the gravitational potential energy that is dissipated goes to raising the body's internal temperature. We will examine this topic again in Chapter 9, in the context of equations of state for fluid planets.

Kelvin–Helmholtz contraction may have been responsible for some fraction of the initial internal energy budgets of the gas and ice giants, but their present day thermodynamic states make this mechanism largely inoperable (Chapter 9). Contraction ceased, probably shortly after formation of the Solar System, when the planets reached density–temperature conditions such that the energies of particle interactions were no longer negligible compared to their thermal energies, and the planetary matter condensed. Present-day heat flows of the fluid planets are thought to derive largely from slow escape of the internal energy stored during the early stage of contraction, although ongoing differentiation may be a significant additional heat source in some of them (Hubbard, 1970, 1980; Flasar, 1973; Graboske *et al.*, 1975; Stevenson, 1982b; Guillot, 2005).

2.6 Differentiation

Differentiation of a homogeneous planet releases gravitational potential energy, as the denser material sinks in the planet's gravitational field to form its core. Because dissipation of mechanical energy takes place deep within the planet, core formation could be a particularly efficient mechanism for raising a planet's internal energy content. Consider an undifferentiated planet of radius R and homogeneous density ρ_0. The planet then differentiates into a core of radius r and density ρ_c, sheathed by a mantle of density ρ_m extending from r to the planet's surface at R (Fig. 2.13). From mass balance we obtain the following relationship between core radius and planetary radius:

$$\frac{r}{R} = \left(\frac{\rho_0 - \rho_m}{\rho_c - \rho_m} \right)^{\frac{1}{3}}. \tag{2.54}$$

We can visualize the core-forming process as follows. Assume that the undifferentiated planet is composed of infinitesimal volume elements that can have a density of either ρ_c or ρ_m. These elements are initially distributed homogeneously throughout the entire planet, in a proportion that yields a mean density ρ_0. Let x be the distance of an arbitrary volume element from the planet's center. During core formation, all volume elements of density ρ_c located at $x > r$ exchange places with elements of density ρ_m located at $x < r$. The result of this process is a core or radius r made up of all volume elements of density ρ_c and a

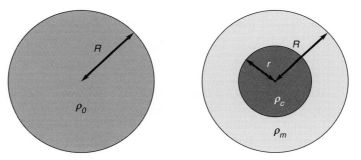

Fig. 2.13   Differentiation of a planet with initial homogeneous density ρ_0 into a core with density ρ_c and a mantle of density ρ_m.

mantle that comprises all elements of density ρ_m. The total mass added to the core is then given by:

$$\Delta m_c = \frac{4}{3}\pi r^3 \left(\rho_c - \rho_m \right). \tag{2.55}$$

The loss in gravitational potential energy of this mass is the amount of mechanical energy dissipated by core formation, and hence the increase in the planet's internal energy, assuming for now that none of it is radiated to space. Each volume element undergoes a different displacement, ranging from R (for an element falling from the planet's surface to its center) to 0 (for an element initially located at r). A characteristic displacement during core formation is thus given by the distance between the center of mass of the mantle, r_m, and the center of mass of the core, r_c. The radii r_m and r_c are defined as the distances from the planet's center such that half of the mantle's mass, and half of the core's mass, are located inside r_m and r_c, respectively (Fig. 2.14). From mass balance we find:

$$r_c = \left(\frac{r^3}{2} \right)^{\frac{1}{3}}; \quad r_m = \left(\frac{R^3 + r^3}{2} \right)^{\frac{1}{3}}. \tag{2.56}$$

A reasonable estimate of the amount of mechanical energy dissipated by core formation is then given by the change in the gravitational potential energy, ΔU_g, of a mass equal to the mass added to the core, Δm_c, falling from r_m to r_c.

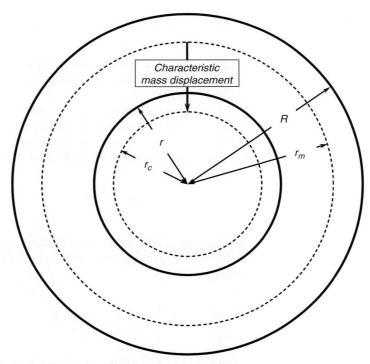

Fig. 2.14   During differentiation, a typical volume element of core material falls from the center of mass of the mantle at r_m to the center of mass of the core at r_c, exchanging places with a volume element of mantle material.

Gravitational potential energy *inside* a self-gravitating body is described by an equation that is different from equation (1.6), because the gravitational force at a point inside the body arises only from the mass internal to that point (this result, which was first derived by Newton, is demonstrated in Box 2.1). The equations are simpler if we work with gravitational potentials. If we call the gravitational potentials of a characteristic volume element (one that falls from r_m to r_c) before and after differentiation $\Phi_{g,i}$ and $\Phi_{g,f}$, respectively, we have:

$$\Delta U_g = \left(\Phi_{g,f} - \Phi_{g,i}\right) \cdot \frac{4}{3}\pi r^3 \left(\rho_c - \rho_m\right). \tag{2.57}$$

In the initial state the characteristic volume element is located at a radius r_m inside a solid sphere of radius R and homogeneous density ρ_0. The gravitational potential at that point is given by (Box 2.1):

$$\Phi_{g,i} = -\frac{GM}{2R^3}\left(3R^2 - r_m{}^2\right) = -\frac{2}{3}\pi G \rho_o \left(3R^2 - r_m{}^2\right). \tag{2.58}$$

In the final state the characteristic element is located at radius r_c inside a homogeneous sphere of radius r and density ρ_c, which is in turn enclosed by a shell of homogeneous density ρ_m extending between radii r and R (Fig. 2.13). The value of $\Phi_{g,f}$ is obtained by adding these two contributions and is given by:

$$\Phi_{g,f} = -\frac{2}{3}\pi G \rho_c \left(3r^2 - r_c{}^2\right) - 2\pi G \rho_m \left(3R^2 - r_c{}^2\right), \tag{2.59}$$

where the first term is the gravitational potential due to the inner sphere (the core), and the second term is the gravitational potential arising from the spherical shell surrounding the core (i.e. the mantle) – see Box 2.1.

The release of gravitational potential energy caused by differentiation increases the (absolute magnitude) of the planet's gravitational binding energy. A useful measure of the amount of energy dissipated during core formation is the non-dimensional ratio that compares this energy to the gravitational binding energy of the undifferentiated planet:

$$D = \frac{\Delta U_g}{U_B}, \tag{2.60}$$

where U_B is the binding energy of the undifferentiated planet, equation (2.8). Substituting equations (2.58) and (2.59) in (2.60) and using (2.54) to eliminate the radii we arrive, after some algebraic manipulations, at:

$$
\begin{aligned}
D = \frac{\Delta U_g}{U_B} = &-\frac{5}{2}\left(\frac{\rho_0 - \rho_m}{\rho_0}\right)^2 \left[1 - \left(\frac{\rho_c - \rho_m}{\rho_0 - \rho_m}\right)^{\frac{1}{3}}\right] \\
&-\frac{5}{6}\frac{\rho_0 - \rho_m}{\rho_0{}^2}\left[\rho_c\left(\frac{\rho_0 - \rho_m}{2\left(\rho_c - \rho_m\right)}\right)^{\frac{2}{3}} - \rho_0\left(\frac{1}{2} + \frac{\rho_0 - \rho_m}{2\left(\rho_c - \rho_m\right)}\right)^{\frac{2}{3}}\right].
\end{aligned}
\tag{2.61}
$$

We can write this equation in terms of the ratios between the densities of the core and mantle and the planet's bulk density, ρ_0. Defining the ratios $\varphi_c = \rho_c/\rho_0$ and $\varphi_m = \rho_m/\rho_0$, we get:

$$
D = \frac{\Delta U_g}{U_B} = -\frac{5}{2}(1 - \varphi_m)^2 \left[1 - \left(\frac{\varphi_c - \varphi_m}{1 - \varphi_m} \right)^{\frac{1}{3}} \right]
$$
$$
-\frac{5}{6}(1 - \varphi_m) \left[\varphi_c \left(\frac{1 - \varphi_m}{2(\varphi_c - \varphi_m)} \right)^{\frac{2}{3}} - \left(\frac{1}{2} + \frac{1 - \varphi_m}{2(\varphi_c - \varphi_m)} \right)^{\frac{2}{3}} \right]
$$

(2.62)

and also:

$$
\frac{r}{R} = \left(\frac{1 - \varphi_m}{\varphi_c - \varphi_m} \right)^{\frac{1}{3}}.
$$

(2.63)

These two non-dimensional equations are plotted in Fig. 2.15. The curves are contours of constant D, as defined by equation (2.62), and the straight lines are contours of constant core radius relative to planetary radius, given by equation (2.63). The graph leads to two conclusions that should be intuitively obvious. First, the thermal contribution that arises from formation of a small core is relatively minor and not very sensitive to the density of the core. The small size of the core limits the value of Δm_c (equation (2.55)) and hence the available amount of gravitational potential energy. Note, however, that even for a planet like Mars, which is likely to have a relatively small core, differentiation may dissipate a fraction of order 10^{-2} of the planet's binding energy, which may be comparable to the internal energy stored by accretion (Worked Example 2.4). Second, in planets with a large core, the thermal effect of core formation is strongly dependent on the density contrast between mantle and core, because this is what ultimately determined the magnitude of Δm_c.

Worked Example 2.4   Thermal effect of core formation in the terrestrial planets

Typical values of specific heat capacity and enthalpy of fusion for rocky materials are 10^3 J kg^{-1} K^{-1} and 5×10^5 J kg^{-1}, respectively. The values for iron are approximately one half of these (see also Section 1.14.3). Density values for Earth are $\rho_0 = 5500$ kg m^{-3}, $\rho_m \approx 4500$ kg m^{-3}, and $\rho_c \approx 10\,700$ kg m^{-3} (you should confirm that these values yield the correct radius for the Earth's core, $\sim$3480 km). Plugging these densities into equation (2.62) we get $D = 0.118$ (see also Fig. 2.15). From equation (2.10) we get the binding energy per unit mass of the Earth, $u_B \approx 3.75 \times 10^7$ J kg^{-1}. We can then conclude that formation of the Earth's core dissipated $\sim$4.43 $\times 10^6$ J kg^{-1}. This is enough thermal energy to raise the temperature of the entire Earth by over 4500 K, or to *melt the entire Earth more than 10 times over!* Upon formation, the core must have been completely molten, and perhaps superheated above its liquidus temperature too. Part of the gravitational energy dissipated was stored as enthalpy of melting. Metal solidification over the age of the Solar System, that continues to this day, releases this enthalpy from the liquid metal thermal reservoir. This is one of the sources of energy that powers convection in the Earth's core and mantle (Chapter 3).

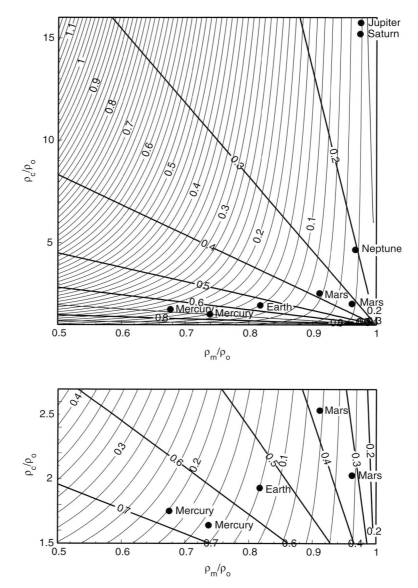

Fig. 2.15 Curves of constant D (equation (2.62)) and constant r/R (equation (2.63), straight lines) as a function of the density ratios φ_c and φ_m. The lower panel magnifies the region of φ_c values characteristic of terrestrial planets. Two different locations are shown for Mercury and Mars, representing different assumed core and mantle density values (see text).

Core and mantle densities are far less well constrained for the other terrestrial planets. Mercury appears to have the largest core among the terrestrial planets, with $r/R \approx 0.6$–0.7 (this can be inferred from the planet's high uncompressed density, see, for example, Strom, 1999). A core of this size allows relatively large D values (Fig. 2.15), which might be an important point in understanding the planet's thermal history given that Mercury appears to still have a partially molten core and a faint magnetic field (Strom, 1999; Purucker *et al.*, 2009). For example, if we assume $\rho_m = 3700 \, \text{kg} \, \text{m}^{-3}$ and $\rho_c = 9500 \, \text{kg} \, \text{m}^{-3}$, and given that for Mercury $\rho_0 = 5430 \, \text{kg} \, \text{m}^{-3}$, we get $r/R = 0.67$ and $D \approx 0.20$. If these values are correct then formation of the Mercurian core may have liberated $\sim 1.1 \times 10^6 \, \text{J} \, \text{kg}^{-1}$, three times what would be needed to melt the entire planet. Because of Mercury's large core this result is sensitive to core and mantle densities. However, even assuming $\rho_m = 4000 \, \text{kg} \, \text{m}^{-3}$ and $\rho_c = 8300 \, \text{kg} \, \text{m}^{-3}$, which are probably extreme permissible values given what we know about Mercury's composition (e.g. a low-Fe crust, Strom, 1999; Taylor & Scott, 2005), we get $D \approx 0.15$ (Fig. 2.15) and an energy dissipation of $\sim 8.4 \times 10^5 \, \text{J} \, \text{kg}^{-1}$, which is still twice the amount of energy needed to melt the entire planet. Data from the *Messenger* mission may allow a refinement of these calculations.

In contrast to Mercury, what we know about the composition of Mars, coupled to its low bulk density, suggests that Mars has a relatively small core, with $r/R \approx 0.3 - 0.4$ (see Wanke & Dreibus, 1988; Lodders & Fegley, 1997; Sanloup *et al.*, 1999). Core and mantle densities may be in the range $\rho_m = 3600 - 3800 \, \text{kg} \, \text{m}^{-3}$ and $\rho_c = 8000 - 10\,000 \, \text{kg} \, \text{m}^{-3}$, respectively (McSween, 2005), which yield $D \approx 0.02 - 0.06$. The largest of these values yields an energy dissipation during core formation of $\sim 4.5 \times 10^5 \, \text{J} \, \text{kg}^{-1}$, which may have been just enough to melt the planet. The energy dissipated at the lower end of the range may have been sufficient to melt the Martian core but not the entire planet.

Application of the results summarized in Fig. 2.15 to the giant planets is less straightforward because, in at least some of them, differentiation may have taken place on more than one level. All giant planets are thought to have very dense cores composed of rocky material and C–H–O–N compounds – "ices" in the terminology of planetary sciences (Guillot, 1999, 2005). These cores, however, are almost certainly quite small ($r/R \approx 0.1$–0.2), so that, even if they formed by differentiation from initially homogeneous bodies, their formation is unlikely to have made a large contribution to the internal energy contents of the giant planets (Fig. 2.15). If the giant planets formed by collapse of gaseous material around earlier accreted dense rocky cores (Thommes *et al.*, 2003) then the rocky cores may have undergone differentiation but the contribution of this process to the heat budget of the full giant planets may still have been relatively minor, compared to the early stage of Kelvin–Helmholtz contraction.

A more important contribution of differentiation to the thermal budget of some of the giant planets may arise from immiscibility between liquid metallic hydrogen and atomic helium (Stevenson & Salpeter, 1977; Hubbard, 1980; Stevenson, 1982a; Hubbard *et al.*, 1999; Guillot, 1999; Fortney & Hubbard, 2004). Hydrogen and helium are the chief constituents of Jupiter and Saturn and are also abundant in Uranus and Neptune, although the bulk compositions of the two ice giants are not as well constrained as those of the two gas giants. In their molecular or atomic forms, hydrogen and helium are miscible in all proportions. Hydrogen, however, undergoes pressure ionization (Section 1.15) at lower pressure than helium. A miscibility gap exists between liquid metallic hydrogen and atomic helium, that closes with increasing temperature (Stevenson & Salpeter, 1977; Chabrier *et al.*, 1992; Saumon *et al.*, 1992, 1995; Morales *et al.*, 2009; Lorenzen *et al.*, 2009). The P–T slope of the boundary that terminates the miscibility gap is poorly known, but it is possible that a planet's thermal gradient may cross it with increasing depth, as shown schematically in Fig. 2.16. The two phases separate at the plasma phase transition (where hydrogen undergoes pressure ionization) and the denser helium "rains out" towards the planet's center, down to a depth at which the temperature is high enough that the solvus disappears, at which depth the helium raindrops redissolve in the liquid metallic hydrogen. Dissipation of gravitational potential energy takes place by friction between the sinking helium drops and the metallic hydrogen medium. This process is thought to be ongoing at the present day in Saturn, and to be an important contributor to its observed heat flux (recall that Saturn's heat flux per unit mass is the highest relative to its binding energy per mass among all fluid planets, Fig. 2.4). Because of Jupiter's higher internal temperatures, the region of H–He immiscibility in it may be much smaller than in Saturn, or altogether absent, so that He differentiation in Jupiter may not be taking place at present.

We now seek an estimate of the helium sinking rate that would be needed to sustain a given planetary heat flow. Let the top and bottom of the two-phase region lie at radii $r_1 = z_1 R$ and $r_2 = z_2 R$, where R is the planet's radius (Fig. 2.16). As the planet ages and cools we expect this region to move towards the planet's center, effectively scavenging He from its outer layers and concentrating it in the planet's interior. As this happens, helium raindrops form at the top of the two-phase layer and redissolve at the bottom, so that each volume element of helium falls from r_1 to r_2, in the process exchanging places with an equal volume of hydrogen. The difference in gravitational potential between the top and bottom of the two-phase layer is given by (Box 2.1 and equation (2.58)):

$$\Delta \Phi_g = -\frac{GM}{2R^3}\left[\left(3R^2 - r_2{}^2\right) - \left(3R^2 - r_1{}^2\right)\right] = -\frac{GM}{2R}\left(z_1{}^2 - z_2{}^2\right). \qquad (2.64)$$

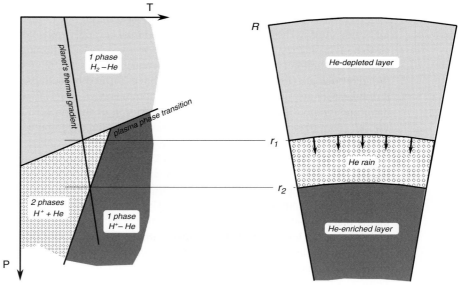

Fig. 2.16   Schematic diagram showing the behavior of H–He mixtures with temperature and pressure. Liquid molecular hydrogen (H_2) ionizes at the plasma phase transition and becomes liquid metallic hydrogen (H^+), which is immiscible with He at low temperature but becomes miscible at high temperature. Depending on the planet's thermal gradient, immiscibility may occur over a limited depth interval, leading to the behavior depicted in the figure, that is thought to be a plausible model for Saturn's interior. Jupiter's interior may be hot enough to miss the miscibility gap. Pressures in Uranus and Neptune are everywhere lower than the plasma phase transition. See also Fig. 1.16.

The helium droplets probably attain their terminal velocity soon after forming, so that we can assume that they fall through the two-phase region with constant velocity dz/dt. Calling the difference in density between atomic helium and liquid metallic hydrogen $\Delta\rho$, the mass settling rate is given by (note that we take the absolute value of dz, as we just seek a mass-flow rate):

$$\frac{dm}{dt} = 4\pi r^2 \Delta\rho \frac{dr}{dt} = 4\pi (zR)^2 \Delta\rho \frac{dr}{dt}, \tag{2.65}$$

where r is some average radius, $r_2 < r < r_1$. This could be the center of mass of the two-phase layer, but if the layer is thin we can take $r = (r_1 + r_2)/2$. From equations (2.64) and (2.65) we can calculate the rate of dissipation of gravitational potential energy:

$$\frac{dU_g}{dt} = \Delta\Phi_g \frac{dm}{dt} = -2\pi GMR\Delta\rho z^2 \left(z_1^2 - z_2^2\right) \frac{dr}{dt}. \tag{2.66}$$

Because dU_g is the loss in gravitational potential energy, the work done by the gravitational force on the planet is $-dU_g$, and the work done by the system (i.e. the planet) is thus $-(-dU_g)$. We seek the value of dr/dt that is needed to sustain a given planetary heat flow, dQ/dt, without change in its internal energy ($dE/dt = 0$). For an average planetary heat flux q the total heat flow is given by $dQ/dt = 4\pi R^2 q$, so that, from the First Law of Thermodynamics with $dE = 0$:

$$\frac{dU_g}{dt} = \frac{dQ}{dt} = 4\pi R^2 q, \tag{2.67}$$

which upon substitution of equation (2.66) yields:

$$\frac{dr}{dt} = -\frac{2Rq}{GM\Delta\rho z^2 \left(z_1^2 - z_2^2\right)}. \tag{2.68}$$

Saturn's heat flux is $q \approx -2$ W m^{-2} (negative because the system loses heat). Part of this heat loss almost certainly derives from a decrease in internal energy ($dE/dt < 0$) that was stored during an early stage of gravitational contraction (Section 2.5). Setting $dE/dt = 0$ thus yields a maximum helium sinking rate. According to Stevenson (1982b) the two-phase region in Saturn extends from $z_1 \approx 0.51$ to $z_2 \approx 0.44$. Plugging in Saturn's mass (5.69×10^{26} kg) and mean radius (58 250 km) and assuming $\Delta\rho = 1000$ kg m^{-3} and $z = 0.47$ we get $dr/dt \approx 4.2 \times 10^{-10}$ m s$^{-1} \approx 1.3$ cm yr^{-1}. This estimate is almost certainly an upper bound, but it translates to a displacement of the order of the planetary radius over the age of the Solar System, which is reasonable because the process appears to be still taking place today. Saturn's atmosphere is depleted in helium, with a H$_2$/He ratio (29.6 by weight) that is higher than that in Jupiter's atmosphere (6.3, data from Lodders & Fegley, 1998). This is qualitatively consistent with helium unmixing as the source of Saturn's high internal energy output (Fig. 2.16).

2.7 Tidal dissipation of mechanical energy

Consider two celestial bodies orbiting their common center of mass. We wish to study the effects of tides on one of them, which we will call the "secondary", caused by the

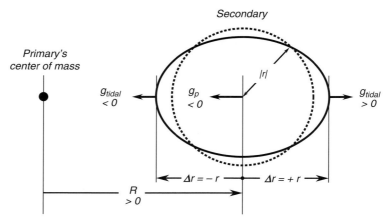

Fig. 2.17 The geometry of tides. Tidal deformation is grossly exaggerated. The secondary's radius has a magnitude | r |, so that the distance from the planet's center to its surface is $\pm r$, as shown.

gravitational pull of the other, which we will call the "primary". The gravitational force between the two bodies is given by equation (1.2), with $x = R$, the distance between their centers of mass (Fig. 2.17). Real objects have finite extent, so that the gravitational attraction of the primary varies over the diameter of the secondary, $2r$, where in general $r \ll R$. If we call the gravitational acceleration of the primary g_p its mass M_p we have from equation (1.7):

$$g_p = -\frac{GM_p}{R^2},$$ (2.69)

where the negative sign expresses the fact that gravitational force is an attractive force. For example, if R is positive towards the right, then g_p, which points towards the left, must be negative (Fig. 2.17). This is the gravitational acceleration at the location of the center of mass of the secondary, but not at its surface. The change in g_p with R is given by:

$$\frac{dg_p}{dR} = \frac{2GM_p}{R^3}.$$ (2.70)

Because the secondary's radius is much smaller than the distance between the objects any distance Δr of order r is equivalent to a very small but finite change in R. We call the small but finite change in gravitational acceleration over this distance the *tidal acceleration*, g_{tidal}, i.e.:

$$g_{tidal} \approx \frac{2GM_p}{R^3}\Delta r.$$ (2.71)

At the point on the surface of the secondary located closest to the primary $\Delta r = -r$ (because we are moving towards the primary), so g_{tidal} is negative, i.e. directed towards the primary. Conversely, at the point furthest away from the primary, $\Delta r = +r$ and in this case g_{tidal} is positive and directed away from the primary. As a result of these tidal accelerations the surface of the secondary is subject to tidal forces that, unless the body is perfectly rigid, give rise to two tidal bulges on opposite sides of the planet, as shown in Fig. 2.17. On the two points on the planet's surface located at 90 degrees from the tidal bulges the tidal force is directed towards the planet's center. This result is not included in equation (2.71) because this is a one-dimensional equation, but it follows immediately from the generalized three-dimensional version of the equation.

If the relative positions of the primary and secondary are fixed, i.e. if the periods of rotation and revolution are the same and if the distance between the two bodies remains fixed, then the tidal deformation of the secondary is constant and there is no energy dissipation. If any of these conditions is relaxed, however, tidal deformation will change with time. Tidal flexing then performs work against frictional forces in the planet's interior, resulting in energy dissipation. The planet heats up as it constantly stretches and squashes.

Consider for instance the case in which rotation of the secondary is faster than revolution of the primary – an example of this situation is motion of the Moon (primary) around the Earth (secondary). The tidal bulge travels over the surface of the Earth as it rotates. If the planet were made up of a perfectly inviscid material (i.e. with no internal friction) then there would be no energy dissipation and, moreover, the tidal bulge would always be exactly aligned with the primary. In reality planets are composed of materials with finite (and commonly large) viscosities. Raising of the tidal bulge takes a finite amount of time, and as a result the tidal bulge is located ahead of the point on the secondary that lies directly below the primary, or in other words at a point that corresponds to an earlier position of the primary, as shown in Fig. 2.18. Because in this case there is internal friction, as the tidal bulge moves over the surface of the planet it performs viscous work, which dissipates kinetic energy of rotation. Angular momentum, however, is conserved, so it must be transferred from the secondary to the primary (if you have trouble visualizing how this happens, note that because the tidal bulge is ahead of the primary it exerts on it a gravitational force that accelerates the primary on its orbit). Over geologic time, the length of the Earth's day increases and the Moon moves faster in its orbit and hence moves away from the Earth. The rotation of the secondary is eventually slowed to become synchronized, or tidally locked,

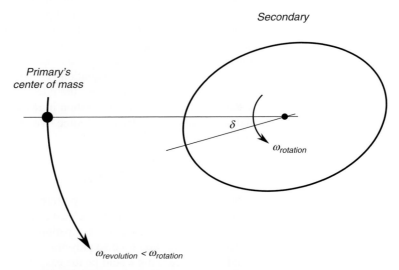

Fig. 2.18 Tidal lag in a body with finite viscosity. The secondary's period of rotation ($2\pi / \omega_{rotation}$) is shorter than the primary's period of revolution ($2\pi / \omega_{revolution}$). The tidal bulge peaks at a point that was located below the primary at an earlier time. The angle δ, greatly exaggerated in the figure, depends on the material properties of the satellite. No scale.

with revolution of the primary. The tidal bulge is now fixed on the surface of the secondary and tidal dissipation due to the secondary's rotation ends. The time needed to achieve tidal locking is called the characteristic time for *tidal despinning* (Kaula, 1964; Hubbard, 1984). For the Earth as secondary and the Moon as primary it is of the order of 10^{11} years, which explains why the Earth's day is still much shorter than its month, although it has lengthened from about 10 hours in the Archean to the current 24 hours. If we consider the Moon as secondary to the Earth as primary the despinning time is of the order of 10^7 years, which means that the Moon's rotation became tidally locked to the Earth very soon after its accretion. The tidal despinning times for the major satellites of the giant planets are much shorter, probably on the order of 10^5 years.

For our purposes there are two important points here. First, when a secondary becomes tidally locked to its primary tidal dissipation from rotation does not take place. At present there is no rotationally induced tidal heating of the Moon but there is a small amount of tidal heating of the Earth. This is, however, insignificant because of the relatively small mass of the Moon, which gives rise to small tidal forces (equation (2.71)). Second, if tidal despinning of the Moon and of the satellites of the giant planets was fast enough, it is possible that they underwent major, but transient, tidal heating events very early in their histories.

If rotation of a secondary is tidally locked to its primary tidal heating is still possible, as long as orbital motion is eccentric. Tidal force varies inversely with the cube of the orbital radius (equation (2.71)) so that as the secondary's distance to the primary changes along an eccentric orbit the tidal bulge grows and shrinks and energy dissipation takes place. The three innermost Jovian satellites, Io, Europa and Ganymede, are locked into orbits with resonant periods in the ratios 1:2:4. This causes their orbits to be eccentric, which could cause potentially important tidal heating in the satellites.

Calculation of the amount of energy dissipated by tidal flexing requires detailed knowledge of the planet's internal structure, chemical composition and physical conditions, and is far from trivial (see Hubbard, 1984, for a detailed discussion). It is possible, however, to derive a simplified equation that contains the essential physics of tidal heating and that yields order of magnitude estimates for the expected energy output. Let the rate at which tidal energy is dissipated by performing work on a planet be $d\boldsymbol{W}_{tidal}/dt$. The work performed by the planet is the negative of this value, so that the rate at which internal energy is added to the planet is $d\boldsymbol{E}/dt = d\boldsymbol{W}_{tidal}/dt$ (equation (1.55)). On physical grounds we can expect that $d\boldsymbol{W}_{tidal}/dt$ must vary directly with: the rate of change of the tidal force, the amount of energy that is stored by tidal deformation in response to the changing tidal force, and the magnitude of the dissipative frictional force. We can express these statements algebraically as follows:

$$\frac{d\boldsymbol{W}_{tidal}}{dt} = \omega \frac{\boldsymbol{U}_{s,o}}{Q_d}. \tag{2.72}$$

In this equation ω is the rate of change of the tidal perturbation and $\boldsymbol{U}_{s,o}$ is the amount of elastic energy stored by tidal deformation. Clearly, the more elastic energy that can be stored, the more energy will be available for dissipation. If the planet were perfectly elastic, however, this elastic energy would constantly travel with the tidal bulge but no energy dissipation would take place. This would be the case if the material that makes up the planet were inviscid, i.e if it had no internal friction. In reality this is not the case and we account for the behavior of real planetary materials with the dimensionless parameter Q_d, called the planet's dissipation factor. For an inviscid material $Q_d = \infty$, and there is no tidal heating, regardless of how much elastic energy is stored. Materials with high internal friction are characterized by small values of the dissipation factor and can thus give rise to high rates of tidal heating. Note very carefully that there is an important difference between the capability of a material for storing elastic energy, ultimately determined by its elastic modulus (equation (1.37)), versus how much energy is dissipated when storing or releasing elastic energy. The latter is a function of the material's internal friction, or viscosity, which in turn depends on several variables, including temperature and the rate at which elastic deformation proceeds. In general, the dissipation factor cannot be calculated from first principles and must be approximated on the basis of the known or inferred internal structure of the planet, including its chemical and physical layering and thermal gradient.

We recall that there are two separate components of tidal heating, one due to rotation (which vanishes if the secondary is tidally locked) and the other one due to orbital eccentricity (which vanishes if orbital motion is circular). We can then break down the tidal work as follows:

$$\frac{dW_{tidal}}{dt} = \frac{dW_{tidal,r}}{dt} + \frac{dW_{tidal,e}}{dt}. \tag{2.73}$$

We seek the equations that describe each of these components, tidal dissipation by rotation ($dW_{tidal,r}/dt$) and tidal dissipation due to eccentricity ($dW_{tidal,e}/dt$).

The elastic energy stored per unit volume by tidally induced shear deformation is given approximately by: $\frac{1}{2}\mu\varepsilon^2$, where μ is the effective shear modulus of the planet, and ε is the strain caused by the tidal forces (equation (1.37)). The elastic energy per unit mass is then given by:

$$u_{s,o} \approx \frac{1}{2}\frac{\mu\varepsilon^2}{\rho}. \tag{2.74}$$

If tidal forces cause the radius of the planet to change by an amount $\delta r \ll r$, then we can approximate the strain as $\varepsilon \approx \delta r/r$. We now seek to relate δr to the secondary's gravitational attraction, which resists deformation, and to the tidal effect of the primary, which drives deformation. Gravitational acceleration is the gravitational potential gradient (equation (1.9)), so we can write the gravitational potential due to tidal acceleration at the surface of the secondary as:

$$\Phi_{tidal} \approx \frac{GM_p}{R^3}r^2. \tag{2.75}$$

Note that the derivative of (2.75) relative to r is (2.71). The ratio between gravitational potential and gravitational acceleration has dimension of length. Thus, on dimensional grounds (see Box 1.2) we can write:

$$\delta r \propto \frac{\Phi_{tidal}}{g_s}, \tag{2.76}$$

where g_s is the gravitational acceleration due to the secondary. The size of the tidal bulge must vary directly with the ratio between the tidal effect of the primary (Φ_{tidal}) and the secondary's own gravitational acceleration, g_s, so this equation is reasonable on physical grounds too. We can then define a non-dimensional parameter, h, such that:

$$\delta r = h\frac{\Phi_{tidal}}{g_s}. \tag{2.77}$$

The parameter h is one among various non-dimesional parameters that are used to describe tides, called Love numbers (in honor of an early twentieth-century British physicist and mathematician who studied tides). The tidal shear strain is then approximately given by:

$$\varepsilon \approx \frac{\delta r}{r} = \frac{h}{r} \frac{\Phi_{tidal}}{g_s}. \tag{2.78}$$

The rotational component of tidal dissipation arises from displacement of the tidal bulge on the secondary's surface, but the magnitude of Φ_{tidal} remains constant. Substituting expressions for Φ_{tidal} and g_s (equations (2.75) and (1.7)) and noticing that we are only interested in absolute values we get:

$$\varepsilon \approx \frac{h M_p r^3}{M_s R^3}, \tag{2.79}$$

where M_s is the mass of the secondary, and all of the other symbols have already been defined. We now substitute this equation for the shear strain in equation (2.74), and the latter in (2.72), and get the following equation for the rate of dissipation of tidal energy by rotation, per unit mass:

$$\frac{dw_{tidal,r}}{dt} \approx \frac{\omega_{tides} \mu h^2}{2\rho Q_d} \left(\frac{M_P}{M_s} \right)^2 \left(\frac{r}{R} \right)^6, \tag{2.80}$$

where ω_{tides} is the characteristic rate with which the tidal bulge travels over the surface of the secondary. For a body whose rotation is tidally locked to the primary, for instance the Moon, $\omega_{tides} = 0$ and there is no tidal heating from rotation. Otherwise $\omega_{tides} = 2\pi/\tau$, where τ is the time elapsed between two consecutive maxima of the tidal bulge on the same point on the secondary and on the same side relative to the primary (i.e. approximately 24 hours and 50 minutes for the Earth's tides).

So far we have considered a situation in which the tidal shear strain remains constant in magnitude, but the locus of this strain moves in response to the secondary's rotation. If the orbit is eccentric we must also consider tidal dissipation as a result of time-dependent variation in the magnitude of the tidal shear strain ε. This arises because the orbital radius changes with time, and hence so does the tidal gravitational potential Φ_{tidal}. The absolute value of the change in Φ_{tidal} with changes in the distance between secondary and primary is given by:

$$\frac{d\Phi_{tidal}}{dR} \approx \frac{3GM_p}{R^4} r^2, \tag{2.81}$$

so that for a small but finite change in the orbital radius, ΔR, we have:

$$\Delta \Phi_{tidal} \approx \frac{3GM_p}{R^4} r^2 \Delta R = 3\Phi_{tidal} \frac{\Delta R}{R} \approx 3\varrho \Phi_{tidal}, \tag{2.82}$$

where $\varrho \approx \Delta R/R$ is the eccentricity of the orbit. The magnitude of the time-dependent component of the tidal strain is then given by:

$$\varepsilon \approx \frac{h}{r} \frac{\Delta \Phi_{tidal}}{g_s} = 3\varrho \frac{h}{r} \frac{\Phi_{tidal}}{g_s} = 3\varrho \frac{h M_p r^3}{M_s R^3}, \tag{2.83}$$

which leads to the following rate of dissipation of tidal energy due to eccentricity, per unit mass (you should work out the algebra to see where this comes from):

$$\frac{dw_{tidal,e}}{dt} \approx \frac{9\omega_{orbital}\varrho^2\mu h^2}{2\rho Q_d}\left(\frac{M_P}{M_s}\right)^2\left(\frac{r}{R}\right)^6. \tag{2.84}$$

The parameter $\omega_{orbital}$ is the rate of orbital motion $= 2\pi/\tau$, where τ is in this case the orbital period of the secondary. This expression of course vanishes for a circular orbit ($\varrho = 0$). Substituting (2.80) and (2.84) in (2.73) we find the total rate of supply of internal energy by tidal dissipation, per unit mass:

$$\frac{de}{dt} \approx \frac{\mu h^2}{2\rho Q_d}\left(\frac{M_P}{M_s}\right)^2\left(\frac{r}{R}\right)^6\left(\omega_{tides} + 9\varrho^2\omega_{orbital}\right). \tag{2.85}$$

In order to obtain numerical estimates for tidal heating we need to substitute appropriate values in this equation. All of the basic planetary properties (M_p, M_s, ρ, r) and orbital parameters ($R, \varrho, \omega_{tides}$ and $\omega_{orbital}$) are well known for the present-day Solar System, although for the early Solar System only educated guesses are possible for the orbital parameters. Much of the uncertainty in the physics of tidal heating is thus contained in the shear modulus, μ, the dissipation factor Q_d and the Love number h. We can get estimates of the possible ranges within which these parameters are likely to vary.

Mean parameter values for the bulk Earth are fairly well known (see Hubbard, 1984; Kaula, 1964; Stacey, 1992; Turcotte & Schubert, 2002; Meyer & Wisdom, 2008). The bulk shear modulus for the Earth is $\approx 7 \times 10^5$ bar ($= 7 \times 10^{10}$ Pa), which can be taken to be a characteristic value for solid rocky planets. Icy satellites, and rocky planets with greater melt contents than the Earth (such as Io ?) are less rigid than the Earth and thus have lower values of μ, of the order of 10^4 bar (10^9 Pa). A range in μ of 10^9 to 10^{11} Pa may cover most solid bodies in the Solar System (rocky and icy planets, satellites and minor bodies). A characteristic value of the dissipation factor for rocky planets is 100, whereas for icy planets the value may be of order 10. The effective Q_d for Earth is also 10, but this low value is due to the fact that much of the Earth's tidal dissipation takes place by turbulent motion in shallow oceans and by friction between ocean water and land. The dissipation factor for the solid Earth is of order 100. We will thus assume that Q_d for solid bodies in the Solar System is in the range 10–100. The dissipation factor for the giant planets, which we do not consider in this section, is much greater, of the order of 10^4–10^5, reflecting their low internal friction.

The Love number h is in fact not independent of a planet's shear modulus. For Earth, $h \approx 0.6$. For rigid bodies smaller than Earth, h is given approximately by:

$$h \approx \frac{10GM_s\rho}{38\mu r}. \tag{2.86}$$

Derivation of this formula from first principles is beyond the scope of this book (see for example Hubbard, 1984, Chapter 4), but we can give a justification of its plausibility *a posteriori*. Substituting this expression into equation (2.77) and then into equation (2.78) we get:

$$\varepsilon \approx \frac{10}{38}\frac{\rho\Phi_{tidal}}{\mu}. \tag{2.87}$$

The product in the numerator, $\rho\Phi_{tidal}$, has dimension of stress (force per unit area) and can thus be interpreted as the stress arising from the tidal acceleration. This stress varies directly with the magnitude of the tidal gravitational potential and, because gravity acts on mass, with the mass contained in unit volume. Equation (2.87) therefore says that shear strain is directly proportional to the tidal stress and inversely proportional to the planet's shear modulus, which agrees with equation (1.33).

Substituting equation (2.86) into (2.85) and simplifying we get:

$$\frac{de}{dt} \approx \frac{0.035 G^2 \rho M_p^2 r^4}{\mu Q_d R^6}\left(\omega_{tides} + 9\varrho^2 \omega_{orbital}\right), \tag{2.88}$$

in which the only uncertain value is the product μQ_d. From our previous discussion of likely values of μ and Q_d we see that this product may be in the range 10^{10}–10^{13} Pa for most solid bodies in the Solar System. Figure 2.19 shows possible ranges of tidal heat generation for various satellites of Jupiter and Saturn, calculated with equation (2.88). The vertical lines, extending over three orders of magnitude, correspond to the three order of magnitude uncertainty in the product μQ_d. Tidal heating values for the Earth and Moon, in contrast, were calculated with equation (2.85) using the better constrained values of μ, Q_d and h for these two bodies. The horizontal dashed lines show measured heat outputs per unit mass for Io and Earth.

What can we conclude from this diagram? Focusing first on the Earth–Moon system, we see that present day tidal heating in these bodies (arising from rotation of the Earth

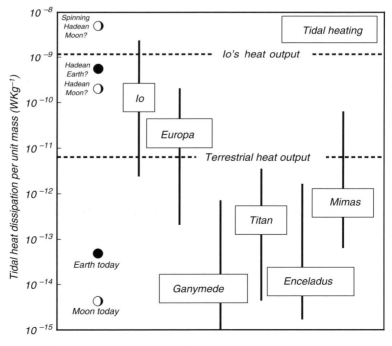

Fig. 2.19 Possible ranges in tidal heating for various planetary bodies. See text for discussion.

and eccentricity of the lunar orbit, respectively) is negligible. The same may not have been true in the early Solar System. The symbols labeled "Hadean Earth" and "Hadean Moon" represent a hypothetical early stage in which the Moon's rotation was already tidally locked to the Earth, but in which its orbit was only 10 times the Earth's radius (compared to the present day's $\sim$60) and the Earth's day was only 10 hours long. Tidal dissipation in the Earth resulted from the planet's rotation, whereas that in the Moon was a result of orbital eccentricity, assumed to have had its present day value of 0.055. The diagram suggests that tidal dissipation may have been an important component of the Hadean internal heat budgets for the two bodies. These calculations, however, assume present-day shear moduli, dissipation factors and Love numbers, all of which may have been radically different if, as seems likely, the Hadean Earth and Moon contained much greater proportions of melt than today. Tidal heat generation in the Moon during tidal despinning may have been horrendous, perhaps one order of magnitude greater than in present day Io (Schultz *et al.*, 1976; Turcotte *et al.*, 1977; Peale & Cassen, 1978; see also Fig. 2.19). The possibility exists that the Moon, and perhaps all satellites of the major planets (and Mercury?) may have had extreme heating episodes of very short duration as their fast initial spins were braked by the tidal pull of their planets (akin to the red-hot glowing brakes in Formula 1 race cars!). The apparently anomalous lunar heat flow relative to its binding energy (Fig. 2.4) may be a relic of this event. Direct observational evidence for this extreme heating event could come if we managed to catch a glimpse of a young extrasolar planetary system at the right evolutionary stage, which may be difficult, given the very short tidal despinning times, $\sim$$10^5$–$10^7$ years.

The Jovian satellites Io, Europa and Ganymede are locked in a 1:2:4 orbital resonance. Io's observed heat output is at the higher end of possible tidal dissipation for the satellite. This means that Io's anomalously large heat flow (Fig. 2.4) may be entirely of tidal origin (Peale *et al.*, 1979; Segatz *et al.*, 1988), but it does not prove that this is the case. Depending on the still poorly known details of Io's interior, tidal heating could be two orders of magnitude less than observed heat output, in which case other explanations would be needed to account for the missing energy flow. The case for Europa, which has a very young and tectonically active crust composed of water ice, is similar. Tidal heating of Europa could provide energy at a rate comparable to the observed terrestrial heat output, which would certainly be sufficient to fuel Europa's tectonism and cryovolcanism (Ruiz, 2005). Whether this would also be possible at the lower end of the calculated range is, however, not clear. Ganymede shows some recent tectonic activity, but the satellite is clearly nowhere near as active as Europa. Is the energy supply at the higher end of Ganymede's range sufficient to explain its relatively young surface?

The moons of Saturn are a deeper enigma than the Jovian satellites. The hydrocarbon content of Titan's atmosphere requires an active supply of methane from the satellite's interior, because the fast rate at which hydrocarbons are destroyed by solar ultraviolet radiation would otherwise render the atmosphere free of hydrocarbons over a time scale of 10^7 years. The Cassini–Huyghens mission has shown that cryovolcanoes may account, at least in part, for the methane supply mechanism and has also revealed that Titan has a young, tectonically active surface (see Lopes *et al.*, 2010). Under the right circumstances, Titan's supply of tidal heat could be comparable to Europa's and perhaps close to the terrestrial heat output. As for Io, however, this does not prove that tidal heating is the chief energy source, given the large uncertainties in the physics of Titan's interior. A bigger problem arises from the contrast between the smaller moons, Mimas and Enceladus. The latter is known to have considerable internal activity, as shown by the presence of active water plumes near the satellite's south pole (Hansen *et al.*, 2008). Mimas, in contrast, appears to be geologically dead. Possible tidal heating of both satellites is comparable, with Mimas actually having a greater potential for tidal heating than Enceladus, because of its greater orbital eccentricity and smaller distance to Saturn (Fig. 2.17). Explanations that have been advanced for the paradox include (see Ross & Schubert, 1988; Nimmo *et al.*, 2007; Meyer & Wisdom, 2007, 2008; Schubert *et al.*, 2007): non-steady state thermal histories (i.e. Enceladus's present-day activity may be driven by heat stored during an earlier epoch of greater tidal dissipation resulting from different orbital parameters), tidal dissipation localized in narrow high-strain domains (i.e. faults), and sources of heat other than tidal.

We can draw the conclusion that tidal dissipation of mechanical (kinetic) energy can be an important, perhaps even dominant, source of planetary internal energy. However, in the absence of accurate knowledge of a body's physical parameters, in particular its rheological properties, it is very hard to be certain that tidal heating is responsible for most of its internal activity. Mercury and Triton could be cases in which more definitive answers appear possible (see end-of-chapter problems).

2.8 Dissipation of electrical energy

A variable magnetic field induces an electric field, and the latter will drive an electric current across a conductive material. Dissipation of electric currents called ohmic heating, see Section 1.8.3) results from friction between moving charges, and takes place in any material with finite electrical conductivity. Because the Solar System is awash with magnetic fields, originating both in the Sun and in the various planets with active internal dynamos, it is in principle possible that ohmic heating may make non-negligible contributions to planetary internal energies. In particular, the process has been invoked in order to explain fast melting and differentiation of asteroids in the early Solar System, but there is no agreement on whether or not ohmic heating was ever a significant planetary heat source. This is so because of large uncertainties in the values of the parameters that control ohmic heating.

There are two ways in which electric currents can be generated in a planetary body by electromagnetic induction. A time-variable magnetic field that permeates a planet generates electric currents that circulate entirely inside the planet. These are called eddy currents. Ohmic heating derived from eddy currents is thought to be negligible for all reasonable ranges of parameters in the Solar System (Colburn, 1980). The other way in which electromagnetic induction can occur is more promising and arises from relative motion between a conductive planet and a steady magnetic field. The currents induced in this case are not confined to the planet, and circulate through the interplanetary medium in order for the electric circuit to close (Fig. 2.20).

The chief contributor to interplanetary magnetic fields is the solar wind, which is a stream of charged particles (protons and electrons, with some heavier ions as well) that move radially away from the Sun at hypersonic speeds (10^2–10^3 km s^{-1}). The origin of the solar wind is expansion of the solar corona. The solar wind is a low-density plasma which is able to conduct currents induced in planetary bodies (Fig. 2.20). An important property of the solar wind for our purposes is that it carries with it the interplanetary magnetic field, which is also rooted in the Sun. As the solar wind streams past a planet the relative velocity between planet and magnetic field is essentially equal to the velocity of the solar wind,

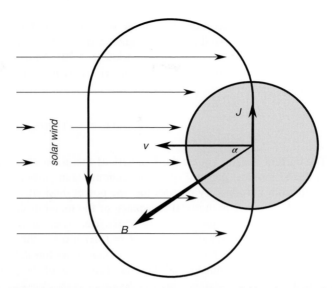

Fig. 2.20 Electromagnetic induction in a planetary body. The interplanetary magnetic field, with intensity B, is directed towards the reader, perpendicular to the page. The solar wind carries the interplanetary magnetic field and moves from left to right, so that the electrical conductor (planet) moves towards the left of the page with velocity v relative to the magnetic field. The induced current J moves upwards across the planet, and the circuit closes via the interplanetary plasma.

given that orbital velocities are typically one order of magnitude smaller. Electromagnetic induction arises because of this relative velocity, but only if the interplanetary magnetic field is able to penetrate the planetary body. This imposes an important limitation on the type of body that can be a candidate for ohmic heating. It must not have an intrinsic magnetic field, for if it does then the solar wind and the interplanetary magnetic field are deflected around the planet. If the planetary body has an atmosphere then there is an additional limitation on ohmic heating, which is that ionization of the outermost atmospheric layers by solar ultraviolet radiation generates an electrically conductive ionosphere. Ionospheric conductivity is typically much greater than that of solid planetary materials, so that most of the induced current is in this case carried by the ionosphere and ohmic heating is shorted out. Given these two limitations – no intrinsic magnetic field and no atmosphere – we can expect that only small airless planetary bodies, such as asteroids, are candidates for electromagnetic induction heating.

An exact solution of the electromagnetic induction equations and the resulting ohmic heating is mathematically quite complex and beyond the scope of this book. A simplified treatment, however, allows us to identify which are the crucial controlling parameters and to analyze the possible magnitude of this heat source. The intensity of the electric field, ϵ, induced by motion with velocity v relative to a magnetic field of intensity B is given by:

$$\epsilon = v\boldsymbol{B}\sin\alpha, \qquad (2.89)$$

where α is the angle between the velocity vector and the magnetic field vector (this is a slightly disguised version of the Lorentz force equation, see Section 1.8.3). The induced electric field is maximum if motion is perpendicular to the magnetic field, and vanishes if motion is parallel to the magnetic field – we will assume for simplicity that the former is always the case ($\alpha = \pi/2$, see Fig. 2.20). An electric field acting across a material with conductivity σ causes an electric current to flow, with the current density, J, given by:

$$J = \sigma \epsilon = \sigma v B \sin \alpha. \tag{2.90}$$

The current density equals the intensity of the electric current (I, measured in Amperes in the SI system) per unit of cross-sectional area perpendicular to the current flow. We recall from introductory physics that the power dissipated by ohmic heating, W, is given by $I^2 R$, where R is the total resistance of the material through which the current flows. The resistance is in turn given by $R = z/a\sigma$, where a is the cross-sectional area of the current-carrying conductor (so that $J = I/a$) and z is the length of conductor over which the current flows. If I is the total current that flows through the planet then ohmic dissipation of this current must equal the rate of increase of the planet's internal energy, dE/dt, so that:

$$\frac{dE}{dt} = I^2 R = a^2 J^2 \frac{z}{a\sigma} = J^2 \frac{az}{\sigma}. \tag{2.91}$$

The product az is the volume of material heated by ohmic dissipation, therefore the rate of increase of internal energy per unit mass, de/dt, is given by:

$$\frac{de}{dt} = \frac{J^2}{\rho\sigma} = \frac{\sigma v^2 B^2 \sin^2 \alpha}{\rho}, \tag{2.92}$$

where ρ is, as usual, the density of the material. The SI unit for the intensity of the magnetic field is called the Tesla (the intensity of the Earth's magnetic field at the planet's equator is approximately 3×10^{-5} Tesla), and conductivity in SI units is measured in mho m^{-1}. Using these units and the standard kg–m–s for density and velocity we obtain de/dt in J s^{-1} kg^{-1}.

The magnitudes of v, the velocity of the solar wind, and B, the strength of the interplanetary magnetic field, are well known in the present-day solar system. Typical values at the radius of the Earth's orbit are $v = 500$ km s^{-1} and $B = 5 \times 10^{-9}$ Tesla. Electrical conductivity of rocks is widely variable and is an exponential function of temperature. For example, the conductivity of carbonaceous chondrites may be given approximately by $\sigma = 10^{-9} \exp(0.014T)$ – see Herbert and Sonett (1979). From 100 K to 1000 K, σ for carbonaceous chondrites varies by almost 6 orders of magnitude, from $\sim 4 \times 10^{-9}$ mho m^{-1} to $\sim 10^{-3}$ mho m^{-1}. Assuming a density of 3000 kg m^{-3}, ohmic heating of a cold chondritic asteroid at the Earth's orbit could account for $\sim 10^{-17}$ J s^{-1} kg^{-1}, which is 6 orders of magnitude less than the present day terrestrial heat output, and 2–3 orders of magnitude less than present day tidal heating in the Earth–Moon system (Fig. 2.19). Heating rate would increase to $\sim 10^{-12}$ J s^{-1} kg^{-1} if the temperature of the asteroid were raised to 1000 K. The conclusion appears to be that, starting with cold material, ohmic dissipation by itself is unlikely to cause a temperature increase of that magnitude.

Worked Example 2.5

It was noticed by Gradie and Tedesco (1982) that the distribution of asteroid types suggests that there was a gradient in maximum temperatures across the asteroid belt. Asteroids that appear to be composed of strongly metamorphosed or melted materials are concentrated in the inner part of the belt (at heliocentric distances smaller than $\sim$2.7 A.U.), whereas those that appear to be composed of undifferentiated early Solar System material, including a significant proportion of ices, dominate the outer part of the belt (distances greater than $\sim$3.4 A.U.). The concentration of asteroids that have undergone relatively low-temperature hydrothermal alteration peaks in between these two distances. Meteorite ages show that thermal processing of asteroidal bodies was a very early event in the history of the Solar System, and was probably completed within the first 10^7 years. Electromagnetic induction heating is in principle consistent with these observations because (a) the interplanetary magnetic field, and hence ohmic heating, varies inversely with heliocentric distance and (b) it is known that, during the transition from gravitational collapse to thermonuclear burning, Sun-like stars go through a brief stage of increased solar wind activity known as the T-Tauri phase. Can early thermal processing of asteroids be a consequence of electromagnetic induction heating during the T-Tauri phase?

Accretionary heating of small planetary bodies is negligible (Section 2.4), so we will assume that the initial temperature of the asteroids was 100 K. We first calculate the mean rate of supply of heat necessary to melt an asteroid, assuming a uniform melting temperature of 1500 K and values for the specific heat capacities and enthalpy of fusion of $c_P = 10^3$ J kg^{-1} K^{-1} and $\Delta h_f = 5 \times 10^5$ J kg^{-1}, respectively. Melting the asteroid in 10 million years then requires a mean heat supply rate of $\sim$6 $\times$ 10^{-9} J s^{-1} kg^{-1}. This is somewhat higher than Io's present day heat output (Fig. 2.19). It is also a minimum estimate because we have ignored heat loss to space, but on the other hand initial temperature in the solar nebula may have been higher than 100 K, so that the two sources of uncertainty may partly cancel each other out.

Let us assume a density of 3000 kg m^{-3} and a conductivity of 10^{-8} mho m^{-1}, consistent with a low initial temperature. If ohmic dissipation was the only source of heat, then it requires that the product $v\boldsymbol{B}$ be $\sim$40 Tesla m s^{-1} (equation (2.92)). Observation of present-day T-Tauri stage stars, and models, suggest that the velocity of the T-Tauri wind was not significantly different from present-day solar wind velocity, 10^2–10^3 km s^{-1} (although the plasma density was of course much higher). The required magnetic field was then of the order of 10^{-4} Tesla at the orbital radius of the asteroid belt. This value is tantalizingly similar to what little is known about the possible intensity of the interplanetary magnetic field in the early Solar System. Magnetic paleointensities measured in a wide range of meteorites are consistent with inducing fields of the order of 10^{-4}–10^{-6} Tesla (Stacey, 1976; Acton et al., 2007). Direct observations suggest magnetic fields of $\sim$0.1 Tesla in the cores of protostellar disks, and comparable values are obtained from proto-star modeling (Donati et al., 2005; Machida et al., 2007). Because the intensity of the magnetic field falls with the square of the distance, the ratio between field intensity at the asteroid belt ($\sim$3 A.U.) and at the proto-solar surface ($\sim$0.01 A.U.) would be $\sim10^{-5}$, which is generally consistent with paleointensities measured in meteorites (see also Herbert, 1989). Electromagnetic induction could have contributed to early heating of asteroids, and might even have been capable of causing large-scale melting, if magnetic field intensities in the early Solar System were at the upper end of the likely ranges and/or if planetesimal conductivities were greater than the 10^{-8} mho m^{-1} assumed in this calculation. This highlights the problem with the hypothesis of planetesimal melting by ohmic heating: it is in principle possible, but the controlling parameters can vary over such wide ranges that it is very difficult to test (see Grimm & McSween, 1989). And there is another nagging problem, which is that the ratio between field intensity at 3.4 A.U. and 2.7 A.U. is $\sim$0.63, i.e. not too different from 1. Explaining the large difference in the degree of thermal processing of asteroids across the asteroid belt would require an uncomfortably fine parameter tuning.

2.9 Radioactive heating

Decay of radioactive isotopes converts nuclear binding energy to kinetic energy, chiefly of subatomic particles, that is dissipated as the particles collide with atoms, slow down and eventually come to a stop. The energy released by decay of one nucleus of radioactive parent, called the decay energy, ϵ_d, is given by the difference between the total rest mass of the decay products (daughter nucleus plus subatomic particles) and the rest mass of the parent nucleus (this difference is also called the Q value of the decay). This is the mass equivalent of the amount of nuclear binding energy liberated by the decay. The decay rate is given by:

$$\frac{dN}{dt} = -\lambda N, \tag{2.93}$$

where N is the number of atoms of parent isotope and λ is the decay constant, which is equal to ln 2 divided by the half life of the isotope. The rate at which decay energy is liberated is then given by:

$$\epsilon_d \frac{dN}{dt} = -\lambda N \epsilon_d, \tag{2.94}$$

so that the rate of heat production per kg of isotope, η_i, is:

$$\eta_i = 6.02 \times 10^{26} \frac{\lambda_i \epsilon_d}{w_i}, \tag{2.95}$$

where w_i is the proper mass of the nucleus of decay constant λ_i, and I have dropped the negative sign because this is heat absorbed by the system (equation (1.55)). In many cases decay from the parent nucleus to the stable daughter nucleus occurs via a number of intermediate steps, called a radioactive decay series. The decay series consists of isotopes with half lives much shorter than that of the initial isotope, so that their concentrations at equilibrium are negligible compared to the concentration of the initial parent isotope. The decay energies of the intermediate decay products, however, are not negligible. The energies liberated by all of the members of the decay series add up so that the general expression for the total rate of radioactive heat production per kg of the initial parent isotope is:

$$\eta_i = 6.02 \times 10^{26} \frac{\lambda_i}{w_i} \sum_k \epsilon_{d,k}, \tag{2.96}$$

Table 2.2 Important heat-generating radioisotopes

Isotope	η_i (J s^{-1} kg^{-1})	λ_i (s^{-1})	Half life (yrs)	Isotopic abundance (present)	Isotopic abundance (4.56 Ga)
^{238}U	9.46×10^{-5}	4.19×10^{-18}	4.47×10^{9}	0.992 75	—
^{235}U	5.69×10^{-4}	3.12×10^{-17}	7.04×10^{8}	0.007 20	—
^{232}Th	2.64×10^{-5}	1.56×10^{-18}	1.41×10^{10}	1	—
^{40}K	2.92×10^{-5}	1.72×10^{-17}	1.28×10^{9}	1.17×10^{-4}	—
^{26}Al	4.55×10^{-1}	3.06×10^{-14}	7.17×10^{5}	0	5.8×10^{-5}
^{60}Fe	7.19×10^{-2}	1.46×10^{-14}	1.50×10^{6}	0	7×10^{-7}
^{53}Mn	6.38×10^{-3}	5.87×10^{-15}	3.74×10^{6}	0	9×10^{-6}

where λ_i and w_i are the decay constant and proper mass of the initial (longer-lived) parent isotope in the series, $\epsilon_{d,k}$ is the decay energy of the kth isotope in the decay series, and we assume that the decay series is in secular equilibrium (Chapter 12). Table 2.2 lists the values of η_i and λ_i (as well as the corresponding half lives) for a number of isotopes that are important planetary heat sources. The isotopes ^{235}U, ^{238}U, ^{232}Th and ^{40}K, with half lives of the order of 10^{9}–10^{10} years, have been important heat producers throughout the entire history of the Solar System, and are still important today. In contrast, ^{26}Al, ^{60}Fe and ^{53}Mn, with half lives of approximately 0.72, 1.5 and 3.75 million years, respectively, have been extinct for most of the age of the Solar System but could have had dramatic thermal effects during the first 5 million years or so. There are two reasons for this. First, owing in part to their short half lives, the rates of heat production per unit mass of ^{26}Al, ^{60}Fe and ^{53}Mn are several orders of magnitude greater than those of the long-lived isotopes (Table 2.2). Second, all three are isotopes of major elements.

If the concentration of radioisotope i in a rock is C_i, then the heat generated by this isotope per kg of rock is $C_i \eta_i$. The concentration of a radioactive isotope in a closed system decays with time according to equation (2.93), and so does the rate of heat production. Let the concentration of isotope i in a rock at the time of formation of the Solar System be C_i^0. Assuming that the rock remains closed to chemical exchange with its environment, the rate of radioactive heat production by isotope i per unit mass of rock at time t after the formation of the Solar System is given by (Chapter 12):

$$\frac{dq_r}{dt} = C_i{}^t \eta_i = C_i{}^0 e^{-\lambda_i t} \eta_i. \tag{2.97}$$

The initial concentrations of ^{235}U, ^{238}U, ^{232}Th and ^{40}K can be determined from their present-day concentrations, C_i^p, as follows (Chapter 12):

$$C_i{}^0 = C_i{}^p e^{\lambda_i t_s} \tag{2.98}$$

where t_s is the age of the Solar System. The present day concentration of active isotopes, in turn, is the product of the corresponding present day isotopic abundance (also given in Table 2.2) and the elemental concentration in the rock. The initial concentrations of the extinct isotopes can obviously not be determined from equation (2.98). The initial isotopic abundances of ^{26}Al, ^{60}Fe and ^{53}Mn at the time of formation of the Solar System (shown in Table 2.2) have been estimated on the basis of minute anomalies in the isotopic abundances of their stable daughter isotopes (^{26}Mg, ^{60}Ni and ^{53}Cr, respectively) measured in meteorites

(Scott, 2007). The values of C_i^0 for the extinct radionuclides are simply the product of these initial isotopic abundances by the corresponding elemental abundances.

We now seek an order of magnitude estimate of the contribution of radioactive decay to present-day terrestrial heat output. This requires knowledge of the Earth's bulk chemical composition, on which there are considerable uncertainties. Estimates of the bulk composition of the Earth are based upon the compositions of chondritic meteorites, complemented by a number of inferences and reasonable assumptions. We know, for example, that the Earth is depleted in volatile elements relative to chondritic meteorites, including in potassium, which is a moderately volatile element. Starting from the present-day bulk earth elemental abundances given by Kargel and Lewis (1993) we use equations (2.98) and (2.97) and the data in Table 2.2 to calculate the temporal evolution of radioactive heat production in the Earth. The results are shown in Fig. 2.21. We can see that at the present time radioactive heating accounts for about one half of the observed terrestrial heat output. The balance represents secular cooling, i.e., slow loss of internal energy that was stored during an earlier epoch. The source of some of this energy may have also been radioactive decay, which generated considerably more heat in the distant geologic past than today, largely as a result of significantly higher abundances of ^{40}K and ^{235}U, with half lives of ~1.3 and 0.7 billion years, respectively (Fig. 2.21). Much of the Earth's secular cooling, however, may still reflect core formation. As we saw in Section 2.6, differentiation of the Earth dissipated enough gravitational potential energy to melt the whole planet several times over. Given that result (Worked Example 2.4) and the temporal evolution of radioactive heat production

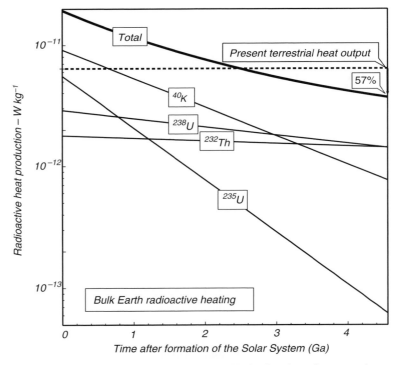

Fig. 2.21 Temporal evolution of bulk Earth radioactive heating per unit mass. Broken line shows the present-day terrestrial heat output. Present radioactive heating accounts for about half of this value.

shown in Fig. 2.19, it is hard to see how one can avoid the conclusion that terrestrial heat output during the Archean and Early Proterozoic must have been much higher than today's. A refined estimate of the history of terrestrial heat flow requires that we also take into account the fact that there are strong geochemical controls on the distribution of radioactive elements, as well as the effects of heat transfer, which we will do in Chapter 3.

A wide range of thermal regimes is known to have affected different early Solar System objects. For instance, some carbonaceous chondrites may have undergone aqueous alteration at liquid water temperatures, whereas basaltic achondrites and iron meteorites indicate that some asteroid-sized bodies underwent essentially complete melting. Electromagnetic induction heating is a possible energy source for thermal processing of early Solar System objects (Section 2.8), but it is not the only one. Decay of short-lived radionuclides may have fueled an early but very short phase of extreme heating of planetesimals. Consider as an example an object with the composition of CI chondrites (this is the group of chondritic meteorites that most closely match the Sun's composition, except for loss of volatile elements). Figure 2.22 shows radioactive heat production in such an object during the first 25 million years after formation of the Solar System, using data from Table 2.2 and chondritic bulk composition from Palme and Beer (1993). Decay of ^{26}Al dominates during the first ~3 million years, and during that time it generates heat at rates 1–2 orders of magnitude greater than Io's present-day extreme heat output. Decay of ^{60}Fe is the chief heat source between ~6 and ~13 million years, after which time the long-lived isotopes take over. The

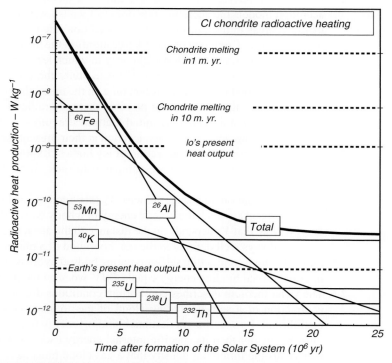

Fig. 2.22 Radioactive heating per unit mass in CI chondrites during the initial 25 million years of the Solar System. The broken horizontal lines labeled "Chondrite melting" are average heating rates required to melt chondritic material in 1 and 10 million years.

most intriguing observation to come out of Fig. 2.22 is that heat output from decay of ^{26}Al during the first million years may have been several times higher than the average rate of supply of heat that would have been required to completely melt the chondritic object in 1 million years. This comparison is of course incomplete, as it ignores the effects of heat loss (Chapter 3), but for a large enough body heat loss over a time scale of 10^6 years, before it melts and starts convecting, is negligible. It is hard to see how objects that accreted while ^{26}Al was still abundant could have escaped wholesale melting. Why, then, are there chondritic meteorites that presumably come from parent bodies that never underwent melting temperatures, and in many cases may have been barely heated? The most likely explanation is that chondritic parent bodies accreted late enough that most short-lived isotopes were well on their way to becoming extinct, whereas differentiated parent bodies may have been the first ones to accrete (Bizzarro *et al.*, 2004; Scott, 2007). A question that in my view has not been satisfactorily answered is how short-lived radioisotope heating explains the thermal gradient across the asteroid belt (see Section 2.8) and whether electromagnetic induction heating also played a role in the origin of this gradient.

Exercises for Chapter 2

2.1 Derive equations (2.20), (2.21) and (2.22). Note that in both the linear and exponential accretion models the accretion rate does not vanish when the planet reaches its final mass, whereas in the sinusoidal model accretion rate is forced to be zero at the final planetary mass. Comment on the relative merits of the three models, and discuss which may be a better representation of the way in which terrestrial planets accrete.

2.2 Use the *Maple* worksheet `accretion` to study the effects of emissivity, nebular temperature, total mass and accretion time on the initial thermal structure of terrestrial planets. Search the literature for possible thermal gradients in the solar nebula, and discuss the extent to which the initial thermal structures of Mercury and Mars could have differed, assuming that they accreted at their present day heliocentric distances.

2.3 Use `accretion` to study accretion of icy bodies. Explore the effects of lower density and lower emissivity of ice compared to rock, and of possibly lower nebular temperatures.

2.4 Derive the equation for escape velocity, V_e (see Section 2.4.1), from the definition that V_e is the speed at which the kinetic energy of a body equals its gravitational potential energy. Note that V_e depends on the distance to the attractor's center of mass, but the value is always well defined in the neighborhood of a planet's surface.

2.5 Show that for a contracting self-gravitating body at hydrostatic equilibrium the change in gravitational binding energy equals $P\,dV$ work on the planet.

2.6 The virial theorem describes equilibrium conditions during contraction. In order for contraction of a self-gravitating cloud of gas to start, however, it must be out of equilibrium. Its specific gravitational binding energy (e.g. given by equation (2.10)) must be greater than its specific internal energy (Section 1.14). Derive an equation that yields the minimum density that a gas cloud must have in order for contraction to start, as a function of its radius and its temperature, assuming that it consists of monatomic ideal gas. Assume that the Sun formed from a cloud of gas with a diameter of 10 light years, at an initial temperature of 50 K. What were the density and the mass of the cloud? How does the mass of the cloud compare to the mass of the Sun? Comment on your results.

2.7 Derive a relationship between the linear contraction rate of a proto-star (dR/dt) and its surface temperature, assuming that it contracts at equilibrium and that it is composed of monatomic ideal gas. Before the discovery of nuclear fusion, one of the explanations proposed to account for solar radiation was Kelvin–Helmholtz contraction. Estimate the necessary contraction rate, assuming that the Sun radiates at 6000 K. Comment on your results.

2.8 Demonstrate that the gravitational acceleration at a point inside a solid sphere is derived only from the mass contained inside the radius of the point (see Box 2.1).

2.9 Prove to yourself that I did not make any mistakes in the tedious derivation of equation (2.61). This is good *Maple* practice.

2.10 Assume that for Mercury $\mu = 5 \times 10^{10}$ Pa, $Q_d = 100$ and $h = 1$. Calculate the likely magnitude of present-day tidal heating in Mercury from Solar-induced tides. Is this a plausible energy source for Mercury's magnetic field? Calculate the possible tidal heating of Mercury before tidal despinning, assuming that it had a 10-hour day. Compare this value with present-day measured planetary heat flows (e.g. Earth, Moon, Io). All data needed to solve this problem not listed here can be found in Lodders and Fegley (1998).

2.11 Assume that for Triton $\mu = 10^{9}$ Pa, $Q_d = 100$ and $h = 1$. Calculate the likely magnitude of present-day tidal heating in Triton (all data needed to solve this problem not listed here can be found in Lodders & Fegley, 1998). Comment on your results, in view of Triton's young and active surface features.

2.12 Discuss how the quasi-static accretion model may be affected if short-lived isotopes such as ^{26}Al, ^{60}Fe and ^{53}Mn are present in the accreted material (it is fairly straightforward to modify `accretion` in order to include this effect). Research the literature to find out what is known about the presence of short-lived isotopes during accretion of the Earth.

3 Energy transfer processes in planetary bodies

We have developed a comprehensive physical description of the processes and pathways by which planetary bodies acquire internal energy. Our next task is to examine how this internal energy drives planetary processes. The hallmark of an active planetary body is that it has surface features, other than impact craters, that have ages that are negligible compared to the age of the Solar System. This is true for any epoch of the Solar System. For instance, the youngest features on the Moon, the immense basaltic plains that we call lunar maria, are about 3 billion years old. This means that the Moon is dead today, but it was active when the age of the Solar System was of the order of 1.5 billion years

Active planetary processes are associated to heat flow, but the causal connection is not always the same. Consider the ascent of magmas. This is a process that transfers mass and heat from the planet's interior towards its surface, and that is made possible by melting, which entails conversion of thermal energy to chemical energy. Ascent of magma and construction of volcanoes, however, are not driven by thermal energy but by gravitational energy. Magmas rise to the surface of a planet if they are buoyant, and magmas are buoyant if melting causes a decrease in density. The essence of the process can be captured by considering a parcel of magma of unit volume and density ρ_m, rising from a depth at which the planetary radius is r to the planet's surface at radius $R > r$, and exchanging places with an equal volume of country rock of density $\rho_c > \rho_m$ (Fig. 3.1). The process takes place inside a solid sphere (Box 2.1). We neglect energy dissipation by friction. The change in gravitational potential energy of the magma is given by:

$$\Delta U_{g,m} = -\frac{GM\rho_m}{2R^3}\left(r^2 - R^2\right) \tag{3.1}$$

and the change in gravitational potential energy of the country rock by:

$$\Delta U_{g,c} = -\frac{GM\rho_c}{2R^3}\left(R^2 - r^2\right). \tag{3.2}$$

The change in gravitational potential energy of the magma ascent process is, therefore:

$$\Delta U_g = \Delta U_{g,m} + \Delta U_{g,c} = \frac{GM}{2R^3}(\rho_c - \rho_m)\left(r^2 - R^2\right) < 0. \tag{3.3}$$

Ascent of magma corresponds to a decrease in gravitational potential energy and, in fact, it is physically and mathematically indistinguishable from planetary differentiation – the distinction is one of scale only. But the First Law of Thermodynamics requires that we

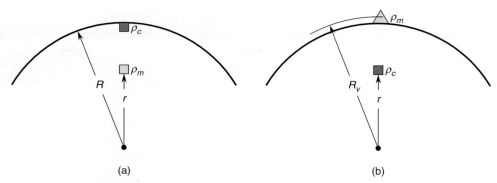

Fig. 3.1 Energy aspects of magma ascent. In (a) a batch of magma of density ρ_m initially located at radius r exchanges places with an equal volume of country rock of density ρ_c, initially located at the geoid, radius $= R$. If the magma ascends no further then there is an "excess" in gravitational potential energy which must be dissipated as heat. In (b) the magma is allowed to rise until its gain in potential energy balances that lost by the country rock, building a volcano with center of mass at radius R_v.

account for the gravitational potential energy that is "lost" in equation (3.3). In order to do so, we must realize that the process depicted in Fig. 3.1a is not the whole story. If the magma stops at or below the planet's geoid (say, it forms a shallow intrusion) then the potential energy that is "missing" in equation (3.3) is dissipated as heat, just as in the case of planetary differentiation. If the magma is able to rise above the geoid then the potential energy released by ascent of magma from its source to the geoid becomes the gravitational potential energy of the magma (i.e. of a volcano) relative to the geoid (Fig. 3.1b). If we neglect frictional losses, the potential energy lost by the sinking country rock must be balanced by the potential energy gained by the magma in rising from its source to the volcano's center of mass (radius R_v in Fig. 3.1b).

Volcanism is not possible in the absence of planetary internal energy, but there is no net conversion of thermal energy to mechanical energy in the process. The thermal energy that is converted to chemical energy during melting is released as heat during crystallization and is eventually radiated to space. There is net transfer of mass in a gravitational field, so that the source of energy for magma ascent is gravity.

Consider now processes such as plate tectonics, atmospheric circulation or planetary dynamos, which are all manifestations of convection. In contrast to volcanism, in all of these processes there is conversion of thermal energy to mechanical energy, but there may not be mass transfer in a gravitational field. This is so because convection is a cyclic process in which there is no accumulation of mass anywhere on the cycle, at least on time scales that are long relative to the time of one convective overturn. Convection is not possible in the absence of a gravitational field, but, since there is no net mass transfer, gravity cannot be the energy source for convection. Convection is driven by conversion of thermal energy (either a planet's internal heat or solar energy) to mechanical energy. There is a remarkable symmetry here with magma ascent, which is not possible in the absence of internal energy, but that is not driven by internal energy. The key to this symmetry is whether or not there is net transfer of mass down a gravitational field. The study of planetary sciences requires a good physical and mathematical understanding of heat-transfer processes and this is the goal of this chapter.

3.1 Transport processes

The First Law of Thermodynamics is the mathematical expression of the principle of conservation of energy. Two other quantities that are conserved are mass and momentum. Each of these three conserved quantities is associated with another quantity, which we may call a potential, such that if there is a potential gradient the conserved quantity will tend to flow down the potential gradient. This is one possible way of stating the Second Law of Thermodynamics, as we shall see in greater detail in Chapters 4 and 12. For example, a temperature gradient drives the flow of internal energy. A gradient in concentration (or, more accurately, in chemical potential, Chapter 5) causes matter to flow, for example in an initially inhomogeneous solution or across an osmotic barrier. Transfer of momentum is more complicated because momentum is a vector quantity, in contrast to energy and mass that are scalar quantities, but the principles are the same. The mathematical law that describes the flow of these three conserved quantities is one and the same and can be written in its most simple form as follows:

$$f = -c\frac{d\Phi}{dx}.$$
(3.4)

In this equation f is the flux of the conserved quantity, i.e. the amount of energy, mass or momentum that is transported per unit of time and per unit of area perpendicular to the direction of transport. The potential that drives the flow is Φ, so that $d\Phi/dx$ is the potential gradient, and the negative sign expresses the fact that the conserved quantity is transported down the potential gradient. The parameter c is a material property which for now we will assume to be a constant, although in general it is a function of temperature and it may also be a function of $d\Phi/dx$. This latter point can introduce immense computational complications. We also note that, in its complete mathematical formulation, f is a vector, $d\Phi/dx$ is a one-form, and c is another geometric object called a *second-order tensor*. In order not to get bogged down in these mathematical complications we restrict our discussion to transport in one spatial dimension, in which case all three geometric objects can be thought of as scalars, as shown in equation (3.4) (in fact, a scalar is a zeroth-order tensor). But one must always keep in mind that flow and flux (flow per unit area) are vectors.

Consider energy conservation first, which is schematized in the drawings at the top of Fig. 3.2. Temperature is the potential that drives heat flow. If there is a difference in temperature across a parcel of matter such that $T_1 > T_2$, then internal energy flows from the high temperature region towards the low temperature one. The magnitude of the heat flux, q, is given by:

$$q = -k\frac{dT}{dx},$$
(3.5)

where k is a material property called thermal conductivity. If we express q in units of J s^{-1} m^{-2} then the units of k are J s^{-1} m^{-1} K^{-1}. Because energy is conserved, in the absence of energy exchanges with the environment the system evolves to a state at a time t_1, later than t_0, in which its internal energy, and therefore its temperature T_3, are uniform and such that $T_1 > T_3 > T_2$ (see Fig. 3.2). Equation (3.5) is also known as Fourier's law of heat conduction, in honor of the French scientist Jean Baptiste Joseph Fourier (1768–1830) who first gave it its mathematical form, and who made many other fundamental contributions to physics and mathematics. In Fourier's time it was not realized that equation (3.5) is

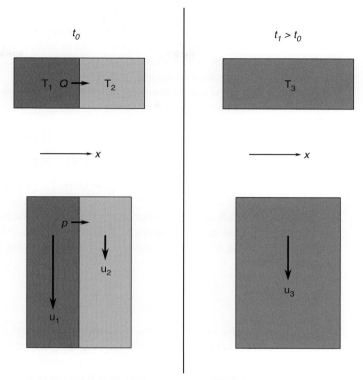

Fig. 3.2 Analogy between diffusive transfer of heat down a temperature gradient dT/dx (top panels, where $T_1 > T_3 > T_2$) and transfer of momentum down a velocity gradient du/dx (bottom panels, with $u_1 > u_3 > u_2$). Time flows from t_0 to a later time t_1.

actually a special case of (3.4). Now it is better to think of equation (3.5) as one of several *constitutive relations*, which describe the physical behavior of a material. The behavior that is described by equation (3.5) is how the material transports heat by *diffusion*, which is the macroscopic expression of propagation of kinetic energy among microscopic particles without bulk macroscopic displacement of matter. Equation (3.5) tells us the rate at which diffusive heat transfer takes place.

Diffusion of momentum is analogous to diffusion of heat. To see this, consider the drawings at the bottom of Fig. 3.2, which represent flow of a fluid in which there is a velocity gradient. The fluid flows downwards, as shown by the velocity vectors $\bar{u}_1 > \bar{u}_2$, but the velocity gradient is in the x direction, which is the direction in which we apply equation 3.4. An important point of this thought experiment is that the boundary of the fluid region drawn in Fig. 3.2 does not represent a physical boundary, but rather a region within an arbitrarily large volume of fluid. In the absence of external forces the momentum inside this region is conserved, just as internal energy is conserved inside a region which undergoes neither heat nor work exchanges with its environment. The result is that at a time t_1, later than t_0, the velocity will be distributed uniformly across the fluid region, and will have a magnitude $\bar{u}_3$, such that $\bar{u}_1 > \bar{u}_3 > \bar{u}_2$, see Fig. 3.2. Momentum has been transported from left to right, driven by the velocity gradient du/dx. One can visualize this process in terms of infinitesimally thin layers of fluid, which initially move with progressively higher velocities towards the left. Because of the velocity difference there is a drag force between the layers. This force transfers momentum from a layer to its neighbor to the right, that moves more slowly, so that the former layer slows down and the latter speeds up. Momentum is thus transported down the velocity gradient.

Momentum flux is the rate of transport of momentum per unit of area. As we can easily see, momentum flux has dimension of stress:

$$[momentum] \times [time]^{-1} \times [distance]^{-2} = [M] \times [L] \times [T]^{-1} \times [T]^{-1} \times [L]^{-2}$$
$$= [M] \times [L] \times [T]^{-2} \times [L]^{-2}$$
$$= [force] \times [area]^{-1}. \qquad (3.6)$$

This dimensional result agrees with our physical image in which the process that transports momentum is friction between layers of different velocity. More precisely, the flux of momentum in a fluid is the shear stress, τ. Equation (3.4) for momentum transport then becomes:

$$\tau = -\mu \frac{du}{dx}, \qquad (3.7)$$

where μ is called the *viscosity* (also, dynamic viscosity) of the fluid and the potential gradient that drives flux of momentum is the velocity gradient taken perpendicularly to the direction of fluid flow (Fig. 3.2). This is another constitutive relationship for a material. Just as the thermal conductivity specifies heat flux, viscosity specifies momentum flux or, in other words, shear stress. For a given velocity gradient, as viscosity increases so does shear stress and, therefore, the magnitude of the drag force between layers. Momentum is then homogenized at a faster rate. This agrees with our intuitive concept of viscosity. For example, if we stir water or motor oil at the same rotational speeds and then stop stirring, the motor oil will come to rest sooner, because momentum flux = shear stress is higher in the more viscous oil than in water. *Note that mechanical energy is not conserved during viscous flow, as viscous drag forces engendered by shear stress are dissipative* (Chapter 1). But momentum is a different quantity and it is conserved. In our example, stirring causes the fluid to heat up by dissipation of mechanical energy. When we stop stirring, momentum is transported across the fluid via the vessel to the Earth.

In this chapter we will use constitutive relations (3.5) (Fourier's law) and (3.7) to study energy transport processes in planetary bodies. The equivalent relation for mass transport will be discussed in Chapter 12.

3.2 Heat transport by diffusion

3.2.1 Overview of heat transfer processes

Heat diffusion (also called *conduction*) is a process in which internal energy flows as a result of the propagation of kinetic energy among microscopic particles (e.g. molecules, atoms, ions, or electrons), without bulk macroscopic displacement of matter. Internal energy is transported as neighboring particles interact with one another and exchange kinetic energy. All microscopic kinetic energy modes: translation, rotation and vibration (Section 1.14), may be involved in the process. Heat *advection* occurs when there is a net macroscopic displacement of matter, which exchanges places with other parcels of matter at a different temperature, so that internal energy is carried by the flow of matter. Although not customary in the engineering literature, in planetary sciences it is convenient to make a distinction between advection, as defined above, and *convection*, which is a process in which heat is advected as a result of a cyclical motion of matter driven by a temperature gradient that engenders buoyancy differences. Engineers call this process natural convection, and distinguish it from forced convection, in which heat advection occurs as a result of motion of fluid propelled by machineries, such as fans or pumps. As an example of the difference between advection and convection, consider heat transfer in the Earth's mantle and heat transfer by magma ascending along a volcanic feeder conduit. The former is heat advection resulting from mantle convection, whereas the latter is heat advection without convection (to an engineer, magma ascent would be an example of forced convection). The processes of diffusion and advection can only occur within matter. *Radiation*, in contrast, is a process in which part of a body's internal energy is transformed to electromagnetic energy that can travel through matter only if it is transparent to that specific range of wavelengths, but that travels unimpeded in the absence of matter. When electromagnetic radiation interacts with an opaque body it is absorbed and converted to internal energy. Radiation is not an important heat transfer process in planetary interiors, but it is important in planetary atmospheres and is the only process by which planetary bodies receive energy from the Sun. We discuss radiation in Chapter 13.

3.2.2 The diffusion equation

Heat diffusion is part of every heat transfer process. Beyond the obvious case, that it is the only form of heat transfer that is possible in a rigid body, heat diffusion takes place between layers of fluid moving at different velocities and at the boundary between a moving fluid and its rigid environment. Heat diffusion is also responsible for carrying internal energy to the interior of a rigid body whose surface receives energy by radiation. We seek to construct an equation that describes heat diffusion in general, allowing for the possibility that both temperature and temperature gradient vary as a function of space and time. We will also include the effects of heat sources (e.g. radioactive, gravitational or electromagnetic heat generation; dissipation of mechanical energy by friction, etc.) and heat sinks (e.g. melting and metamorphic devolatilization reactions).

We begin by considering an infinitesimal volume element inside a material in which there is heat flow (Fig. 3.3). We assume that the material is rigid and at rest relative to a coordinate system fixed to the observer. Our goal is to write the equation for energy conservation in the infinitesimal volume element. Because the process is assumed to take place at constant pressure, any net gain or loss of heat corresponds to a change in the enthalpy content of the volume element, and is reflected in a temperature change given by $c_p \, dT$, where c_p is the specific heat capacity. The lengths of the sides of the volume element are δx, δy and δz, so that the heat flow across the element's left face (amount of heat flowing per unit of time across a surface with area $\delta y \delta z$) is given by (see equation (3.5)):

$$-k\delta y\delta z \frac{\partial T}{\partial x}\bigg|_x . \tag{3.8}$$

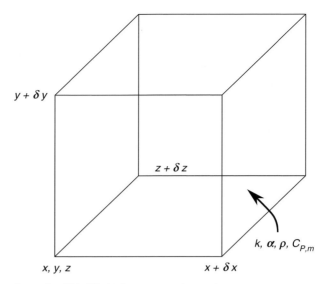

Fig. 3.3 Infinitesimal volume element in which diffusive heat transport is occurring.

We use partial derivatives because temperature is in general a function of the three spatial coordinates and of time. The temperature gradient in equation (3.8) is evaluated at x, which is the coordinate of the element's left face, but the temperature gradient is not necessarily the same on the element's right face. Given that the volume element has infinitesimal dimensions, the heat flow across the right face can be written as:

$$-k\delta y\delta z\frac{\partial T}{\partial x}\bigg|_{x+\delta x} = -k\delta y\delta z\left(\frac{\partial T}{\partial x}\bigg|_{x} + \frac{\partial}{\partial x}\left(\frac{\partial T}{\partial x}\right)\delta x\right)$$

$$= -k\delta y\delta z\left(\frac{\partial T}{\partial x}\bigg|_{x} + \frac{\partial^2 T}{\partial x^2}\delta x\right). \tag{3.9}$$

The net rate of change of enthalpy in the infinitesimal volume element due to heat flow in the x direction is given by expression (3.8) minus equation (3.9) (i.e. "heat entering" minus "heat leaving"):

$$k\delta x\delta y\delta z\frac{\partial^2 T}{\partial x^2}. \tag{3.10}$$

Similar arguments for heat flow along the other two orthogonal directions show that the rate at which the enthalpy of the infinitesimal volume element changes *due only to the effect of heat flow* is:

$$k\delta x\delta y\delta z\left(\frac{\partial^2 T}{\partial x^2} + \frac{\partial^2 T}{\partial y^2} + \frac{\partial^2 T}{\partial z^2}\right) = k\delta x\delta y\delta z\nabla^2 T, \tag{3.11}$$

where the operator ∇^2, called the Laplacian, is shorthand for the sum of the second partial derivatives of a function relative to each of the independent spatial variables (note that the Laplacian is a scalar).

If the material that we are considering contains heat sources or sinks, then their contribution must be added to equation (3.11) in order to obtain the total rate of change of enthalpy. Let the amount of heat added per unit mass and per unit time be α. For a heat source it is $\alpha > 0$, whereas for a heat sink it is $\alpha < 0$. At this point we do not care what the sources or sinks are, or how many of them are present – the value of α is simply the net amount of heat added per unit mass and per unit time, from all sources and sinks combined. The total rate of change of enthalpy of the volume element, from the combined effects of heat flow and heat sources and sinks, is, then:

$$\frac{d\mathbf{H}}{dt} = k\delta x\delta y\delta z\nabla^2 T + \alpha\rho\delta x\delta y\delta z, \tag{3.12}$$

where ρ is the density of the material (recall that α is rate of heat supply *per unit mass*). The rate of change of enthalpy is also given by:

$$\frac{d\mathbf{H}}{dt} = \rho\delta x\delta y\delta z c_P \frac{\partial T}{\partial t}. \tag{3.13}$$

Equating (3.12) and (3.13) and simplifying we arrive at the heat diffusion equation:

$$\frac{\partial T}{\partial t} = \frac{k}{\rho c_P}\nabla^2 T + \frac{\alpha}{c_P} \tag{3.14}$$

or:

$$\frac{\partial T}{\partial t} = \kappa\nabla^2 T + \frac{\alpha}{c_P}, \tag{3.15}$$

where $\kappa = k/(\rho c_P)$ is another material property called the *heat diffusivity*, with dimension $[L]^2\,[T]^{-1}$, e.g., units of $m^2\,s^{-1}$. Equation (3.15) is called the diffusion equation and its mathematical form applies to transport of any quantity that is described by equation (3.4). Momentum diffusivity is described by the *kinematic viscosity* ν given by the ratio $\nu = \mu/\rho$. Chemical diffusivity will be discussed in Chapter 12.

The diffusion equation is an example of a partial differential equation. As all differential equations, it does not tell us directly how to calculate a quantity. Rather, it gives us a mathematical law that describes how the quantity of interest varies. In this case, all that equation (3.15) says is that the rate of change of temperature is a linear function of the curvature (second derivative) of the temperature field. Starting from this information we can construct a precise mathematical description of the spatial and temporal distribution of temperature. At this point you should study Box 3.1 in depth, as it is fundamental for much that follows.

| Box 3.1 | Solution of the heat diffusion equation in a semi-infinite half space |

The heat diffusion equation is an example of a partial differential equation, as it relates the partial derivatives of temperature relative to two independent variables, space and time. In general, solving a partial differential equation requires that we begin by specifying the geometry of the problem and a set of initial and boundary conditions. The geometry of the problem is a precise description of the nature and size of the region of space

Box 3.1   Continued

over which we want to solve the equation, in this case, the region of space for which we wish to know the temperature distribution at any instant of time. The initial condition is the value that the variable of interest (in this case, temperature) takes at some instant chosen as time $= 0$ and at every point in the spatial region of interest. The boundary conditions are the values that the variable (or some of its derivatives) take at some specified points in space (generally, along the boundaries of the region of interest) for all instants of time.

Here we will solve the heat diffusion equation in one dimension over a semi-infinite half space, which is a region of space that has only one boundary. This means that the temperature at a point infinitely far away from the boundary does not change with time. This particular solution is fairly straightforward and leads naturally to the meaning of the characteristic length and time scales of heat diffusion. It is also an important step in understanding heat transfer in terrestrial planets and in inferring their internal structures.

We seek a solution to the heat diffusion equation (3.15) in one dimension and without a heat source:

$$\frac{\partial T}{\partial t} = \kappa \frac{\partial^2 T}{\partial x^2}. \tag{3.1.1}$$

A solution to this equation is a function $T = T(x, t)$. We imagine a semi-infinite half space at a uniform temperature and we impose an instantaneous temperature perturbation on the space's only boundary. We seek the function that describes the propagation of this perturbation in space and time. We need to specify the appropriate initial and boundary conditions. This is illustrated in Fig. 3.4. The initial condition (t_0 in Fig. 3.4) is a semi-infinite half space at a uniform temperature T_1, except at the boundary ($x = 0$) where the temperature is T_0. This is the temperature perturbation that we impose at time t_0. If $T_0 < T_1$ then the half-space cools by losing heat across the boundary, as shown in Fig. 3.4. The temperature at the boundary is held constant at T_0 for all times – this is the boundary condition. Our initial and boundary conditions are, respectively:

$$T(x > 0, 0) = T_1$$

$$T(0, t) = T_0. \tag{3.1.2}$$

As the half space loses heat its temperature distribution changes as shown in Figure 3.4, where t_1, t_2, t_3 are three progressively longer intervals of time.

We wish to find the function that describes temperature distribution in space and time, i.e. the function that describes the curves in Figure 3.4. These curves can be thought of as sections of a surface extending perpendicular to the page, where the third coordinate represents time. One technique for solving partial differential equations (PDEs) is to convert them into ordinary differential equations (ODEs). These are equations that involve derivatives (possibly of different degrees) relative to a single independent variable, and are generally easier to solve. One way of accomplishing this is by combining the two independent variables in the PDE into a single one. Because diffusivity has dimension of $[L]^2 \, [T]^{-1}$ we can define a non-dimensional variable ζ (also called a similarity variable) by combining x and t as follows:

$$\zeta = \frac{x}{2\sqrt{\kappa t}} \tag{3.1.3}$$

(the factor 2 in the denominator has no physical significance, and is justified *a posteriori* because it simplifies the algebraic manipulations). Besides simplifying the solution of the partial differential equation, the similarity variable gives us the scaling factor for length vs. time. Once we know what the temperature is at any given ζ then we can immediately calculate all of the $x - t$ combinations at which temperature has

Box 3.1   Continued

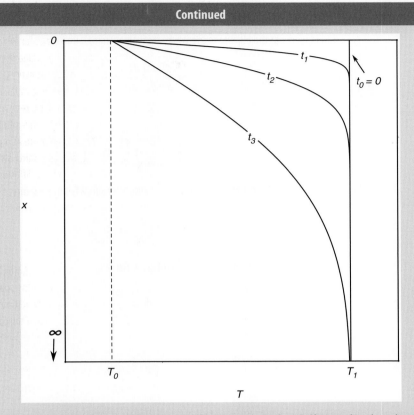

Fig. 3.4   Temporal evolution of the diffusive temperature distribution in a semi-infinite half space whose boundary is kept at a constant temperature $T_0 < T_1$ (the temperature at infinity). At the initial time ($t_0 = 0$) the half space has uniform temperature T_1, except at the boundary, where $T = T_0$. The curves for longer intervals, $t_3 > t_2 > t_1$, are differently-scaled versions of the error function (see text).

this same value. If we seek a function for temperature in terms of this non-dimensional variable, however, then we should also express temperature as a non-dimensional variable. This is accomplished by defining the variable θ:

$$\theta = \frac{T - T_0}{T_1 - T_0},$$ (3.1.4)

which scales temperature to the temperature range characteristic of the problem. In terms of the two non-dimensional variables, our initial and boundary conditions become:

$$\theta = 1 \quad as \quad \zeta \to \infty$$
$$\theta = 0 \quad at \quad \zeta = 0.$$ (3.1.5)

We now use the chain rule to find expressions for $\partial T/\partial t$ and $\partial^2 T/\partial x^2$ in terms of $d\theta/d\zeta$ and $d^2\theta/d\zeta^2$ (partial derivative symbols are obviously no longer needed). Thus:

$$\frac{\partial T}{\partial t} = \frac{dT}{d\theta}\frac{d\theta}{d\zeta}\frac{d\zeta}{dt} = \frac{-\zeta\,(T_1 - T_0)}{2t}\frac{d\theta}{d\zeta}$$ (3.1.6)

Box 3.1	Continued

(you should work out the algebra to see where this comes from). Similarly:

$$\frac{\partial T}{\partial x} = \frac{dT}{d\theta}\frac{d\theta}{d\zeta}\frac{d\zeta}{dx} = \frac{T_1 - T_0}{2\sqrt{\kappa t}}\frac{d\theta}{d\zeta} \tag{3.1.7}$$

and:

$$\frac{\partial^2 T}{\partial x^2} = \frac{d}{d\zeta}\left(\frac{dT}{dx}\right)\frac{d\zeta}{dx} = \frac{T_1 - T_0}{4\kappa t}\frac{d^2\theta}{d\zeta^2}. \tag{3.1.8}$$

We now substitute (3.1.6) and (3.1.8) into (3.1.1) and, after simplifying, we get:

$$-2\zeta\frac{d\theta}{d\zeta} = \frac{\partial^2\theta}{\partial\zeta^2}. \tag{3.1.9}$$

In order to solve this ODE we introduce the auxiliary variable ϕ:

$$\phi = \frac{d\theta}{d\zeta} \tag{3.1.10}$$

and its first derivative:

$$\frac{d\phi}{d\zeta} = \frac{d^2\theta}{d\zeta^2}. \tag{3.1.11}$$

Substituting into (3.1.9) and rearranging:

$$\frac{d\phi}{\phi} = -2\zeta\,d\zeta. \tag{3.1.12}$$

In case you lost track, we have converted the original PDE, equation (3.1.1), into the very simple ordinary differential equation (3.1.12). It may be a good idea to go back and review all of the steps involved in this rather clever trick. The solution to (3.1.12) is, simply:

$$\phi = ce^{-\zeta^2}, \tag{3.1.13}$$

where c is an integration constant. We can now do away with the auxiliary variable ϕ, and we get:

$$\frac{d\theta}{d\zeta} = ce^{-\zeta^2}. \tag{3.1.14}$$

In order to find the value of the constant c we integrate between the two conditions specified by equations (3.1.5):

$$\int_{\theta=0}^{\theta=1} d\theta = c\int_{\zeta=0}^{\zeta=\infty} e^{-\zeta^2}d\zeta. \tag{3.1.15}$$

Box 3.1	Continued

The integral on the right-hand side of this equation appears in a function called the error function (see, for example, Weisstein, 2003, p. 933), symbolized *erf*, and defined by:

$$erf\,(\zeta) = \frac{2}{\sqrt{\pi}} \int_{\eta=0}^{\eta=\zeta} e^{-\eta^2}\,d\eta, \tag{3.1.16}$$

where η is a dummy variable of integration. Equation (3.1.15) then becomes:

$$1 = \frac{c\sqrt{\pi}}{2}\,erf\,(\infty). \tag{3.1.17}$$

The values of the error function are tabulated (see Beyer, 1987) and also available in *Maple*. We find that *erf* $(\infty) = 1$, so that the integration constant is:

$$c = \frac{2}{\sqrt{\pi}}. \tag{3.1.18}$$

Integrating (3.1.14) with this constant:

$$\int_0^\theta d\theta = \frac{2}{\sqrt{\pi}} \int_0^\zeta e^{-\eta^2}\,d\eta \tag{3.1.19}$$

or:

$$\theta = erf\,(\zeta). \tag{3.1.20}$$

This is the solution that we seek: the function that describes temperature distribution in a semi-infinite half space cooled by diffusion and without a heat source. It is straightforward to substitute the definitions of the non-dimensional variables ζ and θ (equations (3.1.3) and (3.1.4)) into equation (3.1.20) and obtain the equation for T as a function of x and t:

$$T = (T_1 - T_0)\,erf\,\left(\frac{x}{2\sqrt{\kappa t}}\right) + T_0. \tag{3.1.21}$$

Solution of the heat diffusion equation in a semi-infinite half space is a relatively simple exercise but it is not necessarily the correct answer to every problem in heat diffusion. The classic text by Carlsaw and Jaeger (1959) develops a large number of solutions of the heat equation with different geometries and boundary conditions, many of them applicable to problems in the Earth and planetary sciences.

3.2.3 The physical meaning of diffusivity

Diffusivity measures the efficiency with which interactions at the microscopic level are able to eliminate the potential gradient that drives the flow. In the expression $\kappa = k/(\rho c_P)$ we can think of the product ρc_P as the *thermal inertia* of the material. The larger the value of this product, the more difficult it is to change the thermal state of the material, thus slowing down heat diffusion. All diffusivities (for heat, momentum and mass) have dimensions of $[L]^2\,[T]^{-1}$, as should be evident from the fact that they have to satisfy the

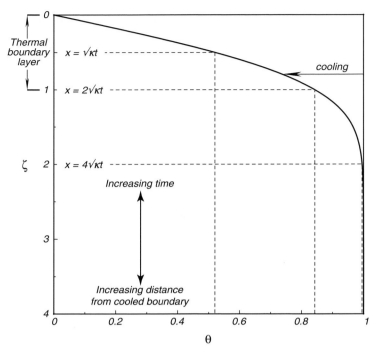

Fig. 3.5 The error function $\theta = erf(\zeta)$. Note that the axes for the independent and dependent variables are switched relative to the mathematical convention, as is common practice in the planetary sciences when representing temperature as a function of depth. The similarity variable $\zeta = x/2\sqrt{\kappa t}$ tracks the propagation of a temperature perturbation in space and time, as shown by the arrows. The magnitude of the perturbation is given by the distance from the non-dimensional temperature $\theta = 1$. The thermal boundary layer can be (arbitrarily) defined at $\zeta = 1$, for which $\theta \approx 0.84$.

same relationship between spatial and temporal derivatives given by equation (3.15). Other physical quantities that appeared in the derivation of the diffusion equation, most notably temperature and energy, do not show up in the dimension of diffusivity. The dimension of diffusivity tells us that in all diffusive processes length scales with the square root of time.

Consider the simple case of diffusion in one dimension (i.e. $\partial T/\partial y = \partial T/\partial z = 0$) without a source ($\alpha = 0$). In Box 3.1 we solve this equation for a semi-infinite half space. Let us look at the general solution in terms of the non-dimensional similarity variable $\zeta = x/2\sqrt{(\kappa t)}$ and non-dimensional temperature, θ. The error function, which is plotted in terms of these variables in Fig. 3.5, describes the propagation of a temperature perturbation ($\theta = 0$) imposed at the boundary of a half space, for which the uniform initial condition is $\theta = 1$. The distance between the curve and the right vertical axis ($\theta = 1$) is the amount of cooling, which increases as ζ decreases. If we fix time, then a displacement towards $\zeta = 0$ represents a displacement towards the boundary of the half space ($x = 0$), which, for finite time, is the only point for which $\theta = 0$. Alternatively, if we consider a fixed point in space, then time, and with it the amount of cooling, increases towards $\zeta = 0$, and cooling to $\theta = 0$ for an arbitrary value of $x \neq 0$ can be achieved only after an infinite amount of time (of course, this is true for a "semi-infinite" half space, as complete cooling of a finite body such as a planet, a dike or lava flow will of course always be accomplished in a finite interval of time).

We can now define a value of ζ that gives a scale estimate for the propagation of a temperature perturbation. As an example, if we set $x = \sqrt{(\kappa t)}$, we get $\zeta = \frac{1}{2}$ and $\theta \approx 0.52$ (Fig. 3.5). Temperature at this point has changed substantially from its initial value, but the effect of the temperature perturbation extends considerably more than this. If we choose $\zeta = 2$, we find that $\theta = erf(2) = 0.995$, which means that the temperature perturbation at this point is insignificant. An intermediate value between these two would give a good estimate of how far the temperature perturbation has propagated. The choice is always arbitrary, and in this book we will choose $\zeta = 1$, for which $\theta = erf(1) = 0.843$ (Fig. 3.5). Although the temperature perturbation can still be seen for $\zeta > 1$, most of the cooling has taken place in the interval $0 < \zeta < 1$. This gives us the following relationship between the characteristic diffusion length, λ, and characteristic diffusion time, τ:

$$\lambda \sim 2\sqrt{\kappa\tau}, \tag{3.16}$$

where the symbol "$\sim$" means "of the same order of magnitude as". Given that θ is non-dimensional temperature, this relationship holds regardless of the absolute magnitude of the initial temperature perturbation.

The physical meaning of equation (3.16) is that a temperature perturbation *of any size* will travel a distance of order λ in a time τ. The numerical value of the diffusivity provides a scale estimate for heat conduction. Consider for example the oceanic lithosphere. It forms by diffusive cooling of mantle asthenosphere as it moves away from spreading centers. We can therefore take the age of the ocean floor as the cooling time of the lithosphere. Using an average age of the ocean floor of 100 million years, and a typical value of heat diffusivity for rocks of 10^{-6} m^2 s^{-1}, we calculate from (3.16) an average thickness for the oceanic lithosphere of $\sim$112 km, which is certainly of the right order of magnitude. Similarly, we can estimate that the characteristic cooling time of a 10 m-thick ash flow (assuming that we can neglect circulation of hydrothermal fluids) is $\sim$10 months, which means that, even if the surface of the ash flow cools to room temperature much sooner than this, we can expect that temperatures deep inside the ash flow will remain noticeably above room temperature for about a year. Recall that in diffusive processes time scales with the square of length. Thus, a 1-m thick lava flow cools 100 times faster than a 10-m thick lava flow – it will have cooled in $\sim$3 days.

These numerical examples illustrate a powerful method for obtaining approximate answers called *scale analysis* or *scaling*. The goal of scaling is to obtain a numerical value that we can be reasonably certain is within an order of magnitude of the exact answer that we seek. In many cases obtaining the exact answer may be computationally very intensive and a scale analysis will return most of the information that we seek for a minimum investment of time and effort. This is especially true in planetary sciences, where the complexity of the processes involved and the large uncertainties in the values of material properties and intensive variables may not always render an "exact" calculation any more informative than a scale analysis. The spirit of scale analysis is to capture the essential physics of the phenomenon that we are studying, even as many of the details may remain uncertain or unknown. One must be careful not to miss some important information, though. In the words of Albert Einstein, "Everything should be made as simple as possible, but not simpler".

3.2.4 The diffusive thermal boundary layer

The thickness of the half space that has been affected by the temperature perturbation (as defined by equation (3.16)) is called the *thermal boundary layer*. Most of the temperature difference between the interior of the half space and the boundary occurs across the thermal boundary layer (Fig. 3.5). A physical interpretation of the diffusive thermal boundary layer is that the material beyond it does not "know" that it will be changing its temperature and, therefore, *does not contribute internal energy to the heat flow across the boundary of the half space*. Heat flow across the boundary comes only from cooling of material inside the thermal boundary layer.

We seek an expression for the rate of heat loss across the thermal boundary layer. We begin from the solution to the heat diffusion equation in a semi-infinite half space (equation (3.1.21)):

$$T = (T_1 - T_0)\, erf\left(\frac{x}{2\sqrt{\kappa t}}\right) + T_0 \tag{3.17}$$

and using the similarity variable ζ defined by equation (3.1.3) we get, from equation 3.17 and the chain rule:

$$\frac{dT}{dx} = (T_1 - T_0)\left[\frac{d}{d\zeta}erf(\zeta)\right]\frac{d\zeta}{dx} = \frac{T_1 - T_0}{2\sqrt{\kappa t}}\frac{d}{d\zeta}erf(\zeta). \tag{3.18}$$

From the definition of the error function, equation (3.1.16), it follows that:

$$\frac{d}{d\zeta}erf(\zeta) = \frac{2}{\sqrt{\pi}}e^{-\zeta^2} \tag{3.19}$$

so that the temperature gradient is given by:

$$\frac{dT}{dx} = \frac{T_1 - T_0}{\sqrt{\pi \kappa t}}e^{-\zeta^2}. \tag{3.20}$$

Since at the boundary $\zeta = x = 0$, the heat flux across the boundary of the half space is given by:

$$q_0 = -k\frac{dT}{dx}\bigg|_0 = -k\frac{T_1 - T_0}{\sqrt{\pi \kappa t}}. \tag{3.21}$$

When applied to surface heat flux in planetary bodies, the negative sign in this equation can be the source of some consternation. If we measure depths in a planet as positive quantities, then heat flux at the planet's surface is a negative quantity, as this equation, which is just Fourier's law, shows. It is customary, however, to express planetary heat flux as a positive quantity, which is what we have been doing throughout this book. In order to follow this near universal practice, which we will continue to do, it is necessary to drop the negative sign from equation (3.21), which thus becomes:

$$q_0 = k\frac{T_1 - T_0}{\sqrt{\pi \kappa t}} \tag{3.22}$$

and it is also necessary to remember that we are inverting the sign of the planetary heat flux, and adjust the sign in other equations, as needed. As the boundary layer cools and thickens the thermally unperturbed material recedes progressively farther away (Fig. 3.5). By Fourier's law (equation (3.5)) heat flux across the boundary of the half space must also decrease with time, as shown by equation (3.22).

3.3 Heat diffusion and cooling of planetary bodies

3.3.1 Lord Kelvin and the thermal structure of the Earth

Students of the Earth and planetary sciences are familiar with the nineteenth-century con-
troversy about the age of the Earth between Lord Kelvin, who calculated an age of the
order of 10^7–10^8 years, and his contemporary geologists, who insisted on much older ages.
Insofar as some geologists advocated an undetermined, perhaps infinitely old age, and by
implication a steady-state Earth, this was an argument about the Second Law of Thermodynamics. Kelvin, who was one of the foundational figures of classical thermodynamics,
correctly pointed out that a steady state Earth would violate the Second Law (Chapter 4).
I believe that the outcome of this argument was split. Kelvin's was a quantitative estimate
based on a rigorous application of physics, and in this sense it was better science than
anything that nineteenth-century geology could offer. Lord Kelvin was also correct that the
Earth is not infinitely old. The age of the planet, however, is one to two orders of magnitude
greater than suggested by his calculations. As it turns out, Lord Kelvin's error provides a
crucial clue about the interior structure of the Earth, which was not fully recognized until
the 1960s (although some scientists had grasped the significance of his error much earlier;
see England *et al.*, 2007, for an account of one such case). Our focus here is to see what
we can learn about the interior of the Earth and other terrestrial planets from the physics of
heat diffusion.

Let us assume that we are ignorant of the interior structure of the Earth and that we
assume that it is a largely solid sphere in which heat moves by diffusion. Because there is
measurable heat flow at the surface, we can also conclude that, in the absence of an active
heat source, the Earth is cooling down from an initially hotter state. This was, more or less,
Lord Kelvin's view of the Earth. Let us now add a crucial piece of information that was
unknown to him: the correct age of the Earth, 4.56 Ga $\approx 1.44 \times 10^{17}$ s. From equation (3.16),
with $\kappa = 10^{-6}$ m^2 s^{-1}, we can conclude that only the outermost 750 km or so of the Earth
have cooled down. Deeper than that the Earth should still preserve temperatures comparable
to its primordial temperature, because there has not been enough time for diffusion to extract
heat from those depths. The observed heat flow at the surface of the planet must therefore
derive exclusively from cooling of the outermost 750-km thick layer. If we know the age of
the Earth then we can use the diffusion equation to do one of two things: assume an initial
temperature for the Earth and calculate the surface heat flux, or use the measured heat flux
and calculate the initial temperature of the Earth. We can also do what Lord Kelvin did:
use measured heat flux and an estimate of the Earth's initial temperature to calculate the
length of time for which the Earth has been cooling down. We will now perform all three
calculations and show that they all lead to erroneous results. More accurately, they lead to
three ways of looking at the same result, namely, that heat diffusion by itself cannot explain
the thermal structure of the Earth.

We model the diffusively cooled Earth as a semi-infinite half space. This is acceptable
because, over the age of the Earth, a layer with thickness of only about one tenth of the
Earth's radius can cool by diffusion, so that most of the planet effectively lies infinitely
distant from the surface in terms of heat transfer. We set $x = 0$ at the surface of the Earth
and $x > 0$ towards the center of the Earth. We assume that the Earth is initially at uniform
temperature T_1, and that temperature at the surface has a constant value T_0 for all times.

Lord Kelvin obtained his estimate for the age of the Earth by solving equation (3.22) for t:

$$t = \frac{1}{\pi \kappa} \left[\frac{k(T_1 - T_0)}{q_0} \right]^2 . \tag{3.23}$$

Box 3.2	Summary of terrestrial heat flow

Throughout this chapter we will use terrestrial heat flow values in numerical examples, but the values appropriate to each example are not necessarily the same. This brief review, based on data discussed by Turcotte and Schubert (2002) and Davies (1999), which are in broad agreement with one another, explains the reasons for the various choices.

The total terrestrial heat output is approximately 4.43×10^{13} W. Divided by the surface area of the Earth (5.1×10^{14} m^2), this yields an average terrestrial heat flux of ≈ 87 mW m^{-2}. This total includes: (i) heat lost from the deep planetary interior by advection (mostly by sea floor spreading, which is the surface expression of mantle convection, plus a small but not negligible contribution from mantle plumes), (ii) heat lost from the deep planetary interior by diffusion across the continental lithosphere, and (iii) heat generated by radioactive decay in the continental crust. Table 10.1 in Davies (1999) summarizes the magnitude of these various contributions. For our purposes, the important concept is that the total average terrestrial heat flux of 87 mW m^{-2} does not discriminate between different sources, depths of origin or heat transport mechanisms. This is the correct value to use, for example, if we wish to model terrestrial heat loss as the product of a single heat transport process that applies indiscriminately to the entire Earth and in which the source of heat is the same for all parts of the Earth. Kelvin's diffusive cooling model is an example of this.

If we subtract radioactive heat generation in the crust, we are left with a total heat loss from the Earth's deep interior (mantle + core) of $\sim 3.6 \times 10^{13}$ W. This includes heat flowing from the mantle across both the oceanic and continental Moho. Divided by the total surface area of the Earth, this yields an average heat flux from the deep Earth of ~ 70 mW m^{-2}. This is not, however, the heat that is transported to the Earth's surface by mantle convection, as part of this heat is transported to the Moho by diffusion across the continental lithosphere. Subtracting the contribution from the subcontinental mantle, including continental shelves, yields a total heat loss across the ocean floor of $\sim 3.1 \times 10^{13}$ W. This is the heat that is transported to the surface by mantle convection (if we ignore radioactive heat generation in the oceanic crust, which is a very small fraction of this total). Divided by the surface area of ocean floor, about 3.1×10^{14} m^2, it yields an average oceanic heat flux of 100 mW m^{-2}. Perhaps 10–20% of this corresponds to heat transport by mantle plumes. The rest, 80–90 mW m^{-2}, is the heat flux that must be accounted for by the moving-plate mode of mantle convection, i.e. formation of oceanic lithosphere at spreading centers and destruction of this lithosphere at subduction zones.

We will discuss mantle convection in other terrestrial planets as well, and in order to do that we need a rough estimate, or at least a bounding value, for the planet's heat flux. If we assume that the bulk compositions and differentiation histories of two planets are the same, then we can tentatively conclude that the heat outputs per unit mass (see Chapter 2) of the two planets are similar. If the sizes of the planets are different, however, the surface heat fluxes will not be the same, as surface area scales with the square of radius and mass with the cube. Calling the two planets 1 and 2, we find:

$$q_2 = q_1 \frac{r_2 \rho_2}{r_1 \rho_1} , \tag{3.2.1}$$

where q is surface heat flux, r radius and ρ density. This formula is strictly applicable *only* if all of the surface heat flux comes from an active heat source, such as radioactive decay. If some of the surface heat flux reflects secular cooling, i.e. slow release of internal energy stored early in the planet's history by processes such as accretion, differentiation, tidal heating or decay of short-lived radioactive isotopes, then smaller planets with a larger surface to mass ratio will cool faster and equation (3.2.1) would not be applicable for the present age of the Solar System. If the planets are not too different in size, however, (3.2.1) may yield a reasonable estimate. This is the case for Venus and Earth. Using (3.2.1) and the average heat flux from the deep Earth (70 mW m^{-2}) yields an estimate for Venusian mantle heat flux of 63 mW m^{-2}. Application of this relationship to smaller terrestrial planets is more contentious. For Mars we calculate an "Earth analog" mantle heat flux of $\sim$30 mW m^{-2}. In Chapter 2 we saw that geochemical estimates suggest a total Martian heat flux of 40 mW m^{-2}, so that, accounting for upwards enrichment of radioactive isotopes, the mantle heat flux estimated from (3.2.1) is not altogether unreasonable. For the Moon we get 12 mW m^{-2}, which is about half of the average value obtained from two measurements by *Apollo* crews. The anomalous lunar heat flow, if real, could be a fossil of an extreme heating event cause by tidal despinning (Chapter 2).

In addition to the material properties k and κ and the surface heat flux, q_0 (see Box 3.2), the solution requires values for the initial temperature of the Earth, T_1, and for the surface temperature T_0. A nice round number such as 300 K is a reasonable enough value for the latter, but what about the initial temperature? As we saw in Chapter 2, this is very poorly constrained, and it was all but unknown in Kelvin's time. He seems to have chosen a value on the order of 4000 K, on the basis of what little was known at that time on melting temperatures of rocks. Given the significant uncertainties in initial temperature, measured geothermal gradients and thermal properties for rocks, Lord Kelvin gave a possible age range of 25–400 million years. Using the value for mean terrestrial heat flux, including continents and oceans, of $\sim$87 mW m^{-2} (Box 3.2), typical values for k and κ of 3 W m^{-1} K^{-1} and 10^{-6} m s^{-2}, respectively, and $T_1 - T_0 = 3700$ K we obtain $t = 5.18 \times 10^{15}$ s $\approx$ 164 million years, within the range of Kelvin's results.

Let us now start from the known age of the Earth, 4.56 Ga $\approx 1.44 \times 10^{17}$ s. If we assume the same initial temperatures, $T_1 - T_0 = 3700$ K , we calculate a mean present-day surface heat flux of $\sim$16.5 mW m^{-2}, about one fifth of the observed value. Alternatively, using the actual heat flux of 87 mW m^{-2}, we arrive at an initial temperature of $\sim$19 800 K for a conductively cooled Earth of the correct age. On the basis of our results from Chapter 2, this temperature is unrealistically high, but there is a bigger problem.

A way of looking at these results is presented in Fig. 3.6 which shows temperature as a function of depth, calculated with equation (3.17), for the three models discussed above. Model 1 is a 4.56 Ga Earth with an initial temperature of 4000 K. This we can discard on the basis of the observed surface heat flux. Model 2 has the same initial temperature but is 164 million years old, which as we saw yields the correct mean terrestrial heat flux of 87 mW m^{-2}. In addition to an erroneous age, this model requires a temperature of 4000 K at a depth of $\sim$300 km, which is above the peridotite liquidus. Model 3, which is based on the correct heat flux (note that dT/dx at the surface coincides with that of model 2) and the correct age of 4.56 Ga, requires a temperature of *19 000 K, or several times the peridotite liquidus,* at a depth of 1500 km, and is obviously unacceptable on geophysical grounds, as it requires a present-day molten mantle.

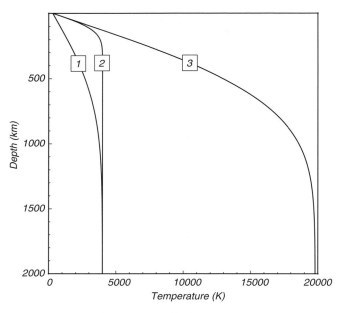

Fig. 3.6 Three diffusive thermal profiles of the Earth, all incorrect. Models 1 and 3 are for a 4.56 billion year old Earth, with initial temperatures of 4000 K and ∼ 20 000 K, respectively. Model 3 yields the correct surface heat flux, as does model 2, which requires a 164 million year old Earth and a 4000 K initial temperature. Note that if the Earth cooled diffusively then most of the planet would still be at its initial temperature, regardless of what this temperature was. Diffusive length and time scales are independent of temperature.

The clue to the failure of the conductive model of terrestrial cooling is that, if the Earth cools by diffusion, then the source of the heat flow observed at the surface is only a small fraction of its internal energy content. The heat flow observed at the surface would be driven by cooling of the thermal boundary layer. For an Earth of the correct age, and even more so for Lord Kelvin's age estimate, most of the planet's interior is located beyond the characteristic diffusion length, so that much of the Earth's primordial internal energy would remain stored and incapable of contributing to the planet's surface heat flow at the present epoch of the Solar System. The importance of Lord Kelvin's mistake is that it shows that conductive cooling of an Earth with no active heat sources fails to explain the observed thermal structure of the planet (which, it must be stressed in his defense, was not known in his time). There are two possible solutions. One, terrestrial heat flow may include a contribution from an active heat source in addition to heat flow derived from cooling of the thermal boundary layer. As we saw in Chapter 2, the only feasible heat source active today is radioactive decay. Alternatively, the planet may have a thin conductive outer layer, below which another heat transfer process is active, that is capable of tapping internal energy from the large proportion of the Earth that lies beyond the diffusive boundary layer, and of delivering this heat to the base of the conductive skin.

3.3.2 Can the conductive model for the Earth be saved?

Let us examine the contribution of radioactive heat production to a conductively cooling Earth. We will assume that the entire terrestrial budget of long-lived radioactive isotopes is concentrated in the thermal boundary layer. This is justified on two grounds. First, it is the best case scenario, for radioactive heating originating beyond the thermal boundary layer cannot contribute to heat flow across the surface. Second, the assumption is sound on geochemical grounds, because all long-lived isotopes are incompatible elements that tend to migrate to the outer layers of the Earth.

If α is the present-day bulk Earth rate of radioactive heat production per unit mass, M is the Earth's mass and R its radius, then the heat flux derived from radioactive decay is $(\alpha M)/(4\pi R^2)$. Using the value of α from Fig. 2.20, $\sim 2.7 \times 10^{-12}$ W kg^{-1}, we get a maximum contribution from present-day radioactive decay of 31 mW m^{-2}. Given a total terrestrial heat flux of 87 mW m^{-2}, it is necessary to derive 56 mW m^{-2} from diffusive cooling. From equation (3.22) we calculate that the initial temperature of the Earth required to yield this heat flux is $\sim 12\,200$ K, and this primordial temperature would still exist today at a depth of ~ 1500 km (see Fig. 3.6), which we know is not possible (it is still several times the peridotite liquidus).

The Earth is certainly still today losing internal energy from an initial hotter state. The fact that the core has not yet completely solidified is one indication of this. Diffusion, however, cannot be the heat transfer process responsible for extracting heat from the Earth's deep interior. The only viable alternative is convection, as silicates are opaque to infrared radiation (Chapter 13).

We should look at Lord Kelvin's mistake in a different light. Unknown to him, his greatest contribution to the Earth Sciences was to demonstrate that convection must be occurring in the interior of our planet. In my view, the realization that this is the case, and the ensuing quantitative understanding of convection in planetary mantles in general, is as pivotal to planetary sciences as the replacement of the Ptolemaic view of the cosmos with the Copernican one. One could say that geology before the understanding of mantle convection was what Aristotelian mechanics is to Newtonian mechanics.

3.4 Convection as a heat engine

Heat engines are thermodynamic cycles that convert thermal energy to mechanical energy continuously and indefinitely, as long as there is a net input of heat (this is the First Law of Thermodynamics), and as long as there are two thermal reservoirs at different temperatures that the engine can exchange heat with (this is mandated by the Second Law of Thermodynamics, which we will study in Chapter 4). The heat engine extracts heat from the high-temperature reservoir and releases heat to the low-temperature reservoir. If the difference between heat absorbed at high temperature and heat released at low temperature is greater than zero then there is the capability of performing work. We see from this description that a heat engine also acts as a heat transport mechanism. Convection fulfills this dual role in planetary bodies. On the one hand, convection is the heat engine that underlies such diverse processes as plate tectonics, mountain building, planetary dynamos, and hurricanes. Convection is also the predominant heat transport process in active planetary mantles and cores, as well as in planetary atmospheres and oceans.

We will construct a "toy model" of convection in a planetary mantle. By "toy model" I mean that I will make a number of simplifying assumptions in order to extract some of the physics of the process. First, it is customary to refer to the convecting material as a fluid, even if, as in the case of planetary mantles, the material is a crystalline solid that is

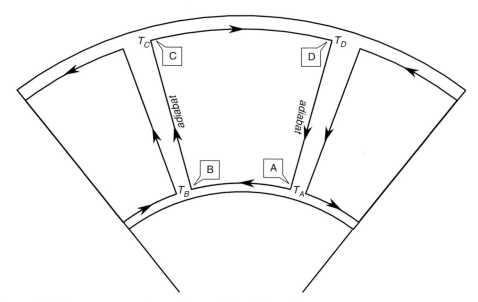

Fig. 3.7 Toy model of planetary convection. A parcel of material absorbs heat isobarically along the bottom boundary of the convective layer, so that its temperature increases from A to B. It then rises adiabatically from B to C, loses heat isobarically from C to D and sinks adiabatically from D to A, completing the cycle.

able to flow at rates of cm to m per year by a combination of microscopic and macroscopic creep mechanisms (see Poirier, 1985). Convection occurs when the temperature difference across a fluid layer is such that, given the magnitudes of material properties of the fluid such as thermal diffusivity, coefficient of thermal expansion and viscosity, the rate of heat transfer by advection exceeds the rate of heat transfer by diffusion. The fluid layer (Fig. 3.7) absorbs heat across its bottom boundary and releases heat across its top. In our toy model we assume that pressure is constant along both boundaries. This is physically reasonable for terrestrial mantle convection. The top of the convecting mantle corresponds to the surface of the Earth (more precisely, the ocean floor). The bottom corresponds to the core–mantle boundary, where the pressure is determined by the lithostatic load of the overlying mantle so that it is essentially constant. Significant lateral temperature variations exist along both boundaries. As the mantle releases heat through the planet's surface, it cools from asthenospheric temperature at mid-ocean ridges to atmospheric temperature at subduction zones. We are less certain about what happens at the bottom boundary, but a reasonable physical model is that cold downwelling mantle hits the core–mantle boundary, moves laterally and its temperature rises as it absorbs heat from the core (see Fig. 3.7). This toy model of convection assumes that heat exchange occurs only at the bottom and top of the convective layer, and that there are neither heat sources nor heat sinks within the convecting fluid. The latter is seldom the case in real planetary systems. For example, radioactive heat is generated in planetary mantles (and perhaps cores as well), frictional dissipation of mechanical energy takes place in planetary cores, mantles and atmospheres, ohmic dissipation is always part of planetary dynamos, and phase changes of condensable species are huge heat sources and sinks in planetary atmospheres (H_2O in Earth, hydrocarbons in Titan, CO_2 in Mars).

It is the nature of convection that it must transfer heat faster than diffusion. This is so because convection is driven by buoyancy forces, which in thermal convection arise from temperature differences. Buoyancy can be maintained only if temperature contrasts between different parcels of fluid can be maintained. The rate of change of the thermodynamic state of a parcel of fluid *not located at the heat exchange boundaries* must therefore be much faster than the rate at which heat can be exchanged by diffusion with neighboring parcels. This means that the interior of the convective layer is adiabatic (Fig. 3.7). Planetary convective systems may differ in some important aspects, though. For example, in contrast to the model in Fig. 3.7 in which both heat absorption and heat loss are isobaric, hurricanes are convective systems in which both heat transfer boundaries are approximately isothermal. The basic structure of two heat-transfer boundaries linked by an adiabatic interior is always present, however.

The convective overturn depicted in Fig. 3.7 maps as a four-leg cycle on a pressure–temperature diagram, shown in Fig. 3.8. A parcel of material located at the bottom of the convective cell, at point **A**, absorbs heat at constant pressure P_b. As its temperature increases from T_A to T_B the material expands until it eventually becomes buoyant at **B**. It then rises and expands adiabatically, from P_b to P_t. Adiabatic expansion causes its temperature to drop from T_B to T_C. At **C** the parcel has reached the top of the convective layer and begins to lose heat. It cools down isobarically from T_C to T_D and contracts, until at **D** the density has increased sufficiently that it sinks and compresses adiabatically. Its temperature increases in response to adiabatic compression until it reaches the bottom boundary at **A**. The function $T = T(P)$ along an adiabatic path such as **BC** or **DA** is called the *adiabat* and is discussed

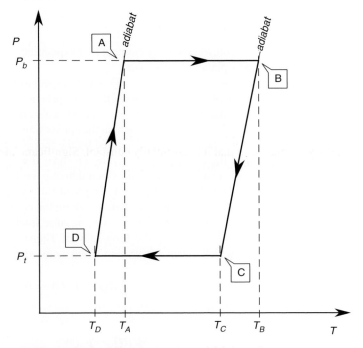

Fig. 3.8   Pressure–temperature diagram for the planetary convection model shown in Fig. 3.7.

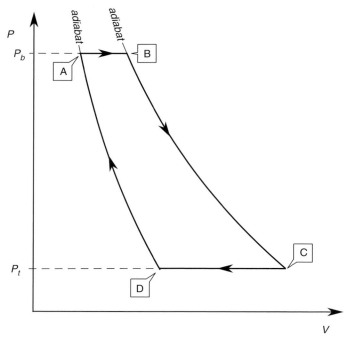

Fig. 3.9 Pressure–volume diagram for the planetary convection model shown in Fig. 3.7. The area bound by the four curves is the net work delivered by the cycle. In a planetary mantle most of this work is dissipated by viscous flow, and a very small fraction is transformed to other types of mechanical energy such as potential energy (topography) and elastic energy in the lithosphere.

fully in the next section, as it is a key element in the quantitative analysis of convection and in the study of many planetary processes.

Because the integral of a function $P = P(V)$ with respect to V represents mechanical work (Section 1.4.1) mapping the convection cycle onto a pressure-volume diagram allows us to explore the energetic aspects of the process. The adiabats are logarithmic curves on a P–V diagram (see end-of-chapter problems) so that the convection cycle maps onto the P–V plane as shown in Fig. 3.9. The net work performed by the thermodynamic cycle is represented by the area bound by the four legs of the cycle. The heat absorbed at high pressure (from **A** to **B**) places the fluid on a hotter adiabat, which delivers work during expansion. The fluid must then return to the colder adiabat, by losing heat from **C** to **D**, and then work must be performed on the system in order to compress it along the colder adiabat, from **D** to **A**. The integral of $d\boldsymbol{E}$ around a full cycle must vanish, because internal energy is a state variable (Section 1.9). Therefore, applying the First Law of Thermodynamics to convection and using the symbol $\oint$ which means integration along a closed path, we see that:

$$\oint d\boldsymbol{E} = \oint d\boldsymbol{Q} - \oint d\boldsymbol{W} = 0 \tag{3.24}$$

or:

$$\oint d\boldsymbol{Q} = \oint d\boldsymbol{W}. \tag{3.25}$$

The net work performed by a thermodynamic cycle equals the net amount of heat absorbed by the cycle. In our toy convection model, this equals the heat absorbed along the high-temperature isobaric path (from **A** to **B**), minus the heat released along the low-temperature isobaric path (from **C** to **D**). The amount of work varies directly with the pressure difference across the convecting layer (i.e. the length of the adiabats) and with the amount of heat that can be extracted from the high-temperature source *or delivered to the low-temperature sink* (i.e. the length of the two heat-exchange paths). Convective overturn is driven by buoyancy, i.e. by gravitational forces, but what makes it possible is the temperature difference between the ascending adiabat (**B** to **C**) and the descending adiabat (**D** to **A**). A temperature difference between adiabatic paths can only be maintained if there is heat exchange at both boundaries (see Fig. 3.9), or, equivalently, if there are at least two non-adiabatic legs in the cycle. Either of them, or both, can exert the dominant control on convection. For instance, the chief driving force for terrestrial mantle convection is cooling at the upper boundary and formation of the negatively buoyant and mobile lithosphere. Terrestrial mantle convection is largely controlled by the length of the cooling path **CD** in Fig. 3.9, i.e. by the temperature difference $T_C - T_D$ (Fig. 3.8).

The convection heat engine absorbs heat at the high-temperature boundary and releases heat at the low-temperature one, thus playing the part of a heat transport mechanism. It can keep operating only as long as it is not outpaced by heat diffusion. As a planet cools down the four legs of the cycle depicted in Fig. 3.9 approach one another and geological activity declines. At some point the material becomes too stiff to be able to convect and the planetary heat engine seizes. The planet may still contain a substantial amount of thermal energy but this is diffused to the surface and radiated to space without performing work. The Moon is an example of a planetary body in this evolutionary stage.

In real convection there are important additional considerations. A fraction of the mechanical energy that convection delivers at the expense of the planet's thermal energy is always dissipated. If the convective layer consists of very viscous and electrically insulating material, such as the Earth's mantle, virtually all of the mechanical energy is dissipated by friction and returned to the mantle as heat. The remainder, a very small fraction of the total energy flow, is transferred to the planet's lithosphere, where it is stored as elastic energy and gravitational potential energy (topography; e.g. Worked Example 1.1). In a convective layer that is electrically conductive some of the mechanical energy appears as electric currents, which are dissipated by ohmic heating while sustaining the planet's magnetic field (Section 1.8.3). In planetary atmospheres, because of their low viscosity, there may be significant conversion of heat to kinetic energy of wind, that is eventually dissipated by friction, both within the atmosphere itself and between the atmosphere and the planet's liquid or solid surface (wind turbines tap a small fraction of this kinetic energy).

3.5 Planetary adiabats

The work performed by an adiabatic process must be balanced by a change in internal energy. Therefore, a function must exist that relates temperature to pressure anywhere along an adiabatic path. Such a function is called the adiabat and allows us, for instance, to calculate temperature as a function of pressure (or depth) in an actively convecting mantle, or atmospheric temperature as a function of elevation in a convecting atmospheric layer (the troposphere). If we know temperature and pressure at one point on the adiabat then we can calculate the temperature at any other pressure. For instance, we can see by reference to Fig. 3.8 that if a planet has a convective interior, then we can calculate the temperature at depth in the planet from temperature measurements at some reference level close to its surface.

We seek a function $T = T(P)$ along an adiabatic path. To begin, we re-state the First Law of Thermodynamics in terms of enthalpy, because enthalpy allows us to write equations in terms of the variables P and T more compactly than internal energy. From the definition of enthalpy (equation (1.59)), an infinitesimal change in H is given by:

$$dH = dE + PdV + VdP \tag{3.26}$$

substituting into the equation for the First Law (equation (1.56)) and simplifying:

$$dH = dQ + VdP. \tag{3.27}$$

Along an adiabatic path $dQ = 0$, so for the adiabat equation (3.27) becomes:

$$dH = VdP. \tag{3.28}$$

If pressure and temperature are the variables of interest we must express dH in terms of infinitesimal T and P increments. This is given by (see equation (1.3.3), Box 1.3):

$$dH = \left(\frac{\partial H}{\partial T}\right)_P dT + \left(\frac{\partial H}{\partial P}\right)_T dP = C_P dT + \left(\frac{\partial H}{\partial P}\right)_T dP. \tag{3.29}$$

Substituting in equation (3.28) and rearranging:

$$C_P dT = \left[V - \left(\frac{\partial H}{\partial P}\right)_T\right] dP. \tag{3.30}$$

To simplify this equation further we must express $(\partial H/\partial P)_T$ in terms of some combination of the state variables: P, V, T and the material properties: K_T, α and C_P (Section 1.11). In order to derive the required expression we must use both the first and second laws of thermodynamics, so we will defer a full discussion until Chapter 4. The derivation is presented in detail in Appendix 2, where it is shown that:

$$\left(\frac{\partial H}{\partial P}\right)_T = V(1 - \alpha T). \tag{3.31}$$

Substituting in equation (3.30) we obtain the equation for the adiabat:

$$\frac{dT}{T} = \frac{\alpha V}{C_P} dP. \tag{3.32}$$

In the derivation of equation (3.32) we have tacitly assumed that the adiabatic transformation is also reversible (Section 1.4.2 and Chapter 4). This happened when in equations (3.26) and (3.27) we equated PdV to the work performed by the adiabatic expansion, which is only true if there is no energy dissipation and therefore, as we shall see in Chapter 4, no entropy generation. Although it is certainly an adiabat, equation (3.32) is more accurately called an *isentrope*, as there are adiabatic transformations that are not isentropic (Chapter 4). It is customary in planetary sciences, however, to call (3.32) "the adiabat" and I will generally follow this custom. This commonly presents no problems, but in cases in which it is necessary to examine the constant entropy assumption more rigorously we will do so (Chapter 10).

We are often interested in how temperature varies with depth in a planetary interior, or elevation in the atmosphere. From the chain rule of partial differentiation:

$$\frac{dT}{dy} = \frac{dT}{dP}\frac{dP}{dy}.$$
(3.33)

The pressure gradient inside a planet is given by the condition of hydrostatic equilibrium (equation (2.34)), but whereas r in equation (2.34) is measured from the center of the planet outwards, y is defined as positive downwards, so we have:

$$\frac{dP}{dy} = g\rho$$
(3.34)

and, using (3.32) and (3.33):

$$\frac{dT}{dy} = \frac{\alpha g T}{c_P},$$
(3.35)

where we recall that $c_P = C_P/(V\rho)$ is the specific heat at constant pressure (= constant pressure heat capacity per unit mass).

Because we have made no assumptions regarding the equation of state of the material that undergoes the adiabatic transformation, equations (3.32) and (3.35) are completely general. In order to integrate them, however, it is necessary to know the equation of state of the material of interest, so as to express α, V and C_P as functions of P and T. It may also be necessary to consider changes in g with depth. We can nevertheless get an order of magnitude estimate of the adiabatic thermal gradient in the Earth's upper mantle by using characteristic values for forsterite. These are: $\alpha \approx 3 \times 10^{-5}$ K^{-1}, $v \approx 3 \times 10^{-4}$ m^3 kg^{-1} and $c_P \approx 1200$ J K^{-1} kg^{-1}. Note that although (3.32) is written in terms of molar properties, it is also valid if we express volume and heat capacity as specific properties, as the unit conversions cancel out. Taking $g = 9.8$ m s^{-2} and a characteristic upper mantle temperature $T = 1700$ K, we get an adiabatic gradient for the asthenosphere of $\sim$0.4 K km^{-1} or $\sim$1.3 K kbar^{-1}. Temperature increases with depth, and the coefficient of thermal expansion and the heat capacity are both functions of temperature and pressure (Chapter 8). The adiabatic temperature gradient decreases at greater depths. It is always much smaller than the temperature gradient in the lithosphere, however, in which heat transfer is by diffusion. For example, the temperature difference across the oceanic lithosphere is of the order of 1400 K. With a lithospheric thickness of 100 km, this yields, assuming no radioactive heat generation, a mean diffusive temperature gradient of 14 K km^{-1}.

Worked Example 3.1   Atmospheric lapse rate

Equation (3.35) can be used to calculate how temperature varies with elevation in the atmosphere, provided we are careful about some points. First and foremost, the thermal gradient will be adiabatic only in those layers in which the atmosphere is convecting. This is true, for example, for the terrestrial troposphere, up to $\sim$10 km above sea level, but not at higher elevations. Second, even in the convective troposphere the adiabatic condition expressed by equation (3.35) is rigorously true only as long as there are no phase changes of any of the atmospheric components. For example, we saw in Section 1.12.1 that the heat associated with phase changes of H_2O in the terrestrial atmosphere is not negligible, rendering $dQ \neq 0$ and equation (3.35) invalid in any atmospheric region in which there is condensation or evaporation of H_2O (or any other condensable species, e.g. CO_2 in Mars or hydrocarbons in Titan). Finally, we need to choose an equation of state. The atmospheres of Earth, Mars and Titan are tenuous enough that the ideal gas equation of state is applicable, but this may not be true for the atmosphere of Venus, nor for the atmospheres of the gas giants below the top few kilometers (Chapter 9).

For a convective atmospheric layer that can be treated as an ideal gas we have $\alpha = 1/T$ (Exercise 1.11), so equation (3.35) simplifies to:

$$\frac{dT}{dy} = \frac{g}{c_P}. \tag{3.36}$$

The temperature gradient given by equation (3.36) is a positive quantity called the *dry adiabatic lapse rate*. It corresponds to y measured as positive downwards, so that if we are interested in how temperature changes with elevation we need to take the negative of equation (3.36). For an ideal gas atmosphere, we can also write the dry lapse rate in terms of the number rotational and translational of degrees of freedom of the gas molecules, f, as follows (see equation (2.45)):

$$\frac{dT}{dy} = \frac{2gw}{(f+2)R}, \tag{3.37}$$

where w is the molecular weight (or proper mass....) of the gas. Clearly, this will work only if the atmosphere can be considered to be composed of a single gas (e.g. Mars), or if all the gas species in the atmosphere have the same number of degrees of freedom (Earth). Calculation of the adiabatic lapse rate in atmospheres in which there are condensable species is suggested as an end-of-chapter exercise.

3.6 Heat advection

3.6.1 The diffusion–advection equation

To study convection as a heat transfer mechanism we must begin by deriving the equation that describes heat advection in general. In our derivation of the diffusion equation (equation (3.15)) we assumed that there was no motion of the material in which heat diffusion takes place. To describe advection we must specify a reference frame relative to which we measure the state of motion. A description of the situation to which equation (3.15) applies is that the material is at rest relative to the observer. If we now allow the material to move relative to the observer, then there will be heat advection relative to him/her/it as a result of motion of material in which there is a gradient of internal energy. This is in addition to the heat that is diffused, and that is described by equation (3.15). We need to add a term to that equation, called the *advective term*, to account for the transport of additional internal energy. Let us assume that the material inside the volume element in Fig. 3.3 is moving relative to an observer with velocity components u_x, u_y, and u_z. The volume element, or equivalently the coordinate axes, remains fixed relative to the observer, but the material inside the volume element moves. The volume of material moving in the x direction per unit time is $\delta y \, \delta z \, u_x$. Because the temperatures at x and $x + \delta x$ are generally different, the net enthalpy advected per unit time into the volume element in the x direction equals the enthalpy of the incoming material at temperature $T(x)$ minus the enthalpy of the outgoing material at $T(x + \delta x)$, as follows:

$$\delta y \delta z u_x \rho c_P T(x) - \delta y \delta z u_x \rho c_P T(x + \delta x) = -\delta y \delta z u_x \rho c_P \frac{\partial T}{\partial x} \delta x$$

$$= -\delta x \delta y \delta z \rho c_P u_x \frac{\partial T}{\partial x}. \qquad (3.38)$$

Identical arguments applied to the other two orthogonal directions yield the following expression for the total amount of enthalpy advected into the infinitesimal volume element per unit time:

$$-\delta x \delta y \delta z \rho c_P \left(u_x \frac{\partial T}{\partial x} + u_y \frac{\partial T}{\partial y} + u_z \frac{\partial T}{\partial z} \right), \qquad (3.39)$$

which can be written more compactly as:

$$-\delta x \delta y \delta z \rho c_P \left(\mathbf{\bar{u}} \cdot \nabla T \right), \qquad (3.40)$$

where $\bar{u}$ is the velocity vector, with components u_x, u_y, u_z and ∇T is the temperature gradient (a one-form), with components $\partial T/\partial x$, $\partial T/\partial y$ and $\partial T/\partial z$ (see also Box 1.1). The total rate of change in the enthalpy content of the volume element is given by adding equation (3.40) to equation (3.12).

The rate of change of enthalpy is related to the rate of change in the temperature of the volume element by equation (3.13). Equating the latter to the sum of (3.12) plus (3.40), simplifying and rearranging we get:

$$\frac{\partial T}{\partial t} + \mathbf{\bar{u}} \cdot \nabla T = \kappa \nabla^2 T + \frac{\alpha}{c_P} \qquad (3.41)$$

or, in one dimension:

$$\frac{\partial T}{\partial t} + u_x \frac{\partial T}{\partial x} = \kappa \frac{\partial^2 T}{\partial x^2} + \frac{\alpha}{c_P}. \qquad (3.42)$$

Equation (3.41) (or (3.42)) is called the diffusion–advection equation, as it accounts for both modes of heat transfer. It is an energy conservation equation that expresses the First Law of Thermodynamics. The diffusion equation, (3.14), is a special case of (3.41) with $\bar{u} = \mathbf{0}$.

3.6.2 A velocity scale for advection

When material in which temperature is not uniform is in motion, heat transfer occurs by both advection and diffusion, as described by equations (3.41) or (3.42). An important question that arises is that of the relative rates at which thermal energy is transported by advection and by diffusion. Consider a system in a steady state ($\partial T/\partial t = 0$) and with no heat generation ($\alpha = 0$). To simplify notation we drop the subscript x from u_x because we consider heat flow in one dimension. Equation (3.42) becomes:

$$u \frac{\partial T}{\partial x} = \kappa \frac{\partial^2 T}{\partial x^2}. \tag{3.43}$$

The two sides of this equation represent the advective and diffusive contributions to heat flow, which in a steady state and with no heat generation must balance one another. Let Λ be the characteristic lengthscale of the problem in the x direction, in which material is moving with velocity u, and such that a finite temperature difference, ΔT, occurs over the distance Λ. We then write equation (3.43) as follows:

$$u \frac{\Delta T}{\Lambda} \approx \kappa \frac{\Delta T}{\Lambda^2}, \tag{3.44}$$

which simplifies to:

$$u \approx \frac{\kappa}{\Lambda}. \tag{3.45}$$

The ratio κ/Λ, which has dimensions of $[L][T]^{-1}$, provides a velocity scale for advective flow. We can define a non-dimensional velocity by forming the ratio $u/(\kappa/\Lambda)$. This non-dimensional velocity is called the *Péclet number* and is symbolized by *Pe*:

$$Pe \equiv \frac{u\Lambda}{\kappa}. \tag{3.46}$$

A value of $Pe > 1$ means that advection outpaces diffusion. The higher the value of *Pe*, the less significant diffusive heat transfer is. We could have also arrived at the definition of *Pe* by noticing that the product of the characteristic velocity of the problem, u, times its characteristic length, Λ, has units of diffusivity, so that the product $u\Lambda$ can be thought of as a measure of the efficiency with which advection diffuses heat. This leads to the same physical interpretation of the Péclet number.

Measuring the velocity of the advective flow with a non-dimensional number such as *Pe* determines the efficiency of advective heat transport independently of the absolute scale of the problem, just as the non-dimensional parameters ζ and θ describe heat diffusion in a scale-free way. As an example, consider convection in the Earth's mantle. If we take a characteristic rate of plate motion of 10 cm yr$^{-1} \approx 3 \times 10^{-9}$ m s^{-1} as the velocity of mantle convection, u, and the mantle thickness of ~ 3000 km as its characteristic lengthscale Λ, then with $\kappa = 10^{-6}$ m^2 s^{-1} we get $Pe \approx 9000$. This says that the velocity of the advective flow in the Earth's mantle is large enough to render heat diffusion negligible, so that the Earth's deep interior loses most of its thermal energy by convection. Compare this statement, and how we arrived at it, with our discussion of diffusive planetary cooling in Section 3.3. The conclusion that we reached there is the same one that follows from calculating the mantle's Péclet number, except that this is a lot simpler and it allows us to make a quantitative statement. The flaw in Lord Kelvin's argument about the age of the Earth becomes obvious, but, of course, calculating the Péclet number requires that we know that the mantle is convecting in the first place, and the rate at which it convects. Lord Kelvin had no way of knowing this nor, I believe, did he have any compelling reasons to assume that the mantle is convecting.

Heat transfer within the Earth's continental crust is primarily by diffusion. There are some obvious exceptions to this statement, such as regions of active magmatic emplacement and regions of permeable fluid flow. In the absence of local perturbations such as these, one could expect that the geothermal gradient in the continental crust can be approximated as a steady-state conductive temperature distribution (see Turcotte & Schubert, 2002, Chapter 4 for an in-depth discussion of continental geotherms). But is this always the case? Consider denudation of metamorphic rocks equilibrated at depth, or burial of surface rocks. Rocks will stay on the steady-state diffusive geotherm only if they are uplifted or buried with a Péclet number of order 1 or less. If $Pe \gg 1$ then their temperature will not be able to adjust fast enough to what the steady-state geotherm would be, and a transient temperature distribution, either hotter or colder than the steady-state geotherm, will develop (Fig. 3.10).

In order to estimate the maximum rate of burial or denudation that can preserve the conductive geotherm we set the non-dimensional velocity $Pe = 1$ and solve equation 3.46 for u. We take the characteristic length of the problem, Λ, as being of the order of the thickness of the continental crust, ~30 km. With $\kappa = 10^{-6}$ m^2 s^{-1}, we obtain $u \sim 3 \times 10^{-11}$ m s^{-1} ~1 mm yr^{-1}. This rate is comparable to the rate of denudation of active mountain

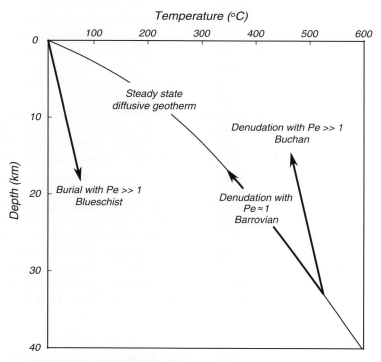

Temperature (°C)

Steady state
diffusive geotherm

Denudation with Pe ≫ 1
Buchan

Burial with Pe ≫ 1
Blueschist

Denudation with
Pe ≈ 1
Barrovian

Depth (km)

Fig. 3.10   Schematic view of three metamorphic field gradients. If metamorphic rocks are unroofed with a Péclet number ~1 their temperature will approximately follow the steady state diffusive geotherm. The resulting assemblages correspond to those of Barrovian metamorphism. High-temperature/low-pressure Buchan assemblages can form if denudation occurs with $Pe \gg 1$. Low-temperature/high-pressure blueschist metamorphism requires burial with $Pe \gg 1$.

belts, but is one to two orders of magnitude slower than characteristic rates of tectonic processes. If we take the latter rates to be in the range 1–10 cm yr^{-1}, then an average estimate for u in tectonically active areas is $\sqrt{(10 \times 1)}$ cm $yr^{-1} \approx 10^{-9}$ m s^{-1}, leading to $Pe \approx 30$. In areas of active tectonic activity, therefore, diffusive heat transfer is modest relative to advection and the geothermal gradient cannot correspond to a steady-state diffusive temperature distribution. It is important to understand that advection in this case has nothing to do with motion of fluids of any kind and occurs entirely as a result of motion of solid rock by some combination of faulting, ductile shearing and folding.

These contrasting modes of heat transfer are expressed in the *metamorphic field gradients* of exposed orogens, which are the thermal gradients preserved in metamorphic assemblages exposed at the surface but equilibrated over a range of pressures (see, for example, the classic text by Miyashiro, 1994). Metamorphic field gradients tend to cluster in three distinct regions of P-T space which, in order of increasing temperature at a fixed pressure, are often labeled Blueschist, Barrovian and Buchan. Metamorphic field gradients are not fossil geotherms, because rocks at different depths do not attain their metamorphic peak temperatures simultaneously, but it can be shown that Barrovian metamorphism develops when rocks are able to stay close to a diffusive geotherm during uplift and denudation (see Patino Douce *et al.*, 1990). This is only likely to happen if the core of an orogenic belt is denuded by erosion. The other two metamorphic field gradients must therefore develop in environments in which advection is the dominant mode of heat transfer, and in which the crust is either heated (Buchan) or cooled (Blueschist) relative to the steady-state diffusive geotherm, in response to tectonic activity. Blueschist metamorphism requires fast tectonic burial of cold material (Fig. 3.10), which characteristically takes place at subduction zones and during the initial stages of continental collisions. Buchan metamorphic conditions may develop during tectonic collapse of orogenic belts, where low-angle detachment faulting leads to tectonic unroofing of deep-seated rocks with $Pe \gg 1$.

In this analysis we ignored the geometry of the environments in which metamorphism takes place, but as we shall see in the following paragraph this can be an important consideration. For instance, there is a minimum thickness that a fault block must have in order to be able to preserve its internal temperature unperturbed during tectonic displacement.

Consider the geometry sketched in Fig. 3.11. Advective flow occurs in the x direction only and heat transport perpendicular to this direction is by diffusion ($u_y = 0$). We ask what is the distance, λ, that a thermal perturbation will propagate by diffusion perpendicular to the direction of advective flow. This length scale is related to the diffusion time scale by equation (3.16): $\lambda \sim 2\sqrt{\kappa\tau}$. Because we want to compare advective and diffusive lengthscales, we make τ equal to the time that the advective flow takes to cover the characteristic advective lengthscale, Λ, i.e. $\tau = \Lambda/u_x$. From this identity and the definition of Pe:

$$Pe = \frac{u_x \Lambda}{\kappa} = \frac{\Lambda^2}{\kappa\tau} \tag{3.47}$$

or:

$$\Lambda = \sqrt{Pe}\sqrt{\kappa\tau} \sim \frac{\lambda}{2}\sqrt{Pe}, \tag{3.48}$$

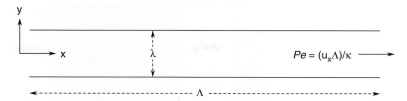

Fig. 3.11 A rigid slab of thickness λ moves with velocity u_x in the x direction, in which temperature changes occur over a characteristic length Λ. Heat transfer in the y direction is by diffusion. The geometry is a reasonable model for the oceanic lithosphere.

which yields the following scaling relationship, *valid only for the geometry sketched in Fig. 3.11*:

$$\frac{\lambda}{\Lambda} \sim 2Pe^{-1/2}. \tag{3.49}$$

The geometry of sea-floor spreading corresponds to that of Fig. 3.11 The mantle flows horizontally away from mid-ocean ridges and as it cools it gives rise to the rigid lithosphere, in which vertical heat transfer is by diffusion. The thermal perturbation that we are interested in is the exposure of hot asthenospheric mantle to the temperature at the surface of the Earth. What we are asking is what is the thickness of mantle ($\pm$ crust) that will cool down significantly while it is being transported laterally by mantle convection. In other words, what is the thickness of the oceanic lithosphere? As we saw previously, for the Earth's mantle $Pe \approx 9000$, so that with $\Lambda = 5000$ km (= characteristic width of an oceanic plate) we get $\lambda \sim 105$ km, which agrees reasonably well with the thickness of the oceanic lithosphere.

3.7 Convection as a heat transport mechanism

We have discussed the energetic underpinnings of convection (Section 3.4) and we have derived a general expression for the adiabat in a convective fluid (Section 3.5). We have also found a general equation that describes advective heat transport (Section 3.6). Our next task is to derive the mathematical equations that describe heat transport by convection. This is not simple. A complete mathematical description of convection in a system such as a planetary mantle, core, atmosphere or magma chamber seeks to describe the precise geometry of the convective flow and its evolution with time. Exact analytical solutions generally do not exist, and the significance of numerical solutions is sometimes hard to assess, owing in part to the fact that solutions tend to be quite sensitive to the choice of initial and boundary conditions, and also because there are significant uncertainties in material properties, chemical composition and phase transformations. Our goal is more modest. We seek to lay out the basic principles of what is called *parametrization* of convection, which consists of developing a set of relatively simple equations that yield order of magnitude estimates for global parameters such as heat flux and convective velocity, while ignoring the local geometry and temporal evolution of the convective flow. It is a remarkable fact that all of these equations are functions of a single parameter that encapsulates the dynamics of convection, i.e. the forces that drive and oppose convective flow.

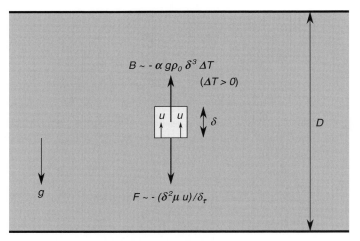

The forces acting on an element of fluid of characteristic linear dimension δ, inside a fluid layer of thickness D, are buoyancy, B, and viscous drag, F. The sign convention is positive upwards, so that gravity, g, is negative.

3.7.1 Dynamics of thermal convection

We wish to derive a criterion that will allow us to decide whether convection will take place in a fluid layer. Consider a fluid layer of depth D, and in it a fluid element of characteristic linear dimension δ, see Fig. 3.12. Convection will take place only if the fluid element is buoyant throughout the thickness of the layer. The buoyancy, $\boldsymbol{B}$, is a gravitational force that acts on the fluid element and that arises from the difference, $\Delta\rho$, between its density and that of the surrounding fluid. The volume of the fluid element is of order δ^3, so:

$$\boldsymbol{B} \sim g\delta^3\Delta\rho. \tag{3.50}$$

Because g, which is directed towards the center of the planet, is negative (equation (1.7)) a mass deficit ($\Delta\rho < 0$) results in a positive buoyancy force, i.e. directed upwards, as expected. In the case of thermal convection the density contrast is a consequence of a difference in temperature, but convection can also be driven by compositional gradients or, in the case of double-diffusive convection, by both temperature and composition gradients. Staying for now with thermal convection, we use the definition of the coefficient of thermal expansion, α (equation (1.66)), to find the effect of temperature on density at constant pressure:

$$\alpha = -\frac{1}{\rho}\left(\frac{d\rho}{dT}\right)_P, \tag{3.51}$$

Integrating this equation between a reference state T_0, ρ_0 and any arbitrary state T, ρ we get:

$$\rho = -\rho_0 e^{-\alpha(T-T_0)}. \tag{3.52}$$

A useful approximation is that, for small x, $\ln(1-x) \approx -x$, or, equivalently, $e^{-x} \approx 1-x$. The coefficient of thermal expansion for solids and liquids is small (typically of order 10^{-5} K^{-1}, see also Chapter 8), so that, for all reasonable temperature contrasts (of order 10^3 K), this approximation is valid. The density contrast is then given by:

$$\Delta\rho = \rho - \rho_0 \approx -\alpha\rho_0\left(T - T_0\right) \approx -\alpha\rho_0\Delta T, \tag{3.53}$$

where ΔT is the temperature contrast that drives convection. This corresponds to the temperature change that takes place at the heat transfer boundaries of the convection cycle, for example $T_B - T_A$ or $T_D - T_C$ in Fig. 3.9. The temperature change along the adiabatic paths ($T_C - T_B$ or $T_A - T_D$) does not generate buoyancy differences, because it affects equally all fluid elements in the layer. Convection can take place only if there is a temperature difference between the boundaries *once the effect of the adiabatic gradient has been subtracted.* This condition is described by saying that the temperature difference across the convecting fluid layer is *superadiabatic.* It is equivalent to our statement in Section 3.4 that convection is only possible if there are at least two heat transfer boundaries that separate the upwelling and downwelling adiabats.

As the fluid element ascends it cools by diffusing heat to its surroundings. Buoyancy will be maintained only as long as a significant temperature contrast exists. We can reasonably postulate that convection will only take place if the time that the fluid element takes to rise through the thickness of the fluid layer, which we can call the convection time t_c, is small compared to the characteristic diffusive cooling time of the fluid element, t_d. The latter is given by equation (3.16):

$$t_d \sim \frac{\delta^2}{4\kappa} \tag{3.54}$$

and t_c by:

$$t_c \sim \frac{D}{u}, \tag{3.55}$$

where u is the characteristic velocity that describes vertical motion of the fluid.

Buoyancy, equation (3.50), is opposed by a frictional force that arises from viscous drag. In all of the examples that we will consider in this book it is acceptable to neglect changes in momentum of the fluid, i.e. we will only consider cases in which fluids move with constant velocity. If this is the case then viscous drag must exactly balance buoyancy, so that there is no net force acting on the fluid element. The frictional force, F, is the product of the shear stress acting on the surface of the fluid element times the surface area of the fluid element. The shear stress is given by equation (3.7). If velocity decays from its value u at the boundary of the fluid element to zero over a characteristic length δ_τ, then we can write:

$$\tau = -\mu\frac{du}{dx} \approx -\mu\frac{u}{\delta_\tau}. \tag{3.56}$$

The surface area of the fluid element is of order δ^2, so that the viscous drag ($=$ total frictional force) is:

$$\boldsymbol{F} \sim -\delta^2\tau \sim -\frac{\delta^2\mu u}{\delta_\tau}. \tag{3.57}$$

This force equals the buoyancy force, equation (3.50), see also Fig. 3.12. Therefore, using (3.53):

$$-\frac{\delta^2 \mu u}{\delta_\tau} \sim g\delta^3 \Delta\rho = -g\delta^3 \alpha\rho_0 \Delta T. \tag{3.58}$$

Solving for u and substituting in (3.55):

$$t_c \sim -\frac{\mu D}{g\Delta\rho\delta\delta_\tau} = \frac{\mu D}{g\alpha\rho_0\Delta T\delta\delta_\tau}. \tag{3.59}$$

Recall that we are interested in the ratio t_d/t_c, which includes the factor $\delta^3\delta_\tau$. The problem with these equations is that they contain three different lengthscales: D, δ and δ_τ. Of these, only the first one is well defined, as it is the thickness of the fluid layer. In order to define the values of δ and δ_τ it is necessary to specify the geometry of the problem, so that it is not possible to specify general values for these two lengthscales. The only way to come up with a result of general validity is to define the timescales for diffusion and convection in terms of the only lengthscale of the problem, which is the thickness of the fluid layer, D. The diffusive and convective time scales defined in this way become:

$$t_{d,D} \equiv \frac{D^2}{4\kappa} \tag{3.60}$$

and:

$$t_{c,D} \equiv \frac{\mu D}{g\alpha\rho_0\Delta T D^2} = \frac{\mu}{g\alpha\rho_0\Delta T D}. \tag{3.61}$$

The ratio between these two time scales, omitting numerical factors, is the definition of a non-dimensional parameter known as the *Rayleigh number*, and symbolized by Ra:

$$Ra \equiv \frac{t_{d,D}}{t_{c,D}} = \frac{g\alpha\rho_0\Delta T D^3}{\mu\kappa}. \tag{3.62}$$

The Rayleigh number gives an indication of the tendency of the fluid layer to convect. The greater the value of Ra, the longer the diffusion timescale is relative to the convection timescale, and therefore the more likely it is that the fluid layer will undergo convective overturn, because parcels of fluid will be able to traverse the thickness of the layer without cooling down significantly. There is a critical value of the Rayleigh number, Ra_c, that marks the onset of convection and hence the transition between diffusive and convective heat transport. The actual critical value depends on the geometry and boundary conditions of the problem, so that there is no universal value of Ra_c. This is to be expected, because in equations (3.60) and (3.61) we substituted the fixed lengthscale D for the two lengthscales that depend on the particular configuration of the fluid layer, δ and δ_τ. Exact analytical solutions for Ra_c are available for some simple cases; in many other instances its value can only be determined numerically or experimentally.

For a laterally extended layer (in which the width is much greater than D) heated from below, which is a reasonable approximation to planetary mantles and tabular igneous intrusions, Ra_c is of order 10^3–10^4. Typical values for the parameters in equation (3.62) for the Earth's mantle are: $D = 3000$ km; $\alpha = 10^{-5}$ K^{-1}; $\rho_0 = 3500$ kg m^{-3}; $\mu = 10^{21}$ Pa s; $\kappa = 10^{-6}$ m^2 s^{-1}; $g = 9.8$ m s^{-2}; $\Delta T = 1400$ K. The value of ΔT is the temperature difference across the oceanic lithosphere, which is the temperature drop that corresponds to the low-pressure cooling leg of the convection cycle (CD in Figs. 3.8 and 3.9). This is the temperature difference that engenders negative buoyancy at the top of the mantle and thus drives convection. With these numbers we find that Ra for the Earth's mantle is $\sim 10^7$. This value is so much higher than the critical value that one must conclude that the Earth's mantle is not only convecting, but doing so vigorously. This is also reflected in its high Péclet number (Section 3.6.2). Note an important conceptual difference between Ra and Pe, however. The Rayleigh number is based on the dynamics of fluid flow and is an *a priori* predictor of whether or not a fluid layer is likely to convect. Calculation of Pe starts from the knowledge that advective heat transfer is taking place and its value is a measure of the efficiency of advection as a heat transfer mechanism.

3.7.2 The thermal boundary layer

In Section 3.6.2 we calculated the thickness of the oceanic lithosphere by specifying that heat flow across it takes place solely by diffusion. In the terminology of heat transfer, the lithosphere is the *thermal boundary layer* of mantle convection. When diffusion is the only heat transfer mechanism the thermal boundary layer is the thickness of material across which most of the thermal gradient occurs (Section 3.2.4, Fig. 3.5). This is also true of advection but in this case the reason why most of the temperature gradient occurs across the thermal boundary layer is that heat transfer across it is by diffusion, whereas motion of matter transports internal energy throughout the remainder of the system, where advection outpaces diffusion (see Section 3.5).

The thermal boundary layer is the main barrier to heat transfer, because diffusion is a much less efficient heat transfer mechanism than advection. As heat transfer across the boundaries is what drives convection (Section 3.4), it follows that the nature and thickness of the thermal boundary layer is what ultimately controls the nature of convection. Figure 3.13a is a schematic view of the thermal boundary layer. Let x be the direction parallel to the thermal boundary layer, and hence to the heat transfer interface, and y be perpendicular to it. Fluid velocity in the y direction, u_y, vanishes in the thermal boundary layer but has a non-zero, and perhaps variable, value in the remainder of the fluid layer. The velocity in the x direction, u_x, varies between some characteristic value inside the actively convecting layer and zero at the boundary. The top diagram in Fig. 3.13 shows a continuously decreasing velocity profile, but this is not always the case, as we shall see. The key point is that there is no heat advection in the y direction across the thermal boundary layer, because u_y vanishes in it. We can think of the thermal boundary layer as being made up of a number of layers moving parallel to one another, like the cards in a deck. Regardless of the actual distribution of velocities throughout the thermal boundary layer, heat transfer between adjacent layers, and hence across the entire thermal boundary layer, can take place only by diffusion. If the system has been convecting long enough to have reached a steady state, and there is no heat generation in the thermal boundary layer, then the temperature gradient across the thermal boundary layer is linear, from T_0 at the heat transfer interface to the temperature of the actively convecting fluid, T_i. This is not the case in the convective interior, where heat is advected in the y direction, with $u_y \sim u_x$. We will consider the temperature in the convecting fluid to be constant, as shown in Fig. 3.13a. An adiabatic gradient can be superimposed on this temperature distribution but this has no effect on the arguments that follow.

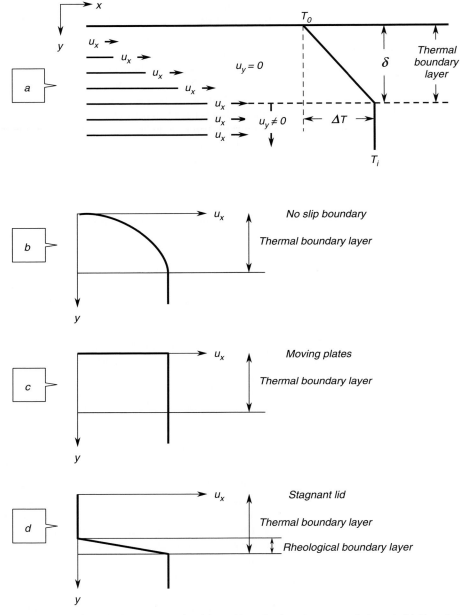

Fig. 3.13 Convective thermal boundary layer. A generalized thermal boundary layer in x–y space is shown in (a). The velocity distribution inside the thermal boundary layer is shown with the arrows of variable length, and the temperature distribution with the thick line. Different types of boundary layers are shown in (b) through (d) in terms of the magnitude of the velocity in the x direction, u_x, as a function of y.

The effectiveness of convection as a heat transport mechanism is controlled to a large extent by the velocity distribution across the thermal boundary layer. This is in turn determined by a number of factors, including material properties, temperature and temperature gradient, and convective velocity. For example, if velocities are high enough that frictional forces at the boundary are non-negligible then the velocity must vanish at the interface. This is called a no slip boundary. The velocity in the boundary layer then varies continuously (but not necessarily linearly) from 0 at the interface to the velocity of the actively convecting fluid (Fig. 3.13b). This is the case, for example, at the interface between ocean and atmosphere. Plate tectonics is different, as in this case friction at the interface (i.e. the ocean floor) vanishes owing to the very small velocity. In this case u_x is constant across the thermal boundary layer (the oceanic lithosphere) and equal to the velocity in the actively convecting mantle, and there is a velocity discontinuity at the interface (Fig. 3.13c). The thermal boundary layer forms because the mantle close to the surface cools to the point where it cannot flow anymore, forming a rigid lid on the ductile asthenosphere. What is special about plate tectonics is that the lid is broken into plates that are able to subduct, allowing fragments of diffusive lid to be carried along by mantle convection.

Formation of the rigid lid takes place because there is a strong inverse relationship between temperature and viscosity (Box 3.3). In the case of plate tectonics, however, the

---

Box 3.3   **Temperature effect on viscosity**

For many materials, including mantle rocks, it is found that viscosity varies with temperature following an inverse exponential law that is called Arrhenius's law (see Chapter 12). A dependency on strain rate (called non-Newtonian rheology) may be superimposed on Arrhenian behavior, but that is beyond the scope of this book. The Arrhenius law for viscosity can be written as follows:

$$\mu = K\, e^{\frac{E_a}{RT}}, \tag{3.3.1}$$

where K and E_a are material properties and R is the gas constant. If we restrict temperature to a relatively narrow interval we can define an (arbitrary) reference temperature T_0 within this interval and, given the viscosity μ_0 at this temperature, we can calculate the viscosity at any other temperature inside the interval by:

$$\mu = \mu_0\, e^{\frac{E_a}{R}\frac{T_0 - T}{T T_0}}, \tag{3.3.2}$$

which, if T and T_0 are not too different, we can approximate as follows:

$$\mu \approx \mu_0\, e^{-\gamma(T - T_0)} \tag{3.3.3}$$

with the parameter γ defined as follows:

$$\gamma = \frac{E_a}{R T_0^{\,2}}. \tag{3.3.4}$$

Equation (3.3.3) is a *linearized approximation* to Arrhenius's law. The material property E_a is called the *activation energy* for viscous flow (see Chapter 2). Typical values for mantle materials are $\mu_0 \approx 10^{21}$ Pa s at $T_0 = 1950$ K, and $E_a \approx 300$ kJ mol^{-1}.

contrast in viscosity between lithosphere and asthenosphere, which spans many orders of magnitude, is masked by the fact that the lithosphere can break and thus move. Consider now the cases of Mars or Venus. A case can be made for both planets that convection takes place below the rigid outer layer (Section 3.8.2), but there is no evidence that this outermost layer is broken up and carried along by the underlying convective flow. For example, there are no subduction zones nor transform faults on either of the planets and volcanoes are not arranged in Hawaii-style volcanic chains. Venus and Mars are examples of what is called *stagnant lid convection*, depicted in Fig. 3.13d. The strong inverse effect of temperature on viscosity generates a rigid lid, but in this case the lid does not break up. The velocity gradient is then concentrated in a relatively thin layer called the *rheological boundary layer*, sandwiched between the stagnant lid and the actively convecting fluid. The y velocity component does vanish in the rheological boundary layer, so that it is part of the thermal boundary layer (see Fig. 3.13d).

The driving force for thermal convection derives from buoyancy engendered by temperature differences which in turn are caused by heat transfer across the thermal boundary layer (Section 3.4). Some part of the thermal boundary layer must always be entrained in the convective flow. It is in this part of the thermal boundary layer that buoyancy is generated. In the case of convection with moving plates (Fig. 3.13c) the entire thermal boundary layer enters the convective flow (e.g., at a subduction zone), so that the full temperature difference ΔT (Fig. 3.13a) is available to drive convection. In contrast, convection with a stagnant lid is driven only by that fraction of ΔT that occurs across the rheological boundary layer, which is the only part of the thermal boundary layer that is entrained in the convective flow and that generates buoyancy. We can expect that, other things being equal, convection with a stagnant lid is less efficient at transporting heat than moving plate convection.

3.7.3 Scaling of heat transport by convection

Consider a convective layer of thickness D capped by a thermal boundary layer of thickness δ, where in general it is $\delta \ll D$. If the system is in a steady state then the heat flux across the convecting fluid layer, q, must equal the heat flux across the thermal boundary layer. By Fourier's law, and with the convention that heat flux q is always a positive quantity (Section 3.2.4) we have:

$$q = k \frac{\Delta T}{\delta},$$

(3.63)

where ΔT is the temperature difference across the thermal boundary layer. We now compare this to the heat flux that would exist across the entire fluid layer *if it was not convecting,* i.e. if heat diffused across the entire thickness D driven by the same temperature difference ΔT. We label this heat flux q^*:

$$q^* = k \frac{\Delta T}{D}.$$

(3.64)

The advective heat flux, q, and the hypothetical diffusive flux, q^*, are driven by the same temperature difference. The ratio between q and q^* is therefore a measure of how much more efficiently advection transports heat relative to diffusion. This ratio defines a non-dimensional parameter called the *Nusselt* number, *Nu*:

$$Nu \equiv \frac{q}{q^*} = \frac{q}{k\frac{\Delta T}{D}} = \frac{D}{\delta}. \tag{3.65}$$

The Nusselt number is the non-dimensional advective heat flux, just as *Pe* is the non-dimensional advective velocity. *Nu* is also the ratio between the thickness of the fluid layer and the thickness of the thermal boundary layer.

Worked Example 3.3   Heat transfer efficiency of terrestrial mantle convection

In the case of convection with moving plates, such as Earth's plate tectonics, the entire thermal boundary layer moves with the same velocity as that of the convecting fluid (Fig. 3.13c). If the thickness of the fluid layer, D, is comparable to its characteristic lateral dimension then D and the thickness of the thermal boundary layer, δ, scale according to equation (3.49), i.e.:

$$\frac{\delta}{D} \sim 2Pe^{-1/2}. \tag{3.66}$$

We can then derive the following scaling relationship between *Nu* and *Pe*:

$$Nu \sim \frac{1}{2}Pe^{1/2}. \tag{3.67}$$

Note that (3.67) applies *only* to convection with moving plates, as we have explicitly used the geometry of this type of thermal boundary layer in the derivation of this equation. Since *Pe* and *Nu* are both non-dimensional numbers, there is no mathematical requirement that they be related by any specific type of function. This is true of any scaling relationship among non-dimensional numbers.

For the Earth's mantle, with $Pe \approx 9000$ (Section 3.6.2), we get $Nu \sim 50$. Mantle convection cools the Earth 50 times faster than diffusion would. It may sound counterintuitive, but this is the explanation for why Kelvin's age estimate was too low. In a diffusive Earth most of the planet would not have began cooling after 4.5 billion years (Fig. 3.6). In contrast, in a convective Earth the deep interior of the planet contributes to surface heat flow from the very inception of the convective regime, most likely immediately after differentiation. A convective planet will cool down completely and meet its thermal demise much faster than if the same planet cooled by diffusion. It is ironic, in view of his argument with geologists, that Kelvin's diffusive model implied that the Earth would have had a much longer life expectancy than what our rapidly cooling planet actually has (disregarding, of course, the fact that the Sun will burn out before diffusive cooling could have had time to reach the center of the Earth).

We can also use equation (3.67) to estimate the rate of mantle convection in the Earth. Let the average mantle heat flux be q_o and assume that heat production in the oceanic lithosphere is negligible. We write equation (3.65) as follows:

$$Nu = \frac{q_o D}{k \Delta T}. \tag{3.68}$$

From equations (3.46) and (3.67), making $\Lambda = D$, we have:

$$Nu \sim \frac{1}{2} Pe^{1/2} = \frac{1}{2} \left(\frac{uD}{\kappa} \right)^{1/2}.$$ (3.69)

We can eliminate Nu between these two equations and solve for u, the velocity of the mantle flow:

$$u \sim 4 \frac{q_o^2 D\kappa}{k^2 (\Delta T)^2}.$$ (3.70)

The value of q_o in this equation is the average oceanic heat flux of 100 mW m^{-2} (Box 3.2). Other characteristic mantle values are: $D = 3000$ km, $\Delta T = 1400$ K, $\kappa = 10^{-6}$ m^2 s^{-1} and $k = 3$ W m^{-1} K^{-1}. We then estimate $u \sim 6.8 \times 10^{-9}$ m s$^{-1} \approx 20$ cm yr^{-1}. This is a factor of 3 higher than typical rates of plate motion. Agreement is not great, but it is within an order of magnitude.

3.7.4 Energy conservation in a convecting fluid

The goal of the parametrized description of convection is to generate estimates for heat flux, convective velocity and thickness of the thermal boundary layer in terms of the Rayleigh number, depth of the convective layer, temperature change at the heat exchange boundaries (e.g. Fig. 3.8), and nature of the thermal boundary layer (i.e. stagnant or moving). If we need to solve for three unknowns: heat flux, q, convective velocity, u and thickness of the boundary layer, δ, we need three independent equations relating these variables. We already have two of these equations: (3.63), which gives the relationship between heat flux and thickness of the thermal boundary layer, and (3.66), which relates thickness of the thermal boundary layer to fluid velocity. The third equation must be one that relates heat flux to velocity. Recall from equation (3.7) that the velocity gradient determines shear stress in the fluid, and that shear stress in turn determines the rate of heating by viscous dissipation (Chapter 1). This suggests that fluid velocity and heat flux must be related by an energy conservation equation, which in a convecting fluid must be the diffusion advection equation. We write equation (3.41) in two dimensions:

$$\frac{\partial T}{\partial t} + u_x \frac{\partial T}{\partial x} + u_y \frac{\partial T}{\partial y} = \kappa \left(\frac{\partial^2 T}{\partial x^2} + \frac{\partial^2 T}{\partial y^2} \right) + \frac{\alpha}{c_p}$$ (3.71)

(do not confuse α in this equation, which is the rate of heat production per unit mass, with the coefficient of thermal expansion). Consider a fluid layer of depth D in the y direction, capped by a boundary layer of thickness δ (Fig. 3.14). We are interested in steady state conditions with no horizontal thermal gradient, so $\partial T/\partial t$ and $\partial T/\partial x$ vanish. We will now substitute a fictional but mathematically equivalent picture for heat transfer inside the convecting fluid. Imagine that the vertical velocity component, u_y, vanishes everywhere in the convective fluid, and that the advective heat flux in the y direction is accounted for by diffusion with a large but fictive value of the thermal conductivity, k^*. It will not be necessary to calculate the value of k^*, all that we need to keep clear is that it is the thermal conductivity that the fluid would have to have in order to transport heat in a steady state at the same rate

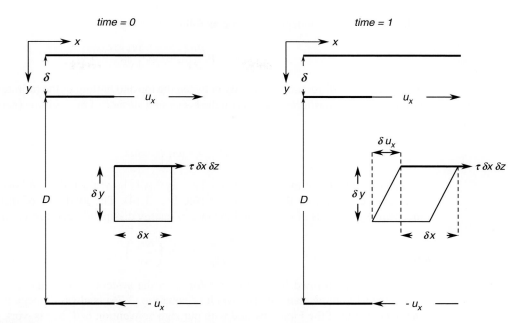

Fig. 3.14 Viscous dissipation in an element of convecting fluid. The dimensions of the fluid element are δx, δy and δz (coming out of the page). The force acting on the horizontal faces of the element is $\tau \, \delta x \, \delta z$. The horizontal velocity changes by an amount δu_x over length δy, so that the top surface of the element moves a distance δu_x in unit time. Horizontal velocity varies from u_x to $-u_x$ over the thickness D of the fluid layer. The thermal boundary layer has thickness δ.

as convection, assuming that the temperature gradient is the same adiabatic temperature gradient present in the convective layer. In contrast to the vertical velocity, we will make no assumptions about the horizontal velocity, u_x, which does not vanish. With these constraints we can drop the partial derivative symbols because $\partial T/\partial t = \partial T/\partial x = \partial^2 T/\partial x^2 = u_y = 0$, and equation (3.71) simplifies to:

$$\kappa^* \frac{d^2 T}{dy^2} + \frac{\alpha}{c_p} = 0, \tag{3.72}$$

where κ^* is the heat diffusivity that corresponds to the fictive conductivity k^*. From the definition of heat diffusivity, $\kappa = k/(\rho \, c_p)$, and following the convention that heat flux is always a positive quantity, this equation can be re-written as follows:

$$-k^* \frac{d^2 T}{dy^2} + \phi = 0, \tag{3.73}$$

where $\phi = \alpha \rho$ is heat production per unit volume. If we consider a simple case in which there are no phase changes nor radioactive heat generation in the fluid layer, and in which the fluid is a perfect electrical insulator, then the only source of heat is viscous dissipation of mechanical work.

Consider an infinitesimal volume of fluid with dimensions δx, δy, δz (Fig. 3.14, in which δz comes out of the page). The viscous stress is given by equation (3.7), which, if horizontal velocity changes with depth, we write as follows:

$$\tau = -\mu \frac{du_x}{dy} \approx -\mu \frac{\delta u_x}{\delta y}, \tag{3.74}$$

where δu_x is the difference in velocity between the top and bottom of the infinitesimal fluid volume. The viscous force that acts on the horizontal surface of this volume (perpendicular to the page) is:

$$\tau \delta x \delta z \approx -\mu \frac{\delta u_x}{\delta y} \delta x \delta z. \tag{3.75}$$

In unit time this force moves a distance δu_x, as this is how much the top horizontal face is displaced relative to the bottom one (see Fig. 3.14). The total rate of dissipation of mechanical energy is $\tau \delta x \delta z \delta u_x$, and the rate of energy dissipation per unit volume is:

$$\frac{\tau \delta x \delta z \delta u_x}{\delta x \delta y \delta z} \approx -\mu \left(\frac{\delta u_x}{\delta y}\right)^2. \tag{3.76}$$

This is work performed by the viscous force on the system, where "the system" is the infinitesimal volume element. The result of equation (3.76) is always a negative quantity. In the language of the First Law and with our sign convention ($dW > 0$ is work performed by the system), this means that "the environment" always performs work on "the system". The system, i.e. our infinitesimal volume element, absorbs this energy as heat. The rate of viscous heating per unit volume is thus the negative of equation (3.76), i.e.:

$$\phi \approx \mu \left(\frac{\delta u_x}{\delta y}\right)^2. \tag{3.77}$$

Taking the small increments to the limit we rewrite equation (3.73) as follows:

$$-k^* \frac{d^2 T}{dy^2} + \mu \left(\frac{du_x}{dy}\right)^2 = 0. \tag{3.78}$$

We now approximate the values of the derivatives by averaging over the depth of the fluid layer, D. If u_x is the characteristic fluid velocity, then the horizontal velocity must change from u_x to $-u_x$ over the distance D (Fig. 3.14), hence we can write:

$$\left(\frac{du_x}{dy}\right)^2 \approx \left(\frac{2u_x}{D}\right)^2. \tag{3.79}$$

Also, from:

$$\frac{d^2 T}{dy^2} = \frac{d}{dy}\left(\frac{dT}{dy}\right) \tag{3.80}$$

we see that we can approximate this term starting from the thermal gradient across the fluid layer. Because the fluid is convecting the thermal gradient is adiabatic, so we can substitute equation (3.35), as follows:

$$k^* \frac{d^2 T}{dy^2} = k^* \frac{d}{dy}\left(\frac{dT}{dy}\right) = k^* \frac{d}{dy}\left(\frac{\alpha g T}{c_P}\right) = \frac{\alpha g}{c_P}\left(k^* \frac{dT}{dy}\right). \tag{3.81}$$

Since we are averaging the derivatives over the depth of the fluid layer the expression in parentheses in the last term in equation (3.81) is the heat flux across the convective layer, q, which is also the heat flux across the thermal boundary layer. Hence (with the convention of q always positive):

$$k^* \frac{d^2 T}{dy^2} \approx \frac{\alpha g}{c_P} q.$$

(3.82)

Substituting equations (3.79) and (3.82) into (3.78) we arrive at the following equation for energy conservation averaged over the thickness of the fluid layer, which relates heat flux to velocity:

$$\frac{\alpha g}{c_P} q \approx 4\mu \left(\frac{u_x}{D} \right)^2.$$

(3.83)

Note that α in this equation is the coefficient of thermal expansion.

3.8 Parametrization of convection in planetary interiors

The interiors of many planetary bodies in the Solar System, present and past, can be discussed on the basis of either moving plate or stagnant lid convection. The goal of this section is to develop simple mathematical descriptions of these two convection modes, that capture some of the essential physics of the processes without getting into any of the details. A very complete and rigorous discussion of mantle convection, including the gory details, is presented in the massive treatise by Schubert *et al.* (2001). It is my hope that the material that I present here will serve as an introduction to that work for the motivated reader.

3.8.1 Convection with moving plates

The parametrized solution for convection with moving plates is a set of values $\{q, u, \delta\}$ that simultaneously satisfy equations (3.63), (3.66) and (3.83). We rewrite the equations in terms of a single characteristic velocity, u, make $\Lambda = D$, and convert Pe to dimensional velocity using (3.46). The three equations become, respectively:

$$q = k \frac{\Delta T}{\delta}$$

(3.84)

$$\delta \sim 2 \left(\frac{\kappa D}{u} \right)^{1/2}$$

(3.85)

$$\frac{\alpha g}{c_P} q \approx 4\mu \left(\frac{u}{D} \right)^2.$$

(3.86)

Combining (3.84) and (3.85) to eliminate δ:

$$q \sim \frac{k \Delta T}{2} \left(\frac{u}{\kappa D} \right)^{1/2}$$

(3.87)

and eliminating u between (3.86) and (3.87):

$$q^3 \sim \frac{\alpha g k^4 (\Delta T)^4}{64 \mu \, c_P \kappa^2}.$$

(3.88)

The right-hand side of this equation can be written in terms of the Rayleigh number (equation (3.62)) and a combination of dimensional parameters. After a bit of algebra and using the definition of heat diffusivity we find that (3.88) can be simplified to:

$$q \sim 0.25 Ra^{1/3} \left(k \frac{\Delta T}{D} \right) \tag{3.89}$$

or in non-dimensional form (see equation (3.65)):

$$Nu \sim 0.25 Ra^{1/3}. \tag{3.90}$$

From equation (3.65) we also see that the thickness of the thermal boundary layer is given by:

$$\delta \sim 4 D Ra^{-1/3}. \tag{3.91}$$

We can now substitute this value for δ in (3.85) to get the velocity:

$$u \sim 0.25 \frac{\kappa}{D} Ra^{2/3}. \tag{3.92}$$

There are two implicit assumptions in this derivation. In the first place, it is assumed that the thermal boundary layer is in steady-state thermal equilibrium with the interior of the convective fluid. This means that the convecting fluid loses heat at the same rate as it is conducted across the thermal boundary layer, as specified by equation (3.84). Second, we assume that there is a source of energy for convection that supplies heat to the convecting fluid at this same rate. Our derivation of the energy conservation equation, (3.83), specifically rules out heat generation in the convecting fluid, so that the solution set given by equations (3.89), (3.91) and (3.92) is strictly applicable only to a fluid layer heated from below.

Worked Example 3.4   Archaean mantle convection

We will apply equations (3.89) through (3.92) to study how mantle convection may have evolved throughout Earth's history. One must keep in mind that these equations yield order of magnitude estimates and not exact results, but because they rely on a single non-dimensional parameter, the Rayleigh number, that encapsulates the dynamics of convection, they can yield a high rate of return in terms of insight gained relative to effort invested.

Mantle convection may have differed in the distant geologic past because mantle temperatures must have been higher than today's (Chapter 2). Temperature enters into this analysis in two ways: directly in the factor ΔT and indirectly as a result of the temperature dependence of viscosity. The viscosity–temperature function is discussed in Box 3.3. A subtle but crucial point is that the temperature that enters into the calculation of viscosity with equation (3.3.3) is a characteristic temperature of the interior of the fluid layer which is *not the same as the temperature at the bottom of the thermal boundary layer*. The latter is the one that defines the temperature contrast with the surface, ΔT, that drives convection by causing negative buoyancy. In a thick fluid layer such as a planetary mantle the two temperatures can be very different, owing to the adiabatic gradient. Moreover, temperature varies across the convecting layer, so what is the characteristic value that we must use in order to calculate viscosity with equation (3.3.3)? There is no unique answer to this question. One possibility, which is what I have chosen to do here, is to use as characteristic mantle temperature the temperature at a depth halfway across the fluid layer. Since the convecting fluid lies on an adiabat, we can set the temperature at the base of the lithosphere, and then calculate the temperature anywhere in the convecting layer by integrating equation (3.35). The result is:

$$T_y = T_0 \, e^{\frac{\alpha g}{c_P}(y - y_0)},$$ (3.93)

where T_y is the temperature at depth y, and T_0 is the temperature at some reference depth y_0 (e.g. the base of the lithosphere, but it could be anywhere else).

The results of parametrization of terrestrial mantle convection with equations (3.89)–(3.92) is shown in Fig. 3.15 as a function of characteristic mantle temperatures (T_y) calculated with equation (3.93). The corresponding temperatures at the base of the lithosphere (T_0) are shown in the bottom panel of the drawing. The temperature of the ocean floor is close enough to $0\,°C$ that we can assume that $T_0 \approx \Delta T$, as suggested in the figure. The characteristic mantle temperature for the present day Earth's mantle according to Schubert *et al.* (2001) is 1950 K, shown with a thick broken vertical line in Fig. 3.15. This corresponds to a temperature at the base of the lithosphere of $\sim 1400\,°C$, which is consistent with petrologic constraints (Chapter 10). The bottom panel of the figure also shows mantle viscosities calculated with equation (3.3.3), using a reference viscosity of 10^{21} Pa s at 1950 K and an activation energy of 300 kJ mol^{-1} (see Box 3.3 and Chapter 12). Other model parameters are listed in the figure caption.

Our first task is to gauge the performance of the model against the terrestrial mantle that we can observe today. The parametrized model predicts a present-day mantle heat flux of ~ 82 mW m^{-2}, a convection velocity of ~ 15 cm yr^{-1}, and a lithospheric thickness of ~ 50 km. The predicted heat flux corresponds almost exactly (and probably fortuitously) to the observed heat loss that can be attributed to the spreading–subduction cycle (Box 3.2). Agreement is less good for the other two parameters, which are off by a factor of approximately 2 (observed values are 5–10 cm yr^{-1} and ~ 100 km), but certainly of the correct order of magnitude.

What can we infer about the Earth's mantle in the distant geologic past? Petrologic evidence (e.g. eruption of ultramafic lavas such as komatiites) and models of the Earth's thermal history suggest that Archaean mantle temperatures may have been 200–400 K higher than today's. The model then suggests that Archaean mantle heat flux may have been about 3–4 times higher than today's, plate thicknesses may have been about half of today's, and convection velocity may have been up to one order of magnitude faster, maybe approaching 1 meter per year! Plate tectonics, in the sense of mantle convection with a mobile thermal boundary layer, almost certainly existed, for the processes that make it possible – breaking and bending of plates – are facilitated by a thinner lithosphere (see next section). But owing to their lesser thickness (and hence rigidity) Archaean plates must have been smaller, and therefore more numerous, than today's, and may have moved up to ten times faster. The total length of plate margins, convergent and divergent, would therefore have been greater than

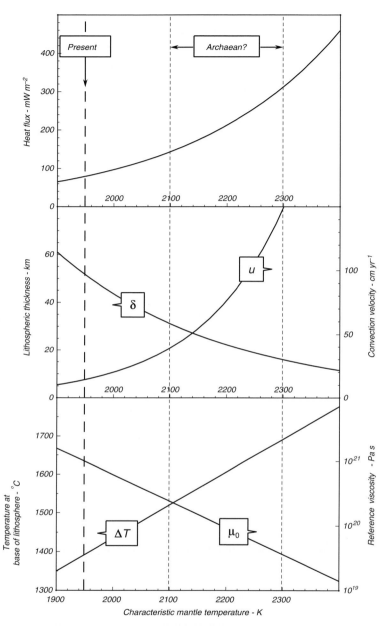

Fig. 3.15 Parametrization of terrestrial mantle convection, as a function of characteristic mantle temperature (calculated with equations (3.89)–(3.92)). The model assumes a reference mantle viscosity of 10^{21} Pa s at 1950 K and an activation energy for mantle viscous flow of 300 kJ mol^{-1}(see Box 3.3). Other parameters are: $\alpha = 10^{-5}$ K^{-1}, $\rho = 3500$ kg m^{-1}, $c_P = 1$ kJ K^{-1} kg^{-1}, $\kappa = 10^{-6}$ m^2 s^{-1}, $k = 3$ W m^{-1} K^{-1}, $D = 3000$ km, $T_{surface} = 300$ K, $g = 9.8$ m s^{-2}. Conditions for the present-day mantle correspond to a characteristic mantle temperature of 1950 K. The range in possible Archaean conditions arises from assumed mantle temperatures of 2100–2300 K.

today and cycles of continental assembly and break up, which since the late Proterozoic have had a characteristic time scale of a few hundred million years, may have operated on time scales of the order of 10^7 years. The fast rate of crustal growth that characterizes the Archaean may reflect, at least in part, the higher rates of mantle convection and greater lengths of plate margins.

3.8.2 Convection with a stagnant lid

An implicit assumption in the derivation of the scaling laws for convection with moving plates is that the viscosity of the fluid does not vary strongly with temperature. This is what allows the thermal boundary layer to move with the same velocity as the underlying fluid. Of course this is not true for mantle rocks nor for any other material, as viscosity is always a strong function of temperature (Box 3.3). This temperature dependency is what makes the rigid lithosphere physically distinct from the asthenosphere. Convection with moving plates is possible on Earth because the lithosphere is broken into fragments that can move relative to one another, and bend to some extent, by a combination of localized brittle and ductile deformation mechanisms. The result of this is that the *apparent* global viscosity of the thermal boundary layer (the lithosphere) is comparable to that of the asthenosphere. Locally, however, the lithosphere preserves its rigid behavior that makes diffusion the only possible heat transport mechanism. Convection with moving plates is a rather special case, as one might surmise from the uniqueness of plate tectonics in the present-day Solar System. A more common situation in solid planets is that of convection with a stagnant lid (Fig. 3.13d).

In order to parametrize this style of convection it is necessary to include the viscosity–temperature relationship of the fluid (this does not show up in the equations for moving plate convection because we implicitly assume that viscosity is constant). Let T_i be the temperature of the convecting fluid, T_r the temperature at the top of the rheological boundary layer (Fig. 3.16) and T_0 the temperature at the top of the thermal boundary layer (see also Fig. 3.13d). Convection with moving plates is driven by the full temperature difference $\Delta T = T_i - T_0$, but if a stagnant lid forms then convection is driven only by the temperature difference across the rheological boundary layer, $\Delta T_r = T_i - T_r$. This is so because this is the only volume of cool rock that is entrained in the convective flow, and the negative buoyancy of the rheological boundary layer is what drives convection. We can then define the rheological boundary layer as that portion of the thermal boundary layer in which the viscosity, μ_r, is of the same order of magnitude as the viscosity of the convecting fluid, μ_i. From equation (3.3.3):

$$\mu_r = \mu_i e^{-\gamma(T_r - T_i)} = \mu_i e^{\gamma \Delta T_r}. \tag{3.94}$$

The condition $\mu_r \sim \mu_i$ is satisfied if $\gamma \Delta T_r = 1$, where γ is defined by equation (3.3.4), Box 3.3. The temperature difference that drives convection is therefore given by:

$$\Delta T_r = \frac{1}{\gamma}. \tag{3.95}$$

The effective Rayleigh number of the fluid corresponds to this temperature difference. We will symbolize it by Ra_e, to distinguish it from Ra (equation (3.62)), which is calculated

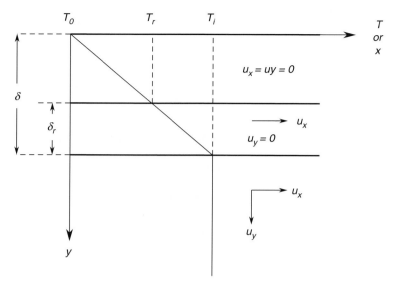

Fig. 3.16 Thermal boundary layer in stagnant lid convection. The diffusive lid has thickness δ, but only a portion of it, the rheological boundary layer of thickness δ_r, participates in convective flow, the remainder is immobile. Vertical heat transport across the rheological boundary layer is by diffusion ($u_y = 0$), so that the temperature gradient is constant across the entire boundary layer (assuming no heat generation). Convection is driven by the temperature drop across the rheological boundary layer, $T_r - T_i$.

with the full temperature difference, ΔT. Thus:

$$Ra_e = \frac{g\alpha\rho_0 \Delta T_r D^3}{\mu\kappa} = \frac{g\alpha\rho_0 D^3}{\mu\kappa\gamma} = Ra\,(\gamma\Delta T)^{-1}. \qquad (3.96)$$

With the Rayleigh number defined in this way the problem of stagnant lid convection becomes the same as that of convection with moving plates, *as long as we ignore the portion of the thermal boundary layer that lies beyond* δ_r (Fig. 3.16) – we will return to this shortly. Convection is driven by the rheological boundary layer becoming negatively buoyant and sinking. We can therefore think of the rheological boundary layer as being made up of plates that move under the stagnant upper portion of the thermal boundary layer, suggesting that we modify equations (3.89), (3.91) and (3.92) as follows:

$$q \sim 0.25\,Ra_e^{1/3}\left(k\frac{\Delta T_r}{D}\right) = 0.25\,Ra_e^{1/3}\left(\frac{k}{\gamma D}\right) \qquad (3.97)$$

$$\delta_r \sim 4D\,Ra_e^{-1/3} \qquad (3.98)$$

$$u \sim 0.25\frac{\kappa}{D}\,Ra_e^{2/3}. \qquad (3.99)$$

We have to be careful with the Nusselt number, however, as it is defined on the basis of the full temperature difference, ΔT and not ΔT_r. From equation (3.65):

$$Nu = \frac{q}{k\frac{\Delta T}{D}} = 0.25\,Ra_e^{1/3}\,(\gamma\Delta T)^{-1} \qquad (3.100)$$

or, using the relationship between Ra_e and Ra (equation (3.96)):

$$Nu = 0.25 Ra^{1/3} (\gamma \Delta T)^{-4/3}. \tag{3.101}$$

These equations highlight an important difference with moving plate convection. In that case the efficiency of convective heat transport scales with the Rayleigh number only (equation (3.90)), whereas in stagnant lid convection the viscosity–temperature relationship of the fluid also enters the scaling law for heat transfer. We should expect this on physical grounds, as the viscosity–temperature law is what determines the thickness of the rheological boundary layer and hence the effective temperature difference that drives convection.

We still need to calculate the total thickness of the thermal boundary layer, δ. If we assume that the stagnant lid is in steady-state thermal equilibrium with the convective fluid and that there is no heat generation in it then the thermal gradient across the entire thermal boundary layer is linear. The thickness of the thermal boundary layer is in that case related to that of the rheological boundary layer as follows (Fig. 3.16):

$$\delta = \frac{\Delta T}{\Delta T_r} \delta_r = \gamma \Delta T \delta_r. \tag{3.102}$$

Worked Example 3.5   Stagnant lid convection vs. moving plate convection

An instructive way of comparing the two styles of convection is by considering what may happen during the transition from moving plate to stagnant lid convection, for example, if plate tectonics on Earth were to seize up. During moving plate convection the mantle is cooled by subduction of cold lithosphere. If subduction stops then convection becomes less efficient at transporting heat. We can consider two end-member situations: either the rate of heating does not change, in which case the internal energy content of the mantle must go up, or the temperature of the mantle does not change, because there is no active heat source. In both cases we can expect the lithosphere to thicken because it ceases to be recycled and is thus no longer heated by being immersed in the hot convecting fluid.

Mantle convection is driven by a combination of radioactive decay and secular cooling, of which core crystallization may be an important component. Radioactive decay delivers energy at the same rate independently of temperature and, therefore, of the rate at which heat is transported away from its source. In contrast, the rate of energy supply by secular cooling generally varies with temperature. For example, if mantle temperature increases the thermal gradient relative to the core will decrease, and so will the rate of delivery of enthalpy of crystallization. It follows that the relative contributions of radioactive decay and secular cooling will determine the evolution of mantle temperature in our hypothetical planet in which plate tectonics comes to a sudden stop.

Figure 3.17 is constructed for a planet identical to the present day Earth. The thick curves labeled "Moving plates" are the same ones for lithospheric thickness and heat flux as in Fig. 3.15, calculated with equations (3.91) and (3.89), respectively. The thick curves labeled "Stagnant lid" show lithospheric thickness and heat flux for a planet with the same characteristics as Earth, but calculated with equations (3.102) and (3.97). We seek to understand how the planet may transition from the moving plate to the stagnant lid regime. If stagnant lid convection had been taking place on Earth since its formation then its mantle temperature would be higher than it is now, because plate tectonics is a more efficient cooling mechanism than stagnant lid convection, but the comparison between the two convection styles would still be valid.

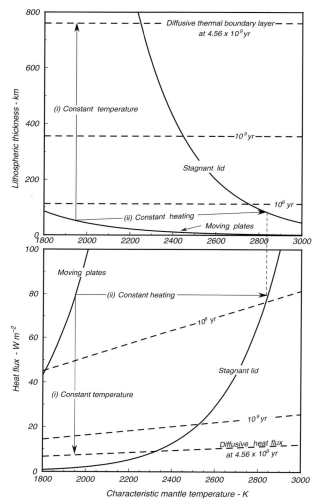

Fig. 3.17 Transition from moving plate to stagnant lid convection for a hypothetical planet with the characteristics of the Earth. The moving plate regime (thick curve labeled "Moving plates") is the same as in Fig. 3.15. The stagnant lid regime (thick curve labeled "Stagnant lid") is calculated with equations 3.97–3.102. If the plates seize up and mantle temperature cannot increase (because there are no active heat sources) surface heat flux and lithospheric thickness will vary along path (i). The dashed lines show the values of these two variables at different times after the inception of stagnant lid convection, as a function of mantle temperature when plate tectonics end (the temperature then remains constant). The thermal boundary layer cannot be thicker than ∼760 km, its thickness for the age of the Solar System. If there is heat generation in the mantle, or heat input from the core, and the rate of heat supply remains constant, then steady state stagnant lid convection will be attained when the mobile-plate heat flux is reestablished, at the end of path (ii). This is accompanied by a substantial increase in mantle temperature, and modest increase in lithospheric thickness.

Suppose that there is no radioactive heat generation, nor any other active source of internal energy (e.g. tidal, electromagnetic). In this case mantle temperature cannot increase when plate tectonics stops and the thermal boundary layer will cool and thicken by diffusion. Assuming that the moving plates before seizing were thin relative to the fluid layer, the thickness of the rigid lid capping the fluid mantle a time t after the plates lock up will be of order $2\sqrt{\kappa t}$, and the surface heat flux at that time will be given by equation (3.22). The broken lines in Fig. 3.17 show heat flux and thickness of the thermal boundary layer for three times, 10^8 years, 10^9 years and the age of the Solar System. If the temperature of the mantle at any given time is such that the convective heat flux is less than the diffusive heat loss across the lid then the fluid mantle will keep convecting (as long as its Rayleigh number is above the critical value) and the lid will continue cooling diffusively and thickening. This is shown by the thin arrows labeled (i) in Fig. 3.17. For example, if the Earth had formed with a mantle temperature of 1950 K and had never had plate tectonics, then convective heat flux under a stagnant lid at the age of the Solar System would still be far lower than diffusive heat loss across the thermal boundary layer (recall that in this model there is no heat generation – this is essentially Kelvin's model). If enough time is available then diffusive heat loss across the lid and convective heat transport may eventually become equal and, if at this point the mantle is still convecting, then subsequent changes in the thickness of the stagnant lid will be controlled by the evolution of the fluid mantle, rather than by diffusive cooling. The maximum possible stagnant lid thickness for this end-member situation is the diffusive cooling length $2\sqrt{\kappa t}$ (e.g. $\sim$760 km for the age of the Solar System, Fig. 3.6).

The other end-member situation is one in which there is an active energy source, such as radioactive decay. In this case mantle temperature must increase, as the rate of energy supply does not change but the rate of heat loss decreases when subduction stops. The evolution is shown by the thin arrows labeled (ii) in Fig. 3.17. Steady-state convection is reestablished once the heat flux in the stagnant lid regime matches the moving plate heat flux. The continuous lithosphere at this point is thicker than it was when it was broken into plates. This may sound counterintuitive, given that mantle temperature is now higher, but the point is that, in order to preserve the same heat flux under a higher temperature gradient, the diffusive lid must become thicker (by Fourier's law, equation (3.5)).

Real planetary bodies with unbroken lithospheres must lie somewhere in between these two end-members. Lithospheric thickness cannot be greater, and mantle heat flux cannot be lower, than the diffusion values corresponding to the time since inception of the stagnant lid regime, because diffusion is the least efficient heat transfer mechanism. Minimum lithospheric thickness or maximum heat flux are not so easily constrained, but either variable can be calculated from the other one with the equations of stagnant lid convection.

3.9 Convection and cooling of solid planetary interiors

I have treated convection as being driven exclusively by heat loss across the cold upper boundary layer. I have ignored the bottom boundary layer, and have implicitly assumed that it is able to transport heat at whatever rate is demanded by the convective system, and otherwise behave in a completely passive way. This is not true in nature. What I have described is only one of the possible modes of heat advection in planetary mantles. The hot lower boundary layer drives another type of advective heat transport process which is expressed as mantle plumes. This mode is superimposed on the convection process driven by the top boundary layer. I will not offer a quantitative description of mantle plumes, but an accessible and lucid explanation can be found in the book by Davies (1999). On the Earth, mantle plumes are responsible for perhaps 20% of the total heat loss from the planet's deep interior (Box 3.2). The proportion may be different in other planets.

At present only the Earth undergoes convection with moving plates. But was this true for all epochs of the Solar System? Differentiation dissipates a large amount of gravitational potential energy, which may be sufficient to at least partially melt a planet (Section 2.6). In addition, there may be other important heat sources during the early stages of planetary evolution, such as tidal heating, decay of short-lived radioisotopes and planet-sized impacts. Vigorous convection must take place in partially molten planetary mantles. If a magma ocean forms it may be reasonable to postulate, from analogy with lava lakes, that when the surface layer first solidifies it will form thin moving plates that subduct. But what if this solid layer consists of low density material, such as lunar anorthosites or water ice? These materials remain buoyant relative to ultramafic melts and liquid water, respectively. The Moon and the icy worlds must have undergone stagnant lid convection from the time of formation of the first continuous solid outer layer. As the mantle solidifies the thermal boundary layer thickens in response to declining heat flux and convection of the solid mantle continues under the stagnant lid as long as the value of Ra_e (equation (3.96)) is above the critical value for convection ($\sim 10^3$–10^4).

This is not the only way in which stagnant lid convection can be established, however. The Earth today convects with a healthy moving-plate regime, but at some time in the future, as heat flux dwindles, it is inevitable that the plates will thicken to the point that they will be unable to break, bend and subduct. The thermal boundary layer will then lock up and display its true (high) viscosity. Could this have happened in the other two large terrestrial planets? Mars may have had something similar to plate tectonics early in its history, as hinted by magnetic anomalies in its ancient southern hemisphere, that resemble stripes of alternately magnetized ocean floor on Earth (Acuña *et al.*, 1999). The case of Venus is more puzzling.

We will attempt to gain some insight about the internal structures of Venus and Mars as an example of the application of the equations of parametrized convection. We will approach each planet with two competing working hypotheses. The first one will be to assume that we know for how long the present-day stagnant lid regime has existed, and that the thermal boundary layer has attained its maximum possible thickness (case (i) in Fig. 3.17). Based on their average surface ages, we will take this time to be ~ 4 billion years for Mars (Carr, 1999), and 600 million years for Venus (Head & Basilevsky, 1999). Using equation 3.16 we estimate maximum lid thicknesses of ~ 250 km for Venus and ~ 700 km for Mars. The other working hypothesis will be that we know the planetary heat flux *and that there is no heat generation in the lithosphere*. This last statement is patently untrue, but we will address that too. Assuming that the full planetary heat flux comes from the deep mantle, and that convection is in a steady state, we can calculate a minimum thickness for the stagnant lid as in case (ii) in Fig. 3.17, and a characteristic mantle temperature. As discussed in Box 3.2, we postulate mantle heat flux values of 30 mW m^{-2} for Mars and 63 mW m^{-2} for Venus.

What viscosity should we use in our calculations? There is no clear nor unique answer to this question. The mantles of Venus and Mars must be made up of Mg-rich silicates, but there may be important differences with the Earth's mantle that may affect viscosity. For example, we know that the Martian mantle is richer in Fe than the terrestrial mantle, and we can surmise that the Venusian mantle is probably drier than the Earth's mantle. The approach that I follow here is to allow the viscosity to vary inside a feasible range, and to study the sensitivity of the results to these variations. As an example, I will allow the reference viscosity μ_0 (equation (3.3.3)) to vary between 10^{20} and 10^{22} Pa s at 1950 K, i.e. one order of magnitude in each direction relative to the likely terrestrial value. I assume that the activation energy is 300 kJ mol^{-1}, as in the Earth's mantle.

Let us look at the results for Venus first (Fig. 3.18). A steady state mantle heat flux of 63 mW m^{-2} requires mantle temperatures 600–1000 K higher than on Earth, depending on the assumed viscosity (shown by the thin dashed line in the center panel of the figure). This appears unlikely, as at such temperatures there should be a continuous layer of partially molten asthenosphere and Venus would be expected to be much more volcanically active than Earth. This is not the case, for although there certainly are young volcanoes on Venus, they do not appear to cover the planet with a density comparable to Earth's. In fact, Venus may be somewhat less volcanically active than Earth. Given the planet's high surface temperature ($\sim$800 K), its lithosphere is likely to be less rigid than Earth's and, moreover, it would rest on a significantly less viscous and partially molten asthenosphere. Gravitational potential energy stored in topography should be dissipated faster on Venus than on Earth. The lithospheric thickness for this model ($\sim$70–80 km, shown by the thin arrow in the top panel) is therefore hard to reconcile with the significant topographic relief of Venus, comparable to that of the Earth. Note that the calculated lithospheric thickness is fairly insensitive to mantle viscosity (interval between the thin dashed lines in the top panel), and that the Rayleigh number required to sustain the assumed heat flux varies by a factor of $\sim$2 for a viscosity contrast of 2 orders of magnitude (bottom panel).

The other end-member possibility is that Venus has a lithosphere $\sim$250 km thick, that has been cooling conductively since the planet was resurfaced about 600 million years ago. The steady-state mantle heat flux in this case would be $\sim$20 mW m^{-2}, varying slightly depending on the assumed mantle viscosity (center panel). This would necessitate that Venus has considerably less radioactive heat generation than Earth, that the Venusian core is not an important heat source, that the planet has undergone a stronger fractionation of incompatible elements towards the surface than Earth, or a combination of all of the above. Interestingly, Venus has no intrinsic magnetic field, suggesting that its core is not convecting, and that it might therefore not be an important energy source for mantle convection. Lack of core convection could result either from a "cold" core in which crystallization is complete or near-complete, or from a "hot" core that is mostly above its liquidus and is thus not liberating enthalpy of crystallization. High mantle temperatures that arise from stagnant lid convection would argue for the latter explanation. Venus could also be in a steady state if the proportion of internal energy that is transported to the surface by mantle plumes is larger than in the Earth, in which case it is required that there be significant heat flux from the core (see Davies, 1999). In this view the planet's notable coronae could be active plume heads.

An alternative is that Venus is not in a steady state, but is rather in the process of following a path such as (ii) in Fig. 3.17. The mantle may be heating up as the rigid lid thickens. Depending on the mechanical behavior of the lithosphere, it may eventually complete the transition to a steady state stagnant lid regime (the end point of path (ii) in Fig. 3.17), or the lithosphere may break up and a regime of convection with moving plates may be established. In this view (suggested by Turcotte, 1995), Venusian plate tectonics would consist of catastrophic global subduction events separated by periods of non-steady-state stagnant lid convection, and the coronae could be incipient subduction zones rather than plume heads. I do not take sides in this argument regarding steady state vs. episodic Venusian convection, but I would like to suggest that they lead to tectonic interpretations of coronae

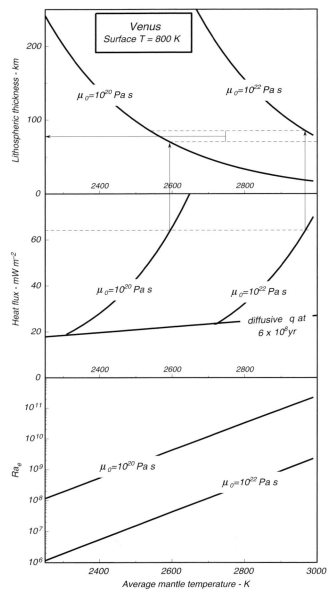

Fig. 3.18 Possible stagnant lid convection regimes in Venus. Two models are shown, calculated as in Fig. 3.17 for reference viscosities at 1950 K of 10^{20} and 10^{22} Pa s. It is assumed that the stagnant lid regime has existed for 600 million years. Lithospheric thickness varies between a maximum of 250 km (corresponding to a constant temperature mantle, case (i) in Fig. 3.17) and a minimum of 70–80 km constrained by an assumed mantle heat flux of 63 mW m^{-2} (thin dashed line in center panel). The minimum mantle heat flux for Venus would be $\sim$ 20 mW m^{-2}, corresponding to a 250-km thick, 600 million year old stagnant lid. Model parameters as in Fig, 3.14, except $T_{surface}$ = 800 K, D = 2700 km and g = 8.9 m s^{-2}.

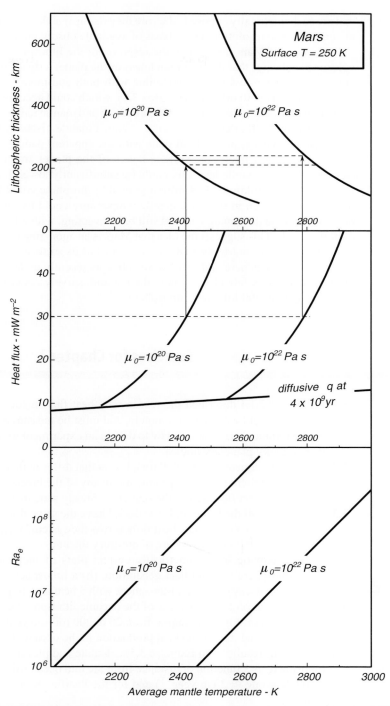

Fig. 3.19   Possible parametrization of stagnant lid convection in Mars. Same as Fig. 3.18, except that the age of the stagnant lid regime is 4 billion years. Other model parameters as in Fig. 3.15, except $T_{surface} = 250$ K, $D = 1600$ km and $g = 3.7\,\mathrm{m\,s^{-2}}$.

that are so diametrically opposed – i.e., are they rising or sinking? – that it should be possible to come to some resolution on the basis of available observations.

In the case of Mars (Fig. 3.19), the response of the lithosphere to the load of the large volcanoes indicates that the mean thickness of the planet's rigid lid is of the order of a few hundred kilometers, which is compatible with both end-member steady-state models. A mantle heat flux of 30 mW m^{-2}(see Box 3.2), which could account for most of the Martian heat output, implies a lithosphere $\sim$200 km thick and mantle temperatures some 500–700 K higher than on Earth, depending on the assumed mantle viscosity (thin dashed line in the center panel). This is probably at odds with the apparent paucity of active volcanism on Mars. If, as seems likely, the surface layers of the planet are enriched in incompatible elements then the mantle heat flux could be significantly lower than 30 mW m^{-2}. Mantle heat flux for a 700-km thick, 4 billion year old lithosphere could be around 10 mW m^{-2} (center panel). Even in this case mantle temperature could be 200–500 K higher than in the Earth and the deep mantle would still be convecting, with a Rayleigh number of $\sim 10^6$ (bottom panel). This suggests the idea that Mars is an agonizing planet that is not quite dead yet (with apologies to Monty Python). Yes, most of its surface is old and heavily cratered, but huge volcanoes here and there were active as recently as 160–200 Ma, perhaps more recently, marking a few places where the hot and actively convecting mantle was able to poke through the 700-km thick lithosphere.

Exercises for Chapter 3

3.1 In Section 2.1 I described the reason why heat flux in solid planetary bodies other than Io cannot be estimated remotely, and must be measured with direct observation on the ground. Use Fourier's law to design an experiment to measure the natural heat flux of rocky planetary bodies.

3.2 Consider a sequence of parallel rock layers that differ in their thermal conductivities. Assume that there is no heat production in any of the layers. Define the condition for steady state heat flow across the sequence. Steady state means $dT/dt = 0$ everywhere.

3.3 Assume that all the layers in Exercise 3.2 have the same thickness. What is the factor that determines steady-state heat flow across the system? Generalize your criterion to a system in which the layers are of arbitrary thickness.

3.4 Draw schematic but geometrically correct plots of the steady-state geotherm for (a) a lithosphere with no heat generation, (b) a lithosphere with a heat source (e.g. radioactive decay) and (c) a lithosphere with a heat sink (e.g. melting).

3.5 Calculate and plot the thickness of the oceanic lithosphere as a function of the age of the ocean floor, for ages ranging from 20 to 200 million years. Let $\kappa = 10^{-6}$ m^2 s^{-1}, and assume that there is no heat production in the oceanic lithosphere.

3.6 Using your results from Exercise 3.5, calculate and plot the thickness of the oceanic lithosphere as a function of distance from the mid ocean ridge, for the Atlantic Ocean (spreading rate $\approx$ 2 cm year^{-1}) and for the Pacific Ocean (spreading rate $\approx$ 10 cm year^{-1}).

3.7 Using your results from Exercise 3.6, plot heat flux through the ocean floor as a function of age of the ocean floor. Assume $k = 3$ J s^{-1} m^{-1} K^{-1}, temperature at the base of the lithosphere = 1400 °C, temperature at the ocean floor = 0 °C.

3.8 Discuss why it is not possible to use the approach in Exercises 3.5 to 3.7 to calculate continental heat flux.

3.9 In the definition of heat diffusivity the product (ρc_P) describes the thermal inertia of the material. What is the dimension of ρc_P? Comment on this.

3.10 How long will it take for each of the following to cool to ambient temperature? (i) A lava flow 10 m thick. (ii) A 10-m thick sill emplaced at a depth of 2 km. (iii) A pyroclastic layer 10 cm thick. (iv) A granitic pluton with a 10 km diameter.

3.11 In Section 2.4 we assumed that the temperature distribution inside an accreting planet is not modified by heat flow during the accretion process. Justify this assumption for accretion times of 10^5 to 10^7 years.

3.12 What is the *maximum* asteroid size for which we can be *certain* that cooling is complete? What can you say about cooling of asteroids larger than this? For an asteroid larger than this size, how would you determine whether or not it is still cooling down?

3.13 In Section 2.9 we saw that decay of short-lived isotopes may liberate enough thermal energy to melt a chondritic asteroid in 1–10 million years. Discuss whether there is some relationship between the size of the asteroid and the likelihood that it will melt by this process. Derive an approximate relationship between accretion rate and half life and concentration of a short-live isotope that may allow you to make semi-quantitative predictions about melting of planetesimals in the very early Solar System.

3.14 Derive the equation for the adiabat on a $P–V$ diagram (Fig. 3.9).

3.15 Calculate the dry adiabatic lapse rates for the five solid planetary bodies with "substantial" atmospheres: Venus, Titan, Earth, Mars and Triton. Look up the necessary data in Lodders and Fegley (1998). For each case, discuss whether or not the dry adiabatic lapse rate is a good approximation to the temperature structure of convective atmospheric layers.

3.16 Modify equations (3.36) or (3.37) to include the effect of condensation of atmospheric species (e.g. H_2O in Earth or CH_4 in Titan). This is known as the *wet adiabatic lapse rate*. (Hint: start from equation (3.27), but now $dQ \neq 0$. Write dQ in terms of latent heat and mass of vapor that changes state).

3.17 Estimate the minimum thickness of a thrust sheet in which Buchan metamorphism can develop, and show why blueschist metamorphism can develop in accretionary wedges. (Hint: start from equation (3.49)).

3.18 Discuss the Gulf Stream and the Hudson current from the point of view of the Péclet number of the ocean.

3.19 A glacier advects enthalpy from the zone of ice accumulation to the zone of ablation. If we consider a steady-state atmosphere then the terminus of the glacier can be thought of as being located at a point where the temperature perturbation caused by exposure to air warmer than that in the zone of accumulation has been able to penetrate the entire thickness of ice. Derive an equation that relates the length of a glacier to its thickness and flow rate. Test your equation by comparing its predictions with observed dimensions and flow rates of alpine and continental glaciers. Discuss possible sources of discrepancies. For ice $\kappa = 1.5 \times 10^{-6}$ m^2 s^{-1}.

3.20 In the accretion model (Section 2.4) we ignored compression of the growing planet. Derive an equation for temperature increase in the planetary interior caused by adiabatic compression. What is the source of this thermal energy?

3.21 Decide whether icy satellites convect. Assume that they are composed of H_2O ice, for which: $\mu = 10^{12}$ Pa s, $\kappa = 1.5 \times 10^{-6}$ m^2 s^{-1}, $\alpha = 2 \times 10^{-4}$ K^{-1} and $\rho_0 = 1000$ kg m^{-3}. Consider and justify a range of likely values for ΔT (temperature drop across the thermal boundary layer). You may want to analyze separately the cases of large satellites (e.g. Callisto) and small ones (e.g. Mimas or Enceladus). Necessary data not listed above can be found in Lodders and Fegley (1998).

3.22 Discuss whether the large asteroid Vesta may have convected, whether it is likely to be convecting now, and how long convection may have lasted. Use material properties for silicate planets given in the text (e.g. Figure 3.15), and other necessary data from, yes, Lodders and Fegley (1998).

3.23 Estimate ranges of possible convective velocities in Venus and Mars. Compare your results with the velocity of terrestrial mantle convection and discuss the physical reasons for the differences.

3.24 Io has an observed heat flux of ~ 2.5 W m^{-2}. Assuming that Io is undergoing steady-state stagnant lid convection, estimate the thickness of, and temperature difference across, the thermal boundary layer, and the velocity of Ionian convection. Are your results likely? If not, how can Io's heat flux be accounted for?

4

The Second Law of Thermodynamics and thermodynamic potentials

The First Law of Thermodynamics, like all conservation laws, is expressed mathematically by an identity relationship. As such, it is incapable of predicting the direction in which a natural process will occur. For example, on the basis of energy conservation alone it is not possible to decide whether heat flows from a hot body to a colder one, or the other way around. We know that heat flows down a temperature gradient, but this does not follow from the First Law. Similarly, energy conservation cannot predict that ice will melt at $20\,^\circ$C, or that water will freeze at $-20\,^\circ$C, and it cannot predict that a gas will expand to fill all of the volume available to it.

Another law of nature is required to predict the direction of spontaneous changes. By "spontaneous" we mean a process that occurs in nature in the direction towards equilibrium and without outside intervention. For example, heat flow down a temperature gradient is a spontaneous process. It is possible to transfer heat from a cold body to a hotter one, but this requires "outside intervention" in the form of a heat pump, which uses mechanical energy to accomplish a process that is not "naturally spontaneous". As soon as the expenditure of mechanical energy ceases the spontaneous process takes over and the cold body heats up at the expense of the hotter one.

The law that we are searching for, which is the Second Law of Thermodynamics, cannot be expressed by a mathematical identity, because identities are unable to determine direction. Therefore, it cannot be a conservation law. Rather, the Second Law of Thermodynamics must be a law that states that some quantity changes when a spontaneous process takes place, or, equivalently that the total content of some quantity in the observable universe varies monotonically with time.

4.1 An intuitive approach to entropy

Consider two bodies at different temperatures, $T_1 > T_2$, enclosed in a container that is a perfect thermal insulator, is also perfectly rigid (i.e. it cannot undergo expansion work) and is impervious to any other imaginable type of energy transfer. The system (in this case, our two bodies) enclosed in such a container is called *isolated* (Fig. 4.1). Recall that adiabatic means impervious to heat transfer, but not to exchanges of mechanical energy. We know that heat will flow from body 1 to body 2 until their temperatures become the same, at which point heat transfer will stop. Heat does not spontaneously flow from a cold body to a hotter one. This is an observation as fundamental, and as impossible to demonstrate from simpler concepts, as any conservation law, such as those for energy, momentum or electric charge. It is, in fact, one possible statement of the Second Law of Thermodynamics.

The First Law of Thermodynamics applied to our system goes as follows. Let us assume that the coefficients of thermal expansion of the two bodies are zero, so that there is no

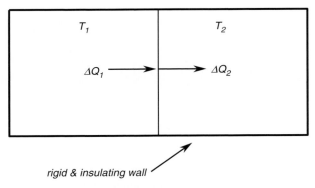

Fig. 4.1 Transfer of heat between two bodies at different temperatures. The two bodies constitute a system that does not interact with its environment.

PdV work. For a finite heat exchange, the changes in internal energy of the two bodies are given by:

$$\Delta E_1 = \Delta Q_1$$
$$\Delta E_2 = \Delta Q_2. \tag{4.1}$$

Since the system is isolated, its total energy content is constant, so:

$$\Delta E_1 + \Delta E_2 = 0 \tag{4.2}$$

from which we conclude what we already knew:

$$\Delta Q_1 = -\Delta Q_2. \tag{4.3}$$

This is as far as the First Law will take us. It is *only* from the knowledge that $T_1 > T_2$ that we can conclude that $\Delta Q_1 < 0$ and $\Delta Q_2 > 0$. Let us now define a new variable, Z, as follows:

$$\Delta Z = \frac{\Delta Q}{T} \tag{4.4}$$

and apply it to the heat transfer process in the isolated system:

$$\Delta Z_{isolated\ system} = \frac{\Delta Q_1}{T_1} + \frac{\Delta Q_2}{T_2} = \frac{\Delta Q_1}{T_1} - \frac{\Delta Q_1}{T_2} = \Delta Q_1 \frac{T_2 - T_1}{T_2 T_1} > 0. \tag{4.5}$$

Note very carefully: ΔQ_1 and $(T_1 - T_2)$ always have opposite signs, regardless of which of the two bodies is hotter. This is a restatement of the fact that heat flows down a temperature gradient, and again, a possible way of expressing the Second Law. We can now draw three conclusions.

(i) ΔZ for heat transfer in an isolated system is always a positive quantity.
(ii) $\Delta Z = 0$ only when $T_1 = T_2$, in which case there is no heat transfer.
(iii) Thermal equilibrium in an isolated system is attained when the variable Z attains its maximum possible value for a given (and fixed) value of the total energy content of the system.

The last point is crucially important in much of what follows, so let me re-phrase it. If the two bodies are initially at different temperatures, then Z will increase as heat flows from the hotter to the colder one. Heat will stop flowing when the two temperatures become equal, at which point we say that the system has attained thermal equilibrium because no further changes take place with time. At this point Z has attained its maximum possible value. If the two bodies were initially at the same temperature then the system was already at equilibrium and the value of Z was already the maximum possible for the total internal energy content of the system. The fact that heat does not spontaneously flow from a cold body to a hotter one is called the Clausius statement of the Second Law of Thermodynamics and is reflected in the fact that ΔZ for an isolated system can never be a negative quantity, so that Z can never decrease *in an isolated system*. The italics are important, as forgetting about this last condition is the source of many misconceptions about the Second Law of Thermodynamics, especially among pseudo-scientists and other assorted charlatans.

This looks very much like the law that we are looking for, that will allow us to predict the direction of a spontaneous natural change. We have defined a quantity that always increases when such a change takes place. There are, however, at least two problems with the variable Z as we have defined it. In the first place, it may allow us to tell the direction in which heat flows (which, by the way, we already knew) but it is not at all clear that it will allow us to predict, for example, whether a chemical reaction will take place, or whether a substance will melt or crystallize at certain conditions. Second, and more subtly, Z cannot possibly be a state variable because we have defined it on the basis of a non-equilibrium process: heat transfer between bodies that differ in temperature by a finite amount is not a quasi-static process (Section 1.4.2). In the next section we see that with just some minor tweaking we can define a state function that allows for completely general predictions about the direction of spontaneous natural processes.

4.2 The entropy postulate and the Second Law of Thermodynamics

Presenting the Second Law of Thermodynamics is the hardest part of any thermodynamics course. It may seem remarkable that this is the case if the only thing that it does is to codify everyday experience, such as the fact that your coffee will become cold if you don't drink it soon enough. It may even seem that conservation of energy is more removed from intuition than this. The problem is that the direction of heat flow is just one aspect of the Second Law, and we seek a statement, and the mathematical law that goes with it, that is completely general. There are several approaches to accomplishing this. The one that came historically first relies on thought experiments and theorems about thermodynamic cycles and heat engines, going back to experimental observations by Carnot and Clausius in the first half of the nineteenth century, and Kelvin's work a bit later. I will discuss some of this material in order to complete the discussion of convection in Chapter 3, but I will not follow this path in order to introduce the Second Law. The primary reason for eschewing this approach is that, at some point, one always comes up against postulates or conclusions that appear to be arbitrary or concocted (although they are not). A better approach, in my view, is to state our postulates at the very beginning and accept their validity later, on the basis of how well the mathematical structure that we build from these postulates agrees with reality. In the words of Callen (1985, p. 27):

> "We . . . formulate . . . a set of postulates depending upon *a posteriori* rather than *a priori* justification. These postulates are, in fact, the most natural guesses that we might make"

This is the spirit of what follows, even though the treatment is much simplified relative to those of Callen (1985) or Guggenheim (1967).

Drawing from our example in Section 4.1, we define an extensive variable called *entropy* and symbolized by S, by means of the following function:

$$dS = \frac{dQ_{rev}}{T}.$$

(4.6)

The meaning of dQ_{rev} is that it is an infinitesimal amount of heat that is transferred reversibly, which means that no energy dissipation takes place, and that the transformation takes place between equilibrium states (it is "quasi-static"). The problem is that this definition of entropy is perfectly circular, because in order to define dissipation (or lack thereof) we need to know what entropy is, or at least how to measure it. But this is the key: I am not telling you what entropy is, only how to calculate it. We still have to deal with the circularity, or at least incompleteness, of (4.6), so we make the following entropy postulate (Callen's "most natural guess"):

> A process in an isolated system takes place spontaneously only if it causes the entropy of the isolated system to increase. Equilibrium is attained when the entropy takes the maximum possible value for the system's energy content (which is constant in an isolated system).

This postulate can be motivated by the discussion in Section 4.1, but now we are not making any mention of heat exchange between different parts of the system. Our postulate is that entropy increases in any isolated system that changes towards equilibrium (i.e. that undergoes a spontaneous change), regardless of what the process that leads to equilibrium is (heat flow, diffusion, chemical reactions, phase transitions, etc.). We express this mathematically as follows:

$$dS_{isolated\ system} \geq 0.$$

(4.7)

We also state that: (i) entropy is a state function, meaning that its value at equilibrium is fully defined by the state of the system (Section 1.5), (ii) that entropy is additive, i.e. that it is an extensive property, and (iii) that it is a continuous and monotonically increasing function of internal energy, E. All of these statements may sound like postulates, but they can actually be formally demonstrated from (4.6) (see for example Callen, 1985).

At this point it is necessary to address the issue of the "entropy of the universe". By definition, the observable universe is an isolated system, so its entropy is constantly increasing as natural processes unfold. Sometimes, when applying the Second Law to a specific problem, one analyzes whether or not the entropy of the universe increases, so as to decide whether or not the process under scrutiny will take place. What we mean with this phrase (which I will try not to use) is that we are considering the entropy change in a system that is large enough to include all of the energy and matter exchanges that we are interested in, so that, as far as our problem is concerned, this system does not interact with anything outside of it. Commonly we do not have to get anywhere near the edge of the observable universe, though.

When a substance melts it absorbs heat. Therefore, by (4.6) its entropy must increase. Consider melting of water ice at a constant pressure of 1 bar and $T = 273$ K. The heat absorbed by melting of ice at constant pressure is the enthalpy of melting, $\Delta H_{melting} \approx$ 6 kJ mol^{-1}. The (finite) increase in entropy that accompanies melting, called the entropy of melting, $\Delta S_{melting}$, is therefore given by:

$$\Delta S_{melting} = \frac{\Delta H_{melting}}{T_{melting}} \approx 22 \, \text{JK}^{-1}\text{mol}^{-1}. \tag{4.8}$$

Note that the units of entropy are the same as those of heat capacity. When water crystallizes it gives off the same amount of heat as it absorbs during melting, so that $\Delta H_{crystallization} = -\Delta H_{melting}$, and thus $\Delta S_{crystallization} = -\Delta S_{melting} \approx -22 \, \text{J K}^{-1} \, \text{mol}^{-1}$. Now, since the entropy of crystallization is negative, doesn't the freezing of water violate the Second Law of Thermodynamics? Of course not. Equation (4.7) applies to an isolated system. Freezing of water cannot be an isolated system by itself, since the enthalpy of crystallization must be absorbed by some other body. The temperature of the absorbing body must be lower than the freezing temperature, because otherwise heat would not flow and water would not freeze. By the First Law the heat absorbed by this body must be the same as the heat liberated by the water, but because its temperature is lower, by (4.6) it follows that its entropy increase is greater than the entropy decrease of freezing water. We can arrange for the freezing water and the heat absorbing body to conform to an isolated system, whose entropy increases as ice melts. But what if the temperature of the freezing water and of the absorbing body are the same? Then ΔS for the isolated system vanishes, which means that the system is at equilibrium. In such an ideal system water at its freezing temperature never freezes, and ice at its melting temperature never melts.

4.3　The First Law of Thermodynamics revisited

When we write the First Law as in (1.55):

$$d\boldsymbol{E} = d\boldsymbol{Q} - d\boldsymbol{W} \tag{4.9}$$

we make no requirements on the kind of process involved. Equation (4.9) is always true, regardless of whether or not the process that it describes is reversible, or, equivalently, of whether or not energy dissipation occurs. Energy conservation cannot be circumvented under any circumstances. We have implicitly used the generality of (4.9) in many of our derivations in Chapters 2 and 3 which describe transformations that are dissipative, or irreversible. For instance, storage of thermal energy in planetary interiors by dissipation of mechanical, electrical or nuclear energy. As we discussed in Section 1.9, $d\boldsymbol{Q}$ and $d\boldsymbol{W}$ do not stand for differentials of functions, and for this reason they are often called "inexact" differentials. Equation (4.9) assures us, however, that their *difference* is the differential of a state function. Let us see what happens then, when we write the First Law as follows:

$$d\boldsymbol{E} = d\boldsymbol{Q} - Pd\boldsymbol{V}. \tag{4.10}$$

Because E, P and V are all state functions, dQ in this case must be the differential of some function that takes on a single well-defined value for each state of the system. This is only possible if (4.10) applies only to transformations between equilibrium states. If this is the case then we can use (4.6) to make the substitution:

$$dQ = T dS \qquad (4.11)$$

from which we get the following expression for the First Law, *applicable only to transformations between equilibrium states:*

$$dE = T dS - P dV. \qquad (4.12)$$

This innocent-looking equation, which combines the First and Second Laws, has several important consequences. We will examine the most momentous ones in Section 4.8. For now, we note that it is written only in terms of state variables, i.e. quantities that are well defined in any equilibrium state. We can therefore integrate equation (4.12) in order to calculate the change in entropy between two arbitrary *equilibrium* states. We can apply the same arguments to the First Law written in terms of enthalpy (equation (3.27)):

$$dH = T dS + V dP. \qquad (4.13)$$

The following two identities follow immediately from equation (4.12):

$$\left(\frac{\partial E}{\partial S}\right)_V = T \qquad (4.14)$$

and:

$$\left(\frac{\partial E}{\partial V}\right)_S = -P. \qquad (4.15)$$

These equations are the thermodynamic definitions of temperature and pressure. They allow a number of algebraic manipulations that are crucial in the formal development of the conditions of chemical equilibrium. We shall have much more to say about this, beginning in Section 4.8.

4.4 Entropy generation and energy dissipation

We will use an example to gain some insight into the nature of entropy, and more specifically into the relationship between entropy generation and energy dissipation. Consider a system undergoing two different types of adiabatic expansion (Fig. 4.2). In case (a) the system undergoes a quasi-static expansion from an initial volume V_i to a final volume V_f. If the system is a gas then we recall from Section 1.4.2 that this means that the expansion is slow enough that molecular velocities preserve an equilibrium statistical distribution, but equivalent statements can be made for a system in any aggregation state (e.g. distribution of vibrational frequencies in a solid). During the quasi-static transformation the system expands against an external pressure that is always infinitesimally close to its own pressure, and that decreases infinitesimally slowly. In case (b) the system undergoes the same change in volume as a result of a free expansion. This means that the substance of interest, for

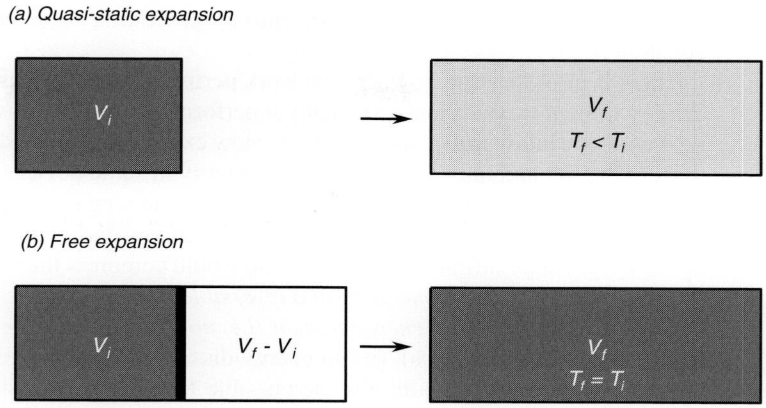

(a) Quasi-static expansion

(b) Free expansion

Fig. 4.2 (a) Quasi-static expansion of an ideal gas performing work on its environment. (b) Free expansion of an ideal gas into a vacuum. Shading represents temperature.

example a gas, initially occupies a volume V_i, and the rest of the system's volume, $V_f - V_i$, is empty and separated from the volume filled with gas by a partition. The partition is removed and the gas expands to fill the entire volume V_f. The transformation is not quasi-static, but we are free to allow enough time to elapse after the expansion, such that the system eventually reaches thermal equilibrium, in the sense of Section 1.4.2.

Because both transformations are adiabatic, we have, for both cases:

$$dQ = dE + dW = 0. \tag{4.16}$$

During the quasi-static transformation (a), the system expands against an external pressure equal to its own pressure, so $dW = PdV$:

$$dE_{quasi\text{-}static} + PdV = 0 \tag{4.17}$$

and, substituting (4.12):

$$TdS_{quasi\text{-}static} - PdV + PdV = 0 \tag{4.18}$$

or:

$$dS_{quasi\text{-}static} = 0. \tag{4.19}$$

Thus, the quasi-static adiabatic expansion that we described is also isentropic. We may ask, if there is no entropy change during a quasi-static expansion, then why does the expansion occur at all? Wouldn't that imply that the system is at equilibrium, and that a spontaneous compression, or better yet, no change at all, are as likely, or unlikely, as a spontaneous expansion? There are several answers to this question, but they all boil down to the fact that *the description that I gave of the system is incomplete, and that the system as described is not isolated*. Expansion occurs within a larger isolated system in which entropy must increase. The condition $dS = 0$ applies *only to the system expanding quasi-statically*. But in order for it to be possible for the system to expand and perform work against an external force the energy must have been previously stored in the system as internal energy, for example by heating the system and causing its temperature to rise. During the heating stage the system is not an adiabatic one, and entropy generation accompanies heat transfer, as in equation (4.5).

There is also the issue of where the work performed by the quasi-static expansion goes. When a system expands quasi-statically it performs work on its environment. This must be so if we are to allow for an infinitesimally slow expansion. It would be in principle possible to store the full amount of this mechanical energy in some device, such as by compressing a spring or lifting a weight suspended from a cord running over a pulley. If we could completely eliminate friction then there would be no energy dissipation and releasing the spring or the weight at an infinitesimally slow rate would compress the substance adiabatically to its initial state. *This is the meaning of a reversible transformation: one in which there is no conversion of mechanical energy to heat (i.e. no dissipation)*. The three conditions: (i) no entropy generation ($dS = 0$), (ii) no energy dissipation and (iii) reversible transformation are thus equivalent. It would also be possible to make the quasi-static expansion in (a) take place by adjusting a frictional force so that expansion proceeds infinitesimally slowly by dissipating energy. Equation (4.19) would still be true for the expansion process, but entropy would be generated by frictional dissipation somewhere else, and the entropy of an isolated system large enough to contain all of the required components ("the Universe") would increase.

Let us now examine transformation (b). By hypothesis the system does not perform external work during free expansion, because it is not expanding against any external force. Let us assume for the sake of argument that the system is an ideal gas, so there are no intermolecular forces and there is no work performed in separating the molecules either. Then, from (4.16):

$$dE_{free\ expansion} = 0.$$
(4.20)

If after the free expansion is completed we wait long enough for the system to attain thermal equilibrium (i.e. for all its molecules to "communicate" with one other and establish the equilibrium velocity distribution) then we can substitute (4.12) in (4.20):

$$TdS_{free\ expansion} - PdV = 0$$
(4.21)

or:

$$dS_{free\ expansion} = \frac{P}{T}dV.$$
(4.22)

For expansion $dV > 0$, and P and T are always positive quantities, so:

$$dS_{free\ expansion} > 0.$$
(4.23)

Equation (4.23) says, for instance, that a gas will always expand to fill any empty space available to it ("nature abhors a vacuum"), and it also says that a gas will never spontaneously contract.

What is the physical meaning of the difference between equations (4.19) and (4.23)? During free expansion the system does not perform mechanical work, as it is not expanding against an external force, but rather just filling empty space (recall that we ignore intermolecular forces). *The entropy increase measures the magnitude of the mechanical work that could have been performed by the expansion, but wasn't*. When the system is in its initial state, occupying volume V_i, it has the capability of performing the amount of work that it performs during a quasi-static expansion. When we allow it to expand freely this capability is *irretrievably lost*. This is equivalent to saying that the system initially stores some amount of "potential for doing work" (careful! do not confuse this with potential energy as defined in Chapter 1) that was dissipated when it expanded without performing any work. From (4.21) we see that the integral of $T\,dS$ is the amount of mechanical energy that was lost, or dissipated, during the free expansion. This also shows that $dW = P\,dV$ only for a reversible transformation. For the irreversible free expansion $dW = 0$, but $P\,dV \neq 0$.

The derivation of equation (3.32) for planetary adiabats assumes that the process is isentropic, i.e. that no energy dissipation takes place. If energy dissipation takes place during adiabatic expansion or compression then additional terms must be added to equation (3.32) to account for this thermal energy. This is the reason why (3.32) is rigorously an isentrope, which is a special case of an adiabat. We return to this in Chapter 10.

4.5 Planetary convection and Carnot cycles

4.5.1 Thermodynamic efficiency

A temperature gradient represents a potential for doing work. If heat is allowed to diffuse without performing work then the loss of this work potential is energy dissipation, which is measured by the amount of entropy that is generated. Consider a system bound by two isothermal boundaries at different temperatures, $T_2 > T_1$, where the temperatures are held constant by a heat source at temperature T_2 and a heat sink at temperature T_1 (Fig. 4.3). This is a reasonable approximation for atmospheric convection, in which the planet's surface is the heat source and space is the heat sink. We can arrange for our idealized atmosphere

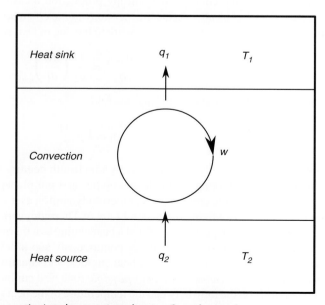

Fig. 4.3   Idealized view of convection in a planetary atmosphere as a Carnot heat engine.

together with its heat source and its heat sink to make up an isolated system (this could be the universe). Let the source at temperature T_2 transfer heat to the atmosphere at a rate $(-q_2) < 0$, where $q = dQ/dt$. Assume initially that heat diffuses across the atmosphere, so that no work is performed. The atmosphere then delivers heat at the same rate rate q_2 to the sink at temperature T_1. Entropy generation is in this case maximum because no work is being performed, so that the maximum rate of entropy generation, $(dS/dt)_{max}$, is given by:

$$\left(\frac{dS}{dt}\right)_{max} = \frac{-q_2}{T_2} + \frac{q_2}{T_1} = q_2 \frac{T_2 - T_1}{T_2 T_1} > 0. \tag{4.24}$$

Now suppose that the atmosphere convects. In this case it performs work, which appears as kinetic energy of moving air. The ideal maximum possible rate at which the atmosphere can perform work must be such that the there is no entropy generation, so we make $(dS/dt)_{min} = 0$, although in nature we never get this close (more on this soon). In this case part of the heat flow q_2 is converted to work at a rate $w_{max} = (dW/dt)_{max}$, so that the rate of heat loss across the top boundary, q_1, is given by:

$$q_1 = q_2 - w_{max}. \tag{4.25}$$

We can then write the minimum rate of entropy production as follows:

$$\left(\frac{dS}{dt}\right)_{min} = \frac{-q_2}{T_2} + \frac{q_1}{T_1} = \frac{-q_2}{T_2} + \frac{q_2}{T_1} - \frac{w_{max}}{T_1} = q_2 \frac{T_2 - T_1}{T_2 T_1} - \frac{w_{max}}{T_1} = 0. \tag{4.26}$$

Using (4.24), the maximum rate at which work can be performed is given by:

$$w_{max} = T_1 \left(\frac{dS}{dt}\right)_{max}, \tag{4.27}$$

which re-states the concept that entropy production measures the loss of work potential (Section 4.4). Let us now define the *thermodynamic efficiency* of convection, η, as the ratio of this maximum rate of performing work to the rate of heat absorption from the source, i.e.:

$$\eta = \frac{w_{max}}{q_2} = \frac{T_1}{q_2} \left(\frac{dS}{dt}\right)_{max}. \tag{4.28}$$

Substituting (4.24) and simplifying, we find:

$$\eta = 1 - \frac{T_1}{T_2}. \tag{4.29}$$

This result is originally due to the early nineteenth century French scientist Sadi Carnot, although he was talking about steam engines and not planetary atmospheres. It is rightly considered one of the cornerstones of thermodynamics, as it is what set Clausius and Kelvin on the path to formalizing the Second Law of Thermodynamics.

Recall that in Chapter 3 we defined a heat engine as a thermodynamic cycle that converts thermal energy to mechanical energy continuously and indefinitely. A planetary mantle or atmosphere, or a steam engine, are heat engines, and equation (4.29) states that the (ideal) *maximum thermodynamic efficiency possible* for a heat engine, called the Carnot efficiency, depends *only* on the ratio between the temperature of the heat sink and the temperature of the heat source. This is true for *any* heat engine, as in deriving (4.29) we made no assumptions whatsoever about the nature of the process by which some of the heat absorbed from the source is transformed to mechanical energy.

There are two corollaries that follow from equation (4.29). First, the thermodynamic efficiency is always less than 1, as $T_1/T_2 = 0$ would require either a sink at absolute zero or an infinitely hot source, and both are impossible (we will return to this when we discuss the Third Law of Thermodynamics later in this chapter). Since $\eta < 1$, even if we could construct a heat engine in which all friction and all other sources of dissipation are eliminated we would still not be able to achieve complete conversion of thermal energy to mechanical energy. The second corollary is that, if $T_1 = T_2$, then $\eta = w_{max} = 0$. This result is called the Kelvin–Planck statement of the Second Law: it is not possible to construct a heat engine that performs work by extracting heat from a single source at a constant temperature. Recall that most of the mechanical energy generated in planetary mantles and cores is dissipated by a combination of viscous and ohmic heating. Consider a hypothetical planet in which there is neither tectonic nor magnetic work. If it were not for the Second Law all that we would need for the mantle or core of such a planet to convect would be an initial supply of heat and perfectly insulating boundaries. Once convection starts it would feed on heat generated by its own dissipation. This is called a perpetual motion machine of the second kind, and is impossible by (4.29) (or, more generally, by the Second Law).

4.5.2 Carnot's cycle

As it turns out, it is possible to get a bit more specific about the type of heat engine that achieves the maximum thermodynamic efficiency given by (4.29), and the result is relevant to planetary convection. There are several ways of demonstrating what follows, which is known as Carnot's theorem. I think that this derivation, modified from Guggenheim (1967), is particularly simple. Suppose that our heat engine absorbs heat at various temperatures, T_i, of which T_2 is the maximum, and releases heat at various temperatures T_o, of which T_1 is the minimum. Let the various rates of heat absorption and emission be q_i and q_o, respectively, where, as in Section 4.5.1, all of the qs are positive quantities. The maximum thermodynamic efficiency is always obtained when $(dS/dt)_{min} = 0$. We now write this condition with the following general equation ((4.26) is a special case of this equation, for a single input and a single output temperature):

$$\left(\frac{dS}{dt}\right)_{min} = -\sum \frac{q_i}{T_i} + \sum \frac{q_o}{T_o} = 0, \qquad (4.30)$$

which we rewrite as follows:

$$\sum \frac{q_i}{T_i} = \sum \frac{q_o}{T_o}. \qquad (4.31)$$

The rate at which work is performed is given by (compare (4.25)):

$$w = \sum q_i - \sum q_o \qquad (4.32)$$

and the thermodynamic efficiency by (see also (4.28)):

$$\eta = \frac{w}{\sum q_i} = 1 - \frac{\sum q_o}{\sum q_i}. \qquad (4.33)$$

The thermodynamic efficiency η is maximum when the ratio $\sum q_o / \sum q_i$ is minimum. In order to minimize this ratio we have to make each q_o as small as possible, and each q_i as large as possible. By (4.31), we will accomplish this by making all T_o equal to the minimum value, T_1, and all T_i equal to the maximum value, T_2. Thus, η takes its maximum possible value, given by equation (4.29), when the heat engine absorbs heat at a single maximum temperature, and releases heat at a single minimum temperature. The most efficient cycle is one that works between two isothermal boundaries, with no heat exchanged at any other temperature. Changes in the system that take place between the two isothermal boundaries are therefore adiabatic.

A thermodynamic cycle that is bound by two adiabats and two isotherms is called a *Carnot cycle*. It is sketched in terms of its pressure–volume evolution in Fig. 4.4, where it is compared to the isobaric–adiabatic cycle that I proposed as a model for mantle convection in Section 3.4. Carnot's cycle operates between two temperatures, $T_2 > T_1$, which are re-labeled T_B and T_D, respectively, for comparison with Fig. 3.9. Starting from the highest pressure point in the diagram (P_1), the fluid in Carnot's cycle absorbs heat at constant temperature T_B, expanding (performing work) and decompressing to P_2. From this point the fluid expands adiabatically and cools down, until it reaches temperature T_D at pressure P_3. It then releases heat at constant temperature T_D, so that it contracts and its pressure increases to P_4, from which point the fluid is compressed adiabatically until it reaches the

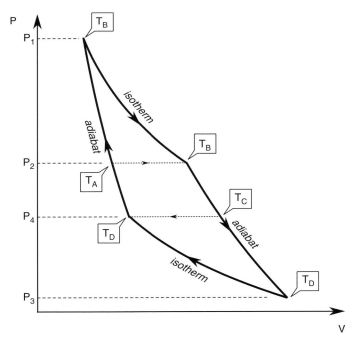

Fig. 4.4   The mantle convection cycle proposed in Chapter 3: isobaric heating at the core–mantle boundary from T_A to T_B, adiabatic decompression during mantle upwelling from T_B to T_C, isobaric cooling at the surface from T_C to T_D, and adiabatic compression during mantle downwelling from T_D to T_A (the isobaric legs are shown with thin dotted lines). Mantle convection approximates a Brayton cycle, that is compared in the figure to a Carnot cycle working between the same extreme temperatures, T_B and T_D.

initial state, at temperature T_B and pressure P_1. If the cycle is completed in unit time then the area bound by the four curves is the working rate, w. According to equation (4.29), as the two isotherms become further apart the efficiency of conversion of thermal energy to mechanical energy (η) increases. Convesely, if $T_B = T_D$ then it must be $\eta = 0$.

Worked Example 4.2   Efficiency of planetary convection systems

In contrast to Carnot's cycle, the convection cycle in Section 3.4 absorbs enthalpy at constant pressure, between temperatures T_A and T_B, and releases enthalpy at constant pressure, cooling from from T_C to T_D at the planet's surface (see Fig. 4.4). Such an isobaric–isothermal cycle is called a *Brayton cycle* or *Joule cycle*, and, interestingly, it is also an approximate description of the working of jet engines. We have just proved that, given that in this case heat is absorbed and released over a range of temperatures, the thermodynamic efficiency of convection in planetary interiors must be less than the efficiency of a Carnot cycle working between the same extreme temperatures, T_B and T_D, given by (4.29) (see Figs. 3.8, 3.9 and 4.4). We will now calculate the thermodynamic efficiency of the Brayton cycle. Let q_H be the rate of heat absorption at high pressure (from T_A to T_B) and q_L be the rate of heat release at low pressure (from T_C to T_D). Work is performed at a rate $w = q_H - q_L$, so that the thermodynamic efficiency of the Brayton cycle, η_B, is given by (compare equation (4.28)):

$$\eta_B = \frac{w}{q_H} = 1 - \frac{q_L}{q_H}. \tag{4.34}$$

For notation simplicity we will take the unit of time as being equal to the cycle's period. Then, given that the two heat transfer legs of the cycle are isobaric, we have $q_H = c_P(T_B - T_A)$ and $q_L = c_P(T_C - T_D)$, where we have arranged for all the qs to be positive quantities. Substituting into (4.34):

$$\eta_B = 1 - \frac{T_C - T_D}{T_B - T_A}. \tag{4.35}$$

Note that T_C and T_B lie on the same adiabat, and so do T_D and T_A. Assuming that the adiabats are also isentropes we can write the temperature ratio as a function of the thickness of the convecting layer, D, by integrating (3.35), as follows:

$$\frac{T_C}{T_B} = \frac{T_D}{T_A} = e^{-\frac{\alpha g D}{c_P}}. \tag{4.36}$$

Substituting in (4.35) we get an estimate for the thermodynamic efficiency of convection in planetary interiors:

$$\eta_B = 1 - e^{-\frac{\alpha g D}{c_P}}. \tag{4.37}$$

The exponent $(\alpha g D)/c_P$ is a non-dimensional parameter that is a function of material properties, gravitational acceleration and thickness of the convecting layer, but *not* of temperature, except indirectly through the effect of T on α.

We can use equation (4.37) to compare the thermodynamic efficiency of convection in planetary interiors. Figure 4.5 shows the function $\eta_B = 1 - e^{-x}$ (equation (4.37)) and rough estimates of the value of the exponent $x = (\alpha g D)/c_P$ for various planetary bodies. A small

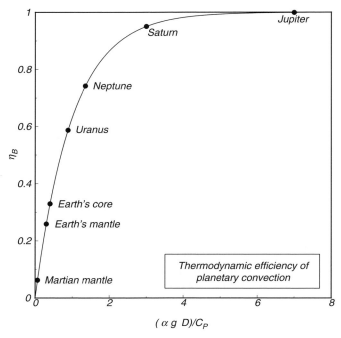

Fig. 4.5 Thermodynamic efficiency of the Bryton cycle as a model for convection in planetary interiors. Values for giant planets are very rough approximations, calculated with the assumption of very simple internal structures and equations of state, and assuming constant g. The correlation of greater efficiency with larger size is likely to be qualitatively correct.

planet such as Mars is not efficient at transforming thermal energy into mechanical energy. Even at an early age, when its internal heat flow may have been significantly higher than today's terrestrial heat flow, it is unlikely that Mars could have supported a rate of tectonic activity comparable to Earth's plate tectonics. In contrast, the giant planets, by virtue of their large sizes, can be very efficient heat engines. Convection of ionized fluids in their interiors is thought to be responsible for their magnetic fields (Section 1.8.3). Efficient conversion of their high heat flows to mechanical energy may be an important factor in explaining the strong magnetic fields of the giant planets. Of course, Fig. 4.5 shows only thermodynamic efficiencies, and only a very small fraction of the work performed by convection leaves the planetary interior as mechanical energy, for example, in the form of lithospheric deformation or planetary magnetic fields. Most of the mechanical energy is dissipated by viscous flow and electric currents (see also Section 1.8.3), and leaves the planetary interior as heat. The actual efficiency of convection in planetary interiors is close to 0, and the rate of entropy generation is close to the maximum given by equation (4.26).

We can also show that the thermodynamic efficiency of the convection cycle, given by equation (4.37), is less than that of a Carnot cycle working between the same extreme temperatures (Fig. 4.4) and given by equation (4.29):

$$\eta_C = 1 - \frac{T_D}{T_B}. \tag{4.38}$$

We can write this temperature ratio as follows (see equation (4.36)):

$$\frac{T_D}{T_B} = \frac{T_D}{T_C}\frac{T_C}{T_B} = \frac{T_D}{T_C}e^{-\frac{\alpha g D}{c_P}}. \tag{4.39}$$

Calling $\Delta T = T_C - T_D$ we write the efficiency of the Carnot cycle as:

$$\eta_C = 1 - e^{-\frac{\alpha g D}{c_P}} + \frac{\Delta T}{T_C} e^{-\frac{\alpha g D}{c_P}}, \qquad (4.40)$$

which leads to:

$$\eta_B = \eta_C - \frac{\Delta T}{T_C} e^{-\frac{\alpha g D}{c_P}} \qquad (4.41)$$

or, in other words, $\eta_B < \eta_C$, as required by Carnot's theorem.

Equation (4.41) has an interesting implication. This is that, as ΔT becomes smaller, the thermodynamic efficiency of the Brayton cycle approaches that of the Carnot cycle. This is immediately obvious from Fig. 4.4: as ΔT becomes smaller, the adiabatic excursions needed to construct a Carnot cycle between the same extreme temperatures become smaller and the two cycles become more similar. However, from Fig. 3.9 and our discussion of convection in Chapter 3, we recall that ΔT is the temperature contrast that drives convection. We must therefore conclude that the thermodynamic efficiency of stagnant lid convection is closer to the maximum Carnot efficiency than that of convection with moving plates. This may seem counterintuitive, for example in view of the fact that Earth has a more active lithosphere than Venus. But one must not confuse efficiency with the actual rate of performing work. The rate at which the convection cycle performs work is given by:

$$w = q_H - q_L = c_P \left[(T_B - T_A) - (T_C - T_D) \right] = c_P (T_C - T_D) \left(e^{\frac{\alpha g D}{c_P}} - 1 \right), \qquad (4.42)$$

which, for a given planetary size (constant g and D), yields:

$$w \propto \Delta T. \qquad (4.43)$$

Plate tectonics is driven by a greater temperature contrast than stagnant lid convection, so it performs mechanical works at a greater rate than the latter.

Atmospheric circulation over tropical and subtropical oceans approaches a Carnot cycle more closely than convection in planetary interiors. This is particularly striking in the case of hurricane formation (see Emanuel, 1986, 2006), but we can also see Carnot at work in Earth's global climate patterns. Because of the immense heat capacity of the oceans, air absorbs heat and humidity from tropical oceans at their (nearly) constant surface temperature, T_O, causing atmospheric pressure to decrease from P_1 to P_2 (Fig. 4.6). When the density has decreased sufficiently it rises and cools adiabatically from T_O to T_A, causing condensation and precipitation in the tropics. As air moves towards higher latitudes it radiates heat to space at the approximately constant temperature of the tropopause (troposphere–stratosphere boundary), T_A. Air density thus increases isothermally from P_3 to P_4, at which point it becomes negatively buoyant and sinks. Adiabatic compression raises its temperature from T_A to T_O. This hot and dry sinking air gives rise to the belts of subtropical deserts on both sides of the equator.

To calculate the Carnot efficiency of atmospheric convection we integrate (3.36) as follows:

$$\frac{T_A}{T_O} = 1 - \frac{gh}{T_O c_P}, \qquad (4.44)$$

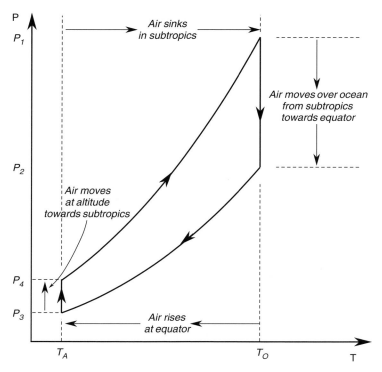

Fig. 4.6 Highly idealized view of atmospheric convection in the tropics as a Carnot engine. Air absorbs heat isothermally from the oceans and releases heat isothermally to space. The large thermal effect of condensation during adiabatic rise near the equator is ignored in the figure. For a rigorous discussion see Emanuel (1986, 2006).

where h is the thickness of the troposphere, $\sim$10 km and T_O is a characteristic surface temperature, $\sim$300 K. With these values we obtain $\eta \approx 0.3$ for the terrestrial atmosphere. This would appear to suggest that atmospheric convection is able to convert about a third of the thermal energy that it absorbs from the Earth's oceans into mechanical energy. But of course this is not quite so, because the thermodynamic efficiency given by equation (4.44) does not take into account dissipation in the engine itself, i.e. the finite viscosity of air (see Section 3.7.4) and friction at the interface between the atmosphere and the Earth's surface. The effective efficiency of atmospheric convection is much lower than 0.3.

4.6 A microscopic view of entropy

4.6.1 Microstates and macrostates

Entropy, like internal energy, is a macroscopic variable. It allows us to make accurate predictions about natural processes and, in regards to much of the material that we will cover in this book, the macroscopic understanding of entropy is all that we need. There are, however, some topics, such as solid solutions (discussed in Chapter 5) and equations of state (discussed in Chapters 8 and 9) for which it is useful to have an understanding of the microscopic underpinnings of entropy. In addition to this practical reason, there is a more fundamental one. We saw in Section 1.14 that internal energy, the macroscopic concept that forms the basis for the First Law, has a well-defined microscopic interpretation. It would be deeply unsatisfactory if the same were not true of entropy, which is at the core of the Second Law. There is also much misuse and abuse of the word entropy by non-scientists and pseudo-scientists, and for this reason too it is important to have a deeper understanding of what it is.

The microscopic interpretation of entropy begins with the concepts of the *microstate* of a system vs. its *macrostate.* The macrostate of a system is defined solely on the basis of macroscopic variables and does not require knowledge of the underlying microscopic configuration of the system, which is what we call its microstate. For example, the macrostate of an ideal gas made up of a single chemical species is fully specified by its temperature and its pressure, from which we can calculate all of its other macroscopic parameters, such as molar volume, internal energy, enthalpy and entropy (we shall see how). Although this macroscopic specification of the sate of a system is independent of any knowledge of its underlying microscopic structure, in Section 1.14 we saw that we can also define the internal energy of an ideal gas in terms of a microscopic model called the kinetic theory of gases. In order to describe the microstate of a system made up of ideal gas we would need to specify the position and velocity of each of the individual molecules. Here lies the crucial difference between macrostates and microstates: a given macrostate may arise from more than one different microstate. For example, we know that the distribution of molecular speeds is as given by Fig. 1.12, but there are many ways in which 10^{23} individual molecules may arrange themselves to give this distribution (see Box 4.1). Each of these different ways is a microstate. As far as the macrostate is concerned we don't care what the velocity of each individual molecule is, i.e. which specific equivalent microstate we have, as long as the velocity distribution is the one that corresponds to that specific macrostate.

Box 4.1　　　　　　　　　　　　　**Counting microstates**

The microscopic interpretation of entropy, and indeed much of statistical mechanics, relies on being able to count the number of equivalent microstates. We define all microstates that give rise to the same macrostate as being equivalent. Consider a situation in which there are N objects of the same kind, each of which can exist in i different states, with $N \geq i$. For example, it could be a group of N coins, each of which can be in one of two states (heads or tails), or a group of N octahedral sites in a crystal of olivine, each of which can be in one of two states (filled with Mg or Fe^{2+}), or a group of N particles in a gas, each of which can be in one of i different energy levels. We want to know how many possible microstates these systems have. A microstate is defined by the number of objects that are in each possible state. All microstates with the same number of objects in each state are equivalent and correspond to the same macrostate. For example, if we have 5 coins, then all microstates with 3 heads and 2 tails are equivalent, and correspond to a single macrostate that is different from the one that arises from all microstates in which there are 4 heads and 1 tail. We wish to calculate the number of equivalent microstates that underlie a given macrostate.

The total number of arrangements of the N objects, which you can think of as the number of ways in which we can choose them one at a time, is $N!$: we can choose the first in N ways, the next in $(N-1)$ ways, the next one in $(N-2)$ ways, and so. You can visualize this process as arranging the objects in a row, but this is just a help in visualization and has nothing to do with any putative spatial distribution of the objects. We thus have $N!$ possible arrangements of objects (e.g. coins), and each object can be in one of i different states (e.g. heads or tails). Let n_i be the number of objects that have the property i, i.e. that are in

Box 4.1	Continued

state i. Obviously, $\sum_i n_i = N$. We subdivide the set of N objects into i subsets of n_i objects. Two microstates are said to be equivalent if n_i is the same for all i. If two microstates fulfill this requirement but, additionally, each of the individual objects in each subset n_i are the same except that they are arranged in a different order, then they are the same microstate, rather than equivalent microstates. By the same argument that we used for N, each subset of n_i objects can be arranged in $n_i!$ different ways. This is true for all n_i, so the product $\prod (n_i!)$ is the total number of repeated (i.e. identical) microstates. The product of the number of equivalent microstates, which we symbolize with Ω, times the number of repeated microstates, $\prod (n_i!)$ must be equal to the total number of possible arrangements of the N objects, $N!$, as there is no other possibility that we have not considered. The number of equivalent microstates is therefore given by:

$$\Omega = \frac{N!}{\prod n_i!},\qquad(4.1.1)$$

which in the case of $i = 2$ (that we will use frequently), simplifies to:

$$\Omega = \frac{N!}{n!\,(N-n)!},\qquad(4.1.2)$$

where n objects are in state 1, and $(N-n)$ objects in state 2.

Consider again the example of 5 coins. The number of microstates defined by 3 heads and 2 tails is $5!/(3!\,2!) = 10$. All of these microstates are equivalent and correspond to the same macrostate. In contrast, the macrostate corresponding to 4 heads and 1 tail has $5!/(4!\,1!) = 5$ microstates.

The numbers of objects (molecules, atoms, ions, crystalline sites, etc.) in systems of physical interest are much larger than this, typically of order 10^{23}. The factorial of 10^{23} is a very large number, beyond the computational range of everyday calculators and computers. Luckily, however, what we are interested in is the logarithm of Ω, rather than in Ω itself, and this we can easily calculate by means of Stirling's approximation, which states that, for N very large:

$$\ln N! \approx N \ln N - N.\qquad(4.1.3)$$

Where does this come from? First, we can write $\ln N!$ as a sum of logarithms:

$$\ln N! = \sum_{i=1}^{i=N} \ln i.\qquad(4.1.4)$$

Although the factorial function is defined only for integers, which are discrete variables, when N becomes very large we can approximate the right-hand side of (4.1.4) as an integral, i.e.

$$\sum_{i=1}^{i=N} \ln i \approx \int_1^N \ln i\, di = N \ln N - N + 1 \approx N \ln N - N,\qquad(4.1.5)$$

which is (4.1.3).

A given macrostate can arise from more than one microstate. We shall say that these microstates are equivalent. We will also postulate that: (i) All individual microstates are equally probable and (ii) the observed macrostate, which we will call the equilibrium

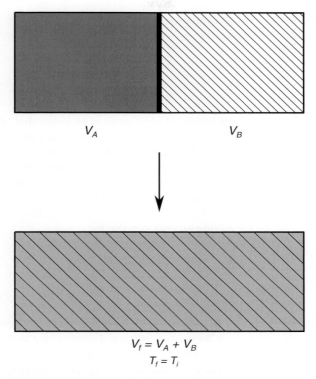

Fig. 4.7 Mixing of two different ideal gases in an isolated container. Each of the gases can be thought of as undergoing a free expansion from its initial volume to the same final volume, V_f.

macrostate, is the one that corresponds to the greatest number of equivalent microstates. An example will clarify the meaning and implications of these statements and show their plausibility, although it must be borne in mind that their acceptance as a valid description of nature relies on the much larger theoretical framework of statistical mechanics, and on an enormous amount of observations.

Let us consider the process of spontaneous mixing of two different ideal gases at uniform and constant temperature and pressure. Suppose that we have two containers separated by a wall, with volumes V_A and V_B, as shown in Fig. 4.7. Container A is filled with a moles of ideal gas A, and container B is filled with b moles of ideal gas B. When the partition is removed the gases mix by diffusion, so that after some time the entire volume, $V_f = V_A + V_B$ is filled with a homogeneous mixture in which the mol fractions of the two gases are $X_A = a/(a+b) = V_A/V_f$ and $X_B = b/(a+b) = V_B/V_f$ (use the ideal gas EOS to prove these identities). Pressure and temperature were the same for the two gases before mixing and remain unchanged during mixing.

Diffusive homogenization of the two gases in an isolated system is a spontaneous process, so it must be accompanied by an increase in entropy, which we calculate as follows. From equation (4.12) we write the entropy change ΔS from an initial state i to a final state f as follows:

$$\Delta S = \int_i^f \frac{dE}{T} + \int_i^f \frac{P}{T} dV. \tag{4.45}$$

Using the ideal gas equation of state and the fact that internal energy of an ideal gas is a function of temperature only (Section 1.14), we re-write equation (4.45) for an ideal gas as follows:

$$\Delta S = \int_i^f \frac{C_V}{T} dT + R \int_i^f \frac{dV}{V} = C_V \ln\left(\frac{T_f}{T_i}\right) + R \ln\left(\frac{V_f}{V_i}\right). \tag{4.46}$$

Diffusive homogenization of ideal gases is an isothermal process, because ideal gases are made up of non-interacting particles (i.e. there are no forces between the particles, and hence no work is performed when particles of different gases mix). However, the volume occupied by each of the gases increases during mixing, from the volume of the respective initial container to the total final volume, V_f. The entropy of each gas therefore increase during mixing by the same amount that it would increase if it was undergoing a free expansion (compare equation (4.22)). Because entropy is additive (Section 4.2), the total entropy increase for the mixing process is given by:

$$\Delta S_{mixing} = a\Delta S_A + b\Delta S_B = aR \ln\left(\frac{V_f}{V_a}\right) + bR \ln\left(\frac{V_f}{V_b}\right)$$
$$= -aR \ln(X_A) - bR \ln(X_B) > 0. \tag{4.47}$$

As mol fractions in a mixture are always less than 1, spontaneous mixing of gases is always accompanied by an increase in entropy. I never cease to marvel at this result. We defined entropy on the basis of heat transfer (equation (4.6)), and yet it can correctly predict the behavior of a system in which there is no heat transfer involved! As you will soon see, the importance of equation (4.47) goes beyond merely telling us that ideal gases will mix spontaneously.

Let us look at this same process microscopically. The two vessels initially contain aN and bN molecules of different ideal gases, where $N \approx 6.02 \times 10^{23}$ is Avogadro's number. We will now count microstates, as described in Box 4.1. You can think of the "objects" that I refer to in Box 4.1 as being little boxes into which the containers are subdivided, such that there is exactly one box per molecule. Averaging over a sufficiently long time, there is always exactly one molecule inside each box, and there are no empty boxes. The boxes can be in two different states: occupied by either a molecule of gas A or a molecule of gas B. Before the partition is removed all boxes in side A are filled with molecules of A, and all boxes in side B are filled with molecules of B. According to equation (4.1.2), the number of microstates in each of the two vessels, Ω_A and Ω_B, is:

$$\Omega_A = \frac{(aN)!}{(aN)!0!} = \frac{(bN)!}{(bN)!0!} = \Omega_B = 1. \tag{4.48}$$

The total number of microstates for the system before the gases are allowed to mix, Ω_i, is the product of these two numbers, as any microstate in A can be combined with any microstate in B, i.e.:

$$\Omega_i = \Omega_A \Omega_B = 1. \tag{4.49}$$

When the gases are allowed to mix there are $(a+b)N$ boxes in a single vessel of volume V_F, of which aN are in one state (occupied by an A molecule) and bN in the other (occupied by a B molecule). Using equation (4.1.2), the number of microstates is given by Ω_f, as follows:

$$\Omega_f = \frac{[(a+b)\,N]!}{(aN)!\,(bN)!}. \tag{4.50}$$

We now use Stirling's approximation (equation (4.1.3)):

$$\ln \Omega_f = (a+b)\,N \ln [(a+b)\,N] - (a+b)\,N$$
$$- [(aN) \ln (aN) - (aN)] - [(bN) \ln (bN) - (bN)], \tag{4.51}$$

which simplifies to:

$$\ln \Omega_f = aN \ln \left(\frac{a+b}{a} \right) + bN \ln \left(\frac{a+b}{b} \right) \tag{4.52}$$

or:

$$\ln \Omega_f = -aN \ln (X_A) - bN \ln (X_B). \tag{4.53}$$

Consider these results in the light of the two postulates about microstates and macrostates. According to postulate (i) all microstates are equally probable. Equation (4.49) says that there is only one microstate that corresponds to the initial macrostate: all boxes on each side filled with only one kind of molecule. We know that, if the partition is not there, this is not the equilibrium macrostate. Rather, the equilibrium macrostate is one in which the gases are mixed uniformly over both containers. According to equation (4.53), for a system size of order 1 mol this macrostate corresponds to $\sim e^N$ equivalent microstates ($N \approx 6.02 \times 10^{23}$). By postulate (i) *each of these equivalent microstates has the same probability of occurring as the single microstate that describes the un-mixed gases.* There are, however, e^N times more equivalent microstates than in the first case, so it is e^N times more likely that we will observe a homogeneous mixture than separate gases (as long as the partition is not there). This is what postulate (ii) says, that the equilibrium macrostate, i.e. the final state that the system will spontaneously evolve to, is the one that corresponds to the most microstates. Note that we have not proved that equation (4.53) is indeed the maximum, only that it is (much) larger than (4.49). It happens to be the maximum, though (see, for example, Glazer & Wark, 2001). The probability that the two gases will spontaneously un-mix and return to the state they were in before the partition was removed is not zero (this is what postulate (i) says), but it is so vanishingly small that we can be certain that we will not observe this phenomenon in the lifetime of the Solar System (quite a bit longer than that, actually). There is an additional point in this argument. This is the fact that fluctuations take place in all natural systems, during which the system visits microstates that are very close, but not identical, to an equilibrium microstate. These fluctuations are so small and swift that they are not reflected in the macrostate of the system, except when a system is close to a critical phase transition (more on this later). In principle, another exception could occur when the system of interest is so small that $1/e^N$ becomes an "observable" number. In Chapter 14 we will see that this may be a limiting factor for the minimum size of living organisms.

4.6.2 Boltzmann's postulate

The attentive reader might have noticed a striking similarity between equations (4.47) and (4.53). The two equations are identical, but for a constant multiplicative factor. They suggest, *but do not prove*, the following relationship between entropy and number of microstates:

$$S = k_B \ln \Omega, \tag{4.54}$$

where $k_B = R/N = 1.38 \times 10^{-23}$ J K^{-1} molecule^{-1} is known as Boltzmann's constant. Equation (4.54), due to Ludwig Boltzmann and reputedly inscribed on his tombstone, is known as *Boltzmann's postulate*. As with the laws of thermodynamics, it remains a postulate because it cannot be shown to be valid from simpler statements or observations. We accept its validity a posteriori, on the basis of the agreement of the theory built upon it with the observed behavior of nature. Boltzmann's postulate provides a microscopic interpretation of entropy that is intuitively satisfying. It says that the highest entropy macrostate is the one that corresponds to the greatest number of equivalent microstates. The increase in entropy that accompanies all spontaneous processes in an isolated system is simply a reflection of the system "falling onto" the most likely set of equivalent microstates. The impossibility of entropy decreasing in an isolated system should actually be seen as a vanishingly small probability – "practically impossible" for any conceivable macroscopic system. As to the relationship between entropy and disorder, which obsesses philosophers, social scientists, post-modernist writers and other commentators, we shall have more to say about it shortly.

Let us go back to the relationship between (4.53) and (4.47). Using (4.54), we can write the following equation:

$$S_{configurational} = k_B \left(\ln \Omega_f - \ln \Omega_i \right) = k_B \ln \left(\frac{\Omega_f}{\Omega_i} \right). \tag{4.55}$$

The increase in entropy that accompanies mixing of the gases is a function only of the *ratio* between the number of microstates in the final and initial states. This increase in entropy is called the *configurational entropy*, as it reflects a change in the microscopic configuration of the system. It is important to realize that equation (4.54) gives the functional relationship between S and Ω, but it says nothing about the factors that enter into the definition of the microstates that are counted by Ω. The number of microstates in our example was defined solely on the basis of the identity of the molecules, but in reality other parameters must be considered too, most importantly temperature. As the temperature increases so does the number of quantized energy levels available for atomic vibrations, and therefore the number of accessible microstates and, by (4.54), the entropy. Because we consider an isothermal transformation the effect of temperature on the value of Ω cancels out in equation (4.55), leaving the change in configuration as the only contribution to the entropy of mixing.

There is an important question here: how does the microscopic interpretation of entropy relate to the concept of energy dissipation and the fact that an entropy increase reflects an irreversibly lost opportunity to perform mechanical work? The following is a possible way of looking at this, which traces its origin to James Clerk Maxwell (who was perhaps the greatest scientist of the nineteenth century). Suppose that rather than a removable partition the two vessels with different gases in Fig. 4.7 are separated by a fixed wall with a small opening covered by a revolving trapdoor that will turn only when it is simultaneously hit by a pair of different molecules coming from opposite directions. The door is just large enough to allow a single molecule through in each direction at any one time (Maxwell used imaginary demons for this sort of thought experiment, and his demons were actually asked to perform a task opposite to the one we describe here, see, for example, Baylin, 1994). The door is connected to an external reservoir of mechanical energy, such as a spring or a weight, so that rotation in one direction compresses the spring (or raises the weight) by a small amount, and rotation in the opposite direction releases the spring by the same amount. As long as the door keeps revolving in the same direction there is net storage of mechanical energy, but if the door flip-flops then there is no net storage nor release of mechanical energy. When we first free the revolving door with different gases in each container it will only be hit by A molecules coming from the left and B molecules coming from the right, so it will spin in only one direction and store mechanical energy. As the gases mix the door may start flip-flopping, but as long as there is a gradient in composition across the partition there will be a net storage of mechanical energy, because the door will rotate more often in one direction than in the other. When the gas compositions become identical on both sides there will be a net amount of mechanical energy stored in the spring or weight, which came exclusively from diffusive mixing of the two gases. If we use a removable partition rather than our little trapdoor this work potential is lost, and entropy is generated, or equivalently, energy is dissipated.

But the gases were allowed to mix anyway, even if they perform work, so didn't their entropy increase, as calculated by (4.47) or (4.53)? No, as one must still abide by the First Law. The mechanical energy stored in the spring came from somewhere: it must have come from the kinetic energy of the molecules. If the container is adiabatic then the gas cooled down. Equation (4.53) is now not a complete description of the number of microstates in the final state, as it does not account for the fact that, as temperature decreases, the number of accessible energy levels of the molecules decreases too, and so does the number of microstates (see Glazer & Wark, 2001). Does the trapdoor example carry over to entropy generation by diffusive heat flow? Yes, but it is less obvious how. The key is that Ω changes with temperature, so we need to imagine some contraption that is able to store mechanical energy by intercepting packages of molecular kinetic energy exchanged between bodies at different temperatures.

Worked Example 4.3   Configurational entropy in crystals

Configurational entropy plays an important role in the thermodynamic characterization of minerals and melts, that we will study in later chapters. Here we focus on the configurational entropy of minerals of constant composition and variable crystalline structure. Consider the case of potassium feldspar. A unit cell of potassium feldspar consists of four formula units, so that it has the composition: $K_4Al_4Si_{12}O_{32}$. Silicon and aluminum occupy tetrahedral sites in the crystalline structure. There are a total of sixteen tetrahedral sites in a potassium feldspar unit cell. In all potassium feldspar polymorphs it is possible to distinguish between two types of tetrahedral sites, on the basis of their sizes and symmetries, which we can call T1 and T2 sites, and such that there are eight T1 sites and eight T2 sites. In addition, in some of the crystal forms it is also possible to discriminate between two subsets of the T1 sites, which we will label T1a and T1b. The four Al and twelve Si atoms can occupy the sixteen tetrahedral sites in a number of different ways, giving rise to minerals with different crystal symmetries and different amounts of configurational entropy.

Consider microcline first. In this mineral the T1a and T1b sites are distinguishable. The four T1a sites are occupied by Al, whereas the four T1b sites and the eight T2 sites are occupied by Si. This regular distribution of Si and Al atoms is said to constitute a structure with *long-range order*. It is also a structure with a high information content: we have total certainty of what we are going to find in each type of site. In sanidine, on the other hand,

Al and Si distribute themselves randomly over all sixteen tetrahedral sites, such that, one average (i.e. in a macrsocopic crystal with $\sim N$ atoms in it) there are six Si and two Al atoms in the T1 sites and six Si and two Al atoms in the T2 sites. This structure lacks long-range order, and has a low information content: all we know is that we have a one in three probability of finding an Al atom in any one tetrahedral site, compared to the certainty that we had in the case of the ordered structure of microcline.

We will now count the number of equivalent microstates that give rise to each of the two macrostates, microcline and sanidine. One "unit cell mol" of potassium feldspar contains N unit cells, i.e. $4N$ Al atoms and $12N$ Si atoms. The objects that we will count are crystalline sites, each of which can be in two possible states: filled with Si or filled with Al. In the following equations, the numerator is the number of crystalline sites, the first term in the denominator is the number of those sites occupied by Si, and the second term in the denominator is the number of sites occupied by Al. For microcline we must consider the number of microstates of each of the three types of distinguishable sites, so that we have:

$$\Omega_{T1a} = \frac{(4N)!}{0!\,(4N)!}$$

$$\Omega_{T1b} = \frac{(4N)!}{(4N)!0!} \tag{4.56}$$

$$\Omega_{T2} = \frac{(8N)!}{(8N)!0!}.$$

The total number of microstates for the microcline macrostate is:

$$\Omega_{microcline} = \Omega_{T1a}\Omega_{T1b}\Omega_{T2} = 1 \tag{4.57}$$

or:

$$\ln \Omega_{microcline} = 0. \tag{4.58}$$

For sanidine:

$$\Omega_{T1} = \frac{(8N)!}{(6N)!\,(2N)!}$$

$$\Omega_{T2} = \frac{(8N)!}{(6N)!\,(2N)!}, \tag{4.59}$$

using Stirling's approximation we find:

$$\ln \Omega_{T1} = \ln \Omega_{T2} = 6N \ln\left(\frac{4}{3}\right) + 2N \ln 4 \tag{4.60}$$

and:

$$\ln \Omega_{sanidine} = \ln \Omega_{T1} + \ln \Omega_{T2} = 2N\left[6\ln\left(\frac{4}{3}\right) + 2\ln 4\right]. \tag{4.61}$$

The sanidine macrostate corresponds to a (much) greater number of microstates than the microcline macrostate, so by Boltzmann's postulate it must have a higher entropy. This configurational entropy does not reflect a change in the composition of the phase, as in the example of the mixing gases, but rather a change in the microscopic configuration of the crystal. We use Bolztmann's equation, (4.54), to calculate the configurational entropies of the two polymorphs. Noting that equations (4.58) and (4.61) are based on four formula units of potassium feldspar, the molar configurational entropies (i.e. per mol of $KAlSi_3O_8$) are:

$$S_{configurational,\ microcline} = \frac{1}{4}k_B \ln \Omega_{microcline} = 0 \tag{4.62}$$

and:

$$S_{configurational,\ sanidine} = \frac{1}{4}k_B \ln \Omega_{sanidine} = R\left[3\ln\left(\frac{4}{3}\right) + \ln 4\right] = R\ln\left(\frac{256}{27}\right)$$
$$\approx 18.7 \mathrm{J\ K^{-1}mol^{-1}} \tag{4.63}$$

The total entropy of a crystal, S_{total}, is in general made up of two contributions:

$$S_{total} = S_{thermal} + S_{configurational}. \tag{4.64}$$

One contribution, $S_{thermal}$, arises from the distribution of (quantized) energy levels of the constituent atoms, and is called the thermal or vibrational component of entropy. Thermal entropy increases with temperature, as the vibrational energy of the atoms gets "dispersed" over a greater number of possible energy levels, which increases the number of accessible microstates. Configurational entropy, $S_{configurational}$, arises from the distribution of atoms in the crystalline structure and remains constant with temperature as long as the distribution of atoms between different crystalline sites does not change (e.g. as long as sanidine does not invert to microcline). In a crystal with perfect long-range order there is no configurational entropy, but at all finite temperatures the ordered crystal still has thermal entropy (see Section 4.7).

The lower "information content" of the crystalline structure with greater configurational entropy (sanidine in this example) is at the core of the relationship between "entropy and disorder": greater order implies greater certainty of what kind of atom is in what kind of crystalline site, and therefore a lower number of microstates and lower configurational entropy. It is a tragedy that this concept, that has a strict and unambiguous meaning in the context of the microscopic structure of matter, has been misunderstood and misappropriated by non-scientists, who then proceed to apply it erroneously in a wide range of contexts where it does not belong. Entropy as a measure of disorder is a strictly microscopic concept and is meaningless as a description of "macroscopic order". Perhaps the most egregious example of the misuse of the concept of entropy is its erroneous application to support the spurious claim that biological evolution, and indeed the existence of life itself, requires a supernatural explanation. I will dispel these detestable notions in Chapter 14.

There is one final point that you may be wondering about. If sanidine has higher entropy than microcline (i.e. it is a macrostate that is more likely because it corresponds to a larger number of equivalent microstates), then why does microcline exist at all? Shouldn't it spontaneously invert to sanidine? If the crystal were an isolated system (constant internal energy and volume) then it would, subject to the removal of constraints (interatomic forces, which would be the equivalent of the partition in Fig. 4.7). But the direction of a spontaneous change in a crystal that is not an isolated system (which is the common situation in nature) is not necessarily determined by an increase in entropy, but rather by a decrease in the *thermodynamic potential* appropriate to the constraints on the system. We return to this in Sections 4.8 and 4.9.

4.7 The Third Law of Thermodynamics

4.7.1 Statement of the Third Law of Thermodynamics

There is an additional principle, called the Third Law of Thermodynamics, that is independent of the First and Second Laws. It is essential in the development of chemical thermodynamics, although much of classical thermodynamics and its applications to heat engines and other engineering processes do not require it. We introduce the Third Law by stating that experimental evidence shows that, as temperature approaches 0 K, heat capacities (C_P and C_V) approach zero faster than temperature, i.e.:

$$\lim_{T \to 0} \left(\frac{C_P}{T} \right) = 0. \tag{4.65}$$

A few examples are shown in Fig. 4.8. A consequence of (4.65) is that the entropy difference of a substance between 0 K and any other temperature, T, is a finite value. Assuming that heating takes place at constant pressure, and that there are no phase transitions between 0 and T, then $dQ = C_p dT$, and we have:

$$S(T) - S(0) = \int_0^T dS = \int_0^T \frac{C_P}{T} dT \tag{4.66}$$

with condition (4.65) guaranteeing that the integral does not blow up. An unknown integration constant remains, however, which is the entropy at 0 K.

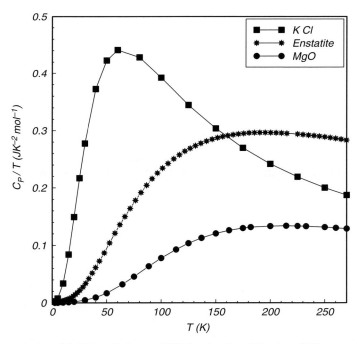

Fig. 4.8 Low temperature behaviors of C_P / T for an ionic crystal (KCl, data from Berg & Morrison, 1957), a crystalline oxide (MgO, data from Barron *et al.*, 1959) and a crystalline silicate (enstatite, data from Krupka *et al.*, 1985).

It is found that the *entropy change* associated with any transition between equilibrium states of perfect crystalline substances vanishes as well as $T \to 0$. The meaning of "perfect crystalline substance" is that the crystal has no configurational entropy, i.e. that its atoms are perfectly ordered in the sense that we discussed in Worked Example 4.3. Consider a transformation $A \to B$ between equilibrium states of perfect crystals. This could be, for instance, a chemical reaction. Using the notation for changes in state variables that we laid out in Section 1.13.1, we have:

$$\Delta_r S_{P,T} = (S_{P,T})_B - (S_{P,T})_A$$
$$= (S_{P,0})_B - (S_{P,0})_A + \int_0^T \left(\frac{C_P}{T}\right)_B dT - \int_0^T \left(\frac{C_P}{T}\right)_A dT \qquad (4.67)$$

and $\Delta_r S_{P,T} \to 0$ when $T \to 0$. The low-temperature behavior of heat capacities, equation (4.65), means that the two integrals in (4.67) also vanish as $T \to 0$. One must therefore conclude that:

$$\lim_{T \to 0} \left[(S_{P,0})_B - (S_{P,0})_A \right] = 0. \qquad (4.68)$$

The integration constant in (4.66), which is the entropy at absolute zero, is therefore the same for all perfect crystalline substances. This was the original statement of the Third Law, put forward by the physical chemist W. Nernst in 1906. Note that (4.68) does not fix the absolute value of S at $0\,K$, only that the value is the same for all crystalline substances. Max Planck gave a stronger statement of the Third Law in 1916, by postulating that entropy vanishes at $0\,K$. Planck's statement of the Third Law, which is the one that is widely accepted today (and which subsumes Nernst's original statement) is that *the entropy of all perfect crystalline substances vanishes at $0\,K$*. The emphasis on "perfect crystalline substance" is important. For example, Planck's statement of the Third Law implies that the entropy of microcline vanishes at $0\,K$. The entropy of sanidine, however, does not vanish, because its configurational entropy remains unchanged with temperature (as long as the atoms don't order themselves spontaneously, in which case it is no longer sanidine).

The Third Law remains a postulate (just as the First and Second Laws), but its plausibility can be demonstrated from quantum mechanical arguments. There is another statement of the Third Law, which can be shown to be equivalent to Planck's statement, and which is also due to Nernst. This is that it is impossible to reach $T = 0$ in a finite number of reversible steps. We will not be concerned with this statement in this book, but the interested reader can see Baylin (1994) for a derivation.

4.7.2 An absolute entropy scale

The Third Law of Thermodynamics makes it possible to define absolute values of entropy for all substances. This is in contrast to energy functions, such as enthalpy. Recall that in Section 1.13.1 we constructed a scale for enthalpy by setting to zero the enthalpies of all pure elements in their stable forms at 298.15 K and 1 bar, and then defining enthalpies of formation for compounds relative to this arbitrary zero. The Third Law says that we cannot do this for entropy. From equation (4.66) it follows that the entropy of all substances, including pure elements, at 298.15 K and 1 bar is *not* zero. Using Planck's postulate, we define the reference state entropy, also called the reference state Third Law entropy, and commonly symbolized by S^0_{298}, as:

$$S^0{}_{298} = \int_0^{298.15} \frac{dQ_{rev}}{T} dT \tag{4.69}$$

with $S(0) = 0$ if the substance is a perfect crystal at absolute zero. If configurational entropy is not zero at 0 K then it must be added to (4.69) (more on this in Chapter 7). If there are no phase transitions between 0 K and 298.15 K (which is the case for the three examples shown in Fig. 4.8) then this integral is equal to (4.66) and is simply the area under the curves in Fig. 4.8. If there are phase transitions between 0 K and 298.15 K then the entropy changes associated to the phase transitions must be added to (4.69), as discussed in Worked Example 4.4.

Values of reference state Third Law entropies are listed in thermodynamic data bases. It is important to reiterate this point: reference states entropies are *not* "entropies of formation". An entropy of formation could be defined as the difference between the Third Law entropy of a compound and those of its constituent elements. If we call this difference $\Delta(S^0_{298})$, then:

$$\Delta\left(S^0_{298}\right) = S^0_{298} - \sum_{elements} S^0_{298, \, elements} \tag{4.70}$$

but, by (4.66), $S^0_{298, \, elements} \neq 0$ for all elements, so $\Delta(S^0_{298}) \neq S^0_{298}$ for all substances. Entropies of formation are seldom, if ever, used. The point of 4.70 is to make clear the difference between *reference state entropy* and *reference state enthalpy of formation*.

Worked Example 4.4   Entropies of phase transitions

Heat capacity becomes undetermined at phase transitions, because the system absorbs heat without experiencing a temperature change. In Chapter 1 we discussed this concept in terms of sensible heat and latent heat, and saw that the enthalpy associated with a phase transition must be treated separately from the heat capacity integrals that track enthalpy changes associated with sensible heat (see Section 1.13.2). The same is true of entropy: if phase transitions occur within the temperature range that one is integrating over (e.g. equation (4.69)), then the entropy of the phase transitions, calculated as discussed in Worked Example 4.1, must be added separately. A simple example is the element chlorine, which melts at 172.12 K and vaporizes at 239.05 K. Figure 4.9a shows values of C_P/T for chlorine between 0 K and 270 K, measured by Giauque and Powell (1939). There are three different curves, separated by two discontinuities that correspond to the two phase transitions, melting and vaporization. The area under each of the curves corresponds to the contribution of each of the phases to the Third Law entropy of chlorine at 270 K, but the entropies associated with the phase transitions are not accounted for in this figure.

Third Law entropy of chlorine is plotted as a function of temperature in Fig. 4.9b. The entropy for the solid at any temperature between 0 K and 172.12 K is equal to the area under the first curve in Fig. 4.9a between 0 K and that temperature. The entropy of liquid chlorine at the melting point of 172.12 K is equal to the entropy of the solid at the melting point plus the entropy of melting, which is equal to the enthalpy of melting divided by the melting temperature (Worked Example 4.1). We then add the area under the second curve in Fig. 4.9a to get the entropy of liquid chlorine at the vaporization point, 239.05K. This, plus the entropy of vaporization (= enthalpy of vaporization divided by temperature of vaporization), yields the entropy of chlorine gas at 239.05 K, and so on.

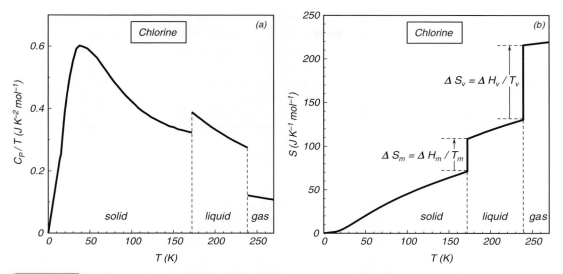

Fig. 4.9 Heat capacity, phase transitions and Third Law entropy of chlorine. Data from Giauque and Powell (1939).

The existence of a non-zero enthalpy of transition causes the entropy to be discontinuous at the phase transition, as this example shows. For reasons that we discuss in Chapter 7, a phase transition of this type, in which there is an entropy discontinuity and $\Delta H_{transition} \neq 0$, is called a *first-order phase transition*. Melting, sublimation, vaporization and polymorphic transformations (such as graphite $\rightleftarrows$ diamond or kyanite $\rightleftarrows$ sillimanite) are common examples of first-order phase transitions. Continuous (or "second-order") phase transitions, in which entropy is continuous (and thus $\Delta H_{transition} = 0$) are important in the study of fluids (Chapter 9) and also occur in some minerals (Chapter 8).

The entropy discontinuity associated with first-order phase transitions reflects a discontinuity in the number of accessible microstates. A simplified view is as follows. In the solid there are vibrational degrees of freedom only. At the melting point rotational degrees of freedom appear, and so the energy of the molecules can be distributed over a greater number of microstates. Even more microstates appear when the substance vaporizes and translational degrees of freedom become available.

4.8 Thermodynamic potentials

4.8.1 The entropy maximum principle revisited

The goal of the rest of this chapter is to develop a formal thermodynamic definition of chemical equilibrium, that will allow us to calculate phase equilibrium and phase compositions in planetary environments. In order to reach that goal and to understand where each result comes from we must follow a somewhat tortuous path that I will try to simplify as much as possible without losing any of the necessary physical and mathematical justifications. We begin by re-stating the Second Law of Thermodynamics in a formal mathematical language. The Second Law says that the entropy of an isolated system can only increase, or, equivalently, that the state of equilibrium in an *isolated* system is the one in which entropy takes its maximum possible value for the (constant) total internal energy content of the system (we will ignore forms of energy other than E). By definition, an isolated system does not interact with anything else, so we see that the condition $dE = 0$ (i.e constant internal energy) must be true of an isolated system. It must also be true that $dV = 0$, because otherwise the system would exchange work with its environment. Moreover, an isolated system must also be closed to the exchange of chemical components. If we call the mol number (= number of mols) of the ith component n_i, then for an isolated system it must be $dn_i = 0$, for all i.

In the mathematical notation of thermodynamics, the maximum entropy statement of the Second Law is often written as follows:

$$dS_{E,V,n_i} = 0 \tag{4.71}$$

$$d^2 S_{E,V,n_i} < 0. \tag{4.72}$$

These are mangled versions of the conditions for the maximum of a function: the first equation says that the first derivative vanishes at an extremum, whereas the second one says that, if the second derivative is negative at the extremum, then the extremum is a maximum. The problem is that equations (4.71) and (4.72) are not written in terms of derivatives, but rather in terms of the total differentials of entropy. This notation is mathematically sloppy, as loudly pointed out by Truesdell (1984), but, regrettably, the use of equations such as (4.71) and (4.72) is so deeply ingrained in thermodynamics that it is difficult to get away from it. It is important, however, to understand what these equations are actually saying. In particular, if E, V and n_i are all constant, exactly what variable are we differentiating entropy relative to, so as to find an extremum for the function? Which begs the question: what (else) is entropy a function of? Or, in physical terms, why is entropy changing in the first place?

The way to think about this is by imagining that, initially, there are restrictions, or *constraints*, that prevent the system from changing towards equilibrium. For instance, a partition separating two different gases that can mix by diffusion, or two different electrolyte solutions that will precipitate a solid when they mix, or a perfect thermal insulator separating two bodies at different temperatures. When we remove the restriction the system changes towards equilibrium, and as it does so entropy changes as a function of some physical quantity that drives the displacement towards equilibrium. This could be, for example, exchange of gas molecules between the two sides of the container in Fig. 4.7, or exchange of ions between the electrolyte solutions, or exchange of internal energy between two bodies at different temperatures. We will analyze the latter example in formal mathematical language so as to clarify the meaning of equations (4.71) and (4.72).

At constant E, V and n_i, the entropy of the isolated system varies as a function of the amount of internal energy exchanged between the two bodies. Let the internal energy, entropy and temperature of body j be E_j, S_j and T_j, respectively, and the corresponding properties of the isolated system composed of the two bodies be E, S and T. We will assume that there is no exchange of matter between the bodies, and that they are incompressible. We then have:

$$E_1 + E_2 = E \tag{4.73}$$

and, because the system is isolated (E constant):

$$\frac{dE_1}{dE_2} = -1. \tag{4.74}$$

We will track the change in entropy of the system relative to the internal energy content of one of the bodies, say 1, at constant internal energy of the system. For an extremum we need the first derivative to vanish, so that:

$$\left(\frac{\partial S}{\partial E_1}\right)_{E,V,n_i} = 0. \tag{4.75}$$

This is the same equation as (4.71), except that now we are explicitly stating how to calculate the derivative. Using the fact that entropy is an extensive variable, we expand (4.75) as follows:

$$\left(\frac{\partial S}{\partial E_1}\right)_{E,V,n_i} = \left(\frac{\partial S_1}{\partial E_1}\right)_{E,V,n_i} + \left(\frac{\partial S_2}{\partial E_1}\right)_{E,V,n_i} = 0 \tag{4.76}$$

and using (4.74):

$$\left(\frac{\partial S}{\partial E_1}\right)_{E,V,n_i} = \left(\frac{\partial S_1}{\partial E_1}\right)_{E,V,n_i} - \left(\frac{\partial S_2}{\partial E_2}\right)_{E,V,n_i} = 0. \tag{4.77}$$

From (4.14) and the properties of partial derivatives (Box 1.3, equation (1.3.18)), this becomes:

$$\left(\frac{\partial S}{\partial E_1}\right)_{E,V,n_i} = \frac{1}{T_1} - \frac{1}{T_2} = 0, \tag{4.78}$$

which says that the entropy of the closed system takes an extremum value when $T_1 = T_2$. We know, of course, that this is the equilibrium condition, *but equation (4.78) by itself does not prove it*, because it does not tell us whether the extremum is a maximum or a minimum. In order to test for this we need to find the sign of the second derivative, so we write the formal equation equivalent to (4.72):

$$\left(\frac{\partial^2 S}{\partial E_1^2}\right)_{E,V,n_i} = \frac{\partial}{\partial E_1}\left(\frac{\partial S}{\partial E_1}\right)_{E,V,n_i} = \frac{d}{dE_1}\left(\frac{1}{T_1} - \frac{1}{T_2}\right) \tag{4.79}$$

or, using (4.74):

$$\left(\frac{\partial^2 S}{\partial E_1^2}\right)_{E,V,n_i} = \frac{d}{dE_1}\left(\frac{1}{T_1}\right) + \frac{d}{dE_2}\left(\frac{1}{T_2}\right). \tag{4.80}$$

From the chain rule:

$$\frac{d}{dE}\left(\frac{1}{T}\right) = \frac{d}{dT}\left(\frac{1}{T}\right)\frac{dT}{dE} = -\frac{1}{T^2}\frac{dT}{dE} \tag{4.81}$$

and, since we assume that the bodies are incompressible, heat transfer is at constant volume, so, using (1.13.18):

$$\frac{dT}{dE} = \frac{1}{C_V}. \tag{4.82}$$

Putting it all together, and noting that C_V is always a positive quantity, we arrive at:

$$\left(\frac{\partial^2 S}{\partial E_1^2}\right)_{E,V,n_i} = -\frac{1}{C_V T_1^2} - \frac{1}{C_V T_2^2} < 0, \tag{4.83}$$

which is the equivalent of (4.72) and shows that entropy is indeed maximum when $T_1 = T_2$. As an aside, because the isolated system consists of the two bodies only, this result also shows that the equilibrium temperature of the system, T, is uniform and equal to that of the two bodies.

Exchange of internal energy is the only restriction that we are allowed to impose and release in our example, so that it is the only variable relative to which we can track entropy changes in the isolated system. Other restrictions are in general possible, such as exchanges of chemical components, PdV work, electric work or radiant energy between different parts of an isolated system. There may also be more than one independent restriction, a point that will become important when we study chemical equilibrium, so that we may need more than one variable to track the total entropy change. We will use the symbol Z_k to represent the independent variables that represent the quantities that are exchanged between parts of an isolated system when restrictions are released and the system evolves spontaneously towards equilibrium. The equilibrium condition, or, if you wish, the Second Law of Thermodynamics, is then written as follows:

$$dS_{E,V,n_i} = \sum_k \left(\frac{\partial S}{\partial Z_k}\right)_{E,V,n_i} dZ_k = 0; \quad dZ_k \neq 0 \tag{4.84}$$

and:

$$d^2 S_{E,V,n_i} = \sum_{k,j} \frac{\partial}{\partial Z_j}\left(\frac{\partial S}{\partial Z_k}\right)_{E,V,n_i} dZ_j dZ_k < 0; \quad dZ_j \neq 0, dZ_k \neq 0. \tag{4.85}$$

In our example there is only one restriction, because changes in E_1 and E_2 are not independent (equation (4.74)), and the quantity whose flow drives the system towards equilibrium is internal energy. These equations then become:

$$dS_{E,V,n_i} = \left(\frac{\partial S}{\partial E_1}\right)_{E,V,n_i} dE_1 = 0; \quad dE_1 \neq 0 \tag{4.86}$$

and:

$$d^2 S_{E,V,n_i} = \frac{\partial}{\partial E_1}\left(\frac{\partial S}{\partial E_1}\right)_{E,V,n_i} (dE_1)^2 < 0; \quad dE_1 \neq 0, \tag{4.87}$$

which, given the condition $dE_1 \neq 0$ are identical to (4.78) and (4.83), respectively. Throughout the rest of this book I will use both notations. I think that it is important to become familiar not only with the one that is widely used in thermodynamics (e.g. $dS = 0$) but also with the one that is mathematically complete (e.g. $\partial S/\partial Z = 0$). This is a point that is unfortunately not made in most thermodynamics textbooks. You can always refer to equations (4.84) and (4.85) to clarify the relationship between the two notations.

4.8.2 The energy minimum principle

We now have a formal mathematical statement that defines thermodynamic equilibrium (equations (4.84) and (4.85)), but this definition is unwieldy on two accounts. In the first place, it is based on entropy, which is neither an intuitively simple concept nor a quantity that is directly observable. Second, it only applies to isolated systems, and in nature we frequently have to deal with systems that are not isolated. We will now take care of these difficulties. The first step is to recast the Second Law of Thermodynamics in terms of an extremum in internal energy rather than entropy. The full justification for doing this will become apparent a posteriori. At this stage let us view it as a simple exercise in calculus: can we start from (4.84) and (4.85) and obtain equivalent (*NOT identical* !!) expressions for $\partial E / \partial Z$ and $\partial^2 E / \partial Z^2$? In particular, what can we say about the internal energy of a constant-entropy system at equilibrium?

We first seek an expression for $(\partial E / \partial Z)_S$ in terms of $(\partial S / \partial Z)_E$. Using equation (1.3.19) (Box 1.3), and omitting the subscripts V and n_i because these variables stay constant throughout, we write:

$$\left(\frac{\partial E}{\partial Z} \right)_S = -\left(\frac{\partial E}{\partial S} \right)_Z \left(\frac{\partial S}{\partial Z} \right)_E = -T \left(\frac{\partial S}{\partial Z} \right)_E. \qquad (4.88)$$

If $(\partial S / \partial Z)_E$ vanishes, then so does $(\partial E / \partial Z)_S$. Thus, *an extremum for entropy at constant internal energy is also an extremum for internal energy at constant entropy*. We now have to decide whether it is a minimum or a maximum, so we take the second derivative. To simplify the notation, we make $(\partial E / \partial Z)_S = Y$. Then, using (1.3.12) (Box 1.3):

$$\left(\frac{\partial^2 E}{\partial Z^2} \right)_S = \left(\frac{\partial Y}{\partial Z} \right)_S = \left(\frac{\partial Y}{\partial Z} \right)_E + \left(\frac{\partial Y}{\partial E} \right)_Z \left(\frac{\partial E}{\partial Z} \right)_S, \qquad (4.89)$$

which, using (4.88) and the fact that at an extremum it is $(\partial E / \partial Z)_S = 0$, simplifies to:

$$\left(\frac{\partial^2 E}{\partial Z^2} \right)_S = \left(\frac{\partial Y}{\partial Z} \right)_E = \left[\frac{\partial}{\partial Z} \left(\frac{\partial E}{\partial Z} \right)_S \right]_E = \left[\frac{\partial}{\partial Z} (-T) \left(\frac{\partial S}{\partial Z} \right)_E \right]_E. \qquad (4.90)$$

If you are wondering whether it is licit to use the condition $(\partial E / \partial Z)_S = 0$ to simplify (4.89), but ignore this simplification in (4.90), it is. The reason is that in (4.90) we are evaluating $(\partial^2 E / \partial Z^2)_S$ in general, and $(\partial E / \partial Z)_S = 0$ is only a special point. We could carry over the last term in (4.89) to the end of the calculation, and drop it there. The result would be the same, because we are not operating on this term any further, but why make the equations any more messy than they have to be?

Applying the product rule to the right-hand side of (4.90), and using the extremum condition $(\partial S / \partial Z)_E = 0$, we have:

$$\left(\frac{\partial^2 E}{\partial Z^2} \right)_S = -\left(\frac{\partial T}{\partial Z} \right)_E \left(\frac{\partial S}{\partial Z} \right)_E - T \left(\frac{\partial^2 S}{\partial Z^2} \right)_E = -T \left(\frac{\partial^2 S}{\partial Z^2} \right)_E > 0, \qquad (4.91)$$

where the inequality, which shows that this extremum is a minimum, follows directly from (4.85), i.e. the fact that S at constant E is a maximum at equilibrium: $(\partial^2 S / \partial Z^2)_E < 0$. The Second Law of Thermodynamics therefore leads to a minimum energy principle: at equilibrium in a **constant entropy** system the internal energy takes it minimum possible value. In the notation of thermodynamics we write the minimum energy principle as follows:

$$dE_{S,V,n_i} = 0 \qquad (4.92)$$

$$d^2 E_{S,V,n_i} > 0, \qquad (4.93)$$

which can also be expanded in a way analogous to (4.84) and (4.85) (exercise left for the reader).

The entropy maximum and the energy minimum principles are both valid definitions of thermodynamic equilibrium. To understand what this means, you should now do Exercise 4.7, in which you are asked to prove the condition of thermal equilibrium ($T_1 = T_2$) in a constant entropy system, by developing a set of equations for the derivatives of internal energy that parallels equations (4.73) to (4.83). It is necessary to be extremely careful here, though. ***The equilibrium state defined by equations (4.92) and (4.93) is not the same as the one defined by equations (4.71) and (4.72).*** It cannot possibly be, as in the first case we are dealing with an isolated system and in the second case we are not. If entropy were kept constant in an isolated system then the system would not change and the minimum energy principle would be meaningless, because internal energy would not be able to change. ***In order for constant entropy minimization of the internal energy to be possible, the system of interest must be part of a larger isolated system in which entropy does increase***. Consider again the example of the two bodies at different temperatures, $T_2 > T_1$, shown now in Fig. 4.10. If we want to achieve thermal equilibrium at constant internal energy (case I) then the two bodies must make up an isolated system, A in the figure. This is the situation considered in Section 4.8.1. If, on the other hand, we want to achieve thermal equilibrium at constant entropy, which is the situation considered in this section (case II), then system A cannot be isolated. It must be able to exchange energy with its surroundings, which conform an isolated system, labeled B in the figure, *and the entropy of B must increase.* In order for the entropy of B to increase there must be heat transfer between A and the rest of B. Because equilibrium of A at constant entropy requires that its internal energy be minimized, heat must be transferred from A to its surroundings. Therefore, the final temperature of A if it is allowed to reach equilibrium at constant entropy must be lower than its final temperature if it reaches equilibrium at constant internal energy.

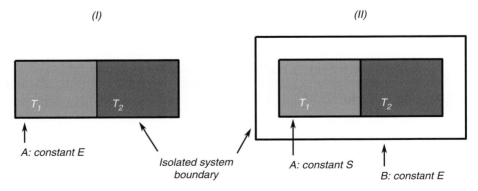

Fig. 4.10 (I) Thermal equilibrium at constant energy maximizes entropy in isolated system A. (II) Thermal equilibrium at constant entropy minimizes internal energy in isentropic but non-adiabatic system A, and maximizes entropy in a larger isolated system B.

We reached the conclusion in the last sentence on the basis of physical arguments only. We will now derive it formally. Consider Fig. 4.10 again. Let the initial temperatures of the two bodies, $T_2 > T_1$, and their masses, m_1 and m_2, be the same in both thought experiments, I and II in the figure. We will call the final equilibrium temperatures reached at constant internal energy (case I) and constant entropy (case II) T_E and T_S, respectively. The constant energy condition is:

$$\Delta E = \Delta E_1 + \Delta E_2 = m_1 C_V (T_E - T_1) + m_2 C_V (T_E - T_2) = 0 \qquad (4.94)$$

from which we get:

$$T_E = \frac{m_1 T_1 + m_2 T_2}{m_1 + m_2}. \qquad (4.95)$$

From the constant entropy condition, and assuming that heat capacities are constant:

$$\Delta S = \Delta S_1 + \Delta S_2 = m_1 C_V \int_{T_1}^{T_s} \frac{dT}{T} + m_2 C_V \int_{T_2}^{T_s} \frac{dT}{T} = 0 \qquad (4.96)$$

we get:

$$\ln T_S = \frac{m_1 \ln T_1 + m_2 \ln T_2}{m_1 + m_2}. \qquad (4.97)$$

We can now make $m_2 = km_1$, with $k > 0$, and $T_2 = hT_1$, with $h > 1$. These two conditions cover all possible situations: T_2 is always greater than T_1, but either of the two bodies can be larger, and by any factor that we wish. Substituting into (4.95) and (4.97) we get:

$$T_E = \frac{T_1 (1 + kh)}{1 + k} \qquad (4.98)$$

and:

$$T_S = T_1 h^{\frac{k}{1+k}}. \qquad (4.99)$$

Take the ratio T_S / T_E:

$$R = \frac{T_S}{T_E} = \frac{(1+k)h^{\frac{k}{1+k}}}{1 + kh}. \qquad (4.100)$$

For $k > 0$ and $h > 1$, we find that $R < 1$, always. This is shown in Fig. 4.11, that shows that the equilibrium temperature at constant entropy (case II) is always lower than the equilibrium temperature at constant internal energy (case I). If one of the bodies becomes much larger than the other one ($k \to 0$ or $k \to \infty$) then R approaches 1, which is also what one should expect. The effect of varying initial temperature contrast (parametrized by h) is, however, very asymmetric.

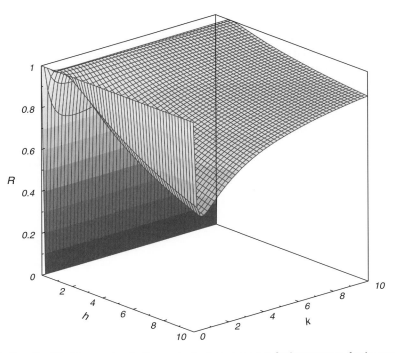

Fig. 4.11   The $R = T_S/T_E$, ratio of final temperature for the constant entropy process to final temperature for the constant energy process, for the systems in Fig. 4.10, as a function of the ratio between the masses of the (h) and the ratio between their initial temperatures (k).

4.8.3 Internal energy as a thermodynamic potential

Internal energy is one of several *thermodynamic potentials*. The name comes from analogy with potentials in mechanics. Thermodynamic equilibrium is attained when a thermodynamic potential is minimized subject to specific constraints. In the case of E these are: constant entropy, volume and chemical composition, and the release of some internal restrictions of the system, for example, removal of a partition between different gases or solutions, or of an insulating wall between bodies at different temperatures. Mechanical equilibrium is attained when a mechanical potential (= potential energy per unit mass, Section 1.3.1) is minimized. For example, a mountain range is not in mechanical equilibrium because it has potential energy relative to the geoid (equipotential surface), which sets the minimum potential level. When restrictions are released, for example by weathering, erosion drives the surface of the planet towards mechanical equilibrium by lowering its gravitational potential. Equilibrium would be attained if the surface of the planet were to become everywhere identical to its equipotential surface. This state of mechanical equilibrium is easily achieved in fluid planets but not in solid bodies.

When we discussed the entropy maximum and energy minimum principles we explicitly stated that they applied to systems that are closed to changes in chemical composition, which we specified with the subscript n_i, meaning that the mol number of component i remains constant, for all is. When studying natural systems we have to make allowance for the possibility that changes in chemical composition may occur. In order to do this we add a term to equation (4.12) that accounts for changes in internal energy that arise from possible changes in chemical composition. The equation becomes:

$$dE = TdS - PdV + \sum_i \mu^i dn_i. \tag{4.101}$$

The extensive variable n_i is the mol number (= number of mols) of chemical component i, and μ^i is an intensive variable called the *chemical potential* of i. The definition of chemical potential follows immediately from (4.101):

$$\mu^i = \left(\frac{\partial E}{\partial n_i} \right)_{S,V,n_{j \neq i}}. \tag{4.102}$$

The chemical potential of component i in a thermodynamic system is the first derivative of the internal energy of the system relative to the mol number of i, taken while keeping S, V and the mole numbers of all other components, $j \neq i$, constant. If you prefer physical terms, the chemical potential of a chemical species i keeps track of how the internal energy of the system varies when the amount of i varies by an infinitesimal amount, everything else being held constant. Chemical potential will play a central role in much of the remainder of this book. We shall soon see that there are alternative definitions of μ^i that are better suited to solving problems in chemical equilibrium. It must be absolutely clear, however, that no matter how one defines it, the chemical potential is always one and the same variable and, everything else being the same, it has a single and well-defined value. It is no different from the other two intensive variables in equation (4.102), temperature and pressure: no matter how one chooses to define or measure them, their physical meaning and magnitude are the same.

Equation (4.101) is called a *fundamental equation*, and the variables S, V and n_i, are called the *natural variables* for the thermodynamic potential E. There is more to these definitions than just attaching labels to things. In the first place, the natural variables are the ones that are held constant in order to specify the type of system or process for which the thermodynamic potential E takes on a minimum value at equilibrium. Second, the derivatives of E relative to its natural variables are physically significant properties. One of these properties is the chemical potential, defined by (4.102). Others are:

$$\left(\frac{\partial E}{\partial S} \right)_{V,n_i} = T; \quad \left(\frac{\partial^2 E}{\partial S^2} \right)_{V,n_i} = \left(\frac{\partial T}{\partial S} \right)_{V,n_i} = \frac{T}{C_V} \tag{4.103}$$

$$\left(\frac{\partial E}{\partial V} \right)_{S,n_i} = -P; \quad \left(\frac{\partial^2 E}{\partial V^2} \right)_{S,n_i} = -\left(\frac{\partial P}{\partial V} \right)_{S,n_i} = -\frac{1}{V\beta_S}. \tag{4.104}$$

If we know the fundamental equation for a system, $E = E(S, V, n_i)$ then we can know its thermodynamic state by using (4.102), (4.103) and (4.104) to calculate its pressure and its temperature, in addition to the chemical potentials of its chemical components. In the planetary sciences this result is much more than a clever-sounding mathematical gimmick. In general, we cannot measure intensive variables such as pressure and temperature in deep planetary interiors by dropping instruments into holes. The fundamental equation for the thermodynamic potential suggests that there may be alternative ways of doing this.

4.8.4 Conjugate variables, extensions and transformations of the thermodynamic potential

Each of the pairs of variables that appear in the fundamental equation (4.101): P–V, T–S, μ–n, are called *conjugate variables* and have some important regularities. First, and obviously, their product in every case has dimension of energy. Second, each conjugate pair consists of an intensive variable (P, T, μ) and an extensive variable (V, S, n). Third, in every case the intensive variable is a "driving potential" for "displacement" of the extensive variable: a pressure gradient causes a change in volume, a temperature gradient causes heat flow and hence a change of entropys and a gradient in chemical potential causes diffusion of chemical components and hence a change in mol numbers. Each of the three products of conjugate variables tracks a separate contribution to the thermodynamic potential: expansion work, heat transfer and mass transfer. We now see that adding the compositional term $\sum(\mu\,dn)$ to (4.12) in order to obtain (4.101) consisted of adding the product of a pair of conjugate variables that accounts for the energetic contribution of chemical composition to the thermodynamic potential E. There is no reason why we should stop here. Suppose that we are interested in a system in which mechanical work is also performed. We then need to add the product of an intensive variable that measures a "driving potential", in this case a force F, times its extensive conjugate, which is the distance over which the force acts, dx. The fundamental equation for E then becomes:

$$dE = T\,dS - P\,dV + \sum_i \mu^i\,dn_i + F\,dx. \qquad (4.105)$$

Other terms could be added if we were also interested in the work of, for example, gravitational forces, elastic forces, Lorentz forces or nuclear forces. Once we identify all of the energetic contributions to the thermodynamic potential we are free to add or remove products of conjugate variables as needed.

Each of the terms in equations (4.101) or (4.105) is the product of an intensive variable times the differential of its extensive conjugate. In other words, if we restrict ourselves to equation (4.101) then internal energy is a function of the form $E = E\,(S, V, n_i)$. This raises the question, is this the only way of writing a thermodynamic potential? Could we define a thermodynamic potential in which the natural variables are intensive variables, so that the minimum energy condition can be applied to a system in which intensive variables, such as P and T, are held constant? On dimensional grounds all we can expect is that a thermodynamic potential must be a sum of products of conjugate variables. In each of the products, however, either of the two variables could be the one that appears in differential form. Consider some other function, say $Y = Y(T, V, n_i)$, or any other combination of variables that are not conjugate among themselves. Do such functions exist? And do they have the properties of thermodynamic potentials?

4.8.5 The Legendre transform

The answer to the questions in the last paragraph is yes, but we have to be careful with how we construct these other functions. In order to ensure that any new function that we construct is a thermodynamic potential we must begin with a thermodynamic potential, such as $E = E\,(S, V, n_i)$, and transform this function in such a way that all of its properties are inherited by the new function. This is accomplished by means of a mathematical procedure called the *Legendre transform*. The Legendre transform is in essence a change of coordinates. We choose any (or all) of the coordinates of the original function (its natural variables) that are not convenient for our purposes, and find a new function that contains the same information in terms of the conjugates of the original coordinates, which become the natural variables of the new function.

Suppose you have a function $\phi(x)$, and that its natural variable x is a physical quantity that is hard to measure and/or that is difficult to control. The first derivative of ϕ: $d\phi/dx$, is, however, a physical quantity that is easy to handle. By this I mean that it is easy to come up with natural situations in which $d\phi/dx$ stays constant, and also that $d\phi/dx$ is easy to measure. Call this quantity $d\phi/dx = y$. It would be useful if we could construct a new function, $\psi(y)$, that contains the same information as ϕ, but that is a function of y rather than x. An example is internal energy. Entropy is one of the natural variables of E, but entropy is a physical quantity that cannot be measured directly. We thus have the easy-to-define but hard-to-work-with function $E(S)$. Its first derivative, $(\partial E/\partial S)_V = T$ (equation (4.103)) is, however, a physical quantity that is easy to measure and is intuitively easy to grasp. We seek another function that contains the same thermodynamic information as E, but that is written in terms of T as one of its natural variables, instead of its conjugate S.

The Legendre transform allows us to do this for certain types of function. The concept of the Legendre transform is best approached geometrically (Fig. 4.12). The key idea is

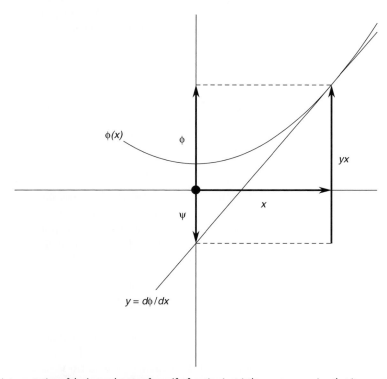

Fig. 4.12 Geometric interpretation of the Legendre transform. If a function is strictly convex, meaning that its second derivative does not change sign such as $\phi(x)$ in the figure, then the function can also be described by the set of all of its tangent lines. The value of the primitive function is given by $\phi = yx + \psi$, where y is the slope of the tangent line at x. Note that $\psi(y)$ is the Legendre transform of $\phi(x)$, given by $\psi(y) = \phi(x) - yx$.

that the set of all tangent lines to the curve $\phi(x)$ describes the function $\phi(x)$ as well as the function itself. This only works if there is a one-to-one relationship between x and y or, in other words, if each value of the slope y occurs at one and only one value of x. This is the case for the function shown in the figure. It means that x is a function of y, i.e. $x = x(y)$. The converse is necessarily true if ϕ is a function of x. A function for which there is a one-to-one correspondence between the independent variable, x, and the slope, y, is called *strictly convex* and is characterized by the fact that its second derivative never changes sign. Equations (4.103) and (4.104) show that this is the case for the natural variables of internal energy, so the Legendre transform can be applied to E.

From Fig. 4.12 it is clear that we can write ψ as follows:

$$\psi = \phi - yx. \tag{4.106}$$

So as to keep track of the fact that the function that we seek is a function of y, we write (4.106) more completely as follows:

$$\psi(y) = \phi(x(y)) - yx(y), \tag{4.107}$$

where the convexity of ϕ guarantees that the function $x(y)$ exists. Equation (4.107) is the Legendre transform of ϕ: it transforms the function $\phi(x)$ into the function $\psi(y)$, that contains the same information as ϕ, but now in terms of our preferred variable y. In particular, *the Legendre transform switches the roles of the conjugate variables in thermodynamic functions*. From the definition of $y = d\phi/dx$ we have:

$$d\phi = ydx. \tag{4.108}$$

Differentiating ψ relative to y in equation (4.106):

$$\frac{d\psi}{dy} = \frac{d\phi}{dx}\frac{dx}{dy} - y\frac{dx}{dy} - x = y\frac{dx}{dy} - y\frac{dx}{dy} - x = -x \tag{4.109}$$

or:

$$d\psi = -xdy. \tag{4.110}$$

If ϕ is a function which has x as one of its natural variables, the ψ is another function for which y, which is the conjugate of x, is a natural variable.

We need to demonstrate that if ϕ is a thermodynamic potential, then ψ is also a thermodynamic potential. First we note that the Legendre transform conserves the convexity of the function, as we can easily see from:

$$\frac{d^2\psi}{dy^2} = \frac{d}{dy}\left(\frac{d\psi}{dy}\right) = -\frac{dx}{dy} = -\frac{1}{\frac{d^2\phi}{dx^2}}. \tag{4.111}$$

This says that, if the second derivative of ϕ does not change sign (ϕ is strictly convex), then the second derivative of ψ does not change sign either, and it is strictly convex too. From here it is simple to show that the Legendre transform is invertible, and that the Legendre transform of ψ is ϕ (exercise left to the reader).

Suppose now that ϕ is a thermodynamic potential. This means that in a system in which its natural variables are held constant ϕ takes a minimum value at equilibrium. As usual we track the behavior of the systems relative to an unspecified variable Z that can vary when some restriction is released. Recall that Z is not one of the natural variables of ϕ, but rather a quantity that can vary while the natural variables of ϕ are held constant. For example, this could be the progress of a chemical reaction towards equilibrium measured by the amount of matter that reacts, or the progress of a system towards thermal equilibrium measured by the amount of heat transferred. It is not necessary to specify what the variable Z actually is. We write the equilibrium condition for the thermodynamic potential ϕ as follows (see equations (4.84), (4.85) and (4.92), (4.93)):

$$\left(\frac{\partial \phi}{\partial Z}\right)_x = 0 \tag{4.112}$$

$$\left(\frac{\partial^2 \phi}{\partial Z^2}\right)_x > 0, \tag{4.113}$$

where for simplicity we consider only one natural variable, x. From equation (4.106) we find the derivative of ψ at constant value of its natural variable, y:

$$\left(\frac{\partial \psi}{\partial Z}\right)_y = \left(\frac{\partial \phi}{\partial Z}\right)_y - y\left(\frac{\partial x}{\partial Z}\right)_y = \left(\frac{\partial \phi}{\partial Z}\right)_y. \tag{4.114}$$

The last identity in (4.114) comes from the fact that the following must always be true:

$$\left(\frac{\partial y}{\partial Z}\right)_x = \left(\frac{\partial x}{\partial Z}\right)_y = 0 \tag{4.115}$$

given that, as $x = x(y)$, if y is constant then x is constant, and conversely if x is constant then y is constant. Using identities (1.3.12) and (4.115):

$$\left(\frac{\partial \phi}{\partial Z}\right)_y = \left(\frac{\partial \phi}{\partial Z}\right)_x + \left(\frac{\partial \phi}{\partial x}\right)_Z\left(\frac{\partial x}{\partial Z}\right)_y = \left(\frac{\partial \phi}{\partial Z}\right)_x, \tag{4.116}$$

which substituting in (4.114) shows that:

$$\left(\frac{\partial \psi}{\partial Z}\right)_y = \left(\frac{\partial \phi}{\partial Z}\right)_x. \tag{4.117}$$

This is true in general, and therefore also at an extremum at which (4.112) is valid. Therefore, an extremum of the function ϕ is also an extremum of the function ψ. In order to prove that ψ is a thermodynamic potential all we need to do is prove that this extremum is a minimum. We take the second derivative of (4.106):

$$\left(\frac{\partial^2 \psi}{\partial Z^2}\right)_y = \left(\frac{\partial^2 \phi}{\partial Z^2}\right)_y - y\left(\frac{\partial^2 x}{\partial Z^2}\right)_y = \left(\frac{\partial^2 \phi}{\partial Z^2}\right)_y, \tag{4.118}$$

where the second identity follows from (4.115): derivatives *of any order* of y at constant x, and of x at constant y, vanish. Switching the order of differentiation and applying identity

(4.116) repeatedly:

$$\left(\frac{\partial^2 \phi}{\partial Z^2}\right)_y = \frac{\partial}{\partial Z}\left[\left(\frac{\partial \phi}{\partial Z}\right)_y\right]_y = \frac{\partial}{\partial Z}\left[\left(\frac{\partial \phi}{\partial Z}\right)_x\right]_y$$

$$= \frac{\partial}{\partial Z}\left[\left(\frac{\partial \phi}{\partial Z}\right)_y\right]_x = \frac{\partial}{\partial Z}\left[\left(\frac{\partial \phi}{\partial Z}\right)_x\right]_x \tag{4.119}$$

$$= \left(\frac{\partial^2 \phi}{\partial Z^2}\right)_x$$

from which we conclude that:

$$\left(\frac{\partial^2 \psi}{\partial Z^2}\right)_y = \left(\frac{\partial^2 \phi}{\partial Z^2}\right)_x \tag{4.120}$$

so, from (4.113), a minimum for ϕ is also a minimum for ψ. This completes the proof that the Legendre transform constructs a new function that inherits its properties from the primitive function. *The Legendre transform of a thermodynamic potential is another thermodynamic potential in which the roles of the conjugate variables have been switched*. The advantage of deriving this relationship in terms of a generic variable with the properties of a thermodynamic potential is that the results are valid for any and all of the natural variables of any thermodynamic potential. For functions of several variables, such as $E = E(S,V,n_i)$, we can transform as many or as few of the natural variables as we wish. Relations (4.117) and (4.120) are true for every variable that we transform.

4.8.6 The four fundamental thermodynamic potentials

By means of the Legendre transform we can construct three other thermodynamic potentials starting from E, by transforming each of its natural variables, S and V, separately, or both of them simultaneously. Let H be the potential in which pressure replaces its conjugate, volume, as a natural variable. By (4.107):

$$H(S,P,n_i) = E(S,V(P),n_i) - V(P)\left(\frac{\partial E}{\partial V}\right)_{S,n_i} \tag{4.121}$$

or, more compactly and using (4.104):

$$H(S,P,n_i) = E(S,V,n_i) + PV, \tag{4.122}$$

which is identical to (1.59). Enthalpy is the thermodynamic potential which has entropy and pressure as its natural variables. Using the notation of thermodynamics and substituting (4.101) (compare equation (4.13)):

$$dH = dE + PdV + VdP = TdS + VdP + \sum_i \mu^i dn_i. \tag{4.123}$$

Enthalpy is the thermodynamic potential that allows us to define equilibrium at constant pressure and entropy.

In the study of equations of state (Chapters 8 and 9) it is convenient to work with a thermodynamic potential that is a function of temperature and volume. This function is called the *Helmholtz free energy* and is symbolized by $F = F(T, V, n_i)$. Some authors use A for this potential but I reserve A for a different function called the affinity (Chapter 12). The Helmholtz free energy is the Legendre transform of internal energy relative to entropy, i.e.:

$$F(T, V, n_i) = E(S(T), V, n_i) - S(T) \left(\frac{\partial E}{\partial S} \right)_{V, n_i} \tag{4.124}$$

or, using (4.103):

$$F(T, V, n_i) = E(S, V, n_i) - TS, \tag{4.125}$$

which results in:

$$dF = dE - TdS - SdT = -SdT - PdV + \sum_i \mu^i dn_i. \tag{4.126}$$

A function that is central to much of chemical thermodynamics is the *Gibbs free energy*, $G = G(T, P, n_i)$ which is the double Legendre transform of internal energy relative to entropy and volume:

$$G(T, P, n_i) = E(S, V, n_i) - S \left(\frac{\partial E}{\partial S} \right)_{V, n_i} - V \left(\frac{\partial E}{\partial V} \right)_{S, n_i}. \tag{4.127}$$

Using (4.103), (4.104), (4.125) and (1.59) we find the following identities:

$$G = E - TS + PV$$
$$G = F + PV \tag{4.128}$$
$$G = H - TS.$$

In thermodynamics notation the total differential of Gibbs free energy in terms of its natural variables is:

$$dG = -SdT + VdP + \sum_i \mu^i dn_i. \tag{4.129}$$

Other thermodynamic potentials are possible, by transforming any of the other conjugate variables in (4.105), or any additional ones that we may need to add for any specific problem. In particular, potentials in which mol numbers n_i are switched with their conjugates, chemical potentials μ^i, are known as *grand thermodynamic potentials*.

4.9 Gibbs free energy

Planetary bodies act as large reservoirs of thermal energy and mass that tend to buffer P and T while chemical reactions take place, so that it is commonly a good approximation to assume that natural chemical reactions occur at constant temperature and pressure. The thermodynamic potential appropriate to studying equilibrium under these conditions is therefore the Gibbs free energy, which has pressure and temperature as its natural variables (equations (4.127) and (4.129)). *Equilibrium in a system at constant pressure and constant temperature is defined by the minimum value of its Gibbs free energy.* In the differential notation of thermodynamics we express this as follows:

$$dG_{T,P,n_i} = 0 \tag{4.130}$$

$$d^2G_{T,P,n_i} > 0. \tag{4.131}$$

A significant portion of the remainder of this book is focused on calculating Gibbs free energy for a wide range of planetary materials over a wide range of pressures and temperatures. Minimization of Gibbs free energy is the fundamental mathematical tool used to study chemical equilibrium. Here we focus on some general properties of the Gibbs free energy function.

4.9.1 Derivatives of the Gibbs free energy

The first and second derivatives of G relative to its natural variables, P, T and n_i, show up repeatedly in chemical equilibrium calculations, so we review them here. From (4.129) we have the following first derivatives:

$$\left(\frac{\partial G}{\partial T}\right)_{P,n_i} = -S \tag{4.132}$$

$$\left(\frac{\partial G}{\partial P}\right)_{T,n_i} = V \tag{4.133}$$

$$\left(\frac{\partial G}{\partial n_i}\right)_{T,P,n_{j\neq i}} = \mu^i. \tag{4.134}$$

The chemical potential of component i is, thus, the first derivative of G relative to n_i, taken at constant T, P and mol numbers of all the other system components, $n_{j\neq i}$. This definition, which is the one that we will most commonly use in this book, is different from (4.102), but *the chemical potential is the same as the one defined there, and has the same value.* The second derivative relative to T is:

$$\left(\frac{\partial^2 G}{\partial T^2}\right)_{P,n_i} = -\left(\frac{\partial S}{\partial T}\right)_{P,n_i} = -\frac{C_P}{T}, \tag{4.135}$$

which relies on the definitions of entropy (equation (4.6)) and heat capacity at constant pressure (equation (1.53)). We also have:

$$\left(\frac{\partial^2 G}{\partial P^2}\right)_{T,n_i} = \left(\frac{\partial V}{\partial P}\right)_{T,n_i} = -V\beta_T, \tag{4.136}$$

where we have used the definition of isothermal compressibility, equation (1.65). Using the definition of coefficient of thermal expansion, equation (1.66), the mixed second derivative yields:

$$\left(\frac{\partial^2 G}{\partial P \partial T}\right)_{n_i} = \left[\frac{\partial}{\partial T}\left(\frac{\partial G}{\partial P}\right)_{T,n_i}\right]_{P,n_i} = \left(\frac{\partial V}{\partial T}\right)_{P,n_i} = V\alpha. \tag{4.137}$$

The three material properties that I defined in section 1.11, C_P, β_T and α, are simple functions of the three second derivatives of Gibbs free energy. Other material properties such as heat capacity at constant volume, adiabatic compressibility and the Grüneisen parameter are similarly related to the second derivatives of other thermodynamic potentials (see Exercise 4.9).

A fundamental result of multivariable calculus says that, if a function complies with certain requirements about continuity and differentiability (and thermodynamic potentials do), then the order of differentiation in mixed second- and higher-order derivatives does not matter. This means that we could have also written (4.137) as follows:

$$\left(\frac{\partial^2 G}{\partial P \partial T}\right)_{n_i} = \left[\frac{\partial}{\partial P}\left(\frac{\partial G}{\partial T}\right)_{P,n_i}\right]_{T,n_i} = -\left(\frac{\partial S}{\partial P}\right)_{T,n_i}, \tag{4.138}$$

which results in:

$$\left(\frac{\partial S}{\partial P}\right)_{T,n_i} = -\left(\frac{\partial V}{\partial T}\right)_{P,n_i} = -V\alpha. \tag{4.139}$$

Equation (4.139) is one example of a large number of analogous relationships, called *Maxwell relations*, that can be obtained from the identity of mixed second derivatives of any thermodynamic potential. These relations are sometimes tabulated, but I don't see the point of that, as they are very easy to derive starting from the definition of the thermodynamic potential appropriate to the problem at hand (see Exercise 4.10). Maxwell's relations are often useful when solving problems in thermodynamics (e.g. Chapters 8, 9 and 10). They are also used to derive thermodynamic identities as discussed in Appendix 2.

Worked Example 4.6   Isentropes and adiabats revisited

An example of the application of the derivatives of thermodynamic potentials is to find the equation that describes P and T along an isentropic transformation. Consider an isentropic transformation in a system in which the only intensive variables are P and T. This could be, for instance, compression or expansion of a convecting fluid in a planetary mantle, ocean or atmosphere. For a constant entropy transformation we have:

$$dS = \left(\frac{\partial S}{\partial T}\right)_P dT + \left(\frac{\partial S}{\partial P}\right)_T dP = 0. \tag{4.140}$$

Substituting (4.135) and (4.139) in (4.140):

$$\frac{C_P}{T}dT - V\alpha dP = 0 \tag{4.141}$$

or:

$$\left(\frac{\partial T}{\partial P}\right)_S = \frac{\alpha V T}{C_P}, \tag{4.142}$$

which is the equation for the adiabat, (3.32), that we derived in Chapter 3. We have now rigorously demonstrated that this equation describes a process that is isentropic, hence the partial derivative notation in (4.142). As we saw in Section 4.4, a transformation can be adiabatic but not isentropic. In such a case the entropy of the system **must** increase during the adiabatic transformation, and the relationship between P and T given by equation (4.142) is no longer valid, because (4.140) is no longer valid. When we apply equation (3.32), or (4.142), to calculate temperature in a convecting fluid (e.g. a planetary mantle) we are *assuming* that the process is isentropic. Although this is commonly a good approximation, it may not always be the case. In Chapter 3 we examined, and discarded, heat diffusion as a possible source of entropy. We did not call heat diffusion an entropy-generating process, but now we known that that is what it is. There may be other sources of entropy, however. Examples include dissipation of gravitational potential energy by separation of phases of different densities, radioactive heating and viscous deformation. We return to this topic in Chapter 10.

4.9.2 The Gibbs free energy surface

Calculating and analyzing chemical equilibrium is in essence an exercise in computing and comparing Gibbs free energy surfaces, and seeking the conditions under which different surfaces intersect one another. We will perform these tasks algebraically rather than geometrically (Chapter 5), but having a mental image of what the Gibbs free energy surfaces look like is always a powerful aid in understanding what one is doing. The Gibbs free energy of a system of constant composition at equilibrium varies as a function of pressure and temperature as described by equations (4.132), (4.133) and (4.135) to (4.137). The function $G = G(P, T)$ for a system at equilibrium is therefore always a surface with the geometrical properties sketched in Fig. 4.13. The entropy of a system is always a positive

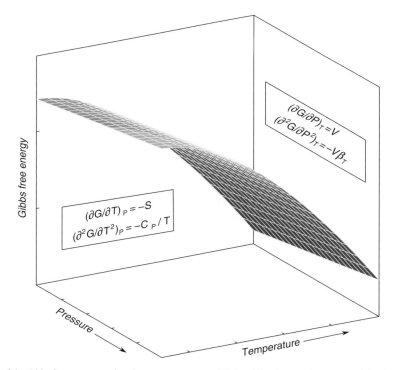

$$(\partial G/\partial P)_T = V$$
$$(\partial^2 G/\partial P^2)_T = -V\beta_T$$

$$(\partial G/\partial T)_P = -S$$
$$(\partial^2 G/\partial T^2)_P = -C_P/T$$

Fig. 4.13 Geometry of the Gibbs free energy surface for any system at equilibrium. The slope and curvature of the G surface on the T and P sections are given by the first and second derivatives of G, as shown in the figure, and always have the same signs.

quantity. (Careful! Do not confuse this statement with the fact that the *entropy change* for a process can be negative.) Therefore, the first temperature derivative of G is always negative, whereas the first pressure derivative (V) is always positive, as shown in the figure. The corresponding second derivatives are always negative (equations (4.135) and (4.136)), so that the Gibbs free energy surface has negative curvature, as shown in Fig. 4.13. The magnitudes of the slopes and curvatures may vary greatly, depending on the nature of the system. For example, the curvature of the G surface relative to pressure at constant temperature (equation (4.136)) is greater for gases than for solids and liquids, reflecting the greater compressibility of gases. The first derivative relative to temperature at constant pressure is also commonly greater for gases than for other substances, because gases have higher entropy. The geometric properties of the G surface, however, are always as shown in Fig. 4.13.

Exercises for Chapter 4

4.1 Show that isothermal expansion is accompanied by an increase in entropy.

4.2 Calculate the rate of entropy production by frictional heating during fault motion, Section 1.12.3. This is an example of a process that is adiabatic but not isentropic. Can you give other examples among planetary processes?

4.3 Formation of a mineral solid solution is accompanied by an increase in configurational entropy, which as we shall see in Chapter 5 is an important aspect in the calculation of equilibrium in natural systems. As a first example, derive a general expression for the configurational entropy of orthopyroxene solid solution along the binary join $MgSiO_3$–$FeSiO_3$, as a function of composition (i.e., X_{Mg} or X_{Fe}). What are the configurational entropies of pure enstatite and of pure ferrosilite? What orthopyroxene composition has maximum configurational entropy? How does configurational entropy change with temperature?

4.4 Repeat Exercise 4.3 for olivine solid solution, between the end-members forsterite: Mg_2SiO_4 and fayalite: Fe_2SiO_4. How does you expression compare to the one for orthopyroxene?

4.5 Consider a homogeneous mixture of two gases, 1 and 2, with mol fractions X_1 and X_2. Calculate the value of Ω for the mixed gas phase. Now consider the possibility that there is a small inhomogeneity in the composition of the gas between different parts of the container, such that the left hand side is richer in gas 1 and the right hand side is richer in gas 2. Call this difference δ. Call the number of microstates of the inhomogeneous gas mixture Ω'. Derive an equation for the ratio (Ω'/Ω) as a function of δ and of the total number of molecules of gas, N. What is the meaning of the ratio (Ω'/Ω)? What can you say about the magnitude of the largest random concentration fluctuations that are likely to appear spontaneously in a homogeneous gas phase? How does the magnitude of likely random concentration fluctuations relate to the size of the system (i.e., the value of N)?

4.6 The following is known as *Gibbs paradox*. Consider three different experiments based on the setup in Fig. 4.7. (i) Side A is filled with He at 1 bar and 298 K, assumed to behave as an ideal gas, and side B is empty. The partition is removed and He expands to fill the entire volume. Calculate the increase in entropy. (ii) Side B is filled with

He at 1 bar and 298 K, assumed to behave as an ideal gas, and side A is empty. The partition is removed and He expands to fill the entire volume. Calculate the increase in entropy. (iii) Both sides are filled with He at 1 bar and 298 K, assumed to behave as an ideal gas. The partition is removed and the gases are allowed to mix. Does entropy increase in this case? Why or why not? Resolve the paradox. The microscopic view is essential.

4.7 Prove that the minimum energy principle leads to the condition of thermal equilibrium (uniform temperature distribution) in a system evolving at constant entropy.

4.8 Modify the fundamental equation for Gibbs free energy, (4.129), to account for the effect of mechanical work on Gibbs free energy. Derive equations for the rate of change of Gibbs free energy of the Earth's mantle during (i) the isobaric cooling leg and (ii) the adiabatic decompression leg. Assume that the mantle is composed of pure forsterite and ignore phase transitions of magnesium silicates. Estimate the rate of change of Gibbs free energy of the mantle during the isobaric and adiabatic legs. Thermodynamic properties can be found in Holland and Powell (1998) – use values at 298 K and 1 bar. Assume that the mantle supports a shear stress of order 100 bar. Comment on the relative contributions of the various terms in the fundamental equation to the rate of change of Gibbs free energy, and reconsider Exercise 4.3 in the light of your results. Comment on the relative magnitudes and signs of the rate of change of Gibbs free energy during adiabatic decompression and isobaric cooling, and discuss how your results relate to the driving mechanism for mantle convection.

4.9 Find expressions for the heat capacities at constant volume and constant pressure, the isothermal and adiabatic bulk moduli, the isobaric and adiabatic coefficients of thermal expansion, and the Grüneisen parameter in terms of the second derivatives of the thermodynamic potentials E, H, F and G.

4.10 Derive the relationships among all the mixed second derivatives of the thermodynamic potentials E, H, F and G (Maxwell's relations). Simplify these relationships as much as possible using the expressions for material properties from Exercise 4.8.

5

Chemical equilibrium. Using composition as a thermodynamic variable

A comprehensive understanding of planetary bodies requires that we study how changes in physical conditions give rise to chemical phenomena. Physical conditions may be determined, for example, by the intensive variables P and T, where possible P–T combinations are in turn determined by the nature of heat sources and heat transfer mechanisms (Chapters 2 and 3). Chemical phenomena are transformations that entail redistribution of matter among and within phases. Some examples are: mineral transformations and melting in solid planets, changes in the relative amounts of molecular species that make up a gas or a supercritical fluid phase, and changes in the ionic constituents in an electrolyte solution such as seawater. The study of phenomena such as these is based on a mathematical description of chemical equilibrium, even in those cases in which departures from equilibrium cannot be ignored (Chapter 12). In this chapter we lay the foundations for the study of chemical equilibrium, including a comprehensive discussion of the use of composition as a thermodynamic variable. The principles and mathematical formalisms that we develop here are general, but important differences in implementation for different types of systems exist. These are dealt with in subsequent chapters.

5.1 Chemical equilibrium

5.1.1 Fundamental concepts

We begin by distinguishing between homogeneous and heterogenous systems. A homogeneous system consists of a single phase. Some examples are: gas in a container with nothing else in it, a planetary atmosphere with no clouds nor suspended particulate matter, or a mineral. A heterogeneous system is one in which we can identify more than one phase, for example, the contents of a liquefied gas cylinder (liquid + gas), a planetary atmosphere with clouds, or a polymineralic rock. The chemical composition of a system is specified by the relative amounts of a minimum number of independently variable chemical components. There are two requirements on this set of components, which are properly called *system components:* (i) the components must be linearly independent, and (ii) they must span the full compositional range of the system of interest. The system components may or may not correspond to actual chemical species present in the system (e.g. molecules, ions, etc.). For example, the composition of a homogeneous gas phase that contains the chemical species H_2O, H_2, CO_2, CO, CH_4 and NH_3 can be described in terms of the relative amounts of the four system components C, H, O and N. We shall define these concepts more rigorously in Chapter 6, but for now this intuitive introduction will suffice.

The state of chemical equilibrium in a system at constant pressure and temperature is the one in which the Gibbs free energy of the system takes its minimum possible value.

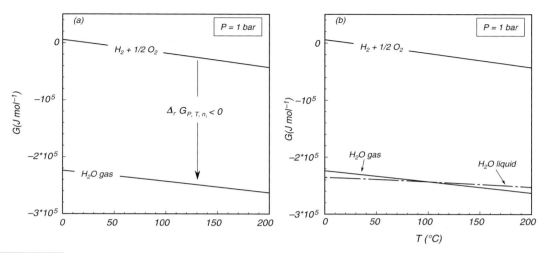

Isobaric sections across the Gibbs free energy surfaces for a 2:1 mixture of molecular hydrogen and oxygen, H_2O gas and H_2O liquid. The equilibrium state at any P and T is the one for which G takes its minimum value.

Consider a homogeneous system consisting of a gas composed of hydrogen and oxygen in an atomic ratio 2 H : 1 O. There are at least two possible states for this homogeneous system: a mixture of molecular H_2 and O_2 in a ratio of 2:1 or gaseous H_2O (other possibilities that we are not considering would be mixtures of atomic and/or ionic species). Which of the two is the equilibrium state? If we specify the temperature and pressure for which we seek an answer, then all we need to do is calculate the Gibbs free energy of each of the two possible states of the system, and determine which of the two is the smallest value.

Figure 5.1a shows Gibbs free energy for a 2:1 hydrogen–oxygen mixture, and for H_2O gas, as a function of temperature and at a constant pressure of 1 bar. The theoretical framework for these calculations is discussed in Box 5.1, and a calculation procedure using *Maple* is explained in Software Box 5.1. The figure shows that the Gibbs free energy of gaseous H_2O is everywhere lower, within this temperature range, than that of a physical mixture of hydrogen and oxygen with the same chemical composition. What this means is that the spontaneous process at constant temperature and pressure, i.e. the one that will minimize the thermodynamic potential G, is the combination of hydrogen and oxygen to form H_2O. The difference in Gibbs free energy that accompanies the chemical reaction is called the *Gibbs free energy of reaction*, $\Delta_r G$ (see Fig. 5.1a). For a spontaneous chemical reaction, it must be $\Delta_r G < 0$. Thus, H_2O does not spontaneously break up into $H_2 + O_2$ at these conditions because $\Delta_r G$ for that process is greater than zero. Water can break up into hydrogen and oxygen at room temperature, for example by electrolysis, but in that case electrical work is being performed on the system so the process is not "spontaneous".

Box 5.1   Calculation of Gibbs free energy

We seek an explicit expression for the Gibbs free energy of a chemical species at any arbitrary temperature and pressure. For now let us define a species as an entity (compound, pure element or ion) whose composition in the system of interest remains fixed. For example, O_2 and H_2O are two of the chemical species in the terrestrial atmosphere, H_2O and Na^+ are two of the chemical species in seawater, and forsterite and fayalite are chemical species in olivine.

Box 5.1   **Continued**

In this book I symbolize the standard state molar Gibbs free energy of a chemical species at P and T by $G_{P,T}^0$. Symbols for thermodynamic variables in all chemical equilibrium equations will be in uppercase non-bold font, representing molar quantities (Section 1.9). The superscript ^{0} means "standard state" and indicates that the value of G corresponds to the pure chemical species at those conditions. We will see that this is different from G of the chemical species in a solution. For example, the standard state molar Gibbs free energy of oxygen at 1 bar and 20°C, symbolized by $G_{1,298}^0$, is the molar Gibbs free energy of pure oxygen at those conditions, but it is *not* equal to the Gibbs free energy of oxygen in air at 1 bar and 20°C. Similarly, G^0 of forsterite is not the same as the Gibbs free energy of forsterite in an olivine solid solution at the same P and T.

The reference state for Gibbs free energy is 298.15 K and 1 bar, and we symbolize G^0 at these conditions by $G_{1,298}^0$. This notation is not standard, but in my experience it is the clearest one, because it states explicitly the conditions at which one is evaluating the thermodynamic function. Using (4.132) and (4.133), and recalling that S and V are functions of temperature and pressure, we see that the Gibbs free energy for the chemical species at any other P and T is given by:

$$G_{P,T}^0 = G_{1,298}^0 - \int_{298}^T S(P,T)\, dT + \int_1^P V(P,T)\, dP. \tag{5.1.1}$$

We need to find explicit expressions for the three terms in the right hand side of equation (5.1.1). The two integrals can be evaluated in any order, but the simplest, and standard, way is the following: evaluate the temperature integral along an isobaric path at 1 bar, from 298 K to T, and evaluate the pressure integral along an isothermal path at T, from 1 bar to P. Alternative ways of doing this, which lead to the same result, are proposed in end-of-chapter problems.

(i) *Gibbs free energy at the reference state.* The value of $G_{1,298}^0$ is listed in some thermodynamic data bases as the reference state Gibbs free energy of formation, $\Delta_f G_{1,298}^0$. This is defined as the Gibbs free energy of a chemical species measured relative to G of its constituent elements, which are arbitrarily set to zero for all pure elements in their stable configurations at 298.15 K and 1 bar. This is the same as the definition of enthalpy of formation (Section 1.13.1), but *recall that the entropy of pure elements at these conditions is not zero* (Section 4.7.2). Thus, whereas $\Delta_f H_{1,298}^0$ is a measured quantity (heat evolved at constant pressure), $\Delta_f G_{1,298}^0$ is not. It includes a contribution from the difference between the entropy of the compound and the (non-zero) entropies of the elements at 298 K. The reference state Gibbs free energy of formation is commonly (but not always!) defined as follows:

$$\Delta_f G_{1,298}^0 = \Delta_f H_{1,298}^0 - 298\Delta\left(S_{298}^0\right), \tag{5.1.2}$$

where $\Delta\left(S_{298}^0\right)$ is the difference between the Third Law entropy of the species of interest and those of its constituent elements in their stable configuration, taken at 298.15 K and 1 bar (equation (4.70)). All thermodynamic data bases list values of $\Delta_f H_{1,298}^0$ and S_{298}^0, so that it is always possible to calculate $\Delta_f G_{1,298}^0$ (which is the value of $G_{1,298}^0$ in equation (5.1.1)) with (5.1.2), if its value is not listed. We will see, however, that although the value of $G_{1,298}^0$ is needed in order to calculate the Gibbs free energy of a chemical species, it is not required when calculating the *change* in G associated with chemical reactions, called the Gibbs free energy of reaction, $\Delta_r G$.

Box 5.1   **Continued**

(ii) *The temperature integral*. We expand the temperature integral in equation (5.1.1) by using (4.135), as
follows:

$$\int_{298}^{T} S(1,T)\, dT = \int_{298}^{T} \left(S_{298}^0 + \int_{298}^{T} \frac{C_P}{T}\, dT \right) dT$$

$$= S_{298}^0 (T - 298) + \int_{298}^{T} \int_{298}^{T} \frac{C_P}{T}\, dT\, dT, \tag{5.1.3}$$

where S_{298}^0 is the Third Law entropy of the chemical species at 298 K (*not* the "entropy of formation", see
Section 4.7.2). The double integral is easily solved by parts. Defining:

$$u = \int_{298}^{T} \frac{C_P}{T}\, dT$$

$$v = T \tag{5.1.4}$$

we get:

$$T \int_{298}^{T} \frac{C_P}{T}\, dT = \int_{298}^{T} \int_{298}^{T} \frac{C_P}{T}\, dT\, dT + \int_{298}^{T} C_P\, dT. \tag{5.1.5}$$

Rearranging and substituting in (5.1.3):

$$\int_{298}^{T} S(1,T)\, dT = S_{298}^0 (T - 298) + T \int_{298}^{T} \frac{C_P}{T}\, dT - \int_{298}^{T} C_P\, dT. \tag{5.1.6}$$

In order to evaluate the heat capacity integrals one must substitute the appropriate heat capacity
equations, see Software Box 1.1. These equations are polynomials in T that are empirical fits to measured
C_P values and may not have a strong physical foundation, beyond being able to reproduce the weak
temperature dependence of heat capacity above the Debye temperature (Section 1.14.3 and Chapter 8).
The choice of $C_p(T)$ equation is generally dictated by the choice of data base. Throughout most of this
book we use thermodynamic data from Holland and Powell (1998), who rely on the Shomate heat capacity
equation. An important exception is the calculation of phase equilibria at very high pressures, for which a
different heat capacity equation will be used (Chapter 8).

It is useful to start collecting terms incrementally, as we find expressions for each of the components of
equation (5.1.1). Using (5.1.6) we can write an expression for the Gibbs free energy of a chemical species
at T and P as follows:

$$G_{P,T}^0 = \Delta_f G_{1,298}^0 - S_{298}^0 (T - 298) - T \int_{298}^{T} \frac{C_P}{T}\, dT + \int_{298}^{T} C_P dT + \int_{1}^{P} V(P,T)\, dP. \tag{5.1.7}$$

(iii) *The pressure integral*. Evaluation of the pressure integral requires that we substitute an explicit equation of
state for the material, which is a function $V = V(P,T)$. As we discussed in Chapter 1, condensed phases
(solids and liquids) and non-condensed phases (gases) respond very differently to changes in pressure.
Their equations of state are different enough that we need to consider each case separately.

(iii)(a) *Condensed phases*. The simplest approximation for condensed phases is to assume that their volume
does not change with pressure or temperature. The pressure integral in equation (5.1.7) is evaluated

Box 5.1   **Continued**

at constant temperature T. Let $V_{1,T}$ be the volume of the chemical species at 1 bar and T. If V is constant, then the pressure integral is simply:

$$\int_1^P V(P,T)dP = V_{1,T} \int_1^P dP = (P-1)V_{1,T} \approx PV_{1,T}, \qquad (5.1.8)$$

where for pressures of more than a few tens of bars we can assume that the lower limit of integration vanishes. Note that equation (5.1.8) is written in terms of the volume at the temperature of interest, T, but molar volumes of chemical species are typically tabulated at a reference temperature of 298 K. Thermal expansion is not insignificant, but, again, we can ignore it as a first approximation and calculate (5.1.8) by using the reference state volume, $V_{1,298}$. These approximations (incompressible phases that undergo no thermal expansion) are generally acceptable for near-surface conditions, but accurate calculation of Gibbs free energy requires that we account for changes in molar volume with pressure and temperature, as well as for the temperature dependence of the coefficient of thermal expansion and the temperature and pressure dependencies of the bulk modulus. We discuss this in Chapter 8.

(iii)(b) Gases. The simplest treatment for gases results from assuming ideal gas behavior. Using the ideal gas EOS, the pressure integral is simply:

$$\int_1^P V(P,T)dP = RT \int_1^P \frac{dP}{P} = RT \ln P. \qquad (5.1.9)$$

Real gases approach ideal gas behavior if their temperature is much higher than the critical temperature, and at very low pressure (Chapter 9). Equation (5.1.9) is applicable to surface environments in the terrestrial planets (except perhaps Venus). Equations of state for fluids in planetary interiors are discussed in Chapter 9.

Summary

We summarize our results so far. For pressure and temperature conditions characteristic of the surface and shallow crust of the large terrestrial planets (and perhaps much of the interior of small solid bodies), the Gibbs free energy of a species in a condensed phase can be approximated to first order by combining equations (5.1.7) and (5.1.8):

$$G_{P,T}^0 = \Delta_f G_{1,298}^0 - S_{298}^0 (T - 298) - T \int_{298}^T \frac{C_P}{T} dT + \int_{298}^T C_P dT + PV_{1,298}. \qquad (5.1.10)$$

For gases at P–T conditions that are far removed from their critical point (low pressure, high temperature) the pressure term is replaced with (5.1.9), and we obtain:

$$G_{P,T}^0 = \Delta_f G_{1,298}^0 - S_{298}^0 (T - 298) - T \int_{298}^T \frac{C_P}{T} dT + \int_{298}^T C_P dT + RT \ln P. \qquad (5.1.11)$$

Software Box 5.1   Calculation of Gibbs free energy of ideal gas and incompressible phase

The file `th_template_2.mw` contains the following two new *Maple* procedures:

`Gsurf_idealgas`: calculates the Gibbs free energy of an ideal gas, or of a mixture of ideal gases, with equation (5.1.11). Thermodynamic properties are entered in the spreadsheet `RefStateData`, and the gas or mixture of gases are specified in a table with two columns, as explained in Software Box 1.1. The procedure calls on procedures in the package `th_shomate.mw` to perform the heat capacity integrals, and then adds the pressure integral. It performs the calculations over a *P–T* range and with *P–T* increments that are specified in the procedure call. Output is sent to a text file whose name is specified in the procedure call.

`Gsurf_Vconst`: works as `Gsurf_idealgas` but assumes that the volume of the phase or assemblage is constant (pressure integral as in equation (5.1.10)).

The data for hydrogen, oxygen and water are stored in tab-delimited format in a file named `waterprops`.

There is, of course, more to this story. In the first place, we know that if the temperature is lower than $100\,°C$ at 1 bar, then the equilibrium condition is not H_2O gas but liquid H_2O. The reason why this does not show up in Fig. 5.1a is that I have not included the curve that represents the Gibbs free energy function for liquid H_2O. This situation is rectified in panel (b) of the figure. We can now compare Gibbs free energies for three different possible states of a system with the same chemical composition. At any given pressure and temperature, the equilibrium state, which we also call the *stable state*, is the one in which Gibbs free energy is lowest. At 1 bar pressure molecular H_2O is stable relative to $H_2 + O_2$ everywhere in the temperature interval 0–$200\,°C$. At $T < 100\,°C$ the Gibbs free energy of liquid water is lower than that of H_2O gas, so that the stable state is liquid water. The converse is true at $T > 100\,°C$. At $T = 100\,°C$ and $P = 1$ bar the Gibbs free energies of liquid and gaseous H_2O are the same, so that both phases are stable.

There is another aspect that we must consider. If we mix hydrogen and oxygen in a container at room temperature we know from Fig. 5.1a that the system is not at equilibrium and that a spontaneous reaction that lowers the system's Gibbs free energy by forming molecular H_2O should take place. But does this happen? Hydrogen-filled airships were built and flown in Germany for many years with no apparent inconveniences, until the *Hindenburg* disaster in 1937. The answer is that in order for the reaction to take place we need to "excite" the system, for example, by supplying heat in the form of an open flame or an electrical spark (which is what apparently doomed the *Hindenburg*). The effect of this excitation is to break the bonds in the H_2 and O_2 molecules, so that the atoms are able to recombine as H_2O molecules. This is the chemical equivalent of removing a restriction in a system, such as a wall separating two different gases or an insulating layer between bodies at different temperatures. We saw that, when such restrictions are removed, isolated systems evolve spontaneously in the direction mandated by the Second Law of Thermodynamics. In the case of a chemical reaction the restrictions are the chemical bonds in the molecules of the reactant species. We remove that restriction by supplying the amount of energy required to break those bonds. This energy is called the *activation energy* for the chemical reaction, which we will discuss in Chapter 12.

5.1.2 Gibbs free energy surfaces and phase boundaries

The curves in Fig. 5.1 are isobaric sections across Gibbs free energy surfaces such as that in Fig. 4.13. We can represent Gibbs free energy surfaces by means of contour lines of constant G projected on a P–T plane, as shown in Fig. 5.2. The solid curves are contours on the G surface for H_2O gas, and the dotted lines are contours on the G surface for liquid H_2O. The different shapes of the two sets of curves reflect the different effects of pressure on the Gibbs free energy of condensed and non-condensed phases, as discussed in Box 5.1 and in more detail in Chapters 8 and 9. The Gibbs free energy of a gas is a strong function of pressure, whereas G of a liquid or solid is much less sensitive to pressure. Within the P–T range of the figure, and for the contour interval chosen (2 kJ), there are four intersections between contour lines, marked with the solid circles. These are four points on the intersection between the two surfaces, which is shown by the thick solid curve. Along this curve the Gibbs free energy of the liquid and of the gas are the same, so that the two phases are in *heterogeneous chemical equilibrium*. The curve is a *phase boundary*. We can see that the thermodynamic condition for a phase boundary is simply:

$$\Delta_r G_{P,T,n_i} = 0, \tag{5.1}$$

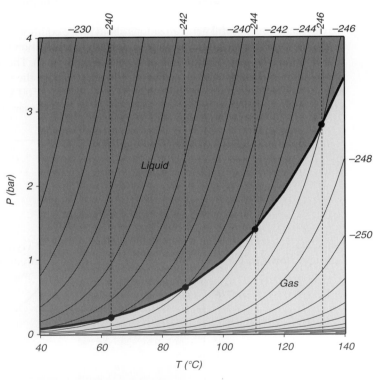

Fig. 5.2 Phase diagram for H_2O. Data from Robie et al. (1995) were used to calculate the Gibbs free energy surfaces for liquid and gaseous H_2O as explained in Software Box 5.1, assuming ideal gas behavior for H_2O gas. Solid curves are contours on the G surface for gas, dashed lines are contours on the G surface for liquid. Numbers on the right and top side of the diagram are Gibbs free energies in kJ mol^{-1}. Intersection between the two surfaces yields the phase boundary ($\Delta_r G = 0$), shown by the thick curve. Four intersection points between the contour lines are shown with big dots.

where the Gibbs free energy of reaction, $\Delta_r G$ is defined in the same way as $\Delta_r H$ (equation (1.86)). A phase boundary is the locus of points for which $\Delta_r G = 0$.

The diagram in Fig. 5.2 is an example of a *phase diagram*. It shows which is the stable phase for different combinations of intensive variables. For each combination of pressure and temperature that is not on the phase boundary there is one phase that has lower Gibbs free energy. Equilibrium requires that the phase with lower Gibbs free energy, called the *stable phase,* forms at the expense of the other one. In this example, two phases (liquid and gaseous H_2O) are stable (i.e. they are at equilibrium) along the phase boundary. On the high pressure - low temperature side of the phase boundary, shown in dark grey, $G_{liquid} < G_{gas}$, so the stable phase is liquid H_2O. On the other side of the phase boundary, shown in light grey, we have $G_{gas} < G_{liquid}$, so the stable phase is H_2O gas. Note that these relations refer to *molar* Gibbs free energy, i.e. an intensive property.

Phase diagrams can become more complex in systems composed of more chemical species than this one, and phase diagrams can be constructed so as to show phase equilibrium as a function of other intensive variables besides pressure and temperature. In every case, however, it is of great help to keep in mind that a phase diagram is simply a set of curves defined by intersections among Gibbs free energy surfaces.

5.1.3 Properties of phase boundaries and phase transitions

We can make some useful generalizations about phase equilibrium by examining the geometry of the G surfaces in Fig 5.2 along an isobaric cross section (e.g. at $P = 1$ bar, Fig. 5.3) and an isothermal cross section (e.g. at $T = 100\,°C$, Fig. 5.4). The intersections of the curves in these two diagrams represent one and the same point on the phase boundary in Fig. 5.2: equilibrium between liquid and gaseous H_2O at 1 bar and 100°C. This example deals with a phase boundary that separates two distinct phases with the same chemical composition. A

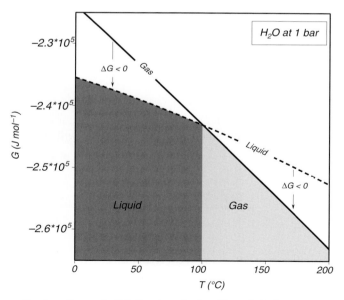

Fig. 5.3 Isobaric section across the phase diagram for H_2O. The phase that is stable on the high-temperature side of the phase transition must have a steeper G–T curve, so its entropy must be higher than that of the low-temperature phase.

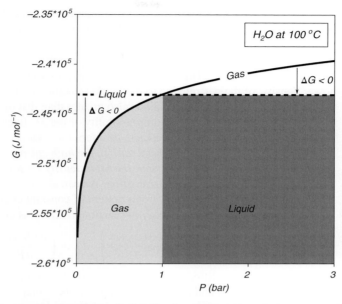

Fig. 5.4 Isothermal section across the phase diagram for H_2O. The phase that is stable on the low-pressure side of the phase transition must have a steeper G–P curve, so its molar volume must be higher (its density lower) than that of the high-pressure phase. The stronger curvature of the G surface for the gas reflects the higher compressibility of gases relative to condensed phases.

phase boundary in a one-component system is called a discontinuous, or first-order, *phase transition*. The Gibbs free energy is continuous at a discontinuous phase transition, as it must be given that this is the definition of equilibrium between the two phases. The first derivatives of G (S and V) are, however, discontinuous at the phase transition (this is where the name comes from). This also implies that there is a finite enthalpy change associated with first-order phase transitions (see Worked Example 1.1). Moreover, if the first derivatives of G are discontinuous, then the second derivatives, heat capacity and compressibility (equations (4.135) and (4.136)), are undefined at the phase transition. This means that as long as the two phases coexist at equilibrium the system water–steam can absorb indefinite amounts of thermal and mechanical energy without its temperature or pressure changing, although the relative amounts of the two phases will change.

The isobaric section (Fig. 5.3) shows that the slope of the G surface for H_2O gas is steeper than that for liquid H_2O. By equation (4.132), the interpretation of this is that the entropy of the gas is higher than the entropy of the liquid. This conclusion is general and expected on physical grounds. The entropy of a non-condensed phase must be higher than that of a condensed phase with the same chemical composition because transition from a condensed to a non-condensed state (boiling or sublimation) requires absorption of thermal energy in order to break intermolecular bonds, and heat absorption means an increase in entropy (equation (4.6)). The microscopic interpretation is that there is an increase in the number of accessible microstates for molecular vibrations.

From the geometry of the isobaric section it follows that, among condensed phases, the entropy of the solid must be lower than the entropy of the liquid, because the intersection that defines the melting point must occur at lower temperature than the one that defines the boiling point. This agrees with the fact that melting requires absorption of heat in order to break atomic bonds in the solid, so that the entropy of melting is always a positive quantity (Worked Example 4.1).

On the isothermal projection, Fig. 5.4, the slope of the G surface for the gas is also steeper than that for the liquid. This is the geometric expression of the higher molar volume, or lower density, of the gas relative to the liquid (equation (4.133)). The curvature of the gas surface on the isothermal projection is also greater than that of the liquid. From (4.136) we see that this reflects the greater compressibility of the gas (a consequence of the molecules being further apart and of the weakness of intermolecular forces).

We can make an important generalization: *the high entropy phase is on the high temperature side of a phase transition, and the high density (= low molar volume) phase is on the high pressure side of a phase transition.* We can also derive an algebraic expression for the slope of the phase transition. For a system of constant chemical composition, the total differential of the Gibbs free energy of reaction $d(\Delta_r G)$, is given by:

$$d(\Delta_r G) = \left(\frac{\partial(\Delta_r G)}{\partial T}\right)_P dT + \left(\frac{\partial(\Delta_r G)}{\partial P}\right)_T dP. \tag{5.2}$$

From the linearity of differentiation (the derivative of a sum is the sum of the derivatives) and equations (4.132) and (4.133) it is simple to show that:

$$\left(\frac{\partial(\Delta_r G)}{\partial T}\right)_P = -\Delta_r S \tag{5.3}$$

and:

$$\left(\frac{\partial(\Delta_r G)}{\partial P}\right)_T = \Delta_r V, \tag{5.4}$$

where $\Delta_r S$ and $\Delta_r V$, the entropy and volume of reaction, are the differences in entropy and volume across the phase transition. Substituting in (5.2) we get:

$$d(\Delta_r G) = -\Delta_r S\, dT + \Delta_r V dP. \tag{5.5}$$

Along a phase transition the Gibbs free energy of reaction is identically zero, i.e. $d(\Delta_r G) = 0$. It follows from (5.5) that the slope of the phase transition is given by:

$$\frac{dP}{dT} = \frac{\Delta_r S}{\Delta_r V}. \tag{5.6}$$

This equation is known as the Clapeyron equation, and the slope of a phase transition is often called the *Clapeyron slope*. It is important to keep in mind that both $\Delta_r S$ and $\Delta_r V$ are generally functions of P and T, so that the Clapeyron slope is not constant.

5.2 Equilibrium among pure chemical species

In the preceding section we used a one-component system to derive the thermodynamic condition of chemical equilibrium and some of the properties of phase transitions. All of these results generalize to equilibrium in multicomponent systems. In multicomponent systems more than one phase is stable on at least one of the sides of a phase boundary (Chapter 6). Properly speaking a phase boundary in a multicomponent system is not a phase transition, but the use of the phrase "phase transition" to refer to multicomponent phase boundaries is common in the planetary sciences, for instance, as in the "spinel–garnet phase transition" or the "perovksite phase transition". In any event, because Gibbs free energy is an extensive property (i.e. it is additive) phase boundaries between multi-phase assemblages in multicomponent systems have the same geometric properties as first order phase transitions. The goal of the rest of this chapter is to build the mathematical framework needed to calculate chemical equilibrium in systems composed of an arbitrary number of components and in which phase compositions may be variable. There are several ways of doing this. In the most general case, one starts with a list of chemical components and a set of possible phases, and seeks the subset of phases (and perhaps their compositions, if they can vary) that minimizes Gibbs free energy for every combination of intensive variables and for a constant bulk composition of the system. A way to visualize this procedure, and indeed to implement it, is to calculate the Gibbs free energy surface for each possible phase and then seek the combination of surfaces that minimizes G subject to the constant bulk composition constraint. Phase boundaries are then defined by intersections between these surfaces, in a manner analogous to what we did in Fig. 5.2. We examine this approach to phase equilibrium in the context of species distribution in homogeneous fluids (Chapters 9 and 14). In this chapter we follow a different route which, if not as general, is pedagogically clearer. We will assume that we know beforehand the phase boundary that we want to locate. The location of the phase boundary will then be found by solving for the set of intensive variable combinations that makes $\Delta_r G = 0$. In other words, we will calculate the intersection between Gibbs free energy surfaces directly, without calculating the surfaces themselves. This is procedurally simpler than finding Gibbs free energy minima, but requires more prior qualitative knowledge of the system of interest. Specifically, we need to know what is the exact phase assemblage or phase boundary that applies to our problem. In most instances this is either known from observations or can be inferred with reasonable certainty (Chapter 6).

We begin by considering chemical equilibrium among chemical species in their *standard state*. The precise meaning of standard state will become clear later on, but for now we note that a pure chemical species at a specified temperature and pressure and in a well-defined physical state is one possible definition of standard state. For example, at a given temperature and pressure, pure liquid H_2O, or pure solid H_2O, or pure H_2O gas are all possible (though different) standard states for the chemical species H_2O.

Consider a balanced chemical reaction involving an arbitrary number of chemical species, each of them in its standard state, where v_i is the stoichiometric coefficient of species i ($v_i < 0$ for reactants, $v_i > 0$ for products). Chemical equilibrium among the species occurs at all coordinate combinations for which $\Delta_r G = 0$. The coordinates can be any intensive variables, for example, pressure and temperature, but other choices of intensive variables are also possible. *Finding the equilibrium position consists simply of solving the equation $\Delta_r G = 0$ for the intensive variables of interest.* If the chemical reaction takes place among chemical species in their standard states, then we symbolize the Gibbs free energy change of the reaction at pressure P and temperature T by $\Delta_r G^0_{P,T}$, where the superscript ^{0} means standard state. For every thermodynamic variable Z, we define the difference operator for a balanced chemical reaction, Δ_r, as follows:

$$\Delta_r Z = \sum_i v_i Z^i, \tag{5.7}$$

where Z^i is the molar value of Z for species i. This is a generalization of equation (1.86). The equilibrium condition for a reaction among chemical species in their standard state is thus given by:

$$\Delta_r G^0_{P,T} = \sum_i v_i G^{0,i}_{P,T} = 0, \tag{5.8}$$

where $G^{0,i}_{P,T}$ is the standard state Gibbs free energy of species i at P and T.

The linearity of differentiation allows us to combine G and any of its derivatives (S, V, Cp, etc.) using (5.7). Therefore, using equation (5.1.7) we write the standard state Gibbs free energy change for a chemical reaction at P and T as follows:

$$\Delta_r G^0_{P,T} = \Delta_r \left(\Delta_f G^0_{1,298} \right) - \Delta_r S^0_{298} (T - 298)$$
$$- T \int_{298}^T \frac{\Delta_r C_P}{T} dT + \int_{298}^T \Delta_r C_P dT + \Delta_r \left(\int_1^P V(P,T) dP \right). \tag{5.9}$$

In this equation we have placed the difference operators for the heat capacity terms under the integral signs. This implies that heat capacities for all of the species can be combined linearly, which is possible if they are all given by the same polynomial function. This is almost always the case (see Box 5.1 and Software Box 1.1), but if it is not then the integral for each species must be performed separately. The pressure integral generally cannot be treated linearly, so equation (5.9) will not be made explicit in pressure for now.

From equation (5.1.2), the difference in Gibbs free energy of formation among chemical species is:

$$\Delta_r \left(\Delta_f G^0_{1,298} \right) = \Delta_r \left(\Delta_f H^0_{1,298} \right) - 298 \Delta_r \left[\Delta \left(S^0_{298} \right) \right]. \tag{5.10}$$

Now, $\Delta(S^0_{298})$, the "entropy of formation" of a species, is the difference between the Third Law entropy of the species and those of its constituent elements, given by equation (4.70). Chemical stoichiometry requires that the sum of chemical *elements* be the same for both sides of a chemical reaction (this is what balancing the reaction is all about!), so that Δ_r for the entropies of the constituent elements must always vanish. This means that:

$$\Delta_r \left[\Delta \left(S^0_{298} \right) \right] = \Delta_r S^0_{298}, \tag{5.11}$$

where $\Delta_r S^0_{298}$ is simply the difference in reference state Third Law entropies between products and reactants, as given by equation (5.7). Following equation (1.98) we also write:

$$\Delta_r \left(\Delta_f H^0_{1,298} \right) = \Delta_r H^0_{1,298}, \tag{5.12}$$

where $\Delta_r H^0_{1,298}$ is the difference in reference state enthalpies of formation between products and reactants. Substituting in (5.9) and simplifying we get the following equilibrium equation:

$$\Delta_r G^0_{P,T} = \Delta_r H^0_{1,298} - T \Delta_r S^0_{298}$$
$$- T \int_{298}^T \frac{\Delta_r C_P}{T} dT + \int_{298}^T \Delta_r C_P dT + \Delta_r \left(\int_1^P V(P,T) dP \right), \tag{5.13}$$

which we can also write as follows, for the sake of keeping track of where the various contributions to the Gibbs free energy of reaction come from:

$$\Delta_r G_{P,T}^0 = \Delta_r H_{1,T}^0 - T \Delta_r S_{1,T}^0 + \Delta_r \left(\int_1^P V(P,T) dP \right)$$

$$= \Delta_r G_{1,T}^0 + \Delta_r \left(\int_1^P V(P,T) dP \right). \tag{5.14}$$

Equation (5.13) (or (5.14)) is the starting point for all our calculations of chemical equilibrium. Setting $\Delta_r G_{P,T}^0 = 0$ (equation (5.8)) allows us to calculate heterogeneous equilibrium among pure phases of fixed composition. The set of intensive variable combinations that satisfy the equation is the phase boundary. This is what we did in Section 5.1. The application of equation (5.13) goes beyond calculating phase diagrams among species in their standard states, however. We will see that calculating heterogeneous equilibrium among phases of variable composition, and calculating homogeneous chemical equilibrium within a phase, are simple extensions of (5.13).

Worked Example 5.1   Calculation of a phase boundary: the spinel–garnet transition in planetary mantles, part (i)

The mineral assemblage of the Earth's upper mantle, and almost certainly of the upper mantles of the other terrestrial planets, consists of olivine, orthopyroxene, clinopyroxene and an aluminous phase. The aluminous phase changes with increasing pressure, from plagioclase to spinel to garnet. Although the modal abundance of the aluminous phase is generally subordinate to those of the other upper mantle phases, its identity exerts a powerful control on the melting relationships of mantle lherzolite, such as melting temperature, melt productivity and major element melt compositions. Each aluminous phase imparts distinct trace element signatures to mafic magmas formed in its stability field, and they can also affect the physical properties of mantle rocks, such as density and elastic parameters. It is thus of interest to know the pressure–temperature conditions under which the plagioclase–spinel and spinel–garnet transitions occur in planetary mantles. Here we focus on the spinel–garnet transition. Plagioclase peridotites are restricted to shallow and high-temperature environments in the suboceanic mantle. It is left as an exercise to the reader to find out the limits of plagioclase stability in planetary mantles.

All of the mantle phases involved in the spinel–garnet transition are Fe–Mg solid solutions. In addition, both pyroxenes dissolve significant amounts of Al, Cr is a major component of spinel and also enters pyroxenes and garnet, and Ca is an important component in garnet. All of these compositional characteristics result in considerable complications when trying to determine the conditions under which the spinel to garnet transition takes place, which have been the subject of many published studies (Green & Ringwood, 1967; Asimow *et al.*, 1995; Robinson & Wood, 1998; Klemme & O'Neill, 2000a; Klemme, 2004). We begin with the simplest possible model for the transition, which is the Mg end-member reaction:

$$MgAl_2O_4 + 2Mg_2Si_2O_6 \rightleftharpoons Mg_2SiO_4 + Mg_3Al_2Si_3O_{12}. \tag{5.15}$$

In this simple model the spinel–garnet transition is the phase boundary defined by equilibrium (5.15). Calculation of the phase boundary consists of finding the set of all P–T combinations for which $\Delta_r G = 0$. We need an explicit function for the pressure integral in equation (5.13), and the simplest one comes from assuming that crystalline solids are incompressible and undergo no thermal expansion. The following equation for the phase boundary then follows directly from equations (5.13) and (5.1.10):

$$\Delta_r H^0_{1,298} + \int_{298}^T \Delta_r C_P dT - T\left(\Delta_r S^0_{298} + \int_{298}^T \frac{\Delta_r C_P}{T} dT\right) + (P-1)\Delta_r V^0_{solids} = 0,$$
(5.16)

where $\Delta_r V^0_{solids}$ is the volume change of the reaction calculated from the molar volumes of the solid phases at the reference state, 1 bar and 298.15 K. We shall relax the constant volume assumption in Chapter 8 – consider equation (5.16) as an "interim rough approximation" only, the goodness of which needs to be determined.

We have one equation, (5.16), in two unknowns, P and T. We say that this is a system with one *degree of freedom*, meaning that we can freely specify one of the two variables, and solve the equation for the other one. The equation is linear in P but not in T, and the heat capacity integrals are messy polynomial functions that do not have analytical roots. Solving for P by hand would be simple, if computationally intensive. Trying to come up with an approximate solution for T by hand is much harder. Either solution, however, is very easy to implement in a symbolic computation system such as *Maple*, as explained in Software Box 5.2.

Software Box 5.2   Calculation of a phase boundary among pure phases. Spinel–garnet equilibrium, part (i)

The file `th_template_3.mw` contains *Maple* procedures that solve for a phase boundary among pure solid phases, assuming constant volume. The procedures in this worksheet solve equation (5.16). Calculation of the Gibbs free energy change of reaction is placed in its own procedure, named `dGPT`. This procedure is called as a function by *Maple*'s equation solver, which makes solving for a phase boundary very straightforward. There are two different procedures that do this.

Pbound: solves for pressure along the phase boundary, at a specified temperature, by making `dGPT`=0. The temperature range and increment are specified in the procedure call, as are the name of the table containing the reaction stoichiometry and the name of the file where output is to be sent. Since the $\Delta_r G = 0$ equation is linear in P the solution is very fast.

Tbound: solves for temperature along the phase boundary, at a specified pressure, by making `dGPT`=0. The pressure range and increment are specified in the procedure call, as are the name of the table containing the reaction stoichiometry and the name of the file where output is to be sent. In this case the $\Delta_r G = 0$ equation is non-linear (and messy) because of the heat capacity integral. An initial temperature guess must therefore be supplied in the procedure call, and this guess is updated at each iteration of the solver (for a different pressure) with the solution for the last iteration. If the procedure fails to find a solution it is almost certainly because the initial temperature guess is in an unfeasible solution region. Enter a different initial guess and try again.

The data for the spinel–garnet phase boundary are stored in tab-delimited format in a file named `spgrt`.

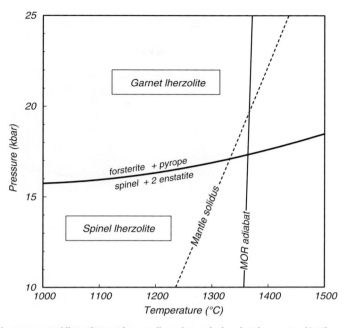

Phase boundary between spinel lherzolites and garnet lherzolites calculated with equation (5.16), assuming (incorrectly) that no Al dissolves in orthopyroxene. The MOR adiabat is a possible adiabatic thermal gradient beneath Earth's mid-ocean ridges, calculated from equation 3.32: $P = P_0 + (C_p/\alpha V)\ln(T/T_0)$, assuming $T_0 = 1623$ K (1350 °C) at $P_0 = 3$ kbar. The solidus is a possible lherzolite solidus (from McKenzie & Bickle, 1988). The calculated phase boundary predicts that initial melting under mid ocean ridges should take place in the stability field of garnet lherzolites, which is thought not to be the case (see also Chapter 10).

The results of the calculations are shown in Fig. 5.5. The figure suggests that, if the terrestrial mantle were composed of the elements Mg, Al, Si and O only, *and under the incorrect assumption that all species remain in their standard states*, garnet would become stable under mid-ocean ridges at $P \sim 17$ kbar. They also suggest that mid-ocean ridge basalts (MORB) should form entirely in the stability field of garnet lherzolites. We will come back to these results and see the extent to which our conclusion will have to be modified once we account for some of the changes in phase compositions that actually take place in the mantle.

The geometric properties of a multicomponent phase boundary are the same as those of a one-component phase transition, but in this case the *assemblage* with higher entropy is the one that is on the high-temperature side of the phase boundary, and the assemblage with lower molar volume is on the high-pressure side. The Clapeyron slope of the phase boundary is the ratio of the change in entropy to the change in volume between reactant and product *assemblages*.

5.2.1 An important digression: uncertainties in calculated equilibrium conditions

There is a question that we need to address. This is: how sensitive is a calculated phase boundary, such as the one shown in Fig. 5.5, to uncertainties in the values of reference state thermodynamic properties? We begin our analysis of this question with Fig. 5.6, which

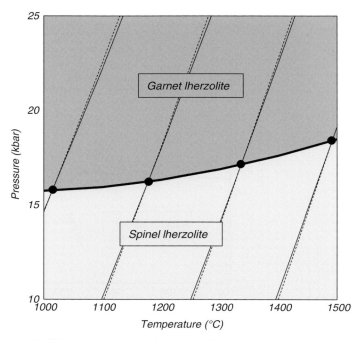

Intersection between the Gibbs free energy surfaces for the assemblages spinel +2 enstatite (solid contour lines) and garnet + forsterite (dashed contour lines). Dots show four intersection points between the surfaces. Phase boundary, the same one as in Fig. 5.5, is shown by thick curve. Note the very shallow intersection between the Gibbs free energy surfaces of the two condensed assemblages.

shows contours on the Gibbs free energy surfaces for the assemblages *spinel + 2 enstatite* (solid lines) and *forsterite + pyrope* (dashed lines), together with the phase boundary for reaction (5.15) (this diagram is equivalent to Fig. 5.2). There is a crucial message here, the importance of which I cannot overemphasize: *the two Gibbs free energy surfaces are almost exactly parallel to one another.* Clearly, even a small relative displacement of the G surfaces can have a large effect on their intersection, which is what defines the phase boundary. This is in stark contrast to the phase diagram for H_2O (Figure 5.2), in which the intersection between the G surfaces for liquid and gas is sharp. It is important to understand the reason for the difference. The local slope of the Gibbs free energy surface is given by the entropy and volume of the assemblage (see Fig. 4.13). Entropy and volume are significantly different for a liquid and a gas of the same chemical composition, leading to a sharp intersection such as that in Fig. 5.2. In contrast, in a reaction in which only solid phases participate the entropies and volumes of the reactant and product assemblages are typically so close that the two surfaces meet almost tangentially.

We can attach some numbers to this conclusion. From equations (4.132) and (4.133) and the linearity of differentiation the temperature and pressure derivatives of $\Delta_r G$ are $\Delta_r S$ and $\Delta_r V$, respectively. The values of these functions for reaction 5.15 at a characteristic temperature of $1400\,°C$ are ~ 6.9 J $K^{-1}mol^{-1}$ and 0.8 J $bar^{-1}mol^{-1}$. This means that a relative displacement of the G surfaces of, say, 1 kJ corresponds to a displacement in their intersection of $\sim 150\,°C$ or ~ 1.25 kbar. These are certainly not negligible values, but their significance becomes even more crucial when they are compared to the *absolute* Gibbs free energies of the assemblages, i.e. to the actual values of the contours in Fig. 5.6. For reaction (5.15), the Gibbs free energies of the assemblages *spinel + 2 enstatite* or *forsterite + pyrope* at 1400 °C are $\sim$8800 kJ mol^{-1}. A displacement of 1 kJ mol^{-1} corresponds to approximately 0.01% of this value.

Although this analysis is based on a specific example, the conclusion is general: even small errors in the determination of reference thermodynamic properties can have enormous effects on calculated phase diagrams. This is true despite the fact that enthalpies and Gibbs free energies are referenced to an arbitrary zero, because their measurement nevertheless entails measuring energy transfers of the order of thousands of kilojoules. And, in any case, entropy values *are* absolute values, by the Third Law. A description of the experimental and mathematical procedures used to determine thermodynamic properties is beyond the scope of this book; see, for example, Anderson, 2005; Berman, 1988; Holloway and Wood, 1988; Holland and Powell, 1998; Anderson, 1995.

In addition to errors in reference state thermodynamic properties, ignoring the compressibility and thermal expansion of the solid phase can have important energetic implications, with large effects on the positions of calculated phase boundaries. We discuss this in Chapter 8, where we will see that the constant volume assumption is not generally an acceptable approximation, except for surface and near-surface conditions. Uncertainties in solution properties (discussed later in this chapter) and in the energetics of higher-order phase transitions (discussed in Chapter 7) must be considered too.

5.3 Phases of variable composition: chemical potential revisited

5.3.1 Equilibrium among chemical species in an arbitrary state

Most planetary materials, whether solids, liquids or gases, are phases of variable composition. We refer to such phases as *solutions*. The chemical species that make up a solution, such that the amount of any of them can be varied independently of all the other species, are called *solution components* or *phase components*. We will often refer to them simply as components, but they must not be confused with the system components defined in Section 5.1. For example, consider a system made up of olivine (forsterite–fayalite) and orthopyroxene (enstatite–ferrosilite) solid solutions. This system can be described with the three system components FeO, MgO and SiO_2, as they constitute a linearly independent set that spans the composition of the system. None of these is, however, a phase component, as their amounts cannot be varied independently of the others while preserving the integrity of the phases olivine and orthopyroxene. The appropriate phase components in this case are $Mg_2Si_2O_6$, $Fe_2Si_2O_6$, Mg_2SiO_4 and Fe_2SiO_4. This set does not constitute a set of system components, as they are not linearly independent. It is generally clear from the context whether one is referring to system components or phase components, but the type of component will be specified if there is any possibility of confusion. We shall return to this topic in Chapter 6.

We now seek an equation that describes chemical equilibrium among phase components in solutions of variable composition, or, equivalently, among chemical species that are not necessarily in their standard states. Consider a system made up of h phases, in which there are k phase components among which it is possible to write a balanced chemical reaction. We make no claims as to the relative values of h and k, nor as to whether or not other phase components that are not part of the chemical reaction between the k components are also present. For instance, if we consider a system made up of the solid solutions spinel, orthopyroxene, olivine and garnet ($h = 4$), we can write a balanced chemical reaction among the phase components $MgAl_2O_4$, $Mg_2Si_2O_6$, Mg_2SiO_4 and $Mg_3Al_2Si_3O_{12}$ ($k = 4$), regardless of whether or not there are other phase components that do not participate of this chemical reaction. If we consider a homogeneous gas phase composed of the system components C, H and O ($h = 1$) we can write a balanced chemical reaction among the phase components CH_4, H_2O, CO_2 and H_2 ($k = 4$), even if other phase components are present (e.g. O_2, CO, C_2H_6, etc.). Other chemical reactions involving these components are of course possible.

For now we will identify each phase by a number j ($0 < j \leq h$) and each component by another number i ($0 < i \leq k$). We arrange the is so that all components in each phase are identified by consecutive numbers. Thus, phase 1 consists of components 1 through ϕ_1, phase 2 of components $\phi_1 + 1$ to ϕ_2, and so on where ϕ_i is the identity of the last of the phase components of interest that is present in phase i. We study the behavior of the system when infinitesimal amounts of matter are transferred among phase components. Let dn_i be an infinitesimal change in the amount of phase component i, and v_i the stoichiometric coefficient of component i in the balanced chemical reaction among the k phase components. If we choose to make $v_1 = 1$, which we can do without loss of generality, then we have $dn_i = v_i dn_1$ for all k phase components.

We use equation (4.129) to write the change in Gibbs free energy for each phase, dG_j, when matter is exchanged at constant temperature and pressure. For phase 1 we have:

$$dG_1 = \sum_{i=1}^{i=\phi_1} \mu^i dn_i, \tag{5.17}$$

where μ^i is the chemical potential of component i. Similarly,

$$dG_2 = \sum_{i=\phi_1+1}^{i=\phi_2} \mu^i dn_i \tag{5.18}$$

and so on until we get to the last phase:

$$dG_h = \sum_{i=\phi_{h-1}+1}^{i=\phi_h} \mu^i dn_i, \tag{5.19}$$

where obviously it must be $\phi_h = k$. The total change in the Gibbs free energy of the system when matter is exchanged at constant pressure and temperature is given by:

$$
\begin{aligned}
dG &= \sum_{j=1}^{j=h} dG_j \\
&= \sum_{i=1}^{i=\phi_1} \mu^i dn_i + \sum_{i=\phi_1+1}^{i=\phi_2} \mu^i dn_i + \cdots + \sum_{i=\phi_{h-1}+1}^{i=\phi_h} \mu^i dn_i \\
&= \sum_{i=1}^{i=k} \mu^i dn_i.
\end{aligned}
\tag{5.20}
$$

Applying the stoichiometry constraint ($dn_i = v_i dn_1$) this equation becomes:

$$dG = dn_1 \sum_{i=1}^{i=k} v_i \mu^i.$$ (5.21)

At equilibrium it is $dG = 0$ (equation (4.130)), and because we have specified that transfer of matter does take place, $dn_1 \neq 0$. We then arrive at the following fundamental equation that expresses the condition of isothermal and isobaric chemical equilibrium among chemical species *in any state* related by a balanced chemical reaction:

$$\sum_i v_i \mu^i = 0.$$ (5.22)

This equation is formally identical to (5.8), except that standard state Gibbs free energy has been replaced by chemical potential. All explicit references to specific phases have disappeared from equation (5.22), so that it is valid for heterogeneous equilibrium among solutions, homogeneous equilibrium within a solution, or a combination of these situations.

I chose to identify phases and components with numbers because it simplifies the notation in the algebraic steps that lead to the compact equation (5.22). This notation is not convenient in practice, however, as it is better to state explicitly the identity of each chemical species and of the phase in which it is present. In explicit applications I shall therefore replace μ^i by μ_p^c, where c is the component and p is the phase, and similarly, if needed, v_i by v_c^p. For example, if we consider equilibrium among the Mg-bearing species in spinel, orthopyroxene, garnet and olivine solid solution, then equation (5.22) becomes:

$$\mu_{spinel}^{MgAl_2O_4} + 2\mu_{opx}^{Mg_2Si_2O_6} = \mu_{garnet}^{Mg_3Al_2Si_3O_{12}} + \mu_{olivine}^{Mg_2SiO_4}.$$ (5.23)

Even though (5.22) (or a specific application such as (5.23)) is the rigorous thermodynamic definition of chemical equilibrium among chemical species in any arbitrary state, it does not by itself allow us to calculate an equilibrium condition. In order to do so we need to find a function that relates chemical potential to phase composition and standard state Gibbs free energy. We tackle this problem in subsequent sections, but before that we explore some other important properties of the chemical potential.

5.3.2 Chemical potential as a driving force for mass transfer

Consider two solutions, call them A and B, and a chemical species σ that is a component in both phases and that can be transferred between them. Simple examples could be the species $NaAlSi_3O_8$, which can be a component of the phases plagioclase, alkali feldspar and silicate melt, or the species H_2O contained in the phases seawater, moist air and ice. By equation (5.22) equilibrium between the phases requires that the following condition be met:

$$\mu_A^\sigma = \mu_B^\sigma.$$ (5.24)

Equilibrium requires that the chemical potential of a component be the same in all phases in which the component is present. The importance of this conclusion is that it is always true, and in particular it is independent of how one chooses to define the standard state. For example, in order for sanidine to coexist at equilibrium with a silicate melt the chemical potential of the species $KAlSi_3O_8$ must be the same in the crystal as in the melt, and in order for water ice to be stable on the Martian surface the chemical potential of H_2O must be the same in surface ice as in the Martian atmosphere. This seemingly trivial corollary of (5.22) is a fundamental concept that we will rely on many times.

Suppose now that the system composed of the two phases A and B is not at equilibrium. Then a spontaneous transfer of component σ between the phases at constant temperature and pressure, say from A to B, can take place only if it leads to a decrease in the Gibbs free energy of the system. Expanding (5.21) we have:

$$dG = \left(\mu_A^\sigma - \mu_B^\sigma\right) dn_\sigma{}^A < 0. \tag{5.25}$$

But if matter is transferred from A to B then it must be $dn_\sigma^A < 0$, which implies that:

$$\mu_A^\sigma > \mu_B^\sigma. \tag{5.26}$$

In other words, chemical species are transferred down chemical potential gradients. This is what we should have expected given that μ and n are conjugate variables (Section 4.8.4) and also justifies the name chemical potential for the thermodynamic variable μ. Equilibrium is reached once mass transfer among phases eliminates chemical potential gradients. The proof that equation (5.26) generalizes to systems with any number of phases and chemical species is left as an exercise.

5.4 Partial molar properties

Finding the function that relates the chemical potential of a phase component to its standard state Gibbs free energy and to the composition of the solution requires several steps. We begin by noting that we can express any extensive thermodynamic property of a solution, Z^{sol}, as a sum of products of intensive quantities z^i for each of the i components of the solution, times the amount (an extensive quantity, for example the number of mols, n_i) of each component, as follows:

$$Z^{sol} = \sum_i z^i n_i. \tag{5.27}$$

We make no claims as to the functional form of the z^is, and in particular, as to whether or not the various z^is are a function of each other. Equation (5.27) is simply an algebraic identity. We do require, however, that the amount of each component, n_i, be independently variable, so these must be phase components (for example, we can add forsterite to olivine without changing the amount of fayalite). Z^{sol} being a thermodynamic variable, it is also a function of temperature and pressure. It then follows from (5.27) that the quantities z^i, called *partial molar quantities*, are defined by the following derivative:

$$z^i \equiv \left(\frac{\partial Z^{sol}}{\partial n_i}\right)_{P,T,n_{j\neq i}}. \tag{5.28}$$

It is important to understand what this equation means: *partial molar z of component i*, an intensive variable symbolized z^i, equals the *rate of change of Z of the solution* (an extensive variable) relative to a change in the amount of component i (also extensive), while keeping pressure, temperature and the amounts of all other components constant. Dividing (5.27) by the total number of mols in the system we obtain molar Z of the solution, Z^{sol}, as follows:

$$Z^{sol} = \frac{\sum_i z^i n_i}{\sum_i n_i} = \sum_i z^i X_i, \tag{5.29}$$

where X_i is the mol fraction of component i.

Obviously, the thermodynamic property Z also takes a definite value for each of the solution components when they are in their standard states. For a component i we symbolize this by $Z^{0,i}$, reserving the subscript slot to state the pressure and temperature of the standard state, if needed (e.g. Section 1.13.1). On dimensional grounds it is evident that z^i and $Z^{0,i}$ refer to the same type of thermodynamic property (e.g. they must both have dimensions of volume per mol, energy per mol or entropy per mol). Note, however, a very important point: we have made no assumptions about the relationship between z^i and $Z^{0,i}$. In particular, the definitions of the two variables are not the same. Whereas the standard state property, $Z^{0,i}$, is defined independently of the properties of a solution (for example, for a pure chemical species), the definition of the corresponding partial molar property z^i is based on the behavior of a particular solution that contains i. There is no expectation that the two variables take on the same value, so that we can write the following equation:

$$Z^{sol} = \sum_i z^i X_i = \sum_i Z^{0,i} X_i + \Delta Z_{mixing}. \tag{5.30}$$

We take (5.30) as the definition of ΔZ_{mixing}, noting that this quantity is defined for the solution (not for the components). For a pure chemical species (equivalently, a phase of fixed composition), there is only one component, for which $X_i = 1$, so that in this case z^i is always equal to $Z^{0,i}$, for any thermodynamic variable, and ΔZ_{mixing} is obviously zero.

There is an additional relationship among partial molar properties and the molar property of the solution that plays a central role in the study of solutions. For a solution of k components, the partial molar property of the ith components is given by:

$$z^i = Z^{sol} + \sum_{j=2}^{j=k} (\delta_{ij} - X_j) \frac{\partial Z^{sol}}{\partial X_j}, \tag{5.31}$$

where the symbol δ_{ij}, called the Kronecker delta, takes the value 1 if $i = j$, and 0 otherwise. The proof and geometric interpretation of (5.31) are given in Box 5.2.

Box 5.2	Proof of equation 5.31

Let Z be any extensive thermodynamic variable. We can write Z for a solution of k components as a function of the form:

$$Z^{sol} = Z^{sol} (P, T, n_1, \dots, n_k) \tag{5.2.1}$$

or, equivalently, as a function of the total number of mols, $n = \sum_i n_i$ and the mol fractions of any $k - 1$ components (the kth is not linearly independent):

$$Z^{sol} = Z^{sol} (P, T, n, X_2, \dots, X_k). \tag{5.2.2}$$

Box 5.2   Continued

This is a simple change of coordinates: note that the functions (5.2.1) and (5.2.2) have the same number of variables, and that, as long as we know n, it is always possible to convert the n_is to X_is and vice versa. Molar Z of the solution, Z^{sol}, is obtained by dividing (5.2.2) by n, and, because n is the only extensive independent variable in (5.2.2), this results in:

$$Z^{sol} = \frac{Z^{sol}\left(P, T, n, X_2, \ldots, X_k\right)}{n} = Z^{sol}\left(P, T, 1, X_2, \ldots, X_k\right). \tag{5.2.3}$$

We now use (5.2.3) to re-write the definition of the partial molar property z^i (equation (5.28)) as follows:

$$z^i = \left(\frac{\partial \left(n Z^{sol}\right)}{\partial n_i}\right)_{P, T, n_{j\neq i}} = Z^{sol}\left(\frac{\partial n}{\partial n_i}\right)_{P, T, n_{j\neq i}} + n\left(\frac{\partial Z^{sol}}{\partial n_i}\right)_{P, T, n_{j\neq i}}. \tag{5.2.4}$$

Because all $n_{j\neq i}$ are kept constant, it is:

$$\left(\frac{\partial n}{\partial n_i}\right)_{P, T, n_{j\neq i}} = 1. \tag{5.2.5}$$

Using (5.2.3) we write the partial derivative in the second term of the right-hand side of (5.2.4) as follows:

$$\left(\frac{\partial Z^{sol}}{\partial n_i}\right)_{P, T, n_{j\neq i}} = \sum_{j=2}^{j=k}\left(\frac{\partial Z^{sol}}{\partial X_j}\frac{\partial X_j}{\partial n_i}\right)_{P, T, n_{j\neq i}} \tag{5.2.6}$$

and, using (5.2.5):

$$\left(\frac{\partial X_j}{\partial n_i}\right)_{P, T, n_{j\neq i}} = \left(\frac{\partial X_j}{\partial n}\frac{\partial n}{\partial n_i}\right)_{P, T, n_{j\neq i}} = \left(\frac{\partial X_j}{\partial n}\right)_{P, T, n_{j\neq i}} = \left(\frac{\partial X_j}{\partial n}\right)_{P, T}. \tag{5.2.7}$$

In the last identity we have dropped the constraint $n_{j\neq i}$ because we are now differentiating relative to the total number of mols. Each mol fraction is given by:

$$X_j = \frac{n_j}{n} \tag{5.2.8}$$

so:

$$\left(\frac{\partial X_j}{\partial n}\right)_{P, T} = \frac{\partial}{\partial n}\left(\frac{n_j}{n}\right)_{P, T} = \frac{1}{n}\left(\frac{\partial n_j}{\partial n} - X_j\right)_{P, T}. \tag{5.2.9}$$

Component i can be any arbitrary component, either component 1 or any one of the $k - 1$ components between $j = 2$ and $j = k$. We need to account for all possibilities. For $n_{j\neq i}$ it is $\partial n_j / \partial n = 0$, whereas for $n_{j=i}$ we have $\partial n_j / \partial n = 1$. We can then write (5.2.9) using the Kronecker delta: $\delta_{ij} = 1$ if $i = j$, $\delta_{ij} = 0$ if $i \neq j$, as follows:

$$\left(\frac{\partial X_j}{\partial n}\right)_{P, T} = \frac{1}{n}\left(\delta_{ij} - X_j\right). \tag{5.2.10}$$

Box 5.2   **Continued**

Substituting into (5.2.6):

$$\left(\frac{\partial Z^{sol}}{\partial n_i}\right)_{P,T,n_{j\neq i}} = \frac{1}{n}\sum_{j=2}^{j=k}\left(\delta_{ij}-X_j\right)\frac{\partial Z^{sol}}{\partial X_j}.$$

(5.2.11)

Substituting (5.2.11) in (5.2.4) and using (5.2.5):

$$z^i = Z^{sol} + \sum_{j=2}^{j=k}\left(\delta_{ij}-X_j\right)\frac{\partial Z^{sol}}{\partial X_j},$$

(5.2.12)

which is (5.31).

This equation comes up repeatedly in the study of solutions, and it is helpful to visualize its geometry. This is best done by considering a system of two components, a and b, in which case (5.2.12) becomes:

$$z^a = Z^{sol} - X_b\frac{\partial Z^{sol}}{\partial X_b}.$$

(5.2.13)

Rearranging as follows:

$$\frac{\partial Z^{sol}}{\partial X_b} = \frac{Z^{sol} - z^a}{X_b}$$

(5.2.14)

it is easy to see (Fig. 5.7) that the partial molar properties z^a and z^b for a given solution composition, say at mole fractions ξ_a, and $\xi_b = 1 - \xi_a$, are the intercepts of the tangent line to Z^{sol} at that composition with

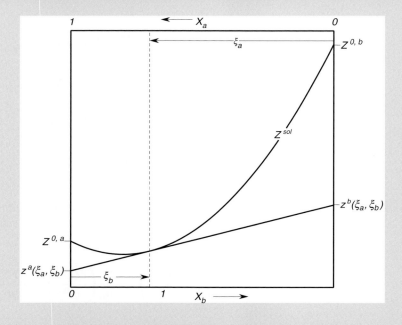

Fig. 5.7   Geometric interpretation of the relationships between the molar property of a solution, Z^{sol}, standard state molar properties of end-member species, $Z^{0,a}$ and $Z^{0,b}$, and partial molar properties of the species in solution, z^a and z^b. Partial molar properties, shown in the figure for a specific solution composition given by ξ_a and $\xi_b = 1 - \xi_a$, are functions of composition.

Box 5.2   Continued

the Z axis at $X_a = 1$ and $X_b = 1$, respectively. This is true for any extensive thermodynamic function that can be written as (5.27), and regardless of the functional form of Z^{sol}. The geometric point of view (Fig. 5.7) is particularly helpful in emphasizing that, whereas $Z^{0,i}$ is a constant at a given P and T, z^i varies with the composition of the solution at constant P and T (except for a property for which Z^{sol} is a linear function). The geometry is the same for multicomponent solutions, except that the tangent line is replaced by a tangent $n - 1$ plane embedded in an n-dimensional space, where n is the number of components in the solution.

Equations (5.27) through (5.31) are algebraic identities that can be applied to any extensive thermodynamic state variable, for instance, enthalpy, entropy, volume and Gibbs free energy. With one exception, the partial molar property is symbolized with the same letter used for the standard state molar property but written in lowercase (e.g. partial molar entropy, s) rather than uppercase (e.g. standard state molar entropy, S^0). The exception is Gibbs free energy. Comparison of (5.28) with (4.134) shows that the partial molar Gibbs free energy is the chemical potential, μ. It is customary to symbolize the standard state molar Gibbs free energy of a pure species by μ^0, and to call it the standard state chemical potential. This is physically appealing given that the interpretation of chemical potential as a driving force for transfer of matter (equation (5.26)) is also true for pure chemical species. In this book we will use both $\mu^{0,i}$ and $G^{0,i}$, depending on the context, but you must keep in mind that the identity $\mu^{0,i} = G^{0,i}$ is always true. Summarizing, we have:

$$G^{sol} = \sum_i \mu^i X_i = \sum_i \mu^{0,i} X_i + \Delta G_{mixing} \tag{5.32}$$

$$S^{sol} = \sum_i s^i X_i = \sum_i S^{0,i} X_i + \Delta S_{mixing} \tag{5.33}$$

$$H^{sol} = \sum_i h^i X_i = \sum_i H^{0,i} X_i + \Delta H_{mixing} \tag{5.34}$$

$$V^{sol} = \sum_i v^i X_i = \sum_i V^{0,i} X_i + \Delta V_{mixing}. \tag{5.35}$$

The pressure and temperature derivatives of chemical potential follow the same rules as those of Gibbs free energy. Thus (omitting the constant variable subscripts for clarity, see Box 1.3):

$$\frac{\partial \mu^i}{\partial P} = \frac{\partial}{\partial P}\left(\frac{\partial \boldsymbol{G}^{sol}}{\partial \boldsymbol{n}_i}\right) = \frac{\partial}{\partial \boldsymbol{n}_i}\left(\frac{\partial \boldsymbol{G}^{sol}}{\partial P}\right) = \frac{\partial \boldsymbol{V}^{sol}}{\partial \boldsymbol{n}_i} \tag{5.36}$$

or:

$$\left(\frac{\partial \mu^i}{\partial P}\right)_T = v^i. \tag{5.37}$$

And by an identical procedure:

$$\left(\frac{\partial \mu^i}{\partial T}\right)_P = -s^i. \tag{5.38}$$

5.5 Generalized equilibrium condition. Activity and the equilibrium constant

5.5.1 Definition of activity

We now have the background needed to derive the algebraic expression that describes equilibrium among chemical species in any state, as a function of temperature, pressure, composition and thermodynamic properties in the standard state. We begin by defining a new thermodynamic variable called *activity*. Using equation (4.128) (last line) we write the following equation for the molar properties of a solution:

$$G^{sol} = H^{sol} - T S^{sol}. \tag{5.39}$$

We can re-write this in terms of partial molar properties, by using (5.32), (5.33) and (5.34):

$$\sum_i \mu^i X_i = \sum_i h^i X_i - T \sum_i s^i X_i \tag{5.40}$$

from which, given that the mol fractions can vary independently, we derive an identity valid for each component i:

$$\mu^i = h^i - T s^i. \tag{5.41}$$

Equation (4.128) also yields the following relationship between standard state molar properties:

$$\mu^{0,i} = H^{0,i} - T S^{0,i}. \tag{5.42}$$

Subtracting (5.42) from (5.41) and rearranging:

$$\mu^i = \mu^{0,i} + \left(h^i - H^{0,i}\right) - T \left(s^i - S^{0,i}\right). \tag{5.43}$$

The terms in this equation have dimension of energy. We can therefore choose to write the sum of the last two terms on the right-hand side as the product of the gas constant, R, times temperature, T, times a non-dimensional variable. We define this non-dimensional variable, called *activity* and symbolized by a, as follows:

$$\ln a^i \equiv \frac{1}{RT}\left(h^i - H^{0,i}\right) - \frac{1}{R}\left(s^i - S^{0,i}\right) \tag{5.44}$$

from which we get:

$$\mu^i = \mu^{0,i} + RT \ln a^i. \tag{5.45}$$

There are different ways of defining activity (see, for example, Guggenheim, 1967). Equation (5.44) is not orthodox, but it is elegant. All definitions converge on equation (5.45). It is very important to understand what this equation is saying. Activity is a non-dimensional parameter that measures the chemical potential of a component in a solution relative to the chemical potential of the same chemical species in a specified standard state. *Activity is a relative quantity that has meaning only if the standard state that it refers to is given explicitly*. This standard state could be, for example, the pure chemical species in the same physical state as the solution (e.g. solid in a certain crystalline state, liquid, gas) and at the pressure and temperature of interest. We could also specify a different standard state, and if we do so then, as is readily apparent from (5.45), the value of the activity must also change, because it now measures the same chemical potential relative to a different reference level.

Worked Example 5.2   Activity of components in silicate melts: an example of changing the standard state

Consider a silicate melt saturated in quartz, for instance, a granitic melt. By "saturated" we mean that crystalline quartz is at equilibrium with the melt, so that we have:

$$\mu_{quartz}^{0,SiO_2} = \mu_{quartz}^{SiO_2} = \mu_{melt}^{SiO_2} \qquad (5.46)$$

The identity $\mu_{quartz}^{0,SiO_2} = \mu_{quartz}^{SiO_2}$ comes from (5.30), noting that for pure quartz it is $X_{SiO_2} = 1$. We wish to specify the activity of SiO_2 in the melt, but in order to do so we must first define the standard state relative to which the activity is to be measured. One possibility is to define the standard state as pure crystalline quartz at the pressure and temperature of interest. As with chemical potential, it is convenient to label activity with the identities of the chemical species and the phase, thus: a_p^c. If we call the activity of SiO_2 in the melt relative to a standard state of pure crystalline quartz at the pressure and temperature of interest a_1 we have, from (5.45):

$$\mu_{melt}^{SiO_2} = \mu_{quartz}^{0,SiO_2} + RT \ln \left(a_{melt}^{SiO_2} \right)_1 \qquad (5.47)$$

and, using (5.46):

$$\left(a_{melt}^{SiO_2} \right)_1 = 1. \qquad (5.48)$$

Another possibility is to define the standard state as pure liquid SiO_2 at the pressure and temperature of interest. This standard state, μ_{liquid}^{0,SiO_2}, is related to the pure quartz standard state by:

$$\mu_{liquid}^{0,SiO_2} = \mu_{quartz}^{0,SiO_2} + \Delta_m G_{P,T}^0, \qquad (5.49)$$

where $\Delta_m G_{P,T}^0$ is the Gibbs free energy of melting of quartz at the pressure and temperature of interest, i.e. $\Delta_r G$ for the reaction *quartz* $\rightarrow$ *liquid* SiO_2 at P and T. Using liquid SiO_2 as the standard state and calling the activity relative to it a_2 we have:

$$\mu_{melt}^{SiO_2} = \mu_{liquid}^{0,SiO_2} + RT \ln \left(a_{melt}^{SiO_2} \right)_2 \qquad (5.50)$$

and, using (5.46) and (5.49):

$$\left(a_{melt}^{SiO_2}\right)_2 = \exp\left(-\frac{\Delta_m G_{P,T}^0}{RT}\right). \tag{5.51}$$

Now, $\Delta_m G_{P,T}^0$ vanishes only if P and T are on the melting curve of pure quartz, in which case the two standard state chemical potentials are equal (equation (5.49)), and $a_1 = a_2$. Otherwise, it is $\Delta_m G_{P,T}^0 \neq 0$ and $a_1 \neq a_2$. Note very carefully that $\mu_{melt}^{SiO_2}$ is the same in (5.47) and (5.50). What has changed is the standard state, and therefore the activity scale that measures this fixed chemical potential.

5.5.2 The equilibrium constant

Substituting (5.45) in (5.22), the condition of equilibrium among chemical species in an arbitrary state and at temperature T can be written as follows:

$$\sum_i v_i \mu^{0,i} + RT \sum_i \ln\left(a^i\right)^{v_i} = 0. \tag{5.52}$$

The first term of the sum in (5.52) is the standard state Gibbs free energy of reaction (equation (5.7)):

$$\sum_i v_i \mu^{0,i} = \sum_i v_i G_{P,T}^{0,i} = \Delta_r G_{P,T}^0, \tag{5.53}$$

but this is now not necessarily zero, as equation (5.8) is true only for species in their standard state, and we have relaxed this restriction. We define another non-dimensional variable, called the *equilibrium constant* of the chemical reaction, and symbolized by K, as follows:

$$K = \prod_i \left(a^i\right)^{v_i}. \tag{5.54}$$

Using (5.53) and (5.54) in (5.52) we get the following generalized equation of chemical equilibrium:

$$\Delta_r G_{P,T}^0 + RT \ln K = 0. \tag{5.55}$$

Finding the equilibrium position for a chemical reaction among species in any state entails solving equation (5.55). This equation is a generalization of (5.8), and collapses to (5.8) if all species are in their standard states, in which case, according to (5.45), it is $a^i = 1$ for all is, and, therefore $\ln K = 0$. The following are two examples of equilibrium calculations among solutions in heterogeneous and homogeneous systems.

In order to study phase changes in real planetary mantles we need to calculate the equilibrium position of reaction (5.15) for the general case in which the Mg-bearing species are not pure phases but components in mineral solid solutions. We define the standard state of each Mg-bearing species as the pure crystalline solid (pyrope, forsterite, spinel or enstatite) at the pressure and temperature of interest. Applying (5.54), we get the following expression for the equilibrium constant of reaction (5.15):

$$K = \frac{a_{garnet}^{Mg_3Al_2Si_3O_{12}} \cdot a_{olivine}^{Mg_2SiO_4}}{a_{spinel}^{MgAl_2O_4} \cdot \left(a_{opx}^{Mg_2Si_2O_6}\right)^2}. \tag{5.56}$$

We then write the equilibrium condition (5.55) like this:

$$\Delta_r G_{P,T}^0 + RT \ln \left[\frac{a_{garnet}^{Mg_3Al_2Si_3O_{12}} \cdot a_{olivine}^{Mg_2SiO_4}}{a_{spinel}^{MgAl_2O_4} \cdot \left(a_{opx}^{Mg_2Si_2O_6}\right)^2} \right] = 0. \tag{5.57}$$

Keeping in mind that $\Delta_r G_{P,T}^0$ is a function of P and T, given by (5.13), we see that (5.57) contains six unknowns: P, T and the four activities, which, as we shall discuss beginning in the next section, are functions of composition. Therefore, equation (5.57) by itself has five degrees of freedom, which means that in order to solve it we must specify the values of five of the variables. We can ask (i) which five, and (ii) how? The answer depends on what we are trying to accomplish. Let us look at two distinct possibilities.

Suppose first that we have a sample of lherzolite that contains the four-phase assemblage olivine–orthopyroxene–garnet–spinel, and that we have good reasons (petrographic or other) to be confident that the four phases crystallized at equilibrium. We can then calculate the four activities from the respective mineral compositions (more on this in subsequent sections), leaving us with two unknowns in (5.57), and thus one degree of freedom. We are now in the same situation that we discussed in Worked Example 5.1: we can calculate a curve in P–T space that maps all the possible P–T combinations at which minerals with the observed compositions could have crystallized. The location of this curve will in general be different from the one in Fig. 5.5 (unless, by chance, activities are such that $K = 1$) and its physical meaning is also subtly different (more on this in Chapter 6). The curve nonetheless constrains the possible crystallization conditions of our lherzolite. If we have an independent way of fixing one of the two variables, for instance temperature, then equation (5.57) can be solved for the pressure of crystallization (subject to the uncertainties discussed in Section 5.2.1, plus others that arise from activity–composition relationships, as we shall see). We may label this procedure *inverting* mineral compositions in order to determine the values of intensive variables during crystallization. It is also called "geothermometry" or "geobarometry", depending on which is the target intensive variable.

Alternatively, we may wish to do *forward modeling* in order to predict mineral compositions as a function of pressure and temperature. In this case we specify arbitrary values of P and T, leaving us with three degrees of freedom in equation (5.57). In order to solve the problem we need to come up with three additional equations, or some other way of constraining three of the compositional variables (Worked Example 5.6).

Carbon–oxygen equilibrium plays important roles in many planetary processes, from determining the composition of volcanic gases (Chapter 9) to determining ice compositions in the outer reaches of the solar nebula, and perhaps the compositions of early post-nebular atmospheres in terrestrial planets (Chapter 14). One often wishes to know the *speciation* in the gas phase, i.e. the relative amounts of chemical species present in the gas phase, which in this case would be CO_2, CO and O_2. In almost every realistic speciation calculation in planetary fluids a meaningful calculation requires that equilibria involving other system components, chiefly hydrogen, nitrogen and sulfur, be included as well (more on this in Chapters 9 and 14). The C–O example, however, allows us a simple first look into how the calculations are carried out.

The following two equilibria describe the chemical reaction between carbon and oxygen:

$$Reaction\ 1:\quad C + \frac{1}{2}O_2 \leftrightharpoons CO,$$

$$Reaction\ 2:\quad C + O_2 \leftrightharpoons CO_2. \tag{5.58}$$

Let us assume that equilibrium is established in the presence of excess graphite, i.e. there is not enough oxygen to oxidize all of the carbon. The assemblage then consists of graphite and a gas phase composed of CO, CO_2 and O_2. It is convenient in this case to define pure graphite at the temperature and pressure of interest as the standard state for C, as this results in: $a^C_{graphite} = 1$. We can now write one version of (5.55) for each reaction, as follows:

$$K_1 = \frac{a^{CO}_{gas}}{\left(a^{O_2}_{gas}\right)^{1/2}} = \exp\left(-\frac{\Delta_r G^{0,1}_{P,T}}{RT}\right) \tag{5.59}$$

and:

$$K_2 = \frac{a^{CO_2}_{gas}}{a^{O_2}_{gas}} = \exp\left(-\frac{\Delta_r G^{0,2}_{P,T}}{RT}\right), \tag{5.60}$$

where the 1 or 2 superscripts identify the reaction. If we specify pressure and temperature then these are two equations in three unknowns (the three activities). The system of equations appears to have one degree of freedom. We shall see, however, that we can recast the activities as functions of the concentration, or mol fraction, of each of the three gas species in the gas phase, X_{CO_2}, X_{CO} and X_{O_2}. Solving for these concentrations is what we call determining the *speciation* of the gas phase. If we assume that no other components are present then we can write an additional equation which states that the gas contains only CO_2, CO and O_2: $X_{CO_2} + X_{CO} + X_{O_2} = 1$. Together with (5.59) and (5.60) this constitutes a system of equations with no degrees of freedom, that we can solve for the three mol fractions (at given P and T). We have not defined the standard states for the three gas species: O_2, CO and CO_2. Neither the activities nor the Gibbs free energies of reaction in equations (5.59) and (5.60) have any meaning unless standard states are specified. The usual way of defining standard states for gases is different from condensed phases, and will be discussed briefly in Worked Example 5.5, where we will solve equations (5.59) and (5.60), and in detail in Chapter 9.

5.6 Introduction to solution theory: ideal solutions

The remainder of this chapter is devoted to discussing the nature of the functions that relate the concentration of a component in a solution to its activity. These are called *activity–composition relationships*. The study of these relationships is a physically and mathematically rich theory, called solution theory. We will cover some of the fundamental concepts in these sections, and some specific applications to solids, liquids and gases in subsequent chapters, but a thorough coverage of the theory would exceed the available space and the intent of this book (see, for example, Guggenheim, 1952; Darken & Gurry, 1953; Kerrick & Darken, 1975; Wood & Nicholls, 1978; Ganguly & Saxena, 1987; Navrotsky, 1987; Wood, 1987; Pitzer, 1995; Kress, 2003).

An important concept of solution theory is that of *ideal solution*. We define an ideal solution as one in which the activity of every component, which we shall term its *ideal activity* (sometimes also called *thermodynamic mol fraction*) follows a strict and simple rule, which is specified in the next section. As with ideal gases, ideal solutions and ideal activities are a simplification of the behavior of natural systems, which under some circumstances may be approached more or less closely, but never perfectly, and often not even closely. The importance of the concept of ideal solution rests, as for ideal gases, on the fact that its mathematical definition is straightforward and it captures fundamental aspects of the behavior of natural systems. For this reason the mathematical formalism of ideal solutions provides a suitable starting point to describe the behavior of real, or also called *non-ideal*, solutions.

5.6.1 Definition of ideal solution and ideal activity

An ideal solution is defined as one that complies strictly with two conditions. The *first condition is that the enthalpy of mixing vanishes for all solution compositions*, i.e. $\Delta H_{mixing} = 0$. From (5.34) it follows that $(h^i - H^{0,i})$ is identically zero, and hence, from (5.44), that the activity of any component in an ideal solution, called its *ideal activity* and symbolized $a^{i,ideal}$ is:

$$\ln a^{i,ideal} = -\frac{1}{R}\left(s^i - S^{0,i}\right). \tag{5.61}$$

The *second condition* focuses on the definition of the quantity $(s^i - S^{0,i})$. From (5.33) we find the following relationship:

$$\sum_i X_i \left(s^i - S^{0,i}\right) = -R \sum_i X_i \ln a^{i,ideal} = \Delta S_{mixing}. \tag{5.62}$$

The entropy of mixing, ΔS_{mixing}, must be related in some way to the increase in configurational entropy (= increase in the number of allowed microstates) that arises from mixing of distinguishable particles. The particles could be, for example, ions in a crystal or in an electrolyte liquid solution, molecules in a multicomponent molecular liquid, or molecules in a mixed gas phase (the latter is the example discussed in Section 4.6.1). We can justify on physical grounds a relationship between ΔS_{mixing} and $S_{configurational}$ as defined in Chapter 4 (equation (4.55)), but we have no reason to assume that the two quantities are in general identical. This is so because end-member components may contain configurational entropy that will also be present in the solution but that does not arise from mixing with other components. Hence the *second condition* in the definition of an ideal solution, which is that *the entropy of mixing of an ideal solution equals the configurational entropy of the solution minus a constant entropy contribution from each end-member component*. We can write this additional entropy term as $R \ln C_i$ (Boltzmann's postulate), and:

$$\Delta S_{mixing} = S_{configurational} - \sum_i X_i R \ln C_i. \tag{5.63}$$

The constant $\ln C_i$ may take different values for each solution component, and may possibly vanish for some or all components, but it is important to emphasize that it is a constant in terms of every thermodynamic variable, in particular, temperature, pressure and composition of the solution. From (5.62) we write:

$$\sum_i X_i \left(s^i - S^{0,i} + R \ln C_i \right) = S_{configurational} \tag{5.64}$$

and note that, because $S_{configurational}$ is a molar property of the solution, the quantity $(s^i - S^{0,i} + R ln C_i)$ behaves algebraically as a partial molar property (see (5.29)). Using the definition of partial molar properties (5.28) and rearranging:

$$\left(s^i - S^{0,i} \right) = -R \ln C_i + \left(\frac{\partial \left(S_{configurational} \right)}{\partial n_i} \right)_{P,T,n_{j \neq i}}. \tag{5.65}$$

The definition of ideal activity then becomes:

$$\ln a^{i,ideal} \equiv \ln C_i - \frac{1}{R} \left(\frac{\partial \left(S_{configurational} \right)}{\partial n_i} \right)_{P,T,n_{j \neq i}}. \tag{5.66}$$

Again, unorthodox but elegant.

5.6.2 Ideal activity and configurational entropy

We can approximate the partial derivative in (5.66) by calculating the change in the number of microstates that arises when we add one molecule of component i to a solution. Say that the solution contains m_i molecules of component i. It may contain any number of molecules of any number of additional components, the only requirement being that all of these other quantities stay fixed. We will symbolize the number of microstates of this solution with Ω_{m_i}. The configurational entropy of this solution, $S_{configurational(m_i)}$, is given by equation (4.55):

$$S_{configurational(m_i)} = k_B \ln \left(\frac{\Omega_{m_i}}{\Omega_i} \right), \tag{5.67}$$

where Ω_i is the number of microstates of the end-member configuration that corresponds to the chosen standard state. Adding one molecule of component i changes the number of mols of i by N^{-1} ($N = $ Avogadro's number), which is an infinitesimally small increment for all practical purposes. The configurational entropy changes to:

$$S_{configurational(m_i+1)} = k_B \ln \left(\frac{\Omega_{m_i+1}}{\Omega_i} \right). \tag{5.68}$$

The partial derivative in (5.66) can be approximated as follows:

$$\left(\frac{\partial \left(S_{configurational} \right)}{\partial \boldsymbol{n}_i} \right)_{P,T,n_{j \neq i}} \approx \frac{S_{configurational(m_i+1)} - S_{configurational(m_i)}}{N^{-1}}. \tag{5.69}$$

Substituting (5.67) and (5.68) in (5.69), and then in (5.66), and simplifying we arrive at:

$$a^{i,ideal} = C_i \frac{\Omega_{m_i}}{\Omega_{m_i+1}}. \tag{5.70}$$

Equation (5.70) is general, as we have made no assumptions regarding the nature of the solution or its components, beyond the fact that it is an ideal solution. The reason why I stated that (5.44) and (5.66) are "elegant" is because they lead naturally to (5.70). In order to evaluate $a^{i,ideal}$ for a specific component in a specific solution, however, it is necessary to know the rules under which mixing takes place, such as whether there is ordering over different types of crystallographic sites and whether or not there are coupled substitutions. We examine some specific cases in the next sections that will exemplify most of the common calculation techniques that are involved in finding ideal activities. Given space constraints, the coverage cannot possibly be exhaustive, but using (5.70) as a starting point it is always possible to calculate an ideal activity in any solution that one may come across.

5.6.3 Ideal activity–composition relationships in fluids and in simple crystalline solids

Consider a crystalline solid solution in which mixing takes place in only one type of crystallographic site, that may however appear repeated in the standard formula unit. We call the number of repetitions of the crystallographic site its *site multiplicity, u*. For example, a simplistic view of Ca-free feldspar is that it is a solid solution of the components $KAlSi_3O_8$ and $NaAlSi_3O_8$, in which the cations K and Na mix in only one site, so that in this case $u = 1$. Olivine is a solid solution of Mg_2SiO_4 and Fe_2SiO_4, so in this case Mg and Fe mix in two crystallographic sites, and $u = 2$. If we disregard complications arising from substitution of other cations in the Si and Al sites (we will get to these later), garnet is a solid solution of the four components: $Mg_3Al_2Si_3O_{12}$, $Fe_3Al_2Si_3O_{12}$, $Mn_3Al_2Si_3O_{12}$ and $Ca_3Al_2Si_3O_{12}$, and in this case $u = 3$ for mixing of Mg, Fe, Mn and Ca.

We expand equation (4.1.1) and write the number of microstates Ω_{m_i} as follows:

$$\Omega_{m_i} = \frac{N!}{n_i! \prod n_{j \neq i}!}. \tag{5.71}$$

In this equation N, typically $\sim$ Avogadro's number, is the total number of atoms involved in mixing (e.g. total number of Mg + Fe + Mn + Ca in garnet, Mg + Fe in olivine or orthopyroxene, and so on), n_i is the number of atoms of the component of interest, and the $n_{j \neq i}$s are the numbers of atoms of all other kinds (I have simply factored out $n_i!$ from the product in the denominator of equation (4.1.1)). We now add one molecule of the component of interest, which means that we add u atoms of i while leaving the amounts of every $j \neq i$ unchanged (e.g. if we want to calculate the activity of pyrope we add one molecule of pyrope to garnet, which adds 3 Mg atoms). The number of microstates is now given by:

$$\Omega_{m_i+1} = \frac{(N+u)!}{(n_i+u)! \prod n_{j \neq i}!}. \tag{5.72}$$

Noting that for $N \gg u$ we can write $(N + u)! \approx N^u N!$, we get:

$$\frac{\Omega_{m_i} + 1}{\Omega_{m_{i1}}} = \frac{(N+u)! n_i!}{(n_i + u)! N!} \approx \frac{N! N^u n_i!}{n_i! n_i^u N!} = \left(\frac{1}{X_i}\right)^u \tag{5.73}$$

since $n_i / N = X_i$. Substituting in (5.70) we arrive at:

$$a^{i, ideal} = C_i (X_i)^u . \tag{5.74}$$

In order to calculate the value of the constant C_i we must specify the standard state. If we choose as standard state the pure chemical species at the temperature and pressure of interest then it follows from (5.45) that it must be $a^{i, ideal} = 1$ for $X_i = 1$, and hence $C_i = 1$. With this choice of standard state, which is the norm, we have:

$$a^{i, ideal} = (X_i)^u . \tag{5.75}$$

You can see from (5.63) that in this case ΔS_{mixing} and $S_{configurational}$ are identical, which recovers our discussion in Section 4.6.2. The condition $\Delta S_{mixing} = S_{configurational}$ is not an *a priori* assumption, but follows from a specific rule for mixing in the solution and a specific choice of standard state. Equation (5.75) is deceptively simple, but must be applied with care. In particular, one must give careful consideration to how the configurational entropy of the entire solution is affected by compositional changes, as more than one distinct crystallographic site may be involved. The following example, subsequent sections and end-of-chapter problems will make this clear.

> **Worked Example 5.5   Chemical potential and activity in a mixture of ideal gases. Carbon–oxygen equilibrium, part (ii)**

Mixing of ideal gases is the simplest possible type of solution, as all locations in the gas are equivalent (there is only one possible "site" on which mixing takes place). We seek an expression for the chemical potential of an ideal gas in a gas phase composed of a mixture of ideal gases. Let the gas phase occupy a volume V at pressure P and temperature T. The total number of mols of gas in the mixture is $N = \sum_i n_i$, where n_i is the number of mols of component i. Because ideal gases are made up of dimensionless and non-interacting particles (Section 1.14) each component exerts a pressure, p_i, equal to the pressure that the same amount of gas would exert if it occupied by itself a volume V at temperature T. Thus:

$$p_i V = n_i RT . \tag{5.76}$$

Dividing by the equation applied to the full mixture, $PV = NRT$:

$$\frac{p_i}{P} = \frac{n_i}{N} \tag{5.77}$$

or:

$$p_i = X_i P . \tag{5.78}$$

We call p_i the *partial pressure* of component i, and it is obvious from (5.78) that $P = \sum_i p_i$. From (5.36) and (5.37) we see that:

$$v^i = \left(\frac{\partial \mu^i}{\partial P}\right)_T = \left(\frac{\partial V}{\partial n_i}\right)_{P,T,n_{j\neq i}} = \frac{\partial}{\partial n_i}\left(\frac{NRT}{P}\right)_{P,T,n_{j\neq i}}$$

$$= \frac{RT}{P}\left(\frac{\partial \sum_i n_i}{\partial n_i}\right)_{P,T,n_{j\neq i}} = \frac{RT}{P} = V^{0,i}, \tag{5.79}$$

where $V^{0,i}$ is the molar volume of *any* ideal gas at P and T. Thus, the partial molar volume of an ideal gas equals the molar volume of the pure component, $v^i = V^{0,i}$, and therefore $\Delta V_{mixing} = 0$ for mixtures of ideal gases (see equation (5.35)).

In order to find the chemical potential of component i in the mixed gas phase at P and T, $\mu^i_{P,T}$ we can integrate $d\mu$ at constant temperature T, from a standard state $\mu^{0,i}_{P,T}$ defined as pure i at P and T, to a state in which the component is at partial pressure p_i in the gas mixture. Thus:

$$\int_P^{p_i} d\mu^i = \mu^i_{P,T} - \mu^{0,i}_{P,T}. \tag{5.80}$$

We re-write this equation using (5.79), as follows:

$$\mu^i_{P,T} = \mu^{0,i}_{P,T} + \int_P^{p_i}\left(\frac{\partial \mu^i}{\partial P}\right)_T dP = \mu^{0,i}_{P,T} + RT\int_P^{p_i}\frac{dP}{P}, \tag{5.81}$$

which, integrating and substituting (5.78), yields:

$$\mu^i_{P,T} = \mu^{0,i}_{P,T} + RT\ln X^i. \tag{5.82}$$

Comparing (5.82) with (5.45) shows that, if we define the standard state of an ideal gas as the pure gas *at the temperature and pressure of interest*, then:

$$a^i = X_i. \tag{5.83}$$

From (5.75) it also follows that, with the standard state define in this way, a mixture of ideal gases behaves as an ideal solution with site multiplicity $u = 1$. Note that we have not assumed this to be the case – we have proved it.

For reasons that we will discuss in Chapter 9, it is more convenient to define the standard state for gases as the pure gas at the temperature of interest *and 1 bar*. Calling the standard state chemical potential in this case $\mu^{0,i}_{1,T}$, we can relate it to the standard state chemical potential of pure i at P and T as follows:

$$\mu^{0,i}_{P,T} = \mu^{0,i}_{1,T} + \int_1^P\left(\frac{\partial \mu^{0,i}}{\partial P}\right)_T dP = \mu^{0,i}_{1,T} + \int_1^P V^{0,i}dP = \mu^{0,i}_{1,T} + RT\int_1^P\frac{dP}{P} \tag{5.84}$$

or:

$$\mu^{0,i}_{P,T} = \mu^{0,i}_{1,T} + RT\ln\left(\frac{P}{1}\right). \tag{5.85}$$

In equation (5.85) I have written the division by 1 explicitly, in order to make it clear that the argument of the logarithm is not pressure, but rather a non-dimensional parameter that equals the ratio between the pressures in one standard state, P bars, and the pressure in the other standard state, 1 bar. The explicit division will generally be absent from equations such as (5.85) if the denominator is 1, but it must always be remembered that this is an abuse of notation, and that the argument of the logarithm function is a non-dimensional variable, which in this case is numerically equal to pressure.

Substituting (5.85) in (5.82) and using (5.78) we get:

$$\mu^i_{P,T} = \mu^{0,i}_{1,T} + RT \ln \left(\frac{p_i}{1} \right), \tag{5.86}$$

where again the division by 1 will be used only here, as a reminder that the argument of the logarithm is a non-dimensional variable. Although we could call the ratio $p_i/1$ the activity of an ideal gas relative to a standard state of 1 bar and the temperature of interest, a different terminology is used for gases, that we will examine in Chapter 9. Equations (5.82) and (5.86) are equally valid representations of the chemical potential of an ideal gas in a mixture of ideal gases, but (5.86) is preferred because it leads to a simpler generalization to real gases. The standard state for gas species, real as well as ideal, is taken at 1 bar and the temperature of interest.

We can now complete the speciation calculation for a C–O fluid at 1 bar that we began in Worked Example 5.4. At 1 bar the ideal gas approximation is generally valid and there is no ambiguity about the standard state, as this is simply the temperature of interest and 1 bar (which is also the pressure of interest). We use (5.83) and re-write (5.59) and (5.60) as follows:

$$\frac{X_{CO}}{(X_{O_2})^{1/2}} = \exp \left(-\frac{\Delta_r G^{0,1}_{1,T}}{RT} \right) \tag{5.87}$$

and:

$$\frac{X_{CO_2}}{X_{O_2}} = \exp \left(-\frac{\Delta_r G^{0,2}_{1,T}}{RT} \right). \tag{5.88}$$

Together with the condition $X_{CO_2} + X_{CO} + X_{O_2} = 1$ we have three equations that can be solved for the three mol fractions at any specified temperature and 1 bar. An additional simplification is possible in this case. The exponential functions in (5.87) and (5.88) (which are the equilibrium constants K_1 and K_2, compare with (5.59) and (5.60)) are very large numbers. For example, we find that at 500 °C, K_1 and K_2 are of order 10^{12} and 10^{26}, respectively. This means that the mol fraction of oxygen in the gas phase is vanishingly small, which of course reflects the fact that oxygen is an extremely reactive chemical species (more on this in Chapters 13 and 14). We can then make $X_{CO_2} + X_{CO} \approx 1$ and eliminate X_{O_2} between (5.87) and (5.88) by squaring (5.87) and dividing by (5.88), leaving us with two equations in X_{CO_2} and X_{CO}. We eliminate one of these variables between these equations (say, X_{CO_2}) and with some manipulation end up with the following quadratic equation:

$$(X_{CO})^2 + K_r X_{CO} - K_r = 0, \tag{5.89}$$

where:

$$K_r = \frac{(K_1)^2}{K_2} = \exp \left[\frac{1}{RT} \left(\Delta_r G^{0,2}_{1,T} - 2\Delta_r G^{0,1}_{1,T} \right) \right]. \tag{5.90}$$

The mol fraction of CO_2 is simply $1 - X_{CO}$, and X_{O_2} ($\ll 1$) is calculated from either (5.87) or (5.88). It is very straightforward to write a *Maple*© procedure to perform all of these calculations, using previously written procedures that handle the thermodynamic calculations (Software Box 5.3). The results are plotted in Fig. 5.8 for temperatures between $25\,°C$ and $1000\,°C$. These are probably of limited applicability to real planetary environments, as hydrogen-bearing species are unlikely to be absent from the gas phase. Recall that the calculations assume that graphite is at equilibrium with the gas phase. They show that carbon is a powerful reducing agent at high temperature, as the gas phase, which is almost pure CO_2 at low temperature, becomes almost pure carbon monoxide at $T \sim 1000\,°C$. Calculated oxygen mol fractions pose an interesting question: what is the physical meaning of

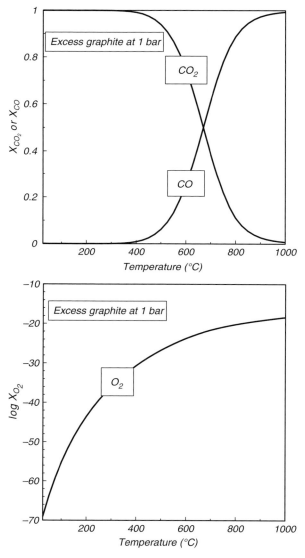

Fig. 5.8   Speciation in a carbon–oxygen gas phase in equilibrium with graphite at $P = 1$ bar, assuming ideal gas behavior.

numbers such as $X_{O_2} = 10^{-40}$ (Fig. 5.8)? This represents a concentration of the order of *one molecule* of oxygen per 10^{16} *mols* of gas. The concept of *fugacity*, that we develop in Chapter 9, allows us to circumvent this physically unappealing interpretation and attach to numbers such as these a rigorous thermodynamic significance.

Software Box 5.3   Calculation of species distribution in a mixture of ideal gases

The *Maple* worksheet `gas_mix_1.mw` contains a procedure named `COeq1` that solves for species distribution in a C–O gas phase in equilibrium with graphite at 1 bar assuming ideal gas behavior (Worked Example 5.4 and 5.5). The procedure uses `dGPT` (see Software Box 5.2) to calculate the values of equilibrium constants K_1 and K_2 (equations (5.59) and (5.60)), and solves for the mol fraction of CO with equation (5.89). Oxygen concentration is calculated from equation (5.88). Output for a specified temperature range is sent to a file, name must be provided in the procedure call. The data for CO–CO_2–graphite equilibrium are stored in tab-delimited format in a file named `gasdata1`. The procedure is easily modified to calculate other gas phase equilibria.

5.7 The geometric view of activity and Gibbs free energy of mixing

The concept of an ideal solution is that the only change in Gibbs free energy that takes place when the solution forms arises from a change in the configurational entropy of the solution (see equation (5.61) and associated discussion). The attentive reader may have noticed, however, that we never examined whether an ideal solution in fact forms. The term $\sum \mu^{0,i} X^i$ in equation (5.32) is the Gibbs free energy of a system that consists of a macroscopic aggregate of components in the same relative proportions as they are present in the solution, except that in the latter mixing occurs at a microscopic scale. A solution will form only if $\Delta G_{mixing} < 0$ (equation (5.32)). Starting from the relationship $G^{sol} = H^{sol} - TS^{sol}$ (this is simply the definition of Gibbs free energy) it is straightforward to show using (5.32) to (5.34) that:

$$\Delta G_{mixing} = \Delta H_{mixing} - T \Delta S_{mixing}. \tag{5.91}$$

From the definition of ideal solution ($\Delta H_{mixing} = 0$) and the fact that mixing at the microscopic scale always results in an increase in configurational entropy (a corollary of Boltzmann's postulate, see Section 4.6.2) we conclude that:

$$\Delta G_{mixing}^{ideal} = -T \Delta S_{mixing}^{ideal} < 0. \tag{5.92}$$

An example of the behavior described by (5.92) is shown in Fig. 5.9, where we plot ΔG_{mixing} for a simple one-site ideal binary solution (equation (5.83)), using (5.62) to calculate ΔS_{mixing}. From (5.62) it also follows that:

$$\Delta G_{mixing}^{ideal} = \sum_i X_i RT \ln a^{i,ideal} \tag{5.93}$$

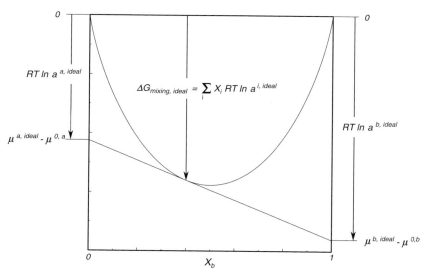

Fig. 5.9 Geometric representation of equations (5.93) and (5.94). Compare with Fig. 5.7. The curve represents the Gibbs free energy of mixing, which vanishes for pure species. The intersections of the tangent line with the vertical axes give the differences between chemical potentials in the solution and standard state chemical potentials $= RT \ln a^{i,ideal}$. The actual values of the standard state properties do not affect this relationship, so they can be set to zero, as in the figure.

and comparison with (5.29) shows that $RT \ln a^{i,ideal}$ is a partial molar property. Using (5.45) and (5.2.13) we find that, for an ideal binary solution such as that depicted in Fig. 5.9:

$$\mu^a - \mu^{0,a} = RT \ln a^{a,ideal} = \Delta G_{mixing}^{ideal} - X_b \frac{\partial \left(\Delta G_{mixing}^{ideal} \right)}{\partial X_b} \tag{5.94}$$

leading to the geometric interpretation of $\mu^a - \mu^{0,a}$ $(= RT \ln a^{ideal})$ shown in Fig. 5.9. There are two additional considerations. First, for solutions with more than two components the function $\Delta G_{mixing\ ideal}$ is a higher-dimensional equivalent of the bowl-shaped curve in Fig. 5.9, the tangent line becomes a tangent plane, and the geometric meaning of $\mu^a - \mu^{0,a}$ stays the same. Second, (5.93) and (5.94) remain true for non-ideal solutions, by simply removing the superscript "ideal" from the variables. We will expand on this shortly.

5.8 More complex ideal activity–composition relationships

5.8.1 Crystalline solutions with multiple ionic sites

In order to apply equation (5.74) to crystalline solids it is necessary to know how ionic substitution in the solid generates configurational entropy. Let us look first at a simple case like olivine, in which substitution takes place only in two octahedral sites that can be considered to be equivalent. Configurational entropy of olivine solid solutions arises only from mixing of cations in these two sites, so we have:

$$a_{ol}^{Fo,ideal} = \left(X_{\text{Mg}}\right)^2$$
$$a_{ol}^{Fa,ideal} = (X_{\text{Fe}})^2 . \tag{5.95}$$

Consider now a mineral in which ionic substitution takes place in two distinct ionic sites, A and B, such that the mineral's structural formula is $A_x B_y O_z$, and suppose that we are interested in calculating the ideal activity of an end-member component consisting of elements α and β, $\alpha_x \beta_y O_z$. We need to consider two possibilities. Let us assume first that substitution in the two sites is uncoupled, because only cations with the same charge mix in each site. In this case each site makes its own independent contribution to configurational entropy. One way of seeing why this is the case is to consider mixing on one site while leaving the composition of the other one fixed at one of its end-members, and then perform the converse operation. The total configurational entropy is the sum of the configurational entropy contribution from each distinct crystallographic site, each of which has its own site multiplicity. It follows from Boltzmann's postulate that the ideal activity of the end-member phase component is the product of activity contributions from each crystallographic site. In our example we have:

$$a^{\alpha_x \beta_y O_z, ideal} = (X_\alpha)^x \left(X_\beta\right)^y . \tag{5.96}$$

Garnets within the compositional range $(\text{Mg, Fe, Mn, Ca})_3(\text{Al, Cr})_2 \text{Si}_3 \text{O}_{12}$ can be treated in this way. For example:

$$a_{grt}^{Prp,ideal} = \left(X_{\text{Mg}}\right)^3 (X_{\text{Al}})^2$$
$$a_{grt}^{Alm,ideal} = (X_{\text{Fe}})^3 (X_{\text{Al}})^2 \tag{5.97}$$
$$a_{grt}^{Kno,ideal} = \left(X_{\text{Mg}}\right)^3 (X_{\text{Cr}})^2$$

where Prp = pyrope, Alm = almandine and Kno = knorringite $(\text{Mg}_3\text{Cr}_2\text{Si}_3\text{O}_{12})$. Conceivably, a small tetravalent cation could substitute for Si in the tetrahedral sites of garnet, in which case the expressions in (5.97) would have to be multiplied by $(X_{Si})^3$ to obtain the correct ideal activities (but I am not aware of this ever being necessary).

A second possibility is that, owing to charge-balance constraints, solid solution forms by coupled substitution. In this case a substitutes for α in site A only if b substitutes simultaneously for β in site B. The ionic charges are such that [charge $(a+b)$] = [charge $(\alpha + \beta)$]. We shall assume for now that in a "perfect" crystal this substitution takes place while preserving local charge balance, such that adjoining A and B sites are substituted simultaneously. If this is the case then the "particles" that mix are not the individual ions but the ionic pairs, ab and $\alpha\beta$. In virtually all minerals in which coupled substitution is important the site multiplicity of the sites that undergo substitution is 1, even if the actual multiplicities of A and B may be different. We then have, from (5.74) with $u = 1$:

$$a^{\alpha\beta, ideal} = X_{\alpha\beta} = X_\alpha = X_\beta. \tag{5.98}$$

An important example of (5.98) is the calculation of activities of Tschermak's components in pyroxenes. The Tschermak's substitution can be written in general as an exchange of the form $\text{Al}^O \text{Al}^T R_{-1}^O \text{Si}_{-1}^T$, where the superscripts O and T signify octahedral and tetrahedral crystallographic sites, and R stands for a divalent cation. For example, in orthopyroxene taking R$=$Mg gives us the Mg-Tschermak's end-member component: MgAlAlSiO$_6$. The two Al cations are kept separate in the formula to emphasize that they occupy different crystallographic sites. In order to calculate the ideal activity of Mg-Tschermak's we begin by noting that the two octahedral sites in pyroxenes are distinct. One of them, labeled M1, is smaller than the other one, M2. In a "perfect" orthopyroxene M2 sites are occupied by Mg and Fe only. If such orthopyroxene dissolves Al then half of the Al cations enter the M1 site, where they substitute for Mg or Fe cations, and the other half enter a neighboring tetrahedral (T) site, where they substitute for Si. Because we assume that this coupled substitution preserves local electrical neutrality the couples AlAl and MgSi behave as mixing units. In a binary Mg–Al orthopyroxene (i.e., no Fe, Ca, etc.) the ideal activities of enstatite and Mg-Tschermak's in orthopyroxene are then given by equation (5.98):

$$a_{opx}^{MgTs,ideal} = X_{Al,M1} = X_{Al,T} = \frac{1}{2}n_{Al}$$

$$a_{opx}^{En,ideal} = X_{Mg,M1} = 1 - X_{Al,M1} = 1 - \frac{1}{2}n_{Al}$$

(5.99)

where n_{Al} is the total number of Al cations per orthopyroxene formula unit: $A_2B_2O_6$.

In the case of an Fe–Mg–Al orthopyroxene one possibility is that the coupled M1–T substitution is independent of Fe–Mg substitution, and that Fe and Mg do not order between M1 and M2 sites. The latter condition implies that $(X_{Mg,M1})/(X_{Fe,M1}) = (X_{Mg,M2})/(X_{Fe,M2}) = n_{Mg}/n_{Fe}$, i.e. the ratio of Mg/Fe occupancy of M1 and M2 sites equals the ratio between the number of Mg and Fe cations per orthopyroxene formula unit. The following activity–composition expressions then follow from (5.96) and (5.98):

$$a_{opx}^{MgTs,ideal} = X_{Al,M1} \cdot X_{Mg,M2}$$

$$a_{opx}^{En,ideal} = X_{Mg,M1} \cdot X_{Mg,M2}$$

$$a_{opx}^{Fs,ideal} = X_{Fe,M1} \cdot X_{Fe,M2}.$$

(5.100)

As an exercise, you should derive equations (5.100) formally starting from (5.70).

Worked Example 5.6   The spinel–garnet transition in planetary mantles, part (iii)

In Worked Example 5.1 we calculated a hypothetical location of the spinel–garnet transition by calculating the phase boundary for the Mg end-member reaction (5.15). This phase boundary cannot be correct, as examination of natural peridotite samples shows that orthopyroxene always dissolves some amount of Al. By (5.99), the activity of enstatite in orthopyroxene must be less than 1, so that the equilibrium constant for reaction (5.15), given by (5.56), does not vanish. Calling reaction (5.15) equilibrium (i), and using (5.99) for the activity of enstatite, we re-write (5.57) as follows:

$$\Delta_r G_{P,T}^{0,(i)} - 2RT \ln X_{Mg,M1} = 0$$

(5.101)

since the activities of the other species remain unity. We can now write another reaction, involving the Mg-Tschermak's component in orthopyroxene, which we shall call equilibrium (ii):

$$MgAl_2O_4 + Mg_2Si_2O_6 \rightleftharpoons MgAlAlSiO_6 + Mg_2SiO_4$$

(5.102)

the equilibrium condition for which is (see (5.99)):

$$\Delta_r G_{P,T}^{0,(ii)} + RT \ln \left(\frac{X_{Al,M1}}{X_{Mg,M1}} \right) = 0. \tag{5.103}$$

Equations (5.101) and (5.103), together with the crystallographic constraint for a binary Al–Mg orthopyroxene, $X_{Mg,M1} + X_{Al,M1} = 1$, constitute a system of three equations with one degree of freedom, as there are four unknowns: $P, T, X_{Mg,M1}, X_{Al,M1}$. We can specify one of the variables, which in this case will be T, and solve the system of equations for the other three variables, so as to obtain the location of the spinel–garnet phase boundary and the composition of orthopyroxene as a function of temperature and pressure.

We begin by eliminating one of the mol fractions, say $X_{Al,M1}$, and re-write (5.103) as follows:

$$\Delta_r G_{P,T}^{0,(ii)} + RT \ln \left(\frac{1 - X_{Mg,M1}}{X_{Mg,M1}} \right) = 0. \tag{5.104}$$

We now have a system of two equations, (5.101) and (5.104), in three unknowns $(P, T, X_{Mg,M1})$, which we solve by specifying temperature. The *Maple* implementation is described in Software Box 5.4, and the results are shown in Fig. 5.10.

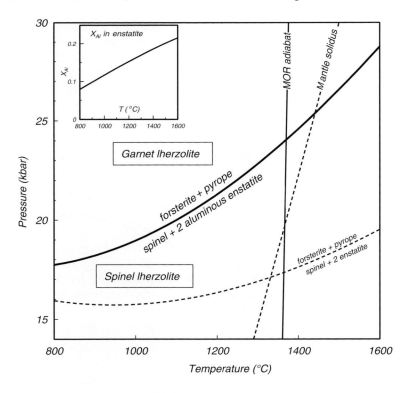

Fig. 5.10   Phase boundary between spinel lherzolite and garnet lherzolite calculated taking into account the Mg-Tschermak's component in orthopyroxene. The incorrect phase boundary from Fig. 5.5 (dashed line) is shown for comparison. Predicted initial melting under mid ocean ridges shifts from the garnet to the spinel stability field. Inset shows calculated mol fraction of Al in orthopyroxene as a function of temperature (pressure is not constant in this graph).

Software Box 5.4   Calculation of spinel–garnet equilibrium, assuming ideal Al–Mg mixing in orthopyroxene

The *Maple* worksheet sp_grt_MAS.mw contains a procedure named spgrAlMAS that solves the two simultaneous equations, (5.101) and (5.104), for the two unknowns, P and $X_{Mg,M1}$ in orthopyroxene. The two equations are labeled R1 and R2 in the *Maple* procedure, which calls dGPT (see Software Box 5.2) to calculate the values of the standard state Gibbs free energy change of equilibria (5.15) and (5.102). *Maple*'s numerical solver, fsolve, is quite powerful and is able to find solutions to systems of non-linear equations with relative ease, as this example shows and as we shall have plenty of opportunity to test in later chapters.

Because spinel–garnet equilibrium is relatively "flat" in P–T space the procedure solves for pressure and orthopyroxene composition at a given temperature, for temperatures over a range that is specified in the procedure call. The name of the file for output is also specified in the procedure call. Output is: $T - P - X_{Al,M1}$. Thermodynamic data are stored in tab-delimited format in a file named spgrt. The procedure can be modified to calculate other equilibria, by changing and/or adding tables defining the required reactions (e.g. reaction1, reaction2, etc.).

The calculated phase boundary is located at pressures 2–10 kbar higher than those obtained assuming that there is no Al in opx (Fig. 5.5). Solubility of Al in orthopyroxene increases with temperature (inset in Fig. 5.10), which agrees with the fact that the right-hand side of reaction (5.102) has the higher entropy (exercise left for the reader). The pressure shift arises because dissolution of Al causes the activity of enstatite to decrease (equation (5.99)), so that its chemical potential decreases as well (equation (5.45)). Looking

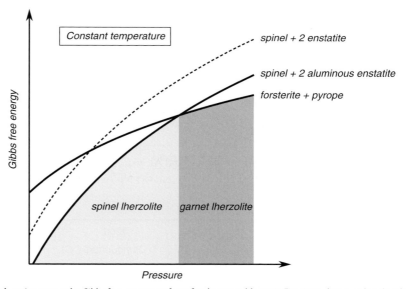

Fig. 5.11   Isothermal section across the Gibbs free energy surfaces for the assemblages in Fig. 5.10, showing that dissolution of Al in orthopyroxene lowers the Gibbs free energy of the assemblage orthopyroxene + spinel, and must therefore shift the equilibrium to higher pressure.

at a schematic G–P diagram (Fig. 5.11), we see that lowering the Gibbs free energy of the low-pressure assemblage in a reaction forces the equilibrium position to higher pressure. This is a necessary consequence of the geometry of Gibbs free energy surfaces (Fig. 4.13).

The phase boundary in Fig. 5.10 is a better approximation than that in Fig. 5.5 to the conditions under which the spinel–garnet transition takes place in planetary mantles, because it accounts for the non-negligible solubility of Al in orthopyroxene. In contrast to what we would have predicted if we had ignored the Al content of orthopyroxene, the more complete phase diagram predicts that garnet becomes stable under mid-ocean ridges at pressures of $\sim$24 kbar, and that MORB magmas form in the stability field of spinel lherzolites (Chapter 10). Our phase boundary in Fig. 5.10 is, however, still deficient on two accounts. First, there are at least three other elements – Ca, Cr and Fe – that can be expected to have non-trivial effects on the location of the spinel–garnet equilibrium. Second, our calculations have been based on the assumption that crystalline solid solutions behave ideally, but we have not examined the validity of this assumption.

5.8.2 Crystalline solutions with non-vanishing configurational entropy in the end-member species

Because they can exist in several different states of cation ordering, plagioclase feldspars are an excellent example of the effect of variable degrees of configurational entropy on ideal activity. They also serve to illustrate further the use of equation (5.70). Using the nomenclature from Worked Example 4.3, we recall that we can distinguish two T1 sites and two T2 sites among the four T (tetrahedral) sites in one formula unit of feldspar. We will consider mixing along the albite–anorthite join using three different assumptions about cation ordering in the tetrahedral sites, as follows.

Model (i) assumes perfect local charge balance, such that Al–Si substitution only occurs on T sites adjoining an octahedral site in which Ca–Na substitution occurs simultaneously.

Model (ii) assumes that Al and Si mix randomly over the two T1 sites, so that the two T2 sites are occupied by Si only. Mixing in tetrahedral and octahedral sites is independent, so that there is no local charge balance.

Model (iii) assumes that Al and Si mix randomly over the four T sites, so that, again, octahedral and tetrahedral mixing are independent and local charge balance is not preserved.

The number of atoms of Na, Ca and Al, n_{Na}, n_{Ca} and n_{Al}, per formula unit of plagioclase and the mol fractions of Na and Ca, are related by the following equations:

$$n_{Na} + n_{Ca} = 1$$
$$n_{Al} = 1 + X_{Ca} = 2 - X_{Na}. \tag{5.105}$$

The local charge balance model, case (i), corresponds to the simple example of coupled substitution discussed in Section (5.8.1). For this model we therefore have:

$$a_{plg}^{Ab, ideal(i)} = X_{Na}$$
$$a_{plg}^{An, ideal(i)} = X_{Ca}. \tag{5.106}$$

In order to calculate the activities for model (ii) we must go back to equation (5.70) and write out the number of microstates explicitly. Because octahedral and tetrahedral mixing are now independent of one another the total number of microstates equals the product of the numbers of octahedral and tetrahedral microstates:

$$\Omega_{m_{Ab}} = \Omega_{m_{An}} = \frac{(n_{Na} + n_{Ca})!}{n_{Na}! n_{Ca}!} \cdot \frac{(n_{Al} + n_{Si,T1})!}{n_{Al}! n_{Si,T1}!}. \tag{5.107}$$

The number of Si atoms in the T1 sites is given by $n_{Si,T1} = 2 - n_{Al}$. Adding one molecule of albite we find:

$$\Omega_{m_{Ab+1}} = \frac{(n_{Na} + n_{Ca} + 1)!}{(n_{Na} + 1)! n_{Ca}!} \cdot \frac{(n_{Al} + n_{Si,T1} + 2)!}{(n_{Al} + 1)!(n_{Si,T1} + 1)!} \tag{5.108}$$

and similarly for one molecule of anorthite:

$$\Omega_{m_{An+1}} = \frac{(n_{Na} + n_{Ca} + 1)!}{n_{Na}!(n_{Ca} + 1)!} \cdot \frac{(n_{Al} + n_{Si,T1} + 2)!}{(n_{Al} + 2)! n_{Si,T1}!}. \tag{5.109}$$

Substituting these equations in (5.70) and simplifying:

$$a_{plg}^{Ab, ideal(ii)} = C_{Ab(ii)} X_{Na} X_{Al,T1} X_{Si,T1} \tag{5.110}$$

and:

$$a_{plg}^{An, ideal(ii)} = C_{An(ii)} X_{Ca} \left(X_{Al,T1}\right)^2. \tag{5.111}$$

The mol fractions of Al and Si in the T1 sites are given by:

$$X_{Al,T1} = \frac{n_{Al}}{2} = \frac{2 - X_{Na}}{2} = \frac{1 + X_{Ca}}{2}$$
$$X_{Si,T1} = \frac{n_{Si,T1}}{2} = \frac{X_{Na}}{2} = \frac{1 - X_{Ca}}{2}. \tag{5.112}$$

We now note that for pure albite it is $X_{Al,T1} = X_{Si,T1} = \frac{1}{2}$, whereas for pure anorthite $X_{Al,T1} = 1$. Substituting these values in (5.110) and (5.111), respectively, and recalling that we require activities to be unity at the standard state (pure end-member species), we find $C_{Ab(ii)} = 4$ and $C_{An(ii)} = 1$. With these values for the constants and mol fractions as in (5.112) we obtain:

$$a_{plg}^{Ab, ideal(ii)} = (X_{Na})^2 (2 - X_{Na})$$
$$a_{plg}^{An, ideal(ii)} = \frac{1}{4} X_{Ca} (1 + X_{Ca})^2. \tag{5.113}$$

We follow the same procedure in order to calculate ideal activities for model (iii). Equation (5.107) needs to be modified slightly, in order to reflect the fact that we must now count all four silicon atoms, with $n_{Si} = 4 - n_{Al}$:

$$\Omega_{m_{Ab}} = \Omega_{m_{An}} = \frac{(n_{Na} + n_{Ca})!}{n_{Na}! n_{Ca}!} \cdot \frac{(n_{Al} + n_{Si})!}{n_{Al}! n_{Si}!}. \tag{5.114}$$

Adding one molecule of each end-member:

$$\Omega_{m_{Ab+1}} = \frac{(n_{Na} + n_{Ca} + 1)!}{(n_{Na} + 1)! n_{Ca}!} \cdot \frac{(n_{Al} + n_{Si} + 4)!}{(n_{Al} + 1)!(n_{Si} + 3)!} \tag{5.115}$$

and:

$$\Omega_{m_{An+1}} = \frac{(n_{Na} + n_{Ca} + 1)!}{n_{Na}!(n_{Ca} + 1)!} \cdot \frac{(n_{Al} + n_{Si} + 4)!}{(n_{Al} + 2)!(n_{Si} + 2)!}, \tag{5.116}$$

which, using (5.70) and simplifying, leads to:

$$a_{plg}^{\mathrm{Ab},ideal(iii)} = C_{\mathrm{Ab}(iii)} X_{\mathrm{Na}} X_{\mathrm{Al,T}} \left(X_{\mathrm{Si,T}} \right)^3 \tag{5.117}$$

and:

$$a_{plg}^{\mathrm{An},ideal(iii)} = C_{\mathrm{An}(iii)} X_{\mathrm{Ca}} \left(X_{\mathrm{Al,T}} \right)^2 \left(X_{\mathrm{Si,T}} \right)^2. \tag{5.118}$$

The mol fractions of Al and Si over the four T sites are:

$$X_{\mathrm{Al,T}} = \frac{n_{\mathrm{Al}}}{4} = \frac{2 - X_{\mathrm{Na}}}{4} = \frac{1 + X_{\mathrm{Ca}}}{4}$$
$$X_{\mathrm{Si,T}} = \frac{n_{\mathrm{Si}}}{4} = \frac{2 + X_{\mathrm{Na}}}{4} = \frac{3 - X_{\mathrm{Ca}}}{4}. \tag{5.119}$$

Unit activities at the standard state require that $C_{\mathrm{Ab}(iii)} = 256/27$ and $C_{\mathrm{An}(iii)} = 16$, so that we finally arrive at:

$$a_{plg}^{\mathrm{Ab},ideal(iii)} = \frac{1}{27} X_{\mathrm{Na}} \left(2 - X_{\mathrm{Na}} \right) \left(2 + X_{\mathrm{Na}} \right)^3$$
$$a_{plg}^{\mathrm{An},ideal(iii)} = \frac{1}{16} X_{\mathrm{Ca}} \left(1 + X_{\mathrm{Ca}} \right)^2 \left(3 - X_{\mathrm{Ca}} \right)^2. \tag{5.120}$$

It is now clear that $C_i \neq 1$ if different cations mix on the same site in an end-member species, as in this instance their mol fraction in the end-member species is not unity, but the activity for the chosen standard state must be unity. From (5.63) we see that $C_i \neq 1$ implies that the entropy of mixing and configurational entropy of the solution are not the same, and we can now understand why. The end-member plagioclase species may contain configurational entropy that arises from Si–Al mixing in the tetrahedral sites (see also Worked Example 4.3), and this entropy is "carried over" into the solution, so that it must be subtracted from the total configurational entropy of the solution in order to obtain the *net increase* in entropy generated by mixing.

The three sets of equations, (5.106), (5.113) and (5.120), yield significantly different ideal activities for plagioclase of a given composition, but none of them is more "correct" than any other. Choosing among the three models requires that one knows the structural state of the particular plagioclase of interest. This is seldom trivial, and is complicated by the fact that the two end-members may show different Al–Si ordering in their standard states, as anorthite always has full long-range ordering (except perhaps near its melting point) whereas the ordering state of end-member albite changes with temperature (see, for example, Putnis, 1992). We will not discuss the details of configurational entropy in plagioclase any further, but it is important to gain some insight into the energetic effects of the different ideal activity models and their likely geological significance. This is shown in Fig. 5.12, where we plot values of $RT \ln a^{i,ideal}$ using $T = 1000$ K as an example. The calculated chemical potentials shift by the same amount as the energy difference between models (equation (5.45)), and the effect is far from trivial. For example, a typical anorthite content in plagioclase in medium to high-grade metamorphic rocks is An_{20-40}. Figure 5.12 shows that the calculated chemical potential for anorthite would vary by 5–10 kJ mol^{-1}, depending on the ideal activity model used. Plagioclase in mantle rocks and in mafic igneous rocks may contain $\sim$10–20% albite, and in this case the shift in calculated chemical potential could be 20 kJ mol^{-1} or more, as the temperature in such rocks is likely to be greater than the 1000 K assumed in the figure. Energy differences of this magnitude would translate to displacements in calculated equilibrium positions of hundreds of degrees or of kilobars to perhaps tens of kilobars (see Section 5.2.1).

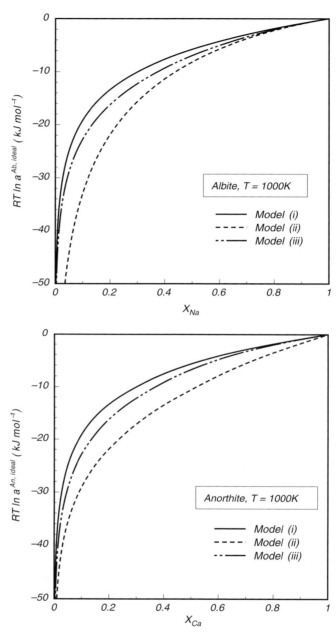

Fig. 5.12 Comparison of three ideal activity models for plagioclase, at a constant temperature of 1000 K.

5.9 Non-ideal solutions

We defined an ideal solution as one for which $\Delta H_{mixing} = 0$. Ideal solutions have another important property, which is that ΔV_{mixing} in them vanishes too, as is easy to prove starting from (5.37) and (5.40). Ideal solutions are an algebraic construct based upon Boltzmann's postulate, but is it reasonable to expect that real substances will follow the ideal behavior that is defined by the condition $\Delta H_{mixing} = \Delta V_{mixing} = 0$? There are physical reasons to believe that in general this will not be the case. For example, molecules in some fluids are polar (e.g. H_2O), whereas in others they are not (CO_2). When the two fluids mix CO_2 molecules screen electrostatic interactions among H_2O molecules, and this should be observed macroscopically as a non-zero ΔV_{mixing}. Different ions mixing in a crystalline lattice generally differ in size, even if slightly, and therefore their surface charge densities will differ too. Ionic substitution should then be accompanied by absorption or liberation of energy ($\Delta H_{mixing} \neq 0$), and perhaps by a change in volume relative to the equivalent macroscopic aggregate ($\Delta V_{mixing} \neq 0$). More subtly, mixing of different ions may have effects on microscopic ordering that are not accounted for by the ideal mixing model, so that the actual configurational entropy of the solution may be different from the value given by (5.63). For example, ions may arrange themselves as if they are forming compounds (at a microscopic level) and this will affect the configurational entropy of a crystal. Or molecules may react to some extent in a mixed gas phase.

There are different ways of treating the behavior of real solutions. Here I focus on what is perhaps the most widely used, and certainly the simplest, approach to real solutions. This consists of assuming a reasonable microscopic mixing model from which one calculates ideal activities, and then approximating the departure of the real (= observed) behavior of the solution from this ideal model by fitting an empirical or semi-empirical function with a variable number of free parameters. This is not the most elegant approach, as the function has no strong physical justification, but there are a number of arguments that can be made in its defense. Above all, it is simple and makes it possible to construct at least a rough description of the behavior of real solutions on the basis of limited experimental observations. Calibration of non-ideal solution models that have a better physical basis often require experimental observations that for many phases and chemical species of planetary interest do not exist. Their application to complex multicomponent phases under very high temperatures and pressures may become computationally unwieldy, yet they carry uncertainties that may make their results indistinguishable from those of simpler empirical models, especially when compounded with possibly large uncertainties about physical conditions in planetary interiors.

5.9.1 Excess mixing functions

Equation (5.91) ($\Delta G_{mixing} = \Delta H_{mixing} - T \Delta S_{mixing}$) is valid for any solution, ideal or non-ideal, as no assumptions about the nature of the solution were made in its derivation. If a solution is non-ideal then in general it must be:

$$\Delta G_{mixing} \neq \Delta G_{mixing}^{ideal}. \tag{5.121}$$

The inequality may arise from a combination of enthalpy and entropy contributions to Gibbs free energy, but there may not be an a priori way of discriminating between them. We therefore convert (5.121) into an identity by adding a Gibbs free energy contribution to the Gibbs free energy of ideal mixing. This contribution is called *excess* Gibbs free energy, G^{excess}, and is defined by the following equation:

$$G^{excess} \equiv \Delta G_{mixing} - \Delta G_{mixing}^{ideal}. \tag{5.122}$$

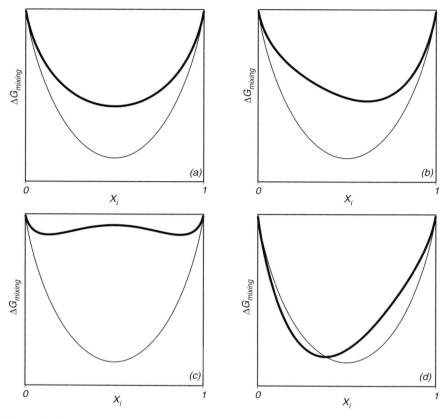

Fig. 5.13 Possible non-ideal behaviors of a binary one-site solution, shown by the thick curves. The thin curve is the same in all four graphs and shows ideal one-site mixing. The examples shown do not exhaust all possibilities. For example, the inflected behavior shown in (c) is symmetric, but it can also be asymmetric, as in (b). Negative deviation from ideality, as in the left side of (d), can also be symmetric (at least in principle).

Figure 5.13 exemplifies possible behaviors of real solutions. The thick curves in the figure represent the Gibbs free energies of four possible real solutions. The thin curves are the Gibbs free energies of the same solutions, assumed to be ideal. The excess Gibbs free energy is the distance between the curves. We see that G^{excess} may be symmetric (Fig. 5.13a) or asymmetric (Fig. 5. 13b) relative to composition, it may be such that the curvature of the ΔG_{mixing} function changes with composition (Fig. 5.13c), or such that the sign of G^{excess} changes with composition (Fig. 5.13d), or some combination of these behaviors. Because it is a Gibbs free energy, G^{excess} is related to excess enthaply, entropy and volume functions:

$$G^{excess} = H^{excess} - T S^{excess} \qquad (5.123)$$

$$\left(\frac{\partial G^{excess}}{\partial T} \right)_{P, n_i} = -S^{excess} \qquad (5.124)$$

$$\left(\frac{\partial G^{excess}}{\partial P} \right)_{T, n_i} = V^{excess}. \qquad (5.125)$$

We use (5.2.12) to define a partial molar property, called excess chemical potential, μ^{excess}, as follows:

$$\mu^{i,excess} \equiv G^{excess} + \sum_{j=2}^{j=k} (\delta_{ij} - X_j) \frac{\partial G^{excess}}{\partial X_j} \qquad (5.126)$$

so that, by equation (5.29), $G^{excess} = \sum_i X_i \mu^i$. From (5.126), (5.122) and (5.93) we get:

$$\Delta G_{mixing} = \sum_i X_i RT \ln a^{i,ideal} + \sum_i X_i \mu^{i,excess} \qquad (5.127)$$

and, by using (5.45):

$$\mu^i - \mu^{0,i} = RT \ln a^i = RT \ln a^{i,ideal} + \mu^{i,excess}. \qquad (5.128)$$

In order to construct a compact equation for the activity of a component in a non-ideal solution we define a new non-dimensional parameter, called the *activity coefficient* and symbolized by γ, as follows:

$$\gamma^i \equiv \exp\left(\frac{\mu^{i,excess}}{RT}\right) \qquad (5.129)$$

so that:

$$a^i = \gamma^i \cdot a^{i,ideal}. \qquad (5.130)$$

Equations (5.122) through (5.130) are the foundations of the mathematical treatment of non-ideal solutions. They have a simple geometric interpretation which helps to visualize the behavior of real solutions (Fig. 5.14). The excess chemical potential of a species is the distance between the intersects of the tangents to the ideal and real Gibbs free energy of mixing curves with the G axis for that species (compare Fig. 5.9). This distance vanishes as the mol fraction of the species of interest approaches unity, and approaches a constant finite value as the species becomes infinitely dilute (Fig. 5.14).

5.9.2 Raoult's law and Henry's law

If we consider a simple one-site binary solution then (5.130) becomes:

$$a^i = \gamma^i X_i \qquad (5.131)$$

and, calling the two components 1 and 2, we get from (5.126):

$$\mu^{1,excess} = G^{excess} - X_2 \frac{\partial G^{excess}}{\partial X_2}. \qquad (5.132)$$

For $X_1 \to 1$ we have $X_2 \to 0$ and also $G^{excess} \to 0$ (see Fig. 5.14), so that, as long as $\partial G^{excess}/\partial X_2$ stays bound, $\mu^{1,excess} \to 0$. We then have, for a nearly pure component:

$$a^i = X_i, \quad \text{for } X_i \to 1. \qquad (5.133)$$

On the other hand, for $X_1 \to 0$, $X_2 \to 1$ but it is still $G^{excess} \to 0$. Assuming again that $\partial G^{excess}/\partial X_2$ stays bound, we have:

$$\mu^{1,excess}_{X_1 \to 0} = -\left(\frac{\partial G^{excess}}{\partial X_2}\right)_{X_2=1}. \qquad (5.134)$$

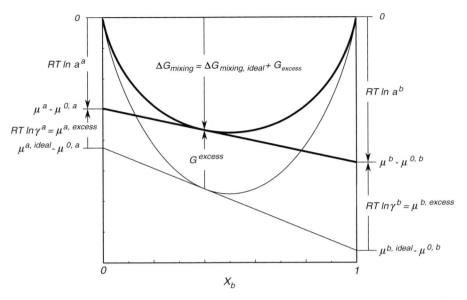

Geometric interpretation of excess mixing properties. The thin curve and tangent line correspond to an ideal solution, the thick ones to a (symmetric) non-ideal solution. Compare with Fig. 5.9.

We define a constant, K_H^1, as follows:

$$K_H^1 = \gamma_{X_1 \to 0}^1 = \exp\left[-\frac{1}{RT}\left(\frac{\partial G^{excess}}{\partial X_2}\right)_{X_2=1}\right], \qquad (5.135)$$

which, using (5.131), yields the following relation for an infinitesimally dilute component:

$$a^i = K_H^i X_i, \quad \text{for } X_i \to 0. \qquad (5.136)$$

Equations (5.133) and (5.136) are known as Raoult's and Henry's laws, respectively, and were found empirically by the eponymous scientists during the nineteenth century. Raoult's law is perhaps self-evident, but Henry's law is not. It requires that $\partial G^{excess}/\partial X_2$ approach a constant finite value as X_1 becomes infinitesimally small.

Raoult's and Henry's laws were originally derived for binary mixtures of liquid and gases, but they are applicable to any type of solution. For complex multi-site solid solutions we can re-state them as follows:

$$a^i = a^{i,ideal}, \quad \text{for } X_i \to 1 \qquad (5.137)$$

and:

$$a^i = K_H^i a^{i,ideal}, \quad \text{for } X_i \to 0. \qquad (5.138)$$

Henry's law is the thermodynamic basis of trace element geochemistry (Chapter 10) and is also important in the thermodynamic treatment of electrolyte solutions (Chapter 11).

5.9.3 Polynomial expansion of excess mixing functions

The empirical approach to describing non-ideal solutions consists of finding a function of T, P and composition that can reproduce observed values of G^{excess} (i.e., the varying distance between the curves in Fig. 5.13) as closely as possible. This function must subsume Raoult's and Henry's laws. Thus, G^{excess} must vanish as the end-member compositions are approached, and all partial derivatives $\partial G^{excess} / \partial X_i$ must be finite over the entire compositional range, and in particular at the end-members. In addition to these two properties, the G^{excess} function must be such that it is able to reproduce a range of possible non-ideal behaviors, as depicted in Fig. 5.13. A polynomial in powers of mol fractions can, subject to some constraints, be made to do all of these things. The most general way of constructing such a polynomial is to begin with the following series, in which the w coefficients are constant with respect to composition but are in general functions of temperature and pressure:

$$
\begin{aligned}
G^{excess} = \sum_i \sum_j w_{ij} X_i X_j + \sum_i \sum_j \sum_k w_{ijk} X_i X_j X_k \\
+ \sum_i \sum_j \sum_k \sum_l w_{ijkl} X_i X_j X_k X_l + \cdots,
\end{aligned}
\tag{5.139}
$$

where $i, j, k, l, \ldots$ are solution components. Higher-order terms can be added as needed, but it is often sufficient, or necessary owing to lack of data, to truncate the series after the second term, in cubes of mol fractions.

A number of simplifications ensue. First, it is convenient to set coefficients with different permutations of the same subindices equal to one another, i.e. $w_{ij} = w_{ji}$, $w_{ijk} = w_{ikj} = w_{jik} = w_{jki} = w_{kij} = w_{kji}$ and so on. Second, in order for G^{excess} to vanish for all pure end-member components every coefficient in which all subindices are the same must vanish too, i.e. $w_{ii} = w_{iii} = w_{iiii} = 0$, and so on if higher-order terms are included. Third, because of the closure condition ($\sum_i X_i = 1$), there are dependencies among the non-zero coefficients, which cuts down on the number of free parameters needed to fit G^{excess}. This is easily seen in a binary solution, but is true, although algebraically more complex, for solutions of any number of components. Applying (5.139) to a binary solution, truncating the series at the second term, collecting coefficients with different permutations of the same subindices and omitting those coefficients that are zero we get:

$$
G^{excess} = 2w_{12} X_1 X_2 + 3w_{112} X_1^2 X_2 + 3w_{122} X_1 X_2^2,
\tag{5.140}
$$

which, noting that $X_1 + X_2 = 1$, we write as:

$$
G^{excess} = X_1 X_2 \left[3w_{112} X_1 + 3w_{122} X_2 + 2w_{12} (X_1 + X_2) \right]
\tag{5.141}
$$

or:

$$
G^{excess} = X_1 X_2 \left[(3w_{112} + 2w_{12}) X_1 + (3w_{122} + 2w_{12}) X_2 \right].
\tag{5.142}
$$

In equation (5.142) there are only two independent coefficients, which we are free to rename as follows: $W_{21}^G = 3w_{112} + 2w_{12}$ and $W_{12}^G = 3w_{122} + 2w_{12}$. These parameters, written with a capital W, are commonly called *Margules parameters* or *interaction parameters*. Equation (5.142) simplifies to:

$$
G^{excess} = \left(W_{21}^G X_1 + W_{12}^G X_2 \right) X_1 X_2.
\tag{5.143}
$$

The superscript G is added to the parameters to denote the fact that they measure excess Gibbs free energy. It is important to realize that Margules parameters are functions of temperature and pressure *but not of composition*.

We can verify that (5.143) is an acceptable algebraic representation of the excess Gibbs free energy of a binary solution. First, G^{excess} vanishes if the mol fraction of any of the two end members becomes zero (Raoult's law). Second, using infinitely dilute component 1 as an example, we have:

$$\left(\frac{\partial G^{excess}}{\partial X_2} \right)_{X_2=1} = -W_{12}^G, \tag{5.144}$$

i.e. $\partial G^{excess}/\partial X_i$ at the limiting end-member composition has a finite non-zero value and is related to the Henry law constant by:

$$K_H^1 = \gamma_{X_1 \to 0}^1 = \exp \left[-\frac{1}{RT} \left(\frac{\partial G^{excess}}{\partial X_2} \right)_{X_2=1} \right] = \exp \left(\frac{W_{12}^G}{RT} \right). \tag{5.145}$$

There are no *a priori* conditions on the signs and relative magnitudes of the Margules parameters. An inflected ΔG_{mixing} curve (Fig. 5.13c) results if the two interaction parameters are positive and large, whereas a G^{excess} function that changes sign (Fig. 5.13d) arises if the two parameters have different signs. An important simplification ensues if the two interaction parameters are equal, in which case we have $W_{21}^G = W_{12}^G$ and:

$$G^{excess} = W_{12}^G X_1 X_2. \tag{5.146}$$

Equation (5.146) corresponds to the symmetric behavior depicted in Figs. 5.13a and 5.13c, whereas the general asymmetric case in Figs. 5.13b and 5.13d is reproduced by (5.143) with $W_{21}^G \neq W_{12}^G$. A commonly used terminology refers to the symmetric solution described by (5.146) as a *simple mixture*, and to the asymmetric solution given by (5.143) as a *subregular solution*. To make things more confusing, a symmetric solution ($=$ simple mixture) for which $S^{excess} = V^{excess} = 0$ (i.e. for which G^{excess} is independent of P and T) is called a *regular solution*. I find this terminology, which is used primarily for historical reasons, unnecessarily confusing, and will eschew it in favor of the more descriptive terms *symmetric solution* (e.g. Fig. 5.13a and c) and *asymmetric solution* (e.g. Fig. 5.12b and d), with or without T and/or P dependencies.

Because G^{excess} is a linear function of interaction parameters, identities (5.123) to (5.125) carry over to equivalent identities among interaction parameters, so that we have:

$$W_{ij}^G = W_{ij}^H - T W_{ij}^S \tag{5.147}$$

$$\left(\frac{\partial W_{ij}^G}{\partial T} \right)_P = -W_{ij}^S \tag{5.148}$$

$$\left(\frac{\partial W_{ij}^G}{\partial P} \right)_T = W_{ij}^V. \tag{5.149}$$

Note that the subscript n_i is no longer needed in the partial derivatives, because the interaction parameters as defined by (5.139) are not functions of composition. Excess enthalpies, entropies and volumes are calculated with (5.143) or (5.146), using W^H, W^S, or W^V, as needed.

Excess chemical potentials and activity coefficients are obtained by substituting the polynomial expansion for G^{excess} (i.e. equations (5.143) or (5.146), depending on whether the solution is asymmetric or not) in (5.126) and (5.129). For example, for an asymmetric binary solution we have:

$$\mu^{1,excess} = \left[W_{12}^{G} + 2X_1 \left(W_{21}^{G} - W_{12}^{G} \right) \right] X_2^2. \tag{5.150}$$

which for a symmetric solution simplifies to:

$$\mu^{1,excess} = W_{12}^{G} X_2^2. \tag{5.151}$$

The attentive reader must have noticed that, beginning with equation (5.140), the discussion has focused exclusively on binary solutions. This is because the corresponding polynomial expansions for solutions of three or more components quickly become much more cumbersome, and are best dealt with by means of a symbolic algebra package such as *Maple* (see end-of-chapter exercises).

5.9.4 Perils and tribulations of excess mixing functions

Whenever one uses non-ideal mixing models the fact must be kept in mind that the values of excess thermodynamic properties are not independent of the values of standard state properties. In order to understand what this means, and the perils that ensue, it is necessary to sketch out how excess mixing properties are measured. One way of doing this is by means of phase equilibrium experiments. The method is based on equilibrating a phase of variable composition in an assemblage in which all the other phases are end-member species that do not change composition. An example is the determination of Al–Mg excess mixing properties in orthopyroxene using reaction (5.102). The experimental data that one seeks are orthopyroxene compositions coexisting at equilibrium with end-member spinel and forsterite. The equilibrium condition for this reaction is equation (5.103), which contains three free parameters: P, T and $X_{Mg,M1}$ in orthopyroxene (e.g. equation (5.104)). If we wish to determine the excess mixing properties of Al–Mg in orthopyroxene then we can perform a series of phase equilibrium experiments at controlled pressures and temperatures, and measure the composition of orthopyroxene that crystallizes in each experiment. Equation (5.103) assures us that at each P and T there is a unique equilibrium orthopyroxene composition, since neither spinel nor olivine will depart from their Mg end-member compositions. We will not go into the details of how the experiments are performed, what are the uncertainties in experimental temperatures and pressures, how we can ascertain whether equilibrium was attained in the experiments, or what are the likely uncertainties in orthopyroxene compositions arising from analytical techniques (see for example Holloway & Wood, 1988; Berman, 1988; Holland & Powell, 1998; Anderson, 2005). Rather, we will put ourselves in the somewhat optimistic position that none of these is a concern, and see that there is a more fundamental issue in play.

Allowing for the possibility that Al–Mg mixing in orthopyroxene may not be ideal, we rewrite equation (5.103) as follows (see (5.128)):

$$\Delta_r G_{P,T}^{0,(ii)} + RT \ln \left(\frac{X_{Al,M1}}{X_{Mg,M1}} \right) + \mu_{opx}^{MgTs,excess} - \mu_{opx}^{En,excess} = 0. \tag{5.152}$$

If we know the pressure and temperature of the experiment then we can calculate the standard state Gibbs free energy of reaction (the first term in the equation), and we get the second term (the ratio of ideal activities) from the measured orthopyroxene composition. One experiment thus gives us the difference between the excess chemical potentials of Mg–Tschermakite and enstatite. We will see in a second that with multiple experiments we can find the absolute value of each of the excess chemical potentials, but before getting into that you must understand that equation (5.152) is the crux of the issue: *the values of the excess chemical potentials are anchored to the values of the standard state thermodynamic properties.* The problem is that standard state properties for many species of interest in the planetary sciences are not known with high accuracy, and there may be significant differences among values given in different data bases. *Excess mixing properties measured by phase equilibria are relative values. They are determined relative to standard state properties from a specific data base, and they can ONLY be used together with standard state data from that same data base.* Of course, equation (5.152) shows that the values of excess mixing properties are also anchored to our choice of ideal activity model, but accounting for this is less of a problem, as it requires only that one be consistent when calculating ideal activities. In contrast, combining excess properties with standard state properties different from those used in the derivation of the excess properties renders the results questionable at best. Excess mixing properties can also be derived from calorimetric measurements (see Navrotsky, 1986), and although these can be absolute values, the same caveat applies: they should not be used in conjunction with standard state properties derived by some other method.

Extracting Margules parameters from (5.152) is straightforward but requires that we perform multiple experiments over a range of pressures and temperatures, so as to obtain a range of orthopyroxene compositions. Assuming that Mg–Al non-ideal mixing is asymmetric, we use (5.150) to re-write (5.152) as follows:

$$
\begin{aligned}
\Delta_r G_{P,T}^{0,(ii)} &+ RT \ln \left(\frac{X_{\mathrm{Al,M1}}}{X_{\mathrm{Mg,M1}}} \right) \\
&+ W_{\mathrm{AlMg}}^G \left(X_{\mathrm{Mg,M1}}^2 - 2 X_{\mathrm{Al,M1}} X_{\mathrm{Mg,M1}}^2 - 2 X_{\mathrm{Mg,M1}} X_{\mathrm{Al,M1}}^2 \right) \\
&- W_{\mathrm{MgAl}}^G \left(X_{\mathrm{Al,M1}}^2 - 2 X_{\mathrm{Al,M1}} X_{\mathrm{Mg,M1}}^2 - 2 X_{\mathrm{Mg,M1}} X_{\mathrm{Al,M1}}^2 \right) \\
&= 0.
\end{aligned}
\tag{5.153}
$$

This is one equation in two unknowns (the two Margules parameters). In principle, and assuming that there is no temperature nor pressure dependency of the W^G parameters (i.e. $W^S = W^V = 0$, see equations (5.148) and (5.149)), with two experiments at different conditions, in which the values of $\Delta_r G^0$, X_{Mg} and X_{Al} are different, we could solve for the two parameters. In practice many experiments are required so as to have an overdetermined system of equations that allows us to analyze experimental errors, and detect possible P and T dependencies of the Margules parameters (i.e. discriminate W^G into W^H, W^S and W^V, see equation (5.147)). Assuming a symmetric solution simplifies 5.153 considerably, to:

$$
\Delta_r G_{P,T}^{0,(ii)} + RT \ln \left(\frac{X_{\mathrm{Al,M1}}}{X_{\mathrm{Mg,M1}}} \right) + W^G \left(X_{\mathrm{Mg,M1}}^2 - X_{\mathrm{Al,M1}}^2 \right) = 0.
\tag{5.154}
$$

One commonly uses experimental data to test for symmetry vs. asymmetry, for example by determining whether the difference between the two parameters fitted to an asymmetric model is statistically significant.

The following example is designed to show that, no matter what model one chooses to use to represent the behavior of a real solution, unless consistency between the various sources of thermodynamic data is observed, the results can be spectacularly incorrect.

Worked Example 5.7   The spinel–garnet transition in planetary mantles, part (iv)

The spinel–garnet transition can be used as an example of the perils of combining mixing and standard state properties that are not consistent with one another. The phase boundary and Al contents in orthopyroxene in Fig. 5.10 were calculated using standard state properties from Holland and Powell (1998) and ideal Al–Mg mixing in orthopyroxene. Relative to their standard state properties, Holland and Powell find that $W^G_{\text{AlMg}} = 0$. In other words, Al and Mg appear to mix ideally in orthopyroxene. In truth what probably happens is that the available experimental data are too sparse and do not make it possible to accurately discriminate between standard state and excess mixing properties for this particular binary join. The safest course of action in such case is to set the excess chemical potential equal to zero, and let the standard state properties of Mg–Tschermakite "absorb" any excess mixing properties – this is, I believe, what Holland and Powell have done. Whether or not Al–Mg mixing in orthopyroxene is ideal, and it almost certainly is not, if we are going to use Holland and Powell's standard state properties to calculate the phase boundary then we *must* use ideal mixing. The curves in Fig. 5.10 are therefore the correct ones.

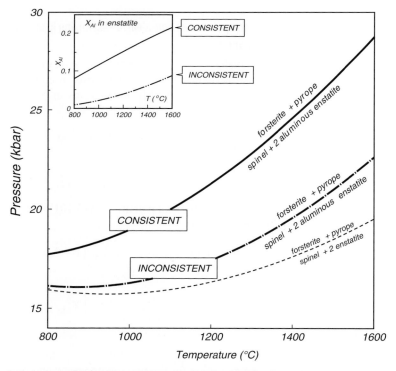

Fig. 5.15   The effect of calculating a phase boundary using excess mixing properties that are inconsistent with the standard state properties. Curves labeled "CONSISTENT" were calculated using standard state properties from Holland and Powell (1998), that require that Al–Mg mixing in orthopyroxene be considered ideal. Curves labeled "INCONSISTENT" were calculated with the same standard state properties, and Al–Mg excess mixing properties from Klemme and O'Neill (2000b). This ***does not*** mean that Klemme and O'Neill's mixing properties are incorrect, only that they are anchored to a different set of standard state properties. The purpose of the diagram is to show that using standard state and mixing properties that are inconsistent with one another can engender very significant errors.

Perusing the literature, however, one finds other studies that report significant non-ideality for Al–Mg mixing in orthopyroxene. One such paper is the one by Klemme and O'Neill (2000b), who find a symmetric non-ideal behavior with $W^G = 20$ kJ mol^{-1}, *but relative to a different set of standard state values* (which unfortunately they do not make explicit). Let us ignore this "but" and recalculate the phase diagram using Holland and Powell's standard state data and Klemme and O'Neill's non-ideal mixing model. The calculation is very easily implemented in *Maple* (Software Box 5.5) and the results are shown in Fig. 5.15. The calculated phase boundary shifts by $\sim$5 kbar, and the calculated Al mol fraction in orthopyroxene drops by about 0.1 to 0.15. These are non-negligible displacements, and the new curves (labeled "INCONSISTENT" in the figure) are incorrect. Note that this *does not* mean that Klemme and O'Neill's non-ideal mixing model is incorrect. Their model is a different way of allocating experimental measurements among standard state and excess properties, as equation (5.152) should make clear. This exercise is designed to demonstrate the fallacy of combining excess properties and standard state properties that are not mutually consistent. *I emphasize: I made this error on purpose, with purely didactic goals.*

Software Box 5.5   Incorporation of non-ideal Al–Mg mixing in orthopyroxene to the calculation of spinel–garnet equilibrium

The *Maple* worksheet `sp_grt_MAS.mw` contains a procedure named `spgrAlMASni` that adds non-ideal mixing terms to equations (5.101) and (5.104). Assuming symmetric non-ideal Al–Mg interactions in orthopyroxene, we find from equation (5.151):

$$\mu_{\text{opx}}^{\text{En}, excess} = W_{\text{AlMg}}^G X_{\text{Al}}^2 \tag{5.155}$$

and:

$$\mu_{\text{opx}}^{\text{MgTs}, excess} = W_{\text{AlMg}}^G X_{\text{Mg}}^2. \tag{5.156}$$

Incorporating these excess chemical potentials in (5.101) and (5.104), and recalling that $X_{\text{Al,M1}} + X_{\text{Mg,M1}} = 1$, we get:

$$\Delta_r G_{P,T}^{0,(i)} - 2RT \ln X_{\text{Mg,M1}} - 2W_{\text{AlMg}}^G \left(1 - X_{\text{Mg,M1}}\right)^2 = 0 \tag{5.157}$$

and:

$$\Delta_r G_{P,T}^{0,(ii)} + RT \ln \left(\frac{1 - X_{\text{Mg,M1}}}{X_{\text{Mg,M1}}}\right) + W_{\text{AlMg}}^G \left[X_{\text{Mg,M1}}^2 - \left(1 - X_{\text{Mg,M1}}\right)^2\right] = 0. \tag{5.158}$$

As in the ideal solution case (Software Box 5.4) we have two equations that, if we fix temperature, we can solve for the two unknowns, P and $X_{\text{Mg,M1}}$ in orthopyroxene. Procedure `spgrAlMASni` is thus otherwise identical to `spgrAlMAS`.

In the example discussed in Worked Example 5.7 I have purposely used values of standard state and excess mixing properties that are inconsistent with one another. Of course, the *Maple* procedure can be used with any combination of thermodynamic properties, by modifying the standard state values stored in the spreadsheet `RefStateData` and/or the value of the excess mixing parameter stored in the variable `WAlMg`.

Exercises for Chapter 5

5.1   Calculate the Clapeyron slope for the reaction *spinel + 2 enstatite ⇄ forsterite + pyrope* at 298 K and 1 bar, and compare your result to the slope of the reaction in Figure 5.5.

5.2   Using the reaction *spinel + 2 enstatite ⇄ forsterite + pyrope*, verify that the *assemblage* with higher entropy is the one that is on the high temperature side of the phase boundary, and the assemblage with higher density is on the high pressure side of the phase boundary.

5.3   Calculate the phase boundary for the transition between plagioclase lherzolite and spinel lherzolite for the end-member Ca–Mg system, assuming that both diopside and enstatite remain as pure end-member phases, and ignoring the order–disorder transition in anorthite (more on this in Chapter 7). You need to write a balanced reaction among anorthite, forsterite, diopside, enstatite and spinel, and program this reaction in the *Maple* worksheet `th_template_3.mw` (see Software Box 5.2). Use standard state properties from Holland and Powell (1998). You will refine this calculation in Exercises 5.15 through 5.17.

5.4   Calculate the activity of diamond relative to the standard state graphite, and the activity of graphite relative to the standard state diamond, at 298 K and 1 bar. What is the physical meaning of $a > 1$? Of $a < 1$?

5.5   Show that the conclusion that chemical species are transferred down chemical potential gradients (i.e. equation (5.26)) is valid in systems with any arbitrary number of phases and components.

5.6   Prove equations (5.95), starting from (5.70).

5.7   Prove equations (5.97), starting from (5.70).

5.8   Prove equations (5.100), starting from (5.70).

5.9   Write equations for the ideal activities of eastonite ($KMg_2AlAl_2Si_2O_{10}(OH)_2$) and muscovite ($KAl_2AlSi_3O_{10}(OH)_2$) in a trioctahedral mica.

5.10   Prove equation (5.91), i.e.:

$$\Delta G_{mixing} = \Delta H_{mixing} - T \Delta S_{mixing}.$$

5.11   Prove that ΔV_{mixing} for an ideal solution is zero.

5.12   Use *Maple*'s plotting capabilities to explore the conditions under which a non-ideal symmetric solution develops an inflected ΔG_{mixing} curve, as in Fig. 5.13c. (Hint: vary the relative values of temperature and the interaction parameter.)

5.13   Use *Maple*'s plotting capabilities to show that a G^{excess} function for an asymmetric solution that changes sign (as in Fig. 5.13d) arises only if the two interaction parameters have different signs.

5.14   Prove equation (5.150).

5.15   A first step in refining the anorthite–spinel phase boundary in lherzolites is to include the Mg–Tschermak's component in orthopyroxene (using reaction (5.102)) and the Ca–Tschermak's component in clinopyroxene: $CaAlAlSiO_6$. Assume that Ca fills the M2 site in cpx, and that Al and Mg mix in the M1 site of cpx, as in opx. You need to come up with an additional balanced chemical reaction that includes the Ca–Tschermak's species in cpx. There are several possibilities, but the simplest one is a reaction among the two Tschermak's components, diopside and enstatite. You will end up with three linearly independent equations ((5.102), the one that you derived in Exercise 5.3, and your new reaction) in four unknowns: P, T, $(X_{Al,M1})_{opx}$ and $(X_{Al,M1})_{cpx}$. The system has one degree of freedom, and can be solved by fixing one variable, for example temperature. Use the discussion in Software Box 5.4, and the worksheet described there, to help you program the solution of this system of equations in *Maple*, assuming ideal mixing in both pyroxenes. Plot the new plagioclase–spinel phase boundary and compare it with the one you generated in Exercise 5.3. Use standard state properties from Holland and Powell (1998).

5.16 In order to further refine the plagioclase–spinel phase boundary you should include excess mixing properties in pyroxenes. According to Holland and Powell, Al–Mg mixing in the M1 site of both pyroxenes is symmetric, with $(W_{AlMg})_{opx} = 0$ and $(W_{AlMg})_{cpx} = 7 \text{ kJ mol}^{-1}$. With the discussion in Software Box 5.5 as a guide, include the corresponding excess chemical potentials, recalculate the plagioclase–spinel phase boundary and compare it with the ones you generated in Exercises 5.3 and 5.15.

5.17 In reality, Ca and Mg also mix in the M2 site of both clinopyroxene and orthopyroxene. You can then add the following two reactions to your system of equations:

$$(CaMgSi_2O_6)_{cpx} \rightleftarrows (CaMgSi_2O_6)_{opx}$$

$$(Mg_2Si_2O_6)_{cpx} \rightleftarrows (Mg_2Si_2O_6)_{opx}$$

and end up with a system of five equations and six unknowns: P, T, $(X_{Al,M1})_{opx}$, $(X_{Al,M1})_{cpx}$, $(X_{Mg,M2})_{opx}$, and $(X_{Mg,M2})_{cpx}$. The system still has one degree of freedom, and can thus be solved by fixing one variable, e.g., temperature. According to Holland and Powell, Ca–Mg mixing in the M2 site of pyroxenes is symmetric, with $(W_{CaMg})_{opx} = (W_{CaMg})_{cpx} = 30 \text{ kJ mol}^{-1}$. Recalculate the plagioclase–spinel phase boundary and compare it with the ones you generated in Exercises 5.3, 5.15 and 5.16.

6

Phase equilibrium and phase diagrams

A phase diagram is a graph that shows the distribution of stable phase assemblages as a function of the values of the intensive variables used to describe the thermodynamic system. If pressure and temperature are the variables of interest then the stable assemblage is the one with the lowest Gibbs free energy, although other thermodynamic potentials may be used if the variables of interest are different, for example, Helmholtz free energy if one is interested in temperature and volume (more on this in Chapter 9). Phase diagrams are powerful analytical tools in many branches of planetary sciences, as they provide a way to quickly visualize how phase assemblages change in response to changes in pressure, temperature, chemical potentials, or any other combination of intensive variables. There are many different types of phase diagrams and it is not the purpose of this book to offer a comprehensive review of all of them. Rather, in this chapter I will focus on the fundamental rules that phase diagrams must abide by in order to be thermodynamically valid. We therefore begin with a discussion of the thermodynamic underpinnings of phase diagrams.

6.1 The foundations of phase equilibrium

6.1.1 The Gibbs–Duhem equation

Any extensive thermodynamic property can be written as a sum of products of partial molar properties times mol numbers (equation (5.27)). Let us re-write equation (5.27) specifying the identity of the phase that it applies to with the index α (α can be a solution or a pure phase, in which case identity (5.27) is trivial):

$$\mathbf{Z}^{\alpha} = \sum_i z_{\alpha}^i \mathbf{n}_i^{\alpha}. \tag{6.1}$$

Taking derivatives:

$$d\mathbf{Z}^{\alpha} = \sum_i z_{\alpha}^i d\mathbf{n}_i^{\alpha} + \sum_i \mathbf{n}_i^{\alpha} dz_{\alpha}^i. \tag{6.2}$$

Now, since $\mathbf{Z}$ is a state variable it is a function of temperature and pressure, i.e. the following function exists:

$$\mathbf{Z}^{\alpha} = \mathbf{Z}^{\alpha}\left(P, T, \mathbf{n}_i^{\alpha}\right). \tag{6.3}$$

Note that (6.1) and (6.3) are two different functions that yield the value of the same extensive property, $\mathbf{Z}$. Taking the derivative of (6.3):

$$d\mathbf{Z}^{\alpha} = \left(\frac{\partial \mathbf{Z}^{\alpha}}{\partial T}\right)_{P,\mathbf{n}_i^{\alpha}} dT + \left(\frac{\partial \mathbf{Z}^{\alpha}}{\partial P}\right)_{T,\mathbf{n}_i^{\alpha}} dP + \sum_i \left(\frac{\partial \mathbf{Z}^{\alpha}}{\partial \mathbf{n}_i^{\alpha}}\right)_{P,T,\mathbf{n}_{j\neq i}^{\alpha}} d\mathbf{n}_i^{\alpha}. \tag{6.4}$$

Substituting the definition of partial molar property, equation (5.28), in (6.4), equating with (6.2) and simplifying we arrive at the following, known as the *Gibbs–Duhem equation*:

$$\left(\frac{\partial \mathbf{Z}^{\alpha}}{\partial T}\right)_{P,\mathbf{n}_i^{\alpha}} dT + \left(\frac{\partial \mathbf{Z}^{\alpha}}{\partial P}\right)_{T,\mathbf{n}_i^{\alpha}} dP - \sum_i \mathbf{n}_i^{\alpha} dz_{\alpha}^i = 0. \tag{6.5}$$

This equation is valid for any thermodynamic extensive variable, but its most common application is to Gibbs free energy, in which case, by using (4.132) and (4.133), it becomes:

$$\mathbf{S}^{\alpha} dT - \mathbf{V}^{\alpha} dP + \sum_i \mathbf{n}_i^{\alpha} d\mu_{\alpha}^i = 0. \tag{6.6}$$

In this form equation (6.6) is also known as *Gibbs'equation 97*. Dividing by the total number of mols, $\mathbf{n}^{\alpha} = \Sigma_i \mathbf{n}_i^{\alpha}$ we can also write the Gibbs–Duhem equation as follows:

$$S^{\alpha} dT - V^{\alpha} dP + \sum_i X_i^{\alpha} d\mu_{\alpha}^i = 0. \tag{6.7}$$

The Gibbs–Duhem equation specifies the number of intensive variables that can vary independently in a homogeneous phase in thermodynamic equilibrium. For example, suppose that we can describe the composition of a phase in terms of the amounts of s different chemical species. There are then $s + 2$ intensive variables in (6.6): pressure, temperature and the chemical potential of each of the s species. The equation has $s + 1$ degrees of freedom, which is the number of intensive variables that must be specified in order to solve it.

You may have noticed that, whereas in Chapter 5 we distinguished between system components and phase components, in the previous paragraph we ignored that distinction and referred simply to the chemical species that make up the phase. The reason for this is that the Gibbs–Duhem equation makes no distinction between phase components and system components. We imposed no restrictions on the type of components when writing equation (6.1), except that whichever components we choose must allow us to fully describe the composition of the phase. Thus, the Gibbs–Duhem equation is equally true whether we choose to describe the composition of the phase in terms of system components, phase components or any combination of the two. We will exploit this flexibility on many occasions.

6.1.2 Gibbs' phase rule

Consider now a system made up of Φ different phases, and let us describe the composition of each of the phases in terms of the same set of system components. Recall from Section 5.1.1 that this is a set of chemical species whose compositions are linearly independent and such that they span the composition space of the full system (and hence of each of the phases that make up the system). In the language of linear algebra the system components conform a *coordinate basis* or *minimal spanning set* for the composition space of the system. The coordinate basis, i.e. the specific set of system components that we choose, is not unique but, importantly, whatever basis we choose it always contains the same number of system components, which we shall call c (in linear algebra, c is the dimension of the composition space). We can construct a system of Φ Gibbs–Duhem equations, as follows:

$$S^1 dT - V^1 dP + \sum_{i=1}^{c} n_i^1 d\mu_1^i = 0$$

$$\cdots\cdots\cdots\cdots\cdots\cdots\cdots\cdots\cdots\cdots\cdots\cdots\cdots \tag{6.8}$$

$$S^\Phi dT - V^\Phi dP + \sum_{i=1}^{c} n_i^\Phi d\mu_\Phi^i = 0.$$

For a system at equilibrium, and given that we have chosen to write all of the Gibbs–Duhem equations in terms of the same set of system components, it must be $\mu_1^i = \cdots = \mu_\Phi^i$, for every one of the c components, $1 \le i \le c$. At equilibrium there must also be identity among the differentials of the chemical potentials, i.e. $d\mu_1^i = \cdots = d\mu_\Phi^i$. There are then Φ equations in $c+2$ independent variables: dT, dP and c $d\mu$s. Two fundamental results follow immediately from the elementary algebraic properties of systems of linear equations.

(i) The maximum possible number of phases at equilibrium in a system composed of c *linearly independent chemical species* (= *system components*) is $c + 2$. If the number of phases, Φ, were greater than this then there would be more Gibbs–Duhem equations than independent variables and the system of equations would be inconsistent.

(ii) The number of degrees of freedom of the system, f, is given by:

$$f = c + 2 - \Phi. \tag{6.9}$$

The number of degrees of freedom, $f \ge 0$, also called the *variance* of the system, is the number of intensive variables that must be specified, or constrained independently, in order to be able to solve the system of equations (6.8) and hence have a complete description of the thermodynamic state of the system.

These two statements constitute *Gibbs' phase rule*. It is a simple algebraic result with profound implications for understanding the physicochemical constitution of planetary bodies.

The phase rule is commonly derived along the lines that I followed here, but I find it interesting that a subtle yet obvious question is seldom addressed explicitly: where does the 2 in equation (6.9) come from? Perhaps the clearest justification for the 2 can be found in the very first two sections of Guggenheim's *Thermodynamics* (1967), which I rephrase as follows:

> *In order to specify the thermodynamic state of a system we need a variable that keeps track of changes in thermal energy – temperature or its conjugate, entropy – and another variable that keeps track of changes in mechanical energy – pressure or its conjugate, volume. These are the two variables that show up in equation (6.9) in addition to the chemical potentials of each of the linearly independent chemical species.*

If additional products of conjugate variables need to be considered (see Section 4.8.4), such as would be the case if the gravitational potential cannot be ignored (Chapter 13), then the constant term in equation (6.9) becomes greater than 2.

6.1.3 Choosing and switching components

The solution of a system of equations such as (6.8) yields a "complete thermodynamic description" of the system of interest. By this we mean that we can calculate its pressure, its temperature and the chemical potential of each of the chemical species that compose the system. Equivalently, since there is a functional relationship between chemical potential and composition (e.g., equation (5.2.12)), the "complete thermodynamic description" could be T, P and the composition of each phase. This is the most useful description in planetary sciences, and the chief goal of many thermodynamic calculations. There are alternative ways of arriving at this description. One possibility is to integrate each of the Gibbs–Duhem equations in (6.8) and then solve for the combination of intensive variables that satisfies the system of integrated equations. This is seldom easy, especially for systems of more than two components. Fortunately, it is also seldom necessary, as several shortcuts are possible. In fact, we have already used some of these shortcuts in the numerical examples in Chapter 5.

In the derivation of the phase rule we specified that the Gibbs–Duhem equations for all phases in the system be written in terms of system components. Because system components constitute a minimal spanning set for the composition of the system, this requirement assures that the number of unknown variables in the system of equations (6.8) is the minimum possible. The number of equations relating these variables in (6.8) is also the minimum possible, as there is one and only one Gibbs–Duhem equation per phase. These two statements signify that the mathematical description of the thermodynamic state of the system is complete and cannot be simplified any further. *In particular, regardless of the way in which we choose to solve for the thermodynamic state of the system, the number of degrees of freedom of our system of equations must be the one given by the phase rule.* This is perhaps the most important consequence of Gibbs' phase rule, and the first key to finding algebraic shortcuts for the thermodynamic description of a system.

The second key is more a question of intuition, experience and the specific goal that one has, rather than a rigorous mathematical rule, although such rule exists, as we shall see in a moment. We begin by choosing the set of chemical species that we are actually interested in. This is determined by the nature of the system that we are investigating, and by what we know about the possible compositional range of each of the phases that make up the system. We must choose at least as many chemical species as the number of system components, and we must choose a set of chemical species that allows us to write the composition of all of the phases in the system. Beyond these restrictions, however, there can be any number of species, and their nature (system components, phase components or a combination of both) is not important. Let us say that we choose a total of s chemical species, with $s \geq c$ (c is the number of system components). The Φ Gibbs–Duhem equations must be re-written in terms of these s components (some of the mol numbers or mol fractions may be zero, but this is not a problem).

Clearly, if $s > c$ then the compositions of all s species cannot be linearly independent. More precisely, there must exist $s - c$ linearly independent equations relating the compositions of the s species. Each of these equations corresponds to a balanced chemical reaction among some or all of the s chemical species. But if we can write a balanced chemical reaction among chemical species then we can also write an equation among their chemical potentials that describes a condition of heterogeneous chemical equilibrium. Each of these equations is a version of (5.22):

$$\sum_i \nu_i \mu^i = 0 \qquad (6.10)$$

and each one can be differentiated to obtain an equation of the form:

$$\sum_i v_i d\mu^i = 0. \tag{6.11}$$

There are $s - c$ equations like (6.11), which, together with the Φ Gibbs–Duhem equations, gives a total of $\Phi + s - c$ equations. These equations contain $s + 2$ unknown variables: pressure, temperature and the chemical potential of each of the s chemical species. The number of degrees of freedom of the augmented system of equations is still $c + 2 - \Phi$, as required by the phase rule. This system of equations is an alternative description of the thermodynamic state of the system that is equivalent to a set of Gibbs–Duhem equations such as (6.8).

The underlying algebraic rule is simple: we must add one equation for each chemical species that we wish to consider beyond the number of system components. Generally, each of these additional equations is an equation of heterogeneous equilibrium of the form (6.10). *And the shortcut that we seek is to solve only these equations, and stay away from the Gibbs–Duhem equations if at all possible.* Equations of the form (6.10) are a lot easier to solve because they are already given in integral form, and all that is required is a function that gives chemical potential in terms of P, T and phase composition. This is exactly what we have been doing throughout Chapter 5.

By now you may be thoroughly confused. The best way of clearing the air is with an example. I encourage you to go over the following example carefully, and return to the preceding discussion often as you do so. The example also demonstrates additional thermodynamic possibilities that we will exploit further in subsequent sections.

Worked Example 6.1   The spinel–garnet transition in planetary mantles revisited

The model for the spinel–garnet transition that we discussed in Chapter 5 consists of the four phases: spinel, orthopyroxene, olivine and garnet. The Mg end-member system is spanned by three system components, which we can choose as SiO_2, Al_2O_3 and MgO (but see Exercise 6.1). Thus, $\Phi = 4$, $c = 3$ and, from (6.9), $f = 1$. This is the same number of degrees of freedom that we found when we solved the equations that describe the phase boundaries in Worked Examples 5.1 and 5.6. Where did those equations come from, and how can we be sure that they are a complete thermodynamic description of the system?

Let us start by writing out the Gibbs–Duhem equations in terms of our chosen system components. The explicit form of (6.8) for this system is:

$$S^{sp} dT - V^{sp} dP + n^{sp}_{SiO_2} d\mu^{SiO_2} + n^{sp}_{Al_2O_3} d\mu^{Al_2O_3} + n^{sp}_{MgO} d\mu^{MgO} = 0$$

$$S^{opx} dT - V^{opx} dP + n^{opx}_{SiO_2} d\mu^{SiO_2} + n^{opx}_{Al_2O_3} d\mu^{Al_2O_3} + n^{opx}_{MgO} d\mu^{MgO} = 0$$

$$S^{ol} dT - V^{ol} dP + n^{ol}_{SiO_2} d\mu^{SiO_2} + n^{ol}_{Al_2O_3} d\mu^{Al_2O_3} + n^{ol}_{MgO} d\mu^{MgO} = 0$$

$$S^{grt} dT - V^{grt} dP + n^{grt}_{SiO_2} d\mu^{SiO_2} + n^{grt}_{Al_2O_3} d\mu^{Al_2O_3} + n^{grt}_{MgO} d\mu^{MgO} = 0. \tag{6.12}$$

I have omitted the phase identification subscripts in the chemical potential terms because at equilibrium the chemical potential of each component is the same in all phases. Some of the mol numbers, such as $n^{sp}_{SiO_2}$ and $n^{ol}_{Al_2O_3}$, may be zero, but there is no harm in keeping the corresponding terms in (6.12). This is a (minimal) system of four equations with five unknowns: dP, dT, $d\mu^{SiO_2}$, $d\mu^{Al_2O_3}$ and $d\mu^{MgO}$. We could solve it by integrating the

Gibbs–Duhem equations and specifying the value of any one of the intensive variables. As an aside, a system of equations such as (6.12), in which all equations have a zero constant term, is called a *homogeneous system* and its only solutions are zeroes. Integrating the equations converts (6.12) into a heterogeneous system with non-zero solutions, thanks to non-zero integration constants. This is an important concept from linear algebra, but need not concern us too much at this point for, if all we are interested in is the location of the phase boundary, there is a much easier solution that does not require any integration.

We begin by re-writing the Gibbs–Duhem equations in terms of an appropriately chosen set of phase components. Suppose first that we did not know that Al can enter orthopyroxene. The compositions of the four phases can in that case be written out in terms of the four phase components: $MgAl_2O_4$, $Mg_2Si_2O_6$, Mg_2SiO_4 and $Mg_3Al_2Si_3O_{12}$. This is one more than the number of system components, so in order to preserve the number of degrees of freedom we must add an equation. Let us re-write the Gibbs–Duhem equations in terms of our new set of components:

$$S^{sp}dT - V^{sp}dP + n^{sp}_{MgAl_2O_4}d\mu^{MgAl_2O_4}_{sp} = 0$$

$$S^{opx}dT - V^{opx}dP + n^{opx}_{Mg_2Si_2O_6}d\mu^{Mg_2Si_2O_6}_{opx} = 0$$

$$S^{ol}dT - V^{ol}dP + n^{ol}_{Mg_2SiO_4}d\mu^{Mg_2SiO_4}_{ol} = 0$$

$$S^{grt}dT - V^{grt}dP + n^{grt}_{Mg_3Al_2Si_3O_{12}}d\mu^{Mg_3Al_2Si_3O_{12}}_{grt} = 0. \qquad (6.13)$$

In (6.13) it is convenient to include the phase identification subscripts in the chemical potentials because the additional equation that we seek is a heterogeneous equilibrium equation among these species, obtained by differentiation of equation (5.23):

$$\mu^{MgAl_2O_4}_{sp} + 2\mu^{Mg_2Si_2O_6}_{opx} = \mu^{Mg_3Al_2Si_3O_{12}}_{grt} + \mu^{Mg_2SiO_4}_{ol}, \qquad (6.14)$$

which results in:

$$d\mu^{MgAl_2O_4}_{sp} + 2d\mu^{Mg_2Si_2O_6}_{opx} - d\mu^{Mg_3Al_2Si_3O_{12}}_{grt} - d\mu^{Mg_2SiO_4}_{ol} = 0. \qquad (6.15)$$

Equations (6.13) and (6.15) constitute a system of five equations in six unknowns: dP, dT and the four $d\mu$s. It therefore preserves the one degree of freedom required by the phase rule, as shown by the system (6.12). *We do not need to solve the full system of equations, however, because we can write each of the chemical potentials in (6.14) as a function of temperature and pressure.* For phases in their standard state (e.g. pure phase at the temperature and pressure of interest) the required function $\mu^0 = \mu^0(P, T)$ is equation (5.1.1). Substituting a function of this kind for each of the four phases in (6.14) reduces this equation to two unknowns: P and T. The resulting equation is (5.16), which we solved in Worked Example 5.1 by specifying temperature.

Of course, we know from Worked Example (5.6) that, given that Al dissolves in orthopyroxene, this solution is not correct. In order to obtain a better solution we need to consider an additional chemical species, the Mg–Tschermak's component in orthopyroxene: $MgAlAlSiO_6$. The Gibbs–Duhem equations are now as follows:

$$S^{sp}dT - V^{sp}dP + n^{sp}_{MgAl_2O_4}d\mu^{MgAl_2O_4}_{sp} = 0$$

$$S^{opx}dT - V^{opx}dP + n^{opx}_{Mg_2Si_2O_6}d\mu^{Mg_2Si_2O_6}_{opx} + n^{opx}_{MgAlAlSiO_6}d\mu^{MgAlAlSiO_6}_{opx} = 0$$

$$S^{\text{ol}}dT - V^{\text{ol}}dP + n^{\text{ol}}_{\text{Mg}_2\text{SiO}_4}d\mu^{\text{Mg}_2\text{SiO}_4}_{\text{ol}} = 0$$

$$S^{\text{grt}}dT - V^{\text{grt}}dP + n^{\text{grt}}_{\text{Mg}_3\text{Al}_2\text{Si}_3\text{O}_{12}}d\mu^{\text{Mg}_3\text{Al}_2\text{Si}_3\text{O}_{12}}_{\text{grt}} = 0. \tag{6.16}$$

But, in order to preserve the number of degrees of freedom we need an additional equation, which is the heterogeneous equilibrium condition for reaction (5.102):

$$\mu^{\text{MgAl}_2\text{O}_4}_{\text{spinel}} + \mu^{\text{Mg}_2\text{Si}_2\text{O}_6}_{\text{opx}} = \mu^{\text{MgAlAlSiO}_6}_{\text{opx}} + \mu^{\text{Mg}_2\text{SiO}_4}_{\text{olivine}}. \tag{6.17}$$

or, equivalently:

$$d\mu^{\text{MgAl}_2\text{O}_4}_{\text{spinel}} + d\mu^{\text{Mg}_2\text{Si}_2\text{O}_6}_{\text{opx}} - d\mu^{\text{MgAlAlSiO}_6}_{\text{opx}} + d\mu^{\text{Mg}_2\text{SiO}_4}_{\text{olivine}} = 0. \tag{6.18}$$

Equations (6.16), (6.15) and (6.18) constitute a system of six equations in seven unknowns: dP, dT and five $d\mu$s. Hence, one degree of freedom (and J. Willard Gibbs stays happy). But if we now focus on the integral forms of (6.15) and (6.18), which are equations (6.14) and (6.17), respectively, we see that these are two equations in three unknowns, as we can write the chemical potentials of $\text{Mg}_2\text{Si}_2\text{O}_6$ and $\text{MgAl}_2\text{SiO}_6$ as a function of P, T and mol fraction of Mg (or Al) in orthopyroxene, and all the other chemical potentials as functions of P and T only. With the appropriate substitutions, and assuming ideal mixing in orthopyroxene, the resulting equations are (5.101) and (5.103), which we solved in Worked Example 5.6. Note that if we choose to treat orthopyroxene as a non-ideal solution there is still one degree of freedom, as the excess chemical potentials of $\text{Mg}_2\text{Si}_2\text{O}_6$ and $\text{MgAl}_2\text{SiO}_6$ are functions of P, T and orthopyroxene composition only (e.g. equations (5.153) or (5.154)) so that no additional variables are introduced (this is true in general, not just for this specific example).

There is no need to stop here, however. For example, we may be interested in the chemical potentials of the species SiO_2, Al_2O_3 and MgO. Even though the amounts of these components cannot vary independently in any of the four phases that we are considering, equations (6.12) assure us that the chemical potentials of these species are well defined, and can be calculated. It is important to understand that this is always true, even if none of these chemical species exist as "free" or "stoichiometric" components in the system of interest. We will have many uses for this fact. For now, we motivate the following calculation by noting that the chemical potentials of the oxide species may be important in order to understand how mantle phases interact with supercritical hydrous fluids (Chapter 9), or in order to understand their melting relationships (Chapter 10).

Since we have already determined that equations (6.14) and (6.17) by themselves constitute a valid thermodynamic description of our system we can start from these two equations only and ignore the Gibbs–Duhem equations. The problem of determining the chemical potentials of the three oxides is simply one of adding three new variables: μ^{SiO_2}, $\mu^{\text{Al}_2\text{O}_3}$ and μ^{MgO} and three new equations of heterogeneous equilibrium, so as to preserve the number of degrees of freedom. There are several possible sets of equations, but as long as we choose three equations that are linearly independent which particular three we choose is immaterial (Exercise 6.3). Here I choose the following three:

$$\text{Mg}_2\text{SiO}_4 + \text{SiO}_2 \rightleftharpoons \text{Mg}_2\text{Si}_2\text{O}_6 \tag{6.19}$$

$$2\text{MgAl}_2\text{O}_4 + \text{SiO}_2 \rightleftharpoons \text{Mg}_2\text{SiO}_4 + 2\text{Al}_2\text{O}_3 \tag{6.20}$$

$$\text{MgAl}_2\text{O}_4 \rightleftharpoons \text{MgO} + \text{Al}_2\text{O}_3. \tag{6.21}$$

We can write the chemical potential of each of the three oxides as the sum of its standard state chemical potential, chosen, for example, as pure crystalline solid at the temperature and pressure of interest, plus an activity term (equation (5.45)). The conditions of heterogeneous chemical equilibrium for the three reactions are then as follows:

$$\Delta_r G_{P,T}^{0,(6.19)} + RT \ln \left(X_{\mathrm{Mg,M1}} \right)_{\mathrm{opx}} - RT \ln a^{\mathrm{SiO_2}} = 0 \qquad (6.22)$$

$$\Delta_r G_{P,T}^{0,(6.20)} + 2RT \ln a^{\mathrm{Al_2O_3}} - RT \ln a^{\mathrm{SiO_2}} = 0 \qquad (6.23)$$

$$\Delta_r G_{P,T}^{0,(6.21)} + RT \ln a^{\mathrm{Al_2O_3}} + RT \ln a^{\mathrm{MgO}} = 0. \qquad (6.24)$$

In these equations the standard state chemical potentials of the oxides are included in the $\Delta_r G$ terms, as usual. I have assumed ideal mixing in orthopyroxene and used (5.99) for the activity of $\mathrm{Mg_2Si_2O_6}$. We now have five equations: (6.14), (6.17) and (6.22) through (6.24), with six independent variables: P, T, $X_{\mathrm{Mg,M_1}}$, $a^{\mathrm{SiO_2}}$, $a^{\mathrm{Al_2O_3}}$ and a^{MgO}. The system of equations preserves the single degree of freedom and can be solved if we specify the value of one of the intensive variables. It is straightforward to add the three new equations to the *Maple* worksheet that we developed in Software Box 5.4 to solve for the spinel–garnet phase boundary. We can then specify temperature (for example) and solve simultaneously for the other five variables. You should verify (Exercises 6.2 and 6.3) that the resulting values of pressure and Al content in orthopyroxene are the same ones that we obtained in Chapter 5 (Fig. 5.10).

The activities of the three oxides calculated in this way are plotted in Fig. 6.1 as a function of temperature. In the ternary $\mathrm{SiO_2}$–$\mathrm{Al_2O_3}$–MgO system, the four-phase assemblage spinel–orthopyroxene–olivine–garnet fixes, or *buffers*, the chemical potentials (or activities) of the three oxide components along the curves shown in Fig. 6.1. This means that, as long

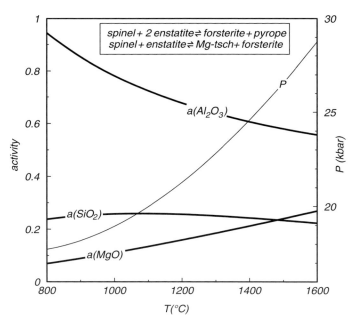

Fig. 6.1   Activities of $\mathrm{Al_2O_3}$, $\mathrm{SiO_2}$ and MgO buffered by the orthopyroxene–spinel–forsterite–pyrope univariant assemblage. Pressure at the univariant equilibrium from Fig. 5.10.

as the four phases are present, the chemical potential of each oxide takes one and only one value at any given temperature. Note that neither pressure nor $(X_{Mg,M1})_{opx}$ are constant along these curves. Rather, their values are the equilibrium values for each temperature, as shown in Fig. 5.10.

The three oxides behave differently with temperature. In particular, the activity of Al_2O_3 increases with decreasing temperature, and reaches a value of 1 at $T \sim 750\,°C$ and $P \sim 17.5$ kbar. Since the standard state that we chose is pure crystalline Al_2O_3 at the temperature and pressure of interest, this means that at those $P–T$ conditions the four-phase system becomes saturated in an additional phase, namely, corundum. According to the phase rule the number of degrees of freedom now becomes zero, as the number of system components has not changed, but there is now an additional phase. A system of equations with no degrees of freedom has a single solution and does not allow us to fix the value of any of the variables independently. Thus, in the ternary $SiO_2–Al_2O_3–MgO$ system there is a single $P–T$ combination at which the five phases spinel–orthopyroxene–olivine–garnet–corundum coexist at equilibrium. We discuss these concepts further beginning in Section 6.2.

You may be wondering, why do we need to go to all the trouble discussed in excruciating detail in this example? In practice we seldom do, as we tacitly skip over the Gibbs–Duhem equations (unless there are reasons to use them, more on this later). However, before immersing oneself in any thermodynamic calculation it is always necessary to do a "phase rule check" of number of phases, components and degrees of freedom (and of course the Gibbs–Duhem equations are implicit in this). *If the number of degrees of freedom of whatever system of equations we set up does not agree with the number predicted by the phase rule then our proposed thermodynamic description cannot be correct.*

As another example, consider the gas speciation calculation from Worked Examples 5.4 and 5.5. Our model system consists of two phases (graphite and gas) and is described by two system components (we can choose, for example, C and O_2). Thus, Gibbs' phase rule assures us that the system has two degrees of freedom. When we solved the system of equations consisting of (5.87), (5.88) and the composition of the gas phase ($X_{CO_2} + X_{CO} + X_{O_2} = 1$) I stated that this is a system of three equations in three unknowns (the three mol fractions), that can therefore be solved exactly. This does not mean, however, that the *thermodynamic* system has no degrees of freedom, as in order to solve this system of equations we had to specify two intensive variables: temperature and pressure (which we fixed at 1 bar). The thermodynamic system has two degrees of freedom, as required by the phase rule. The species distribution can *only* be calculated if we specify both P and T.

6.2 Analysis of phase equilibrium among phases of fixed composition

The phase rule is the foundation for the study of phase equilibrium and phase diagrams. For the sake of clarity it is convenient to break up the discussion of phase diagrams into different parts, focusing first on equilibria among phases of fixed composition and then on equilibria among phases of variable composition. Nature, of course, does not fall neatly into one or the other of these categories, so that we generally have to keep both sets of concepts in mind.

6.2.1 Some fundamental concepts and terminology

A phase assemblage with a single degree of freedom (such as the four-phase assemblage spinel–orthopyroxene–olivine–garnet in the ternary system SiO_2–Al_2O_3–MgO) is called a *univariant* assemblage. Recall that this means that, if the four phases exist at equilibrium, then only one intensive variable can be independently specified. The geometric representation of a univariant assemblage is a segment of a curve, that we also call a phase boundary (e.g. Figs. 5.2 and 5.5). It should be immediately apparent that an assemblage with no degrees of freedom, called an *invariant* assemblage, is expressed geometrically by a point, and an assemblage with two degrees of freedom, termed *divariant*, is represented geometrically by a sector of a two-dimensional surface. We could keep going, and note that a trivariant assemblage corresponds to a portion of a three-dimensional volume, a quadrivariant assemblage to a portion of a four-dimensional hypervolume, and so on. Algebraic descriptions of assemblages with any number of degrees of freedom are not a problem, but phase diagrams limit us to representing information in two dimensions. This means that geometric representations of phase equilibria commonly do not extend beyond divariant assemblages. This would appear to be a serious limitation on the usefulness of phase diagrams but, as we shall soon see, in most cases it is not, as it allows us to focus on those variables that are particularly relevant to the problem at hand. Moreover, low-variance assemblages (say, those with $f \leq 2$) are particularly useful, as they often make it possible to place fairly tight brackets on the values of intensive variables.

An equilibrium invariant assemblage ($f = 0$) in a system of c components consists of $c + 2$ phases (equation (6.9)). This assemblage is represented by a point on a plane, *in which the coordinates are any two intensive variables*. The emphasis is crucial: the variables that we use to track phase equilibrium can be any combination of intensive variables, and *the principles that rule the construction of phase diagrams are the same regardless of which combination of intensive variables we use.* The variables can be P and T, or two chemical potentials, or a chemical potential and T or P, or some other combination. In order to emphasize the fact that the rules that govern phase diagrams are completely general I will use the names Y and Z for the intensive variables, unless the specific example calls for a particular set of intensive variables. Going back now to our system of c components we can see that there are $c + 2$ different univariant assemblages that converge at the invariant point, each of them consisting of $c + 1$ phases. We obtain these univariant assemblages by eliminating each of the $c + 2$ phases that exist at the invariant point, one at a time. Each of these univariant assemblages is represented by a different curve on the Y–Z plane, and all of the curves must have a common intersection at the invariant point.

Consider now a system of one component, in which there are three phases: A, B and C, that exist at equilibrium at an invariant point. There are three univariant curves that intersect at the invariant point, each of them representing univariant equilibrium of one of the three possible two-phase assemblages. This situation is sketched in the left hand side of Fig. 6.2. The labels next to the univariant curves indicate which phases are stable on each side of each phase boundary, and the two phases are of course stable along the corresponding curve. This particular arrangement of phases is not random. It is an arrangement that is thermodynamically possible. In order to see what this means, and to derive some fundamental properties of phase diagrams, we begin by noting that Gibbs free energy is a monotonic function of all intensive variables or, more precisely, that the first and second derivatives of Gibbs free energy relative to any intensive variable never change sign. We have seen that this is the case for temperature and pressure (equations (4.132), (4.133), (4.135) and (4.136)), and it is

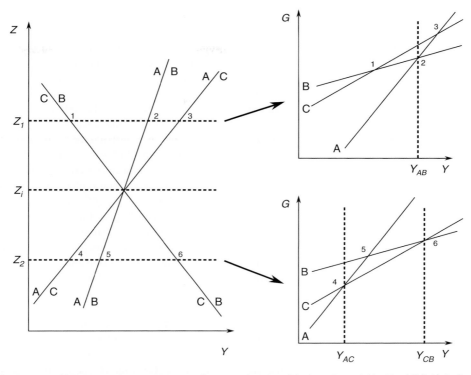

Fig. 6.2 Univariant equilibria in a one-component system shown as a function of the intensive variables Z and Y (left). Each point on the univariant curves corresponds to an intersection between two $G = G(Y)$ curves at constant Z (right diagrams). For example, intersection 2 between the Gibbs free energy curves for phases A and B corresponds to the point on the A–B univariant curve at $Z = Z_1$ and $Y = Y_{AB}$.

trivial to show (e.g. from equation (6.1)) that it is also true for the derivatives of G relative to chemical potential. This property of the Gibbs free energy function means that, if Y and Z are any two intensive variables and Z is kept constant, then the curves $G = G(Y)$ for different univariant assemblages intersect only once. This is shown on the right-hand side of Fig. 6.2, for two different values of Z, greater than and less than the value of Z at the invariant point, Z_i.

Clearly, the three $G(Y)$ curves have a common intersection point at the (Y, Z) coordinates of the invariant point (not shown in the figure). It is also evident that, because G is monotonic relative to all intensive variables, the order in which the curves intersect relative to the value of Y must change as the value of Z becomes greater or less than that at the invariant point. For example, for $Z_1 > Z_i$ the order of intersection may be as shown in the top right of Fig. 6.2. If this is the case, then there must exist a finite interval in the neighborhood of the invariant point, $Z_2 < Z_i$, in which the curves intersect as shown in the bottom right of Fig. 6.2. It follows that some arrangements of univariant and divariant assemblages around the invariant point are possible, whereas others are not thermodynamically permissible. The monotonicity of the Gibbs free energy function assures us of this. Note that it would be equally valid to have the sequence of intersections be 1–2–3 at Z_2 and 4–5–6 at Z_1. However, given the properties of the Gibbs free energy function, other arrangements are not possible (see also Exercise 6.4).

An examination of the diagrams on the right-hand side of Fig. 6.2 shows that not all intersections between Gibbs free energy functions correspond to stable univariant assemblages. For example, at $Z = Z_1$, intersection 1 corresponds to the univariant assemblage BC, but this assemblage has higher Gibbs free energy than A at the same conditions, and intersection 3 corresponds to the univariant assemblage AC, but this has higher Gibbs free energy than B at the same conditions. The two intersections, 1 and 3, are said to correspond to *metastable* univariant equilibria. What this means is that univariant equilibrium among the corresponding phases is possible, and may in some cases persist for extended periods of time, but, given the appropriate activation energy (Chapter 12, recall also our discussion of the *Hindenburg* in Chapter 5) the system will collapse to the stable lowest free energy state. The lowest Gibbs free energy state in this example corresponds to stable divariant equilibrium assemblages, which are in this case the single-phase assemblages A and B relative to the metastable univariant assemblages BC and AC, respectively. We can also see that at $Z > Z_i$ phase C is never stable, and the only possible stable univariant equilibrium assemblage is AB. The same line of argument shows that at $Z < Z_i$ all three phases can be stable, univariant assemblages AC and BC can both be stable, but univariant assemblage AB is always metastable. Most importantly for the construction and analysis of phase diagrams, note that all three univariant equilibria switch from stable to metastable when they cross the invariant point.

A phase diagram such as that on Fig. 6.2 is not satisfactory, as it does not convey the fundamental difference between stable and metastable equilibria. A better representation of the phase relations in this system is given by the phase diagram in Fig. 6.3. The stable univariant equilibria are shown with long solid lines and their *metastable extensions* across the invariant point with short broken lines. The importance of showing these metastable extensions will become clear in the next section. The divariant fields can in this case be uniquely identified with the name of the single phase that is stable in each of them but, as we shall see, this is not true of systems of more than one component. In contrast, there is a simple way of labeling the univariant curves, by using the identity of the single phase that is absent along each univariant curve. This carries over to systems of any number of components. For example, the curve that represents univariant equilibrium between phases A and B will be called the *C absent curve*, and is symbolized by placing C inside parentheses at the end of the curve, as shown in Fig. 6.3. This apparently innocent labeling system constitutes a powerful tool in the construction of phase diagrams, which we will examine in the next section. Here we note that the metastable extension of each univariant curve enters the divariant stability field of the phase that is absent along that curve. This is a necessary consequence of the monotonicity of the Gibbs free energy function, as you can verify for yourself by comparing the Z–Y diagram with the G–Y diagrams in Fig. 6.3. There is a more general way of stating this property of metastatble extensions, easily applicable to systems of any number of components, which is that the metastable extension of each univariant curve enters the only divariant field in which the phase absent along the univariant curve appears as a reactant on the two univariant curves that bound the field. Thus, the metastable extension of the (A) curve enters the only divariant field which is bound by univariant curves in which A appears as a reactant (see Fig. 6.3).

6.2.2 Schreinemakers' rule

Although it is possible to carry out an analysis such as that in the preceding section for systems of more than one component, the number of assemblages and Gibbs free energy

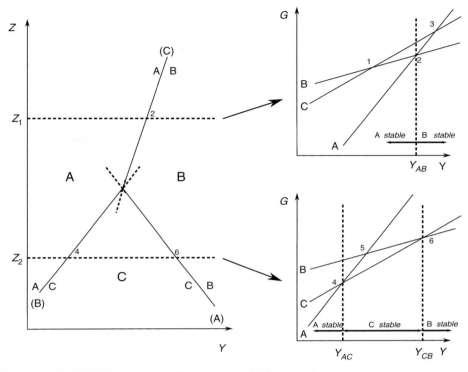

Fig. 6.3 Same as Fig. 6.2, distinguishing between stable and metastable segments of the univariant curves. Intersections 2, 4 and 6 are stable univariant equilibria, because they have the lowest Gibbs free energy for the given values of Z and Y. Intersections 1, 3 and 5 are metastable univariant equilibria. In every case a divariant assemblage with lower Gibbs free energy exists.

curves becomes quickly unmanageable. Notice, for example, that although in the neighborhood of an invariant point the plane is always divided by the $c + 2$ univariant curves into $c + 2$ divariant fields, this does not mean that there are $c + 2$ different divariant assemblages. Except for the case $c = 1$, there are more. To see why, we note that in the same way that we generate a univariant assemblage by omitting one phase from those present at an invariant point, we can generate a divariant assemblage by omitting one phase from those present along a univariant curve. Since there are $c + 1$ phases in a univariant assemblage, there must be $c + 1$ divariant assemblages "radiating" away from each of the $c + 2$ univariant curves. This leads to a total of $(c + 1)(c + 2)$ divariant assemblages, but each of these assemblages is duplicated, as interchanging the order in which the two phases are omitted leads to the same exact assemblage. Thus, the number of divariant assemblages is $1/2\,(c + 1)(c + 2)$. For $c = 1$ this yields three divariant assemblages, in agreement with what we saw in the previous section. In that case there is one and only one assemblage in each divariant field, but for $c = 2$ there are six divariant assemblages distributed over four divariant fields, for $c = 3$ there are ten divariant assemblages distributed over five divariant fields, and so on.

How are we to represent a system with such complexities on a two-dimensional diagram? Fortunately, there is a simple and clever way, which arises from the work, almost a century ago, of the Dutch physical chemist F.A.H. Schreinemakers (see, for example, the 1965

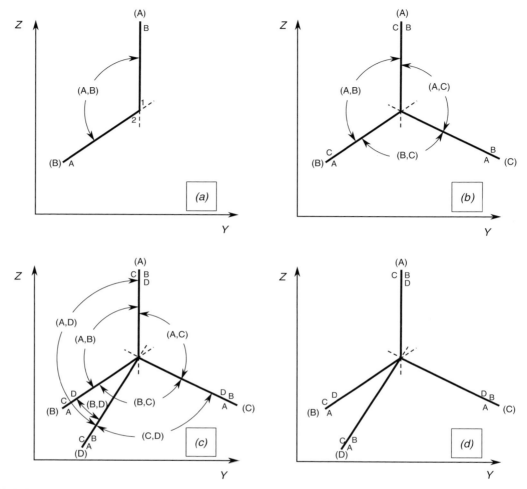

Fig. 6.4 Justification of Schreinemakers' rule. Phases in parentheses are absent along the corresponding univariant curve or inside the divariant field (or fields) spanned by the arc segments with arrow heads. See text for complete explanation.

English translation of his collected papers). The method is remarkably sophisticated and leads to many possible variants of phase diagrams, which we will not review in detail as there are clear and complete treatments available in the literature (especially those by Korzhinskii, 1959, and Zen, 1984). The fundamental principle of Schreinemakers' method is, however, very simple and readily explained. We will use it repeatedly in the construction of different types of phase diagrams. In fact, we have already deduced an important part of the method in the preceding section, as we shall now see.

Let us consider a system with $c > 1$ (the actual value of c is not important for now), and two univariant curves, which we shall label (A) and (B), see Fig. 6.4a. Divariant fields radiate from both univariant curves. In particular, the (unique) divariant field in which both of these phases are absent, which we can label (A, B), radiates from both of these univariant curves. Without a formal demonstration, I will make the intuitively acceptable statement that, just as a univariant curve switches from stable to metastable when it crosses the invariant point, a divariant field switches from stable to metastable when it crosses a univariant curve (see, for example, Zen, 1984). Thus, the divariant field (A, B) is stable on one side of each of the univariant curves and metastable on the other. We can see from Fig. 6.4a that it must be stable on the side of the intersection in which the angle between the two curves is less than 180°, which is shown in the figure by the arc labeled (A, B). To see why this must be the case let us assume that (A, B) was stable on the other side, i.e. to the right of (A) and below (B). If this were the case then we would run into the contradiction that between each stable curve and the metastable extension of the other one, in regions 1 and 2, (A, B) would have to be simultaneously stable and metastable. This is, essentially, Schreinemakers' rule: a stable divariant field always extends between the two univariant curves that each lack one of the two phases absent from the divariant assemblage, and on the side in which the angle between the curves is less than 180° (there is a special case in which the angle equals 180°, which we discuss later). This simple rule makes it possible to construct phase diagrams of arbitrary complexity. Remarkably, it can be stated in even simpler terms. To see how, we note that, if (A, B) is stable to the left of (A) and above (B), then all divariant fields that are stable to the right of (A) must contain B, and all divariant fields that are stable below (B) must contain A. Thus, A and B must show up as reactants on the sides of the univariant curves opposite to the (A, B) field, as shown in Fig. 6.4a – this is of course the same conclusion that we reached in Section 6.1.1 for a one-component system.

Let us now add a third univariant curve, labeled (C). Following the same arguments as above, we label the stable divariant fields and add the reactant phases on the opposite sides of the univariant curves, as shown in Fig. 6.4b. It appears that we have recovered the phase diagram from Fig. 6.3, and if this was a one component system that would be the case. In general, however, this may be only the beginning in the construction of a phase diagram for a system with $c > 1$. We can add a fourth univariant curve, labeled (D), for example as in Fig. 6.4c. Repeating the same arguments we conclude that the identities of the stable divariant fields and the sides on which the reactant phases plot must be as shown in this figure (note that if we had placed the fourth curve anywhere else the diagram would be identical, except for a rotation and some switched divariant field labels). We could keep going, but there is little reason to do so, as Schreinemakers' method already emerges from this example.

Figure 6.4c is a thermodynamically feasible phase diagram for a system of two components. It shows consistent relative locations of all four univariant assemblages and all six divariant assemblages. Two of the latter, (A, D) and (B, C), extend over more than one divariant field, and there are three divariant fields that can contain more than one equilibrium divariant assemblage. Which is the actual equilibrium assemblage in each divariant field depends on the bulk composition of the system, as we shall see in the following section. The labels for the divariant assemblages, and the arcs showing their extents, are not normally shown in phase diagrams, as there is a more compact way of showing this information (Section 6.2.3). We can thus "clean" the diagram and obtain Fig. 6.4d, which shows how Schreinemakers' "180° rule" transforms into a rule that is much simpler to apply, regardless of how many components the system has. This is simply a restatement of the rule that we inferred for the case $c = 1$, which we now see is true in general: *the metastable extension of each univariant curve enters the only divariant field in which the phase absent along the univariant curve appears as a reactant on the two univariant curves that bound the field*. Thus, the metastable extensions of (B) and (D) enter the same field, as B and D are reactant phases on the two boundaries of this field. The metastable extensions of (A) and (C) enter different divariant fields, with A and C as reactant phases on their respective boundaries. In the divariant field bound by (B) and (D) there is no phase that appears as a reactant on both boundaries and, correspondingly, there are no metastable extensions going into this field.

Schreinemaker's rule can be stated in several equivalent ways, but I find the statement written in italics in the preceding paragraph the one that is most generally applicable and simplest to use. The rule determines the order in which univariant curves and divariant fields succeed one another around an invariant point, and is all that one needs to place them in the correct sequence. Schreinemakers' rule is a powerful aid in the construction of phase diagrams for phases of fixed composition, but it is equally important to understand what it is that this method *does not* do. In particular, Schreinemakers' rule *does not* yield: (a) the actual slopes of the univariant phase boundaries and (b) the side of the phase boundary on which the reactant phases appear. This is shown in Fig. 6.5. The first phase diagram in the figure is the one that we constructed in Fig. 6.4. The second one is a rotation of the first, and shows that changing the slopes (case (a)) does not affect the thermodynamic validity of the phase diagram. The third phase diagram is a mirror reflection of the first one, and exemplifies situation (b): reactant phases appear on the opposite side of each phase boundary, and the univariant curves and divariant fields follow one another in the opposite direction around the invariant point, relative to Fig. 6.4, yet the phase diagram is equally feasible from a thermodynamic point of view. All three phase diagrams comply with Schreinemakers' rule, so they are all thermodynamically feasible, yet only one can be correct. Deciding which is the correct phase diagram requires thermodynamic data, as we shall see (review Exercise 6.4).

6.2.3 Chemography

We saw that, except for the case $c = 1$, there are always more possible divariant assemblages than there are divariant fields in the neighborhood of an invariant point. This means that in at least some of the divariant fields more than one assemblage is possible. If we examine the univariant reactions in the phase diagram in Fig. 6.4 or 6.5, we see that these phase relations are possible only if there are some specific compositional relationships among the phases. In particular, whereas phases A and C can be formed by combining subsets of the other phases, this is not true of phases B and D. For example, we can write reactions of the form $C + D \rightarrow A$ and $B + D \rightarrow C$, but similar reactions in which B or D appear by themselves on one side of the reaction do no occur. This means that B and D must be at the ends of the compositional range of the system of interest, and that between the two of them they span this compositional range. The chemical compositions of B and D can thus be taken as system components (Section 6.1.2), which we will label 1 and 2, respectively.

It is often helpful to show the compositional information of a system graphically (obviously, in the case of phase diagrams with phases of variable composition, it is mandatory). This is known as *chemography,* and two examples are shown in Fig. 6.6. Compositional information in a system of two components, also known as a binary system, can be represented on a line segment, in which the two system components correspond to the endpoints. We shall write the names, or chemical compositions, of the system components below the line (top part of Fig. 6.6). In binary systems the compositions of two of the phases of interest coincide with those of the system components, and other phases are located along the line segment. We show phase compositions with black circles, and the name of the phase above the circle. The top part of Fig. 6.6 is an example of a *chemographic diagram* for a binary system. Compositional relations in three-component, or ternary, systems are shown on triangular chemographic diagrams such as that in the lower part of Fig. 6.6. In this case

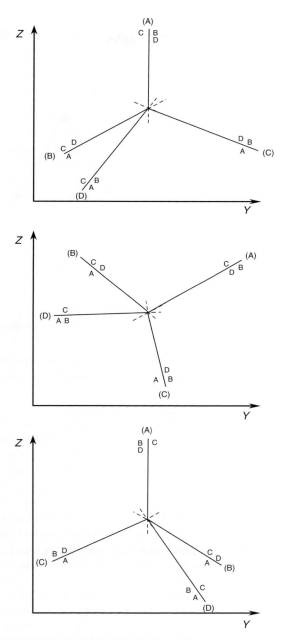

Fig. 6.5 Three possible phase diagrams for a binary system. The middle diagram is a rotation of the top one, the bottom one is the mirror image of the top one. All three are thermodynamically valid.

the system components are generally located at the vertices. There may be phases whose compositions correspond to those of some system components (e.g. E in the lower part of Fig. 6.6). Other phases plot either on the sides of the triangle or inside it. There are many possible variants in the arrangement of phases in three-component systems, which we will

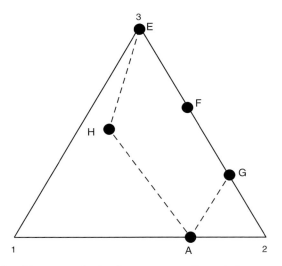

Fig. 6.6 Chemographic diagrams for a binary system (top) and a ternary system (bottom). Components are labeled with numbers, phase compositions are shown with dots, and labeled with letters.

not review systematically here but will rather discuss as the need arises (see Zen, 1984, for a systematic discussion). The arrangement of phases in the lower part of Fig. 6.6 is just an example. The bulk composition of a system in which the phases of interest are A E F G H must plot inside the area defined by the dotted line and the segment EG of the side 32 of the triangle. In this case there is no set of three phase compositions that can serve as system components, but such may be the case in other systems. Chemographic diagrams for systems of more than three components require multiple projections from three or more dimensions onto two. They are generally confusing and, in my view, tend to hinder understanding rather than help it (in other words, I will not use them).

We now revisit the phase diagram in Fig. 6.4 armed with the appropriate chemographic diagrams (Fig. 6.7). Within each divariant field we place a version of the diagram that shows which are the phases that are stable in that field. The univariant reactions mediate the changes between chemographic diagrams in adjacent divariant fields. For example, univariant reaction (C), which the phase diagram shows to correspond to $B + D \rightarrow A$, causes A to appear between B and D in the chemographic diagrams, as the curve is crossed in the direction of decreasing Z. The reason for using chemographic diagrams now becomes clear: they unequivocally show which are the possible divariant assemblages in each divariant field. For example, in the field bound by the curves (C) and (D) the two possible assemblages are AB and AD. This information is also available in Fig. 6.4c, as assemblage (B, C) is the same as AD, and (C, D) is the same as AB, but using a chemographic diagram shows this information more compactly. More importantly, the chemographic diagram shows that the identity of the divariant assemblage is determined by the bulk composition of the system,

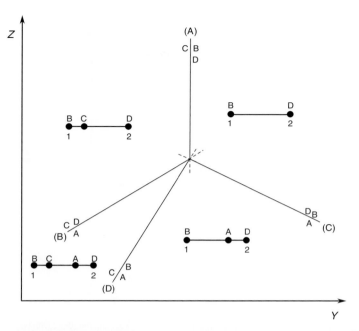

Fig. 6.7 Phase diagram for the binary system shown in Fig. 6.4. The chemographic diagrams show the possible stable divariant assemblages in each divariant field. For example, the possible divariant assemblages in the field bound by the curves (D) and (C) are BA and AD, depending on bulk composition. The chemographic diagrams contain the same information as the arc segments in Fig. 6.4, and in addition they show the relationship between stable assemblage and bulk composition.

as AB or AD will form depending on whether the bulk composition is on one or the other side of A. This information is not present in Fig. 6.4c.

There is some redundancy of information in Fig. 6.7, and as we shall see in some cases it is not necessary, or even meaningful, to include the chemographic diagrams. If chemographic diagrams are used, then there is an alternative way of expressing Schreinemakers's rule, which is that *the metastable extension of each univariant curve enters the divariant field in which the phase absent along the curve is stable for all bulk compositions.* You can verify the validity of this statement in Fig. 6.7. This version of the rule is easy to apply in systems of two and three components, but can be (very) tricky when $c > 3$. In all cases I prefer the statement given in Section 6.2.2, which is equally easy to use in systems of any number of components.

Worked Example 6.2   Evaporites, part (i)

Terrestrial evaporites display a remarkable mineralogical diversity, and, as we come to know and understand evaporites in other planets (chiefly, Mars and Titan) the diversity of evaporites is likely to become greater. Formation of evaporites entails equilibrium among solid, liquid and gas phases, and in this sense thermodynamic description of evaporite-forming processes is fundamentally the same as that of igneous rocks – the differences (and assorted devils) are in the details. Let us look at a simple example, which focuses on the formation of two sodium sulfate minerals that are found in terrestrial evaporites, and could

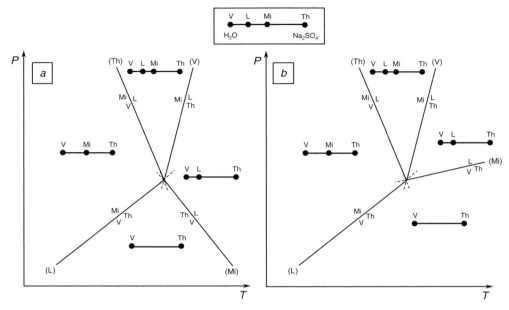

Fig. 6.8 Phase diagram for crystallization of thenardite (Na_2SO_4) and mirabilite ($Na_2SO_4.10H_2O$). The slopes of the (Th), (V) and (L) curves are unambiguous. The (Mi) curve is harder to pin down and two alternatives are shown. The correct one is (b), see text.

conceivably be present on the Martian surface as well: thenardite (Na_2SO_4) and mirabilite ($Na_2SO_4.10H_2O$). These minerals are known to precipitate from marine brines, which constitute a liquid phase. The process occurs at the planet's surface, and therefore in the presence of a gas phase. If we ignore for now the fact that the planet's atmosphere (i.e. the gas phase) contains various other components, we can consider this to be a binary system, and use Na_2SO_4 and H_2O as the system components – we shall return to the issue of atmospheric composition in a moment. In the binary system the four phases thenardite–mirabilite–liquid–vapor constitute an invariant assemblage (we call a gas phase in equilibrium with a condensed phase a "vapor", more on this in Chapter 9). Four univariant curves radiate from the invariant point. Using Schreinemakers' rule and general properties about univariant phase boundaries that we derived in Section 5.1.3, such as their Clapeyron slopes and how $\Delta_r S$ and $\Delta_r V$ determine the side of a reaction that a given assemblage is on, we can construct the P–T phase diagrams shown in Fig. 6.8.

The orientation of the phase diagram is obtained from the (L) and (V) curves. In the first place, both liquid and vapor are higher entropy phases than the solids, so they must be placed on the high temperature side of these equilibria, establishing the mirror orientation of the diagram (see Fig. 6.5). They are also higher volume (lower density) phases than the solids, so the Clapeyron slopes of both curves must be positive. Finally, because the volume change associated with reaction (L), in which a gas is evolved, is greater than that associated with reaction (V), in which a liquid forms in the absence of a gas phase, equilibrium (V) must have a steeper Clapeyron slope than (L), as shown in Fig. 6.8. There is a tacit assumption in the last statement, which is that $\Delta_r S$ varies less than $\Delta_r V$ between reactions (L) and (V). This is true but I have not justified it; I will return to this issue in Chapters 9 and 10.

Schreinemakers' rule determines the location of the (Th) and (Mi) curves, but their slopes are not as easy to ascertain as those of the other two univariant curves. This is so because in both of these equilibria the liquid and gas phases appear on opposite sides of the phase boundary, and the signs of $\Delta_r S$ and $\Delta_r V$ are not always evident. The sign of the Clapeyron slope becomes strongly dependent on two factors: pressure and the nature and stoichiometry of the reaction. At low pressure, such as at and near a planet's surface, molar volumes of gases are orders of magnitude greater than those of liquids, so that it is a safe assumption that reactions that consume vapor and produce liquid have $\Delta_r V < 0$. This is certainly the case at the conditions at which evaporites form. Gases are much more compressible than liquids, however, and at pressures significantly greater than that of the critical point $\Delta_r V$ for reactions that involve aqueous or carbonic fluids on one side and silicate melts on the other may change sign. This becomes important when studying the formation of magmas in the deep crust and mantle of the Earth and other planets. Staying for now with evaporites we note that, although in general gases also have higher entropy than liquids, the difference is not as large as that for molar volume (Chapter 9). The sign of $\Delta_r S$ then becomes dependent on reaction stoichiometry, which determines the relative amounts of vapor and liquid in a balanced reaction. The liquid along the (Th) and (Mi) curves is an aqueous solution saturated in mirabilite and thenardite, respectively. Given the very large H_2O content of mirabilite, it appears reasonable to assume that, when this mineral decomposes and liberates H_2O, little additional H_2O may be required to dissolve any excess sulfate, so that the stoichiometric coefficient of the vapor is probably small. The dominant contribution to the entropy of reaction arises from the increase in configurational entropy that occurs when crystalline mirabilite becomes aqueous sodium sulfate. The resulting $\Delta_r S$ for the (Th) reaction is positive, and hence the phase boundary has a negative Clapeyron slope, as shown in Fig. 6.8.

Reactions such as (Mi) are in general harder to pin down, chiefly because of greater uncertainty about the sign of $\Delta_r S$. Two alternatives are shown in Fig. 6.8. A negatively sloping reaction, case (a), requires that, with increasing temperature, thenardite reacts with vapor to yield a saturated sodium sulfate solution. If the slope is positive, case (b), increasing temperature causes boiling of the liquid (sodium sulfate solution) and crystallization of thenardite. Worded like this (which I did on purpose) it appears obvious that the correct phase diagram must be (b), and this is indeed the case for this particular system, and in general for crystallization of evaporites from liquid H_2O solutions. One can intuitively justify this result by noting that thenardite is an anhydrous mineral, so that all of the H_2O needed to form the saturated sodium sulfate solution at the (Mi) curve must come from condensation of vapor. In contrast to the (Th) curve, the stoichiometric coefficient of vapor must be quite large, and the entropy contribution from the phase change of H_2O overrides the contribution from the breakdown of the crystalline structure of thenardite. We then have $\Delta_r S < 0$, and a positive Clapeyron slope, as in Fig. 6.8b. This conclusion is, however, not general, *and in particular is incorrect for equivalent reactions in igneous systems* (Worked Example 6.3).

We now have a thermodynamically consistent phase diagram for crystallization of sodium sulfate phases (Fig. 6.8b), but how meaningful is it in relation to natural evaporites? There are three issues that must be addressed.

First, this is a purely schematic phase diagram. In order for it to be useful it must be made quantitative, by calculating the actual locations of the phase boundaries in terms of the intensive variables of interest. We have established the principles for doing so in Chapter 5, but we have not yet discussed how to calculate equilibria involving gases at pressures other than 1 bar (Chapter 9) nor how to calculate equilibria involving aqueous solutions (Chapter 11). We can nonetheless see the importance of Schreinemakers' analysis. It allows us to identify erroneous thermodynamic data, for instance if the calculated position of a phase boundary is inconsistent with Schreinemaker's rule. It also allows us to work with only a partial data set, as being able to calculate only a couple of the reactions places constraints on the locations of the others.

The second issue is that the information contained in a phase diagram such as that in Fig. 6.8b, or even in a quantitative version of it, is incomplete. The composition of the liquid phase is not fixed, and its location in the chemographic diagrams is only schematic. Liquid composition is fixed at each point on each univariant curve, because the chemical potentials of each of the liquid components are fixed at each point on each curve, but the liquid composition is not constant along each curve, nor is it equal from one curve to another. In the divariant fields the liquid composition also depends on bulk composition. For example, consider a liquid in the field bound by the reactions (Th) and (V). If the bulk composition of the system is such that there is excess mirabilite, then the liquid is saturated in mirabilite and its composition will be that of the saturated solution at any given P and T (which are the two degrees of freedom). If, on the other hand, all mirabilite dissolves then the liquid will be in equilibrium with a gas phase rather than with mirabilite, and its sodium sulfate content will be less than that of a saturated solution at the same pressure and temperature. This shortcoming is inherent to this type of phase diagram. Recall that in this section we set out to construct phase diagrams among phases of fixed composition. By including a liquid phase we have gone beyond that specification, even if the resulting phase diagram is still thermodynamically valid.

The third, and most important, issue, is whether the choice of intensive variables in Fig. 6.8 is the most appropriate one for this particular example. We must now return to our decision to ignore the fact that the Earth's atmosphere consists chiefly of components other than H_2O. By considering a system of two components and using pressure as an intensive variable we are assuming that the gas phase is made up exclusively of H_2O, so that the pressure on the system is the same as the partial pressure of H_2O (see Worked Example 5.5). This would be true in a hypothetical planet in which the atmosphere consisted of H_2O vapor only, but it is certainly not true of the present-day Earth. Using pressure as an intensive variable to represent the phase relations of sodium sulfate phases is not incorrect, but it leads to unphysical interpretations if the diagram is used to study natural evaporites. For instance, consider a system whose bulk composition is between Mi and L, and assume that conditions are initially within the divariant field bound by (L) and (Th). If conditions change such that reaction (Th) is crossed then all the vapor will be consumed and the assemblage will be liquid + mirabilite. This description may work in a closed vessel in a laboratory, but it does not work on a planetary surface. A better alternative would be to use μ^{H_2O} as an intensive variable. By doing so we can track the evolution of the system as a function of changes in atmospheric humidity at constant atmospheric pressure, as there is a simple relationship between the mol fraction of H_2O in air and the chemical potential of H_2O (e.g., equation (5.82)). We return to this in Worked Example 6.6.

Worked Example 6.3   Hydrous melting of silicate rocks

Melting of hydrous silicate mineral assemblages is an important process in the origin of felsic terrestrial magmas, and in the origin and evolution of the continental crust. Whether or not such processes ever took place in Mars or Venus is as yet unknown, and figuring this out will be important in piecing together the crustal evolution of our sister planets. The fundamental thermodynamic relations of hydrous melting of silicate rocks are a close parallel to those

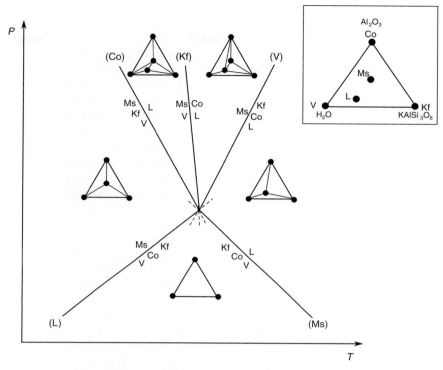

Fig. 6.9　Phase diagram for melting of silicates in the system KAlSi$_3$O$_8$–Al$_2$O$_3$–H$_2$O. The tie lines joining phases in the chemographic diagrams define the possible divariant assemblages. For example, in the diavariant field bound by the curves (V) and (Ms) there are three possible stable divariant assemblages: Co–L–V, Kf–L–V and Co–Kf–L. The only phase stable in all assemblages is L, and the metastable extension of the (L) curve enters this field. This is one possible statement of Schreinemakers' rule. The other one (which I prefer) is that the metastable extension of (L) enters the field in which L appears as a reactant on the two bounding curves, (V) and (Ms).

of evaporite–brine equilibria. A complete analysis of hydrous melting of rocks can only be carried out in systems of at least four or five components, but a simple analog that exhibits some of the key thermodynamic aspects can be based on the ternary system: Al$_2$O$_3$–KAlSi$_3$O$_8$–H$_2$O. Let us consider the five-phase invariant assemblage: muscovite–sanidine–corundum–melt–vapor in this ternary system (Figure 6.9). In this simple example we will assume that the vapor is pure H$_2$O. The melt composition will be taken to correspond to a hydrous and slightly peraluminous syenite (Fig. 6.9) – rocks that approximately match this description are rare in nature, but not unknown.

A schematic P–T phase diagram for this system is shown in Fig. 6.9. Comparison with Fig. 6.8 reveals some remarkable similarities and one important difference. Except for the fact that there is one additional phase in each assemblage (because this is a ternary system), reactions (Co), (L) and (V) are analogous to (Th), (L) and (V), respectively, in the sodium sulfate–H$_2$O system. Thus, the hydrous mineral assemblage (mirabilite, or muscovite + sanidine) reacts with vapor to yield a liquid at (Th) and (Co), respectively. The hydrous mineral by itself breaks down to an anhydrous solid assemblage (thenardite, or sanidine + corundum) plus either vapor or liquid at the (L) and (V) curves in both systems. For reasons that we will discuss in Chapter 10, in igneous systems we typically call the liquid phase a melt, but, as these diagrams suggest, there are important thermodynamic similarities between both systems. The signs of the Clapeyron slopes of these three reactions at low pressure are the same in both systems, for the reasons that we discussed in Worked Example 6.2, but at high pressure, such as in the deep crust or upper mantle of the Earth, the Clapeyron slopes of the (Co) and (V) curves may change signs, in response to the higher compressibility of hydrous fluids relative to silicate melts, and of silicate melts relative to crystalline phases.

In the hydrous silicate system ($c = 3$) there is an additional univariant curve, (Kf), that has no analog in the sodium sulfate–H_2O system ($c = 2$). This reaction is similar to (Co) in the sense that in both of them muscovite reacts with vapor to produce liquid, except that an anhydrous solid phase is produced together with the liquid along the (Kf) reaction, but not along (Co). The one important difference between both systems is in the Clapeyron slope of the reaction in which an anhydrous assemblage, thenardite or sanidine + corundum, reacts with vapor to produce liquid. In the evaporite example we justified the positive slope of (Mi) on the basis of the large value expected for the stoichiometric coefficient of vapor. We can think of the liquid that forms at (Mi) as the result of condensation of vapor and dissolution of the anhydrous solid in the resulting condensate. At the high temperature at which the (Ms) reaction takes place, in contrast, the liquid forms by melting of the crystalline solids and dissolution of vapor in the melt. The consequence is that the stochiometric coefficient of vapor in the (Ms) reaction is characteristically quite small, and $\Delta_r S$ is dominated by the entropy of melting of the silicates, which is of course positive. The volume change of reaction, and therefore the Clapeyron slope, are negative.

In the discussion of igneous phase relations alternative names are used for some of these reactions. Thus, (Co) is called the *vapor-saturated solidus*. This reaction maps the minimum temperature at which melt can form. Because melt forms along this reaction only if an aqueous vapor phase is present the melt at the solidus is saturated in H_2O. Note that for certain bulk compositions (e.g. inside the triangle defined by the phases vapor, muscovite and corundum, see Fig. 6.9) the (Kf) reaction, rather than (Co), is the vapor-saturated solidus. The liquid-absent curve is called the *subsolidus dehydration* reaction, as the assemblage becomes anhydrous without melting. Finally, the (V) curve, where melt forms in response to breakdown of the hydrous mineral without formation of a vapor phase, is called the *dehydration-melting* or *vapor-absent melting* reaction.

Whether we call the liquid phase a liquid (or solution) or a melt depends on the composition of the coexisting solids. A melt is a liquid at equilibrium with a solid of its same composition (Chapter 10). The composition of the liquid that forms in the silicate rock example is close enough to the composition of the solid phases that it is properly called a melt. In the evaporite example, on the other hand, the liquid composition is close to that of a condensed gas and it contains dissolved ions, so we call it a solution. Liquid-forming reactions such as (Co), (Th), (Mi) and (Ms), in which the only product of the reaction is a liquid, are called *congruent melting* or *congruent dissolution* reactions. In contrast, along equilibria such as (Kf) and (V) a solid phase crystallizes on the high-temperature side of the reaction, together with formation of a liquid. Such reactions are called *incongruent melting* or *incongruent dissolution* reactions.

6.2.4 Compositional degeneracy

A detailed discussion of the many possible chemographic relations in systems of two or more components, and of the different topological varieties of phase diagrams that they give rise to, exceeds the space available here (but see Zen, 1984). There is, however, one

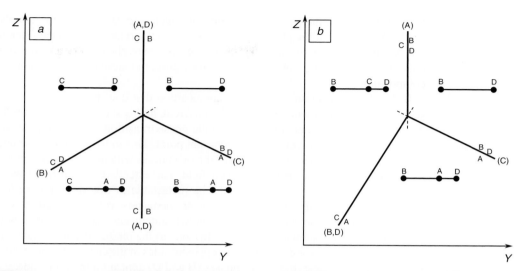

Fig. 6.10 Two types of compositional degeneracy in a binary system. If the degenerate reaction occurs at one of the ends of the compositional range, as in (a), the univariant curve does not become metastable when it crosses the invariant point. If the degeneracy is internal, as in (b), the univariant curve becomes metastable (see text).

instance that needs attention, as it occurs very commonly in natural systems. This is the case in which a subset of phases can be described in terms of a smaller number of components than the system as a whole. Phase transitions (including polymorphic transformations) are an example of this situation, as they are described in terms of only one component (the composition of the substance that undergoes the phase transition). Other examples are: a set of three phases that plot along a straight line in a ternary chemographic diagram, so that equilibria among these phases are described in terms of two components only; a set of four co-planar phases in a four-component system, and so on. Such cases are said to be *compositionally degenerate*.

Consider a binary system in which two phases have the composition of one of the system components (Fig. 6.10a). The phase transition (B ⇌ C in this example) is a univariant equilibrium, which is represented by a curve, labeled (A, D), along which two phases are missing. Two divariant fields, (A, D, C) and (A, D, B), radiate from this univariant curve, and each of these divariant fields must switch from stable to metastable as it crosses the univariant curve (A, D). The only way in which this can be accommodated is if the curve (A, D) crosses the invariant point without becoming metastable, as shown in Fig. 6.10a. Another way of seeing why (A, D) does not become metastable is by thinking of it as a phase boundary in a one-component system in which phases A and D do not exist. The phase transition B ⇌ C takes place regardless of the presence of additional components that make phases A and D possible. Alternatively, comparing Fig. 6.10a with Fig. 6.7, one may imagine that, as C approaches B on the chemographic diagram, the slope of reaction (B) approaches that of reaction (C) on the phase diagram. Coincidence in both diagrams occurs simultaneously, so that when the reaction becomes degenerate each stable curve coincides with the metastable extension of the other (this is the basis of the method developed by G. W. Morey to derive Schreinemakers' rule; see, for example, Williamson & Morey, 1918; Morey & Williamson, 1918). The important point here is to understand how Schreinemakers' rule applies to this degenerate case, which is easily seen in Fig. 6.10a. In my preferred versions of the rule, each of the ends of the (A, D) curve, none of which is metastable, *splits* the only divariant field in which one of the phases absent along the univariant curve appears as a reactant on the two univariant curves bounding that field. Alternatively, each of the ends of the (A, D) curve splits a divariant field in which a phase absent along the curve is stable for all bulk compositions.

If the phase transition occurs for a composition that does not correspond to one of the system components then the behavior is different, and is shown in Fig. 6.10b. In this case the univariant curve that corresponds to the phase transition, (B, D) in the example, *must* become metastable as it crosses the invariant point. This must be so because reactions (A) and (C) correspond to the breakdown of the two phases with the same composition. Neither of the phases is stable in the divariant field that extends between these two reactions, and therefore the phase transition must be metastable in this field. Figure 6.10b shows that the application of either version of Schreinemakers' rule is immediately obvious. It is important to see the subtle difference in Schreinemakers' rule applied to both types of degenerate systems shown in Fig. 6.10. In case (a) the phases absent along the degenerate univariant reaction, A and D, appear on opposite sides of the curves that bound the divariant fields, whereas in case (b) the two phases (B and D) appear on the same side.

The fascination of petrologists with carbonatites is out of proportion to their scarcity in the terrestrial igneous rock record. This is justified, however, as the processes that lead to the formation of carbonate melts in a silicate planet are not as straightforward as those that generate silicate melts. The simplest possible model for carbonate melting can be constructed in the binary system CaO–CO_2 (Fig. 6.11). If we assume that calcite melts congruently to a liquid with the same composition then the melting reaction is degenerate, as shown in the figure. When this reaction, labeled lime- and vapor-absent, crosses the invariant point it becomes metastable – this is the case illustrated in Fig. 6.10b. The invariant point therefore marks the minimum pressure at which calcite melts or, equivalently, the minimum pressure at which calcite can crystallize from a carbonate melt. The pressure of the invariant point is higher than 1 bar, as at atmospheric pressure calcite decarbonates to lime plus vapor along the liquid-absent curve.

The parallels with the examples discussed in Worked Examples 6.2 and 6.3 should be clear. We could call the (L) and (Lm, V) reactions the "subsolidus decarbonation" and "decarbonation-melting" reactions, respectively, and we see that they are close analogs of the corresponding reactions in the H_2O-bearing examples. The vapor-saturated solidus is missing in the carbonate system because we assumed a degenerate melting reaction. Physically, this means that we assumed that the carbonate liquid is not able to dissolve excess CO_2. If this were not the case then the liquid phase would be richer in CO_2 than calcite and the melting reaction would not be degenerate. The (Lm,V) reaction would then split into two different reactions, with the (Lm) reaction: Cc + V = L becoming an exact analog of the (Co) or (Th) reactions in the previous examples. As with the (Mi) and (Ms) reactions, the slope of the (Cc) reaction is the one that is most uncertain. I emphasize this in Fig. 6.11 by plotting it parallel to the T axis. We can be certain that liquid is the high-pressure phase, but whether it is the high- or low-temperature phase depends on the entropy of molten $CaCO_3$ relative to that of a stoichiometric mixture of CaO and CO_2. Note that regardless

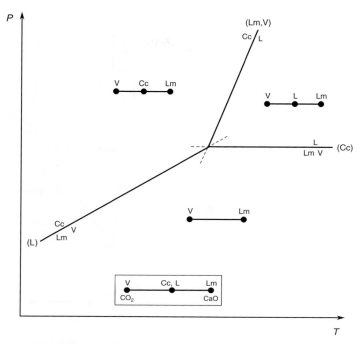

Fig. 6.11 Schematic phase diagram for melting of carbonates. Carbonatitic melts can only exist between the (Cc) and (Lm,V) curves. Calcite cannot crystallize from a melt at a pressure lower than that of the invariant point. Compare Fig. 6.9.

of the slope of the (Cc) reaction the pressure of the invariant point is the minimum pressure at which calcite can crystallize from a melt.

The phase diagram in Fig. 6.11 suggests that one way in which carbonatite melts can form is by subduction of limestones. The "cold" conditions that prevail at subduction zones may be able to keep limestones on the low temperature side of the (L) reaction until the pressure of the invariant point is exceeded. Heating of subducted limestones at high pressure would then produce carbonatite melts. Alternatively, carbonatite melts could be produced by infiltration of CO_2-rich fluids into clinopyroxene- or garnet-bearing peridotites.

Worked Example 6.5   Freezing of brines. Liquid water on the Martian surface?

Consider the system NaCl–H_2O, and the four phases: ice–vapor–halite–liquid (Fig. 6.12). The liquid phase in this system is brine, which we shall label L_2, to distinguish it from pure liquid H_2O, which we label L_1. The four phases I–V–Ha–L_2 exist at equilibrium at an invariant point, O_2 in Fig. 6.12. Because neither ice nor vapor dissolve NaCl the phase transition I $\rightleftharpoons$ V is a degenerate reaction, (L_2, Ha), that in this case crosses the invariant point without becoming metastable (this is the case depicted in Fig. 6.10a). The other two univariant equilibria, (V) and (I), correspond to freezing and boiling of brine, respectively. The negative slope of the (V) reaction arises from the fact that H_2O expands when it freezes, which is of course unusual.

Because sublimation of ice is a liquid-absent reaction, and the composition of both ice and vapor is H_2O, the (L_2, Ha) reaction must be the same liquid-absent reaction that appears in the one component system H_2O, and which we can label (L_1). This reaction meets the

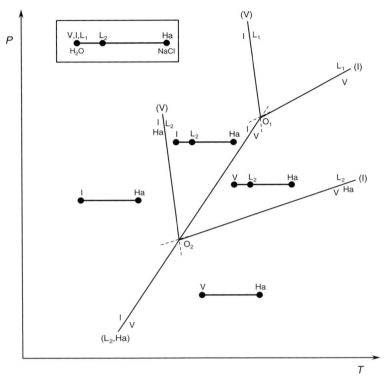

Fig. 6.12 Effect of a solute (e.g. halite) on the phase diagram of H_2O. The triple point for pure H_2O (O_1) shifts to lower pressure and temperature (O_2). The freezing temperature drops and the boiling temperature rises.

freezing and boiling curves of pure water at the invariant point O_1, which is the triple point of H_2O. Reaction (L_1) must become metastable when it crosses O_1, as no other behavior is possible in a one-component system. The univariant curve for ice–vapor equilibrium joins the two invariant points, even as its name changes (by convention) in the neighborhood of each of the invariant points. Schreinemakers' rule also tells us that O_1 must be located at higher pressure than O_2, because O_2 must lie on the stable part of the (L_1) curve. Given vapor's higher volume and higher entropy relative to ice, the slope of the (L_2, Ha) reaction is positive. The two invariant points must therefore be located relative to one another as shown in Fig. 6.12.

Schreinemakers' rule generates a qualitative result that we know well: addition of a solute to water lowers its freezing point and raises its boiling point. The magnitude of the effect of course depends on each particular solute, and thermodynamic data are needed in order to calculate it (Chapter 11). But another prediction follows from this analysis, and this is that the triple point of water shifts to lower pressure and temperature by addition of a solute. Consider a planet (for instance, Mars), in which the partial pressure of H_2O at the surface is below O_1 (why we specify partial pressure will become clear in the next section). In the absence of soluble salts liquid water is not stable at the planet's surface, but if appropriate solutes are available then the pressure of invariant point O_2 may be low enough to allow brines to exist. The implications of these phase relations for the possible existence of liquid H_2O on the Martian surface are discussed in Chapter 11.

I used a binary system to discuss the concept of compositional degeneracy, but application to systems with more components presents no difficulty, as further examples will show (see end-of-chapter Exercises).

6.3 Phase diagrams in open systems

We can define a closed system as one in which the only intensive variables that can be controlled independently and externally (or, equivalently, imposed on the system) are pressure and temperature. An open system is one in which, in addition to pressure and temperature, some chemical potentials can be controlled externally. Let the number of externally controlled chemical potentials be v. Then the total number of independent intensive variables in an open system is $v + 2$. Consider the evaporite system discussed in Worked Example 6.2. If the gas phase is not pure H_2O, and the H_2O content of the atmosphere is variable, then $v = 1$ and there are three independent intensive variables: P, T and μ^{H_2O}. These additional independently variable thermodynamic quantities allow extra flexibility in the construction of phase diagrams for open systems.

6.3.1 Externally controlled chemical potentials

Recall that Schreinemakers' rule is general, as we derived it on the basis of a set of arbitrary and unspecified intensive variables. Pressure and temperature are not always the most convenient combination of variables, particularly for open systems. Evaporites, for instance, are most accurately described as open systems that exchange a chemical component (H_2O) with their environment (the atmosphere). In Worked Example 6.2 I suggested that the chemical potential of H_2O is a better intensive variable than pressure to study the behavior of evaporites. This is so because, whereas pressure at the Earth's surface is approximately constant, atmospheric humidity is not. By using μ^{H_2O} as one of the intensive variables we can track how changes in atmospheric humidity cause H_2O to be transferred between the atmosphere and the evaporite + brine assemblage, and predict changes in the evaporite phase assemblage. Recall that equation (5.86) relates μ^{H_2O} to the partial pressure of H_2O, p_{H_2O}, and that p_{H_2O} is a convenient way of measuring atmospheric humidity.

Consider an equilibrium assemblage with f degrees of freedom, and subject to v externally controlled (or imposed) chemical potentials. In such a system there are $v + 2$ independent intensive variables (the v chemical potentials, pressure and temperature). If we fix the values of *any* f of the $v + 2$ intensive variables then the assemblage appears to behave as an invariant assemblage, *as long as we keep the values of the f chosen intensive variables fixed*. Our goal is to map the f-variant assemblage onto a *pseudo-invariant point* in a two-dimensional graph in which the coordinates are any two intensive variables, (Z, Y), taken from the $v + 2$ available independent variables. In order for this to be possible it must be $v = f$. The variance of the assemblage is of course still f, so that it is stable over an f-dimensional region of intensive-variable space. However, for each point in this region there is a unique combination of the variables Z and Y for which the assemblage is stable. This combination maps as a pseudo-invariant point on the Z–Y plane. Suppose further that the pseudo-invariant assemblage consists of $\Phi = c + 2 - f$ phases. Then, removing each of the Φ phases one at a time generates Φ *pseudo-univariant curves* radiating from the pseudo-invariant point, along each of which $\Phi - 1$ phases are stable. These curves separate Φ *pseudo-divariant fields*, inside each of which $\Phi - 2$ phases are stable. In general, a "pseudo-q-variant assemblage" means that the assemblage actually has $q + v$ degrees of freedom, but the values of v intensive variables are held constant.

For example, suppose that we have a divariant assemblage ($f = 2$) in a four-component open system ($\Phi = 4$). In order for it to be possible to map this assemblage as a pseudo-invariant point there must be two chemical potentials that are controlled externally ($v = 2$). If we fix any two intensive variables, say pressure and temperature, then there is a unique combination of the two chemical potentials for which the assemblage is stable. This combination of values defines a pseudo-invariant point in an isobaric and isothermal phase diagram in which the two coordinates are the two externally controlled chemical potentials. Four phases are stable at the pseudo-invariant point, four pseudo-univariant three-phase curves radiate from the point, and there are four pseudo-divariant two-phase fields between the curves. Note that we can choose to fix any two intensive variables. In this example we chose P and T, but if we had chosen, say, T and one of the chemical potentials then the four-phase assemblage will be pseudo-invariant in a phase diagram in which the coordinates are P and the other chemical potential. All of this is best seen in examples that will make guest appearances throughout this and subsequent chapters.

Worked Example 6.6   Evaporites, part (ii)

We wish to recast the phase diagram for sodium sulfate evaporites shown in Fig. 6.8 as a function of the intensive variables μ^{H_2O} and T (Fig. 6.13). Note that, whereas Fig. 6.8 assumes a closed system (constant bulk composition), in Fig. 6.13 we consider the system mirabilite–thenardite–liquid to be an open system that exchanges H_2O with its environment.

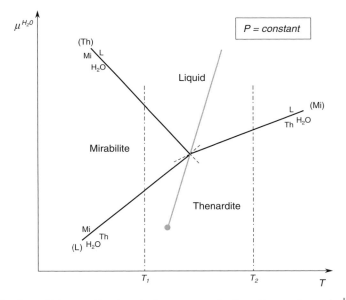

Fig. 6.13 Crystallization of sodium sulfate evaporites at constant pressure, as a function of temperature and μ^{H_2O} (e.g. atmospheric humidity). The faded line shows the path of the psudo-invariant point with changing pressure (it is **not** a phase boundary). The path of the pseudo-invariant point terminates at the dot, which corresponds to the temperature of the invariant point for the binary system (Fig. 6.8). P–T along the path of the pseudo-invariant point correspond to the (V) curve for the binary system.

In the former case the chemical potential of H_2O is fixed by the assemblage that is stable for each combination of pressure and temperature. In contrast, the chemical potential of H_2O in the open system depicted in Fig. 6.13 is an externally imposed variable, just as pressure and temperature. In this case there is one externally controlled chemical potential ($v = 1$), so that a univariant assemblage in the P–T diagram maps to a pseudo-invariant assemblage if we fix one intensive variable. We choose to fix pressure, so that the three-phase assemblage mirabilite + thenardite + liquid is stable at a pseudo-invariant point in the isobaric μ^{H_2O}–T phase diagram. Three pseudo-univariant curves, along each of which two phases coexist at equilibrium, radiate from the pseudo-invariant point. As usual, we label each curve with the name of the absent phase. *Vapor is no longer a phase in our analysis, however*. Rather, we are interested in how H_2O is transferred between our system and its environment (the atmosphere). We therefore substitute the name of this mobile species for V along the pseudo-univariant curves (Fig. 6.13), which makes it clear that an increase in μ^{H_2O} always causes H_2O to move from the environment (atmosphere) to the system (evaporite + brine), and vice versa. This is of course the same conclusion that we reached in Section 5.3.2. A single phase is stable inside each of the pseudo-divariant fields. This is the phase that is absent along the curve whose metastable extension enters that field. Recall that by "pseudo-divariant field" we mean that there are $2 + v$ degrees of freedom: the two coordinates in the phase diagram plus the variables that we chose to fix. In this case $v = 1$, so that the three degrees of freedom required by the phase rule for a single phase in a binary system are preserved. The same is of course true for the pseudo-univariant and pseudo-invariant assemblages. The slopes of the phase boundaries are discussed in the next section.

What about the vapor-absent reaction? The assemblage along this reaction is the same one as in the pseudo-invariant point, so that it cannot exist along a curve on the isobaric μ^{H_2O} – T phase diagram. The location of the pseudo-invariant point shifts with pressure, along the (V) curve for the binary system in Fig. 6.8. It defines a path on the μ^{H_2O} – T plane which is the locus of all points at which the pseudo-invariant assemblage mirabilite + thenardite + liquid is stable. This is shown in Fig. 6.13 with a "ghost" line, to emphasize that it is not a phase boundary but rather the path along which the pseudo-invariant point slides with changing pressure. This path has a lower extremum, which corresponds to the P–T conditions of the invariant point for the binary system (i.e. for the situation in which the vapor phase is pure H_2O, Fig. 6.8).

Note the similarities between the phase diagram in Fig. 6.13 and the phase diagram of H_2O (e.g. Fig. 6.12). We can think of the (Mi) curve as a "boiling" reaction, in the sense that H_2O in the liquid phase becomes vapor. Following this analogy the (Th) curve corresponds to a freezing reaction, in which (most of) the H_2O in the liquid phase "crystallizes" as it is incorporated in the structure of mirabilite. The (L) curve behaves like a sublimation reaction, in which H_2O of crystallization of mirabilite becomes vapor.

Consider two evaporite basins in different locations, such that their characteristic temperatures are below and above the temperature of the pseudo-invariant point, for instance T_1 and T_2 in Fig. 6.13. Say that evaporation occurs at constant temperature and pressure, in response to a decrease in atmospheric humidity (i.e. p^{H_2O}) and hence in μ^{H_2O}. Recall that, at equilibrium, μ^{H_2O} is the same in the gas phase (atmosphere) as in the liquid phase (brine). In the cooler climate the brine crystallizes mirabilite. If atmospheric humidity never drops below the value at the (L) curve then the evaporite bed will consist of mirabilite only. On the other hand, if dessication of the atmosphere continues then the (L) curve, where mirabilite dehydrates to thenardite plus vapor, may be reached, and perhaps crossed. In this case mirabilite will be partially or totally replaced by thenardite. Under conditions hotter than the invariant point, such as T_2, evaporite formation occurs by crystallization of thenardite only and mirabilite never forms. If, however, the climate becomes cooler and the (L) curve is reached from the right, then mirabilite would crystallize at the expense of thenardite. Note that in every case the label "H_2O" on the phase boundaries indicates the direction in which this component is being exchanged between system (evaporite basin) and environment (atmosphere).

There are four paragenetic sequences which, at least in principle, can be identified in evaporite sequences and used as paleoclimatic indicators: mirabilite only, thenardite only, mirabilite followed by thenardite, and thenardite followed by mirabilite. Of course, real evaporites are much more complex than this, but some of the key thermodynamic principles that govern their formation can be seen in this example.

6.3.2 Slopes of pseudo-univariant phase boundaries

Phase diagrams in which the coordinates are chemical potentials are also known as *chemical potential diagrams*. There are simple equations that yield the slopes of phase boundaries in them, analogous to Clapeyron's equation for *P–T* phase diagrams. Let A and B be phase assemblages in an open system, such that they are at equilibrium along a pseudo-univariant curve by exchanging species X and Y with the environment, according to the following balanced chemical reaction:

$$A + v_x X = B + v_y Y. \tag{6.25}$$

The stoichiometric coefficients of X and Y are v_x and v_y, whereas those of the phases that constitute the open system are subsumed in the symbols A and B. We can write the equilibrium condition for this reaction as follows:

$$\Delta_r G_{phases} + v_y \mu^y - v_x \mu^x = 0, \tag{6.26}$$

where $\Delta_r G_{phases}$ is the difference in Gibbs free energy between assemblages B and A. We seek the response of the system to infinitesimal changes in pressure, temperature and the chemical potentials of X and Y. In order for equilibrium to be maintained between the system and its environment the following identity must hold:

$$\left(\frac{\partial \Delta_r G_{phases}}{\partial T} + v_y \frac{\partial \mu^y}{\partial T} - v_x \frac{\partial \mu^x}{\partial T} \right) dT$$
$$+ \left(\frac{\partial \Delta_r G_{phases}}{\partial P} + v_y \frac{\partial \mu^y}{\partial P} - v_x \frac{\partial \mu^x}{\partial P} \right) dP$$
$$+ v_y d\mu^y - v_x d\mu^x = 0. \tag{6.27}$$

Because the chemical potentials of X and Y are controlled externally they can be varied independently of temperature and pressure, which requires that we include the differentials of these chemical potentials in the last line of the equation. The terms in parentheses in the first two lines of (6.27) are simply the temperature and pressure derivatives of the Gibbs free energy change for the complete chemical reaction (6.25), which we will symbolize by $\Delta_r G$. *Note very carefully that this is not*

$\Delta_r G^0$, *as the phases may not be in their standard states.* We can then simplify equation (6.27) to:

$$-\Delta_r S dT + \Delta_r V dP + v_y d\mu^y - v_x d\mu^x = 0. \tag{6.28}$$

The slopes of phase boundaries in various types of diagrams follow immediately from this equation. First, setting $d\mu^y = d\mu^x = 0$ we recover Clapeyron's equation (5.6). For an isobaric and isothermal phase diagram we have:

$$\frac{d\mu^y}{d\mu^x} = \frac{v_x}{v_y}, \tag{6.29}$$

i.e. the slope is the ratio between the stoichiometric coefficients of the externally-controlled species. For a $T - \mu^y$ diagram at constant P and μ^x:

$$\frac{d\mu^y}{dT} = \frac{\Delta_r S}{v_y} \tag{6.30}$$

and for a $P - \mu^y$ diagram at constant T and μ^x:

$$\frac{d\mu^y}{dP} = -\frac{\Delta_r V}{v_y}. \tag{6.31}$$

The signs in equations (6.30) and (6.31) are of course reversed if we interchange μ^y and μ^x. It is sometimes convenient to plot phase relations in terms of activities (or concentrations, or fugacities, see Chapter 9) of externally controlled species. Differentiating equation (5.45) and applying the chain rule equations (6.29) to (6.31) become:

$$\frac{d\ln a^y}{d\ln a^x} = \frac{v_x}{v_y} \tag{6.32}$$

$$\frac{d\ln a^y}{dT} = \frac{\Delta_r S}{RT v_y} \tag{6.33}$$

$$\frac{d\ln a^y}{dP} = -\frac{\Delta_r V}{RT v_y} \tag{6.34}$$

where, again, it must be kept in mind that the entropy and volume of reaction in these equations are generally not the standard state values (more on this later).

Worked Example 6.7   Equilibrium among iron compounds in different oxidation states: a key to early terrestrial environments

Iron, one of the most abundant elements in terrestrial planets, has three oxidation states: Fe^0 (metallic iron), Fe^{2+} (ferrous iron) and Fe^{3+} (ferric iron). As a consequence there is a wide range of phase relations involving iron compounds that play important roles in the evolution of rocky planetary bodies. We begin with a simple example, which focuses on equilibria among the three phases: hematite (Fe_2O_3), magnetite (Fe_3O_4) and siderite ($FeCO_3$). These three phases constitute a divariant assemblage in the ternary system: $FeO-Fe_2O_3-CO_2$ (we could have also chosen the system components as $Fe-O_2-CO_2$, it makes no difference). We can think of situations in which the chemical potentials of O_2 and CO_2 are controlled externally, for instance, if an assemblage of oxides and carbonates equilibrates with a planet's atmosphere, or with groundwater, or with hydrothermal fluids. The two chemical potentials ($v = 2$) allow us to map the divariant assemblage onto a pseudo-invariant point in an isobaric and isothermal phase diagram. Such diagram is shown in the top panel of Fig. 6.14. The three pseudo-univariant curves that radiate from the pseudo-invariant point correspond to the following three reactions:

$$4mt + O_2 \rightleftharpoons 6hm \quad (sd) \tag{6.35}$$

$$4sd + O_2 \rightleftharpoons 2hm + 4CO_2 \quad (mt) \tag{6.36}$$

$$6sd + O_2 \rightleftharpoons 2mt + 6CO_2 \quad (hm). \tag{6.37}$$

The slopes of these reactions in the isobaric and isothermal phase diagram (also called a chemical potential diagram) are given by the stoichiometric coefficients of the two externally controlled chemical potentials (equation (6.29)). Thus, the (sd) reaction is parallel to the μ^{CO_2} axis, and the (mt) and (hm) reactions have $\partial\mu^{O_2}/\partial\mu^{CO_2}$ slopes equal to 4 and 6, respectively. The units on the coordinate axes in Fig. 6.14 are arbitrary, as it shows only the relative position of the reactions and not their absolute locations (we will get to this in Chapter 11). The relative slopes scale with a change in coordinates, however, and could also have been derived from Schreinemakers' rule, as the figure shows. It is important to remember that slopes derived from equation (6.29) must be consistent with Schreinemakers' rule – if they are not then you made a mistake somewhere. The mirror orientation of the phase diagram can be obtained in several different ways, for example, from the fact that hematite is the most oxidized phase, or that siderite must become stable with increasing μ^{CO_2}. Each pseudo-divariant field contains a single phase which, at fixed pressure and temperature, is stable over a $\mu^{O_2}-\mu^{CO_2}$ region (of course, a single phase in a ternary system has four degrees of freedom, but two of these are "used up" by fixing P and T). The diagram confirms what we should intuitively expect: that hematite forms by oxidation of magnetite, that siderite forms by carbonation of the oxides, or the oxides by oxidation of siderite, and that reduction of ferric iron in the oxides to ferrous iron in siderite requires higher μ^{CO_2} the higher μ^{O_2} is. It may seem that we have not gained much information that we did not already know, and this is indeed a very simple example. But a schematic phase diagram such as Fig. 6.14 is the starting point for construction of the rigorous quantitative version, which we will do in Chapter 11.

We can choose to fix different subsets of the four externally controlled intensive variables. The phase diagrams at the bottom of Fig. 6.14 show two possibilities: $\mu^{O_2}-T$ and $\mu^{O_2}-P$ diagrams. Although the slopes can be derived from equations (6.30) and (6.31), this is not always as straightforward as for the $\mu-\mu$ diagram. Only one gas species participates in the (sd) reaction, which immediately identifies the high-entropy and high-volume side of the reaction, and the signs of $\Delta_r S$ and $\Delta_r V$ in (6.30) and (6.31). In the (mt) and (hm) reactions there are four and six times more gas, respectively, on one side of the reaction than on the other. These relationships define the signs of $\Delta_r S^0$ and $\Delta_r V^0$ but not of $\Delta_r S$ and $\Delta_r V$, as the gas species are generally not in their standard states. Schreinemakers' rule would allow for the three lines to have positive slopes in the $\mu^{O_2}-T$ diagram. The justification for the negative slopes of the (mt) and (hm) reactions will be given in Chapter 9. The

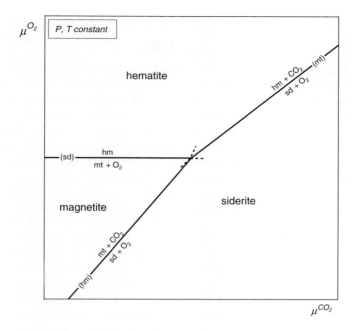

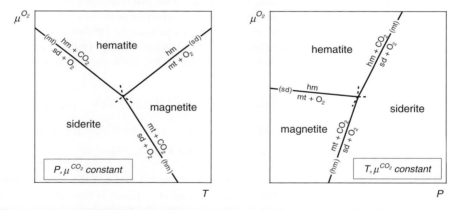

Fig. 6.14 Magnetite–hematite–siderite phase relations in three different projections. See text for discussion, but note that Schreinemakers' rule is followed in all three projections.

μ^{O_2}–T and μ^{O_2}–P phase diagrams show that siderite breaks down to iron oxide plus CO_2 with increasing temperature, and that the resulting oxide is either hematite or magnetite, depending on the chemical potential of oxygen. They also tell us something that may not have been so obvious: that at constant μ^{O_2} hematite is the low-temperature and high-pressure phase relative to magnetite. In fact, the positive slope of the hematite–magnetite phase boundary in the isobaric μ^{O_2}–T diagram is characteristic of all oxidation reactions (Chapter 9).

We can now ask some additional questions. For instance: (i) under what conditions can an aqueous solution contain a significant concentration of dissolved iron (we shall see in Chapter 11 that the relevant aqueous ion is Fe^{2+}) and (ii) what are the identities of the iron phases that crystallize from such solutions under different conditions? Such questions are relevant to understanding changes in the oxidation state of the early terrestrial atmosphere, as well as the possible conditions under which liquid water may have existed on the Martian surface in the geological past. In order to address them we can consider the quaternary system $Fe–O_2–CO_2–H_2O$, and four phases in this system: siderite, magnetite, hematite and an aqueous liquid phase that may or may not contain a significant concentration of Fe^{2+} (I will clarify the meaning of "significant" in the next paragraph). Subsets of three of these phases constitute trivariant assemblages in the quaternary system, that we can map to pseudo-invariant points by finding three externally controlled chemical potentials ($v = 3$), and fixing any three of the resulting five independent intensive variables (P, T and the three μs). Two of the chemical potentials are μ^{O_2} and μ^{CO_2}, as before. The third one is the chemical potential of the hydrogen ion, which can be measured in terms of the pH of the aqueous solution. Recalling that pH is the negative of the logarithm of the H^+ concentration (Chapter 11), we see that we can apply equations (6.32)–(6.34) with the signs switched to determine the slopes of phase boundaries in diagrams in which pH is one of the coordinates.

The phase diagram in Figure 6.14 is valid in the quaternary system, with the proviso that: (a) we indicate that there are three intensive variables that are kept constant: P, T and pH, and (b) we indicate that one of the phases, the aqueous solution, is absent at the pseudo-invariant point and along all reactions that emanate from it. A common convention is to indicate the phase that is absent at an invariant point by enclosing it in square brackets. The pseudo-invariant point in Fig. 6.14 can then be labeled [aq], and the three pseudo-univariant phase boundaries could be re-labeled (sd, aq), (mt, aq) and (hm, aq). Take now the pseudo-invariant point [hm]. The following three pseudo-univariant reactions radiate from it:

$$2mt + 12H^+ \rightleftharpoons 6Fe^{2+} + 6H_2O + O_2 \quad (sd, hm) \tag{6.38}$$

$$sd + 2H^+ \rightleftharpoons Fe^{2+} + H_2O + CO_2 \quad (mt, hm) \tag{6.39}$$

$$6sd + O_2 \rightleftharpoons 2mt + 6CO_2 \quad (hm, aq). \tag{6.40}$$

The aqueous solution phase appears in these reactions as the combination ($Fe^{2+} + H_2O$), but what exactly does this mean? A liquid H_2O phase is typically always present on both sides of reactions such as (6.38) and (6.39). The "appearance" or "disappearance" of the aqueous Fe^{2+} species is more accurately described as a change in the concentration of Fe^{2+}. In order to plot these reactions as pseudo-univariant phase boundaries it is necessary to (arbitrarily) fix some concentration of Fe^{2+} that we will use as the boundary between "dissolved Fe" and "precipitated Fe". This is what I meant by "significant concentration of Fe^{2+}". If, in addition, we note that the chemical potential of H_2O stays approximately constant (because its mol fraction is very close to 1, regardless of how much Fe is dissolved in it), then we see that reactions (6.38) and (6.39) have one degree of freedom (μ^{O_2} or μ^{CO_2}, respectively) at constant P, T and pH. With these restrictions it is possible to treat them as pseudo-univariant reactions, and Schreinemakers' rule applies. In Chapter 11 we will see how to deal with aqueous solutions in a less restrictive way.

Reactions (6.40) and (6.37) are the same one, so this reaction must be the one that joins the [hm] and [aq] pseudo-invariant points (see Worked Example 6.5). Because hematite is not stable at the [hm] point this must be located on the stable side of reaction (6.37). The resulting phase diagram is shown in Fig. 6.15a. The labels of the pseudo-univariant lines have been omitted for clarity but you should convince yourself that their relative locations abide by Schreinemakers' rule. Just as there is a hematite-absent pseudo-invariant point, there is also a magnetite-absent pseudo-invariant point, [mt], where the following three reactions meet:

$$2\text{hm} + 8\text{H}^+ \rightleftharpoons 4\text{Fe}^{2+} + 4\text{H}_2\text{O} + \text{O}_2 \quad (\text{sd}, \text{mt}) \tag{6.41}$$

$$\text{sd} + 2H^+ \rightleftharpoons \text{Fe}^{2+} + \text{H}_2\text{O} + \text{CO}_2 \quad (\text{hm}, \text{mt}) \tag{6.42}$$

$$4\text{sd} + \text{O}_2 \rightleftharpoons 2\text{hm} + 4\text{CO}_2 \quad (\text{mt}, \text{aq}). \tag{6.43}$$

Reaction (6.43) is now the same as (6.36), so this must be the reaction that joins the [mt] and [aq] pseudo-invariant points. Since magnetite is not stable at [mt], this point must be located on the stable side of reaction (6.36). The only way in which this is possible is if both the [hm] and [aq] pseudo-invariant points are located on the metastable side of the (aq, mt) reaction, as shown in Fig. 6.15b. The two phase diagrams in Fig. 6.15 represent two different sets of phase relations for the same system. We will get to that in a moment, but first note that there is an important difference between the two sets of phase relations. When the [mt] point becomes stable magnetite is never stable and both the [hm] and [aq] pseudo-invariant points are metastable. In contrast, when the [hm] point is stable hematite exists around the stable [aq] pseudo-invariant point.

How can two different sets of phase relations be possible for the same system? Because each of the phase diagrams in Fig. 6.15 is valid for a different combination of the three intensive variables that are being held constant, P, T and pH. In particular, they may correspond to different pH values, labeled pH_I and pH_{II} in the figure, at the same pressure and temperature.

We can determine the relative values of pH_I and pH_{II} by drawing the phase relations on the μ^{O_2}–pH plane at constant P, T and μ^{CO_2}. These are shown in Fig. 6.16 – you should demonstrate to yourself that this is the correct diagram. We can now see that magnetite becomes unstable with increasing acidity (decreasing pH). Let the two pH values be as shown in Fig 6.16 and recall that μ^{CO_2} is now being held constant, for example at the value μ_c in Figure 6.15. Under less acidic conditions, pH_I, reduction of hematite at constant μ^{CO_2} produces magnetite first and dissolved Fe^{2+} at lower μ^{O_2}, as in Fig. 6.15a. If conditions are more acidic, pH_{II}, then hematite dissolves in response to a decrease in μ^{O_2} without forming magnetite as an intermediate product, as in Fig. 6.15a.

Banded iron formations (BIF) are chemical sediments that are found in Earth's stratigraphic record beginning in the Early Archaean, reaching their maximum extent in the interval 2.5–2.0 Ga, and tapering off after that. Their major development coincides in time with what is known as the Great Oxidation Event, when μ^{O_2} in the terrestrial atmosphere increased sharply and relatively rapidly (geologically speaking). The most likely explanation for the origin of BIFs is that they formed by precipitation of Fe^{2+} dissolved in seawater. What was the ultimate source of this iron is a different, and controversial, question. There is considerable variability in the mineralogy of BIFs but some consistent patterns exist. Some are dominated by hematite, which in some cases is accompanied by subordinate magnetite. Siderite is rare in hematite-dominated BIFs. Other BIFs are dominated by magnetite, and these commonly also contain large quantities of siderite. Yet a third kind is composed predominantly of siderite, with no oxides. All BIFs contain chert and/or iron silicates.

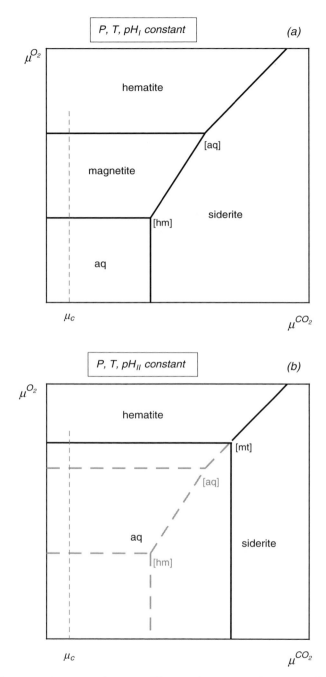

Phase relations of hematite–magnetite–siderite in equilibrium with an aqueous phase at two different values of pH. The boundaries of the aqueous phase field correspond to an arbitrarily chosen concentration of dissolved Fe^{2+}, not to the disappearance of the phase (see text and also Chapter 11). The phase absent at each invariant point is enclosed in square brackets. The phase boundaries shown in grey broken lines in the bottom diagram are metastable relative to dissolved Fe^{2+}. Hence, magnetite is not stable at pH_{II}. Note that μ_c is the chemical potential of CO_2 in Fig. 6.16.

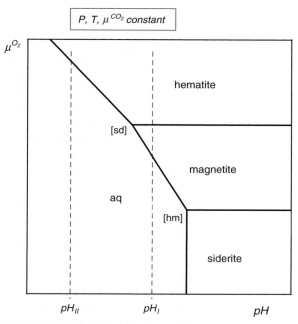

Fig. 6.16 Phase relations for the same system as Fig. 6.15, but at constant μ^{CO_2}.

We can construct a qualitative interpretation of these observations on the basis of the phase relations in Figs. 6.15 and 6.16. Oxide-dominated BIFs are most likely to represent precipitation of Fe^{2+} in response to an increase in μ^{O_2}, but precipitation caused by an increase in the pH of seawater cannot be ruled out. Precipitation of hematite may generally indicate oxidation under more acidic conditions than precipitation of magnetite, but there is also a minimum value of the chemical potential of oxygen that is needed to stabilize hematite. Oxidation under yet less acidic conditions may have caused precipitation of magnetite + siderite BIFs. Carbonate-dominated BIFs with little or no magnetite, in contrast, are more likely to represent an increase in pH or μ^{CO_2} under relatively reducing conditions, at which neither of the two oxides is stable. BIFs preserve a priceless record of the chemical evolution of the atmosphere and ocean in the early Earth, that we will study in a more quantitative fashion in Chapter 11. Widespread deposits of hematite have also been identified on the Martian surface, both by remote sensing and in-situ observations by landers. Magnetite and siderite, in contrast, appear to be absent in Mars, or at the very least to be far less common than in terrestrial sedimentary iron formations. One possible explanation is that Martian hematite formed by precipitation from bodies of consistently acidic water, hinting at an important difference with early terrestrial environments.

6.4 Equilibrium among phases of variable composition

When we consider chemical equilibrium among phases of variable composition we are interested in tracking phase compositions as a function of the intensive variables that control the system, most commonly temperature and pressure. The following discussion focuses exclusively on binary systems but the principles are valid, if considerably more cumbersome to implement, for systems of any number of components. We need to consider two distinct situations.

In the first case each of the compositional end-members undergoes a *discontinuous phase transition* between two phases that have the same chemical composition but different structural states. This could be solid–liquid, solid–gas, liquid–gas or a phase transition between two isochemical solids with different crystallographic structures. Although the concept of a discontinuous phase transition is intuitively easy to grasp, we can give it a precise thermodynamic meaning by noting that they are accompanied by a non-zero enthalpy change (e.g. the enthalpy of fusion, vaporization, sublimation, etc.) and therefore also a finite entropy change, as well as a finite volume change. At a discontinuous phase transition $\Delta_r G$ vanishes but its first derivatives, $\Delta_r S$ and $\Delta_r V$, do not. Thermodynamic analysis of discontinuous phase transitions is based on a set of equations that describe the equilibrium of a chemical species between coexisting phases. These equations are derived in Section 6.5, and applied to the study of phase transitions in Section 6.6.

The second case consists of equilibrium between two phases of different composition that are in the same, or very similar, structural states. These could be two liquids or two isostructural solids (e.g. two feldspars, or two pyroxenes). It is an empirical observation, that we will also justify from thermodynamic considerations, that the compositions of the two phases at equilibrium converge with increasing temperature, until the two phases become indistinguishable at a well-defined temperature called the *critical mixing temperature*. At the critical temperature the system undergoes a phase transition between a sub-critical state in which two phases coexist at equilibrium, and a super-critical state in which only one phase exists at equilibrium. Such phase transitions are called *continuous, or critical, phase transitions*. They are discussed in the next chapter.

6.5 Chemical equilibrium at first-order phase transitions

6.5.1 Condensed phases

Consider a chemical species, A, contained in two condensed phases, 1 and 2, that are at equilibrium at a discontinuous phase transition. We will follow the convention that phase 2 has higher entropy and higher volume than phase 1, so that phase 2 occurs on the high-temperature and low-pressure side of the phase transition, and we will always write the reaction with the high entropy and high-volume phase as a product. The equilibrium condition can be written as follows (see equation (5.55)):

$$\ln\left(\frac{a_2^A}{a_1^A}\right) = \ln K = -\frac{\Delta_r G_{P,T}^0}{RT}. \tag{6.44}$$

If the phases on both sides of the phase transition are condensed phases (i.e., the phase transition is solid–liquid or solid–solid) then the standard states in (6.44) are taken to be pure A in phase 1 and pure A in phase 2 *at the temperature and pressure of the phase transition for the pure substance*. Equation (6.44) vanishes at the pressure and temperature at which pure A undergoes the phase transition, but at any other P–T combination it is *not* zero. We now wish to find how the ratio of equilibrium activities at the phase transition changes with temperature and pressure. Thus, at constant pressure:

$$\frac{\partial \ln K}{\partial T} = -\frac{1}{R}\frac{\partial}{\partial T}\left(\frac{\Delta_r G_{P,T}^0}{T}\right) = -\frac{1}{R}\left(\frac{1}{T}\frac{\partial \Delta_r G_{P,T}^0}{\partial T} - \frac{\Delta_r G_{P,T}^0}{T^2}\right)$$

$$= -\frac{1}{R}\left(-\frac{\Delta_r S_{P,T}^0}{T} - \frac{\Delta_r G_{P,T}^0}{T^2}\right). \tag{6.45}$$

Now, from the definition of Gibbs free energy (equation (4.128)) we can write:

$$\frac{\Delta_r H_{P,T}^0}{T^2} = -\frac{\Delta_r G_{P,T}^0}{T^2} - \frac{\Delta_r S_{P,T}^0}{T} \tag{6.46}$$

which, substituting in (6.45), yields:

$$\frac{\partial \ln K}{\partial T} = \frac{\Delta_r H_{P,T}^0}{RT^2}. \tag{6.47}$$

We can now integrate this expression, at constant pressure, between the temperature of the phase transition for pure A, T_A, and any other arbitrary temperature T:

$$\ln K_T = \int_{T_A}^{T} \frac{\Delta_r H_{P,T}^0}{RT^2} dT + \ln K_{T_A}. \tag{6.48}$$

We will assume that the integration interval is narrow enough that the enthalpy change associated with the phase transition can be considered to be constant, and equal to that for the phase transition for pure A at T_A. Calling this enthalpy change $\Delta_r H^0{}_A$ and given that our choice of standard states makes $\ln K_{T_A} = 0$, we get the following expression for the activity ratio at temperature T:

$$\left(\frac{a_2^A}{a_1^A}\right)_T = K_T = \exp\left[\frac{-\Delta_r H_A^0}{R}\left(\frac{1}{T} - \frac{1}{T_A}\right)\right]. \tag{6.49}$$

If we write the chemical formula of species A in such a way that site multiplicity equals one then the activities are the products of mol fractions times activity coefficients, and we can recast (6.49) as follows:

$$\left(\frac{X_2^A}{X_1^A}\right) = \alpha\,(T)\left(\frac{\gamma_1^A}{\gamma_2^A}\right) \tag{6.50}$$

where, in order to simplify subsequent equations, we have made:

$$\alpha\,(T) = \exp\left[\frac{-\Delta_r H_A^0}{R}\left(\frac{1}{T} - \frac{1}{T_A}\right)\right]. \tag{6.51}$$

Equation (6.49), or its equivalent (6.50), is sometimes called the *freezing point depression* equation, as it tracks how the melting point T of a substance changes relative to the melting point of the pure substance, T_A, as a function of the addition of "impurities" that lower the activity of the substance in the melt phase, a_2, in equilibrium with pure solid (a_1 constant and equal to 1). Equation (6.49) is, however, completely general and applicable to any first-order phase transition, even if some additional considerations are needed when applying it to phase transitions involving a gas phase (next section).

We are also interested in how the activity ratio varies with pressure at constant temperature. In this case it is straightforward to see that:

$$\frac{\partial \ln K}{\partial P} = -\frac{\Delta_r V^0_{P,T}}{RT}. \tag{6.52}$$

If the pressure of the phase transition for pure A is P_A, so that $\ln K_{P_A} = 0$, we get:

$$\ln K_P = -\frac{1}{RT} \int_{P_A}^{P} \Delta_r V^0_{P,T} dP. \tag{6.53}$$

If the phases on both sides of the phase transition are condensed phases then it may be acceptable as a first approximation to treat the volume change as a constant, in which case we have:

$$\left(\frac{a^A_2}{a^A_1}\right)_P = K_P = \exp\left[\frac{\Delta_r V^0_A}{RT}(P_A - P)\right], \tag{6.54}$$

where $\Delta_r V^0_A$ is the volume change associated with the phase transition of pure A at P_A. Separating ideal and excess contributions to activity and defining:

$$\alpha(P) = \exp\left[\frac{\Delta_r V^0_A}{RT}(P_A - P)\right] \tag{6.55}$$

we get:

$$\left(\frac{X^A_2}{X^A_1}\right) = \alpha(P)\left(\frac{\gamma^A_1}{\gamma^A_2}\right). \tag{6.56}$$

The integrals in (6.48) and (6.53) can be refined by considering the temperature and pressure dependencies of the enthalpy and volume changes of the phase transition. It is also possible to write the freezing point depressions equation (6.49) as a function of the entropy of the phase transition, something that we will do and justify in Chapter 10.

6.5.2 Phase transitions involving a gas at low pressure

If the phase transition involves a gas then it is necessary to define the standard state in such a way that the energy contribution that arises from expansion of the gas is properly accounted for. Here we will restrict the discussion to gases at pressures of order 1 bar or less, such that the ideal gas approximation is acceptable. The phase transitions in question are boiling and sublimation, that we will call in general vaporization, and their inverse, condensation, also called deposition in the case that a gas condenses to a solid.

Recall that according to our labeling convention phase 2 is the high entropy and high volume phase, so that it is the gas. Phase 1 can be either a liquid or a solid. We will choose the standard states as pure gaseous A (phase 2) and pure A in phase 1 *at 1 bar and the temperature of the phase transition for the pure substance at 1 bar*. With this definition of standard states the activity of A in the gas is numerically equal to the partial pressure of component A in the gas, p_A (see equation (5.86)). The equilibrium constant in equation (6.48) is therefore the ratio p_A/a_1^A, and given our choice of standard states $\ln K_{TA}$ vanishes, as before. We now write (6.49) as follows:

$$\ln\left(\frac{p_A}{a_1^A}\right)_T = -\frac{\Delta_r H_A^0}{RT} + \frac{\Delta_r H_A^0}{RT_A} \qquad (6.57)$$

For moderate temperature excursions relative to T_A (the temperature of the phase transition for pure A at 1 bar) $\Delta_r H_A^0$ can be considered to be constant. Equation (6.57) can then be re-written as follows:

$$\log_{10}\left(\frac{p_A}{a_1^A}\right) = a + \frac{b}{T} \qquad (6.58)$$

with a and b constants, and the natural logarithm conventionally replaced by decimal logarithm. This is the version of equation (6.49) that is widely used for phase transitions involving a gas at low pressure. The partial pressure of gas species A, p_A, in equilibrium with a condensed phase in which the activity of species A is a_1^A is called the *vapor pressure* of A. We will return to this in Chapter 9, but it is important to understand what it means. First, the definition of vapor is a gas in equilibrium with a condensed phase, solid or liquid, of the same composition. If a condensed phase containing species A exists inside a system whose volume is greater than that of the condensed phase, there is "empty space" if you wish, then thermodynamic equilibrium requires that the chemical potential of A in the "empty space" must be the same as the chemical potential of A in the condensed phase. This means that there must be a vapor of A molecules in the "empty space". The vapor pressure of A is the partial pressure of A that makes the chemical potential of A in the vapor (see equation (5.86)) equal to that of A in the condensed phase. If no other substances are present then this partial pressure equals the total pressure on the system (equation (5.78)), but at a given temperature the vapor pressure of A is fixed, regardless of the total pressure – this is what equation (6.58) stands for. The relation is equally valid whether the condensed phase is a liquid or a solid, although the vapor pressures of a solid and a liquid of the same composition and at the same temperature are different, and so are the a and b parameters in equation (6.58). Substances are said to be more volatile the higher their vapor pressure is at a given temperature. The boiling point of a liquid is the temperature at which its vapor pressure equals the total pressure, so that at least locally (at the liquid–gas interface) the gas is made up exclusively of molecules with the same composition as the liquid. This concept also works for solid–gas equilibrium, even if we don't generally think of a boiling point for solids.

The usefulness of equation (6.58) is that values of the parameters a and b are tabulated for liquid–vapor and solid–vapor equilibria for many substances (see, for example, Lodders & Fegley, 1998, Table 1.20), which makes calculation of vaporization equilibria very straightforward (see below). I emphasize, however, that (6.49) and (6.58) are the same equation. A refinement of (6.58) that accounts for changes in $\Delta_r H_A^0$ with temperature is also widely used and is known as the *Antoine equation* (Antoine equation parameters for many substances can be found in the NIST Chemistry WebBook).

Suppose that we wish to use (6.58) to track changes in the equilibrium compositions of gas and condensed phase with temperature, at a constant pressure P. We recall from equation (5.78) that the mol fraction of species A in the gas phase is given by $X_A = p_A/P$. Separating ideal and excess contributions to activity in the condensed phase we can rewrite (6.58) as follows:

$$\left(\frac{X_2^A}{X_1^A}\right) = \frac{\alpha(T)}{P} \cdot \gamma_1^A,$$
(6.59)

where the exponential function in this case is:

$$\alpha(T) = 10^{\left(a+\frac{b}{T}\right)}.$$
(6.60)

Equation (6.58) can also be recast so as to track changes in phase composition with pressure, at constant temperature. If we fix the temperature at some value of interest then the right-hand side of (6.58), and hence $\alpha(T)$, is a constant. In particular, $\alpha(T)$ is equal to the partial pressure of A in a gas in equilibrium with a condensed phase composed of pure A at the temperature of interest. Let us call these values of partial pressure and activity $p_{A,0}$ and $a_1^{A,0} = 1$, respectively. From (6.58) it follows that for any other combination of equilibrium values it must be:

$$\frac{p^A}{a_1^A} = \frac{p_{A,0}}{a_1^{A,0}} = p_{A,0}.$$
(6.61)

Let the total pressure on the system be P. Dividing (6.61) by P and separating the activity coefficient in the condensed phase we get:

$$\left(\frac{X_2^A}{X_1^A}\right) = \frac{p_{A,0}}{P} \cdot \gamma_1^A,$$
(6.62)

which is the equivalent of 6.56 for vaporization reactions.

6.6 Discontinuous phase transitions in phases of variable composition

We need to consider two distinct behaviors. In one instance there is complete miscibility between two components, which we shall label A and B, in two phases, labeled 1 and 2, separated by a discontinuous phase transition. By *complete miscibility* we mean that all phase compositions between pure A and pure B are stable. Ideal solutions always exhibit this behavior, as do non-ideal solutions above their critical point (Chapter 7). At the opposite end of the spectrum is the case in which there is complete miscibility between A and B in one of the phases, whereas on the other side of the phase transition A and B are present in perfectly immiscible phases of constant composition, which we shall label a (= pure A) and b (= pure B). Intermediate behaviors, in which limited miscibility exists on one or both sides of a discontinuous phase transition, are common but will not be discussed here.

6.6.1 Complete miscibility in both phases

We discuss first the case in which the phase transition occurs between two phases in both of which the two components, A and B, are fully miscible. We will abide by the following conventions regarding phases and components (see Fig. 6.17). First, phase 2 has higher entropy and volume than phase 1. Thus, phase 2 occurs on the high temperature and low pressure side of the phase transition. Second, the transitions for the pure phases occur at different temperatures and pressures, such that, at constant pressure $T_A < T_B$, and at constant temperature $P_A > P_B$. You can think of component B as being more refractory and less volatile than A, but these definitions are not entirely clear if applied, for example, to a solid–solid phase transition, or to the effect of pressure on a melting reaction.

Let us look first at the behavior of the phase transition at constant pressure. For a given temperature we have two versions of equations (6.50) or (6.59), depending on whether only condensed phases or a gas phase is present:

$$\left(\frac{X_2^A}{X_1^A}\right) = \alpha\,(T)\left(\frac{\gamma_1^A}{\gamma_2^A}\right)$$

$$\left(\frac{X_2^B}{X_1^B}\right) = \beta\,(T)\left(\frac{\gamma_1^B}{\gamma_2^B}\right) \tag{6.63}$$

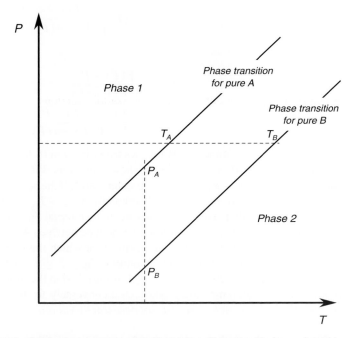

Fig. 6.17 Labeling convention for phases and components in binary phase diagrams with phases of variable composition – see text.

or, if a gas is present:

$$\left(\frac{X_2^A}{X_1^A}\right) = \frac{\alpha(T)}{P} \cdot \gamma_1^A$$

$$\left(\frac{X_2^B}{X_1^B}\right) = \frac{\beta(T)}{P} \cdot \gamma_1^B \qquad (6.64)$$

where I have used $\beta(T)$ to represent the exponential function (6.51) or (6.60) for phase B. In addition, because the system is binary, we have the conditions:

$$X_1^A + X_1^B = 1$$

$$X_2^A + X_2^B = 1. \qquad (6.65)$$

The activity coefficients are functions of composition and, perhaps, temperature. Thus, at constant temperature we have a system of four equations that we can solve for the four mol fractions, i.e. for the compositions of the two phases at equilibrium along the phase transition. The general solution can be quite messy because of the compositional dependency of activity coefficients, and must generally be obtained numerically. However, if the phases can be considered to be ideal mixtures, or if the departures from ideality are comparable in the two phases so that the ratio of activity coefficients in (6.63) is of order 1, then the general solutions are very simple. For phase transitions among condensed phases:

$$X_1^A = \frac{\beta(T) - 1}{\beta(T) - \alpha(T)}$$

$$X_2^A = \alpha(T)\frac{\beta(T) - 1}{\beta(T) - \alpha(T)} \qquad (6.66)$$

and for vaporization transitions:

$$X_1^A = \frac{\beta(T) - P}{\beta(T) - \alpha(T)}$$

$$X_2^A = \frac{\alpha(T)}{P} \cdot \frac{\beta(T) - P}{\beta(T) - \alpha(T)}. \qquad (6.67)$$

The distribution of component B follows trivially from (6.65).

Note some important properties of these equations. First, for $T > T_B > T_A$ both $\alpha(T)$ (or $\alpha(T)/p$) and $\beta(T)$ (or $\beta(T)/p$) are greater than 1. If the activity coefficients are unity then this is impossible by (6.63) or (6.64). The condition $T_B > T_A > T$ leads to another impossible result. Thus, the only feasible solutions are in the interval $T_B \geq T \geq T_A$. This means that the phase transition for any composition intermediate between A and B occurs at a temperature that is intermediate between the temperatures at which the two end-members undergo the phase transition. Second, from the condition $T_B \geq T \geq T_A$ it follows that $X_2^A > X_1^A$ and $X_1^B > X_2^B$. The low-temperature phase, 1, is enriched in the more refractory component, B, relative to the high temperature phase, and conversely for the less refractory component, A. *Note that these conclusions are not necessarily valid if excess mixing properties are not negligible.*

The solution for the behavior of the phase transition with pressure at constant temperature is analogous. Assuming that activity coefficients can be neglected we obtain, for condensed phases:

$$X_1^A = \frac{\beta(P) - 1}{\beta(P) - \alpha(P)}$$

$$X_2^A = \alpha(P) \frac{\beta(P) - 1}{\beta(P) - \alpha(P)} \tag{6.68}$$

and for vaporization:

$$X_1^A = \frac{p_{B,0} - P}{p_{B,0} - p_{A,0}}$$

$$X_2^A = \frac{p_{A,0}}{P} \cdot \frac{p_{B,0} - P}{p_{B,0} - p_{A,0}}. \tag{6.69}$$

In this case we find that, if activity coefficients can be neglected, then the physical solutions are in the interval $P_B \leq P \leq P_A$, and from this condition we find that it must be $X_2^A > X_1^A$ and $X_1^B > X_2^B$. The low-pressure phase, 2, is enriched in the more volatile component, relative to the high-pressure phase, and conversely for the less volatile component, B (see also Fig. 6.17).

Figure 6.18 shows the topology of the phase diagrams calculated with these equations. Each diagram consists of two curves that map the equilibrium compositions of phase 1 and phase 2, expressed as a function of the mol fraction of component A. For example, at T_Z or P_Z, phase 1 of composition X_I is in equilibrium with phase 2 of composition X_{II}.

Phase 1 is stable at lower T and higher P than phase 2. A one-phase assemblage in a binary system has three degrees of freedom, which in this case we choose as P, T and a

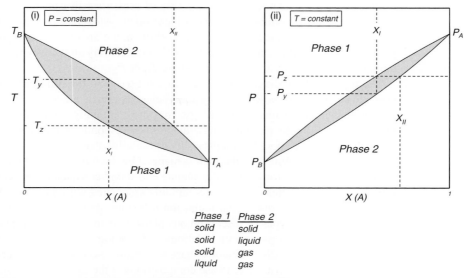

Fig. 6.18 T–X and P–X diagrams for phase transitions between two binary phases, in both of which there is complete miscibility between components A and B. Phase 2 has higher entropy and molar volume than phase 1 (possible combinations of phases are listed at the bottom of the figure). The two phases exist at equilibrium inside the shaded regions. The compositions of the coexisting phases are given by the intersections of the T or P coordinate with the respective bounding curves.

chemical potential that is represented by a compositional variable, $X(A)$. Pressure is fixed in diagram (i) and temperature is fixed in diagram (ii), so each of the one-phase assemblages is stable over a two-dimensional region (T–X or P–X, respectively) in the phase diagrams. Each of these regions terminates at a curve (calculated with the corresponding equation for X_1^A or X_2^A) that represents the location of the phase transition. Inside the shaded area bound by the two curves the two phases are stable, so that there are two degrees of freedom. In diagram (i) these are pressure, which is fixed, and either temperature or composition, whereas in diagram (ii) they are temperature (fixed) and either pressure or composition. At a given temperature in (i), say T_Z, and the pressure chosen to construct the diagram, the compositions of the coexisting phases, X_I and X_{II}, and hence all the chemical potentials, are determined by the solutions to the systems of equations that we derived above, with no possibility of arbitrarily choosing the value of any other variable. For any combination of bulk composition and temperature in diagram (i), or bulk composition and pressure in (ii), that plots inside the shaded regions there are two phases at equilibrium, whose compositions are given by the intersections of the two bounding curves with the temperature or pressure coordinate. The two phases can exist at equilibrium at T_Z, or P_Z, only if the bulk composition of the system lies between X_I and X_{II}. If the system is richer in B relative to this interval then only phase 1 is stable at these conditions, and conversely only phase 2 is stable in a system whose bulk composition is richer in A than X_{II}.

The phase transition in the binary system occurs over a divariant region, rather than along a univariant curve as in the case of a one-component system. For example, as temperature rises or pressure falls, phase 1 of composition X_I will begin to undergo the transition to phase 2 at T_Z, or P_Z, respectively, at which conditions phase 2 of composition X_{II} will form. In a closed system the two phases may remain at equilibrium, while changing composition, until T_y, or P_y, are reached, at which point phase 2 will have attained composition X_I. Further increase of temperature or decrease of pressure will cause phase 1 to disappear. More interesting behaviors become possible in open systems. For instance, if phase 2 is lost from the system then the bulk composition shifts in the direction of component B, and the phase transition will extend beyond T_y, or P_y, conceivably all the way to the values for the univariant equilibrium in the pure B system, T_B or P_B.

The type of phase diagram shown in Fig. 6.18(i), and the different behaviors that are possible depending on whether the system is closed or open, are of course familiar to geologists from elementary igneous petrology. They are commonly exemplified by the olivine or plagioclase melting loops. If the phase diagram corresponds to a melting reaction, then the curve that maps the upper thermal stability of the solid phase is called the *solidus*, whereas the lower thermal stability of the melt is mapped by the *liquidus*. Note that the solidus and the liquidus coincide for the two degenerate end-member systems. I wish to present these results in a different light from that commonly associated with igneous petrology, emphasizing their generality and in particular the following three points. First, melting is only one of the possible types of discontinuous phase transitions involving phases of variable composition. The thermodynamic relations and topology of the resulting phase diagrams are the same for other discontinuous phase transitions. Second, temperature has no special status as an intensive variable. Phase compositions at the transition can also be tracked as a function of pressure, if it is more convenient. Third, the topology of the phase diagram is determined by the solutions to sets of equations such as (6.63) and (6.65), (6.64) and (6.65), and so on. All that these equations require is that the curve that maps the composition of the high-entropy, or high-volume, phase be located at higher temperature, or lower pressure, than the curve that maps the composition of the low-entropy or low-volume phase.

The equations per se do not determine neither the separation between the curves nor their curvature. These quantities depend on the relative values of temperature, pressure, enthalpy change and volume change of the phase transitions for the pure end-member phases, as we shall see in the following examples. Also important are the excess mixing properties of the two phases, which we have ignored in this discussion. It is generally far from trivial to account for these in calculations and we shall not attempt it here. Many phase diagrams are, however, constructed empirically, so that they conflate all of these effects from experimental measurements.

Worked Example 6.8   Crystallization of planetary mantles and cores

At temperatures close to their melting points, olivine forms a complete solid solution between the Fe and Mg end-members, fayalite and forsterite, and so do Fe and Ni metals. Solid–liquid equilibria in these two systems constitute very simplified models for the crystallization of the mantles and cores of the terrestrial planets. These simple models capture some of the important physicochemical aspects of those processes, however. Assuming that excess mixing properties can be ignored all that we need in order to construct phase diagrams for melting of olivine or Fe–Ni alloys using equations (6.66) are the melting points and enthalpies of fusion of each of the end-members – see equation (6.49). The required values are: $T_{Fo} = 2163$ K, $T_{Fa} = 1490$ K, $\Delta_f H_{Fo} = 71.1$ kJ mol^{-1}, $\Delta_f H_{Fa} = 92.2$ kJ mol^{-1} and $T_{Fe} = 1809$ K, $T_{Ni} = 1726$ K, $\Delta_f H_{Fe} = 14.2$ kJ mol^{-1}, $\Delta_f H_{Ni} = 18.2$ kJ mol^{-1}. With these data we calculate $\alpha(T)$ and $\beta(T)$ with equation (6.51), and the compositions of the coexisting phases, for a series of temperatures, $T_B \geq T \geq T_A$ (a spreadsheet program such as *QuattroPro* will do this effortlessly). The resulting phase diagrams (at 1 bar pressure) are shown in Fig. 6.19. If you compare the olivine phase diagram in the figure with the one shown in igneous petrology textbooks you will notice that they differ. This is so because the diagram shown in textbooks is constructed from experimental results, and thus includes the effects of non-ideal mixing in olivine and melt. That diagram is the correct one, but the simplified version shown in Fig. 6.19 will do for our purposes.

There is a striking difference between the olivine and metal phase diagrams. First, the solidus and liquidus curves have curvatures of opposite signs in the olivine phase diagram, whereas both have positive curvature in the metal phase diagram. Second, and more importantly, the curves for olivine are widely separated, whereas they almost coincide with one another in the metal diagram (see inset). Melting, or crystallization, strongly fractionates olivine compositions, but that is not the case for Fe–Ni metal alloys. The different behaviors are rooted in the different enthalpies (and entropies) of fusion of the two systems, which are about five times greater in the silicates compared to the metals (see also Chapter 10). Equation (6.49) shows that, for a given temperature, the ratio between the activity of a component in phase 1 relative to its activity in phase 2 varies exponentially with the enthalpy of transition.

We can think of two ways in which these contrasting behaviors may have affected the evolution of terrestrial planets. First, mantle crystallization from a magma ocean of olivine composition produces olivine crystals considerably richer in Mg than the liquid. If the densities of the solid and liquid phases differ significantly then the two phases would become separated and crystallization would take place in an open system. The residual liquids would become progressively enriched in Fe, resulting in a planetary mantle with

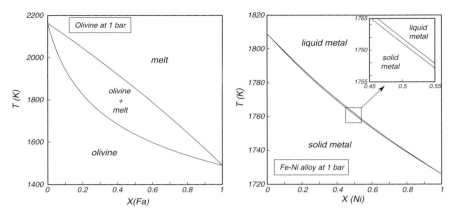

Fig. 6.19 Schematic melting relations for olivine solid solution and Fe–Ni alloy, calculated with the simplifying assumption that the solid and liquid phases are ideal solutions. The enthalpy of fusion of the silicates is ∼5 times greater than that of the metal, causing the wider separation between the solidus and liquidus curves. A section of the metal phase diagram is magnified, showing the extent of solid–liquid fractionation.

primary compositional stratification. Such stratification is less likely to develop in metallic cores, given that the Fe–Ni alloy that crystallizes from a molten metal mixture has virtually the same composition as the liquid. Second, in a crystallizing silicate mantle there is a wide temperature interval over which solid and liquid coexist, which may lead to a sizable depth interval over which the melt fraction varies between 1 and 0. In contrast, solidification of a metal core can be expected to take place along a sharp front.

Worked Example 6.9   Hydrocarbons in Titan's atmosphere–hydrosphere cycle

Titan's atmosphere is composed predominantly of nitrogen, but it has been known for a long time that non-trivial amounts of hydrocarbons are also present. The *Cassini–Huyghens* mission has revealed remarkably Earth-like landforms, including dendritic drainage networks, lakes and dry lake beds. It is thought that Titan's rocky surface consists chiefly of water ice, and that surface liquids are hydrocarbons. Hydrocarbons in Titan play a role comparable to that of water in the terrestrial atmosphere. The dominant hydrocarbon species present in the atmosphere is methane, accompanied by lesser amounts of ethane. The two hydrocarbons are fully miscible in both the liquid and gas phases. Condensation of atmospheric gases and evaporation of liquid hydrocarbons from lakes should lead to methane–ethane fractionation in Titan's climate cycle.

Titan's surface atmospheric pressure is about 1.5 bar, with a hydrocarbon mol fraction of ∼0.05, so that the partial pressure of hydrocarbons in Titan's atmosphere is ∼75 mbar. Using equation (6.58) and *a* and *b* parameters from Lodders and Fegley (1998) and NIST Chemistry WebBook we find that the temperature at which the end-members attain a saturation vapor pressure of 75 mbar are 88 K for methane and 147 K for ethane. Given that the average surface temperature is ∼95 K, there must exist a range of bulk compositions over which liquid and vapor coexist.

We can use equations (6.67) to calculate the compositions of the coexisting phases at $P = 75$ mbar. This calculation assumes that the sum of the partial pressures of the two

Hydrocarbons in Titan's atmosphere and hydrosphere

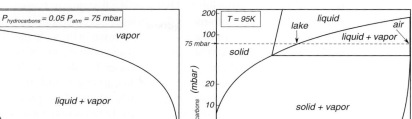

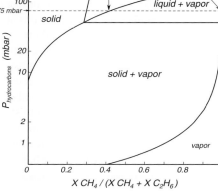

Fig. 6.20 Methane–ethane liquid–vapor equilibrium at the conditions of Titan's surface (95 K and 1.5 bar), assuming a constant hydrocarbon partial pressure of 75 mbar. Lakes are likely to be strongly enriched in ethane relative to methane, compared to atmosphere. The solid–vapor section of the P–X diagram is only schematic, as the saturation vapor pressure of ethane over solid ethane is not well known (see text). Note logarithmic scale on pressure axis.

hydrocarbons does not change with temperature, which is a somewhat artificial constraint that we will address later. As in the previous example, we calculate $\alpha(T)$ and $\beta(T)$ for temperatures in the range 88–147 K, but in this case with equation (6.60). The resulting phase diagram is shown in the left panel of Fig. 6.20. Because hydrocarbons are not the only components present in the gas phase, the horizontal coordinate in the phase diagram is not the absolute mol fraction of methane, but rather the ratio of mols of methane to mols of methane + ethane, which are the two species that contribute to the partial pressure of hydrocarbons in our model.

At the conditions of Titan's surface methane–ethane mixtures with a molar proportion of methane greater than ~0.4 exist as liquids at equilibrium with atmospheric hydrocarbons. One can therefore expect evaporation from hydrocarbon lakes and hydrocarbon rain. There is, however, strong fractionation between the two phases. Hydrocarbon vapor is almost pure methane, but the lakes are about 60% ethane. Note that although it is possible to predict the composition of Titan's lakes from hydrocarbon partial pressure in the atmosphere and temperature, it is *not* possible to derive a bulk methane–ethane ratio from these data alone. The bulk composition can lie anywhere between about 60% ethane and almost pure methane, and would be reflected in the relative masses of liquid and gaseous hydrocarbons.

Some other issues need to be addressed. First, is there a chance that hydrocarbons will solidify on Titan' surface? The melting points of pure methane and ethane are 85 and 101 K, and the enthalpies of fusion are 0.94 and 0.58 kJ mol^{-1}, respectively. Assuming that they mix ideally in the solid state (this is probably incorrect, but it is OK for this discussion, and is explored further in the end-of-chapter exercises), we calculate a binary melting loop as in Worked Example 6.8. The melting loop is also shown in the left panel of Fig. 6.20. Given the very low enthalpies of fusion the solidus and liquidus curves are virtually indistinguishable from one another. For the compositional interval of possible liquid–vapor equilibria the freezing curve is below Titan's average surface temperature. Mixtures with more than about 70% ethane would freeze, however.

What would happen if the bulk proportion of ethane in Titan's hydrocarbons was between 60% and 70%, i.e. in the interval between the freezing and vaporization curves? There would still be liquid hydrocarbons on the surface, but in equilibrium with a lower hydrocarbon partial pressure in the atmosphere. This is shown in the right panel of Fig. 6.20, which shows isothermal phase relations at 95 K, calculated with equations (6.69). The bounding curve of the vapor field (called the *vaporus*, by analogy to solidus and liquidus; see Ricci, 1966) is so steep in this region that atmospheric hydrocarbons would still be essentially pure methane, even if lakes were composed of up to 70% ethane. At approximately this composition the freezing curve is intersected, such that for bulk compositions with more than 70% ethane one would find hydrocarbon icefields and frozen lakes in equilibrium with atmospheric hydrocarbons that would still be dominated by methane, unless the bulk composition was almost pure ethane. The solid–vapor curves are shown (schematically) in the figure, and could be precisely calculated with (6.69) if the vapor pressure of pure solid ethane at 95 K were accurately known, which as far as I can tell is not the case.

6.6.2 Complete immiscibility in the low temperature or high pressure phase

We now turn to the case in which there is complete miscibility between the two components on one side of the phase transition, typically in a liquid or gas phase, and complete immiscibility on the other side, typically between two solid phases. We will assume that the high-entropy and high-volume phase, labeled 2 as before, is the one in which A and B mix without restrictions. Phase 1 is replaced by two distinct phases in the same aggregation state, a and b, composed of pure A and B, respectively. As in the previous case, the phase transitions for pure end-member phases are such that, at constant pressure $T_A < T_B$ and at constant temperature $P_A > P_B$. Consider the isobaric relations for a phase transition between condensed phases first. This is almost always a melting reaction, although in principle it could also take place between two immiscible solids with different crystalline structures on one side and a third solid on the other. Since now a and b are pure phases the activities of A and B in them are unity and equations (6.63) become:

$$X_2^A = \alpha\,(T) \left(\frac{1}{\gamma_2^A} \right)$$

$$X_2^B = \beta\,(T) \left(\frac{1}{\gamma_2^B} \right). \tag{6.70}$$

We have only one mol fraction equation, for the high-entropy phase:

$$X_2^A + X_2^B = 1. \tag{6.71}$$

The situation is rather different from the one that arises if there is full miscibility in both phases. We cannot choose any arbitrary temperature and solve for the composition of phase 2. Rather, 6.70 and 6.71 constitute a system of three equations in three unknowns: the two mol fractions in phase 2 and temperature. The system of equations has zero degrees of freedom, meaning that there is only one temperature at which the three phases co-exist at equilibrium. This is of course the phase rule result: three phases in a binary system have one degree of freedom, that is taken up by pressure, which we fix arbitrarily. Temperature is therefore fixed by the stable co-existence of the three phases. This temperature, which is the solution to the system of simultaneous equations (6.70) and (6.71), is called the *eutectic temperature*, T_e.

In this case the parameters $\alpha(T)$ and $\beta(T)$ equal the activities of each of the components in phase 2, because the phases a and b remain at their standard states (equation (6.70)). Therefore, if phase 2 is a stable solution (we will return to this in the next chapter), then by equation (5.31) both $\alpha(T)$ and $\beta(T)$ must be less than one (see also Fig. 5.7). Given that phase 2, typically a liquid or gas, is the high entropy phase, $\Delta_r H$ is always positive. It then follows from equation (6.51) that the eutectic temperature must be lower than both of the end-member phase transition temperatures: $T_e < T_A < T_B$. This relationship applies most commonly to melting. In order for melting to occur at equilibrium at a point other than the eutectic, one of the solid phases must disappear – the phase rule assures us of this. Say that the phase that disappears is a. Then the chemical potential of A in the liquid must decrease relative to that at the eutectic and, by equation 5.31, the chemical potential of B must increase (Fig. 5.7). If the temperature did not change then the chemical potential of B would be lower in the solid than in the liquid. Because the liquid is the high-entropy phase equilibrium between solid and liquid can be restored only by increasing the temperature. The algebraic expression of this is that, when one of the solids disappears, we are left with only one of the equations in (6.70), which represents equilibrium between liquid and the other solid phase. This equation can be solved for the mol fraction of one component, A or B, in the liquid, in equilibrium with its pure solid, a or b, for any temperature between T_e and T_A or T_B, respectively. There are two physical solutions for $T_e < T < T_A$, and only one for $T_A < T < T_B$. The liquid must disappear at $T < T_e$. In contrast to the binary loops discussed in the previous section there is no analytical solution for this system of equations, because temperature appears in an exponential function, equation (6.51) (i.e. $\alpha(T)$ and $\beta(T)$ in (6.70)). It is, however, very easy to write a *Maple* routine that solves for the eutectic temperature and the composition of phase 2 at the eutectic (Software Box 6.1).

Software Box 6.1   Calculation of melting loop and eutectic melting and vaporization

The worksheet **phasediags1.mw** contains two *Maple* procedures.

Ibin_Tloop: Calculates a melting loop assuming full miscibility and ideal one-site solution behavior in the two coexisting phases. It solves the system of equations (6.63) and (6.65). Note that these calculations can also be implemented in a spreadsheet.

Ibin_eutec: Calculates eutectic melting relations assuming complete immiscibility in the solid and full miscibility and ideal solution in the liquid. It first solves equations (6.70) and (6.71) for the eutectic temperature, and then calculates each branch of the liquidus with the corresponding equation from (6.70). Each branch of the liquidus is stored in a separate file, with "a" or "b" appended at the end of the name. Both branches are stored in terms of mol fraction of species A.

The eutectic behavior for vaporization reactions is obtained in the same way, by solving the system of equations:

$$X_2^A = \frac{\alpha(T)}{P}$$

$$X_2^B = \frac{\beta(T)}{P} \tag{6.72}$$

together with equation (6.71), and recalling that in this case $\alpha(T)$ and $\beta(T)$ are given by equation (6.60) (see Software Box 6.1).

Of considerable interest are isothermal phase transitions involving eutectic vaporization or condensation at low pressure. In this case the system of equations consists of two versions of (6.62):

$$X_2^A = \frac{p_{A,0}}{P}$$
$$X_2^B = \frac{p_{B,0}}{P} \tag{6.73}$$

together with equation (6.71). The total pressure P is the combined partial pressure of the two chemical species. An analytical solution is in this case trivial, with the pressure at the eutectic given by:

$$P_e = p_{A,0} + p_{B,0}. \tag{6.74}$$

The eutectic composition follows immediately from substitution of P_e in (6.73). In this case the combined partial pressure at the eutectic is greater than the partial pressures at the end-member phase transitions. The gas compositions in equilibrium with each of the condensed phases at any pressure P, $p_{B,0} < P < P_e$ or $p_{A,0} < P < P_e$ is also obtained directly from (6.73).

Isothermal phase relations for condensed phases are obtained by substituting $\alpha(P)$ and $\beta(P)$ – see equations (6.55) and (6.56) – for $\alpha(T)$ and $\beta(T)$ in (6.70). They also predict a eutectic pressure, P_e, which must be higher than the phase transition pressures for the two end-members, i.e. $P_e > P_A > P_B$ (exercise left for the reader).

The key aspect of eutectic phase relations is that the melting and boiling points of an assemblage of immiscible phases are lower than the corresponding values for each of the phases in isolation, and the vapor pressure of the assemblage is higher than those of either of the isolated phases. An assemblage of immiscible phases is less refractory and more volatile than each of the phases by themselves. Equally important is the fact that the composition of the liquid or gas that forms at the eutectic point (minimum temperature or maximum vapor pressure) is fixed, and is independent of the bulk composition of the system. These simple facts underlie many fundamental planetary processes. To name just two, they are the reason why terrestrial planets have basaltic crusts (the Earth too, or at least 70% of it) and why granitic rocks have a well-defined and restricted compositional range. We will examine these and other applications of eutectics in later chapters. At this point it is important to build an understanding of how eutectics work, including what are the thermodynamic parameters that determine the magnitude of the displacement of the eutectic temperature or vapor pressure.

Figure 6.21 shows three isobaric phase diagrams for eutectic melting of hypothetical substances that are reasonable models for materials abundant in terrestrial planets. In all cases the pure solids have the same melting point: a (= pure component A) melts at 1600 K, and b (pure B) melts at 1950 K. They also have the same enthalpy of fusion, but this differs among the three diagrams: 100 kJ mol^{-1} in the top panel, 30 kJ mol^{-1} in the middle and 10 kJ mol^{-1} in the bottom. Silicate minerals typically fall in the range between the top and middle panels, and metals between the middle and bottom panels (see Worked Example 6.8). The magnitude of the melting point depression is a strong inverse function of the enthalpy of melting, and it is easy to see why. Fixing the composition of the liquid at any

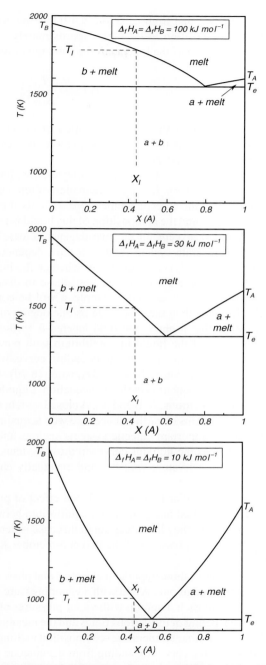

Fig. 6.21 Eutectic melting at constant pressure in three hypothetical systems in which the solids are fully immiscible and the melt is an ideal solution. Enthalpy of melting decreases from top to bottom, causing increased depression of the eutectic temperature.

arbitrary value we find from equation (6.49) that the depression of the melting point for a given liquid composition, ΔT_m, varies approximately as the square of the melting point of the end-member and as the inverse of the enthalpy of melting, i.e.:

$$\Delta T_m \sim \frac{T_A{}^2}{\Delta_m H_A^0}. \tag{6.75}$$

The figure also shows that the composition of the eutectic phase shifts towards that of the refractory phase as the eutectic temperature decreases. This also follows from (6.49) (exercise left to the reader).

It is customary to label eutectic phase diagrams as shown in Fig. 6.21. The labels identify the phases that are stable for any combination of temperature and bulk composition inside each field. Below the eutectic temperature, T_e, only the solid assemblage is stable. Liquid first appears at T_e, and the composition of this liquid is fixed at X_e, for all bulk compositions between A and B, except for the two degenerate cases, pure A and pure B. The solidus of the non-degenerate system (and of a third degenerate system, of composition X_e) is thus the point (X_e, T_e). Depending on whether the bulk composition of the system lies to the left or to the right of X_e the first phase to disappear as the system absorbs heat will be a or b, respectively. For example, a will be consumed first for bulk composition X_I. If this happens then the system gains one degree of freedom allowing its temperature to increase, so that it enters the field labeled $b + liquid$. For any point inside this field the composition of the liquid in equilibrium with pure b is given by the intersection of the temperature coordinate with the bounding curve, which is the liquidus and is also the geometric representation of the second equation (6.70) – the other curve emanating from the eutectic is the representation of the first equation. Liquidus and solidus coincide for the three degenerate compositions, A, B and X_e. A closed system of composition X_I becomes entirely molten at T_I. Students of earth sciences are well acquainted with these relations, commonly exemplified by important eutectic systems such as anorthite–diopside and albite–quartz. Note that whereas the liquidus curves in those systems are convex up, as in the top panel in Fig. 6.21, the curvature vanishes and eventually changes sign as enthalpy of melting decreases.

Perhaps less familiar to geologists is the effect of pressure on eutectic phase relations. These are exemplified in Fig. 6.22, for equilibrium between gas and, say, two solids, a and b. The geometry of the phase diagram in this case is determined exclusively by the relative values of the vapor pressures of the two end-member solids, as shown by equations (6.73) and (6.74).

The figure shows three hypothetical isothermal phase diagrams. Phase a is the same in all of them, composed of pure A and with a vapor pressure of 0.1 bar. Component B and phase b are different in each diagram, with vapor pressures of 0.1, 0.03 and 0.01 bar from top to bottom. The labeling of the various fields and the meaning of the various curves are the same as in the more familiar temperature–composition melting phase diagram of Fig. 6.21, except that in this case the curves extending from the eutectic composition to the vapor pressures of each of the end-members represents the composition of the gas phase and are called vaporus, rather than liquidus. The diagrams make clear what equation (6.74) says: that a mixture of two immiscible solids (or liquids) is more volatile, i.e. has higher vapor pressure, than each condensed phase in isolation. If the two condensed phases have very different vapor pressures the effect is particularly strong for the less volatile substance (e.g. B in the bottom panel of the figure). With increasing disparity in volatilities the composition of the

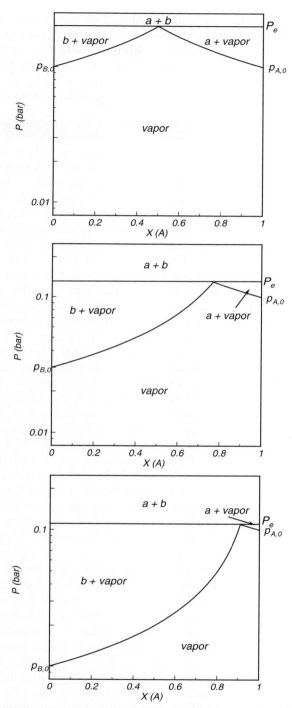

Fig. 6.22 Eutectic vaporization of immiscible solids or liquids at constant temperature. An assemblage of immiscible condensed phases is more volatile (i.e. has a higher saturation vapor pressure) than each phase individually. Note logarithmic scale on pressure axis.

"eutectic" gas shifts in the direction of the more volatile substance. These may have been important factors in the condensation of solids from nebular gas during formation of the Solar System. For example, a solid of relatively low volatility such as B in the bottom panel of Fig. 6.22 may have been prevented from condensing by the presence of more volatile components in the nebular gas, such as A in the figure, until the nebular gas attained a pressure considerably higher than what would have been needed to condense the pure solid.

Worked Example 6.10   Cryolavas in Triton

One of the biggest surprises to emerge from the flyby of the Neptune system by *Voyager 2* in 1989 was the discovery that its large satellite Triton has a tenuous atmosphere and shows signs of current, or at least recent, geological activity. There are stunning pictures of what are almost certainly volcanic calderas and associated lava flows. Icy worlds such as Triton undergo volcanic processes (and, necessarily, plutonic ones) that may resemble those in terrestrial planets in many ways except one: the composition of the magmatic liquids. The temperatures and compositions of icy crusts do not allow the existence of silicate liquids, but melts composed of mixtures of C–H–O–N species, known as cryolavas if they are erupted, are certainly possible. Melting in the binary system CH_4–CO is a simple example of cryomagmatism that may be applicable to Triton (Fig. 6.23). Actual icy magmas are no doubt more complex and varied than this. Using this phase diagram as an introduction to cryomagmatism is akin to introducing the study of basaltic magmatism with the system diopside–anorthite – an informative simplification.

The melting points of pure CH_4 and CO at 1 bar are 90.6 K and 68.1 K, respectively. These values would of course be different at the higher pressures at which the cryomagmas would form in Triton's interior, but we will ignore that. The enthalpies of melting are 0.93 and 0.84 kJ mol^{-1} for CH_4 and CO, respectively. Assuming that the two species crystallize as pure immiscible solids and that they mix ideally in the liquid phase we use the system of equations (6.70) and (6.71) to calculate the phase diagram in Fig. 6.23 (see Software Box 6.1). The system melts at a eutectic temperature of ∼51 K, and the eutectic melt consists of approximately 60 mol% carbon monoxide and 40 mol% methane.

Magmas on Earth seldom reach the surface at temperatures significantly above their liquidus (Chapter 10). If this is also the case in icy satellites then one could expect cryolavas to have temperature–composition combinations along one of the two liquidus curves on Fig. 6.23, and perhaps to be saturated with phenocrysts of either methane ice or carbon monoxide ice, depending on the bulk composition. Say that a cryolava of composition L (Fig. 6.23) is erupted at a temperature of ∼65 K on Triton's surface, where the ambient temperature is about 38 K. The saturation vapor pressures of liquid CH_4 and CO at 65 K are about 1 and 94 mbar, respectively, which are orders of magnitude higher than Triton's atmospheric pressures (tens of μbar, see Worked Example 6.11). The lava will therefore boil upon eruption, but boiling will increase its cooling rate, so that a quenched crust (microcrystalline? glassy?) is likely to form on the surface of the lava flow. Say that this crust cools instantaneously to the ambient 38 K. The saturation vapor pressures of solid CH_4 and CO at 38 K are, respectively, ∼0.006 and 2.6 μbar, i.e., below atmospheric pressure. The solid crust is therefore stable against sublimation in Triton's atmosphere (but see also next Example). If the lava flow is thick enough, or the liquid collects in a lava lake, cooling and crystallization of the interior of the flow may progress slowly enough that the liquid composition will move down the liquidus curve, crystallizing methane ice, as shown by the arrow in Fig. 6.23. An interesting question is what happens to these crystals – do they float or sink? This depends on how the density of methane–carbon monoxide liquids varies with

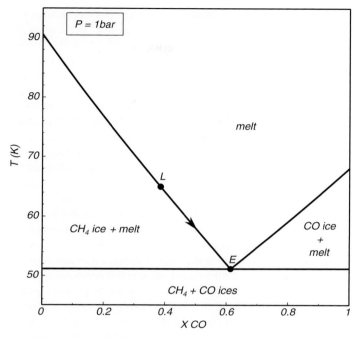

Fig. 6.23 Eutectic melting of a mixture of methane and carbon monoxide ices, assuming ideal mixing in the melt, as a hypothetical model for cryomagmas in Triton. The eutectic temperature ($\sim$51 K) is slightly higher than Triton's average surface temperature ($\sim$38 K), but the saturation vapor pressures of liquid CH_4 and CO at these temperatures are orders of magnitude higher than Triton's atmospheric pressure, suggesting that cryolavas are likely to boil upon eruption.

composition and temperature. In any event, when the liquid reaches the eutectic, point E in the figure, carbon monoxide ice joins the crystallizing assemblage and no further changes in liquid composition nor temperature will take place until solidification is complete. If there was efficient segregation of the liquid from the early-formed methane ice crystals then crystallization of the last pools of liquid may give rise to pods in which the two ices are present in eutectic proportions, perhaps not unlike terrestrial minimum-temperature granites.

An interesting aspect of this model for Tritonian cryomagmatism is that the solidus (eutectic temperature) is only 10–15 K higher than the satellite's surface temperature, making it possible for Triton to be geologically active even if its source of internal heat is feeble.

Cryomagmas in H_2O-rich icy satellites such as Titan and Ganymede are likely to be composed of $H_2O + NH_3 \pm CH_4$ mixtures. This introduces a few complications relative to the simple system that we considered here. First, although water–ammonia mixtures do melt eutectically, several intermediate compounds form (ammonia hydrates) and the system has at least three different eutectics between H_2O and NH_3. Second, for a range of liquid compositions starting at the H_2O end-member the liquid is denser than the solid, so that magmatic ascent and extrusion may be impossible. We return to these issues in Chapter 10.

The dominant species in Triton's atmosphere is N_2, and nitrogen ice is also abundant on its surface. Triton is thus an example of a planetary body in which the chief atmospheric component condenses on the surface. This is distinct from Earth and Titan, in both of which N_2 is also the chief atmospheric gas but the surface temperature is above the boiling point of N_2 ($\sim$78 K). The dominant component of the Martian atmosphere also condenses on the planetary surface, but whereas on Mars there are only relatively minor amounts of frozen CO_2 in the planet's ice caps, much of Triton's surface appears to be composed of frozen nitrogen. We can consider Triton's surface to be a collapsed atmosphere (a situation vividly described in Dan Simmons' *Rise of Endymion*). This may be the norm in the outermost bodies of the Solar System, such as Pluto, Charon and other Kuiper Belt objects.

The following is a simple thermodynamic model for this type of atmosphere. Suppose that the surface of Triton consists of frozen nitrogen and carbon monoxide, two immiscible solids. Triton's surface temperature is $\sim$38 K, the lowest surface temperature so far measured for any body in the Solar System. From equation (6.60) (a and b parameters from Lodders & Fegley, 1998) we calculate vapor pressures at 38 K of $\sim$20.8 and 2.6 μbar for N_2 and CO, respectively. We can then construct the phase diagram in Fig. 6.24, using the system of equations (6.73) and (6.74). The combined vapor pressure of CO and N_2 in equilibrium with the mixture of pure solids at 38 K is $\sim$23.5 μbar. If there are no other components in

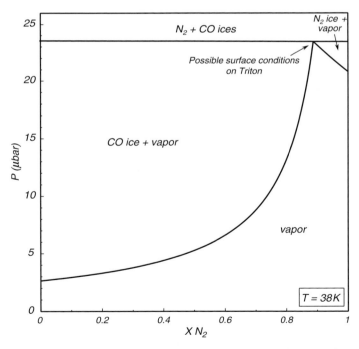

Fig. 6.24   Sublimation of a mixture of nitrogen and carbon monoxide ices at the temperature of Triton's surface. If the surface of the moon consists chiefly of N–C–O ices in equilibrium with the atmosphere then atmospheric pressure and composition should correspond approximately to the eutectic composition.

Triton's atmosphere then this is also the total pressure of the gas, i.e. atmospheric pressure at Triton's surface. The composition of Triton's atmosphere also follows from the phase diagram: it is the composition of the eutectic gas, $\sim$89 mol% N_2 and 11 mol% CO.

Note what we have done here. Beginning with a simple physical model for Triton's atmosphere and the observed surface temperature and composition we calculated atmospheric pressure and composition. The calculation can be refined by including other solid species known to be present on Triton's surface, notably methane (perhaps brought to the surface by lava flows, as discussed in the previous example) and carbon dioxide. The vapor pressures of these two solids at 38 K are, however, many orders of magnitude lower than those of N_2 and CO ($\sim 10^{-9}$ and 10^{-29} bar for CH_4 and CO_2, respectively), so that they are unlikely to be present in the atmosphere in any significant concentrations, and the results are not likely to be much different from those shown in Fig. 6.24.

Exercises for Chapter 6

6.1 Find alternate sets of system components to describe the system spinel–orthopyroxene–olivine–garnet.

6.2 Modify the *Maple* worksheets described in Software Boxes 5.2 and 5.4 to solve the system of five simultaneous equations: (6.14), (6.17), (6.22), (6.23) and (6.24) for the five unknowns: P, $(X_{Mg,M1})_{opx}$, a^{SiO_2}, $a^{Al_2O_3}$ and a^{MgO}, as a function of temperature. Check your results against Figures 5.10 and 6.1.

6.3 Find a set of linearly independent heterogeneous equilibrium equations that fix the chemical potentials of MgO, SiO_2 and Al_2O_3, but different from the set {6.19, 6.20, 6.21}. Repeat Exercise 6.2 and verify that you get the same values of P, $(X_{Mg,M_1})_{opx}$, a^{SiO_2}, $a^{Al_2O_3}$ and a^{MgO}, as a function of temperature.

6.4 Compare the phase diagrams of H_2O and Al_2SiO_5 with Figs. 6.2 and 6.3. Sketch G–P and G–T curves for each of the phases in each system, with the correct relative slopes (recall that the pressure and temperature derivatives G are physically accessible quantities). Convince yourself that the phase diagrams that you are familiar with are the only possible one for each of these systems. What is the only significant difference between the phase diagrams for these two one-component systems? (Hint: the behaviors of andalusite and liquid H_2O are opposite to each other.)

6.5 Construct a schematic phase diagram for the phases jadeite–albite–quartz–melt in the system $NaAlSi_2O_6$–SiO_2. Assume that the melt is silica-saturated.

6.6 Construct schematic phase diagrams in the neighborhoods of each of the following invariant points in the ternary system MgO–SiO_2–Al_2O_3:

   (i) periclase–forsterite–pyrope–sillimanite–spinel
  (ii) forsterite–enstatite–pyrope–sillimanite–spinel
 (iii) periclase–forsterite–pyrope–corundum–spinel
  (iv) periclase–enstatite–quartz–corundum–spinel
   (v) periclase–forsterite–enstatite–quartz–pyrope
  (vi) periclase–quartz–corundum–sillimanite–spinel.

Use thermodynamic properties (e.g. from Holland & Powell, 1998) to decide on the correct orientation of each of the phase diagrams. Compare your phase diagrams and discuss the differences among them.

6.7 Consider the phases forsterite–enstatite–magnesite–vapor (of composition CO_2)–melt (of composition $MgCO_3$) in the ternary system MgO–SiO_2–CO_2. Construct a schematic phase diagram and use it to discuss the generation of carbonatite melts in magnesite-free mantle peridotites. Is it possible, in this simple system, for a Mg-carbonatite melt to crystallize a silicate mineral assemblage, leaving no traces of the existence of a carbonate liquid?

6.8 Consider the four-component system MgO–CaO–SiO_2–CO_2, and the phases forsterite, diopside, enstatite, calcite, vapor (of composition CO_2) and melt (of composition $CaCO_3$). Construct a schematic phase diagram for this system, and connect it to the phase diagram in Fig. 6.11 via the multiply-degenerate reaction (di, fo, en, v). Discuss the similarities and differences between this phase diagram, the one in Exercise 6.7, and the one in Fig. 6.11.

6.9 Construct a schematic P–μ^{O_2} phase diagram for ferroslite–fayalite–magnetite, at constant temperature and μ^{SiO_2}. Discuss the implications of your phase diagram for the effects of pressure and oxidation conditions on the generation of tholeiitic vs. calc-alkaline igneous differentiation trends.

6.10 Construct schematic T–μ^{O_2} and P–μ^{O_2} phase diagrams in the neighborhood of the wustite (FeO)–cohenite (Fe_3C)–metallic Fe–graphite invariant point in the ternary system Fe–C–O. Use your diagrams to discuss whether there is more than one possible explanation for the fact that metallic iron ± cohenite is more characteristic of the interiors of small planetary bodies (e.g. asteroids and the Moon) than of large bodies (e.g. Earth, Mars).

6.11 Construct P–T, P–μ^{O_2}, T–μ^{O_2} and μ^{SiO_2}–μ^{O_2} phase diagrams showing equilibria among the phases: metallic iron, fayalite, schreibersite (Fe_3P), whitlockite ($Ca_3(PO_4)_2$), perovskite and ilmenite, in the six-component system Fe–Si–P–Ca–Ti–O. Discuss the various pathways by which whitlockite microcrystals can exsolve from metal grains in chondritic meteorites.

6.12 In Worked Example 6.9 I calculated a methane–ethane melting loop under the assumption that the two ices are fully miscible. This may not be the case. The opposite "end-member" possibility is that they are perfectly immiscible, and that the ice mixture exhibits eutectic behavior. Calculate the phase diagram that results from this assumption, using the *Maple* worksheet described in Software Box 6.1 (all necessary data are given in Worked Example 6.9). How does this affect the conclusions about the possible nature of Titan's surface hydrocarbons?

6.13 Refine the calculation of Triton's atmospheric composition by calculating the pressure and composition of the vapor in equilibrium with the four ices: N_2, CO, CO_2 and CH_4. All the necessary data can be obtained from Lodders and Fegley (1998), Table 1.20.

The phase transitions that we discussed in Section 6.6 are all discontinuous phase transitions. They are step-wise changes in the structure of matter, for instance, the destruction of the crystalline structure during melting or sublimation, or the breakdown of molecular bonds in a liquid during boiling. These are microscopic changes that are accompanied by a macroscopic exchange of heat with the environment, what we call the enthalpy of transition (melting, vaporization, etc.), or also "latent heat". There is another type of phase transition, which takes place without there being a discontinuity either in the microscopic structure of a substance or in its macroscopic properties, and during which there is no energy exchange with the environment. Such phase transitions are called *continuous* or *critical phase transitions*, and play important roles in many planetary processes. For example, they underlie exsolution phenomena such as are observed in feldspars, pyroxenes, oxides and meteoritic metal, hydrogen–helium unmixing in fluid planets and liquid immiscibility phenomena in magmatic systems. They also explain order–disorder transformations in crystalline substances. Critical phase transitions also play an important role in the study of fluids (Chapter 9).

7.1 An intuitive approach to critical phase transitions

Consider a binary solution between components A and B with unit site multiplicity. If we use X for the mol fraction of component A then the mol fraction of B is $1 - X$, and the Gibbs free energy of ideal mixing is given by equation (5.93):

$$\Delta G_{mixing}^{ideal} = RT\left[X \ln X + (1 - X)\ln(1 - X)\right]. \tag{7.1}$$

This function vanishes at $X = 0$ and $X = 1$, is negative everywhere in between, and becomes more negative with increasing temperature. What this means is that, if two substances mix ideally, then the solution (microscopic mixture) is always stable relative to a mechanical (macroscopic) mixture of the pure substances, and becomes more stable (the tendency to form the solution becomes stronger) with increasing temperature. The curve that represents $\Delta G_{mixing}^{ideal}$ has a minimum and is concave up, as shown in Fig. 5.9. The second derivative of (7.1) must therefore be everywhere positive, which you can verify by differentiating the function. Let us now assume that the solution is not ideal. From (5.122):

$$\Delta G_{mixing} = \Delta G_{mixing}^{ideal} + G^{excess}. \tag{7.2}$$

We begin by considering a hypothetical non-ideal solution that can be described with a symmetric excess Gibbs free energy function (equation (5.146)) that is neither a function of temperature nor of pressure (i.e. $W^S = W^V = 0$). This is the simplest possible description of a non-ideal solution. G^{excess} vanishes at $X = 0$ and $X = 1$ and is positive everywhere in between and independent of temperature. Because G^{excess} and $\Delta G^{ideal}_{mixing}$ have opposite signs and G^{excess} is a function of composition only whereas $\Delta G^{ideal}_{mixing}$ is a function of both composition and temperature, the relative magnitudes of the two functions, and hence of ΔG_{mixing}, vary in interesting ways with temperature and composition. This is shown in Fig. 7.1.

At some high temperature T_1 the excess function is small enough relative to the ideal function that ΔG_{mixing} has the properties of the ideal function, in particular, it is always negative and concave up. This case corresponds to the top diagram on the left of Fig. 7.1.

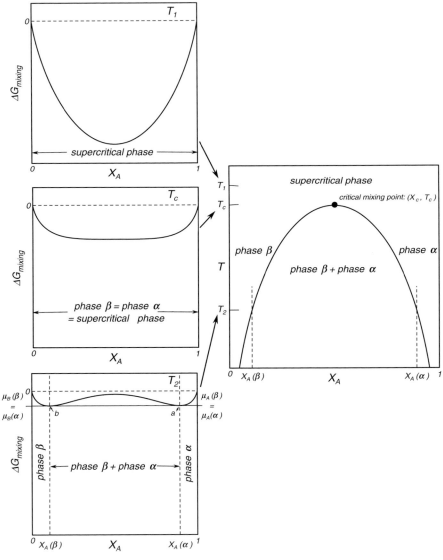

Fig. 7.1 Gibbs free energy of mixing curves curves (left) for a symmetric non-ideal solution at the critical mixing temperature, T_c, and at a higher and a lower temperature. The curve on the T–X diagram (right) is the solvus. See text.

Under these conditions any composition between pure A and pure B forms a stable solution, just as in the case of ideal mixing. As temperature decreases, however, $\Delta G_{mixing}^{ideal}$ becomes less negative but G^{excess} remains unchanged (and positive). The properties of the two functions are such that at some temperature $T_2 < T_1$, ΔG_{mixing} takes the shape shown in the bottom left diagram of Fig. 7.1 (see also Figure 5.13.c). At this temperature the curvature of the ΔG_{mixing} function changes sign twice: there is a central interval with negative curvature separated from two positively-curved intervals by two inflection points. Given a curve with this shape it is possible to find two points, labeled a and b in the figure, which have a common tangent. Note that in Fig. 7.1 these points happen to be minima, but this is only because we are plotting Gibbs free energy of mixing, so that the vertical coordinate of the two end-members is the same. A plot of Gibbs free energy of the solution would not show these points as minima, but they would still have a common tangent. The common tangent is what matters, for it means that there are two compositions in which each of the components, A and B, have chemical potentials that are simultaneously the same (see also Fig. 5.9). These compositions are labeled $X_A(\beta)$ and $X_A(\alpha)$ in Figure 7.1. It is evident from the figure that the Gibbs free energy of any solution that could form in between these two compositions is greater than the Gibbs free energy of a macroscopic mixture of solutions with compositions $X_A(\beta)$ and $X_A(\alpha)$. Such a solution is therefore unstable relative to a macroscopic mixture of phases with compositions $X_A(\beta)$ and $X_A(\alpha)$.

Two distinct phases, which I have labeled phase α and phase β, are therefore stable at T_2. For any bulk composition in the interval $X_A(\beta) - X_A(\alpha)$ the stable assemblage consists of two phases: phase β of composition $X_A(\beta)$ and phase α of composition $X_A(\alpha)$. Two phases in a binary system constitute a divariant assemblage, and since we have chosen arbitrary values of pressure and temperature the compositions of the two phases are fixed. The stable assemblage for bulk composition between $X_A(\beta)$ and $X = 0$ (pure B) consists of phase β only. This can be seen from the fact (Figure 7.1) that any solution within this interval has lower Gibbs free energy than a macroscopic mixture of pure B and $X_A(\beta)$. The one-phase assemblage is trivariant, so that the composition of phase β if it exists at equilibrium by itself is not fixed, but is rather given by the bulk composition of the system. Identical arguments show that phase α alone is stable from $X_A(\alpha)$ to $X = 1$ (pure A), and that its composition within this interval equals the bulk composition of the system.

The two points $X_A(\beta)$ and $X_A(\alpha)$ are compositions of phases that coexist at equilibrium. They correspond to two points on a curve that maps compositions of coexisting phases as a function of temperature, as shown on the right-hand side diagram of Fig. 7.1. This curve is called a *solvus* and is the phase diagram for this system. Bulk compositions inside the solvus consist of two phases at equilibrium, of compositions given by the two intersections of the temperature coordinate with the solvus. Bulk compositions outside the solvus consist of a single phase, of composition equal to the bulk composition. We will get to a formal justification for the shape of the solvus below, but it should be intuitively apparent that the solvus of the system that we are considering must become wider as temperature decreases and the positive contribution of the excess term in (7.2) becomes greater in relative terms.

It is important to realize that the only thing that changed between T_1 and T_2 is temperature. The nature of the phase has not changed, as reflected in the fact that the mathematical description of mixing has not changed. The ideal and non-ideal contributions to Gibbs free energy of mixing are described by the same equations and using the same values for the interaction parameters throughout, and we have also implicitly assumed that the standard state Gibbs free energies always refer to the same two end-members (otherwise leveling ΔG_{mixing} at zero at the end-members for all temperatures would not be valid). The corollary follows that a solvus can exist only between condensed phases that have the same or very similar microscopic structures, such that complete miscibility is possible at some temperature. Examples include: two feldspars, two pyroxenes, two micas, two metals, two liquids, etc. (but obviously not two gases, since gases are always fully miscible).

A temperature must exist at which the nature of the system changes from one in which there is only one phase spanning the entire compositional range between A and B, such as at T_1, to one in which there is a compositional interval within which two phases are stable, such as at T_2. This temperature is called the *critical temperature.* The two branches of the solvus converge on the same composition, X_c, at the critical temperature, defining the *critical mixing point* (X_c, T_c), see Fig. 7.1. This point is also called the *consolute point*, but I prefer critical mixing point because it highlights the deep underlying analogies among all critical phenomena, that we will explore further below and in Chapter 9.

But what exactly happens at the critical mixing point? If there are two distinct phases, α and β, that can exist below the critical temperature, but only one above it, is the high-temperature phase α or β, or something else altogether? The answers to these questions begin with the middle diagram on the left-hand side of Fig. 7.1, which shows Gibbs free energy of mixing at the critical temperature, T_c. As we saw, for $T < T_c$ the Gibbs free energy function has two inflection points, whereas for $T > T_c$ there are no inflection points. Recall that at an inflection point the curvature of a function changes signs, and hence its second derivative vanishes. We can move from $T > T_c$ to $T < T_c$ without any discontinuity, as all that is entailed in doing so is changing the value of T, which is a continuous variable, and monitoring how (7.2), which is a continuous function, responds. Somewhere along this continuous path there must be a temperature – the critical temperature – at which the function goes from having a second derivative that never vanishes to having a second derivative that vanishes at two points. The only way in which this can happen without a discontinuity is if at some temperature, which is the critical temperature, the second derivative vanishes at one point only. Now, if the function vanishes at $X = 0$ and $X = 1$ and has a single inflection point in between, then the sign of the curvature of the function cannot change at this inflection point either. Mathematically this means that the third derivative of the Gibbs free energy function vanishes as well. This special behavior is depicted in the middle left diagram in Fig. 7.1. It corresponds to a curve that is very flat in the neighborhood of X_c. It is, if you wish, almost a straight line but not quite. Generally, the fourth derivative is the lowest order derivative of the Gibbs free energy of mixing that does not vanish at the critical point (Section 7.4).

Consider a phase of composition X_c that forms at $T > T_c$. We shall label any phase that is stable at $T > T_c$ the *supercritical phase*. If the supercritical phase is cooled instantaneously to a temperature $T < T_c$ it will have a finitely higher Gibbs free energy relative to a mixture of phases α and β and this Gibbs free energy will act as a driving potential (see Section 5.3.2) that will cause the single supercritical phase to unmix into a macroscopic mixture of the *subcritical phases* α and β (albeit subject to kinetic constraints, Chapter 12). With an experiment of this kind it would in principle be possible to distinguish between the supercritical phase and the subcritical phases. However, if the final temperature of the experiment is made progressively higher, all the while keeping it below T_c, the driving potential for unmixing becomes smaller (because the ΔG_{mixing} curve becomes flatter), and the compositions and physical properties of the two subcritical phases become closer to one other, and also closer to those of the supercritical phase. *Finally, at T_c it is no longer possible to distinguish between the three phases.* There is no discontinuous phase boundary between the supercritical phase and either of the two subcritical phases. Rather, at the critical point the system changes continuously from a supercritical state in which only one phase is possible to a sub-critical state in which two phases can exist at equilibrium. If the experiment is repeated with a supercritical phase of composition other than X_c then the phase will change continuously and imperceptibly, as temperature is lowered, from the supercritical phase to either of the two subcritical phases (depending on the bulk composition), until the temperature is reached at which this particular bulk composition intersects the solvus and the other subcritical phase will appear, in this case discontinuously. There is a direct analogy with the behavior of fluids: the distinction between liquid and gas, and the discontinuous phase transition that separates the two phases, vanishes continuously as the critical point is approached, and a single super-critical fluid phase is stable above the critical temperature (Chapter 9). Application of the phase rule at a critical point may be confusing, but it shouldn't be (see Box 7.1).

Box 7.1   **The phase rule at a critical point**

Consider a solvus in a binary system such as the one in the left-hand side panel of Fig. 7.2. The solvus is a divariant phase boundary, as can be seen from an application of the phase rule. We have $c = 2$, and along the solvus the two phases, α and βs coexist at equilibrium, so $\Phi = 2$ and $f = 2$. The solvus appears as a one-dimensional curve, i.e. as a pseudo-univariant phase boundary, because we hold pressure constant, thus "using up" one of the two degrees of freedom. At a different pressure the solvus will generally be located at a different temperature. This is shown schematically by the two curves in the left-hand panel of Fig. 7.2, where I have arbitrarily chosen the higher temperature solvus to correspond to a pressure, P_2, lower than that of the lower temperature solvus, P_1. It could also be the other way around, but this diagram is applicable, for example, to the system liquid molecular helium–liquid metallic hydrogen (see Fig. 2.16).

   The solvus is also a first-order phase transition. Consider bulk composition X_z, at temperature T_z and pressure P_2. At these conditions the system consists of the two phases, α and β, of compositions given by the intersection of the temperature coordinate with the two branches of the solvus. If we change the

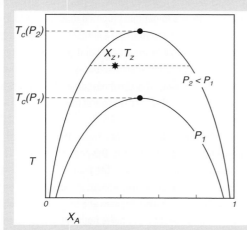
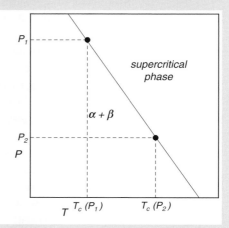

Fig. 7.2   Two solvus curves for a hypothetical system (left) in which the critical mixing temperature varies inversely with pressure. The $P–T$ curve on the right-hand side diagram maps the location of the critical mixing point. It is not a univariant phase boundary, because only one phase is stable along the curve. Following Ricci (1966) we can call it a "singular curve" that separates supercritical conditions (to the right) from subcritical conditions (to the left).

Box 7.1 Continued

temperature while holding the pressure constant one of the phases will form at the expense of the other, and the two phases will change composition. If we raise the temperature then phase α will eventually disappear when the vertical coordinate X_z intersects the solvus. These changes are "first order" because there is a difference in entropy and volume between the two phases, so that there is a non-vanishing enthalpy of reaction associated with the phase changes.

More generally, this non-zero enthalpy change that characterizes first-order phase transitions is what makes it possible for invariant (and univariant, divariant, and so on) assemblages to exist in the first place. We can add or remove heat to an invariant system (e.g. an equilibrium assemblage of the three Al_2SiO_5 polymorphs) and the system will stay at the P and T of the invariant point by forming some of the phases at the expense of the others, until one or more of the phases disappears and the variance of the assemblage increases. With two degrees of freedom rather than zero, this is what happens along the solvus.

Except at the critical point. Critical points have unique coordinates. Thus, in the two component system shown in Fig. 7.2, the critical point at each pressure occurs at a unique temperature, which is the temperature at which coexistence of the two subcritical phases ends. We can show this by a curve in P–T space, as in the right-hand panel of Fig. 7.2, that shows the combination of (T, P) coordinates at which the critical phase transition takes place. The figure would appear to suggest that this curve is a univariant phase boundary **but this is not correct**. By construction, the curve represents the P–T conditions at which phases α and β become identical, and in effect identical to the supercritical phase. It is the locus of all critical points, two of which, corresponding to the diagrams on the left, are shown in the figure. **Along the curve the system consists of a one-phase assemblage**, so according to the phase rule it should be trivariant. Yet it shows as what *looks* like a univariant phase boundary. Similarly, if one considers the critical point of a one-component fluid (Chapter 9), this is the point at which the liquid–vapor equilibrium curve ends, and only one phase is present at the critical point. It *looks* like an invariant point, but according to the phase rule its variance is 2.

What is going on here? Simply, that the variance is indeed what the phase rule says it should be. We will see why in a moment but first a note of caution. It is sometimes proposed that the way to fix this conundrum is by modifying the phase rule at the critical point, by subtracting a number of degrees of freedom equal to the number of phases that become indistinguishable at the critical point (two in both the example of the solvus and that of the fluid critical point). The justification offered for this is having to account for "conditions of criticality", and although it may work arithmetically, it is thermodynamically incorrect. The critical point in a one-component system **is not invariant**, and the P–T locus of critical mixing points in a two-component system (Fig. 7.2) **is not univariant**.

The correct description was formulated by Ricci a long time ago (Ricci, 1966, pages 24–27). He pointed out that a critical point along a univariant phase boundary (e.g. the end of the liquid–vapor coexistence curve, more on this in Chapter 9) is a **singular point**: it is simply the point where the boundary ends. **It is not an invariant point, because exchange of even an infinitesimal amount of energy will move the system away from the critical point, which is distinctly different from what happens at a true invariant point.** Depending on whether one adds (or subtracts) heat or work the system can move both in P and T, so the critical point is indeed divariant. The fact that it is a "point", with unique coordinates, does not make it invariant!

Consider a mixture at its critical mixing point. An infinitesimal exchange of heat, or of mechanical energy, or of matter, will move the system away from the critical point in T, P and X, respectively. The critical point is indeed trivariant. Following Ricci's terminology, we can label the curve on the right-hand diagram of Fig. 7.2 a **singular curve**: it is the boundary of a two-dimensional divariant field, that extends to the left of the curve in the figure. The coexistence of α and β ends at the singular curve because the phases

Box 7.1 **Continued**

become identical. We can also state a general rule as follows: a critical phase transition occurs along an n-dimensional boundary of an $(n + 1)$ dimensional region, over which an assemblage of subcritical phases coexist. The phases become identical at the n-dimensional critical boundary, and only a single supercritical phase exists on the other side of it.

So what makes phase equilibrium at the critical point different? Simply, the fact that there is no latent heat associated with critical (continuous) phase transitions. Because entropy and volume are continuous across a critical phase transition, $\Delta_{transition}H = 0$. It therefore becomes impossible to have, for example, an invariant assemblage, as transfer of even an infinitesimal amount of energy will change the temperature and/or pressure. The phase rule, and the variance that it predicts, are intact and there is no need to invent "conditions of criticality".

I must emphasize that although I used a symmetric Margules-type function to exemplify the concept of critical mixing (because it generates simple diagrams, such as in Fig. 7.1), the behavior is general. It arises from the fact that, if the relative magnitudes of the ideal and excess contributions to Gibbs free energy of mixing change with temperature, then a transition from a Gibbs free energy function with no inflection points to one with two inflection points must happen at some temperature. At this critical temperature both the second and third derivatives of ΔG_{mixing} must vanish, giving rise to a continuous phase transition.

Unmixing of a homogeneous supercritical phase into two or more subcritical phases is an important process in nature. Common examples include exsolution in high-pressure mantle phases (such as pyroxenes from majoritic garnets), Widmanstätten textures in iron meteorites, metal–sulfide–silicate liquid immiscibility, condensation of a supercritical fluid into a liquid and coexisting vapor, and immiscibility between liquid helium and liquid metallic hydrogen in the interiors of some giant planets (Chapter 2).

7.2 Location of the critical mixing point

Finding the critical mixing point of a solution is the first step in determining its solvus. From the preceding discussion it follows that at the critical mixing point the following two conditions must be satisfied:

$$\frac{\partial^2 \Delta G_{mixing}}{\partial X^2} = \frac{\partial^2 \Delta G_{mixing}^{ideal}}{\partial X^2} + \frac{\partial^2 G^{excess}}{\partial X^2} = 0 \tag{7.3}$$

and:

$$\frac{\partial^3 \Delta G_{mixing}}{\partial X^3} = \frac{\partial^3 \Delta G_{mixing}^{ideal}}{\partial X^3} + \frac{\partial^3 G^{excess}}{\partial X^3} = 0 \tag{7.4}$$

which, substituting (7.1), yield:

$$\frac{\partial^2 G^{excess}}{\partial X^2} = -\frac{RT}{X(1-X)} \tag{7.5}$$

and:

$$\frac{\partial^3 G^{excess}}{\partial X^3} = -\frac{RT\,(2X-1)}{X^2\,(1-X)^2}.\tag{7.6}$$

Equations (7.5) and (7.6) are general (for solutions of site multiplicity one). They contain two free parameters, X and T, which means that they can be solved for a unique pair of values that satisfy both equations simultaneously. These are the coordinates of the critical mixing point. In order to solve the equations, however, it is necessary to substitute an explicit expression for the excess Gibbs free energy function. As an example we will assume that the excess Gibbs free energy can be represented by the asymmetric Margules function (5.143), with no temperature dependency. Without loss of generality we can choose to label the components such that $W_{12}^G \geq W_{21}^G$, and make $X = X_2$. We also define the variable K_W, as follows:

$$K_W = \frac{W_{21}^G}{W_{12}^G}\tag{7.7}$$

and note that it must be $K_W \leq 1$, with $K_W = 1$ corresponding to the symmetric function (5.146). Substituting these definitions in (5.143), taking the second and third derivatives of the resulting function, and substituting in (7.5) and (7.6) we obtain, after some simplification:

$$-\frac{RT}{X\,(1-X)} = W_{12}^G[(6X-4)\,K_W - 6X + 2]\tag{7.8}$$

and:

$$-\frac{RT\,(2X-1)}{X^2\,(1-X)^2} = 6W_{12}^G\,(K_W-1).\tag{7.9}$$

Dividing (7.8) by (7.9) and simplifying some more we get:

$$\frac{2X-1}{2X-3X^2} = \frac{3\,(1-K_W)}{2K_W-1}\tag{7.10}$$

which is a quadratic equation in X:

$$9\,(1-K_W)\,X^2 + (10K_W - 8)\,X + (1-2K_W) = 0.\tag{7.11}$$

The physical solution of this equation is X_c, the composition of the critical mixing point. Equation (7.11) shows that the composition of the critical mixture depends only on the relative values of the interaction parameters, and not on their absolute values. In particular, for a symmetric solution $K_W = 1$ and $X_c = 0.5$, which of course we could have inferred from symmetry considerations. Let us now define a non-dimensional temperature, τ, as follows:

$$\tau = \frac{2RT}{W_{12}^G}\tag{7.12}$$

and call the value of τ at the critical temperature τ_c. Substituting this definition in (7.8), and with X_c given by the physical solution of (7.11), we find that τ_c is given by:

$$\tau_c = 4X_c\,(1-X_c)\,[K_W - (1-K_W)\,(1-3X_c)].\tag{7.13}$$

This is a remarkable result: given that X_c depends only on K_W, the non-dimensional critical temperature, τ_c, is also a function of K_W only. But, alas, this concise and rather elegant

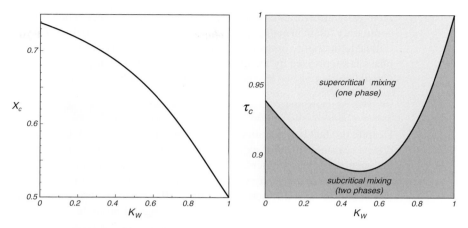

Fig. 7.3 Calculated composition (X_c) and non-dimensional temperature (τ_c) at the critical mixing point, for (possibly hypothetical) solutions with temperature-independent excess Gibbs free energy of mixing.

result is true only because we assumed that G^{excess} is not a function of temperature, and is given by the simple Margules function. Since the Margules approximation is no more than an empirical fit, with little physical fundamentation, we should not expect equation (7.13) to provide more than a rough approximation to the behavior of real solutions.

Even if equations (7.11) and (7.13) are no more than simplified models for the behavior of real solutions, the results, which are summarized in Fig. 7.3, help make the significance of the critical mixing temperature clear. The figure shows X_c and τ_c as functions of K_W. Note that K_W is defined to be ≤ 1, and although it could in principle also be negative, such behavior (which corresponds to Fig. 5.13d) is uncommon and is not included in the figure. The figure shows that, starting from the value $X_c = 0.5$ for a symmetric solution, increasing asymmetry causes the composition of the critical mixture to shift in the direction of the component that is associated with the larger interaction parameter ($X_c > 0.5$), i.e. the component that is responsible for generating the greatest amount of excess Gibbs free energy of mixing. We will see in the next section that this asymmetry carries over into the subcritical region. Symmetric solutions also have the largest value of non-dimensional critical temperature, $\tau_c = 1$. Note that τ_c decreases to a minimum of ~ 0.89 for $K_W \sim 0.5$, but this range may have little practical significance in planetary processes. Figure 7.3 illustrates how equations (7.11) and (7.13) encapsulate a fundamental property of critical mixing. *Namely, all non ideal solutions ($W_{12}{}^G \neq 0$) are in principle capable of exhibiting critical mixing behavior.* This means that all non-ideal solutions can undergo a continuous phase transition between a supercritical one-phase region at $\tau > \tau_c$, and a subcritical region at $\tau < \tau_c$ that, in a binary system, is populated by two phases (Fig. 7.3, right panel). For a given value of K_W (asymmetry) the absolute value of the critical temperature is a linear function of the interaction parameter W_{12}^G, given by (7.12). This leads naturally to a couple of important considerations, that arise from the fact that the stabilities of phases, both supercritical and subcritical, unavoidably terminate at some discontinuous phase boundary. If the excess Gibbs free energy of mixing is small then the critical mixing point may occur at a temperature lower than that at which the supercritical phase breaks down to a lower-entropy assemblage, i.e. the critical mixing point is metastable. The critical temperature may also be sufficiently low that subcritical unmixing, even if thermodynamically favored, is kinetically unfavorable (Chapter 12). In either of these cases only the supercritical phase is commonly found in natural assemblages. A good example of this is olivine. Alternatively, the critical temperature in strongly non-ideal solutions may be high enough that the supercritical phase is suppressed by formation of a stable higher entropy phase, most commonly a melt. Crystallization of two pyroxenes from basaltic melts is an example of this.

Worked Example 7.1   Rock-forming minerals as supercritical vs. subcritical phases

Despite the fact that equation (7.12) is only an approximate description of real solutions, it does a reasonable job of explaining why some minerals display continuous solid solution between two end-members, but others do not. This is summarized in Fig. 7.4. The two diagonal lines in the figure show values of T_c as a function of W_{12}^G, calculated with equation (7.12) and for two different values of τ_c (0.89 and 1), corresponding to K_W values of 0.5 and 1, respectively. These values span the range of possible values for these variables (see Fig. 7.3).

In addition to calculated values of T_c, the figure shows data for a number of rock-forming minerals. The horizontal coordinate of each thick vertical line is an approximate value

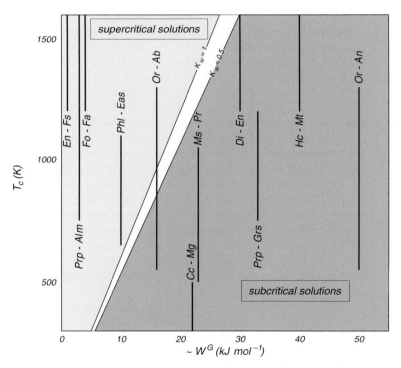

Fig. 7.4   Miscibility relationships in some common rock-forming minerals. The diagonal lines labeled $K_W = 1$ and $K_W = 0.5$ map critical mixing temperature, T_c as a function of W_{12}^G (equations (7.12) and (7.13)). Solid solutions are plotted in terms of approximate values of their largest interaction parameter ($\sim W^G$), and likely temperatures of formation in rocks. En-Fs: orthopyroxene; Prp-Alm: Fe-Mg garnet; Fo-Fa: olivine; Phl-Eas: Mg-Al trioctahedral mica; Or-Ab: K-Na feldspar; Cc-Mg: calcite–magnesite; Ms-Pr: K-Na: dioctahedral mica; Di-En: orthopyroxene–clinopyroxene; Prp-Grs: Mg-Ca garnet; Hc-Mt: Al-Fe^{3+} ferroan spinel; Or-An: K-Ca feldspar.

for the larger interaction parameter (W_{12}^G) measured in each solid solution. The lengths of the lines are very rough estimates of the characteristic temperature ranges in which the minerals form, or at least exist stably. For example, olivine (Fo-Fa) and orthopyroxene (En-Fs) are associated with igneous and solid-state mantle processes, for which characteristic temperatures may be in the range 1200–1700 K. Given the small excess Gibbs free energy of mixing in these minerals their critical mixing temperatures may be close to room temperature, so that they always crystallize as supercritical phases. The same is true of Fe-Mg garnets and Mg-Al micas. On the other hand, mixing between clino- and orthopyroxenes is associated with much higher excess Gibbs free energy of mixing. We could say that two distinct subcritical phases (clino- and orthopyroxene) form even in high-temperature magmatic systems, because the critical point for this solution is metastable relative to (discontinuous) melting reactions. Pyroxenes are actually more complicated than this, undergoing other types of discontinuous phase transitions in the solid state, but the description based on critical mixing is probably correct in general terms, and physically appealing. Other examples of minerals that commonly occur as subcritical phases are K-rich–Ca-rich feldspars, muscovite–paragonite and Mg and Ca carbonates in low-temperature near-surface environments. Note that the critical temperature for calcite-magnesite mixing is ~1000 K, so that a supercritical Ca-Mg carbonate phase might crystallize from carbonatitic magmas. The alkali feldspars are an interesting case in that they can crystallize (e.g. from granitic magmas) as a single supercritical phase that often unmixes into two subcritical phases upon cooling, generating perthitic intergrowths.

7.3 Calculation of non-dimensional solvi

The two branches of the solvus at any temperature $T < T_c$ are the two points $X_A(\alpha)$ and $X_A(\beta)$ that simultaneously satisfy the following two equations (see Fig. 7.1):

$$\left(\frac{\partial \Delta G_{mixing}}{\partial X}\right)_{X=X_A(\alpha)} = \left(\frac{\partial \Delta G_{mixing}}{\partial X}\right)_{X=X_A(\beta)} \tag{7.14}$$

and:

$$\left(\Delta G_{mixing}\right)_{X=X_A(\alpha)} = \left(\Delta G_{mixing}\right)_{X=X_A(\beta)} + [X_A(\alpha) - X_A(\beta)]\left(\frac{\partial \Delta G_{mixing}}{\partial X}\right)_{X=X_A(\beta)}. \tag{7.15}$$

The first equation establishes the common tangent (equal chemical potentials in both phases), the second one simply states that the difference in Gibbs free energy between the two phases equals the slope of the common tangent times the difference between the compositions of the phases at equilibrium. These equations are initially written in terms of $G_{solution}$, rather than ΔG_{mixing}, but all standard state properties drop out, resulting in (7.14) and (7.15) (check this by yourself!). The equations express fundamental thermodynamic relations that are independent of any particular model that one may choose to describe the Gibbs free energy of mixing. They are thus general but, in order to solve them, we must choose an explicit function to describe ΔG_{mixing}. As in the previous section we will choose an asymmetric Margules function with no temperature dependency, with the understanding that we are doing this in order to investigate how solutions in general behave in the subcritical region, and not as a rigorous and accurate description of any specific system.

Since ΔG_{mixing}, or its first derivative, appear on both sides of equations (7.14) and (7.15) we can divide throughout by RT and rewrite these equations in terms of the non-dimensional parameter ($\Delta G_{mixing}/RT$). Introducing the non-dimensional parameters K_W and τ defined in the previous section we have, for an asymmetric Margules excess mixing function:

$$\frac{\Delta G_{mixing}}{RT} = X \ln X + (1 - X) \ln (1 - X) + \frac{2}{\tau} X (1 - X) [K_W (1 - X) + X] \qquad (7.16)$$

and its first derivative:

$$\frac{\partial}{\partial X} \left(\frac{\Delta G_{mixing}}{RT} \right) = \ln X - \ln (1 - X) + \frac{2}{\tau} \left[3 (K_W - 1) X^2 - 2 (2K_W - 1) X + K_W \right].$$
$$(7.17)$$

Substituting (7.16) and (7.17) in (7.14) and (7.15) we end up with two equations in the two unknowns, $X_A(\beta)$ and $X_A(\alpha)$, that can be solved in terms of K_W and τ, for values of $\tau < \tau_c$. The task might be rather unpleasant by hand (I'm not sure it even has an analytical solution) but *Maple* handles it without breaking a sweat (Software Box 7.1).

Software Box 7.1

Procedure `solvi` in worksheet **solvus.mw** solves the non-dimensional solvus specified by equations (7.14) through (7.17). The procedure uses *Maple's* `diff` function to calculate the first derivative of the Gibbs free energy equation (equation (7.17)), so that even though it is written for an asymmetric Margules equation it should be possible to substitute a different Gibbs free energy function and find the solvus for some other description of the excess mixing function. Care must be exercised when specifying the maximum temperature in the input line (parameter `tauhigh`), as if this value is higher than τ_c the procedure will crash. The procedure appears to be stable for `tauhigh` up to $\sim 0.9999 \tau_c$.

The result of these calculations is a solvus that shows the compositions of the coexisting subcritical phases as a function of the non-dimensional temperature τ. Four examples are shown in Fig. 7.5, calculated for different values of K_W. Recall that if G_{excess} is described with a temperature-independent Margules function then τ_c and X_c depend only on K_W (equations (7.11) and (7.13)). With increasing asymmetry the solvus becomes narrower and "leans" in the direction of the component that generates the greatest amount of excess Gibbs free energy of mixing. The effect is that the stability field of the subcritical phase rich in the "more ideal" component becomes wider at the expense of the phase rich in the "less ideal" component (we could have expected this to be the case on purely intuitive grounds). A real solution with a given degree of asymmetry (given by the value of K_W) will exhibit a solvus with the same shape as the corresponding non-dimensional solvus, scaled in its vertical coordinate according to the value of W_{12}^G.

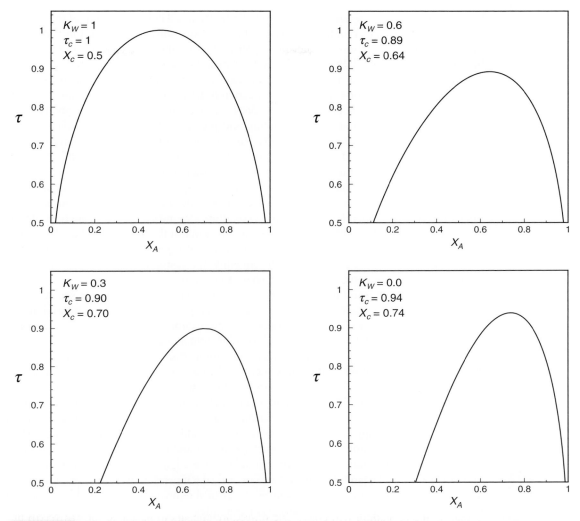

Fig. 7.5 Solvi for hypothetical solutions with temperature-independent interaction parameters, plotted in terms of non-dimensional temperature, τ. In this simple case τ_c, X_c and the width and location of the solvus are determined only by the asymmetry of the solution, given by the non-dimensional parameter K_W.

7.4 Order–disorder phase transitions in crystalline solids

In the preceding sections we discussed critical phase transitions focusing on the example of unmixing of a single-phase solution as its temperature drops below the critical mixing temperature. Critical mixing is only one particular instance of a wide class of phenomena that are governed by the same statistical mechanical principles, and that can be described using the same theoretical framework. We will now begin to see what this means, although our discussion of critical phenomena will not be complete until we discuss fluids, in Chapter 9. A note of caution: many algebraic manipulations in the remainder of this chapter have been omitted owing to space constraints. I believe that a full comprehension of the physics involved requires a clear understanding of where the equations come from. I include some of the important derivations as end-of-chapter problems, and others are left to the reader, but I think that it is crucial that you fill in the gaps as you go over this material.

Minerals can undergo microscopic changes with temperature and pressure, such as changes in their crystalline structures or in cation ordering, that have energy implications. Such changes must be accounted for in order to have a full and accurate thermodynamic description of a mineral at any temperature and pressure. In contrast to Sections 7.1 to 7.3, in this section we will restrict ourselves to transformations at constant composition. It is important to understand that the changes that we are discussing here are distinct from polymorphic transformations. The latter are first order phase transitions: they are step-wise structural changes that take place "in full" at a specific temperature and pressure. For instance, you can have either kyanite or sillimanite, but not something in between. They are different minerals, with distinct crystalline structures, optical properties, densities, entropies, etc. Their Gibbs free energies are the same at the univariant phase boundary, but their entropies and volumes (the first derivatives of G) are not. These discontinuities in $\partial G/\partial T$ and $\partial G/\partial P$ are reflected in the enthalpy of transition, or latent heat, and are the hallmark of discontinuous, or first order, phase transitions.

In critical phase transitions, which for historical reasons are also known as second-order phase transitions, microscopic changes occur gradually over a finite temperature interval and are completed at a well-defined temperature. For example, a mineral may display long-range cation ordering (Worked Example 4.3) at low temperature and become progressively disordered as the temperature increases, until its cation arrangement becomes fully random at a well-defined transition temperature, known, for reasons that we shall presently see, as the *lambda temperature: T_λ*. It then remains in the same state of complete disorder for all temperatures higher than the lambda temperature, until the mineral's demise at a first-order phase transition. Order–disorder phase transitions are observed, for instance, in minerals in which Si and Al occupy similar cation sites, such as feldspars and aluminosilicates. They are also common in compound oxides, such as spinels and ilmenite, and in minerals of Fe and other transition metals, in which they arise from the alignment of magnetic moments of individual atoms. The lambda temperature is a critical temperature with the same properties as those of the critical mixing temperature, the critical temperature of a fluid, or the Curie temperature of a magnet, among many other examples.

Consider a crystalline structure in which there are two types of atoms that can exchange places with one another in two types of crystallographic sites. These could be, for example, Si and Al over different types of tetrahedral sites, or Fe atoms with oppositely pointing magnetic moments over different octahedral sites. Our first task is to define a variable that describes quantitatively the state of long-range order of the crystalline structure. Recall from Chapter 4 that a structure with perfect long-range order is one that has maximum information content, by which we mean that we are absolutely certain of what type of atom we will find in each type of crystallographic site. For example, in microcline, which has long-range order, the four T1a sites are occupied by Al, whereas the four T1b sites and the eight T2 sites are occupied by Si. At the other extreme, in a perfectly random structure we can only define the *probability* of finding a certain kind of atom in a certain kind of site. This probability is equal to the fraction of atoms of the kind we are interested in, relative to the total number of atoms that can enter the site. Thus, in sanidine, which has a fully disordered structure, all we know is that we have a 1 in 4 probability of finding an Al atom, and a 3 in 4 probability of finding a Si atom, in any one tetrahedral site.

Let us call the probability of finding atom X in site ξ, $P_{X\xi}$. We define a parameter ϕ, called the *order parameter*, as follows:

$$\phi = \frac{P_{X\xi} - P_{X\xi, \, random}}{P_{X\xi, \, ordered} - P_{X\xi, \, random}}. \tag{7.18}$$

Clearly, it must be $0 \leq \phi \leq 1$, with $\phi = 1$ for the structure with perfect long-range order, and $\phi = 0$ for a structure with complete disorder, i.e. fully random. This formula is general, but we will apply it to our example of two types of atoms, let us call them A and B, mixing over two crystallographic sites, say α and β. For simplicity we will assume that the stoichiometry and structure of the mineral are such that there are equal numbers of A and B atoms and α and β sites. In principle the relative amounts of A and B could change, but here we are only interested in changes in the crystal at constant composition, AB. We will define the state of perfect long-range order as the one in which A atoms occupy only α sites, and B atoms occupy only β sites. In the disordered state any site, α or β, has the same probability of being occupied by A or B. We then have:

$$\begin{aligned} P_{A\alpha, \, ordered} &= P_{B\beta, \, ordered} = 1 \\ P_{A\alpha, \, random} &= P_{B\beta, \, random} = \frac{1}{2} \end{aligned} \tag{7.19}$$

which, substituting in (7.18), yields:

$$\phi = 2P_{A\alpha} - 1 = 2P_{B\beta} - 1 \tag{7.20}$$

or, equivalently, and, as we shall see, more usefully:

$$\begin{aligned} P_{A\alpha} &= P_{B\beta} = \frac{1}{2}(1 + \phi) \\ P_{A\beta} &= P_{B\alpha} = \frac{1}{2}(1 - \phi). \end{aligned} \tag{7.21}$$

When A and B atoms redistribute themselves in the crystal there is a change in configurational entropy. This change arises not only if A and B are atoms of different elements, but also, for instance, if they are atoms of the same element (e.g. Fe) with differently oriented magnetic moments. *As long as the atoms are distinguishable there is entropy to pay when we shift them around* (Chapter 4). There is also an enthalpy cost, however, that arises from the fact that the energetic interactions between the different types of atoms are not necessarily equal. For example, if A and B are different atoms then the enthalpies of formation of A–A, B–B and A–B bonds may be different. In the case of alignment of magnetic moments, energy is exchanged when the moments flip between parallel (A–A or B–B) and anti-parallel (A–B) arrangements.

To deal with this we split-up the Gibbs free energy of the crystal as follows:

$$G = G(T, P) + G(\phi) = G(T, P) + H(\phi) - TS(\phi). \tag{7.22}$$

In this equation $G(T, P)$ is a function of temperature and pressure (and perhaps composition, but we keep that variable fixed in this case) but not of the state of order of the structure. The free energy contribution that arises from the state of order of the structure, $G(\phi)$, consists of an enthalpy term, $H(\phi)$, and a configurational entropy term, $S(\phi)$, which correspond to the enthalpy and entropy effects that we described in the previous paragraph. *For the crystal* *to be in homogeneous thermodynamic equilibrium at T and P the order parameter, ϕ, must take the value that minimizes $G(\phi)$ at that T and P. Since $G(T, P)$ is not a function of ϕ, minimizing $G(\phi)$ assures that G is also minimized. In general, the value of ϕ that minimizes $G(\phi)$ is a function of T (and P, although initially we will ignore the pressure dependency).* To see why, suppose that the enthalpies of formation of all bonds are negative, but that the enthalpy of formation of A–B bonds is greater in absolute magnitude (more negative) than those of A–A and B–B bonds. One might think naively that the stable configuration, i.e. the one with lowest Gibbs free energy, would be the one that maximizes the number of A–B bonds. Such a structure would, however, have perfect long-range order ($\phi = 1$), and its configurational entropy would vanish. Thus, and given that the enthalpy and entropy contributions to G have opposite signs, maximizing the number of A–B bonds might not minimize $G(\phi)$. Perhaps some value of $\phi < 1$ would produce enough configurational entropy to more than compensate for the loss of some A–B bonds. Moreover, since $S(\phi)$ is multiplied by T, the value of ϕ that minimizes $G(\phi)$ must be a function of temperature. The fact that the Gibbs free energy of a system includes competing enthalpy and entropy contributions is what allows it to undergo a critical phase transition.

In order to construct the function $G(\phi)$ we begin by finding explicit expressions for $H(\phi)$ and $S(\phi)$. The latter is the configurational entropy that we learned to calculate in Chapter 4 (equation (4.55)). Say that in one mol of mineral (N sites) there are $N/2$ each of α and β sites. The total number of microstates, Ω, equals the product of microstates for the α and β sites, Ω_α and Ω_β:

$$\Omega_\alpha = \frac{(N/2)!}{(N_{A\alpha})!\,(N_{B\alpha})!}$$
$$\Omega_\beta = \frac{(N/2)!}{(N_{A\beta})!\,(N_{B\beta})!}. \tag{7.23}$$

Using Stirling's approximation and the definitions of the site occupation probabilities, $P_{X\xi}$, we find after some simplification:

$$\ln \Omega_\alpha = N_{A\alpha} \ln \left(\frac{N/2}{N_{A\alpha}}\right) + N_{B\alpha} \ln \left(\frac{N/2}{N_{B\alpha}}\right) = -(N/2)\, P_{A\alpha} \ln P_{A\alpha} - (N/2)\, P_{B\alpha} \ln P_{B\alpha} \tag{7.24}$$

and an identical expression for $\ln\Omega_\beta$, substituting $P_{B\beta}$ and $P_{A\beta}$ for $P_{A\alpha}$ and $P_{B\alpha}$, respectively. Using (7.21) and simplifying some more we get:

$$\ln \Omega = \ln \Omega_\alpha + \ln \Omega_\beta = N \ln 2 - (N/2)\,[(1+\phi)\ln(1+\phi) + (1-\phi)\ln(1-\phi)] \tag{7.25}$$

finally arriving at:

$$S(\phi) = R \ln 2 - \frac{R}{2}\,[(1+\phi)\ln(1+\phi) + (1-\phi)\ln(1-\phi)]. \tag{7.26}$$

It is a good idea to check what (7.26) is saying. For the fully ordered state $\phi = 1$ and $S(\phi) = 0$ (recall that the limit of $x \ln x$ for x going to zero is 0), whereas for the fully random state, $\phi = 0$ and $S(\phi) = R \ln 2$.

Say now that the formation of each type of nearest-neighbor pair, A–A, B–B and A–B is associated with an enthalpy ε_{AA}, ε_{BB} and ε_{AB}, respectively, and call the number of each type of nearest-neighbor pairs ν_{AA}, ν_{BB} and ν_{AB}. If z is the coordination number of the crystal, then:

$$\nu_{AA} = \frac{zN}{2} \cdot P_{A\alpha} \cdot P_{A\beta} = \frac{zN}{8} \left(1 - \phi^2\right)$$

$$\nu_{BB} = \frac{zN}{2} \cdot P_{B\alpha} \cdot P_{B\beta} = \frac{zN}{8} \left(1 - \phi^2\right) \tag{7.27}$$

$$\nu_{AB} = \frac{zN}{2} \cdot P_{A\alpha} \cdot P_{B\beta} + \frac{zN}{2} \cdot P_{B\alpha} \cdot P_{A\beta} = \frac{zN}{4} \left(1 + \phi^2\right).$$

The enthalpy of formation of all nearest-neighbor pairs in a mol of crystal is then given by:

$$\nu_{AA}\varepsilon_{AA} + \nu_{BB}\varepsilon_{BB} + \nu_{AB}\varepsilon_{AB}$$

$$= \frac{zN}{8} \left(\varepsilon_{AA} + \varepsilon_{BB} + 2\varepsilon_{AB}\right) + \frac{zN}{8} \left(2\varepsilon_{AB} - \varepsilon_{AA} - \varepsilon_{BB}\right) \phi^2. \tag{7.28}$$

Because the first term in the right-hand side of (7.28) is not a function of ϕ, we may consider it to be part of $G(T,P)$ in equation (7.22). We thus get:

$$H(\phi) = \frac{zN}{8} \left(2\varepsilon_{AB} - \varepsilon_{AA} - \varepsilon_{BB}\right) \phi^2 \tag{7.29}$$

or, defining:

$$\eta = \frac{zN}{8} \left(2\varepsilon_{AB} - \varepsilon_{AA} - \varepsilon_{BB}\right) \tag{7.30}$$

we re-write (7.29) as follows:

$$H(\phi) = \eta\phi^2. \tag{7.31}$$

The constant η is the difference between the enthalpy of formation of different nearest neighbors and the enthalpy of formation of same nearest neighbors. Thus, $\eta > 0$ means that the formation of AB nearest neighbors is either less exothermic or more endothermic than the formation of AA and BB nearest neighbors, or, in other words, that there is A–B avoidance in the crystal. Conversely, $\eta < 0$ implies that there is preference for the formation of A–B nearest neighbors.

Collecting terms from (7.22), (7.26) and (7.31) we arrive at:

$$G(\phi) = \eta\phi^2 - RT \ln 2 + \frac{RT}{2} \left[(1+\phi) \ln (1+\phi) + (1-\phi) \ln (1-\phi)\right]. \tag{7.32}$$

We seek the value of ϕ that minimizes this function, so we equate its first derivative to zero:

$$\frac{dG(\phi)}{d\phi} = 2\eta\phi + \frac{RT}{2} \ln \left(\frac{1+\phi}{1-\phi}\right) = 0 \tag{7.33}$$

and note that $\phi = 0$ is always a solution to this equation. But is this particular solution a minimum, and is it the only solution? To answer these questions we begin with the second derivative:

$$\frac{d^2G(\phi)}{d\phi^2} = 2\eta + \frac{RT}{(1-\phi)^2}. \tag{7.34}$$

We now see that, if $\eta \geq 0$, then the second derivative at $\phi = 0$ is always positive, meaning that $G(0)$ is in this case always a minimum. Moreover, if $\eta \geq 0$ then $\phi = 0$ is the only solution to (7.16), as any non-zero solution requires:

$$\ln\left(\frac{1+\phi}{1-\phi}\right) \leq 0, \tag{7.35}$$

which is impossible for $\phi > 0$. What this means is that, if there is either A–B avoidance ($\eta > 0$) or neither preference for nor avoidance of A–B nearest neighbors ($\eta = 0$), then the only stable configuration is the fully disordered state, except for the special case $\eta = 0$ at $T = 0$, where ϕ could be greater than zero (and the Third Law says that, in a perfect crystal, it must be 1). Note that, if $\eta > 0$, then long-range order is not even possible at absolute zero. We could have arrived at these conclusions by simple inspection of (7.22). They are also what one should expect on physical grounds: the only way of spontaneously overcoming disorder is if there is an energetic advantage for formation of A–B pairs, which is not the case if there is A–B avoidance.

The case $\eta \geq 0$ is, then, rather uninteresting, as there is no possibility for the system to undergo a phase transition. Again on physical grounds, we should expect this not to be the case for $\eta < 0$, as in this instance A–B affinity may cause the structure to become ordered at some finite temperature. We now note that, although $\phi = 0$ is a solution to (7.33) with $\eta < 0$, it is not necessarily a minimum. In order for this solution to be a minimum it must be:

$$2\eta + RT > 0 \tag{7.36}$$

or:

$$T > -\frac{2\eta}{R}. \tag{7.37}$$

Thus, the fully disordered structure is the stable one (lowest Gibbs free energy) for temperatures greater than the lambda temperature, T_λ, given by:

$$T_\lambda = -\frac{2\eta}{R}. \tag{7.38}$$

Before we continue, note that this equation for T_λ is formally identical to the one for the critical mixing temperature of a solution: setting $\tau = 1$ in (7.12), we get $T_c = W_{12}^G/2R$. This is no coincidence, as we shall see.

For $T < T_\lambda$ the solution $\phi = 0$ is a maximum, so we need to find whether there is another solution to (7.33) that minimizes $G(\phi)$. We rewrite (7.33) as follows:

$$\frac{1}{2}\ln\left(\frac{1+\phi}{1-\phi}\right) = -\frac{2\eta}{RT}\phi = \frac{T_\lambda}{T}\phi, \tag{7.39}$$

which, making use of the rather obscure hyperbolic trigonometric identity:

$$\tanh^{-1} x = \frac{1}{2}\ln\left(\frac{1+x}{1-x}\right) \tag{7.40}$$

becomes:

$$\frac{T}{T_\lambda} = \frac{\phi}{\tanh^{-1}\phi}. \tag{7.41}$$

Equation (7.41) has non-zero roots over the domain of $\tanh^{-1}\phi$, which is $|\phi| < 1$, so we are assured that for every possible value of the order parameter, $G(\phi)$ will have an extremum at some temperature $T < T_\lambda$. Moreover, since the solution $\phi = 0$ is a maximum and G is a continuous function, the other solution $0 < \phi < 1$ must be a minimum.

Figure 7.6 (top) shows ϕ vs. T/T_λ, calculated with (7.41). If we start from a disordered phase ($\phi = 0$) at high temperature ($T > T_\lambda$) we see that, as the temperature is lowered, ordering begins at $T = T_\lambda$, and ϕ initially increases very rapidly with decreasing temperature below T_λ. The value of the order parameter is very close to 1 at $T \sim 0.5\,T_\lambda$, and changes very little with further cooling to $T = 0$. Figure 7.6 (bottom) shows dimensionless Gibbs free energy: $G(\phi)/RT_\lambda$ (substitute (7.38) in (7.22)), plotted as a function of dimensionless temperature, T/T_λ. As we saw, for $T > T_\lambda$ there is only one possible value of $G(\phi)$, that corresponds to $\phi = 0$. $G(\phi)$ has two possible values below T_λ, a maximum at $\phi = 0$ (shown with a dashed line in Fig. 7.6 (bottom)) and a minimum at the value of $\phi > 0$ given by Fig. 7.6 (top) (shown with the solid curve in 7.6 (bottom)). The disordered phase becomes metastable below T_λ relative to a phase with some degree of long-range order, which increases with decreasing temperature. In nature the disordered high-temperature phase may persist metastably below T_λ, depending on the nature of the cooling process, but that is a different story (Chapter 12). The shape of the function $G(\phi)$ for various values of temperature is shown in Fig. 7.7, plotted as non-dimensional values normalized to $G(0)$: $-[(G(\phi) - G(0))/G(0)]$. Indeed, for $T > T_\lambda$ the minimum occurs at $\phi = 0$, whereas for $T < T_\lambda$ the function goes through a maximum at $\phi = 0$ (corresponding to points on the dashed curve in Fig. 7.6 (bottom)) and a minimum at $\phi > 0$ (points on the solid curve in Fig. 7.6 (bottom)). The distance between the maximum and minimum (i.e. between the two curves in Fig. 7.6 (bottom)) increases with decreasing temperature. Note that at $T = T_\lambda$ the function has a very "flat" minimum, reflecting the fact that the second derivative vanishes (substitute (7.38) in (7.34)), and so does the third derivative (check for yourself). The fourth derivative is positive, though, ensuring that $\phi = 0$ is a minimum at T_λ. The attentive reader will have noticed that this is the same behavior that we found at the critical mixing point (see Section 7.1, and especially Fig. 7.1).

Examining Fig. 7.6 (bottom) we see that $G(\phi)$ is not smooth at T_λ, meaning that its derivatives are discontinuous. The phase transition that occurs at the lambda temperature, called a lambda phase transition is, however, continuous, in the sense that there is no entropy "jump" at the transition. If the lambda phase transition is continuous then the shape of $G(\phi)$ in Fig. 7.6 (bottom) is a problem, as it would imply that there are discontinuities in entropy and volume, which characterize first-order phase transitions. Clearly something is missing from our description of the lambda phase transition. This problem, together with the realization that there are phase transitions that at first sight have no obvious relationship to order–disorder transformations yet behave in the same fundamental way, led to the development of a mathematical formalism to describe phase transitions known as *Landau theory*, named after the great mid twentieth-century Russian physicist Lev Landau. As we shall see, one of the fundamental insights that comes out of Landau theory is the realization that transitional behavior between continuous and discontinuous phase transitions is also possible.

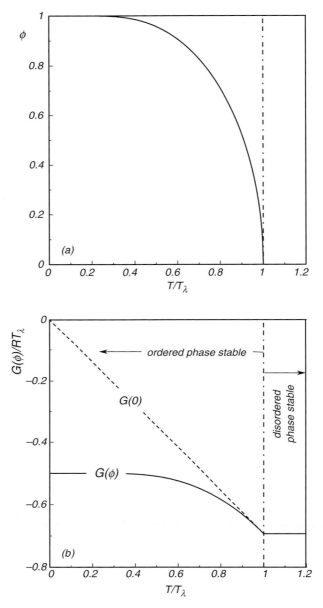

Fig. 7.6 (Top) Order parameter vs. non-dimensional temperature for a phase in which two types of atoms, A and B, occupy two crystallographic sites, α and β, such that $X_A = X_B = X_\alpha = X_\beta = \frac{1}{2}$ and that there is A–B affinity. (Bottom) Ordering contribution to the Gibbs free energy of the phase, shown in non-dimensional units. $G(0)$ is the Gibbs free energy of the disordered phase, which for $T < T_\lambda$ is metastable relative to the ordered phase, shown by $G(\phi)$.

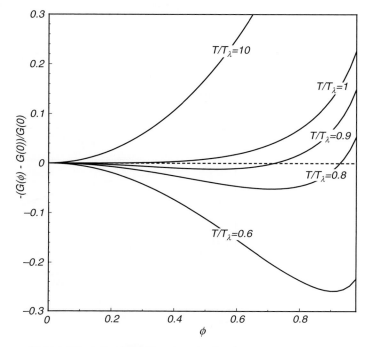

Fig. 7.7 $G(\phi)$ as a function of ϕ at constant temperature. The value of ϕ at each minimum corresponds to the value of ϕ at that temperature in Fig. 7.6, with $\phi = 0$ for $T \geq T_\lambda$.

7.5 Analogies with other phase transitions

There are systems that display behaviors that are remarkably similar to order–disorder phase transitions, even though at first sight they do not appear to bear close similarities to the atomic ordering processes that we discussed in the previous section. The key is to identify a variable in the system that can serve as an order parameter. Figure 7.8 shows some examples. The top panel is a calculated lambda phase transition for atomic ordering – it is in fact the same curve as in Fig 7.6 (top), scaled so as to make comparisons with the other examples easier. The second panel shows the difference in composition between the two branches of a calculated symmetric solvus (e.g. from Figs. 7.1 or 7.5), plotted against the ratio T/T_c, where T_c is the critical mixing temperature. The vertical coordinate is $X_A(\alpha) - X_B(\beta)$. This variable takes values between 0 and 1 (see Fig. 7.1) and it can serve as the order parameter. The curve is identical to that in the top panel, but note that the two curves describe different phenomena. In one case it is atomic ordering in a single phase at constant composition. In the other it is separation of two phases of different composition from a single homogeneous supercritical phase. A way to think of this is as "ordering" of the components A and B into the phases α and β, with perfect order (pure A and pure B) attained only at $T = 0$, but already approached very closely at $\sim 0.5 T_c$ (see Fig. 7.5).

Some minerals acquire higher symmetry with increasing temperature. For example, a mineral may be orthorhombic at low temperature but, owing to anisotropic thermal expansion, the lengths of its a and b crystallographic axes become progressively closer to one

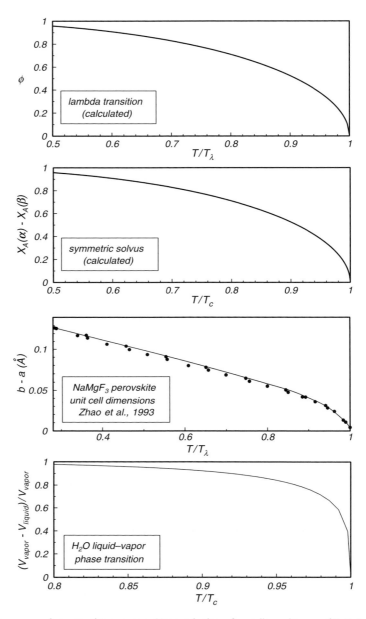

Fig. 7.8 Order parameter vs. non-dimensional temperature. Measured values of crystallographic axes of NaMgF₃ perovskite from Zhao *et al.* (1993). The other three curves are calculated and correspond to: the hypothetical phase in Figs. 7.6 and 7.7, the symmetric solvus in Fig. 7.1, and liquid–vapor equilibrium in H₂O (Chapter 9).

another with increasing temperature, until at a well-defined temperature a and b become equal. When this happens the symmetry changes from orthorhombic to tetragonal. It can also happen that all three axes become equal at a single temperature, and in such case the symmetry of the crystal changes from orthorhombic to cubic. The third panel in Fig. 7.8

shows data for a solid, $NaMgF_3$ with perovskite structure, that behaves in this way. It is orthorhombic at low temperature, but as temperature increases the lengths of its three crystallographic axes approach each other and the solid undergoes a lambda-type phase transition at $T_\lambda \approx 1050\,K$, where it takes on cubic symmetry. The figure shows the difference (in Å) between the lengths of the a and b crystallographic axes (data from Zhao *et al.*, 1993). This difference serves as the order parameter, and could be converted, if one wishes, to a non-dimensional variable that takes the value 1 at $T = 0$ by dividing by $a - b$ at 0 K. Even without doing this, however, the figure shows clearly that $a - b$ goes to 0 at an accelerating rate as T_λ is approached, just as in the case of an order–disorder transition. In fact, the change in crystal symmetry can be understood as an order–disorder transition by considering rotations and distortions in the crystalline framework that affect the entropy of the crystal (see, for example, Putnis, 1992).

The bottom panel in Fig. 7.8 shows the difference between the molar volumes of water vapor and liquid water coexisting at equilibrium (obviously, pressure cannot be constant along this curve – why?), normalized to the volume of the vapor. The molar volumes become equal at the critical temperature (~647.3 K) and a single supercritical phase exists at $T > T_c$. Two phases exist below T_c, and their volumes diverge rapidly immediately below this temperature. We shall return to critical phenomena in fluids in Chapter 9. There are other important critical phenomena that we will not discuss in this book. Examples include: the para-ferromagnetic transition at the Curie temperature, responsible, for example, for the preservation of thermal remanent magnetization in igneous rocks; the appearance of superconductivity in some materials at low temperature; and the existence of a superfluid liquid-helium phase near absolute zero. What is common to all of these processes is that in all of them it is possible to define an order parameter that goes to zero continuously and at an accelerating rate as the critical temperature is approached from below, and that stays at zero above the critical temperature.

It has been known for a long time that systems that behave in this way display anomalies in the second derivatives of the Gibbs free energy in the neighborhood of the critical temperature. This is true of heat capacity, compressibility and thermal expansion (equations (4.135) to (4.137)), but for now we will focus on c_P only. At a first-order phase transition these three quantities become infinite, because as long as a univariant assemblage is present energy transfer (either thermal or mechanical) causes changes in entropy and volume at constant temperature and pressure, by forming some phases at the expense of others. The "infinite heat capacity" occurs only at the first order phase transition, at which S and V are discontinuous, but c_P (and the other second derivatives) behave normally everywhere else. At a lambda phase transition the second derivatives are anomalous in the neighborhood of the transition temperature, and diverge strongly (but may or may not become infinite) at the transition temperature itself. Three examples of this behavior are shown in Fig. 7.9. The top two correspond to crystal symmetry phase transitions, including the example of $NaMgF_3$ perovskite. The bottom panel in Fig. 7.9 corresponds to an order–disorder transition caused by alignment of magnetic moments in ferrosilite. In every case the dashed curve shows what the "normal" heat capacity would be expected to be. The measured values define a strong and steep positive anomaly relative to these expected values, which is characteristically lambda-shaped, hence the name for this type of phase transition. An important point is that, whether or not c_P becomes infinite at the transition temperature, the function is integrable across the transition, i.e., the area under the lambda-shaped anomaly is finite. This area needs to be known in order to be able to calculate the entropy of the solid at $T > T_c$ (Section 4.7.1).

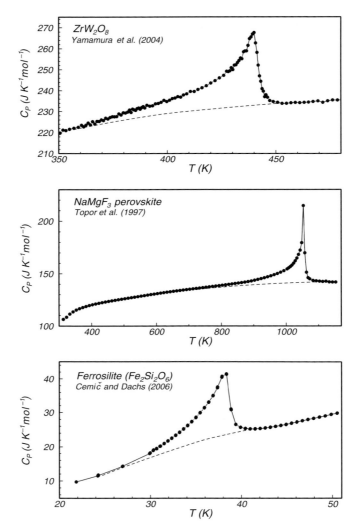

Fig. 7.9 Lambda heat capacity anomalies in three different solid phases. Data from references shown in the figure.

In the simple thermodynamic description of an order–disorder phase transition that we developed in Section 7.4 there is no indication of the behavior depicted in Fig. 7.9. There is need for a theoretical framework that reproduces this observed behavior.

7.6 Landau theory of phase transitions

7.6.1 Critical phase transitions

The key idea of Landau theory of phase transitions is to start with equation (7.22), in which we made G equal to the sum of a function of T and P and a function of ϕ, and expand the function $G(\phi)$ as a power series in ϕ. When this is done $G(\phi)$ is (informally) called the *Landau potential*. As before, we seek the values of ϕ that minimize $G(\phi)$, *but in this case we will make no assumptions about the functional forms of $H(\phi)$ or $S(\phi)$*. Rather, we will derive these, as well as the heat capacity, from the temperature derivatives of the Landau potential. We require that the order parameter be a non-dimensional variable in the interval $0 \le \phi \le 1$ and symmetric relative to 0, so that if we wish to define the order parameter backwards it would be $G(-\phi) = G(\phi)$ (this is a minor point, but simplifies the notation). We also require that at a critical phase transition ϕ goes to zero as the critical temperature is approached from below, and remains at zero for $T > T_c$. Beyond these constraints we attach no specific meaning to the order parameter, which gives us flexibility to define it *a posteriori* in the most convenient way for each specific application. In fact, the requirement that the maximum value of ϕ be 1 can be relaxed trivially, as we shall see.

Symmetry of the order parameter means that we must include only even powers of ϕ, so that, including numerical coefficients that will simplify the derivatives, we write the power series as follows:

$$G(\phi) = \frac{1}{2}g_2\phi^2 + \frac{1}{4}g_4\phi^4 + \frac{1}{6}g_6\phi^6 + \cdots \tag{7.42}$$

The coefficients g_i are empirical macroscopic parameters, as is the order parameter ϕ. The microscopic description that we used to construct (7.32) (number of microstates, energy of nearest neighbor pairs), or any other kind of *a priori* microscopic model, is not a part of Landau theory. Landau theory is an example of a *mean field theory*. The power of this approach is that it is applicable to a large class of phenomena which may have only remote microscopic similarities, and for which our microscopic understanding may be limited. Of course, there is a drawback too, and that is that it provides little insight about the microscopic nature of a specific phase transition.

We will initially assume that we can truncate the power series after the second term. As usual, we require that there be a minimum of $G(\phi)$ for $\phi = 0$ at $T \ge T_c$, and for a value of $\phi > 0$ at $T < T_c$. Taking the first and second derivatives of (7.42), these conditions require the following three relationships (see Exercise 7.2):

$$\begin{aligned} g_2 &> 0, \quad T > T_c \\ g_2 &< 0, \quad T < T_c \\ \phi^2 &= -\frac{g_2}{g_4} > 0, \quad T < T_c. \end{aligned} \tag{7.43}$$

One solution to (7.43) is to make g_4 a positive constant, and g_2 a linear function of $(T - T_c)$:

$$g_2 = \alpha (T - T_c) \tag{7.44}$$

with $\alpha > 0$. We then re-write (7.42) as follows:

$$G(\phi) = \frac{1}{2}\alpha (T - T_c)\phi^2 + \frac{1}{4}g_4\phi^4, \quad \alpha > 0, \quad g_4 > 0. \tag{7.45}$$

This function has the same geometric properties as the function $G(\phi)$ plotted in Fig. 7.7: a sharp minimum at $\phi = 0$ for $T > T_c$, a "flat" minimum for $T = T_c$ and a minimum at $\phi > 0$ and maximum at $\phi = 0$ for $T < T_c$ (Exercise 7.3). Perhaps more interestingly, from the last line in (7.43) we get an equation for ϕ valid for $T < T_c$ (we will use only the positive root, because by construction ϕ is symmetric):

$$\phi = \left[\frac{\alpha}{g_4} (T_c - T) \right]^{1/2}. \tag{7.46}$$

If we scale the order parameter so that $\phi = 1$ at $T = 0$, then we get:

$$T_c = \frac{g_4}{\alpha} \tag{7.47}$$

and:

$$\phi = \left(1 - \frac{T}{T_c} \right)^{1/2}. \tag{7.48}$$

Note that scaling the order parameter to the interval [0,1] is a matter of convenience, not of necessity. Function (7.48) is plotted in Fig. 7.10 (labeled "critical"). Substituting (7.48) in (7.45), and recalling that $\phi = 0$ for $T \geq T_c$ we get:

$$G(\phi) = 0, \quad T \geq T_c$$

$$G(\phi) = -\frac{1}{4} g_4 \left(1 - \frac{T}{T_c} \right)^2, \quad T < T_c \tag{7.49}$$

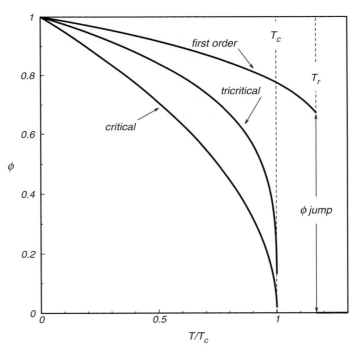

Fig. 7.10 Order parameter vs. non-dimensional temperature predicted by Landau theory for three types of phase transitions. The critical phase transition corresponds to the curve in Fig. 7.6 (top), but the curve in that figure was calculated from a microscopic model of the phase, and the curve in this figure from mean-field Landau theory. The tricritical phase transition is associated with a lambda-shaped heat capacity anomaly, see Fig. 7.11. The first-order phase transition occurs discontinuously at T_r.

and, differentiating (7.49) relative to T:

$$S(\phi) = 0, \quad T \geq T_c$$

$$S(\phi) = -\frac{\alpha}{2}\left(1 - \frac{T}{T_c}\right) = -\frac{\alpha}{2}\phi^2, \quad T < T_c. \tag{7.50}$$

We note that at $\phi = 0$ both $G(\phi)$ and $S(\phi)$ are continuous, and $G(\phi)$ is smooth, confirming that these equations describe a continuous, and not a first-order, phase transition. Calculation of $H(\phi)$ is left as an exercise (Exercise 7.3). Note that spontaneous ordering is always exothermic, so that the entropy of the "universe" increases by a greater amount than the decrease in the entropy of the system given by (7.50), as required by The second Law (more on this in Chapter 14).

We will label the contribution of the ordering process to heat capacity $C_P(\phi)$ and, by analogy with (7.22), we make $C_p = C_p(P,T) + C_p(\phi)$. Differentiating (7.50) and using (4.135):

$$C_P(\phi) = 0, \quad T \geq T_c$$

$$C_P(\phi) = \frac{\alpha}{2}\frac{T}{T_c} = -\frac{\alpha^2}{2g_4}T, \quad T < T_c. \tag{7.51}$$

Heat capacity is thus discontinuous at the phase transition, and this is in fact the origin of the name "second order" – entropy (and volume) are not smooth across a continuous phase transition, and the second derivatives of the free energy are thus discontinuous. The name is discouraged, though, because in some cases discontinuities show up in higher order derivatives. "Critical" or "continuous" are preferable names to encompass all "not first order" phase transitions.

The behaviors of the three thermodynamic functions at a critical phase transition are shown in the top three panels of Fig. 7.11. Comparison of the free energy plot with that in Figure 7.6 (bottom) shows what Landau theory accomplishes: it makes G smooth across the phase transition, while preserving the general relationship between G of the stable ordered phase and G of the metastable disordered phase below T_c. If you rotate the curves to the left of the phase transition in Fig. 7.6 (bottom) counterclockwise until the dashed line (G of metastable disordered phase) becomes the extension of the solid line (G of the ordered phase above T_c) you will get, more or less, the Gibbs free energy diagram in Fig. 7.11. With this transformation G becomes smooth and S becomes continuous at the phase transition.

7.6.2 Lambda phase transitions

Although heat capacity is discontinuous at T_c, the ordering contribution to C_P varies linearly with temperature at $T < T_c$ (Fig. 7.11). Thus, although *equations (7.45) through (7.51) are a correct description of a continuous (or critical) phase transition, they do not describe the behavior observed at lambda phase transitions*, where C_P characteristically diverges strongly (Fig. 7.9). Can Landau theory be made to account for this? Inspection of equations (7.51) suggests that, if g_4 were made a function of $(T - T_c)$, then the heat capacity function would diverge as $T \to T_c$. However, it would not have a finite integral, so this solution is not acceptable. We then do the next best thing, which is to make $g_4 = 0$, immediately invalidating (7.45) through (7.51). We note, however, that if g_4 vanishes, then the only way

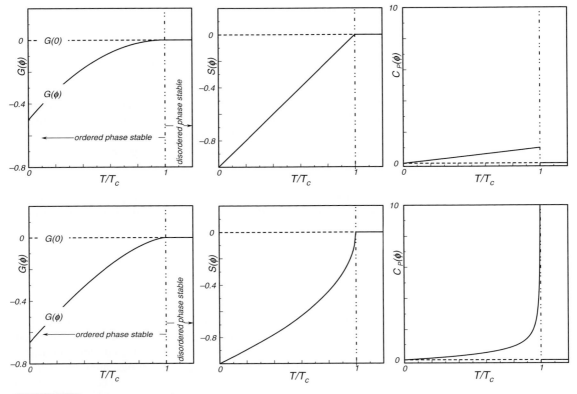

Fig. 7.11   Gibbs free energy, entropy and heat capacity for systems that undergo a critical phase transition (top) or a tricritical phase transition (bottom). The Gibbs free energy diagrams are similar to the one in Fig. 7.6 (bottom), rotated counterclockwise so that the Gibbs free energy function is smooth. The effect of this rotation is to make entropy continuous, but not smooth, at the phase transition.

of allowing for the ordered phase to be stable at $T < T_c$ is to have $g_6 \neq 0$. In effect, we keep (7.44) and the first two lines in (7.43), and from the first derivative of (7.42) with $g_4 = 0$ we replace the third line with (see Exercise 7.5):

$$\phi^4 = \frac{\alpha (T_c - T)}{g_6} > 0, \quad T < T_c, \tag{7.52}$$

which shows that it must be $g_6 > 0$. We then replace (7.45) with:

$$G(\phi) = \frac{1}{2}\alpha (T - T_c) \phi^2 + \frac{1}{6}g_6\phi^6, \quad \alpha > 0, \quad g_6 > 0 \tag{7.53}$$

and note that this function generates a family of curves that have the same general characteristics as those obtained from (7.45) (Exercise 7.6). We now obtain a set of equations that parallel (7.46) through (7.51), but with a crucial difference, as we shall see. Scaling the order parameter as before so that $\phi = 1$ at $T = 0$, we have:

$$T_c = \frac{g_6}{\alpha} \tag{7.54}$$

and, from (7.52):

$$\phi = \left(1 - \frac{T}{T_c}\right)^{1/4}, \quad T < T_c. \tag{7.55}$$

Function (7.55) is labeled "tricritical" in Figure 7.10 (the reason for this name is essentially historical, and need not concern us; see, for example, Griffiths & Wheeler, 1970; Griffiths, 1973). The Landau potential is now given by:

$$G(\phi) = 0, \quad T \geq T_c$$

$$G(\phi) = -\frac{1}{3} g_6 \left(1 - \frac{T}{T_c}\right)^{3/2}, \quad T < T_c \tag{7.56}$$

and the contribution of order–disorder to entropy:

$$S(\phi) = 0, \quad T \geq T_c$$

$$S(\phi) = -\frac{\alpha}{2} \left(1 - \frac{T}{T_c}\right)^{1/2} = -\frac{\alpha}{2} \phi^2, \quad T < T_c. \tag{7.57}$$

The crucial difference is in the contribution of ordering to heat capacity, which now becomes:

$$C_P(\phi) = 0, \quad T \geq T_c$$

$$C_P(\phi) = \frac{\alpha}{4} \frac{T}{\sqrt{T_c}} (T_c - T)^{-1/2}, \quad T < T_c. \tag{7.58}$$

The bottom panels of Fig. 7.11 show plots of the thermodynamic functions for tricritical phase transitions (equations (7.56), (7.57) and (7.58)). The heat capacity diverges at T_c, but it has a finite integral. Comparison of (7.58) with measured heat capacities (Fig. 7.9) suggests that a Landau potential given by (7.53) may provide a good approximation to the behavior of solids that undergo lambda phase transitions.

7.6.3 Discontinuous phase transitions

Beyond the intriguing name, what is a tricritical phase transition? The difference with a critical phase transition becomes clear once we realize that Landau theory can also describe discontinuous phase transitions. We have considered the cases $g_4 > 0$ and $g_4 = 0$, so it is natural to be curious about the case $g_4 < 0$. Let us begin by seeking the equilibrium value of the Landau potential (with $g_4 < 0$) at the critical temperature. Taking the first derivative of (7.42) and equating to zero:

$$\frac{dG(\phi)}{d\phi} = \alpha (T - T_c) \phi + g_4 \phi^3 + g_6 \phi^5 = 0 \tag{7.59}$$

we find that at $T = T_c$ the Landau potential has extrema at $\phi = 0$ and $\phi = (-g_4/g_6)^{\frac{1}{2}}$ — recall that we only use the positive square root. The first non-vanishing derivative for the solution $\phi = 0$ is the fourth derivative:

$$\frac{d^4 G(\phi)}{d\phi^4} = 6g_4 < 0. \tag{7.60}$$

The fact that this derivative is negative means four things. Firstly, that the disordered phase is metastable at the critical temperature, because it corresponds to a maximum of $G(\phi)$. Secondly, that if the other solution must represent the stable phase at T_c, then g_6 must be positive (see equation (7.59)). Thirdly, that the order parameter of the stable phase at the critical temperature is given by (also comes from (7.59)):

$$\phi = \left(-\frac{g_4}{g_6}\right)^{1/2}, \quad T = T_c. \tag{7.61}$$

Plugging this result in the second derivative proves that this is indeed the value of the order parameter that minimizes the Landau potential at T_c. Fourthly, and most momentously, that the phase transition must occur at a different temperature, T_r, which must be $T_r > T_c$, as the disordered phase can only become stable with increasing temperature, and it is not stable at T_c.

The critical point is in effect a metastable critical point, but for reasons that will become clear in a moment we nonetheless wish to know the value of T_c. As always, we scale to $\phi = 1$ at $T = 0$ and, substituting in (7.59) (because the ordered phase must be the stable one at zero temperature), we get:

$$T_c = \frac{g_4 + g_6}{\alpha}. \tag{7.62}$$

It is important to emphasize that (7.62) yields the temperature of a metastable critical point, not of an actual phase transition, but note that if $g_4 = 0$ then the solution collapses to that of the tricritical case, equation (7.54) – you may see where this is going.

At the transition temperature T_r, $G(\phi)$ of the (partially) ordered phase must become zero, because this is the Landau potential of the disordered phase everywhere above T_c. But the ordered phase must also be stable at T_r, so the first derivative of the Landau potential for the ordered phase must also vanish at T_r. We thus have two equations that must vanish simultaneously, one for $G(\phi)$ and the other one for its first derivative. Dividing the first of these equations by ϕ^2 and the second one by ϕ we get:

$$\frac{1}{2}\alpha\,(T_r - T_c) + \frac{1}{4}g_4\phi^2 + \frac{1}{6}g_6\phi^4 = 0$$

$$\alpha\,(T_r - T_c) + g_4\phi^2 + g_6\phi^4 = 0 \tag{7.63}$$

which yield the solutions:

$$\phi = \left(-\frac{3g_4}{4g_6}\right)^{1/2}, \quad T = T_r \tag{7.64}$$

and:

$$T_r = T_c + \frac{3}{16}\frac{g_4^2}{\alpha g_6}. \tag{7.65}$$

Equation (7.64) says that at the temperature T_r given by (7.65) the system undergoes a discontinuous (= first-order) phase transition, as shown by the curve labeled "first order" in Fig. 7.10. This is so because at that temperature the Landau potential of a phase that is still partially ordered (with ϕ given by (7.64)) vanishes and therefore becomes equal to the Landau potential of the disordered phase. The order parameter "jumps" from the value given by (7.64) to 0 at a definite temperature, T_r. The key to understanding the details of this behavior is in the ratio $(-g_4/g_6)$. To see why we first combine (7.62) and (7.65) to obtain:

$$\frac{T_r}{T_c} = 1 + \frac{3}{16}\left(\frac{g_4}{g_6}\right)^2 \frac{1}{1 + \left(\frac{g_4}{g_6}\right)}. \tag{7.66}$$

The second equation in (7.63) is the equilibrium equation (first derivative of the Landau potential equal to zero) so it is valid for $\phi \neq 0$ for all $T < T_r$. Using the quadratic formula and some algebra we find that the non-zero solution is generated by the positive square root (Exercise 7.7). By using (7.62) and some additional algebra, we then arrive at:

$$\phi = \left[-\frac{1}{2}\frac{g_4}{g_6} + \left(\frac{1}{4}\left(\frac{g_4}{g_6}\right)^2 + \left(1 + \frac{g_4}{g_6}\right)\left(1 - \frac{T}{T_c}\right)\right)^{1/2} \right]^{1/2}, \quad T \leq T_r. \tag{7.67}$$

We can see that as $g_4 \to 0$ the solution for the first-order phase transition approaches the tricritical case. Equation (7.66) says that, as $g_4 \to 0$, T_r approaches T_c which, according to (7.62), is the value of T_c for the tricritical case (i.e. for $g_4 \to 0$, (7.62) becomes (7.54)). Moreover, equation (7.67) shows that the solution for the order parameter approaches that of the tricritical case (equation (7.55)) as $g_4 \to 0$.

A tricritical phase transition is therefore a limiting case, between a continuous (critical) and a first-order phase transition. As Fig. 7.11 suggests, it shares characteristics of both: entropy (and volume) are continuous across the tricritical phase transition, but "less smooth" than at a critical phase transition. Heat capacity diverges, but less strongly than in a first-order transition, and the anomaly occurs over a finite temperature interval, rather than only at the transition temperature. Figure 7.12 shows the ratio T_r/T_c and the value of the order parameter at the transition temperature T_r, as a function of $(-g_4/g_6)$. The tricritical phase transition occurs at $(-g_4/g_6) = 0$. As this ratio increases, T_r initially stays infinitesimally close to T_c, but the order parameter at T_r increases rapidly. It is apparent that discriminating experimentally between a tricritical phase transition and one that is first order but associated with a small discontinuity may not always be easy.

7.6.4 Some comments of Landau theory

Landau theory is empirical and provides only an approximate description of actual phase transitions. Still, it is a powerful analytical tool to understand the nature of phase transitions, and in some instances, such as order–disorder transformations in minerals, it does an excellent job of describing the thermodynamic functions associated with these transformations. It is also very flexible – in the preceding section we have barely scratched the surface of what Landau theory is capable of, and although space constraints prevent us from digging deeper into it, there are a couple of points that are worth exploring in some more detail.

First, what is the physical significance of the α and g_i parameters? We begin by noting that α has units of entropy, and g_4 and g_6 have units of energy. We now re-write (7.47), or (7.54), as follows:

$$g_i - \alpha T_c = 0. \tag{7.68}$$

We can interpret (7.68) as meaning that, if the phase transition took place as a first-order transition at T_c, then α would be the entropy of transition, and g_i would be the enthalpy of

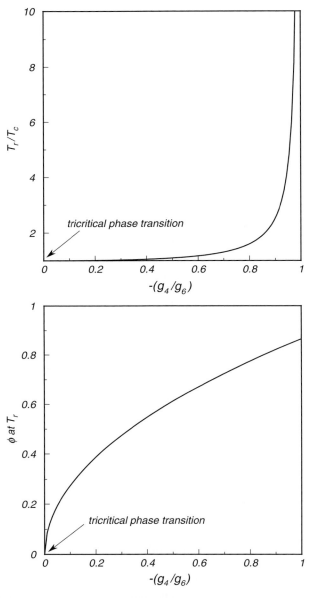

Fig. 7.12   Values of the temperature and order parameter at a discontinuous phase transition, as a function of the ratio between the Landau parameters g_4 and g_6. As $g_4 \rightarrow 0$ the temperature of the phase transition approaches the critical temperature, and the transition becomes smoother. The phase transition is tricritical at $g_4 = 0$.

transition. The phase transition is, however, either critical (7.47) or tricritical (7.54) and in such cases α and g_i may represent the total entropy and total enthalpy associated with the phase transition, "spread out" from absolute zero ($\phi = 1$) to the critical temperature ($\phi = 0$). In the case of a discontinuous phase transition we can manipulate (7.65) to show that α

represents the entropy of the phase transition at T_r, and that the enthalpy of transition is a non-linear function of g_4 and g_6, although the individual meanings of g_4 and g_6 are obscure (Exercise 7.8).

Second, what if we wish to look at the effect of some other intensive variable on the phase transition? This is done by including additional terms in the Landau polynomial expansion. Say that the intensive variable that we are interested in is Z. The simplest case is to add only a quadratic term in Z, so that (7.42) (with g_2 substituted as in (7.44)) becomes:

$$G(\phi) = \frac{1}{2}\alpha\left(T - T_{c,0}\right)\phi^2 + \frac{1}{4}g_4\phi^4 + \frac{1}{6}g_6\phi^6 + \frac{1}{2}\alpha_\zeta Z\phi^2. \tag{7.69}$$

The coefficient α_ζ corresponds to the conjugate variable of Z, so that the product $\alpha_\zeta Z$ has the dimension of energy (recall that α represents entropy, which is the conjugate variable of T). The intensive variables that we are most commonly interested in are either pressure, in which case the coefficient is labeled α_v and represents a volume change associated with the phase transition, or composition, in which case we will have α_μ terms representing chemical potential changes.

By choosing to use only the quadratic term in Z we are coupling α_ζ to α only. This means that the value of α will be the same for any value of Z, but the g_is will vary with Z. I have also added a 0 to the subscript of T_c, to emphasize that this is the critical temperature at a zero value of Z, i.e. setting $Z = 0$ in (7.69) recovers (7.42). We now re-write (7.69) as follows:

$$G(\phi) = \frac{1}{2}\alpha\left[T - \left(T_{c,0} - \frac{\alpha_\zeta}{\alpha}Z\right)\right]\phi^2 + \frac{1}{4}g_4\phi^4 + \frac{1}{6}g_6\phi^6 \tag{7.70}$$

and we define the critical temperature at Z, $T_{c,Z}$ as follows:

$$T_{c,Z} \equiv \left(T_{c,0} - \frac{\alpha_\zeta}{\alpha}Z\right). \tag{7.71}$$

Substituting (7.71) in (7.70) we get an equation in T and ϕ only, which is identical to (7.42) (with g_2 substituted as in (7.44)). This makes all the other equations valid, except that the numerical values of the critical temperature and of the g_i parameters will be different from those obtained with $Z = 0$. From (7.71) we also get the value of critical Z at $T_{c,Z}$:

$$Z_c = \frac{\alpha}{\alpha_\zeta}\left(T_{c,0} - T_{c,Z}\right). \tag{7.72}$$

Note that, in contrast to T_c, there is no *a priori* requirement that the disordered phase be stable on any particular side of Z_c. This depends on the sign of α_ζ, which is not predetermined as in the case of α (by equation (7.44)). For instance, if α_ζ is negative then the critical temperature increases with increasing Z, and the supercritical (disordered) phase is stable on the low Z side of the phase transition (see also Box 7.1).

Worked Example 7.2   Lambda phase transitions and phase equilibrium

Several important rock-forming minerals undergo lambda phase transitions which arise from a variety of microscopic processes. We already mentioned the case of perovskites, in which crystal symmetry changes occur with increasing temperature. More subtle symmetry changes are observed in quartz (transition between α quartz and β quartz at 847 K and 1 bar) and in carbonate minerals. Albite and anorthite display full Si–Al disorder above critical temperatures (at 1 bar) of 950 K and 2300 K, respectively, and progressive ordering below those temperatures. Iron and nickel oxides, as well as the respective metals, undergo order-disorder phase transitions associated with the orientation of magnetic moments. The energetic effects of these phase transitions must be incorporated in the calculation of phase equilibrium. The data set of Holland and Powell (1998 and electronic updates, henceforward HP98) includes lambda phase transition properties that have been regressed simultaneously with standard state properties, so that they are mutually consistent (see Worked Example 5.7 for a refresher of why internal consistency of thermodynamic data is important). Holland and Powell find that the energetic effects of lambda transitions for all minerals in their data base can be accurately modeled by means of a tricritical Landau potential (equation (7.53)), regardless of the microscopic nature of the transformation in each particular mineral. The equations in Section 7.6.2 are applicable to their data, with two important caveats. First, HP98 reverse the sign of $S(\phi)$ in equation (7.57). Second, they use 298 K as the reference point for Landau potentials (i.e. they make $G(\phi) = 0$ at 298 K). The effect of these changes is that the maximum value of entropy, at T_c, is a positive number rather than zero. With their convention $S(\phi)$ is in fact always positive, and would become zero at 0 K (compare equations (7.57)).

Figure 7.13 shows the Landau potentials for several minerals, calculated from HP98 (see Software Box 7.2). The effect of their sign convention is to make the diagrams look like Fig. 7.6 (bottom). As in that case, I also show the Landau potential of the metastable disordered phase below T_c for one mineral (ilmenite) to emphasize that what really matters is the Gibbs free energy *change* that accompanies the ordering process, i.e.

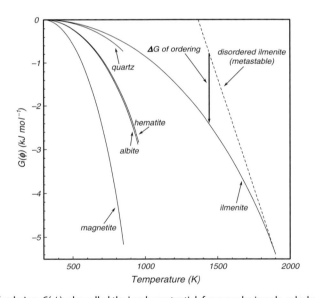

Fig. 7.13 Gibbs free energy of ordering, $G(\phi)$, also called the Landau potential, for several minerals, calculated with data from Holland and Powell (1998). The lambda phase transition for each mineral occurs at the end of the respective curve. Holland and Powell change the convention from $G(\phi) = 0$ for the disordered phase (e.g. Figs. 7.6, 7.10 and 7.11) to $G(\phi) = 0$ at 298.15 K, for consistency with the way in which chemical thermodynamic calculations are usually done. The value of $\triangle G$ of ordering is not affected: compare the example of ilmenite, for which G of the metastable disordered phase is shown with a broken line, with Fig. 7.6 and its rotated equivalent, Fig. 7.11.

the distance between the stable and metastable curves. Restoring Landau's sign and reference conventions rotates the curves as in Fig. 7.11, but the distance between them remains invariant. The figure shows that lambda phase transitions in minerals cause free energy changes of the order of a few kilojoules per mol, which are not insignificant values (see Section 5.2.1).

Software Box 7.2. Incorporation of the Landau potential to Gibbs free energy of minerals

A procedure is included in worksheet **th_shomate.mw** that calculates the Gibbs free energy of order–disorder transitions in minerals, with the formalism of Holland and Powell (1998). The name of the procedure is landau_hp. It calculates the Landau potential of a mineral at any given P and T, using tabulated values of T_c, S_{max} and V_{max}, as defined by Holland and Powell (1998). Recall that their sign convention is different from the one commonly used in physics, but procedure landau_hp takes care of this. Since the procedure is encapsulated in the package **th_shomate.mw** it can be called from any other worksheet.

The strategy to include the Landau potential in mineral equilibrium calculations is to call landau_hp for each of the species that participates in a reaction, multiply by the corresponding stoichiometric coefficient, and add up all of the Landau potentials (they may be zero for some species). The result is added to the Gibbs free energy of reaction as an "excess term". Worksheet **th_template_4.mw** contains a procedure named Gord that does this. The solution for the equilibrium position must now be found iteratively, because there is a conditional statement in landau_hp that checks whether the temperature is above or below the lambda phase transition. This conditional statement is not accepted by *Maple*'s equation solver fsolve. The way around this problem is illustrated in procedures Peq and Teq, that solve for pressure at given temperature, and for temperature at given pressure. The procedures first find a tentative solution assuming that the sum of Landau potentials is zero. They then calculate the Landau potential sum at this pressure and temperature, solve again including this interim Landau potential, and so on, iterating until consecutive solutions converge within a desired interval (set at 1 bar or 1 K). Peq is able to solve for pressure directly, but Teq asks for an initial temperature guess, needed because of the non-linearity of the heat capacity equations. If needed, the procedures can easily be incorporated in a loop to calculate a phase boundary, as in **th_template_3.mw** (e.g. Figure 7.14; exercise left to the reader).

Procedures Peq and Teq also include a call to vdp, which is another procedure, included in the package **th_shomate.mw**, that calculates the pressure integral in equation (5.1.7) with Holland and Powell's compressibilty and thermal expansion data. This is discussed in Chapter 8.

An example of the effect of lambda phase transitions on calculated phase equilibrium is shown in Fig. 7.14. The breakdown of albite to jadeite plus quartz is an important indicator of the transition between medium pressure metamorphism (greenschist–amphibolite facies) and high-pressure metamorphism (blueschist–eclogite facies). Both albite and

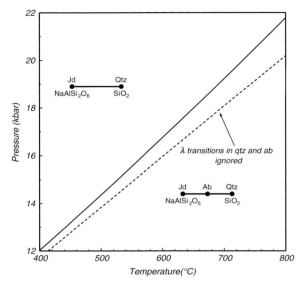

Fig. 7.14 Effect of ignoring ΔG of ordering in albite and *quartz* on the calculated position of the breakdown reaction of albite to jadeite + quartz.

quartz undergo lambda phase transitions. The figure shows the location of the univariant phase boundary calculated with and without the energetic contributions of these phase transitions. Ignoring the Landau potentials displaces the calculated equilibrium position by about 1 kbar. Note that, because albite and quartz are on opposite sides of the reaction, the contributions from the two minerals cancel each other to some extent. The effect would be greater if only one of the phases underwent a lambda phase transition.

Exercises for Chapter 7

7.1 Convert the diagrams on the left of Fig. 7.1 from ΔG_{mixing} to $G_{solution}$, and show that the two points with a common tangent in the subcritical region are no longer minima, but they still have a common tangent.

7.2 Derive the conditions (7.43) for a critical phase transition.

7.3 Show that equation (7.45) generates curves like those in Fig. 7.7. (Use any good scientific plotting program.)

7.4 Find the equation for $H(\phi)$ for a system that undergoes a critical phase transition. Show that spontaneous ordering must be exothermic. Assume that the system that undergoes spontaneous ordering is contained in a larger isolated system. Discuss how the entropy of this isolated system changes during ordering.

7.5 Prove that, if $g_4 = 0$, then ϕ must be given by (7.52).

7.6 Show that (7.53) generates a family of curves with the same general characteristics as (7.45), but with sharper curvature. (Use any good scientific plotting program.)

7.7 Explain why, if the second equation in (7.63) is the equilibrium equation, it must be valid for $\phi \neq 0$ for all $T < T_r$. Show that the non-zero solution of this equation corresponds to the positive square root of the quadratic equation, and derive equation (7.67) for the order parameter.

7.8 Show that for a first-order phase transition α represents the entropy of the phase transition at T_r, and derive an equation for the enthalpy of the phase transition.

8 Equations of state for solids and the internal structure of terrestrial planets

Inferring the internal structure of solid planetary bodies requires that we use thermodynamic theory in order to interpolate and extrapolate often sparse experimental data to very high pressures and temperatures. The constant-volume approximation that we used in Chapter 5 to calculate the Gibbs free energy of solids at high pressure leads to erroneous results at depths greater than a few km. This situation is remedied by introducing a variety of equations of state for condensed phases, that are accurate over progressively greater pressure ranges. The study of equations of state for condensed phases that are valid at very high pressures and temperatures will allow us not only to perform chemical equilibrium calculations relevant to deep planetary interiors but also to predict physical conditions – pressure and temperature – as a function of depth in solid planets.

8.1 An introduction to equations of state for solids

In Chapter 1 we calculated the following integral for solid phases assuming that their molar volumes remain constant:

$$\int_1^P V(P,T)dP. \tag{8.1}$$

This is tantamount to assuming that the second derivatives of the Gibbs free energy vanish (equations (4.136) and (4.137)), which is in general not true. All materials change in volume in response to changes in pressure and temperature, and a change in volume entails a change in free energy. This energy needs to be accounted for, both when studying chemical transformations (i.e. calculation of phase equilibria) and when inferring physical conditions in planetary interiors. For example, equation (3.32) is a general differential equation for the adiabat, based exclusively on thermodynamic relations and thus independent of specific material properties. Integrating this equation so as to find the temperature distribution inside a convective planet requires knowledge of how α, V and C_P (all of which are derivatives of the free energy) vary with temperature and pressure. Throughout Chapter 3 we assumed that these properties are all constant, and although this may be fine as a first approximation it introduces unacceptably large errors in detailed work.

Recall from Chapter 1 that the functional relationship between the three variables, P, V and T, is called an equation of state (EOS). Operationally, the procedure to construct an EOS begins with a reference value for the volume, which we symbolize V_0. Except in surface and near-surface environments, $P \gg 1$ bar, so that the reference pressure is commonly taken to be 0, and the reference volume is called the zero-pressure volume, even if it is customarily measured at 1 bar ($1 \approx 0$, if $P \gg 1$). The reference temperature, T_0, may be taken either at the standard reference temperature for thermodynamics (298.15 K), or at 0 K (in which case we refer to the zero-temperature volume). The most appropriate choice depends on the nature of the EOS, and when stating the value of $V_0 = V(0, T_0)$ I will always explicitly state the value of T_0.

We seek a function that yields either the volume at P and T, $V(P, T)$, which is called a *volume-explicit EOS*, or the pressure at V and T, $P(V, T)$, which is called a *pressure-explicit EOS*. As with other thermodynamic calculations, we construct the function in steps, allowing only one intensive variable to change at a time. Two different approaches are summarized in Fig. 8.1. One possibility is to calculate the effect of isobaric thermal expansion at zero pressure, from T_0 to the temperature of interest. This takes the volume from V_0 to $V(0,T)$. We then hold the temperature constant and calculate the effect of compression to the pressure of interest (labeled "hot isothermal compression"), so that we obtain the volume at the P and T of interest, $V(P, T)$. Equations of state for materials at very high pressures and temperatures, such as in deep planetary mantles and cores, are often constructed following the alternate route shown in Fig. 8.1. In this approach we must consider separate contributions to pressure, so that the total pressure, P, is given by:

$$P = P_0 + P_{th} + P_{el}. \tag{8.2}$$

In this equation P_0 is the pressure required to change the volume from the reference volume V_0 to the volume at the pressure and temperature of interest $V(P, T)$, *along an isothermal compression path at the reference temperature*, which is labeled "cold isothermal compression" in the figure. The pressure at the end of the cold isothermal compression leg, P_0 is such that $V(P_0, T_0) = V(P, T)$. This is the pressure associated with elastic compression of the material at the reference temperature. Heating at constant volume (isochoric) from T_0 to T generates additional pressure contributions. One of them, called thermal pressure, P_{th}, arises from increased vibrations of the atoms about their mean equilibrium positions (see Section 1.14). The other term in (8.2), P_{el}, arises from electron vibrations. It is negligible in dielectric materials such as silicate rocks, but becomes important in metals at high temperature (e.g., in the cores of terrestrial planets) and in plasmas (such as liquid metallic hydrogen in the cores of giant planets). Unless otherwise stated we will assume that $P_{el} = 0$, as in Fig. 8.1.

Both approaches to constructing an equation of state require that we find a functional relationship between pressure and volume at constant temperature (to handle the isothermal legs in Fig. 8.1). Such a function is known as an *isothermal EOS*. In the following sections we discuss three different ways of constructing isothermal EOS: from thermodynamics (Murnaghan EOS), from finite strain (Birch–Murnaghan EOS) and from interatomic potentials (Born–Mie EOS). The first two are macroscopic approaches and the resulting isothermal EOS are applicable at any temperature, as long as the material properties at the temperature of interest are known. Isothermal EOS based on interatomic potentials, in contrast, incorporate a microscopic model of the material and are rigorously correct only at 0 K.

The two approaches summarized in Fig. 8.1 differ substantially in the way they handle the effect of temperature. In the "hot isothermal compression" method one adjusts the material properties in the isothermal EOS to the temperature of interest (Section 8.2.2). In the "cold isothermal compression" method a separate term is added, that accounts for thermal pressure (Sections 8.3 and 8.4).

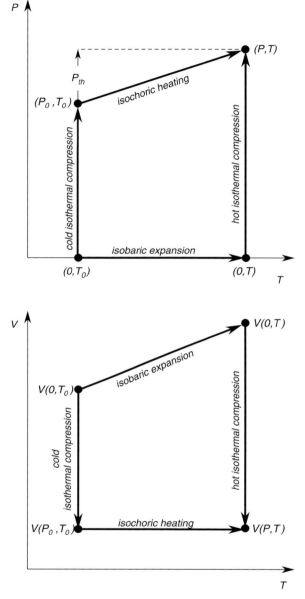

Fig. 8.1 Two ways of calculating $V = V(P,T)$ or $P = P(V,T)$. An isothermal equation of state is needed in both cases.

8.2 Macroscopic equations of state

8.2.1 The Murnaghan isothermal EOS

This equation of state follows directly from thermodynamic identities, even if it was originally derived by Murnaghan (1937) from finite strain considerations. We begin by recalling the definition of the isothermal bulk modulus, equation (1.21):

$$K_T = -V \left(\frac{\partial P}{\partial V} \right)_T.$$

(8.3)

Integrating (8.3) at constant T we get:

$$\ln \left(\frac{V_{P,T}}{V_{0,T}} \right) = - \int_0^P \frac{dP}{K_T}$$

(8.4)

where we use 0 as the lower integration limit because we integrate from "zero pressure". The integral on the right-hand side of (8.4) would be very easy to solve if the isothermal bulk modulus were constant, which would require that all third derivatives of the Gibbs free energy vanish. This is in general not the case, however. Rather, the isothermal bulk modulus is a function of both temperature and pressure. Because equation (8.4) is integrated at constant temperature, however, K_T can be treated as a function of pressure only, at a constant temperature T. The simplest possible assumption now is to make K_T a linear function of P so that, using K_T' for the first pressure derivative of K_T, we write:

$$K_T' = \left(\frac{\partial K_T}{\partial P} \right)_T = constant.$$

(8.5)

This approximation, which is equivalent to assuming that the third pressure derivative of the Gibbs free energy does not vanish but the fourth and higher derivatives do, leads to what is known as the *Murnaghan equation of state*. If we use $K_{0,T}$ to symbolize the bulk modulus at zero pressure and temperature T, then integration of (8.5) yields:

$$K_T = K_{0,T} + K_T' P.$$

(8.6)

Substituting in (8.4):

$$\ln \left(\frac{V_{P,T}}{V_{0,T}} \right) = - \int_0^P \frac{dP}{K_{0,T} + K_T' P}.$$

(8.7)

Making the substitution $u = K_{0,T} + K_T' P$, this integrates easily to:

$$\ln \left(\frac{V_{P,T}}{V_{0,T}} \right) = - \frac{1}{K_T'} \ln \left(\frac{K_{0,T} + K_T' P}{K_{0,T}} \right)$$

(8.8)

or:

$$V_{P,T} = V_{0,T} \left(1 + \frac{K_T' P}{K_{0,T}} \right)^{-\frac{1}{K_T'}},$$

(8.9)

which is the volume-explicit form of Murnaghan's equation of state. The temperature of interest, T, can be any value that is convenient. If $T = T_0$ then we substitute V_0 for $V_{0,T}$ in (8.9) – this is a notation issue, but it is important to keep it straight for consistency with what follows. The Murnaghan EOS works well up to pressures of the order of 200 kbar. More accurate representations of volume are needed at higher pressure.

To apply the Murnaghan EOS (equation (8.9)) or Birch–Murnahgan EOS (see below) at any arbitrary temperature we need to account for thermal expansion at zero pressure, so as to calculate $V_{0,T}$, and we need the values of the bulk modulus and its pressure derivative. Before we proceed with other isothermal EOS we discuss how each of these steps is handled for solid materials.

8.2.2 Thermal expansion at zero pressure

Recalling the definition of the coefficient of thermal expansion, equation (1.66):

$$\alpha = \frac{1}{V}\left(\frac{\partial V}{\partial T}\right)_P \tag{8.10}$$

we see that volume at zero pressure and the temperature of interest is given by:

$$V_{0,T} = V_0 \exp\left(\int_{T_0}^{T} \alpha \, dT\right), \tag{8.11}$$

where V_0 is the volume at zero pressure and the reference temperature T_0 (generally, 298 K). Evaluating 8.11 requires a function $\alpha = \alpha(T)$.

The behavior of the coefficient of thermal expansion with temperature parallels that of heat capacity. In Section 1.14.3 we saw that heat capacity increases strongly with T at low temperature, and that the T dependency becomes weaker as the Debye temperature of the material θ_D is approached. Figure 8.2a shows this relationship, plotted as non-dimensional heat capacity vs. non-dimensional temperature for three minerals, periclase, corundum and forsterite, for which the values of θ_D are approximately 940 K, 1040 K and 760 K, respectively (data from Anderson *et al.*, 1992). In the graph I have normalized temperature to θ_D and heat capacity to the Dulong and Petit values (Section 1.1.4.3) for each of the three minerals. This procedure makes the regularity of the behavior hinted at in Fig. 1.15 strikingly clear. What is perhaps even more remarkable is that thermal expansion behaves in essentially the same way. This is shown in Fig. 8.2b, in which I normalized the coefficients of thermal expansion for the same three minerals to the values measured at or close to the Debye temperature, symbolized by α_{θ_D}. The reason for the similar behavior arises from the fact that heat capacity and thermal expansion are both macroscopic manifestations of changes in atomic vibration modes. The importance of the trends shown in Fig. 8.2 for our present purposes is that they suggest that we can represent thermal expansion coefficients with polynomial functions similar to the ones used to express heat capacity as a function of temperature. There appears to be less common ground regarding the actual function that should be used to represent thermal expansion than there is for heat capacity, though. For instance, Berman (1988) and Chatterjee *et al.* (1998) propose two-parameter polynomials, Saxena *et al.* (1993) a polynomial with four parameters, and Holland and Powell (1998) a single-parameter equation.

8.2.3 Bulk modulus at zero pressure

The second parameter required for the "hot isothermal compression" approach is the bulk modulus at zero pressure and the temperature of interest: $K_{0,T}$. Measurements show that the bulk modulus of minerals decreases linearly with increasing temperature (i.e., minerals become weaker with increasing temperature). This is exemplified in Fig. 8.3, which is drawn for the same minerals shown in Fig. 8.2 (data from Anderson *et al.*, 1992). The behavior of bulk modulus with temperature is quite regular and, in contrast to heat capacity and thermal expansion, does not appear to be significantly affected by whether the mineral is above or below its Debye temperature. One could infer from Fig. 8.3 that our problem is easily solved, as one would just need to obtain data such as those in this figure for all minerals, and values of $K_{0,T}$ would be readily available for any temperature. Alas, this is not the

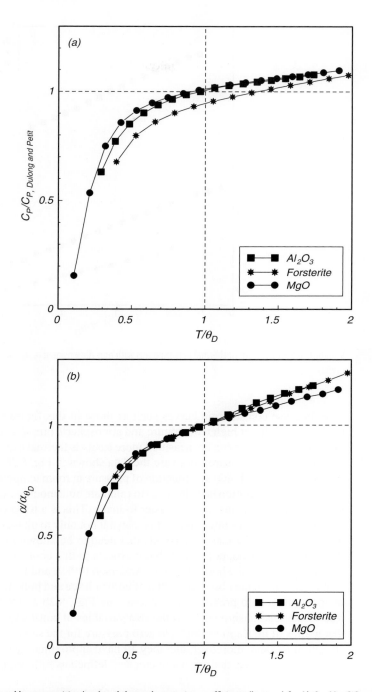

Fig. 8.2 Non-dimensional heat capacities (top) and thermal expansion coefficients (bottom) for Al_2O_3, Mg_2SiO_4 and MgO (data from Anderson *et al.*, 1992), plotted against non-dimensional temperature. Heat capacities are normalized to the Dulong and Petit values (Section 1.14.3). Thermal expansion coefficients are normalized to the values measured at, or close to, the Debye temperature, θ_D.

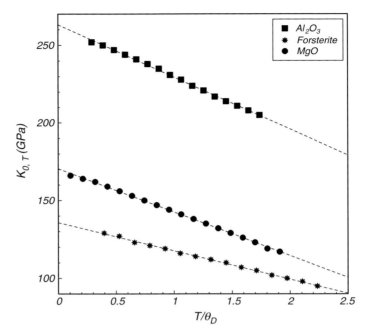

Variation of zero-pressure isothermal bulk modulus with non-dimensional temperature. Data from Anderson *et al.* (1992).

case, because constructing curves such as those in this figure requires measurements of the bulk modulus at high temperature and atmospheric pressure, and this is experimentally rather difficult. What can be measured more easily is thermal expansion at high temperature and atmospheric pressure (these are the data shown in Fig. 8.2b) and compressibility (the inverse of bulk modulus) as a function of pressure at room temperature. We therefore need a thermodynamic relation that allows us to calculate bulk modulus at high temperature starting from the measurements that are easier to make. This is a topic of immense importance in the study of planetary interiors, as it is relevant not only to phase equilibrium but also to the determination of the elastic constants that need to be known in order to interpret seismic data. We can only cover the most basic aspects in this book. Two excellent references for in-depth study are the books by D. L. Anderson (1989) and O. L. Anderson (1995).

We begin from the observation that, if both volume and bulk modulus vary regularly with temperature at zero pressure (as suggested by Figs. 8.2b and 8.3), then the same must be true of the relationship between the two variables at constant pressure. Figure 8.4 shows that graphs of $\ln(K_{0,T})$ vs. $\ln(V)$ at zero pressure for the same three minerals as before are straight lines, and, moreover, that they all have approximately the same slope, ≈ -5. This observation suggests that the parameter δ_T, defined as follows, may be significant:

$$\delta_T \equiv -\left(\frac{\partial \ln K_T}{\partial \ln V} \right)_P, \tag{8.12}$$

where δ_T is called the isothermal Anderson–Grüneisen parameter, and provides the crucial link that makes the calculation of $K_{0,T}$ possible. From the chain rule of

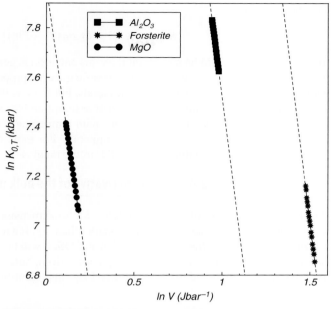

Fig. 8.4 Variation of zero-pressure isothermal bulk modulus with volume. Temperature varies as in Fig. 8.3 (note different units for $K_{0,T}$). Data from Anderson *et al.* (1992).

differentiation we get:

$$\left(\frac{\partial \ln K_T}{\partial \ln V}\right)_P = \frac{\partial \ln K_T}{\partial K_T}\left(\frac{\partial K_T}{\partial V}\right)_P \frac{\partial V}{\partial \ln V} = \frac{1}{K_T}\left(\frac{\partial K_T}{\partial V}\right)_P V \qquad (8.13)$$

or:

$$\delta_T = -\frac{V}{K_T}\left(\frac{\partial K_T}{\partial V}\right)_P. \qquad (8.14)$$

Also from the chain rule, and using the definition of the coefficient of thermal expansion, (8.10):

$$\left(\frac{\partial K_T}{\partial V}\right)_P = \left(\frac{\partial K_T}{\partial T}\right)_P \left(\frac{\partial T}{\partial V}\right)_P = \frac{1}{\alpha V}\left(\frac{\partial K_T}{\partial T}\right)_P. \qquad (8.15)$$

From (8.14) and (8.15) it follows that:

$$\left(\frac{\partial K_T}{\partial T}\right)_P = -\alpha K_T \delta_T. \qquad (8.16)$$

Now, $(\partial K_T/\partial T)_P$ is the slope of the lines in Fig. 8.3. If we can assume that the behavior of these three crystalline solids is true of all minerals, then $(\partial K_T/\partial T)_P$ is a characteristic constant for each mineral. We can approximate this constant by substituting in equation (8.16) values of α and K_T measured at a single temperature, and some characteristic value of δ_T. In particular, we can use the values at the reference state, α_0 and K_0. Integrating equation (8.16) and taking the reference state at 298 K we get the following equation for $K_{0,T}$:

$$K_{0,T} = K_0 [1 - \delta_T \, \alpha_0 \, (T - 298)]. \tag{8.17}$$

If δ_T can be assumed to be a constant then this equation is general, as it follows from thermodynamic identities and not from any specific material properties. The room-temperature values α_0 and K_0 are known for most minerals, but there is the difficulty of the parameter δ_T. Treating it as a constant is a convenient assumption but is not strictly true. In fact, the Anderson-Grüneisen parameter does vary with temperature, although not strongly, and it is also somewhat different for different minerals. For most minerals at terrestrial mantle temperatures δ_T varies within a factor of 2, in the range 4–8.

8.2.4 Pressure derivative of the bulk modulus

The pressure derivative of the bulk modulus, K', is a dimensionless number that is generally within the range 3.5–7 and close to, but smaller than, δ_T. If a measured value of K' exists for the mineral of interest then it can be used in an EOS. It will become important to understand where the relationship between K' and δ_T comes from, however. We begin by forming the product αK_T. Using the definitions of α and K_T, and identity (1.3.19) in Box 1.3, we find that:

$$\alpha K_T = -\left(\frac{\partial V}{\partial T}\right)_P \left(\frac{\partial P}{\partial V}\right)_T = \left(\frac{\partial P}{\partial T}\right)_V. \tag{8.18}$$

Thus, the product of the two material properties α and K_T gives the change in pressure with temperature, at constant volume. This is the thermal pressure that we discussed in Section 8.1 (see Fig. 8.1) and to which we will return later. For now we note that it is an empirical observation that for solids at high temperature, which roughly means above their Debye temperature, αK_T is a very weak function of P and T (see, for example, Anderson et al., 1992). This is of interest for our present discussion. Taking the pressure derivative of αK_T at constant temperature we get:

$$\left(\frac{\partial (\alpha K_T)}{\partial P}\right)_T = K_T \left(\frac{\partial \alpha}{\partial P}\right)_T + \alpha \left(\frac{\partial K_T}{\partial P}\right)_T. \tag{8.19}$$

It is straightforward to prove that (Exercise 8.2):

$$\left(\frac{\partial \alpha}{\partial P}\right)_T = \frac{1}{K_T{}^2} \left(\frac{\partial K_T}{\partial T}\right)_P. \tag{8.20}$$

Substituting in (8.19) and using (8.5) and (8.16), we get:

$$\left(\frac{\partial (\alpha K_T)}{\partial P}\right)_T = \frac{1}{K_T} \left(\frac{\partial K_T}{\partial T}\right)_P + \alpha K' = \alpha \left(K' - \delta_T\right). \tag{8.21}$$

If αK_T can be considered to be constant at high temperature then its pressure derivative vanishes. In the absence of measured values for one or the other of the parameters, it is therefore often assumed that:

$$K' \approx \delta_T \tag{8.22}$$

This remains an approximation, generally valid at high temperature, but it is no substitute for measured values. When the latter do exist it is found that $\delta_T \geq K'$.

Worked Example 8.1   Some mineral reactions in the Earth's upper mantle

Let us look at the effect of ignoring the compressibility and thermal expansion of solid phases on the calculated location of some important phase boundaries in the Earth's upper mantle. As in other examples focusing on crustal and upper mantle conditions, and also because their data set is large, allowing for the formulation of many different examples, I will use standard state thermodynamic properties from Holland and Powell (1998). We must therefore use their treatment for compressibility and thermal expansion, which is based on the isothermal Murnaghan EOS. The *Maple* implementation of the procedures discussed here is given in Software Box 8.1.

Software Box 8.1   Calculation of Gibbs free energy of solid phases at high pressure with Murnaghan equation of state

Software Box 7.2 describes the procedures Peq and Teq in worksheet **th_template_4.mw**. These procedures include a call to the procedure vdp, that is also included in **th_template_4.mw**.

Procedure vdp calculates the total contribution of pressure to the Gibbs free energy of reaction by calling function vdp_hp, which is included in the package **th_shomate.mw**. The function vdp_hp calculates the integral of the Murnaghan equation of state for a phase (equation (8.24)), using Holland and Powell's data and formalism to calculate thermal expansion at 1 bar and the value of the bulk modulus at the temperature of interest. An examination of the lines of code in vdp_hp and comparison with the discussion in Section 8.2 and Worked Example 8.1 is self-explanatory.

Procedure vdp calls vdp_hp for each phase in the reaction, multiplies the Gibbs free energy term by the stoichiometric coefficient, and adds up the contributions of all of the phases in the reaction. In Peq and Teq this Gibbs free energy is then added to the Gibbs free energy of reaction at 1 bar, calculated with DGT_sh, and any contribution from order–disorder, calculated with Gord, and the condition $\Delta_r G = 0$ is then solved for iteratively, as discussed in Software Box 7.2.

Peq and Teq include all of the calculations necessary to calculate the equilibrium of a solid assemblage in which all phases are in their standard states, with Holland and Powell's data base. This includes the heat capacity integral, thermal expansion contribution, Murnaghan EOS pressure integral and Landau potential contribution. Calculation of equilibrium among solid solutions is handled by adding energetic contributions from non-unit activities, as discussed in Software Boxes 5.4 and 5.5 and the *Maple* worksheets referenced there (exercise left for the reader). Calculation of a full phase boundary, e.g. as in Fig. 8.5, is accomplished by including either Peq or Teq in a do loop that iterates over a pressure or temperature range of interest, e.g. as in worksheet **sp_grt_MAS.mw**.

The pressure integral of non-condensed phases is handled differently, and will be discussed in Chapter 9.

Substituting (8.9) in (8.1) and integrating at constant T along a "hot isothermal compression" path from zero pressure (≈ 1 bar) to the pressure of interest we get:

$$\int_0^P V(P,T)dP = V_{0,T} \int_0^P \left(1 + \frac{K'P}{K_{0,T}}\right)^{-\frac{1}{K'}} dP. \tag{8.23}$$

Making the substitution $u = 1 + K'P/K_{0,T}$, we arrive quite easily at:

$$\int_0^P V(P,T)dP = \frac{V_{0,T}K_{0,T}}{K'-1}\left[\left(1 + \frac{K'P}{K_{0,T}}\right)^{\frac{K'-1}{K'}} - 1\right]. \qquad (8.24)$$

There are three parameters in this equation: $V_{0,T}$ and $K_{0,T}$ are functions of temperature only and K' is a constant (by construction of the Murnaghan EOS, see equation (8.5)).

Holland and Powell use a single-parameter equation for thermal expansion, and make the case that trying to fit more complex equations to the high-temperature volume data available for many minerals would amount to "overfitting" – i.e. the more complex empirical equations would not recover any significant information that is not already recovered by the one-parameter equation that they propose. Their equation for α is:

$$\alpha = a_0\left(1 - 10T^{-1/2}\right), \qquad (8.25)$$

where a_0 is an empirical parameter characteristic for each mineral. This equation causes α to approach the limiting value a_0 at very high temperature, which is the behavior suggested by Fig. 8.2b. Substituting in (8.11) and integrating we get:

$$\frac{V_{0,T}}{V_0} = \exp\left(a_0\left[(T - 298) - 20\left(\sqrt{T} - \sqrt{298}\right)\right]\right). \qquad (8.26)$$

For conditions up to those of the terrestrial mantle transition zone the argument of the exponential function in (8.26) is always a small number, as a_0 is characteristically of order 10^{-5}K^{-1}, and temperatures are at most of order 2–3×10^3 K. We can then use the approximation $e^x \approx 1 + x$, for x small (expand e^x as a McLaurin series if you don't remember where this comes from), and we get the following expression for the volume at the temperature of interest and zero pressure:

$$V_{0,T} = V_0\left(1 + a_0\left[(T - 298) - 20\left(\sqrt{T} - \sqrt{298}\right)\right]\right). \qquad (8.27)$$

Bulk modulus at zero pressure and T is given by (8.17). Holland and Powell list values of K_0 in their data base, and set the product $\delta_T\alpha_0$ in equation 8.17 constant and equal to 1.5×10^{-4}. This appears to provide a reasonable approximation for crustal and upper mantle conditions. In any case, because their thermodynamic data base is regressed *assuming* that this is the (constant) value of the product $\delta_T\alpha_0$, calculations using Holland and Powell's data must be carried out with this constant value in equation (8.17) in order to be consistent. Finally, Holland and Powell point out that the amount of data available for most minerals only warrants the use of a single value of K' for all minerals, and choose $K' = 4$ as the value that yields the best fit to all available data. As we shall see below this is also the value of K' that can be expected on theoretical grounds if one assumes that this parameter is constant. Software Box 8.1 shows that numerical implementation of the pressure integral (8.24) is very straightforward.

Figure 8.5 (top) shows the spinel–garnet phase boundary (again!), as defined by equilibrium among spinel, aluminous enstatite, forsterite and pyrope (see Worked Example 5.6), calculated both under the assumption that volumes are constant (dashed curve, which is the same curve as in Fig. 5.9) and using equation (8.24) to account for compressibility and

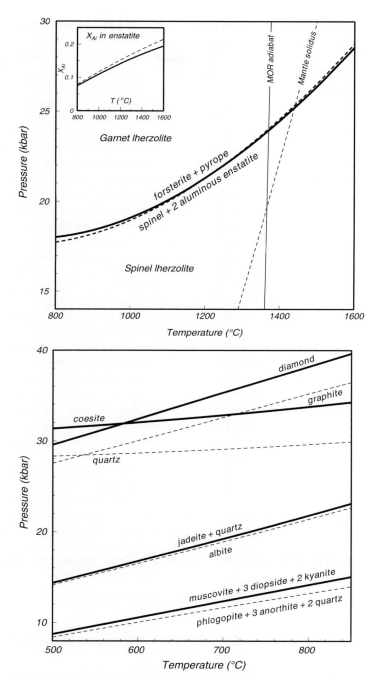

Fig. 8.5 Effect of ignoring compressibility and thermal expansion on the calculated locations of the spinel–garnet phase transition (top) and of some diagnostic ultrahigh-pressure metamorphic reactions (bottom). Note that the error is not systematic. Thermodynamic data from Holland and Powell, 1998. High-pressure volumes calculated with Murnaghan EOS.

thermal expansion (solid curve). You may rightly wonder whether the preceding several pages of formulas and thermodynamic derivations are worth the effort. The difference between the two calculated phase boundaries is of the order of 100 bars at most, and all but vanishes at characteristic mantle temperatures ($\sim$1400°C). The difference in calculated orthopyroxene compositions may barely be outside analytical error. The problem is that this result is not general. It is a lucky coincidence, arising from the fact that compressibilities and thermal expansions on both sides of the reaction essentially cancel each other out. This is not the case if either a particularly "weak" (= small bulk modulus) or "strong" mineral occurs on one side of the phase boundary, or if thermal expansions vary significantly among the participating minerals. It is an even greater problem in mineral reactions that evolve a fluid phase, and in partial melting reactions, as we shall see in the following chapters. The error incurred by ignoring volume changes of solid phases increases with pressure and is always unacceptable at pressures higher than a few tens of kilobars.

Figure 8.5 (bottom) shows calculated phase boundaries for some model mineral reactions relevant to the study of utrahigh-pressure metamorphism (UHPM) of crustal rocks subducted to mantle depths (typically at continental collision zones). In every case the solid line is calculated accounting for the compressibility and thermal expansion of minerals, and is consistent with experimental brackets, whereas calculation of the dashed lines assumes constant volume. Coesite and diamond are perhaps the two chief diagnostic minerals of UHPM. Ignoring volume changes in the reactions graphite $\rightleftarrows$ diamond and quartz $\rightleftarrows$ coesite introduces errors of the order of 3–4 kbar, or about 10% of the actual equilibrium pressure, and equivalent to a depth uncertainty of $\sim$10–12 km. These large discrepancies arise from significant differences in both bulk moduli (diamond = 5800 kbar vs. graphite = 390 kbar, coesite = 1000 kbar vs. quartz = 750 kbar) and coefficients of thermal expansion (a_0 values are: diamond = 1.65×10^{-5} K^{-1} vs. graphite = 4.84×10^{-5} K^{-1} and coesite = 1.8×10^{-5} K^{-1} vs. quartz = 0.65×10^{-5} K^{-1}). Note that the high pressure phase in both cases is the less compressible one, as one would expect from a closer packed crystalline structure. Thermal expansion, in contrast, is greater for graphite relative to diamond, but smaller for coesite relative to quartz.

The two other reactions shown in the figure are the breakdown of albite to jadeite plus quartz, which we discussed in Worked Example 7.2, and a water-conserving reaction that models the formation of the assemblage phengite + clinopyroxene in rocks of tonalitic composition. Reactions such as the latter one, including ones in which amphiboles take part, may be responsible for inhibiting dehydration-melting (see Worked Example 6.3) of metamorphosed crust during continental subduction (Patino Douce, 2005). In both of these cases the effect of ignoring volume changes in solid phases is less severe, but note that the magnitude of the error is not systematic.

8.2.5 The Birch–Murnaghan isothermal EOS

Derivation of the Birch–Murnaghan EOS begins from a description of finite strain (see, for example, Birch, 1952). Strain can be measured relative to either the deformed or undeformed states, leading to different equations. We will describe strain relative to the deformed state, which is known as the *Eulerian convention*. Measurement of strain relative to the undeformed state, known as the *Lagrangian convention*, leads to a more cumbersome EOS.

Let the distance between two points in a strained material be S, given by $S^2 = \sum \Delta X_i^2$, where ΔX_i ($i = 1, 2, 3$) is the distance along each of three orthogonal directions. The distance between the same points in the unstrained state is S_0, where $S_0^2 = \sum (\Delta X_i - dX_i)^2$. The increments dX_i are the elongations along the three orthogonal directions. We are interested in changes in volume only, which is a condition that is called *isotropic strain*. We can thus make $dX_i = k \Delta X_i$, with the same proportionality constant, $k < 0$ (we are interested in compression), for all three directions. The change in distance between the points is then given by:

$$S^2 - S_0^2 = \sum \left[\Delta X_i^2 - (\Delta X_i - dX_i)^2 \right] = \left(2k - k^2 \right) \sum \Delta X_i^2 = 2\varepsilon \sum \Delta X_i^2. \quad (8.28)$$

The coefficient ε, known as the Eulerian strain, is a second-order tensor with nine components ε_{ij}, $i = 1, 2, 3$, $j = 1, 2, 3$. If strain is isotropic then all shear components vanish ($\varepsilon_{ij} = 0$ for $i \neq j$) and all principal components are equal ($\varepsilon_{11} = \varepsilon_{22} = \varepsilon_{33}$). In this special case the Eulerian strain tensor collapses to the scalar ε (a zeroth-order tensor), which can be shown to be $\varepsilon = \frac{1}{2} (2k - k^2)$, as in (8.28) (Exercise 8.3).

If we now call the strained volume $V = \prod \Delta X_i$ and let the unstrained volume be the reference volume $V_0 = \prod (\Delta X_i - dX_i)$, we find that (proof left for the reader):

$$\frac{V_0}{V} = (1 - k)^3. \quad (8.29)$$

We define the parameter $f = -\varepsilon$ and with some manipulations find:

$$f = -\varepsilon = -\frac{1}{2} \left(2k - k^2 \right) = \frac{1}{2} \left[\left(\frac{V_0}{V} \right)^{2/3} - 1 \right]. \quad (8.30)$$

Note that for compression it is always $f > 0$. The idea of the Birch–Murhanghan equation of state is to express the elastic free energy of the material as a power series in the positive quantity f, and then obtain the isothermal pressure, P_0, as a derivative of the elastic free energy. Because we seek to define pressure as a derivative of free energy relative to volume at constant temperature the relevant thermodynamic potential is the Helmholtz free energy, $F = F(V, T)$, rather than the Gibbs free energy (see Section 4.8.6, equations 4.124 – 4.126). We write:

$$F_e = a_1 f + a_2 f^2 + a_3 f^3 + \cdots \quad (8.31)$$

where the subscript e specifies that this is only the elastic contribution to Helmholtz free energy. The power series is commonly truncated at the cubic term, which generates what is known as the *third-order Birch–Murnaghan EOS*, although the fourth-order term is sometimes included too (and the algebra, which is already ungainly at third order, becomes even more so). From equation (4.126) for a constant composition system we have, using the chain rule:

$$P_0 = -\left(\frac{\partial F_e}{\partial V} \right)_T = -\left(\frac{\partial F_e}{\partial f} \right) \left(\frac{\partial f}{\partial V} \right). \quad (8.32)$$

Because:

$$\left(\frac{\partial F_e}{\partial f} \right) = a_1 + 2a_2 f + 3a_3 f^2 \quad (8.33)$$

and P_0 must vanish in the uncompressed state ($f = 0$), we see that it must be $a_1 = 0$. From
(8.30) we find the other partial derivative in (8.32) to be:

$$\left(\frac{\partial f}{\partial V}\right) = -\frac{1}{3V_0}(2f+1)^{5/2} \tag{8.34}$$

so:

$$P_0 = \frac{f}{3V_0}(2a_2 + 3a_3 f)(2f+1)^{5/2}. \tag{8.35}$$

There are two unknown parameters in this equation, a_2 and a_3, which can be evaluated if we
find two linearly independent equations that relate these two parameters to experimentally
observable quantities. For example, we can fix the values of two of the derivatives of P_0
at some reference pressure, such as zero pressure (because (8.35) vanishes at zero pressure
we cannot use this equation). Applying the chain rule to the definition of isothermal bulk
modulus we get:

$$K_T = -V\left(\frac{\partial P_0}{\partial V}\right)_T = -V\left(\frac{\partial P_0}{\partial f}\right)\left(\frac{\partial f}{\partial V}\right). \tag{8.36}$$

We already have $\partial f / \partial V$ (equation (8.34)). Calculation of $\partial P_0 / \partial f$ involves some rather uninteresting algebra. You can do it yourself, or have *Maple* do it for you, and verify the
result:

$$K_T = \frac{(2f+1)^{5/2}}{9V_0}\left[2a_2 + (14a_2 + 6a_3)f + 27a_3 f^2\right]. \tag{8.37}$$

Evaluating (8.37) at $f = 0$ (i.e. zero pressure) we find:

$$K_{0,T} = \frac{2a_2}{9V_{0,T}}. \tag{8.38}$$

We can also take the pressure derivative of K_T, K'_T:

$$K'_T = \frac{\partial K_T}{\partial P} = \frac{\partial K_T}{\partial f}\frac{\partial f}{\partial P} \tag{8.39}$$

and, after some additional uninteresting algebra and evaluation at zero pressure, arrive at:

$$K'_{0,T} = \frac{a_3}{a_2} + 4. \tag{8.40}$$

From (8.38) and (8.40) we can get the values of a_2 and a_3 in terms of three parameters
that are experimentally accessible: $V_{0,T}$, $K_{0,T}$, and $K'_{0,T}$. Substituting these values and
the definition of f (equation 8.30) in the pressure equation (8.35) we finally arrive at the
(Eulerian) Birch–Murnaghan equation of state to third order:

$$P_0 = \frac{3}{2}K_{0,T}\left[\left(\frac{V_{0,T}}{V}\right)^{7/3} - \left(\frac{V_{0,T}}{V}\right)^{5/3}\right]\left[1 + \frac{3}{4}\left[\left(\frac{V_{0,T}}{V}\right)^{2/3} - 1\right](K'_{0,T} - 4)\right].$$

$$\tag{8.41}$$

The subscript T in the three parameters, V, K and K' in (8.41) makes it clear that this is an
isothermal equation of state that can be used at any temperature, *provided that the values of*

the three parameters are known at that temperature. If the equation is applied to calculation of isothermal compression at the reference temperature (cold isothermal compression in Fig. 8.1) then the parameters are labeled V_0, K_0 and K_0', *but it is always necessary to specify whether the reference temperature is 298 K or 0 K*.

The second-order Birch–Murnaghan EOS is obtained by setting $a_3 = 0$ (see equation (8.31)). This makes K_0' constant and equal to 4 (equation (8.40)) which is the value used in the Holland and Powell data set, and is comparable to the value measured for many minerals. The resulting second-order EOS is:

$$P_0 = \frac{3}{2} K_{0,T} \left[\left(\frac{V_{0,T}}{V} \right)^{7/3} - \left(\frac{V_{0,T}}{V} \right)^{5/3} \right]. \tag{8.42}$$

Note that, in contrast to the Murnaghan EOS, both forms of the Birch–Murnaghan EOS are pressure-explicit and have no analytic solutions for V.

Worked Example 8.2   The ringwoodite–perovskite phase transition in planetary mantles

If, as a first approximation, we neglect the relatively minor components Al and Ca, then the chemical composition of the mantles of the terrestrial planets approaches $(Mg,Fe)_2SiO_4$, with $X_{Mg} \sim 0.9$ for the Earth's mantle and ~ 0.8 for the mantles of Mars and the Moon (Mg number for the mantles of Venus, Mercury and other rocky bodies are far less well constrained). At near-surface conditions these components occur as the mineral olivine – in fact, the stability field of olivine is what defines the Earth's upper mantle. Olivine undergoes phase transformations with increasing pressure, which are observed as discontinuities in seismic velocities. In the Earth the first such discontinuity occurs at a depth of ~ 410 km, where olivine transforms to the isochemical phase wadsleyite which has a spinel-like structure. At a depth of ~ 520 km wadsleyite in turn transforms to ringwoodite, also $(Mg,Fe)_2SiO_4$ but with a true spinel structure. Density increases from olivine to wadsleyite to ringwodite, but in all three phases silicon occurs in tetrahedral coordination. At a depth of approximately 660 km in the Earth's mantle a major phase transformation takes place, in which ringwoodite breaks down to a phase with perovskite structure and composition $(Mg,Fe)SiO_3$, and an oxide phase of composition $(Mg,Fe)O$, which is called either magnesiowüstite or, more appropriately in view of its relative Mg and Fe contents, ferropericlase. This transformation, in which a silicate perovskite phase with Si in octahedral coordination becomes stable, marks the top of the Earth's lower mantle. The depth interval 410–660 km, over which the olivine–wadsleyite–ringwoodite–perovksite phase transitions take place, is known as the mantle transition zone. Other reactions involving Al- and Ca-bearing phases such as pyroxenes, majoritic garnets and calcium perovskite also take place in the mantle transition zone. Completing the picture, there is a "final" silicate phase transition that occurs close to the Earth's core–mantle boundary and corresponds to the D'' seismic discontinuity. It has recently been found experimentally that this phase transition gives rise to a dense silicate phase that is isochemical with perovskite, and is at the time of this writing called the "post-perovskite phase".

In this example we will calculate the ringwoodite–perovskite phase boundary, and see what we can learn from it about deep planetary interiors. There are some significant complications, not the least of which is the paucity of well-constrained standard state thermodynamic properties. It is known from experimental results that ringwoodite–perovskite phase relations are shifted significantly in P and T between the Mg and Fe end-member systems. In fact, at P–T conditions such as those of the terrestrial mantle transition zone Fe-perovskite is not stable, and Fe-ringwoodite breaks down to the assemblage wüstite + stishovite. We will thus consider the four phases: perovksite ($pv = MSiO_3$), ringwoodite ($rw = M_2SiO_4$), ferro-periclase ($pc = MO$) and stishovite ($st = SiO_2$), where M stands for Mg or Fe, and calculate the following four univariant reactions in each of the two-component end-member systems.

$$rw \rightleftarrows pv + pc \quad (st)$$
$$pv \rightleftarrows pc + st \quad (rw)$$
$$rw \rightleftarrows 2pc + st \quad (pv)$$
$$2pv \rightleftarrows rw + st \quad (pc)$$

We need a full set of standard state thermodynamic properties, and here is the first problem. As of this writing there appears to be no updated and internally consistent data base for high pressure phases comparable to, for example, those of Holland and Powell or Berman for crustal and upper mantle phases. Part of the problem is that these ultra-high-pressure phases are difficult to synthesize in enough quantity to allow accurate calorimetric measurements, and the error bars in pressure and temperature of phase equilibrium experiments, that can also be used to derive values of thermodynamic functions, can be considerable. I have chosen to harvest thermodynamic data from three sources: Matsuzaka *et al.* (2000), Frost *et al.* (2001) and Mattern *et al.* (2005), as a reasonable compromise between quality and "recentness" of data, on the one hand, and mutual consistency in the treatment of heat capacities, thermal expansion coefficients and bulk moduli on the other. As we shall see, the results of the calculations are generally supportive of this choice. Both C_P and α for these high-pressure phases are expressed by different polynomials from those used by HP98, so that new *Maple* procedures are needed in order to implement the heat capacity and volume integrals (this is of course trivial, see Software Box 8.2). The corresponding equations are:

$$C_P = a_1 + a_2 T + a_3 T^{-2} + a_4 T^2 + a_5 T^{-3} + a_6 T^{-1/2} + a_7 T^{-1} \tag{8.43}$$

and:

$$\alpha = \alpha_0 + \alpha_1 T + \alpha_2 T^{-2}. \tag{8.44}$$

There appears to be some consensus that, at pressures such as those of the mantle transition zone and higher (we will calculate what those pressures are later), the third-order Birch–Murnaghan EOS reproduces mineral volumes reasonably well. In contrast to the Murnaghan EOS, however, the Birch–Murnaghan EOS cannot be written in volume-explicit form, so that VdP must be integrated by parts:

$$\int_0^P VdP = PV_{P,T} - \int_{V_{0,T}}^{V_{P,T}} PdV, \tag{8.45}$$

where $V_{0,T}$ is the zero pressure volume and $V_{P,T}$ is the volume at the pressure of interest, both of them taken at the temperature of interest. If one wishes to calculate thermodynamic functions at a given pressure, for example to locate a phase boundary, it is necessary first to solve (numerically) for $V_{P,T}$, and then integrate (8.45), substituting the desired pressure-explicit EOS. Both of these steps are handled with ease by *Maple* (Software Box 8.2).

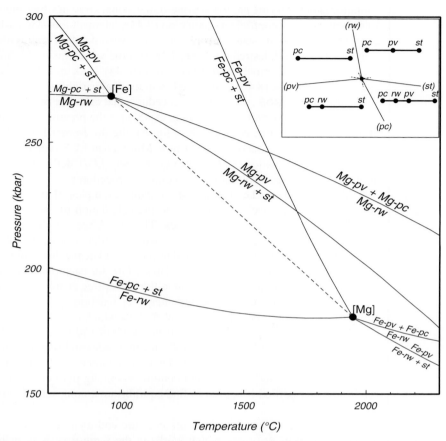

Fig. 8.6 Calculated reactions among ringwoodite (rw), silicate perovskite (pv), ferropericlase (pc) and stishovite (st), for the Mg and Fe end-member systems. The dashed line shows a possible path of the pseudo-invariant point for ternary systems of intermediate composition (it is not a phase boundary!), but note that the actual path does not have to be a straight line. Inset shows schematic phase relations, to emphasize Schreinemakers' legality. Thermodynamic data from Matsuzaka *et al.* (2000), Frost *et al.* (2001) and Mattern *et al.* (2005). High-pressure volumes calculated with third-order Birch–Murnaghan EOS.

Software Box 8.2   *Maple* worksheets for thermodynamic calculations with solid phases at very high pressures.

Holland and Powell estimate that their data base is reliable to pressures of order 100 kbar. Different data, and different equations, must be used to handle thermodynamic calculations at higher pressures. I include many commonly used calculations in two packages, so that they can be called from any other *Maple* worksheet.

The package `th_hiP.mw` parallels `th_shomate.mw` but substitutes heat capacity equation (8.43) for the Holland and Powell (Shomate-style) equation used in `th_shomate.mw`. Package `th_hiP.mw` contains procedures that calculate Cp, $\int Cp$, $\int Cp/T$, H, S and G. There is also a procedure that calculates zero-pressure volume, $V(T)$, by integration of equation 8.11 using equation (8.44) to express $\alpha(T)$, and a procedure that calculates zero pressure bulk modulus $K(T)$ using equation (8.46). Except for bulk modulus, parameters are passed in a one-dimensional array in which elements 1 through 7 are heat capacity coefficients (equation (8.43)), element 8 is standard state enthalpy, element 9 is standard state entropy, element 10 is standard state Gibbs free energy (can be set to zero if desired), element 11 is volume and elements 12 through 14 are the coefficients of the thermal expansion equation (8.44). Bulk modulus at zero temperature and dK/dT (equation (8.46)) are passed as individual variables.

The package **S_EOS.mw** contains procedures that calculate pressure and the value of $\int VdP$ (integrated by parts, see equation (8.45)) for the pressure-explicit version of the Murnaghan EOS (solve equation (8.9) for P), the third-order Birch–Murnaghan EOS (equation (8.41)), the second-order Birch–Murnaghan EOS (equation (8.42)), and the Born–Mie EOS (equation (8.56)). The rather messy $\int PdV$ definite integrals are handled implicitly by *Maple*, resulting in very compact procedures.

Worksheet **delgcalc_hi_p.mw** contains examples of the use of some of the procedures in these packages, applied to the calculation of phase boundaries among pure end-member high pressure phases. This worksheet was used to construct Fig. 8.6. Procedures `load` and `deltareax` work as in previous examples (see Software Box 1.1), but `load` is slightly modified to accommodate the different format of the high pressure data set. Procedure `vdp` calculates $\int VdP$ for each phase in the reaction, and adds up the total pressure contribution to the Gibbs free energy of reaction. Procedure `dGPT` calculates the total Gibbs free energy of reaction at P and T. Solution of the equilibrium condition is implemented in `delg0` which uses an iterative procedure, as the integral in the `vdp` procedure does not allow use of *Maple*'s `fsolve` facility. `delg0` solves for pressure (in kbar) at given T (in centigrade) and requires an initial pressure guess. A value of 500 kbar appears to be a good choice for the initial pressure guess. Procedure `delg0` finds equilibrium conditions among pure end-member phases, but can be easily modified if desired to find equilibrium among solid solutions, by adding an $RT\ln K$ term.

Calculation of a phase boundary among pure end-member phases is accomplished by the `do` loop in `Pbound`, which works in the same way as in many similar prior procedures. Input parameters are the reaction name, the range of temperatures over which to solve, the temperature increment between consecutive solutions, the initial pressure guess, and the name of the output file.

The statement block at the end of the worksheet defines the stoichiometries of all of the reactions included in the phase diagram in Fig. 8.6. Remember that this statement block must be executed before the phase boundary calculation is attempted, or an error message will result.

The third-order Birch–Murnaghan EOS contains three parameters. $V_{0,T}$ is calculated from (8.11), using (8.44). In the data sources used for this example K is assumed to be a linear function of temperature, so that:

$$K_{0,T} = K_0 + \frac{dK}{dT}(T - T_0). \tag{8.46}$$

Finally, K' is also treated as a constant, but with different and characteristic values for each phase (compare Holland and Powell's treatment for lower pressure phases, in which all phases are assigned the same value of K'). Values of K_0, K' and $\partial K/\partial T$, as well as of all the other thermodynamic parameters, for all of the phases used in this exercise are given in Matsuzaka *et al.* (2000), Frost *et al.* (2001) and Mattern *et al.* (2005). The numerical implementation is discussed in Software Box 8.2, and the results are shown in Fig. 8.6.

The calculated reactions are consistent with Schreinemakers's rule, as shown in the inset diagram. The invariant points for each of the two end-member systems are labeled [Fe] and [Mg]. The Fe-absent system, [Fe], is more closely applicable to the mantles of the terrestrial planets. As we shall see later, the temperature in the Earth's mantle at the 660 km discontinuity may be $\sim1600\,^\circ$C, which in the Mg end-member system corresponds to a pressure of ~245 kbar. As we shall also see, this corresponds to a depth of ~700 km, which is close to, but not exactly, the expected value. The calculated phase diagram shows that adding Fe to the system shifts the equilibrium to lower pressures, with the invariant point following a path such as the one suggested by the dashed line (the actual path does not have to be a straight line – the line is for illustrative purposes only). In another numerical example towards the end of this chapter we will incorporate the effect of varying $X_{\rm Mg}$ and see whether a better agreement with the depth of the observed seismic discontinuity can be obtained.

An interesting feature of the phase diagram in Fig. 8.6 is that the ringwoodite–perovskite phase transition has a negative Clapeyron slope (as this is true of both end-member systems it is almost certainly true in general). This means that if the transition is crossed isothermally, or nearly so, in the direction of increasing pressure ($\Delta V < 0$) then the reaction is endothermic ($\Delta H = T\Delta S > 0$) and, conversely, it is exothermic during decompression.

8.3 Isothermal equations of state from interatomic potentials: the Born–Mie EOS

Consider equation (8.2) (with $P_{el} = 0$) and let us define the pressure components more rigorously. We shall require that all of the energy of vibration of atoms about their equilibrium positions be expressed macroscopically as thermal pressure, P_{th}. We will derive expressions for P_{th} in Section 8.4. For now we focus on the fact that from this requirement it follows that P_{th} vanishes at 0 K. The pressure associated with isothermal compression at 0 K arises only from the energy of position of the atoms or, more accurately, from changes in the energy of position that arise from changes in interatomic distances (i.e. $P = -\partial E/\partial V$ if $T = 0$, see equation (4.12)). Now, in order for a crystalline structure to be stable (i.e. neither collapse nor fly apart) there must be both attractive and repulsive forces between atoms. The equilibrium interatomic distance is where the two forces balance each other out. We can then express the energy of position of the atoms, also called the lattice energy, E_L, as the sum of an attractive (negative) potential and a repulsive potential (recall the definition of potential from Section 1.3.1):

$$E_L = -\frac{a}{r^m} + \frac{b}{r^n} = -\frac{a}{V^{m/3}} + \frac{b}{V^{n/3}}, \tag{8.47}$$

where a, b, m and n are positive constants, r is the equilibrium interatomic distance, and V is the equilibrium volume (which goes as the cube of r). At absolute zero we can write:

$$P_0 = -\left(\frac{\partial E_L}{\partial V}\right). \tag{8.48}$$

Note very carefully that, because the natural variables of internal energy are volume and entropy, *the isothermal derivative (8.48) is true only at 0 K* (see equation (4.12)). Isothermal EOS derived from interatomic potentials are therefore only strictly correct at zero temperature, and in this case we must take V_0 at $T_0 = 0$ K, even if the difference in volume between 0 K and 298 K is small and is often ignored.

Applying (8.48) to (8.47) we get:

$$P_0 = -\frac{am}{3}V^{-\left(\frac{m+3}{3}\right)} + \frac{bn}{3}V^{-\left(\frac{n+3}{3}\right)}. \tag{8.49}$$

We now seek the values of two of the constants, a and b, in terms of m and n. Why will become clear soon – for now we note that we need two linearly independent equations in a and b. From (8.49) at zero pressure we easily find:

$$\frac{a}{b} = \frac{n}{m}V_0^{\left(\frac{m-n}{3}\right)}. \tag{8.50}$$

We can also calculate the bulk modulus from (8.49):

$$K_T = -V\left(\frac{\partial P}{\partial V}\right)_T = -\frac{am(m+3)}{9}V^{-\left(\frac{m+3}{3}\right)} + \frac{bn(n+3)}{9}V^{-\left(\frac{n+3}{3}\right)}. \tag{8.51}$$

Evaluating 8.51 at zero pressure ($V = V_0$) and using (8.50) we find:

$$a = \frac{9K_0}{m(n-m)}V_0^{\left(\frac{m+3}{3}\right)}, \quad b = \frac{9K_0}{n(n-m)}V_0^{\left(\frac{n+3}{3}\right)}, \tag{8.52}$$

where K_0 is the isothermal bulk modulus at 0 K and zero pressure. Substituting in (8.49) we finally arrive at:

$$P_0 = \frac{3K_0}{m-n}\left[\left(\frac{V_0}{V}\right)^{\left(\frac{m+3}{3}\right)} - \left(\frac{V_0}{V}\right)^{\left(\frac{n+3}{3}\right)}\right]. \tag{8.53}$$

This equation is known as the Mie equation of state. It is not complete, as the values of the exponents m and n in (8.47) are still undetermined (see also Exercise 8.4). The simplest way of addressing this is to assume that the attractive potential is due only to electrostatic forces, which vary as the inverse square of distance, so that $m = 1$ (compare equation (1.8) for the gravitational potential). We then need an additional equation to get the value of n, which can be estimated from the pressure derivative of K at zero pressure, K_0'. Applying the definition of bulk modulus to (8.53) we find:

$$K = \frac{K_0}{m-n}\left[(m+3)\left(\frac{V_0}{V}\right)^{\left(\frac{m+3}{3}\right)} - (n+3)\left(\frac{V_0}{V}\right)^{\left(\frac{n+3}{3}\right)}\right]. \tag{8.54}$$

Note that in (8.54) K_0 is the bulk modulus at zero pressure and zero temperature, whereas K is the bulk modulus at some other pressure (given by the value of V), but still at zero temperature. Differentiating (8.54) relative to P (more uninteresting algebra) and evaluating at zero pressure we get the pressure derivative of the bulk modulus at zero pressure and zero temperature:

$$K_0' = \frac{1}{3}(n+m+6).$$

(8.55)

Substituting $m = 1$ and n from (8.55) in (8.53) we arrive at the following equation, known as the *Born–Mie EOS*:

$$P_0 = \frac{3K_0}{3K_0' - 8}\left[\left(\frac{V_0}{V}\right)^{\left(\frac{3K_0'-4}{3}\right)} - \left(\frac{V_0}{V}\right)^{\left(\frac{4}{3}\right)}\right].$$

(8.56)

8.4 Thermal pressure

We can argue on the basis of physical intuition that thermal pressure must vary directly with the vibrational energy of atoms about their equilibrium positions, and inversely with volume, i.e.:

$$P_{th} = \gamma \frac{E_{vib}}{V}.$$

(8.57)

This equation states that thermal pressure is proportional to the vibrational energy density. The proportionality factor γ is known as the *Grüneisen ratio* and appears in a number of geophysical applications. Equation (8.57) is the starting point for statistical mechanical approaches to calculating thermal pressure. These entail finding expressions for E_{vib} as sums of individual vibrational modes and allow, in principle, *ab initio* calculations of thermal pressure, i.e. calculations that rely on minimal empirical knowledge of specific material properties. Such calculations are beyond the scope of this book, and I mention them only for completeness. We will calculate thermal pressure following a thermodynamic (i.e. macroscopic) approach.

From (8.2), and neglecting electron pressure, we have:

$$\left(\frac{\partial P}{\partial T}\right)_V = \left(\frac{\partial P_0}{\partial T}\right)_V + \left(\frac{\partial P_{th}}{\partial T}\right)_V.$$

(8.58)

Because in (8.2) we defined P_0 at constant temperature, the first term in the right-hand side of (8.58) vanishes. Using identity (8.18) we infer:

$$\left(\frac{\partial P_{th}}{\partial T}\right)_V = \alpha K_T.$$

(8.59)

Thermal pressure can therefore be calculated by integrating the product of the (macroscopic) material properties αK_T. We shall return to this integral in a moment. First, we note that because solids only have vibrational degrees of freedom, the vibrational energy density equals the internal energy density, so that we can differentiate (8.57) as follows:

$$\left(\frac{\partial P_{th}}{\partial T}\right)_V = \frac{\gamma}{V}\left(\frac{\partial E}{\partial T}\right)_V = \frac{\gamma C_V}{V}. \tag{8.60}$$

Comparing with (8.59) yields a definition of the Grüneisen ratio in terms of experimentally accessible macroscopic quantities:

$$\gamma = \frac{\alpha K_T V}{C_V}. \tag{8.61}$$

We now return to the calculation of thermal pressure. As thermal pressure is calculated along a constant-volume path (see Fig. 8.1) we can find it by integrating (8.59):

$$P_{th} = \int_0^T (\alpha K_T)\,dT. \tag{8.62}$$

The lower limit of integration is 0 because thermal pressure, which arises from atomic vibrations, vanishes only at 0 K. Thus, and as I mentioned previously, in order for the calculation to be rigorously correct isothermal compression must be calculated at 0 K. In order to evaluate the integral (8.62) we not only need to know how the product αK_T varies with temperature but also, and more subtly, how it changes with volume. The latter information is needed even if the integration path is at constant volume because the value of this volume will be different depending on how much the material is compressed at zero temperature (see Fig. 8.1). From our discussion in Sections 8.2.2 and 8.2.3, see in particular Figs. 8.2 and 8.3, we see that K_T decreases linearly with temperature and, above the Debye temperature, α increases approximately linearly with temperature. We can thus expect that their product will not be very sensitive to temperature above θ_D, and this expectation is borne out by experimental measurements. Fig. 8.7 shows non-dimensional plots of αK_T normalized to the value of this product at the Debye temperature, versus non-dimensional temperature, T/θ_D, for the same three minerals as in Figs. 8.2 and 8.3 (data from Anderson *et al.*, 1992). Although in detail it is necessary to account for the small deviations observed above θ_D (see Anderson, 1995), assuming that αK_T is constant above the Debye temperature is commonly an excellent approximation.

For $T > \theta_D$, which is the case for planetary interiors at depths greater than the top few kilometers, we can break up (8.62) as follows:

$$P_{th} = (\alpha K_T)_D\,(T - \theta_D) + \int_0^{\theta_D} (\alpha K_T)\,dT, \tag{8.63}$$

where $(\alpha K_T)_D$ is the approximately constant high temperature value of αK_T, for example measured at some T above the Debye temperature. If the regular non-dimensional behavior depicted in Fig. 8.7 is truly universal (or at least applicable to the chief minerals in deep planetary interiors) then an additional simplification is possible. Let A_D be the area under the (nearly identical) curves in Fig. 8.7 in the interval [0,1], given by:

$$A_D = \int_0^1 \frac{(\alpha K_T)}{(\alpha K_T)_D}\,d\left(\frac{T}{\theta_D}\right) = \frac{1}{\theta_D\,(\alpha K_T)_D}\int_0^{\theta_D} (\alpha K_T)\,dT. \tag{8.64}$$

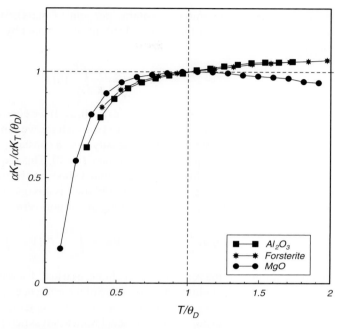

Fig. 8.7 Variation of the product αK_T, normalized to the values of the product at the Debye temperatures, relative to non-dimensional temperature. Data from Anderson *et al.* (1992). The area under the curves between 0 and 1 is approximately 0.77.

Numerical integration of the curves in Fig. 8.7 yields $A_D \approx 0.77$, so:

$$\int_0^{\theta_D} (\alpha K_T)\, dT \approx 0.77 \theta_D \, (\alpha K_T)_D \tag{8.65}$$

and, substituting in (8.63):

$$P_{th} \approx (\alpha K_T)_D \, (T - 0.23 \theta_D), \quad T \geq \theta_D. \tag{8.66}$$

Other thermodynamic approximations to thermal pressure may be found, for example, in Anderson (1995) and Jackson and Rigden (1996), but my approximation, equation (8.66), appears to work reasonably well. Thermal pressures for $T < \theta_D$ can be calculated in a similar manner, by integrating numerically (8.64) to the desired temperature.

Worked Example 8.3   Thermal pressure in planetary mantles

Let us compare the relative magnitudes of thermal pressure and elastic pressure (see equation (8.2)) along mantle adiabats. We re-write the condition of hydrostatic equilibrium (equation (2.34)) as a function of depth, z, as follows:

$$\frac{dP}{dz} = g\rho = \frac{g}{v}, \tag{8.67}$$

where v is specific volume (i.e. volume per unit of mass). We will first calculate zero-temperature pressure as a function of depth inside a planet by integrating this equation as follows:

$$z - z_r = \frac{1}{g} \int_{P_r}^{P} V dP, \tag{8.68}$$

where P_r is the pressure at some reference depth, z_r. If we choose the reference depth as the surface of the planet then $z_r = P_r = 0$, but as we shall see we cannot always do this. Whereas in previous discussions we considered density to be a constant, we will now incorporate the variation of density with pressure (at constant $T = 0$). This requires a function $V = V(P)$, i.e. an isothermal EOS, which for this example I choose to be the Born–Mie EOS, equation (8.56). In order to use a pressure-explicit EOS such as this one we must integrate by parts, so we substitute (8.45) in (8.68) and, accounting for the non-zero lower integration limit, obtain:

$$z = z_r + \frac{1}{g} \left[P V_P - P_r V_{P_r} + \int_{V_0}^{V_{P_r}} P dV - \int_{V_0}^{V_P} P dV \right]. \tag{8.69}$$

In this equation V_0 is the zero-pressure volume, as usual, V_P is the volume at the pressure of depth z, and V_{P_r} is the volume at the pressure of the reference depth, P_r, all taken at the same temperature, 0 K in this case. The integral is easily implemented in *Maple* (see Software Box 8.3), but there is an important point that I want to stress here: equation (8.69) is a function of the form $z = z(V)$. Thus, the calculation procedure requires that we choose a value of V_P, use this value to calculate P from the EOS (e.g. equation (8.56) if we choose the Born–Mie EOS), and then evaluate the integral in (8.69). In other words, the independent variable in the calculations is volume (or density), and both pressure and depth are functions of V. This may sound odd, but it is required by the pressure-explicit nature of the equations of state.

Software Box 8.3   Calculation of zero-temperature pressures in solid planets

The *Maple* worksheet **Z_P.mw** contains two procedures that generate tables of zero-temperature pressure and density as a function of depth, by calling on the procedures in the **S_EOS** package. The two procedures, ZvsP_BoMi and ZvsP_BiMu are identical, except that they use the Born–Mie or Birch–Murnaghan EOS, respectively. The procedures use density as the independent variable, convert density to specific volume, and calculate pressure from the equation of state and depth from equation (8.69). In order to apply these procedures to planets with phase transitions (e.g. Fig. 8.9) the procedure is run separately for each layer, using the pressure and depth at the bottom of each layer as the starting point for the next layer. The material properties for each phase, molar volume and bulk modulus, are always the zero-pressure values (see discussion in Worked Example 8.3).

If we knew nothing about phase transitions in the Earth's mantle, or chose to ignore them, then we could start with olivine at the Earth's surface and solve the equations (the EOS for P and 8.69 for z) all the way to the core–mantle boundary. Using this approach we have $V_0 = 3.006 \times 10^{-4}\,\mathrm{m^3\,kg^{-1}}$ (corresponding to $\rho_0 = 3300\,\mathrm{kg\,m^{-3}}$, which is the zero-pressure density of Fo_{90} olivine), $K_0 = 1.3\,\mathrm{Mbar} = 1.3 \times 10^{11}\,\mathrm{Pa}$, $K' = 4$ and $z_r = P_r = 0$.

The resulting zero-temperature pressure vs. depth curve calculated with the Born-Mie EOS and assuming a constant value of $g = 9.8$ m s^{-2} (see Chapter 2 for a justification of why this is an acceptable approximation) is shown by the thin P_0 curve in Fig. 8.8. This curve can't be much better than an order of magnitude approximation, however, as both density and bulk modulus change at mantle phase transitions.

The correct procedure is, of course, to calculate the curve in various segments, changing the material properties at each phase transition to those of the incoming phase, and setting z_r equal to the depth of the phase transition and P_r equal to the pressure at that depth. *Note very carefully, however, that the material properties of the incoming phase that are used to solve the EOS and the depth equation, (8.69), are always the zero-pressure properties.* If you are not clear on why this is so you should derive equation (8.69) and convince yourself that this is indeed the case.

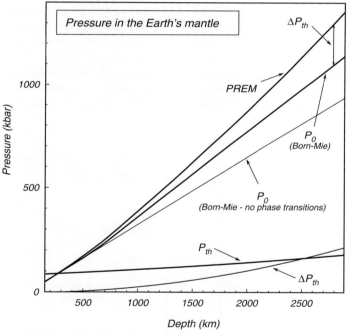

Fig. 8.8   Calculated pressures in the Earth's mantle compared to pressures on the Preliminary Reference Earth Model (PREM) of Dziewonski and Anderson (1981). Chemical composition of the mantle is assumed to be $Mg_{1.8}Fe_{0.2}SiO_4$. P_0 are zero-temperature pressures calculated with the Born-Mie EOS, either ignoring olivine phase transitions (thin curve), or assuming two phase transitions, at depths of 500 km (olivine to ringwoodite) and 670 km (ringwoodite to perovskite + ferropericlase), see also Figure 8.9. P_{th} is the thermal pressure calculated with equation 8.66. Thermodynamic calculation of thermal pressure assumes heating at constant volume (see Figure 8.1). This is not realizable in the Earth's upper mantle and transition zone, because self-compression of the mantle is not sufficient to counter thermal expansion. The actual thermal pressure in the Earth's mantle may be better represented by the difference between PREM and the calculated zero-temperature pressure, shown by the curve labeled ΔP_{th}.

The thick P_0 curve in Fig. 8.8 was calculated in this way, assuming that olivine transforms to ringwoodite at 500 km depth, and ringwoodite transforms to perovskite at 660 km depth. This is still a simplification, as it ignores other phase transitions in the mantle transition zone (olivine transforms to wadsleyite at the top of the transition zone, $\sim$410 km deep), but it is much better than assuming that olivine is stable throughout the entire mantle. Pressures up to 500 km depth were calculated with the olivine properties given above. This yields a pressure of 163 kbar at 500 km depth, and these values are set equal to P_r and z_r for the ringwoodite layer. For ringwoodite we have $\rho_0 = 3700$ kg m^{-3}, $K_0 = 1.9$ Mbar $= 1.9 \times 10^{11}$ Pa, and $K' = 4$. With these values we get $P_r = 222$ kbar at $z_r = 660$ km, which is the top of the perovskite layer, and then use the zero pressure properties of perovskite, $\rho_0 = 4200$ kg m^{-3}, $K_0 = 2.5$ Mbar $= 2.5 \times 10^{11}$ Pa, and $K' = 4$ to calculate pressures all the way to the core–mantle boundary. Figure 8.9 shows zero-temperature densities as a function of depth calculated with this model, as well as the (hypothetical) density of olivine if phase transitions are ignored (i.e. along the thin P_0 curve in Fig. 8.8). Exercise 8.5 asks you to verify these results.

Figure 8.8 shows that ignoring mantle phase transitions underestimates the pressure at the core–mantle boundary by $\sim$200 kbar, or roughly 20% relative. This difference reflects the fact that there are significant density jumps at the phase transitions, as shown by Fig. 8.9. Also shown in Figs. 8.8 and 8.9 are curves labeled PREM. These curves show pressure and density along the Preliminary Reference Earth Model of Dziewonski and Anderson (1981). Details of how these values were calculated are complex and beyond the space available here (see, for example, Anderson, 1989). Suffice it to say that the PREM results are not

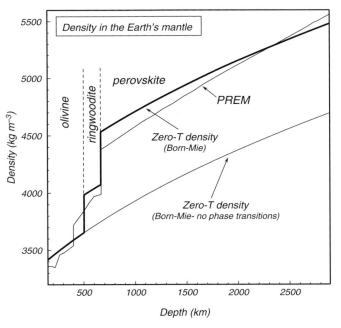

Fig. 8.9   Densities as a function of depth in the Earth's mantle calculated assuming zero-temperature compression of $Mg_{1.8}Fe_{0.2}SiO_4$, with or without phase transitions as in Fig. 8.8. Densities along PREM shown for comparison; note the much more complicated pattern in the upper mantle and transition zone, corresponding to mineral reactions that I ignored in the simplified calculations.

altogether independent of a specific choice of EOS, but they are generally accepted as a good approximation to actual pressures and densities inside the Earth.

We first note that although PREM densities show more "texture" than the ones calculated in this example, reflecting the real phase transitions in the Earth's mantle, they are in generally good agreement with the simple model. PREM pressures and calculated zero temperature pressures are virtually indistinguishable to a depth of ∼500 km, but diverge smoothly as depth increases beyond this value. At the base of the Earth's mantle the calculated zero-temperature pressure is ∼220 kbar lower than the PREM pressure. Clearly, thermal pressure must have a role in this discrepancy, but what exactly is this role? In order to answer this question we calculate thermal pressure along the mantle adiabat.

Temperature along an adiabat is obtained by integrating equation (3.35), as follows:

$$T = T_L \exp\left[\frac{\alpha g}{c_P}(z - z_L)\right], \tag{8.70}$$

where T_L is the temperature at the base of the lithosphere (top of the convective layer) and z_L is the depth to the base of the lithosphere. For Earth we can take $T_L = 1650$ K and $z_L = 150$ km. Substituting in (8.66) we obtain an approximate expression for thermal pressure along a mantle adiabat:

$$P_{th} \approx (\alpha K_T)_D \left(T_L \exp\left[\frac{\alpha g}{c_P}(z - z_L)\right] - 0.23\theta_D\right). \tag{8.71}$$

We now need to choose characteristic values for the parameters in this equation. For many minerals θ_D is of the order of 1000 K and the high-temperature (i.e. above θ_D) values of α and c_P are approximately 3×10^{-5} K^{-1} and 1.2 kJ K^{-1} kg^{-1}, respectively. The high-temperature value of αK_T for closely-packed minerals is of the order of 60 bar K^{-1}. Using these parameter values equation (8.71) generates the curve labeled P_{th} in Fig. 8.8.

A "blind" application of equation (8.2) would lead us to add P_{th} to P_0 in order to obtain the total pressure. Figure 8.8 shows that, whereas this would yield a value comparable to the PREM pressure at the core–mantle boundary, pressures calculated in this way would become progressively more erroneous with decreasing depth, and would be gross overestimates in the upper mantle. What is going on here? Quite simply, that if we had applied equation (8.2) "blindly" we would have been careless with how we applied thermodynamics to the real world. *The definition of thermal pressure* (equations (8.57) or (8.59)) *requires that volume be kept constant as the material is heated, yet the Earth does not behave as a perfectly rigid container*. At shallow depth, where the zero-temperature pressure and the (calculated) thermal pressure are of comparable magnitudes, compression of the mantle under its own weight (often called self-compression) cannot keep the mantle from expanding as its temperature increases. The zero-temperature (elastic) pressure corresponds to the load that is available to keep the volume of the "container" fixed and it is not enough to counteract the thermal expansion of the mantle. In other words, the isochoric heating leg in Fig. 8.1 is not realizable in the Earth's upper mantle, because the upper mantle does not behave as a rigid vessel. In particular, Fig. 8.8 suggests that at depths less than 300 km or so thermal pressure would be higher than zero temperature pressure, which is physically impossible (the material would shatter).

If the material is allowed to expand freely then there can be no thermal pressure. The increase in atomic vibrational energy that occurs with increasing temperature is in such case expressed macroscopically as thermal expansion rather than thermal pressure. The coincidence between PREM and P_0 suggests that this is the case in the Earth's upper mantle

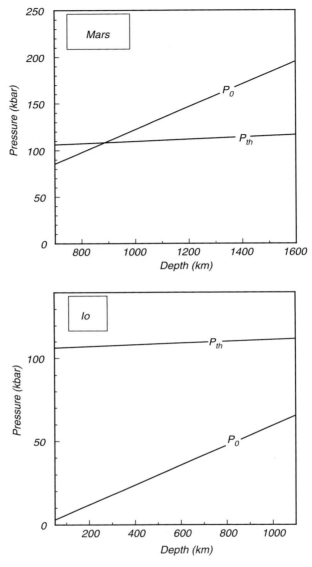

Fig. 8.10 Zero-temperature pressures and thermal pressures (calculated with the constant-volume assumption, Fig. 8.1) in Mars and Io. Thermal pressure is likely to be vanishingly small in Io, and almost negligible in the Martian mantle.

(depths less than ~650 km). The picture at greater depths becomes more blurred. If PREM is an accurate representation of actual pressures inside the Earth and P_0 is the correct zero temperature pressure then the difference between the two values must be thermal pressure. This difference, let us call it the inferred thermal pressure, is shown by the curve labeled ΔP_{th} in Fig. 8.8. The inferred thermal pressure only becomes comparable to the calculated thermal pressure at depths in excess of 1500 km or so. This suggests that at such depths it becomes reasonable to consider the mantle as a "constant volume" vessel, and that a direct application of equation (8.2) to calculate pressure is likely acceptable for the Earth's deep mantle and core. As we move towards the surface there is not enough load to keep the volume perfectly constant, but neither is free expansion possible, so that some thermal pressure builds up. This decreases with decreasing depth, until thermal pressure essentially vanishes in the upper mantle. There is no well-defined ratio between P_0 and calculated P_{th} at which the actual thermal pressure becomes significant, but Figure 8.8 suggests that when P_{th} is less than about $1/3 P_0$, the constant volume approximation, and hence equation (8.2), may become reasonably accurate.

What about application to other planets, for which we do not have PREM as a benchmark? Figure 8.10 shows P_o and P_{th} calculated for the mantles of Mars and Io (core–mantle boundaries assumed to be at 1630 km and 1100 km depth, respectively, after Lodders & Fegley, 1998). In the case of Mars I assumed an adiabatic (convecting) mantle below a 700 km lithosphere, with $T_L = 2000$ K and $g = 3.7$ m s^{-2}. For Io I assumed a thin lithosphere ($z_L = 50$ km), with $T_L = 2000$ K and $g = 1.8$ m s^{-2}. In both cases I assumed that olivine does not undergo phase transitions. The case of Io is unambiguous: temperature is high enough and gravitational acceleration low enough that calculated thermal pressure is everywhere higher than zero temperature pressure, by a large factor. This means that Io's interior must behave as an essentially unconfined solid or, in other words, that pressure in Io's interior is well approximated by zero-temperature pressure, with a vanishingly small thermal component. In the case of Mars we find that at the core–mantle boundary $P_{th} \approx 0.6$ P_0. By comparison, a similar relationship occurs in Earth at a depth of $\sim$600 km, where the inferred thermal pressure, ΔP_{th}, is ~ 10 kbar, or about 5% of the zero-temperature pressure. One could tentatively conclude that thermal pressure is generally negligible in the Martian mantle. Thermal pressures in the Moon and Venus are left as an exercise for the reader.

Worked Example 8.4   The ringwoodite–perovskite phase transition in planetary mantles revisited

In Worked Example 8.2 we calculated phase diagrams for the Mg and Fe end-member systems of mantle silicates. If we focus on the stishovite-absent phase boundary we see that, in an Fe-Mg system, we can write three linearly independent heterogeneous equilibria along this phase boundary:

$$(i) \ Mg_2SiO_4 \rightleftarrows MgSiO_3 + MgO$$
$$(ii) \ Fe_2SiO_4 \rightleftarrows FeSiO_3 + FeO$$
$$(iii) \ MgSiO_3 + FeO \rightleftarrows FeSiO_3 + MgO. \qquad (8.72)$$

Excess mixing properties for perovksite, ringwoodite and periclase are not well known, and in any case a set of mixing properties consistent with the standard state properties that I chose to calculate the phase diagram in Fig. 8.6 does not exist (see Worked Example 5.7 for why this is important). Under these circumstances, and as a first approximation, we will treat the three phases as being ideal solutions. Assuming that all phases are binary Mg–Fe solutions we write the equilibrium conditions for the three reactions as follows:

$$(i) \ K_{(i)} = \exp \left(-\frac{\Delta_r G_{P,T}^{0,(i)}}{RT} \right) = \frac{X_{Mg}^{pv} \cdot X_{Mg}^{pc}}{\left(X_{Mg}^{rw} \right)^2}$$

$$(ii) \ K_{(ii)} = \exp\left(-\frac{\Delta_r G_{P,T}^{0,(ii)}}{RT}\right) = \frac{\left(1 - X_{Mg}^{pv}\right) \cdot \left(1 - X_{Mg}^{pc}\right)}{\left(1 - X_{Mg}^{rw}\right)^2}$$

$$(iii) \ K_{(iii)} = \exp\left(-\frac{\Delta_r G_{P,T}^{0,(iii)}}{RT}\right) = \frac{\left(1 - X_{Mg}^{pv}\right) \cdot X_{Mg}^{pc}}{X_{Mg}^{pv} \cdot \left(1 - X_{Mg}^{pc}\right)}. \tag{8.73}$$

At fixed pressure and temperature these are three equations in three unknowns, so that we can solve for the compositions of the three phases at equilibrium. It is always a good idea to check that this is consistent with the phase rule. In the Fe-Mg system we have $\Phi = 3$ and $c = 3$, so $f = 2$ and everything is fine – fixing two intensive variables fully determines the thermodynamic state of the system.

Let us now calculate the composition of the phases at equilibrium at the 670 km discontinuity, and compare the results to the structure and composition of the terrestrial mantle. In order to do this we use the pressure vs. depth values from PREM (shown in Fig. 8.8), and combine them with temperatures as a function of depth calculated with (8.70), assuming $T_L = 1650$ K at $z_L = 150$ km. This yields a curve of pressure vs. temperature, labeled "Terrestrial adiabat" in Fig. 8.11. Calculated conditions at the 670 km discontinuity are $\sim$238 kbar and $\sim$1600 °C, shown with the dash-dot lines in the figure. Using these values in the solution of the system of equations (8.73) (Software Box 8.4) yields $X_{Mg}^{pv} = 0.86$, $X_{Mg}^{rw} = 0.9$ and $X_{Mg}^{pc} = 0.82$, which compares rather favorably with the Mg number of the terrestrial mantle of $\sim$90. It is hard to say to what extent this agreement is fortuitous, but it is encouraging. If correct, it means that for the composition of the Earth's mantle the stishovite-absent reaction is displaced relative to the Mg end-member so that it goes through 238 kbar at 1600 °C. The calculation also predicts that Fe partitions into the oxide phase relative to the silicates, which agrees with experimental findings.

Software Box 8.4   Fe–Mg exchange among ringwoodite, perovskite and ferropericlase

Maple worksheet **fe_mg_exch_hiP.mw** contains two new procedures that solve the system of equations in Worked Example 8.4 and generate the P–X loop in Fig. 8.12. Procedure FMX calls on dgPT (see Software Box 8.2) to solve the system of non-linear simultaneous equations (8.73), for the three unknowns X_{Mg}^{pv}, X_{Mg}^{rw} and X_{Mg}^{pc}. Executing the statement block following the procedure runs FMX and outputs the three mol fractions to the screen. Procedure FMX_LOOP encloses FMX in a pressure loop and generates a file that lists the three mol fractions as a function of pressure, at constant temperature. This output is used to construct Fig. 8.12. Remember that the statement block at the end of the worksheet, where the reaction stoichiometries are defined, must be executed before the Fe-Mg calculations are attempted, or an error message will result.

The attentive reader should have noticed that, given that ringwoodite–perovskite–periclase equilibrium in the Fe-Mg system is divariant, the three phases must coexist over a pressure range at constant temperature. This is the behavior depicted in the right hand panel of Fig. 6.18. By solving the system of equations (8.73) for a range of pressures at a

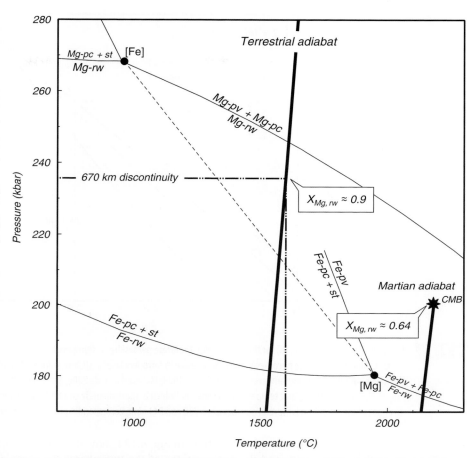

Fig. 8.11 Terrestrial and Martian adiabats superimposed on the phase diagram of Fig. 8.6, with some reactions omitted for clarity. The diagram predicts that stabilization of silicate perovskite in the terrestrial mantle at the 670 km discontinuity would take place for a bulk composition with $X_{Mg} \approx 0.9$, which agrees almost perfectly (and perhaps somewhat fortuitously) with the mantle composition. The Martian adiabat terminates at the conditions of the core–mantle boundary, shown with a star. The ringwoodite–perovskite phase transition would take place at the conditions of the Martian core–mantle boundary for a bulk composition with $X_{Mg} \approx 0.64$, which is more ferroan that any likely Martian mantle composition. Silicate perovksite is therefore unlikely to be present in the Martian mantle. Breakdown of ringwoodite to ferropericlase + stishovite is also unlikely if temperatures in the Martian mantle are of the order of those estimated in Section 3.9, see also Fig. 3.18.

constant temperature (say 1600 °C) we can construct a pressure–composition loop such as that in Fig. 6.18. This is shown in Fig. 8.12, but there are two complications. First, as Fig. 8.6 shows, the stishovite-absent equilibrium must become metastable at a certain X_{Mg} value. Calculation of the exact X_{Mg} and P at which this happens is left as an exercise. In Fig. 8.12 I truncate the diagram at 225 kbar, which is close to the conditions at which the stishovite-absent equilibrium becomes metastable relative to the perovskite-absent one (see Fig. 8.6). The second complication is that, in contrast to the simple phase transitions discussed in Section 6.6, the equilibrium described by equations (8.73) has two phases with

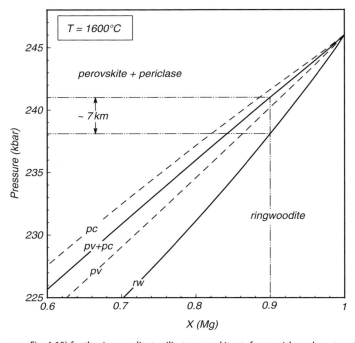

Fig. 8.12 $P–X_{Mg}$ loop (compare Fig. 6.18) for the ringwoodite to silicate perovskite + ferropericlase phase transition at 1600 °C. Silicate perovskite becomes metastable relative to ferropericlase + stishovite for $P–X$ combinations close to the lower bounds of the diagram (see Ito & Takahashi, 1989; Matsuzaka *et al.*, 2000; Frost *et al.*, 2001). For bulk $X_{Mg} = 0.9$, and assuming that temperature is constant (which is approximately correct for a convective mantle), the divariant phase transition is "spread out" between 238 and 241 kbar, corresponding to a depth interval of ∼7 km.

different X_{Mg} on one side of the reaction. In Fig. 8.12 I show the composition of ringwoodite with a solid curve, and the compositions of the two phases that coexist on the other side of the reaction, perovskite and periclase, with broken lines. The bulk composition of the assemblage perovskite + periclase must lie between these two curves, and is shown schematically with a solid curve. The two solid curves, labeled rw and pv + pc, are the boundaries of the divariant phase transition at the 670 km discontinuity. For a mantle with $X_{Mg} = 0.9$, breakdown of ringwoodite at 1600 °C begins at ∼238 kbar and is completed at ∼241 kbar. Thus, the divariant assemblage ringwoodite + perovskite + periclase coexists at equilibrium over a pressure interval of ∼3 kbar, which corresponds to a depth interval of ∼7 km (see, Fig. 8.8). This result compares favorably with estimates based on the sharpness of the seismic discontinuity.

Let us now repeat the exercise for the Martian mantle. In this case we use the zero-temperature Born–Mie pressures shown in Fig. 8.10 and temperatures calculated with (8.70) assuming $T_L = 2000$ K at $z_L = 700$ km (see Section 3.9). Using zero-temperature pressures is justified by our finding that thermal pressure in the Martian mantle is likely to be small. The resulting $P–T$ curve is labeled Martian adiabat in Fig. 8.11, and terminates at the pressure of the Martian core–mantle boundary (CMB), shown with a star in the figure. Calculated conditions at the Martian CMB are ∼200 kbar and ∼2180 K, for which the solution set of equations (8.73) is $X_{Mg}^{pv} = 0.6$, $X_{Mg}^{rw} = 0.64$ and $X_{Mg}^{pc} = 0.53$. Given that the Martian mantle is more magnesian than this (Mg number about 80), and that the perovskite phase transition at any pressure less than that of the CMB would require an even more Fe-rich composition than that calculated for the CMB (see Fig. 8.11), we can infer that perovskite does not exist in the Martian mantle. Moreover, if the temperatures estimated in Section 3.9 are approximately correct, then the Martian adiabat also stays on the high temperature side of the invariant point, and breakdown of ringwoodite to ferropericlase + stishovite is also inhibited. We can thus conclude that the silicate phase at the base of the Martian mantle is likely to be ringwoodite.

Exercises for Chapter 8

8.1 The ringwoodite–perovskite transition, that defines the bottom of the mantle transition zone, has a negative Clapeyron slope (Fig. 8.6). In contrast, the olivine–wadsleyite transition, that defines the top of the transition zone, has a positive Clapeyron slope. Discuss how each of these transitions may affect the mantle adiabat.

8.2 Prove identity 8.20:

$$\left(\frac{\partial \alpha}{\partial P} \right)_T = \frac{1}{K_T^2} \left(\frac{\partial K_T}{\partial T} \right)_P . \tag{8.20}$$

8.3 Prove that if all shear strain components vanish ($\varepsilon_{ij} = 0$ for $i \neq j$) and all diagonal elements of the strain tensor are equal ($\varepsilon_{11} = \varepsilon_{22} = \varepsilon_{33}$) then strain is isotropic, meaning that elongation is proportional to the distance between points only, and independent of orientation. Also, prove that in this special case Eulerian strain is related to the proportionality constant between elongation and distances by $\varepsilon = \frac{1}{2}(2k - k^2)$, see Section 8.2.5.

8.4 Note that the Mie EOS (equation (8.53)) is formally identical to the second-order Birch–Murnaghan EOS (equation (8.42)). To make it identical would require that $m = 4 > n = 2$. Is this physically reasonable? Explain why. (Hint: find the extrema of the function.) How does the Born–Mie EOS get around this problem?

8.5 Reproduce Figs. 8.8 and 8.9, using the *Maple* worksheets described in Software Box 8.3. Plot density vs. pressure in the simplified three-layer mantle described in these figures. Comment on your results.

8.6 On the basis of the discussion in Section 3.9, construct a possible adiabat for Venus. Assuming that the Venusian mantle has the same composition as the terrestrial mantle, estimate the depth of the ringwoodite–perovskite transition in Venus. As a very rough approximation estimate the depth of the olivine-ringwoodite (metastable) transition by scaling your result to Fig. 8.9. Calculate curves of zero-temperature pressure vs. depth and thermal pressure vs. depth for Venus, and comment on your results.

8.7 The phase diagram in Fig. 8.12 shows that the high-pressure assemblage perovskite + ferropericlase is richer in Fe than the low-pressure phase ringwoodite. Yet the mantle transition zone is an isochemical boundary. Explain the apparent contradiction.

Thermodynamics of planetary volatiles

From a physical point of view a fluid is a material that lacks shear strength, i.e. that deforms, or "flows", when subject to shear stress. The strain rate for a given shear stress is of course highly variable, and defines the viscosity of the fluid. From a chemical thermodynamic point of view it is convenient to distinguish between different types of fluids. There are fluids with relatively low densities and low viscosities, which tend to be highly mobile in planetary environments. These are often referred to as "volatiles", and are typically composed of species in the system C–O–H–N–S–F–Cl, with inert gases (particularly He) also important in gas giants. There are also fluids with generally higher densities than volatiles, which also have much higher viscosities, typically by several orders of magnitude. If such fluids exist at equilibrium with solids of broadly similar bulk composition we call them melts (Chapter 10). Melts in terrestrial planets are chiefly silicates, although natural carbonate melts also exist, as do metallic melts in planetary cores. Melts in icy satellites, in contrast, are likely to be composed chiefly of species in the system C–O–H–N. A third type of fluids are liquids at conditions that are far removed from equilibrium with solids of similar bulk composition, but that may contain species in solution that crystallize their own solids. Aqueous solutions (Chapter 11) are an example.

The boundaries between these different types of fluids are not always clearly defined. In this chapter we will discuss the thermodynamics of fluids that can be described as volatiles. These can be gases, i.e. non-condensed phases that expand indefinitely, liquids, which are condensed and strongly incompressible fluids, or supercritical fluids, whose physical properties vary continuously between those of gases and liquids (see also Section 1.15).

9.1 Fugacity and standard state fugacity

In Chapter 5 we saw that the chemical potential of a gas species, i, in a mixture of ideal gases at total pressure P is given by:

$$\mu^i_{P,T} = \mu^{0,i}_{P_{(o)},T} + RT \ln \left(\frac{p^i}{P_{(o)}} \right) = \mu^{0,i}_{P_{(o)},T} + RT \ln \left(\frac{X_i P}{P_{(o)}} \right), \qquad (9.1)$$

where p_i is the partial pressure of species i in the mixture, and $P_{(o)}$ is the pressure of some arbitrarily chosen standard state, in which species i is pure. In Chapter 5 (e.g. equation (5.86)) we chose $P_{(o)} = 1$ bar, and this is indeed the usual practice for gases. We shall return to this convention soon, but for clarity in the following derivations it is best to leave the value of $P_{(o)}$ temporarily unspecified.

We can write the chemical potential of a gas species in a mixture of real gases as a combination of ideal and excess, or residual, contributions:

$$\mu^i_{P,T} = \mu^i_{P,T,id} + \mu^i_{P,T,res} = \mu^{0,i}_{P_{(o)},T} + RT \ln \left(\frac{X_i P}{P_{(o)}} \right) + \mu^i_{P,T,res}. \tag{9.2}$$

We now define a dimensionless parameter ϕ^i, called the *fugacity coefficient* of species i in the mixture, by the expression:

$$RT \ln \phi^i \equiv \mu^i_{P,T,res} \tag{9.3}$$

and re-write (9.2) as follows:

$$\mu^i_{P,T} = \mu^{0,i}_{P_{(o)},T} + RT \ln \left(\frac{X_i \phi^i P}{P_{(o)}} \right). \tag{9.4}$$

We can collect the terms in the numerator of the logarithmic function and define a new thermodynamic function, called *fugacity*, by means of the following two equations:

$$f_i \equiv X_i \phi^i P \tag{9.5}$$

$$\lim_{P \to 0} \frac{f_i}{X_i P} = 1. \tag{9.6}$$

Note the following properties of fugacity: (i) equation (9.5) defines fugacity as a physical quantity with units of pressure, (ii) equation (9.3) makes the fugacity of an ideal gas ($\mu_{res} = 0$) identical to its partial pressure, or its total pressure if $X_i = 1$, and (iii) equation (9.6) states that at vanishingly small *total* pressure the fugacity of *any* gas is equal to its partial pressure. As long as $T \gg 0$ (e.g. $T \geq 298$ K), a pressure of 1 bar is under most circumstances low enough for (9.6) to be valid. Unless otherwise stated, then, we shall assume that (9.6) is always valid at 1 bar and call this condition the *ideal gas limit*. We also note that, although not explicitly stated in (9.3) and (9.5), fugacity and fugacity coefficient are functions of temperature and pressure.

Let us consider now the special case of a pure gas. In this case $X_i = 1$ and we re-write (9.5) as follows:

$$f^0 = \phi^0 P, \tag{9.7}$$

where f^0 and ϕ^0 are called the *standard state fugacity* and *standard state fugacity coefficient*, respectively. Considering (9.4) for a pure substance and substituting (9.7) we find:

$$\mu^0_{P,T} = \mu^0_{P_{(o)},T} + RT \ln \left(\frac{f^0}{P_{(o)}} \right). \tag{9.8}$$

In equation (9.8) I dropped the superscript i to emphasize that we are now dealing with a pure substance. The two chemical potentials in this equation can be taken as two possible standard states, at P and $P_{(o)}$. Using the relationship $V = (\partial G / \partial P)_T$ we can also relate the two standard states as follows:

$$\mu^0_{P,T} = \mu^0_{P_{(o)},T} + \int_{P_{(o)}}^{P} V dP, \tag{9.9}$$

which results in:

$$RT \ln \left(\frac{f^0}{P_{(o)}} \right) = \int_{P_{(o)}}^{P} V \, dP.$$ (9.10)

We now introduce the usual convention for gases, i.e. $P_{(o)} = 1$ bar, and (9.10) becomes:

$$RT \ln f^0 = \int_{1}^{P} V \, dP,$$ (9.11)

which is the definition of standard state fugacity. We will use this equation to calculate f^0 (Section 9.5). Note that, as I will do from now on, I have omitted the denominator in the argument of the logarithmic function in (9.11). It must be tacitly understood that we are dividing by 1 bar, however, so that fugacity retains its pressure dimension.

With the convention $P_{(o)} = 1$ bar equation (9.8) becomes:

$$\mu^0_{P,T} = \mu^0_{1,T} + RT \ln f^0,$$ (9.12)

which shows that standard state fugacity is a function that converts the standard state of a real gas at 1 bar to the standard state at any arbitrary pressure P, and at the same temperature. Physically we can think of the standard state fugacity of a real gas as its *thermodynamic effective pressure*, i.e the pressure that an ideal gas with the same chemical composition would have to be under in order to generate the observed chemical potential of the real gas (compare (9.12) with (9.1), setting $X_i = 1$). Equation (9.12) provides a simple way of calculating the standard state chemical potential of a pure gas species at any pressure and temperature. We first take the gas from the reference temperature (298 K) to T, at a constant pressure of 1 bar, by evaluating the heat capacity integrals for enthalpy and entropy (Box 5.1). This yields $\mu^0_{1,T}$: the standard state chemical potential at 1 bar and T. We then add the standard state fugacity term, in order to obtain the standard state chemical potential at P and T. The standard state fugacity is calculated by integrating (9.11), substituting an appropriate equation of state. The choice of equation of state, and the integration procedure, will be discussed later in this chapter.

If we now consider a gas species in a mixture of real gases then substituting (9.5) in (9.4) we get:

$$\mu^i_{P,T} = \mu^{0,i}_{1,T} + RT \ln f_i$$ (9.13)

where in this case the fugacity of species i in the mixture, f_i, plays the role of a *thermodynamic effective partial pressure*. A more useful interpretation of equation (9.13) is that it defines fugacity as a measure of the difference between the chemical potential of a species in its 1 bar standard state and the chemical potential of the species in the system of interest, *regardless of whether or not the species is actually present in a fluid phase, or even of whether or not a fluid phase is present*. Note that even though we defined fugacity by considering the behavior of chemical species in a gas phase, equation (9.13) as interpreted in the previous sentence is an equally valid definition of fugacity, that makes no *a priori* statement regarding the existence of a fluid phase. In fact, we can also start from (9.13) and work our way backwards to equation (9.1), which is how fugacity was initially defined by G. N. Lewis, about a century ago.

The importance of interpreting fugacity as a difference between chemical potentials is that it makes it possible to attach physical meaning to results such as the one that we obtained in Worked Example 5.5. There we calculated a partial pressure of oxygen of order 10^{-40} bar. It is hard to make sense of this number from a molecular point of view, but in terms of fugacity it is easily understood as a representation of the chemical potential of oxygen, relative to that of pure oxygen at the same temperature and 1 bar. Comparing (9.12) with (9.13) it also follows that the activity of species i in a fluid, a^i is given by:

$$a^i = \frac{f_i}{f^0}.$$

(9.14)

Although (9.14) is rigorously correct we shall not use this relationship, as it is redundant. Because of the condition imposed by equation (9.6) fugacity is defined in absolute terms, relative to the behavior of the ideal gas at vanishingly small pressure, and it is a number that by itself carries all the information that we need.

Worked Example 9.1   Mineral-fluid reactions: hydration of peridotites

Mantle peridotites emplaced in the Earth's continental crust, or incorporated into the ocean floor, typically undergo retrograde metamorphism. The anhydrous high-temperature ultra-mafic assemblage is replaced partially or completely by hydrous minerals, transforming the rock into a serpentinite. A simple Mg end-member model reaction for this process is the formation of chrysotile from forsterite plus enstatite, according to:

$$2Mg_2SiO_4 + Mg_2Si_2O_6 + 4H_2O \rightleftarrows 2Mg_3Si_2O_5(OH)_4.$$

(9.15)

In the presence of a volatile phase that contains no species other than H_2O this is a univariant reaction. If we choose the value of one intensive variable, for example, temperature, then all other intensive variables are fixed by the assemblage. This includes the chemical potential of H_2O or, equivalently, the fugacity of H_2O, which in this case must be the standard state fugacity, $f^0(H_2O)$, at the equilibrium temperature and pressure. Volatiles in planetary interiors are seldom composed of pure H_2O, however, and may contain C, S, N, F and Cl in addition to H and O. The variance of an equilibrium such as (9.15) in nature is thus greater than 1. Regardless of what the actual variance of the system is, however, as long as the three solid phases in (9.15) coexist at equilibrium the chemical potential of H_2O is fixed by this equilibrium. This means that fixing T and P fixes the fugacity of H_2O, $f(H_2O)$. We state this by saying that the assemblage enstatite + forsterite + chrysotile *buffers* the fugacity, or chemical potential, of H_2O. We note from our preceding discussion that it must be $f(H_2O) \le f^0(H_2O)$, with the equality holding only for the special case of univariant equilibrium, in which no other components are present in the fluid phase.

   In order to calculate the fugacity of H_2O buffered by (9.15) we write the equilibrium condition for this reaction as follows:

$$\Delta_r G_{1,T}^{0,(9.15)} + \int_1^P \Delta V_{solids} dP - 4RT \ln f(H_2O) = 0.$$

(9.16)

Note that in (9.16) the standard state Gibbs free energy change is calculated at 1 bar and the temperature of interest. The contribution of pressure to Gibbs free energy change is then split into two components that are calculated separately, one for the solid phases and the other for the gas species. This is necessary because the equations of state for solids and gases are different, as we will see later in this Chapter. We discussed how to calculate the standard state Gibbs free energy change in Chapter 5 (e.g. Box 5.1), and how to calculate the pressure integral for solids in Chapter 8. Hence, calculation of $f(H_2O)$ from (9.16) is easily implemented in *Maple* by calling on previously written procedures (Exercise 9.1).

Fugacities of H_2O calculated from (9.16) are shown in Fig. 9.1, as a function of temperature at constant $P = 6$ kbar (top), and as a function of pressure at constant $T = 500\,^\circ C$ (bottom). By fixing one of these two intensive variables the curves in Fig. (9.1) become pseudo-univariant phase boundaries (Section 6.3.1). Also shown in the figure are curves showing the standard state fugacity of H_2O, obtained by integrating (9.11) and substituting an explicit equation of state; this will be discussed later in this chapter. The standard state fugacity curves separate a "feasible" region from a "prohibited" region, which is shaded in the figure. The standard state fugacity of a fluid species, which is its fugacity in a pure fluid, is the maximum value that fugacity can take at a given P and T. Fugacity can be less than f^0, but not greater. Equivalently, one can state that the chemical potential at equilibrium cannot be greater than the chemical potential of the pure species, for if it

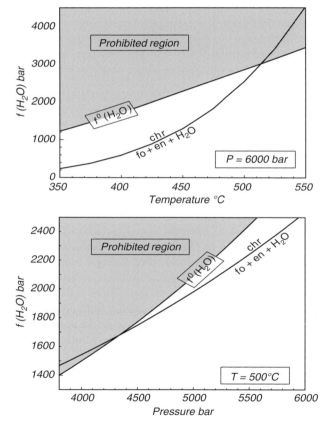

Fig. 9.1 Calculated H_2O fugacity along the forsterite–enstatite–chrysotile–vapor equilibrium, and standard state H_2O fugacity, at a constant pressure of 6 kbar (top) and constant temperature of 500 $^\circ C$ (bottom). A fugacity greater than the standard state fugacity at the same P and T is not an equilibrium condition.

was then the pure phase would form, lowering the Gibbs free energy to that of the pure species.

The curves for $f(H_2O)$ along equilibrium (9.15) intersect the $f^0(H_2O)$ curves. At the intersection $f(H_2O) = f^0(H_2O)$, so these points represent equilibrium in the presence of a fluid composed of pure H_2O. The intersections in Fig. 9.1 correspond to points on the univariant equilibrium phase boundary for reaction (9.15). The segments of the $f(H_2O)$ curves that enter the "prohibited regions" represent impossible conditions. The H_2O fugacity curves must therefore terminate at the $f^0(H_2O)$ curves, but are shown in the figure extending beyond the intersection for the sake of this discussion. The figures can also be interpreted as saying that chrysotile is not stable at temperatures greater than, or pressures less than, those at the intersection, and that these intersections are the ultimate limits of chrysotile stability (because it must be $f(H_2O) \leq f^0(H_2O)$). On the opposite side of the intersections the four-phase assemblage enstatite + forsterite + chrysotile + fluid is stable, and the fugacity of H_2O lies on the pseudounivariant phase boundary. Because this fugacity is lower than the standard state fugacity, however, a pure H_2O fluid cannot be present at equilibrium. If a fluid is present then the distance between the two curves is a measure of the H_2O content of the fluid in equilibrium with the three solid phases.

Worked Example 9.2   Evaporites revisited

Let us take another look at the thernardite–mirabilite equilibrium that we discussed in Chapter 6. In Fig. 6.13 we showed the phase relations as a function of the coordinates μ^{H_2O} and T. Equation (9.13) allows us to represent the chemical potential as a fugacity, but in order to calculate actual fugacity values it is necessary to know standard state thermodynamic properties (see equation (9.16)). These are known for the two crystalline phases but not for the liquid. We can therefore calculate the position of the liquid-absent equilibrium only. Using equation (9.16) and standard state properties from Robie and Hemingway (1995) we calculate the univariant equilibrium curve at 1 bar shown in Fig. 9.2, which is the same as Fig. 6.13. Calculated H_2O fugacities are of order 10^{-2}–10^{-1} bar at $T \sim 20$–$40\,^\circ C$. Because I used standard state properties for H_2O gas, these are the fugacities of H_2O gas. At these P–T conditions (ideal gas limit) fugacity is equivalent to the partial pressure of H_2O. The calculated fugacity can therefore be compared to the vapor pressure of H_2O, i.e. to the partial pressure of H_2O vapor in equilibrium with its liquid (see Section 6.5.2). Saturation vapor pressure is shown in the figure by the thin curve bounding the shaded "prohibited region". Equilibrium fugacities (or partial pressures) inside this region are impossible because liquid H_2O condenses and buffers the chemical potential of H_2O, and $f(H_2O)$, at the saturation vapor pressure. The thenardite–mirabilite phase boundary intersects the vapor saturation curve at a temperature of $\sim 42\,^\circ C$, meaning that the (pseudo)invariant point in Fig. 6.13 must be located at a temperature lower than this. In Fig. 9.2 I show it at $T \sim 32\,^\circ C$, which is its approximate location as determined by phase equilibrium experiments (Rodríguez-Navarro et al., 2000; Marliacy et al., 2000). The metastable extension of the liquid-absent reaction is shown projecting to its intersection with the vapor saturation curve. The other two phase boundaries are shown schematically. The precise conditions for the formation of thenardite vs. mirabilite can be obtained from the diagram, with partial pressure converted to relative humidity if desired.

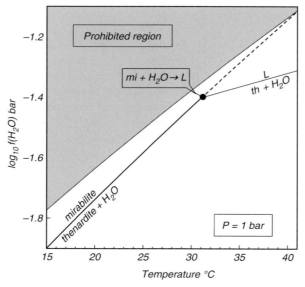

$$Fig. 9.2$$ Calculated H_2O fugacity along the thenardite–mirabilite equilibrium at 1 bar. The locations of the thenardite–liquid and mirabilite–liquid equilibria are only approximate. Condensation of liquid H_2O occurs along the boundary of the "prohibited region", which maps the saturation vapor pressure of H_2O.

Worked Example 9.3   Oxygen fugacity along iron oxidation reactions

The multiple oxidation states of iron: Fe^0, Fe^{2+}, Fe^{3+}, and its very large cosmic abundance give this element the capability of serving as an oxygen buffer in many planetary environments. Increasingly oxidized states of Fe can be related by the following three reactions, conventionally called QFI (quartz–fayalite–iron), QFM (quartz–fayalite–magnetite) and HM (hematite–magnetite), respectively:

$$2Fe + SiO_2 + O_2 \rightleftarrows Fe_2SiO_4 \tag{9.17}$$

$$3Fe_2SiO_4 + O_2 \rightleftarrows 2Fe_3O_4 + 3SiO_2 \tag{9.18}$$

$$4Fe_3O_4 + O_2 \rightleftarrows 6Fe_2O_3. \tag{9.19}$$

Using equation (9.16) it is a simple exercise to calculate $f(O_2) - T$ curves for the three reactions, which are shown for a constant pressure of 1 bar in the top panel of Fig. 9.3 (note that, with the exception of fayalite, all solid phases in these reactions undergo lambda phase transitions, whose free energy contributions must be included in equation (9.16), as we discussed in Section 7.6). The thermodynamic meaning of the $f(O_2) - T$ curves is the same as that of the chrysotile dehydration curves in Fig. 9.1: as long as a given solid phase assemblage is present oxygen fugacity is buffered at the values along the corresponding curve. There are a couple of important differences with the serpentinization example, though. First, because oxygen fugacity values are many orders of magnitude smaller than 1, it is not necessary to worry about the standard state fugacity of oxygen, as the curves stay well within the "feasible" region. This is another way of saying that oxygen has a strong affinity for most other elements, so that one is unlikely to come across oxygen in a state approaching its

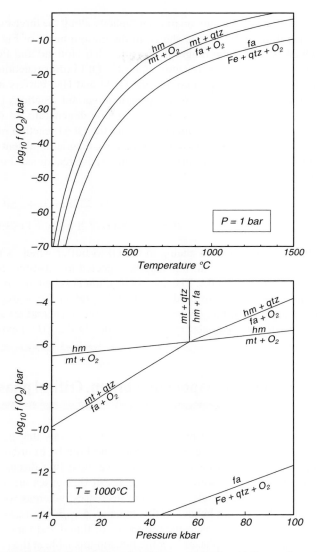

Fig. 9.3   Top: oxygen fugacity as a function of temperature along three successive Fe oxidation reactions, at $P = 1$ bar. Bottom: oxygen fugacities along Fe oxidation reactions as a function of pressure, at $T = 1000\,°C$.

standard state. Oxygen fugacity in the terrestrial atmosphere, ~ 0.2 bar, is not at equilibrium with the Earth's surface, but makes it possible for me to write this and for you to read it. I discuss this topic in Chapter 14. Second, in the case of the chrysotile dehydration reaction it is possible, but not necessary, to give the H_2O fugacity values an interpretation in terms of mol fraction of H_2O in a fluid (more on this later in this chapter). For oxygen fugacities such as those in Fig. 9.3 this interpretation becomes problematic, as we already noted. Oxygen fugacity along buffering reactions such as those in Fig. 9.3 must be seen as a measure of the chemical potential of O_2, independently of whether or not a fluid phase exists at equilibrium with the solid phases. This same interpretation is of course also valid for H_2O fugacities in Fig. 9.1 and 9.2, but in those cases we can also make meaningful inferences about fluid composition, assuming that a fluid phase exists.

The effect of pressure on oxygen fugacity along the three buffering reactions at a constant temperature of $1000\,^\circ$C is shown in the bottom panel of Fig. 9.3. From 1 bar to 100 kbar (which is within the range of pressures that Holland and Powell's data base is expected to be reliable) oxygen fugacity along the QFI buffer reaction stays several orders of magnitude below the other two, but the QFM and HM buffers intersect at $P \approx 57$ kbar. This is an intersection between two pseudounivariant reactions (Section 6.3.1), that defines a pseudoinvariant point. The HM reaction is degenerate, so it crosses the pseudoinvariant point without becoming metastable, but the QFM reaction must become metastable at the pseudoinvariant point. Two other reactions, magnetite-absent and oxygen-absent, must also radiate from the pseudo-invariant point. These reactions, shown in Fig. 9.3 in their correct calculated locations, are:

$$2Fe_2SiO_4 + O_2 \rightleftarrows 2Fe_2O_3 + 2SiO_2 \tag{9.20}$$

$$2Fe_3O_4 + SiO_2 \rightleftarrows 2Fe_2O_3 + Fe_2SiO_4. \tag{9.21}$$

The first reaction is an analog to the QFM buffer which is not stable in the Earth's crust and shallow mantle, but that can be expected to substitute for QFM as the oxide–silicate buffer at greater depths. The second reaction is called an *iron disproportionation reaction*. It entails no change in oxidation state, but rather the segregation of Fe^{2+} and Fe^{3+}, that occur together in magnetite, into a ferric phase (hematite) and a ferrous phase (fayalite). Because it is oxygen-absent it plots parallel to the $f(O_2)$ axis.

9.2 Liquid–vapor equilibrium. Critical phase transitions redux

In Section 1.15, see in particular Fig. 1.16, we saw that non-ideal gases span a range of densities between those of ideal gases and liquids. In order to develop equations of state appropriate for non-ideal gases, then, we must first discuss the nature of the liquid–gas phase transition in some detail. We recall that gases are *non-condensed fluids* because they expand indefinitely as pressure decreases, whereas we call liquids *condensed fluids* because they do not behave in this manner. A *vapor* is a gas in equilibrium with its liquid. As temperature increases, the properties of liquid and vapor at equilibrium approach each other, until the two phases become indistinguishable at the *critical temperature*. Above the critical temperature a single fluid phase is stable, called a *supercritical fluid*. The density of supercritical fluids may vary from values typical of liquids to values typical of ideal gases. The defining property of a supercritical fluid is that in it there is no discontinuous phase transition separating both states. This behavior is reminiscent of that of solid solutions and order–disorder transitions in crystalline solids (Chapter 7).

9.2.1 The van der Waals equation of state

Ideal gases do not condense at any temperature or pressure, as by hypothesis they lack intermolecular forces. Thus, in contrast to the ideal gas EOS, we require that any EOS for a real gas be able to reproduce condensation and the existence of a critical phase transition. The simplest equation of state that can achieve this was proposed by the Dutch physical chemist J. D. van der Waals in 1873. The van der Waals EOS is not quantitatively accurate, but thanks to its simplicity and its correct qualitative behavior it remains a powerful tool with which to gain insight into the behavior of real fluids. Some of the more refined EOS that we will discuss in later sections follow the van der Waals equation in spirit, if not in algebraic detail.

We begin by defining the (poorly named) compressibility factor of a gas, Z, as follows:

$$Z \equiv \frac{PV}{RT}. \tag{9.22}$$

For an ideal gas, $Z = 1$. In a real gas we expect that there are both attractive and repulsive forces between molecules, that are not present in an ideal gas. Intermolecular repulsion will cause Z to increase, whereas attraction will cause it to decrease. We can then write Z for a real gas as follows:

$$Z = 1 + \Delta Z_{repulsion} - \Delta Z_{attraction}. \tag{9.23}$$

The idea behind van der Waals' equation, and in fact behind most EOS for real gases, is to find explicit expressions for the ΔZ terms that match the experimentally measured behavior of fluids over as wide a range of conditions as possible. Many successful approaches, beginning with that of van der Waals himself, entail some combination of intuition, educated guesses and trial and error.

As both the repulsive and attractive terms arise from intermolecular potentials, it is reasonable to postulate that the two ΔZ terms will vary inversely with some power of volume: the closer the molecules are, the stronger they will interact. This is essentially the same argument used to derive the Mie EOS, e.g. equation (8.47). Note that, just as in that case, the qualitative statement about the relationship between volume and intermolecular potentials says nothing about the actual value of the exponents. The van der Waals EOS assumes that both the attractive and repulsive terms go as V^{-1}, but the value of the exponent could be different, and in fact this is one of the problems with this equation.

Real gas molecules have a finite volume, which in the van der Waals equation is defined as the distance at which intermolecular repulsion becomes infinite: no matter how high we make the pressure we cannot squeeze molecules any closer than this value. The total molecular volume defined in this way for a mol of gas is called the excluded volume and symbolized with b. We write the repulsive term as follows:

$$\Delta Z_{repulsion} = \frac{b}{V - b}. \tag{9.24}$$

This equation gives the desired behavior: repulsion becomes stronger with decreasing volume, and diverges as the excluded volume is approached (the term b must appear in the numerator too because Z is a dimensionless number). We can immediately see two pitfalls in this equation: first, molecules are assumed to be perfectly rigid, and second, the effect of temperature, which is likely to affect the b parameter, is ignored. We shall return to this in later sections.

In contrast to repulsion, the effect of attractive forces is assumed to decrease with increasing thermal agitation, so we write:

$$\Delta Z_{attraction} = \frac{a}{RTV}, \tag{9.25}$$

where a is a constant with the units required to make ΔZ non-dimensional. Again there are pitfalls that we will address later: a is assumed to be independent of temperature, and the exponents of T and V are arbitrarily assumed to be -1. Ignoring all of these problems we substitute in (9.23) and obtain:

$$Z = 1 + \frac{b}{V - b} - \frac{a}{RTV},\tag{9.26}$$

which, by using (9.22), we recast as follows:

$$P = \frac{RT}{V - b} - \frac{a}{V^2}.\tag{9.27}$$

Equation (9.27) is the van der Waals EOS. We shall now see what this equation predicts about the behavior of fluids, following a path that parallels our prior discussions of critical phase transitions in Chapter 7.

9.2.2 The critical point of a van der Waals fluid

For a system at equilibrium the derivative $(\partial P/\partial V)_T$ must be negative. This is easy to prove formally from the equilibrium conditions for Helmholtz free energy (Exercise 9.4), but we can also accept it on the grounds of physical intuition: in a system at equilibrium volume can only decrease in response to an increase in pressure. For the ideal gas EOS this condition is always true, but the van der Waals EOS may have a positive $(\partial P/\partial V)_T$ derivative under certain circumstances. To see why, and what these circumstances may be, we differentiate (9.27):

$$\left(\frac{\partial P}{\partial V}\right)_T = -\frac{RT}{(V - b)^2} + \frac{2a}{V^3}.\tag{9.28}$$

It is obvious that (9.28) may take positive or negative values, according to the relative magnitudes of T and V. Inside any $T–V$ region over which the derivative is positive a *single* fluid phase cannot exist at equilibrium (note the emphasis on *single*). In order to identify such regions we note that the derivative vanishes where it changes sign, so we make (9.28) equal to 0 and solve for T:

$$T = \frac{2a}{R} \frac{(V - b)^2}{V^3}.\tag{9.29}$$

The shape of this function is shown in Fig. 9.4. The origin of the coordinate system is fixed at $T = 0$ and $V = b$, as the region $V < b$ is not physically meaningful. By construction, $(\partial P/\partial V)_T$ vanishes along the curve, it is negative for temperatures above the curve and positive for temperatures below it. Thus, the region under the curve in Fig. 9.4 is the $T–V$ region within which a single fluid cannot exist at equilibrium, which I will call the "prohibited region".

The peak of the curve is the maximum temperature at which $(\partial P/\partial V)_T$ may take non-negative values. For temperatures higher than this value, let us call it T_c, $(\partial P/\partial V)_T$ is negative for all values of V. This means that for $T > T_c$ a single fluid phase that obeys van der Waals' EOS (equation (9.27)) is stable for all V and P. By "being stable for all V and P" I mean that we can vary the intensive variables continuously and, as long as we stay inside the region $T > T_c$ there will be no discontinuity in the material properties of the phase. We shall return to this, but for now we notice that the properties of equation (9.27) assure us that this continuity is certainly true of density.

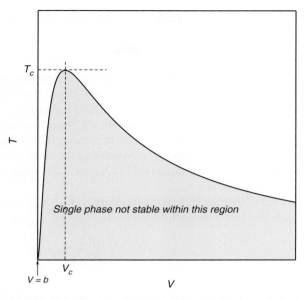

Fig. 9.4 Conditions for reversal of the sign of the derivative $(\partial P/\partial V)_T$ for a van der Waals gas. In the shaded region below the curve the derivative is positive, implying that a single fluid phase cannot be stable. The peak of the curve is the critical temperature: above this temperature a single fluid phase of any density can be stable.

For $T < T_c$ there is a "prohibited" volume (or density) interval within which a single phase is not stable. Any volume inside this region can, however, be obtained algebraically as a linear combination of two arbitrary volumes outside and on opposite sides of the curve – in particular, by two volumes that satisfy equation (9.27) at the same temperature and pressure (the equation is cubic in V, so it must have either one or three real roots). Note that these volumes are *not* the two intersections of a temperature coordinate with the curve in Fig. 9.4, as these intersections simply define the condition $(\partial P/\partial V)_T = 0$. The two volumes that we seek correspond to two stable phases of different densities, so they must lie outside of the curve, and they must be such that they minimize the free energy of the system. We shall return to this crucial point shortly.

At the temperature T_c the system changes from one in which two different fluids are possible (for $T < T_c$) to one in which only one fluid is possible (for $T > T_c$). This looks a lot like a critical temperature, and in fact it is. In order to find its value we need to locate the maximum of the curve in Fig. 9.4, so we differentiate (9.29) and equate to 0:

$$\frac{dT}{dV} = \frac{2a}{R} \left[\frac{2V(V-b) - 3(V-b)^2}{V^4} \right] = 0. \tag{9.30}$$

The solution to (9.30) is the critical volume V_c, i.e. the volume at the critical temperature:

$$V_c = 3b. \tag{9.31}$$

Substituting in (9.29) we get the critical temperature, T_c:

$$T_c = \frac{8a}{27Rb} \tag{9.32}$$

and using (9.31) and (9.32) in (9.27) we get the critical pressure, P_c:

$$P_c = \frac{a}{27b^2}. \tag{9.33}$$

The set $\{T_c, P_c, V_c\}$ gives the coordinates of the critical point of a van der Waals gas with constants a and b. In the next section we will show rigorously that a single supercritical fluid phase is stable at $T > T_c$, whereas two subcritical phases, a liquid and its vapor, can exist at equilibrium for $T < T_c$. At the temperature T_c the properties of the subcritical phases become identical, marking the critical phase transition. Before we demonstrate these conclusions formally it is important to focus on some of the similarities between the critical point of a fluid and those that we discussed in Chapter 7, for solutions and for order–disorder phase transitions in crystals.

We recall that the Helmholtz free energy, F, is a function of the natural variables T and V, such that (see Section 4.8.6):

$$\left(\frac{\partial F}{\partial V}\right)_T = -P. \tag{9.34}$$

Thus:

$$\left(\frac{\partial^2 F}{\partial V^2}\right)_T = -\left(\frac{\partial P}{\partial V}\right)_T = \frac{RT}{(V-b)^2} - \frac{2a}{V^3} \tag{9.35}$$

and:

$$\left(\frac{\partial^3 F}{\partial V^3}\right)_T = -\frac{2RT}{(V-b)^3} + \frac{6a}{V^4}. \tag{9.36}$$

Substituting the values of T_c and V_c we find that both the second and third derivatives of free energy vanish at the critical point. This is the same behavior that we found at the critical point of a solution, and at order–disorder critical phase transitions in general (Chapter 7). We recall that the order parameter is a quantity that vanishes at the critical point and takes non-zero values only for $T < T_c$. For example, the difference in the compositions of coexisting phases or the difference in the dimensions of the crystallographic axes. For a fluid we may take the difference in density, or molar volume, between the two subcritical phases, liquid and vapor, as the order parameter (see Fig. 7.8). Note also that, because the second derivative of free energy ($= -\partial P/\partial V$) vanishes at the critical point, so does the isothermal bulk modulus. Equivalently, the compressibility becomes infinite. Divergence of quantities such as compressibility and heat capacity (Fig. 7.9) is a characteristic of critical phenomena.

9.2.3 The phase diagram of a van der Waals fluid

To continue our study of real fluids it is best to recast van der Waals' EOS into non-dimensional form. We define the non-dimensional pressure, π, volume, ϕ, and temperature, τ, as the ratios P/P_c, V/V_c and T/T_c, respectively. The variables π, ϕ and τ are also known as reduced variables (do not confuse this ϕ with the fugacity coefficient that we defined in Section 9.1). With these coordinate transformations the van der Waals EOS becomes:

$$\pi = \frac{8\tau}{3\phi - 1} - \frac{3}{\phi^2}. \tag{9.37}$$

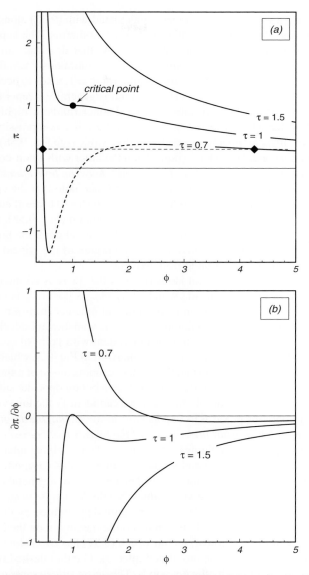

Fig. 9.5 Reduced pressure of a van der Waals gas (a) and its volume derivative (b), plotted as a function of reduced volume. For a given temperature below the critical temperature (e.g. $\tau = 0.7$) the width of the "prohibited" region in Fig. 9.4 corresponds to the dashed segment of the curve in (a). For any pressure between 0 and the local maximum of the isotherm there are two phases at equilibrium. An example is shown by the two diamonds joined by the thin dashed line.

Because the parameters a and b disappear in this equation, this would be a universal EOS if all fluids followed van der Waals behavior (i.e. equation (9.26)) exactly, but unfortunately this is not the case. We will need to develop more complex EOS to deal with real fluids in a quantitative fashion. In the meantime (9.37) is an excellent tool with which to gain physical insight.

Figure 9.5a shows $\pi = \pi(\phi)$ calculated with (9.37) along three isotherms, and Fig. 9.5b shows $(\partial\pi/\partial\phi)_\tau$ along the same three isotherms. The supercritical isotherm, $\tau = 1.5$, is reminiscent of the hyperbola $P \sim 1/V$ that describes an ideal gas. Its slope is always negative, implying that a single phase of continuously variable density is stable everywhere along the isotherm. The critical isotherm, $\tau = 1$, tends to become parallel to the supercritical isotherm at low densities (large ϕ) but is generally steeper than the latter at high densities. This hints at the existence of two distinct fluids: a highly incompressible high density fluid for which $(\partial\pi/\partial\phi)_\tau \to -\infty$ and a fairly compressible low density fluid for which $(\partial\pi/\partial\phi)_\tau$ approaches a small negative value (see Fig. 9.5b). One could identify the former with a condensed phase (liquid) and the latter with a non-condensed phase (gas), *but along the critical isotherm there is continuity between both states, because there is no first-order phase transition between them.* To see this, note that the critical point, $\tau = \pi = \phi = 1$, is the only point on this isotherm for which $(\partial\pi/\partial\phi)_\tau = 0$ and, since this point corresponds to a maximum of the first derivative function (Fig. 9.5b), the second derivative vanishes too, as we expected from (9.36). At the critical point the bulk modulus vanishes, but since it never becomes negative there is no region of the critical isotherm inside which a single fluid phase is prohibited from existing.

Along the subcritical isotherm, $\tau = 0.7$, there is an interval, shown by the broken segment in Fig. 9.5a, inside which $(\partial\pi/\partial\phi)_\tau$ is positive, as seen in the corresponding curve in Fig. 9.5b. This "prohibited" interval extends from a minimum to a maximum on the isotherm, as it must given that $(\partial\pi/\partial\phi)_\tau$ vanishes at both ends of the interval. Of course, it is possible to have an equilibrium system with $\{\tau, \pi, \phi\}$ coordinates inside this region, but the thermodynamic relations summarized in Fig. 9.5, which we will explore in more detail in a moment, mandate that this system cannot consist of a single phase. It must be made up of two phases that lie on the same isotherm but on opposite sides of the prohibited region, and at the same pressure. In this case we can identify the low-volume phase as a liquid and the high-volume phase as its vapor. *Note, however, that the diagrams in this figure by themselves cannot tell us what the volumes of the phases that coexist at equilibrium are.* In particular, these are not the volumes at the ends of the prohibited interval, first because they do not lie at the same pressure and second because, in this case, one of the volumes corresponds to the unphysical condition $\pi < 0$. The latter is not a general constrain, as isotherms closer to $\tau = 1$ never take negative π values, but the former is (as an aside, negative pressures are not altogether impossible, but their magnitude is limited by the cohesiveness of the material). There is a π interval, between 0 and the maximum on the isotherm located at the right end of the broken segment, within which (9.37) has two solutions in ϕ for each value of π, on opposite sides of the prohibited interval. The thin dashed line is an example, with the two solutions shown by the diamonds. *The phase rule assures us that one and only one of these solution pairs along each isotherm represents thermodynamic equilibrium, as we have a system of one component and two phases, and therefore one degree of freedom, which we have chosen to be the temperature.* The phase rule does not tell us, of course, which is the equilibrium pair. What we can be certain about, however, and what characterizes the subcritical region, is that the two phases must be separated by a first-order phase transition, because there is a discontinuity in volume and hence in enthalpy and entropy.

In order to find the volumes of coexisting liquid and vapor at equilibrium we begin by calculating the Helmholtz free energy of the fluid. This we do by integrating (9.34) and substituting the van der Waals EOS (equation (9.27)). The result is (check with *Maple*):

$$F = -\int_{V_0}^{V} P \, dV = -RT \ln (V - b) - \frac{a}{V} + \left[F_0 + RT \ln (V_0 - b) + \frac{a}{V_0} \right], \qquad (9.38)$$

where F_0, V_0 are the Helmholtz free energy and volume at some arbitrary reference state. Let us now define a non-dimensional Helmholtz free energy, $\Psi = F/RT_c$. Substituting (9.31) and (9.32) in (9.38) we find:

$$\Psi = -\tau \ln (3\phi - 1) - \frac{9}{8\phi} + \left[\Psi_0 + \tau \ln (3\phi_0 - 1) + \frac{9}{8\phi_0} \right]. \qquad (9.39)$$

We can, without loss of generality, set the constant term in square brackets equal to zero – this will move the curve of $\Psi(\phi)$ up or down, but will not change its geometric properties. We also find, using the chain rule and required substitutions:

$$\frac{\partial \Psi}{\partial \phi} = \frac{\partial \Psi}{\partial F} \frac{\partial F}{\partial V} \frac{\partial V}{\partial \phi} = -\frac{3}{8} \pi, \qquad (9.40)$$

which is the non-dimensional equivalent of (9.34).

The left-hand-side panels in Fig. 9.6 show $\Psi(\phi)$ calculated for three different values of τ. It is now important to recall that, whereas Gibbs free energy defines the equilibrium condition at constant temperature and pressure, Helmholtz free energy does the same thing at constant temperature and volume (Sections 4.8.5 and 4.8.6). This means that the equilibrium state of a system at constant temperature and volume is the one that minimizes Helmholtz free energy, F, or its non-dimensional avatar, Ψ. We see in Fig. 9.6 that, for $\tau > 1$, i.e. above the critical temperature, the Helmholtz free energy has continuously negative curvature, meaning that at any given volume there is only one possible stable phase. The situation is different below the critical temperature, $\tau < 1$. There are now two inflection points, and two points on the curve that have a common tangent. Because the tangent to the Helmholtz free energy curve is the negative of the pressure (equation (9.40)) the ϕ coordinates of these two points are the volumes of two different fluids at the same pressure. Moreover, for all volumes in between these two the Helmholtz free energy of a combination of these two fluids (i.e. a point on the common tangent) is lower than that of a single fluid with the same volume, implying that the single fluid is unstable relative to formation of liquid + vapor. The interval between ϕ_{liq} and ϕ_{vap} in the bottom left panel of Fig. 9.6 is thus the "prohibited" density interval. Outside of this interval Helmholtz free energy has negative curvature and a single phase is stable, a high-density liquid or a low-density gas.

The transition between the two different behaviors occurs at the critical temperature, $\tau = 1$. As the critical temperature is approached from below the common tangent points approach one another, and merge at a single point at $\tau = 1$. At this temperature the free energy curve becomes "flat" in the neighborhood of the coordinate $\phi = 1$ (the critical volume), reflecting the fact that its second and third derivatives vanish (equations (9.35) and (9.36)). A comparison of Fig. 9.6 with Fig. 7.1 should bring out the similarities between the critical point of a fluid and the critical mixing point of a solution, but there are also differences, most notably the fact that symmetry between the subcritical phases is never possible in a fluid.

The diagram on the right of Fig. 9.6 shows the phase relations of the fluid in terms of the reduced variables π, ϕ and τ, and completes the analogy with critical mixing (Fig. 7.1). The thick curve, called the liquid–vapor loop, bounds the "prohibited region". Calculation of the

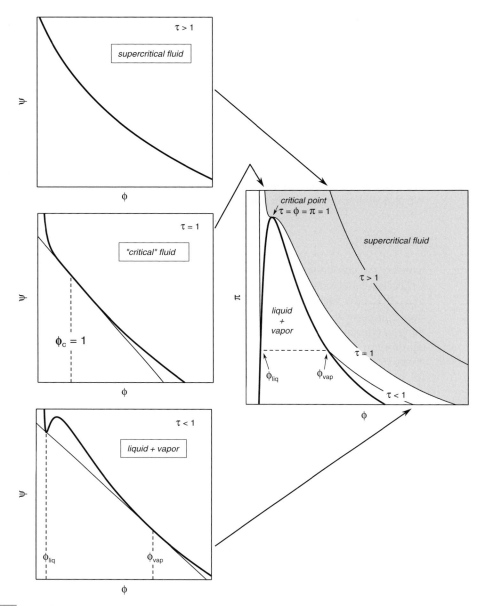

Fig. 9.6 Non-dimensional Helmholtz free energy as a function of non-dimensional volume (left panels, for the critical temperature and temperatures above and below the critical temperature), and the phase diagram of a van der Waals gas (right panel). Compare with Gibbs free energy diagram for a non-ideal solution in Fig. 7.1.

loop is explained in Box 9.1. Any π–ϕ combination inside the loop consists of liquid + vapor, with the equilibrium volumes given by the two intersections of the pressure coordinate, indicated by the dashed line, with the loop. The liquid–vapor loop corresponds to the solvus in Fig. 7.1. The supercritical fluid phase unmixes into two subcritical phases as temperature drops below T_c. The width of the loop, $\phi_{vap} - \phi_{liq}$, is the order parameter. The thin curves are isotherms, with the critical isotherm ($\tau = 1$) separating the supercritical region (shaded) from the subcritical region. The critical isotherm is tangent to the liquid–vapor loop at the critical point.

Box 9.1	Calculation of the liquid–vapor loop

The problem of calculating the liquid–vapor loop, e.g. in Fig. 9.7 and 9.8, consists of finding the volumes of the two subcritical phases at equilibrium. We seek to solve for three variables, the non-dimensional volumes of the liquid and its vapor, ϕ_{liq} and ϕ_{vap}, respectively, and the non-dimensional pressure, π. Hence, we need three independent equations, which in this case are non-linear. The first one is the equality of pressures between the two phases, $\pi_{liq} = \pi_{vap}$, which from (9.37) yields:

$$\frac{8\tau}{3\phi_{liq} - 1} - \frac{3}{\phi_{liq}^2} = \frac{8\tau}{3\phi_{vap} - 1} - \frac{3}{\phi_{vap}^2}. \tag{9.1.1}$$

The second equation is the equality of Gibbs free energy between the two phases. Note that the Helmholtz free energy of the two fluids at equilibrium is not the same – Fig. 9.6 makes this clear. The two fluids have different volumes, so Helmholtz free energy does not provide the correct equilibrium criterion. The two fluids coexist at equilibrium at constant temperature and pressure, however (Fig. 9.6, right panel), so it must be $G_{liq} = G_{vap}$. Now, applying the Legendre transform we find that:

$$G = F + PV \tag{9.1.2}$$

so the equilibrium condition at constant P and T is:

$$\Delta G = \Delta F + P\Delta V = 0 \tag{9.1.3}$$

which in non-dimensional form becomes:

$$\Delta \psi + \frac{3}{8}\pi \, \Delta \phi = 0 \tag{9.1.4}$$

and substituting (9.39):

$$-\tau \ln\left(\frac{3\phi_{liq} - 1}{3\phi_{vap} - 1}\right) - \frac{9}{8}\left(\frac{1}{\phi_{liq}} - \frac{1}{\phi_{vap}}\right) + \frac{3}{8}\pi\left(\phi_{liq} - \phi_{vap}\right) = 0. \tag{9.1.5}$$

The third equation is the equation of state, which we can apply to either the vapor or the liquid, e.g.:

$$\pi = \frac{8\tau}{3\phi_{vap} - 1} - \frac{3}{\phi_{vap}^2}. \tag{9.1.6}$$

At a given temperature equations (9.1.1), (9.1.5) and (9.1.6) are a system of three non-linear equations in the three unknowns ϕ_{liq}, ϕ_{vap}, and π. They are easily solved with a *Maple* procedure that you are asked to write in Exercise 9.5. Solving for the vapor–liquid loop with the dimensional van der Waals EOS, or any other EOS, is in principle no more complicated.

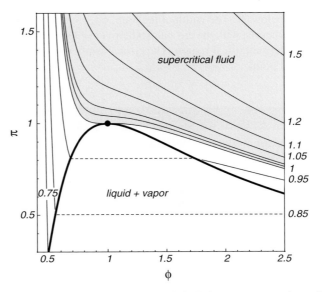

Fig. 9.7 Non-dimensional pressure–volume diagram for a van der Waals fluid. The contours are isotherms (τ given on the right edge of the diagram). The critical isotherm is tangent to the vapor–liquid loop at the critical point. In a supercritical fluid there is no discontinuous phase transition separating "low density fluids" (= gases) from "high density fluids" (= liquids). A discontinuous phase transition between the two phases always exists at temperatures below the critical temperature.

Close-ups of the neighborhood of the critical point are shown in Figs. 9.7–9.9. The first of these figures shows the pressure–volume projection, with temperature depicted by isotherms (the numbers on the right-hand side are the values of τ, the two missing values are 1.01 and 1.02). Figure 9.8 shows the temperature–volume projection, with pressure depicted by isobars (showing values of π), and Fig. 9.9 shows the pressure–temperature projection with contoured isochores (showing values of ϕ). The fat dot in all figures is the critical point. The univariant phase boundary in Fig. 9.9 maps the location of the liquid–vapor discontinuous phase transition, i.e. it tracks the P–T path of the liquid–vapor loop that appears in the other two projections. The density jump across the first-order phase transition decreases towards the critical point, and disappears at $\tau = \pi = 1$. Therefore, the univariant phase boundary necessarily terminates at the critical point, which is a singular point on the curve, *not* an invariant point (see Box 7.1). No discontinuous phase transition is possible beyond the termination of the univariant phase boundary, but note that the critical isochore ($\phi = 1$) is the continuation of the univariant phase boundary (Exercise 9.6).

The distinction between condensed and non-condensed fluids, expressed, for example, by their different densities and compressibilities, is discontinuous below the critical point. The density of the liquid varies relatively little along the vapor–liquid coexistence curve, but the density of the vapor decreases rapidly with decreasing pressure. Above, but in the neighborhood of, the critical point there is a sharp decrease in the compressibility as density increases, but no discontinuity between condensed and non-condensed phases. Where one chooses to place the boundary between condensed and non-condensed fluid is conventional (or unimportant): in the supercritical region, which is shaded in Fig. 9.7 and 9.8, there is no discontinuous phase transition between the two states. Supercritical fluid means that the

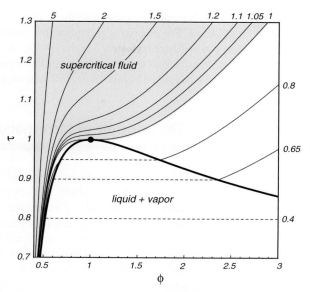

Fig. 9.8 Non-dimensional temperature–volume diagram for a van der Waals fluid. The contours are isobars (π given along the top and right edges). The critical isobar is tangent to the vapor–liquid loop at the critical point. Liquid–vapor unmixing at a discontinuous phase transition is possible only if the pressure is less than the critical pressure.

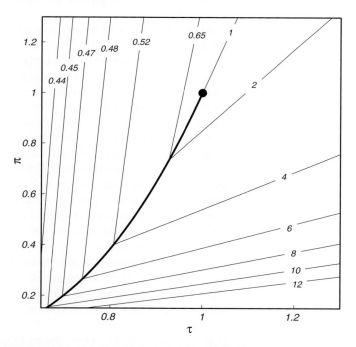

Fig. 9.9 Non-dimensional pressure–temperature diagram for a van der Waals fluid. The contours are isochores. The univariant phase boundary, i.e. the discontinuous phase transition mapped by the liquid–vapor loop in Figs. 9.7 and 9.8, ends at the critical point, which is *not* an invariant point (Box 7.1). The density jump across the discontinuous phase transition vanishes at the critical point, and the critical isochore is continuous with the univariant phase boundary.

fluid's properties change continuously from those of a gas to those of a liquid, but carries no density connotation. For example, given that the critical temperature and pressure for nitrogen and oxygen are approximately 126 K, 34 bar and 154 K, 50 bar, respectively, air cannot be made to condense at a discontinuous phase transition by compressing it at room temperature – it must be cooled below its critical temperature.

9.3 The principle of corresponding states

It is an empirical observation, which was put on solid footing largely by the work of Guggenheim (for example, 1967, pp. 135–140), that some gases follow the same behavior if their pressure, temperature and volume are expressed as the non-dimensional reduced quantities π, τ and ϕ. This is known as the principle of corresponding states. It suggests that a universal equation of state may exist, but it does not tell us what form the universal EOS has, nor, for that matter, whether an analytic function with the properties of a universal EOS even exists. We can quickly discard the non-dimensional van der Waals EOS, however. If gases followed this equation as a universal EOS then their critical compressibility factor, Z_c, would be (using (9.31)–(9.33)):

$$Z_c = \frac{P_c V_c}{R T_c} = \frac{3}{8}.$$

(9.41)

A compilation of critical parameters for eighteen gases of interest in planetary sciences (Table 9.1) shows that this is not the case, and that Z_c is in every case significantly less than 0.375. The gases in Table 9.1 are listed in order of decreasing Z_c. There is a group of gases for which Z_c is ~ 0.3. These gases are sometimes called simple gases and follow the principle of corresponding states more or less closely. As we shall see, a simple EOS exists which, if not completely accurate, is at least acceptable as a first order semi-quantitative approximation to their behavior. At the other extreme there are substances such as H_2O, NH_3, HCN and HF for which Z_c is much less than 0.3 and, more importantly, significantly different among the various gases. This suggests that a universal EOS for these substances does not exist. These gases do not obey the principle of corresponding states. What distinguishes these substances from the "simple gases" is that their molecules are strongly polar. In between there are some gases, most notably CO_2 and SO_2, which depart significantly from the behavior of simple gases but not as much as, say, H_2O. This emphasizes the fact, pointed out by Guggenheim, that there is no sharp boundary that one can draw between gases that follow the principle of corresponding states and those that don't.

One important consequence of the principle of corresponding states is that, if a sufficiently simple universal EOS can be found, then the adjustable parameters of the EOS can be calculated from the critical parameters of the gas. To exemplify this, let us assume for the sake of argument that the van der Waals equation is the universal EOS. We could then pick any two of the equations (9.31)–(9.33) and solve for a and b in terms of any two of the critical parameters. If the gas followed the van der Waals EOS exactly then we would get the same values for a and b regardless of whether we chose to solve in terms of $\{T_c, P_c\}$, $\{T_c, V_c\}$ or $\{P_c, V_c\}$, as the three critical parameters would be related by (9.41). For real gases this is not the case, however, as $Z_c \neq 0.375$. Because the pressure and temperature of the critical point are easier to measure accurately than the critical volume, it is customary

Table 9.1 Critical properties and Redlich–Kwong parameters of some planetary fluids

Species	T_c k	P_c bar	V_c cm^3 mol^{-1}	ρ_c kg m^{-3}	n_c m^{-3}	Z_c	a (R–K)	b (R–K)	ρ_{Moho} kg m^{-3}	ρ_{Moho}/ρ_c
Ne	44.4	27.6	42.00	519.0	1.4×10^{28}	0.3140	1.51×10^{6}	12.0		
H$_2$	33.0	12.9	65.00	30.8	9.3×10^{27}	0.3066	1.81×10^{6}	15.3		
He	5.2	2.3	57.00	70.2	1.1×10^{28}	0.2998	1.05×10^{6}	11.1		
CO	132.9	35.0	93.10	300.8	6.5×10^{27}	0.2948	1.72×10^{7}	27.4	809.1	2.7
Ar	150.9	49.0	75.00	532.6	8.0×10^{27}	0.2929	1.69×10^{7}	22.2		
N$_2$	126.2	34.0	89.20	313.9	6.7×10^{27}	0.2890	1.56×10^{7}	26.7		
O$_2$	154.6	50.5	73.40	436.0	8.2×10^{27}	0.2884	1.74×10^{7}	22.1		
CH$_4$	190.5	46.0	98.40	162.6	6.1×10^{27}	0.2856	3.22×10^{7}	29.9	432.4	2.7
F$_2$	144.1	51.7	66.00	575.8	9.1×10^{27}	0.2849	1.43×10^{7}	20.1		
H$_2$S	373.2	89.4	98.50	345.2	6.1×10^{27}	0.2838	8.89×10^{7}	30.1	925.1	2.7
Cl$_2$	416.9	79.9	123.00	576.4	4.9×10^{27}	0.2836	1.31×10^{8}	37.6		
CO$_2$	304.2	73.8	94.43	466.0	6.4×10^{27}	0.2756	6.46×10^{7}	29.7	1257.7	2.7
SO$_2$	430.8	78.8	122.00	524.6	4.9×10^{27}	0.2684	1.33×10^{8}	37.4	1390.0	2.6
HCl	324.6	83.1	81.00	450.6	7.4×10^{27}	0.2494	6.75×10^{7}	28.1	1047.3	2.3
NH$_3$	405.6	112.8	72.50	234.5	8.3×10^{27}	0.2425	8.68×10^{7}	25.9	525.0	2.2
H$_2$O	647.1	220.5	56.00	321.4	1.1×10^{28}	0.2295	8.80×10^{7}	14.6	944.1	2.9
HCN	456.7	53.9	139.00	194.2	4.3×10^{27}	0.1973	2.44×10^{8}	61.0	398.2	2.0
HF	461.0	64.8	69.00	289.9	8.7×10^{27}	0.1167	7.80×10^{7}	12.8	345.4	1.2

Critical temperature, pressure and density from Mathews (1972), checked for recent corrections against *NIST Chemistry WebBook*.

Critical volume and particle number calculated from critical density.

Z_c calculated from P_c, V_c and T_c (equation (9.41)).

Redlich–Kwong a and b parameters calculated from T_c and P_c (equation (9.50)), except values in italics taken from Holloway (1987, Table 1). Units are: bar cm^6 K$^{1/2}$ for a, cm^3 for b.

Estimated density at the Moho, ρ_{Moho}, calculated with R–K EOS at 950 K, 10.7 kbar.

to solve for a and b in terms of these variables and then we have, from (9.32) and (9.33):

$$a_{vdW} = \frac{27}{64} \frac{R^2 T_c^2}{P_c}, \quad b_{vdW} = \frac{RT_c}{8P_c}. \tag{9.42}$$

Equations (9.42) are of interest primarily for historical and didactic reasons, as the van de Waals EOS, even if qualitatively correct, does not provide an accurate quantitative representation of the behavior of any real gas. For any EOS with two adjustable parameters, however, the same methodology can be applied in order to derive the values of the parameters from measured values of T_c and P_c.

9.4 Equations of state for real fluids at P–T conditions typical of the crusts and upper mantles of the terrestrial planets

There are two basic approaches to constructing EOS applicable over a wide range of temperatures and pressures. The first one, exemplified by the van der Waals equation, is to start from the ideal gas EOS and add adjustable parameters that account for repulsion and attraction between molecules, and for the ways in which these forces vary with temperature and density. These are empirical equations that are cubic in volume and are thus called *cubic equations of state*. We will discuss two equations of this type, in addition to the van der Waals EOS: the Redlich–Kwong EOS (Redlich & Kwong, 1949) and a successful modification to this equation due to Kerrick and Jacobs (1981).

The other approach is philosophically more satisfying because it can be shown to have physical fundamentation, but it unfortunately results in an equation of state that performs very poorly and that can only be improved by empirical tweaks not unlike those used in cubic EOS. As we did for some of the equations of state for solids, we begin with an expression for the Helmholtz free energy of the fluid. In this case we consider two contributions to F, one corresponding to the Helmholtz free energy of the ideal gas and a second one, called the residual free energy, F_{res}, that encapsulates all of the energetic effects that arise from intermolecular interactions:

$$F = F_{ideal} + F_{res}. \tag{9.43}$$

We now write the residual term as a power series in density or, equivalently, in the inverse of volume:

$$F_{res} = RT \left(\frac{a_1}{V} + \frac{a_2}{2V^2} + \frac{a_3}{3V^3} + \cdots \right), \tag{9.44}$$

where the coefficients a_i, called the *virial* coefficients, are functions of temperature. The virial coefficients can be shown to represent the energetic effects arising from interactions between two molecules (a_1), three molecules (a_2) and so on (see, for example, Mason & Spurling, 1969). The name derives from the fact that the distribution of molecular energies in terms of kinetic and potential energy terms is described by the virial theorem that we discussed in Chapter 2.

Differentiating (9.43) we get:

$$P = -\left(\frac{\partial F}{\partial V} \right)_T = -\left(\frac{\partial F_{ideal}}{\partial V} \right)_T - \left(\frac{\partial F_{res}}{\partial V} \right)_T = P_{ideal} - \left(\frac{\partial F_{res}}{\partial V} \right) \tag{9.45}$$

and from (9.44) and the ideal gas EOS:

$$P = RT \left(\frac{1}{V} + \frac{a_1}{V^2} + \frac{a_2}{V^3} + \frac{a_3}{V^4} + \cdots \right) \qquad (9.46)$$

or, equivalently:

$$Z = 1 + \frac{a_1}{V} + \frac{a_2}{V^2} + \frac{a_3}{V^3} + \cdots. \qquad (9.47)$$

Equation (9.46), or (9.47), with $a_i = a_i(T)$, is known as the virial equation of state. Despite its theoretical foundation the equation is unsatisfactory. Many terms are required in order to represent the behavior of real gases at even moderate pressures, and then at high densities (small V) the series may diverge, i.e. terms in progressively higher powers of V become larger rather than smaller. Its one strong point is that it provides an explicit expression for the Helmholtz free energy (equation (9.44)), which is convenient when calculating other thermodynamic functions such as fugacity (Section 9.5.1). We will discuss an equation of state (the Pitzer–Sterner EOS) which is constructed following the same idea as the virial EOS, i.e. beginning from an empirical function for the Helmholtz free energy, but which is not a virial equation because the function is not a power series in V, and is therefore not a physical representation of intermolecular potentials. Finally, we will discuss another empirical EOS (Brodholt–Wood EOS) that combines a cubic EOS with virial-like terms.

9.4.1 Cubic equations: the van der Waals EOS revisited

We begin our discussion with the van der Waals (VDW) EOS, with the sole purpose of understanding why it fails and what can be done about it. We will compare the predictions of VDW and three other EOS with the measured density of H_2O at pressures of 0.1–10 kbar. Figure 9.10 shows compressibility as a function of reduced density, $\rho/\rho_c = V_c/V$. From the definition of compressibility (equation 9.22) we also get $Z = V/V_{ideal}$. We can thus interpret the vertical coordinate either as the ratio of the volume of the gas to that of an ideal gas at the same pressure, or as the ratio of the pressure acting on a given volume of real gas to the pressure that would be required to take an ideal gas to the same volume. In any case an ideal gas would plot in this figure as a horizontal line at $Z = 1$.

Molar volumes of H_2O were measured by Burnham et al. (1969) to 8.9 kbar, and extrapolated by them to 10 kbar. The symbols in Fig. 9.10 represent the density of H_2O from 0.1–10 kbar along two isotherms, circles for the critical isotherm ($T_c = 647.14$ K) and triangles for 700°C, which corresponds closely to $1.5T_c$. Focusing first on the measured behavior of H_2O we see that at low density it is more compressible than an ideal gas ($Z < 1$), and that it becomes less compressible than an ideal gas as density increases. This is the behavior that we should expect from a substance with strongly polar molecules: attraction predominates at low density, but, as the molecules are squeezed more closely together, repulsion (or the finite size of molecules) takes over. Attraction becomes less important at high temperature, as thermal agitation tends to swamp intermolecular potentials, explaining why the data at $1.5T_c$ plot at higher Z than data at the critical temperature.

At low densities, generally much lower than the critical density, the VDW EOS is moderately successful, but as density increases it fails spectacularly (Fig. 9.10). This behavior arises because the repulsive term is a constant. As density increases and V approaches b the

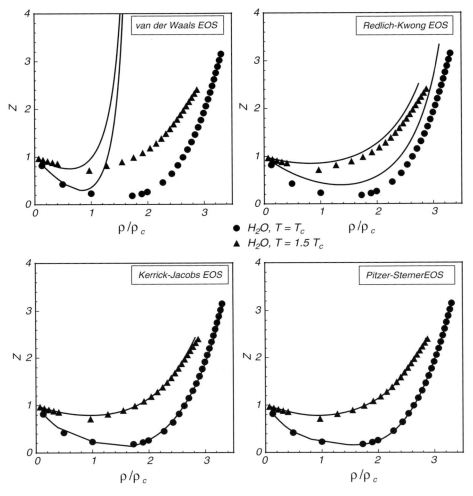

Fig. 9.10 Comparison of four equations of state for H_2O against measured densities at 1 and $\sim$1.5 T_C (experimental data from Burnham *et al.*, 1969).

pressure required to accomplish further compression diverges (equation (9.27)). An important reason for the inadequacy of the VDW EOS is its failure to take into account the fact that molecules are not rigid.

9.4.2 Cubic equations: the Redlich–Kwong EOS

The first successful modification to the VDW EOS was proposed by Redlich and Kwong in 1949. In this equation the repulsive term (equation (9.24)) is left unchanged, but the attractive term (equation (9.25)) is modified as follows:

$$\Delta Z_{attraction} = \frac{a}{RTV} \cdot \frac{1}{T^{1/2}} \cdot \frac{1}{1 + \frac{b}{V}}. \qquad (9.48)$$

The additional factors accomplish two things. The first factor causes the attractive force to fall off faster with increasing temperature than in the VDW equation. The second factor makes the attractive force less dependent on density. Redlich and Kwong stated that there is no theoretical justification for these particular correction terms. Rather, they just happen to vastly improve the agreement between the EOS and measured densities for many fluids, especially the simple gases discussed in Section 9.3. As in the VDW equation, a and b in (9.48) are constants with specific values for each gas. Substituting (9.24) and (9.48) in (9.23) we get the Redlich–Kwong (RK) EOS:

$$P = \frac{RT}{V - b} - \frac{a}{T^{1/2} V (V + b)}. \tag{9.49}$$

Because this equation has only two adjustable parameters it is possible to derive their values from the critical properties of the gas, by using the condition that the second and third derivatives of the Helmholtz free energy vanish at the critical point. Applying these conditions to (9.49) we get:

$$a_{RK} = 0.427\,480\,233\,7 \frac{R^2 T_c^{5/2}}{P_c}, \quad b_{RK} = 0.086\,640\,349\,97 \frac{RT_c}{P_c} \tag{9.50}$$

and $Z_c = 1/3$. This value for the critical compressibility factor is closer to the value of ~ 0.3 measured for the simple gases in Table 9.1. One could then expect that the RK EOS with parameters calculated from (9.50) may do a reasonable job of representing the properties of these gases, at least up to moderate pressures. For substances that depart significantly from corresponding state behavior, however, it is better to obtain the values of the a and b parameters by fitting the equation to P–V–T measurements spanning as wide a range of conditions as possible. The parameters for H_2O in Table 9.1 were obtained in this way (data from Holloway, 1987).

The RK EOS does not solve the problem of the divergence of pressure at a finite volume that is a characteristic of the VDW EOS. However, comparing (9.50) with (9.42) we see that the value of the b parameter in the RK EOS is about half of that in the VDW EOS. This means that the useful range of the RK EOS can be expected to extend to higher densities than the VDW EOS. This is borne out by a comparison of its predictions with the measured density of H_2O up to 10 kbar (Fig. 9.10). The agreement is far from perfect, and generally inadequate for accurate thermodynamic calculations, but is much better than for the VDW EOS. The fit improves with increasing temperature. However, at the highest densities shown in the diagram the tendency for pressure to diverge clearly insinuates itself. This is a problem with the EOS, not with H_2O in particular, which means that its application to simple gases is also limited to moderate pressures, such that the molar volume remains significantly greater than b.

9.4.3 Cubic equations: the Kerrick–Jacobs modified Redlich–Kwong EOS

Since Redlich and Kwong's proposal of equation (9.49) many modifications have been proposed in order to overcome the limitations of that equation. All of these EOS come under the general label of modified Redlich–Kwong (MRK) EOS, and they are all empirical fits of varied complexity. Among the most successful ones for conditions in planetary interiors is the MRK equation proposed by Kerrick and Jacobs in 1981. This equation leaves the attractive term in the RK EOS unchanged but: (1) it makes the parameter a a function of temperature and volume (and hence pressure) and (2) it modifies substantially the repulsive term. The Kerrick and Jacobs (KJ) EOS is:

$$P = \frac{RT\left(1 + y + y^2 - y^3\right)}{V\left(1 - y\right)^3} - \frac{a\left(V, T\right)}{T^{1/2} V\left(V + b\right)} \tag{9.51}$$

where:

$$y = \frac{b}{4V} \tag{9.52}$$

and:

$$a = a_1\left(T\right) + \frac{a_2\left(T\right)}{V} + \frac{a_3\left(T\right)}{V^2} \tag{9.53}$$

and:

$$a_i\left(T\right) = c_{i,1} + c_{i,2}T + c_{i,3}T^2. \tag{9.54}$$

The KJ EOS has ten adjustable parameters, which obviously makes it much easier for it to reproduce measured volumes over a wide pressure–temperature range. Figure 9.10 shows that, up to ~ 10 kbar, it reproduces the measured behavior of H_2O almost perfectly, although at $\tau = 1.5$ there is a hint of the KJ EOS beginning to show divergence in the pressure, as V approaches $b/4$.

In contrast to two-parameter equations, it is not possible to estimate the values of the adjustable parameters of the KJ EOS from the critical properties. The equation can only be calibrated by fitting it to measured volumes over a pressure–temperature range. Values for b and the nine $c_{i,j}$ parameters are given for H_2O and CO_2 by Kerrick and Jacobs (1981) and for CH_4 by Jacobs and Kerrick (1981). Because the KJ EOS is an empirical equation its performance beyond the range of conditions used to fit the parameters (~ 0–10 kbar) is uncertain, and application to such conditions must be done with caution.

9.4.4 Expansion of the residual Helmholtz free energy: the Pitzer–Sterner EOS

Pitzer and Sterner (1994) proposed an equation of state based on an empirical formulation for the residual Helmholtz free energy. Their expression is most easily written in terms of density, $\rho = 1/V$:

$$\begin{aligned} \frac{F_{res}}{RT} = a_1\rho &+ \left(\frac{1}{a_2 + a_3\rho + a_4\rho^2 + a_5\rho^3 + a_6\rho^4} - \frac{1}{a_2}\right) \\ &- \left(\frac{a_7}{a_8}\right)\left(e^{-a_8\rho} - 1\right) - \left(\frac{a_9}{a_{10}}\right)\left(e^{-a_{10}\rho} - 1\right) \end{aligned} \tag{9.55}$$

where each of the ten adjustable parameters a_i is a function of temperature parametrized by the following polynomial:

$$a_i(T) = c_{i,1}T^{-4} + c_{i,2}T^{-2} + c_{i,3}T^{-1} + c_{i,4} + c_{i,5}T + c_{i,6}T^2. \tag{9.56}$$

The equation has 60 adjustable parameters, although many of them are found to be zero. Using (9.45) the Pitzer–Sterner (PS) EOS is:

$$\frac{P}{RT} = \rho + a_1\rho^2 - \rho^2\left(\frac{a_3 + 2a_4\rho + 3a_5\rho^2 + 4a_6\rho^3}{(a_2 + a_3\rho + a_4\rho^2 + a_5\rho^3 + a_6\rho^4)^2}\right)$$
$$+ a_7\rho^2 e^{-a_8\rho} + a_9\rho^2 e^{-a_{10}\rho}. \tag{9.57}$$

The equation has no physical justification but is found to reproduce the volumetric properties of H_2O and CO_2 to very high pressure with a remarkable degree of accuracy. Figure 9.10 shows that the agreement with the measured properties of H_2O to 10 kbar is essentially perfect, but the agreement is also excellent to much higher pressures (of order 10^2–10^3 kbar), for which scattered volume measurements obtained by a variety of experimental methods are available. The form of equation (9.57) keeps the EOS from blowing up at high densities, as happens with cubic EOS when the molar volume approaches the value of the excluded volume, b. The equation is calibrated for H_2O and CO_2 only. The coefficients for the two gases are given by Pitzer and Sterner (1994). Given the large number of adjustable parameters one would expect that the PS EOS would also work well for other fluid species, but as of this writing, and to the best of my knowledge, no attempt has been made to perform the necessary calibrations. It would be interesting to know whether the PS EOS can be applied successfully to other important fluid species such as CH_4, CO, H_2S and NH_3.

9.4.5 The Brodholt–Wood EOS for H_2O at high pressure

Brodholt and Wood (1993) performed molecular dynamics simulations of the properties of H_2O to 2500 K and 350 kbar. They then fit the following EOS to the results of their simulations:

$$P = \frac{RT}{V-b} - \frac{a}{T^{1/2}V(V+b)} + \frac{c}{V} + \frac{d}{V^2} + \frac{e}{V^3} + \frac{f}{V^4} \tag{9.58}$$

with:

$$a = a_0 + a_1T + a_2T^2 + a_3T^{-2} \tag{9.59}$$

and:

$$b = b_0 + b_1V < 0 \tag{9.60}$$

and c, d, e and f constants (values of all coefficients are given by Brodholt & Wood, 1993). The Brodholt–Wood (BW) EOS has been described as an MRK equation with virial terms, but I would argue that this is not correct. In the first place, the b term is always negative, as both b_0 and b_1 are negative constants. The physical meaning of the b parameter is thus not the excluded volume. As a consequence the first term in (9.58) is not a repulsive term, which is the intention in cubic equations of state such as MRK EOS. Second, the terms in inverse powers of V may look like virial terms, but rigorously they are not, as virial coefficients are defined on the basis of a series expansion of residual Helmholtz free energy (equation (9.44)), which is not the case in equation (9.58). The BW EOS is an empirical equation that fits the results of molecular dynamics calculations very well, and also reproduces isolated data points for H_2O volumes at extreme pressures (350 kbar), but it has no obvious physical justification. Its one drawback is that it does not work below 10 kbar, which is a minor (but not fatal) inconvenience in the calculation of thermodynamic functions such as fugacity. It is not known whether an equation of this type works for other fluids too. EOS based on molecular dynamics simulations for H_2O as well as for other species of geological interest have also been proposed by Duan and co-workers (see, for example, Duan *et al.*, 1992, 1996, 2000; Zhang & Duan, 2009).

9.4.6 Summary

The preceding list does not come close to being an exhaustive compilation of available equations of state. I have chosen to discuss these particular EOS either because they are important from a historical point of view (VDW), because they have been shown to perform adequately up to temperatures and pressures corresponding to the Earth's upper mantle (KJ) or the Earth's mantle transition zone (PS and BW), or because they are a reasonable (or only) "stopgap" EOS for some species, that can at least provide semi-quantitative predictions (RK). Figure 9.11 compares four of these EOS for H_2O and three of them for CO_2 (the BW EOS is not calibrated for CO_2) from 1 bar to 300 kbar, and at 1000 K and 2000 K. The RK and KJ EOS are not expected to yield accurate results over most of this pressure range but are included for comparison. Both of them predict strongly incompressible fluids. This arises from the repulsive term in cubic EOS, which becomes unphysical at high density. The BW and PS EOS avoid this problem and are in generally good agreement with each other and with high pressure experimental data (see Brodholt & Wood, 1993 and Pitzer & Sterner, 1994). The PS EOS has two advantages, however: (1) it works continuously from the ideal gas limit to very high pressure, whereas the BW EOS breaks down below 10 kbar, and (2) it is calibrated for both H_2O and CO_2.

The plots in Fig. 9.11 suggest that at densities of up to $\sim$2.5–3 times the critical density the departure of the RK EOS from the PS EOS is rather small. One could then, tentatively and rather carefully, suggest that the RK EOS might be used for other fluid species too at conditions such that $\rho < 3\rho_c$. For those gases listed in Table 9.1 that may conceivably exist as free species in planetary interiors this possible upper validity bound for the RK EOS corresponds approximately with the P–T conditions of the base of the Earth's continental crust. The two last columns in the table show densities calculated at 10.7 kbar and 950 K with the RK EOS, and the ratio of these densities to the corresponding critical densities. The latter values generally fall in the region where the RK EOS for H_2O and CO_2 begin to depart significantly from the corresponding PS EOS. The RK EOS is by no means accurate, but it may provide a reasonable first-order approximation to the behavior of fluids at $P < 10$ kbar in those cases in which other EOS are not available.

The densities listed in Table 9.1 offer a glimpse into the nature of deep planetary volatiles. These are dense supercritical fluids. The density of supercritical H_2O at the base of the Earth's continental crust is approximately the same as that of liquid H_2O on the Earth's

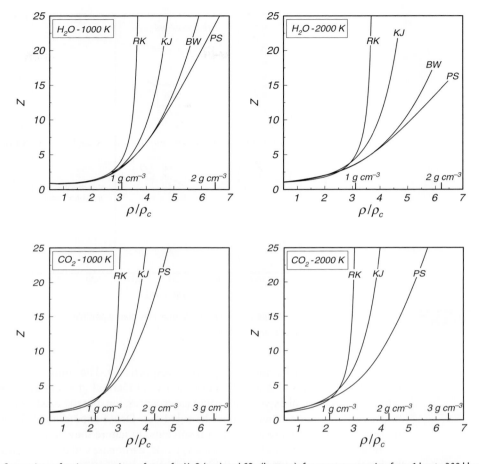

Fig. 9.11 Comparison of various equations of state for H_2O (top) and CO_2 (bottom), for pressures ranging from 1 bar to 300 kbar (coordinate axes as in Fig. 9.10, actual fluid densities added for comparison). The RK EOS is not accurate above a few kbar, and the KJ EOS may be valid to a few tens of kbar. Both the PS and BW EOS are supposed to be valid to ~300 kbar, but at densities above ~5 times the critical density they predict significantly different densities for H_2O (the BW EOS is calibrated for H_2O only).

surface. Supercritical CO_2 at those conditions is even denser. A more complete view is presented in Fig. 9.12, in which I have plotted the densities of H_2O and CO_2 along a possible terrestrial continental geotherm, also shown in the figure. The conductive portion of the geotherm is constructed by fixing the temperature at 300 K at the surface, 950 K at the Moho (40 km depth) and 1650 K at the base of the lithosphere (150 km), with pressure calculated as in Worked Example 8.3. This is physically unrealistic because it causes a discontinuity in thermal gradient at the Moho, but the curve as a whole approximately mimics the curved steady-state geotherm of a crust with radioactive heat production (see Turcotte & Schubert, 2002, Chapter 4). The adiabatic geotherm at depths greater than 150 km was calculated with (8.70). The densities of H_2O and CO_2 were calculated with the PS EOS which is thought to be reliable at least to pressures of the base of the mantle transition zone (~240 kbar, Fig. 8.11). The densities of H_2O and CO_2 at those conditions

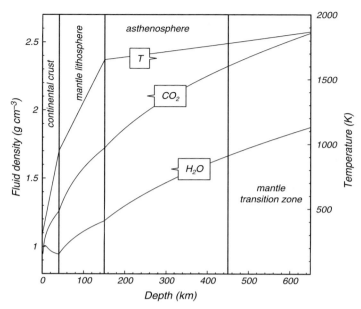

Fig. 9.12 Densities of H_2O and CO_2 calculated with the Pitzer–Sterner EOS along a possible terrestrial geotherm, from the surface to the top of the Earth's lower mantle.

are approximately 1.85 and 2.55 g cm^{-3}, respectively. The minimum in H_2O density at the Moho may simply be an artifact of the strong thermal gradient assumed for the crust, which causes thermal expansion to outpace compression, but the existence of this minimum under certain circumstances cannot be ruled out. Dense supercritical fluids such as those represented in this figure are powerful solvents. In nature they do not exist as pure fluids, but rather as complex solutions of silicate, oxide, carbonate and other species, with densities higher than those shown in Fig. 9.12.

9.5 Calculation of fugacity in fluid phases

In Section 9.1 we defined fugacity and we saw that it is straightforward to calculate the fugacity of a fluid species in equilibrium with a buffering solid assemblage. Those values by themselves do not tell us much about the nature of a putative fluid in equilibrium with the solids. They are best understood as a measure of the chemical potential of the fluid species relative to its 1 bar standard state. A different problem is that of calculating the fugacities of chemical species in a fluid phase, and the distribution of chemical species in the fluid. We now address these questions.

9.5.1 Standard state fugacity of pure fluids

Standard state fugacity is calculated by substituting an equation of state in (9.11). Because equations of state are generally pressure explicit we cannot integrate directly, so we follow the same procedure that we used to calculate the pressure integral for solid phases (equation (8.45)). Integrating (9.11) by parts we get:

$$RT \ln f^0 = PV_{P,T} - V_{1,T} - \int_{V_{1,T}}^{V_{P,T}} P dV. \tag{9.61}$$

Note that, in contrast to the integral for solid phases, we cannot drop the $V_{1,T}$ term (which is multiplied by 1 bar) because for a gas the difference between 1 bar and zero pressure is not negligible. Once a particular EOS is chosen the integral is easily implemented in *Maple* (Software Box 9.1). The procedure entails, first, solving the EOS for $V_{P,T}$ and $V_{1,T}$ and then using these values to calculate the definite integral. We can also make the substitution, valid at the ideal gas limit, $V_{1,T} \approx RT$; the result is numerically indistinguishable from calculating $V_{1,T}$ explicitly with the EOS. If the EOS is written in terms of density, ρ, rather than molar volume then 9.61 is easily converted to:

$$RT \ln f^0 = \frac{P}{\rho_{P,T}} - \frac{1}{\rho_{1,T}} + \int_{\rho_{1,T}}^{\rho_{P,T}} \frac{P}{\rho^2} d\rho. \tag{9.62}$$

Software Box 9.1   Calculation of standard state fugacities

The package **fluideos.mw** contains procedures that calculate pressure, volume and standard state fugacity with the equations of state discussed in Section 9.4. Pressure is calculated directly, volume is calculated by invoking *Maple*'s numerical solver, fsolve, to solve the corresponding EOS for V, and standard state fugacity is calculated either by performing the integral (9.61), or, for the Pitzer–Sterner EOS, with (9.62) and (9.63). The units that must be used in every case are $cm^3 mol^{-1}$ for molar volume, Kelvin for temperature and bar for pressure and fugacity. Conversion to or from other units must be handled by the calling procedure. Particular care must be exercised with the volume units, which are different from the ones used in other thermodynamic calculations in this book (i.e. $J bar^{-1} mol^{-1}$). The reason for this is that use of $cm^3 mol^{-1}$ is deeply ingrained in the literature of thermodynamics of fluids, and the parameters for the various EOS are almost always given in these units.

There are slight differences in the way each procedure must be invoked, which are explained as comments in the *Maple* code. Some additional notes follow.

For the Van der Waals EOS there are procedures to calculate pressure and volume only (no fugacity calculation). The values of the a and b parameters must be provided by the calling procedure.

For the Redlich–Kwong EOS the calling procedure must provide the values of the a and b parameters and, for calculation of volume and fugacity, an initial volume guess. This is required in order to force the solution to the liquid or gas branch of the loop. The actual value of the volume guess varies with each gas species and with the specific P–T conditions. Some trial and error may be necessary (an inappropriate volume guess will generate an error message). The *Maple* code listing includes some suggestions.

Two sets of procedures are provided for the Kerrick and Jacobs EOS, for H_2O and CO_2. The EOS parameters are included in the package, so they are not passed by the calling procedure. An initial volume guess is required for calculation of volume and fugacity (see above).

Software Box 9.1   Continued

The procedures for the Brodholt–Wood EOS (for water only) should only be used for $P > 10$ kbar, as the equation is not valid at lower pressures. Standard state fugacity is calculated by adding the Kerrick–Jacobs fugacity at 10 kbar to the Brodholt–Wood integral from 10 kbar to the pressure of interest. The EOS parameters are included in the package, so they are not passed by the calling procedure. The initial volume guess is included in the procedure and is not passed by the calling procedure either.

Two sets of procedures are provided for the Pitzer–Sterber EOS, for H_2O and CO_2. The EOS parameters are included in the package, so they are not passed by the calling procedure. Pressure is calculated as a function of density, not volume, but this does not affect the procedures that calculate volume and standard state fugacity. An initial volume guess is required for calculation of volume and fugacity (see above).

An accompanying *Maple* worksheet named **fluidtest.mw** includes examples of the use of the procedures in **fluideos.mw**.

If an explicit expression for Helmholtz free energy is available (as in the case of the PS EOS) then a further simplification is possible, as the PdV integral can be evaluated as follows:

$$\int_{\rho_{1,T}}^{\rho_{P,T}} \frac{P}{\rho^2} d\rho = -\int_{V_{1,T}}^{V_{P,T}} P\,dV = \int_{F_{1,T}^0}^{F_{P,T}^0} dF = F_{P,T}^0 - F_{1,T}^0$$

$$= \left(F_{P,T}^0 - F_{1,T}^0 \right)_{ideal} + \left(F_{P,T}^0 - F_{1,T}^0 \right)_{res} \tag{9.63}$$

$$= RT \ln \frac{\rho_{P,T}}{\rho_{1,T}} + \left(F_{P,T}^0 - F_{1,T}^0 \right)_{res}.$$

At the ideal gas limit we have $\rho_{1,T} = 1/RT$ and $(F_{1,T}^0)_{res} = 0$, so this expression simplifies to:

$$-\int_{V_{1,T}}^{V_{P,T}} P\,dV = RT \ln \frac{\rho_{P,T}}{RT} + \left(F_{P,T}^0 \right)_{res} \tag{9.64}$$

and, substituting in (9.62):

$$\ln f^0 = \frac{P}{RT\rho_{P,T}} + \ln \rho_{P,T} + \frac{\left(F_{P,T}^0 \right)_{res}}{RT} + \ln RT - 1, \tag{9.65}$$

where $(F_{P,T}^0)_{res}$ is given by an equation such as (9.55).

Other procedures to calculate standard state fugacities, which rely on calculating the fugacity coefficient rather than the fugacity itself, are commonly found in the literature. I believe that such procedures are largely a product of the time before easily accessible computers, because their chief advantage is that they facilitate graphic integration. I find them to be unnecessarily confusing, as implementing (9.61) or (9.65) in a symbolic algebra software package is straightforward.

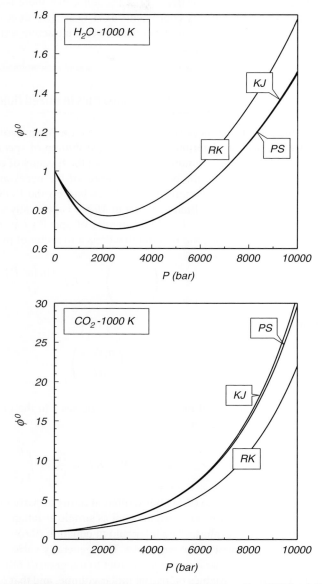

Fig. 9.13   Standard state fugacity coefficients for H_2O and CO_2 as a function of pressure at $T = 1000$ K, calculated with the Redlich–Kwong, Kerrick–Jacobs and Pitzer–Sterner EOS.

Figure 9.13 shows standard state fugacity coefficients for H_2O and CO_2 at 1000 K, calculated from 1 bar to 10 kbar with the RK, KJ and PS EOS. Over this pressure range the results for the last two EOS are for all practical purposes identical. The RK EOS predicts noticeably different fugacities beginning at pressures of $\sim$1 kbar, and, relative to the other two equations, its performance is generally better for H_2O than for CO_2. Note that fugacity coefficients are about one order of magnitude greater for CO_2 than for H_2O, so that the thermodynamic effective pressure of CO_2 is many times higher than its actual pressure. The two gases also behave differently in the sense that there is a pressure range, below about 7 kbar, in which fugacity coefficients of H_2O are less than 1, but this is never the case for CO_2. This difference arises from the stronger attractive term of H_2O, which overwhelms repulsion at low densities.

9.5.2 Fugacities in mixed fluids

Calculation of chemical equilibrium between a mixed volatile phase and solid or melt phases, as well as determination of the distribution of species in a homogeneous volatile phase at equilibrium, requires knowledge of the fugacity of each of the individual chemical species that make up the phase. Derivation of the necessary equations is rather tedious, and can become quite cumbersome depending on the EOS that one chooses to use, but implementation of the final equation in *Maple* is generally straightforward. We begin with equation (9.4), for the chemical potential of species i in a mixed fluid. Differentiating relative to P, and keeping in mind that ϕ^i is a function of pressure but X_i is not, we get:

$$\left(\frac{\partial \mu^i_{P,T}}{\partial P}\right)_T = RT \frac{\partial \ln\left(\phi^i P\right)}{\partial P}. \tag{9.66}$$

But:

$$\left(\frac{\partial \mu^i_{P,T}}{\partial P}\right)_T = v^i \tag{9.67}$$

where v^i is the partial molar volume of species i in the mixture (equation (5.37)). From (9.66) and (9.67) we get:

$$RT \ln \phi^i P = RT \ln \frac{f_i}{X_i} = \int_1^P v^i dP, \tag{9.68}$$

which is equivalent to (9.11) but written in terms of partial molar properties. The problem is that integrating this expression is not as simple as integrating (9.11). In that case V is the volume of the gas phase. The integrand in equation (9.68) is the partial molar volume of a specific component of the gas phase. This generally varies with pressure, temperature and composition, so what is required in order to integrate (9.68) is an EOS that accounts for the effects of these variables on partial molar volume, and that can do so continuously over the integration interval $[1, P]$.

As usual when working with partial molar properties, the simplest way to include the effect of phase composition is by converting to extensive properties. From the definition of partial molar properties, equation (5.28), we write:

$$v^i = \left(\frac{\partial V^m}{\partial n_i}\right)_{P,T,n_{j\neq i}} \tag{9.69}$$

where V^m is the (extensive) volume of the gas mixture. Now, from partial derivative identities (equation 1.3.19) we find that:

$$\left(\frac{\partial V^m}{\partial n_i}\right)_{P,T,n_{j\neq i}} = -\left(\frac{\partial V^m}{\partial P}\right)_{T,n_i}\left(\frac{\partial P}{\partial n_i}\right)_{V,T,n_{j\neq i}} \tag{9.70}$$

which allows us to change the variable of integration in (9.68), as follows:

$$RT\ln\phi^i P = -\int_{V_{1,T}^m}^{V_{P,T}^m}\left(\frac{\partial P}{\partial n_i}\right)_{V,T,n_{j\neq i}} dV^m. \tag{9.71}$$

This equation is general and rigorously complete , but in order to evaluate the integral we need an explicit expression for $(\partial P/\partial n_i)$, and no general expression of such type exists, as it is obviously dependent on a particular EOS. Therefore, a specific expansion of the derivative must be found for each particular equation of state. In the following example I illustrate the procedure with the Redlich–Kwong EOS. The procedure for other EOS is analogous, but can be more time consuming and fill many more pages.

Worked Example 9.5   Species fugacities in a mixed Redlich–Kwong fluid

We begin by re-writing the Redlich–Kwong EOS, (9.49), in terms of the (extensive) total volume of the gas mixture $V^m = nV$, where V is the molar volume of the mixture, and $n = \sum n_i$ is the total number of mols in the mixture:

$$P = \frac{nRT}{V^m - nb} - \frac{n^2 a}{T^{1/2}V^m(V^m + nb)}. \tag{9.72}$$

The RK parameters a and b in this case describe the mixed gas phase, not a specific species. As we shall see, determining reliable values for the mixture parameters is the main difficulty in calculating fugacities in mixed fluids. Using (9.72) we calculate the integrand in (9.71), which is:

$$\left(\frac{\partial P}{\partial n_i}\right)_{V,T,n_{j\neq i}} = \frac{RT(V^m - nb) + nRT\left(\frac{\partial(nb)}{\partial n_i}\right)}{(V^m - nb)^2}$$

$$- \frac{1}{T^{1/2}V^m}\left[\frac{(V^m + nb)\left(\frac{\partial(n^2 a)}{\partial n_i}\right) - n^2 a\left(\frac{\partial(nb)}{\partial n_i}\right)}{(V^m + nb)^2}\right]. \tag{9.73}$$

This expression allows us to evaluate the definite integral in (9.71). Below I present the final result, recast in terms of the molar volume of the mixture, $V = V^m/n$, and including several simplifications that arise from the fact that, at the lower limit of integration (ideal gas at 1 bar), the following approximations are true to an excellent degree of accuracy:

$$V_{1,T} + b \approx V_{1,T} - b \approx V_{1,T} \approx RT, \quad \frac{1}{V_{1,T}} \approx 0. \tag{9.74}$$

The definite integral in (8.71) is, then:

$$RT \ln \phi^i P = RT \ln \left(\frac{RT}{V_{P,T} - b} \right) + \frac{RT \left(\frac{\partial (nb)}{\partial n_i} \right)}{V_{P,T} - b}$$

$$+ \frac{\left(\frac{\partial (n^2 a)}{\partial n_i} \right)}{nbT^{1/2}} \ln \left(\frac{V_{P,T}}{V_{P,T} + b} \right) \tag{9.75}$$

$$- \frac{a \left(\frac{\partial (nb)}{\partial n_i} \right)}{b^2 T^{1/2}} \ln \left(\frac{V_{P,T}}{V_{P,T} + b} \right) - \frac{a \left(\frac{\partial (nb)}{\partial n_i} \right)}{bT^{1/2} \left(V_{P,T} + b \right)}.$$

Equation (9.75) is specific to the RK EOS, and it still contains derivatives for which we must find explicit values. These values depend on how the parameters a_i and b_i for each of the individual species in the gas mixture combine to yield the a and b parameters for the mixture. From a physical point of view the question that we are asking is how attraction and repulsion among molecules of different gases depend on the properties and concentration of each of the gases in the mixture. This is akin to finding an expression for excess mixing properties in a crystalline solution (e.g. Section 5.9.3). For gases the problem is often expressed algebraically by means of *mixing rules*, which are functions of the form $a = f(a_i, n_i)$ and $b = f(b_i, n_i)$. Many mixing rules have been proposed, with greater or lesser amounts of theoretical justification and experimental verification. In general, microscopic mixing rules that describe interatomic potentials tend to be the ones with greater theoretical justification, but they can only be used with EOS constructed on the basis of molecular dynamics simulations, such as those of Brodholt and Wood and Duan and collaborators. Mixing rules for macroscopic EOS are largely empirical. For two-parameter cubic EOS such as the RK EOS the following mixing rules, due originally to van der Waals, are widely used and reasonably successful:

$$b = \sum_i X_i b_i \tag{9.76}$$

$$a = \sum_i \sum_j X_i X_j \left(a_i a_j \right)^{1/2}. \tag{9.77}$$

The intuitive justification for these mixing rules is easy to see. Equation (9.76) states that the excluded volume of the mixture is the weighted average of the excluded volumes of the individual components, whereas (9.77) states that the attractive term is the weighted average of attractions over all types of molecular pairs. Both rules recover the corresponding parameters for pure species. The fact that they are intuitively reasonable, however, does not mean that they are theoretically justified, nor that they are the only rules that yield empirically acceptable results. We will use (9.76) and (9.77) in this example, but it must be understood that if any other set of mixing rules is to be used then the equations that follow must be modified accordingly (as (9.72) through (9.75) must be modified for any EOS other than the RK EOS). From (9.76) and (9.77) we obtain the following values for the derivatives:

$$\frac{\partial (nb)}{\partial n_i} = b_i \tag{9.78}$$

$$\frac{\partial (n^2 a)}{\partial n_i} = 2 (a_i)^{1/2} \sum_j n_j (a_j)^{1/2}, \tag{9.79}$$

where the summation in (9.79) is over all components in the mixture, i.e. including component i (it is a good idea to work out the derivatives on your own to see why this is so). Substituting in (9.75) we arrive at:

$$RT \ln \frac{f_i}{X_i} = RT \ln \phi^i P = RT \ln \left(\frac{RT}{V_{P,T} - b} \right) + \frac{b_i RT}{V_{P,T} - b}$$
$$+ \frac{2 (a_i)^{1/2} \sum_j X_j (a_j)^{1/2}}{b T^{1/2}} \ln \left(\frac{V_{P,T}}{V_{P,T} + b} \right) \qquad (9.80)$$
$$- \frac{ab_i}{b^2 T^{1/2}} \ln \left(\frac{V_{P,T}}{V_{P,T} + b} \right) - \frac{ab_i}{b T^{1/2} (V_{P,T} + b)}.$$

It is important to understand how this equation must be used. The calculation procedure starts with the composition of the mixed gas phase (i.e. the values of X_i for all component species), uses these mol fractions to calculate a and b of the mixture with (9.76) and (9.77), then these parameters are used to calculate the molar volume *of the mixture* at P and T, $V_{P,T}$, with the RK EOS, equation (9.49), and finally the values for the mixture $(a, b, V_{P,T})$ are combined with a_i, b_i of the species of interest in (9.80), to calculate the fugacity coefficient and fugacity of species i. A *Maple* implementation is discussed in Software Box 9.2.

Software Box 9.2   Calculation of fugacities and fluid speciation in mixed fluids

The *Maple* package **RKmixing.mw** contains procedures that calculate species fugacities in a mixed fluid containing an arbitrary number of species, by using the Redlich–Kwong EOS and mixing rules given by equations (9.76) and (9.77). The mixing rules are implemented in procedures bmix and amix, and the derivative of the a mixing rule (equation (9.79), converted to mol fraction as in (9.80)) in procedure damix. The package also contains procedures that calculate volume and fugacity with the RK EOS, identical to those in **fluideos.mw** and included here for convenience only.

Procedure rkphimix is the key to this package. It calculates the fugacity coefficient of the ith species using equation (9.80). The call to this procedure is rkphimix(X, a, b, i, P, T, vguess), where:

X is a one-dimensional array containing the mol fractions of all species in the mixture
a is a one-dimensional array containing the RK a parameters, in the same order as X
b is a one-dimensional array containing the RK b parameters, in the same order as X
i is the index of the species of interest in the arrays (i.e. the identity of the species for which the fugacity coefficient is to be calculated)
P and T are pressure in bar and temperature in Kelvin
vguess is an initial guess for the volume of the mixture (trial and error may be necessary).

The procedure returns the fugacity coefficient of species i.

The *Maple* worksheet **phimixcalc.mw** shows an example of the use of procedure rkphimix. It calculates fugacity coefficients and fugacities, and the ratio of each of these variables to the corresponding standard state value (as plotted in Fig. 9.14) for a binary join. The example is for the H_2O–CH_4 join in the figure. In order to calculate some other join the RK parameters must be changed in the procedure itself, as explained in the *Maple* code.

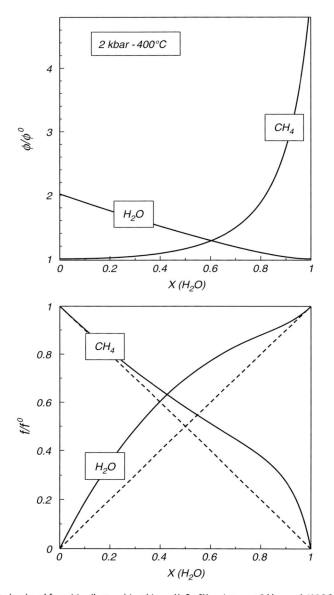

Fig. 9.14   Fugacity coefficients (top) and fugacities (bottom) in a binary H_2O–CH_4 mixture at 2 kbar and 400 °C, calculated with the RK EOS, and normalized to the values of the corresponding standard state functions.

An example of fugacity-composition relationships for a binary fluid, calculated with the RK EOS (equation (9.80)), is shown in Fig. 9.14. The fluid is assumed to consist of the species H_2O and CH_4 only, although as we shall see in the following section this may be an unrealistic constraint. The calculations are done at 2 kbar and 400 °C, at which conditions the RK EOS may still yield reasonably accurate results (e.g. Fig. 9.11). The top panel shows the ratio of the fugacity coefficient of each species in the mixture to the corresponding standard state fugacity coefficient. In a mixture of ideal gases the two curves would be straight lines at a constant value of 1. The bottom panel shows the ratio of fugacity of each species in the mixture to its standard state fugacity. This ratio must vanish as the concentration of the component goes to zero, and approach one as the fluid becomes pure. A mixture of ideal gases would follow the dashed lines in the figure. The H_2O–CH_4 mixture displays a positive departure from ideal behavior, which reflects a positive excess Gibbs free energy of mixing (Section 5.9.1). In principle, all of the information describing non-ideal mixing behavior of real gases is contained in the equation of state and the mixing rules. This of course does not mean that Fig. 9.14 is an accurate representation of an H_2O–CH_4 mixture, first because, as we shall see in the next section, other species are also likely to be present in non-negligible amounts in this system, and second because the RK EOS is not accurate above pressures of a few hundred bars. The latter point becomes particularly serious when dealing with mixtures of polar and non-polar species, as in this case. Multiparameter equations, such as the PS EOS, are better suited to the task, but as of this writing they have not been calibrated for species other than H_2O and CO_2, thus greatly limiting their usefulness for modeling many planetary processes.

9.6 Speciation in multicomponent volatile phases

In a multicomponent volatile phase there is virtually unrestricted freedom of formation of chemical species. Distinct chemical species can be considered to exist in a crystal, but the type and number of species is limited by the stoichiometry of the compound and the crystal chemical constraints of the crystalline structure. In a gas there are no such constraints. Calculating the distribution of species in a homogeneous fluid is important for understanding its chemical properties (e.g. its oxidizing potential) and physical behavior (e.g. stratification of fluid composition in a gravitational field).

We will discuss two different approaches to calculating fluid speciation. The chemical equilibrium approach is based on finding a linearly independent set of homogeneous equilibrium equations among the fluid species. It is best suited to studying species distribution in those cases in which there may be mineral assemblages that buffer the fugacities of some of the fluid species. An alternate approach is to seek the species distribution that minimizes the Gibbs free energy of the fluid. This is an efficient way of studying species distribution as a function of the bulk composition of the fluid, and of identifying the conditions under which saturation of condensed phases may take place.

9.6.1 The chemical equilibrium approach to fluid speciation

Consider a homogeneous gas phase composed of c system components, in which there are s distinct chemical species (or phase components). The speciation problem consists of finding the equilibrium mol fraction, X_i, $i = 1 \ldots s$, of each of the s chemical species. We therefore need s linearly independent equations relating these variables. There must be $s - c$ linearly independent conditions of homogeneous equilibrium in the gas phase (Section 6.1.3.). Each of these equations corresponds to a balanced chemical reaction among a subset of chemical species. The homogeneous gas phase is a system with $f = c + 2 - 1 = c + 1$ degrees of freedom. Therefore, at constant temperature and pressure it is necessary to specify the value of $c - 1$ intensive variables in order to fully determine the thermodynamic state of the system. These could be, for instance, the values of the fugacities of $c - 1$ of the fluid species. The fluid could exist at equilibrium with a mineral assemblage that buffers these fugacities, or we could just specify their values arbitrarily and see how fluid speciation changes in response to changes in these variables. Alternatively, the $c - 1$ intensive variables could be mol fractions of $c - 1$ system components, which fully specify the bulk composition of the fluid phase. In either case we have so far identified a total of: $s - c + c - 1 = s - 1$ linearly independent equations. The final equation required to solve for the species distribution is the closure condition, $\sum_i X_i = 1$. Finding the species distribution consists of writing each of these equations, and solving the system.

Worked Example 9.6   Species distribution in C–O–H fluids, (i): general relations

Fluids in which the only, or at least major, components are C, O and H are important in many planetary environments. In this case we have $c = 3$, so in order to fully specify the thermodynamic state of the system, and thus be able to calculate the species distribution in the fluid, we need to specify the values of $c - 1 = 2$ intensive variables in addition to pressure and temperature. These could be the fugacities of two of the species in the fluid, or the bulk contents of two of the system components, say C and O (H content then follows by difference). We shall see examples of both approaches, and why we would choose one or the other.

In order to calculate species distribution we must begin by deciding which species are present, or likely to be present, in the fluid phase. There are no general rules on how to do this, and the approach is largely a combination of intuition and trial and error. For example, under most planetary conditions (the terrestrial atmosphere being the one notable exception) oxygen fugacity is so low that it is fine to assume $X_{O_2} = 0$. This *does not* mean that we will also make $f(O_2) = 0$, however. We will use oxygen fugacity as a measure of the chemical potential of O_2, while assuming that there is no free molecular oxygen. For other species the decision may not be that simple. In this example we will assume that the C–O–H fluid contains five species: H_2O, H_2, CH_4, CO_2 and CO and we will calculate the mol fractions of these five species over a range of conditions. Other species could also conceivably be present in non-negligible amounts, for instance: C_2H_6, CH_2O, CH_3OH. If one suspects that this may be the case, then the fugacities of these other species can be estimated a posteriori from the calculated species distribution, and if some of the fugacities are found to suggest non-negligible mol fractions then the offending species can be added to the list of species, and the species distribution calculated again.

For now let us stay with the five species, H_2O, H_2, CH_4, CO_2 and CO. This makes $s = 5$, so that there must be $s - c = 2$ linearly independent homogeneous equilibrium reactions, which we can choose as follows:

$$CH_4 + 3CO_2 \rightleftarrows 4CO + 2H_2O \quad (\textit{reaction 1})$$

$$CH_4 + H_2O \rightleftarrows CO + 3H_2. \quad (\textit{reaction 2})$$

Together with the condition: $X(CO_2) + X(CO) + X(CH_4) + X(H_2) + X(H_2O) = 1$ and the values of two externally controlled intensive variables, this gives us five equations that we can solve for the five mol fractions.

Homogeneous chemical equilibrium is described by equation (5.22):

$$\sum_i \nu_i \mu^i = 0 \tag{9.81}$$

where ν_i is the stoichiometric coefficient of the ith species in the reaction. Substituting (9.13) and simplifying we get the following equation for equilibrium in a homogeneous gas phase:

$$\Delta_r G_{1,T}^0 + RT \sum_i \nu_i \ln f_i = 0 \tag{9.82}$$

and substituting (9.5) and rearranging:

$$\prod_i (X_i)^{\nu_i} = \frac{\exp\left(\dfrac{-\Delta_r G_{1,T}^0}{RT}\right)}{\left[P^{(\sum_i \nu_i)}\right] \prod_i (\phi_i)^{\nu_i}}. \tag{9.83}$$

At fixed P and T the values of all of the variables on the right-hand side of (9.83) are known, except for the fact that the ϕ_i are functions of X_i (Section 9.5.2). This requires an iterative solution of the system of equations, but except for this procedural detail (9.83) is an equation in a subset of the unknowns, X_i. Each of the $(s - c)$ homogeneous equilibrium conditions gives rise to an equation of this form. Incorporation of the externally controlled intensive variables is accomplished by some modification of (9.83), as we shall now see.

Worked Example 9.7 Species distribution in C–O–H fluids, (ii): inorganic methane production by serpentinization of ultramafic rocks

The Martian atmosphere contains trace amounts of methane (Krasnopolsky *et al.*, 2004). Because atmospheric methane decomposes by photodissociation over times many orders of magnitude shorter than the age of the Solar System, the observation implies that there is active methane outgassing from the Martian surface or subsurface. One possibility is microbial activity (Chapter 14). Another possibility is that methane is produced by inorganic reactions involving hydration of ultramafic rocks in the presence of a carbon-bearing fluid. The process consists of reduction of CO_2 to CH_4 by H_2 generated by serpentine-producing reactions, and is thought to be important in the Earth's ocean floor. We will calculate the species distribution in C–O–H fluids in equilibrium with ultramafic rocks undergoing serpentinization. The output of the exercise will be plots of fluid species distribution as a function of $f(O_2)$ and $f(H_2O)$, at constant temperature and pressure.

We choose to specify the values of the two intensive variables $f(O_2)$ and $f(H_2O)$ because during serpentinization $f(H_2O)$ may be buffered by reactions such as (9.15), whereas $f(O_2)$ may be buffered by silicate-oxide reactions such as (9.18). If $f(H_2O)$ is specified arbitrarily then $X(H_2O)$ is given by equation (9.5):

$$X_{H_2O} = \frac{f_{H_2O}}{\phi^{H_2O} P}. \tag{9.84}$$

In order to incorporate $f(O_2)$ we write the equilibrium:

$$2CO + O_2 \rightleftarrows 2CO_2 \qquad\qquad (reaction\ 3)$$

from which, by using (9.83) and simplifying, we get:

$$\frac{X_{CO_2}}{X_{CO}} = \frac{\phi^{CO}}{\phi^{CO_2}} \left(f_{O_2}\right)^{1/2} \exp\left(\frac{-\Delta_r G_{1,T}^{0,3}}{RT}\right). \qquad (9.85)$$

Applying (9.83) to each of the two homogeneous equilibrium reactions, 1 and 2 in Worked Example 9.6, we get:

$$\frac{(X_{CO})^4}{(X_{CO_2})^3 X_{CH_4}} = \frac{(\phi^{CO_2})^3 \phi^{CH_4}}{(\phi^{CO})^4} \left(f_{H_2O}\right)^{-2} \exp\left(\frac{-\Delta_r G_{1,T}^{0,1}}{RT}\right) \qquad (9.86)$$

and:

$$\frac{X_{CH_4}}{X_{CO} (X_{H_2})^3} = \frac{\phi^{CO} (\phi^{H_2})^3}{\phi^{CH_4}} P^3 \left(f_{H_2O}\right)^{-1} \exp\left(\frac{-\Delta_r G_{1,T}^{0,2}}{RT}\right). \qquad (9.87)$$

Finally, the closure condition is:

$$X_{H_2O} + X_{H_2} + X_{CH_4} + X_{CO_2} + X_{CO} = 1. \qquad (9.88)$$

Equations (9.84)–(9.88) are our five equations in the five unknown mol fractions. The system of equations does not have an analytical solution, and the fugacity coefficients are functions of the fluid composition. The strategy to solve the system of equations is to calculate first the *standard state* fugacity coefficients, use these to solve for an initial set of mol fractions, use these mol fractions to calculate fugacity coefficients in the mixture (e.g. with (9.80) if the RK EOS is chosen), recalculate mol fractions, and iterate between these last two steps until consecutive solutions converge within a desired interval. An implementation in *Maple* is discussed in Software Box 9.3.

Software Box 9.3   Calculation of fluid speciation by chemical equilibrium

The *Maple* worksheet `eq_fluid_species.mw` contains procedures that calculate fluid speciation in a C–O–H fluid as a function of $f(O_2)$ and $f(H_2O)$, at constant temperature and pressure. The procedures are written specifically to solve the problem in Worked Example 9.7. Because in this case we choose $f(O_2)$ and $f(H_2O)$ as independent variables the system of equations can be simplified considerably. First, the mol fraction of H_2O is calculated from (9.84), and the mol fraction of H_2 from:

$$X_{H_2} = \frac{f_{H_2O}}{\phi^{H_2} P} \left(\frac{K_I}{f_{O_2}}\right)^{1/2}, \qquad (S9.3.1)$$

where K_I is the equilibrium constant for the reaction:

$$2H_2O \rightleftarrows 2H_2 + O_2. \qquad (I)$$

This reduces the problem to a system of three equations in three unknowns:

$$\frac{X_{CO_2}}{X_{CO}} = \frac{\phi^{CO}}{\phi^{CO_2}} \left(K_{II} \cdot f_{O_2} \right)^{1/2}$$

$$\frac{X_{CO}}{X_{CH_4}} = \frac{\phi^{CH_4}}{\phi^{CO}} \frac{K_{III} \left(f_{O_2} \right)^{3/2}}{\left(f_{H_2O} \right)^2} \qquad \text{(S9.3.2)}$$

$$X_{CO_2} + X_{CO} + X_{CH_4} = 1 - X_{H_2O} - X_{H_2}$$

where K_{II} and K_{III} are the equilibrium constants for the reactions:

$$2CO + O_2 \rightleftarrows 2CO_2 \qquad \text{(II)}$$

and:

$$CH_4 + 3/2 O_2 \rightleftarrows CO + 2H_2O. \qquad \text{(III)}$$

Note that the equations appear to be linear but they are not, because the fugacity coefficients are functions of composition. The system is solved with *Maple*'s fsolve command. There are two procedures.

COH1 solves the fluid speciation as a function of log $f(O_2)$, at constant temperature, pressure and H_2O fugacity. It calculates an arbitrary number of points within a specified log $f(O_2)$ interval. The procedure is invoked with the following parameters, in this order: (pressure in bar, temperature in centigrade, H2O fugacity in bar, initial log oxygen fugacity, final log oxygen fugacity, number of points to be calculated, name of output file).

COH2 solves the fluid speciation as a function of $f(H_2O)$, at constant temperature, pressure and oxygen fugacity. It calculates an arbitrary number of points within a specified $f(H_2O)$ interval. The procedure is invoked with the following parameters, in this order: (pressure in bar, temperature in centigrade, initial H2O fugacity in bar, log oxygen fugacity, final H2O fugacity in bar, number of points to be calculated, name of output file).

Thermodynamic properties of the gas species are entered in the spreadsheet RefStateData, with a Shomate-type heat capacity equation, as explained in Software Box 1.1. The properties used in this example can be imported in tab-delimited format from the file agaspeciesdata, or they can be copied from a spreadsheet.

An example of the results generated by the *Maple* procedure is shown in Fig. 9.15. All calculations were done at 2 kbar and 400 °C. I chose these conditions as representative of serpentinization at shallow depth in the Earth's oceanic crust. Similar conditions may exist in the Martian crust if some type of magmatic activity survives to this day. The top diagram shows species distribution as a function of oxygen fugacity, calculated at constant $f(H_2O) = 230$ bar. The serpentinization reaction (9.15) buffers the fugacity of H_2O at values ranging from 215 bar for $X_{Fo} = 0.9$ to 250 bar for $X_{Fo} = 0.8$, corresponding to possible

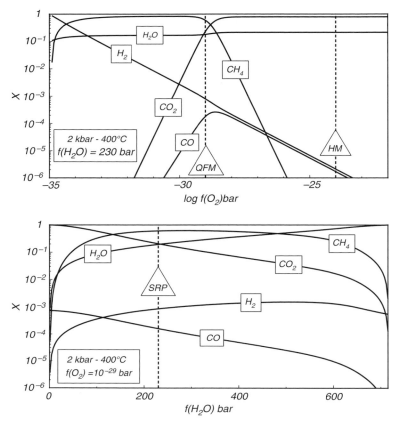

Fig. 9.15 Species distribution in a C–O–H fluid at 2 kbar and 400°C. Top: as a function of $f(O_2)$ at $f(H_2O)$ buffered by serpentinization reaction (9.15) with $X_{Mg} \approx 0.9$. The QFM and HM oxygen buffers are shown for reference. Bottom: as a function of $f(H_2O)$ at the $f(O_2)$ of the QFM buffer. H_2O fugacity at the serpentinization reaction shown by "SRP". Standard state fugacity of H_2O at these conditions is 717 bar. Voluminous production of inorganic methane is possible as long as oxygen fugacity stays below that of the QFM buffer.

terrestrial and Martian compositions, respectively (Exercise 9.10) – 230 bar is in the middle. The approximate locations of the QFM and HM oxygen buffers are shown by the dashed vertical lines. Under oxidizing conditions the fluid is a rather dull binary mixture of CO_2 and H_2O, with vanishingly small contents of other species. At very reducing conditions the fluid can be considered a ternary mixture of CH_4, H_2O and H_2. It is an empirical observation that most terrestrial upper mantle rocks crystallized within a relatively narrow range of oxygen fugacity, roughly between 0.01–1 times QFM. The Martian mantle appears to be slightly more reducing than this. Fluid speciation undergoes interesting and rapid changes in the neighborhood of these conditions. Assuming that mantle assemblages buffer oxygen fugacity during serpentinization, the fluids would be dominated by CH_4, with a mol fraction of H_2O of ~0.15 and much smaller concentrations of H_2, CO_2 and CO. However, within one order of magnitude of the QFM buffer, towards more oxidizing conditions, CH_4 concentration drops rapidly, and is replaced by CO_2 as the dominant carbon species.

The effect of H_2O fugacity at constant $f(O_2) = 10^{-29}$ bar ($\approx$ QFM) is shown in the bottom diagram of Fig. 9.15. The standard state H_2O fugacity at 2 kbar and 400 °C is 717 bar (calculated with the RK EOS), and this is the upper bound of the horizontal coordinate axis in the figure (for which $X_{H_2O} \to 1$). The assemblage forsterite + enstatite + chrysotile buffers $f(H_2O)$ at $\sim$230 bar, shown by the vertical line labeled SRP. H_2O fugacity can be greater than this if one or both of the anhydrous silicates are absent, whereas chrysotile would not be present at equilibrium if $f(H_2O)$ was less than this value (see also Fig. 9.1). At the oxygen fugacity of the QFM buffer CH_4 is the dominant species over a wide range of H_2O fugacities. It is replaced by H_2O as $f(H_2O) \to f^\circ(H_2O)$, i.e. as the fluid approaches pure H_2O, and by CO_2 as $f(H_2O) \to 0$. At these oxidation conditions H_2 and CO are always minor species. In fact, CO concentration is always vanishingly small at this pressure and temperature, and ignoring this species would have a negligible effect on the results (Exercise 9.11).

The calculations summarized in Fig. 9.15 suggest that inorganic methane production in the shallow Martian crust is feasible, and perhaps capable of explaining the observed methane concentration in the Martian atmosphere. This conclusion, however, rests on five premises: (1) that ultramafic rocks are present in the shallow Martian crust, (2) that a shallow heat source exists, (3) that a source of H_2O exists, (4) that a source of CO_2 exists and (5) that the ultramafic assemblage buffers oxygen fugacity to conditions not much more oxidizing than QFM. Of these, we can be reasonably confident of (3) and, by analogy to terrestrial examples, perhaps of (5) as well. Sedimentary carbonates could be a source of CO_2, but confirming their presence in Mars has been elusive; (1) and (2) are likely to remain unknown until geological and geophysical work on the surface of Mars become possible, which, in the humble opinion of this pundit, will require the development of highly mobile robots capable of autonomous intelligent decision making.

9.6.2 Calculation of fluid speciation by Gibbs free energy minimization

There are instances in which one wishes to know how speciation in a fluid depends on its bulk composition, rather than on externally buffered chemical potentials of some of the species. Although it is in principle possible to use the chemical equilibrium approach to do this, it rapidly becomes unwieldy and a much simpler method is available. It relies on the fact that the thermodynamic potential of a system takes its minimum value at equilibrium. As in the previous example, we consider a fluid composed of c system components. In this instance it is convenient to choose elements as system components, for example, C, H, O, N, S, Cl, F, etc. The total mol number (number of gram atoms) of the jth system component will be represented by N_j, $j = 1 \ldots c$. There are also s distinct chemical species (phase components), with the mol number of species i labeled $n_i, i = 1 \ldots s$. If we call the total number of mols of phase components in the fluid phase n_t, then:

$$n_t = \sum_{i=1}^{s} n_i, \quad X_i = \frac{n_i}{n_t}. \tag{9.89}$$

Let us write the chemical potential of the ith species as follows:

$$\mu^i = \mu_{1,T}^{0,i} + RT \ln \left(\phi^i P \right) + RT \ln \left(\frac{n_i}{n_t} \right). \tag{9.90}$$

The total Gibbs free energy of the fluid phase is given by equation (5.27):

$$G_{fluid} = \sum_{i=1}^{s} n_i \mu_{1,T}^{0,i} + RT \sum_{i=1}^{s} n_i \ln\left(\phi^i P\right) + RT \sum_{i=1}^{s} n_i \ln\left(\frac{n_i}{n_t}\right). \tag{9.91}$$

The equilibrium fluid distribution at constant temperature and pressure is given by the set of n_is that minimizes its Gibbs free energy.

The problem of determining the fluid speciation consists of finding the minimum of the function (9.91). Since this is a function of s variables it has s partial derivatives, relative to each of the n_is. If an absolute minimum of the function exists it would correspond to a point where all of the partial derivatives vanish. We do not seek the absolute minimum of (9.91), however, because there are constraints on the values that the n_is can take. These constraints are given by the bulk composition of the system. Recalling that N_j is the total number of atoms of the jth system component we can define C new functions that describe the bulk composition constraints, as follows:

$$\varphi_j = RT \left[\sum_{i=1}^{s} \eta_j^i n_i - N_j \right] = 0, \tag{9.92}$$

where η_j^i is the number of atoms of component j in species i, and I have multiplied by RT because it leads to a more compact final equation. The problem now consists of finding the set of n_is that minimize (9.91), subject to the n_is also satisfying (9.92). A simple yet powerful technique exists to solve this problem, known as the method of *Lagrange multipliers*. The method consists of multiplying each of the C functions (9.92) by a constant, λ_j (the Lagrange multiplier), adding these C new functions to (9.91) to construct a new function, Γ, as follows:

$$\Gamma = G_{fluid} + \sum_{j=1}^{c} \lambda_j \varphi_j \tag{9.93}$$

and then minimizing Γ. The mathematical justification of the method is not complicated but is beyond the scope of this book – a particularly clear explanation can be found in the textbook by Sokolnikoff and Redheffer (1966).

The function Γ has i partial derivatives relative to the n_is, which must all vanish simultaneously at a minimum. Performing the algebra (which you should verify for yourself) we find that each of these equations is of the form:

$$\frac{\partial \Gamma}{\partial n_i} = \frac{\mu_{1,T}^{0,i}}{RT} + 1 + \ln\left(\phi^i P\right) + \ln\left(\frac{n_i}{n_t}\right) - \frac{n_i}{n_t} + \sum_{j=1}^{c} \lambda_j \eta_j^i = 0 \tag{9.94}$$

where I have divided by RT in order to make the equation more readable (since we are equating to zero this is always legal). At constant temperature and pressure we have a total of s equations like (9.94) plus c equations (9.92), with a total of $s + c$ unknowns: the s n_is and the c λ_js (n_t is given by (9.89)). The remarkable fact is that this system of equations is generally quite easy to solve numerically, as we shall see in the following example. The Lagrange multipliers have mathematical meaning, but they do not have physical meaning in this instance and we will just discard those values. We are only interested in the n_is.

The Gibbs free energy minimization method does not require that we find homogeneous equilibrium equations among the fluid species, but it does require that we decide beforehand on which species are likely to be present. However, adding or removing species simply consists of adding or removing equations like (9.94), and making the necessary modifications to the constraining equations (9.92). Of course, the phase rule must always be obeyed and it is necessary to check on this before attempting to solve the system of equations. If we restrict the system of interest to a homogeneous fluid phase with c components then in order to completely specify the thermodynamic state of this system we must specify the values of $c + 1$ intensive variables. Two of these are pressure and temperature, leaving $c - 1$ intensive variables to be specified. These typically would be bulk composition parameters of the fluid. Alternatively, we may be interested in speciation in a fluid phase saturated with some specific condensed phases, in which case we must minimize the Gibbs free energy of a heterogeneous system, including the Gibbs free energy contribution of the condensed phases. The best way to understand all of this is with an example, as follows.

> ### Worked Example 9.8   Species distribution in C–O–H fluids, (iii): fluid composition during core formation in planetesimals

Meteorites preserve evidence that planetesimals with radii of order 10^2 km underwent core formation very early in the history of the Solar System. The chondritic precursors of differentiated planetesimals may have been rich in carbon and H_2O, so that C–O–H fluids may have been present during core formation. The bulk compositions of these putative fluids are unknown, but we are interested in understanding how species distribution in them may have varied over a range of bulk compositions, and also as a function of temperature and pressure. The Gibbs free energy minimization method is well suited to address these questions.

As in previous examples, we shall assume that the fluid consists of the five species: CO_2, CO, CH_4, H_2 and H_2O. If we consider only a homogeneous C–O–H fluid phase then this is a ternary system with four degrees of freedom, of which two are temperature and pressure. We can specify the other two as bulk composition variables, for example the ratios N_O/N_H and N_O/N_C. Note that these two ratios completely define the bulk composition of the fluid. Because graphite is present in many differentiated and undifferentiated meteorites, however, it is of interest to study the fluid distribution in C–O–H fluids saturated in this phase. In this case we have a ternary system with two phases and three degrees of freedom, in which we can specify only one compositional variable. Given that the system is saturated in graphite, carbon is present in excess, so that we must choose the ratio N_O/N_H as the compositional variable. Molecular O_2 will be assumed to be present in vanishingly small amounts, which leads to a simple way of calculating its chemical potential (or fugacity) as part of the Gibbs free energy minimization exercise.

We begin by writing the three constraining bulk composition equations, (9.92):

$$\varphi_1 = RT \left[n_{CO_2} + n_{CO} + n_{CH_4} + n_{graphite} - N_C \right] = 0$$
$$\varphi_2 = RT \left[4n_{CH_4} + 2n_{H_2O} + 2n_{H_2} - N_H \right] = 0$$
$$\varphi_3 = RT \left[2n_{CO_2} + n_{CO} + n_{H_2O} + 2n_{O_2} - N_O \right]$$
$$\approx RT \left[2n_{CO_2} + n_{CO} + n_{H_2O} - N_O \right] = 0.$$

(9.95)

Note that I wrote two versions of the oxygen equation (φ_3). Because n_{O_2} is vanishingly small it is permissible to ignore its contribution to the function φ_3, *but not to the derivative* $\partial\varphi_3/\partial n_{O_2} = 2$, which enters into the function $\partial\Gamma/\partial n_{O_2}$. This will become clear in a second.

The Gibbs free energy of the system fluid + graphite, G_{system}, is made up of three contributions: G_{fluid} as given by (9.91), G_{O_2} and $G_{graphite}$. Even if molecular oxygen is present in vanishingly small quantities there is a finite chemical potential of oxygen that contributes to the Gibbs free energy of the system and that must therefore be accounted for in the minimization calculation. We write the contribution of O_2 to the total Gibbs free energy of the system as follows:

$$G_{O_2} = n_{O_2}\mu_{1,T}^{0,O_2} + n_{O_2}RT\ln fO_2. \tag{9.96}$$

Because we are considering a graphite saturated system we must also have an equation for the Gibbs free energy of graphite. As this is a pure phase it is simply:

$$G_{graphite} = n_{graphite}\,\mu_{1,T}^{0,graphite} + n_{graphite}\int_1^P V_{graphite}dP. \tag{9.97}$$

The function Γ in this case is as follows:

$$\Gamma = G_{fluid} + G_{O_2} + G_{graphite} + \sum_{j=1}^3 \lambda_j\varphi_j \tag{9.98}$$

with G_{fluid} given by (9.91). This function has seven partial derivatives, all of which must vanish at the minimum. Five of the partial derivatives are of the form of equation (9.94), one each for CO_2, CO, CH_4, H_2 and H_2O. As an example, the CO_2 equation is:

$$\frac{\partial\Gamma}{\partial n_{CO_2}} = \frac{\mu_{1,T}^{0,CO_2}}{RT} + 1 + \ln\left(\phi^{CO_2}P\right) + \ln\left(\frac{n_{CO_2}}{n_t}\right) - \frac{n_{CO_2}}{n_t} + \lambda_1 + 2\lambda_3 = 0. \tag{9.99}$$

In this equation $n_t = n_{CO_2} + n_{CO} + n_{CH_4} + n_{H_2} + n_{H_2O}$, as n_{O_2} can be ignored in this context. The other four partial derivatives, for CO, CH_4, H_2 and H_2O, are analogous, and you should write them out yourself. The partial derivative relative to n_{O_2} is:

$$\frac{\partial\Gamma}{\partial n_{O_2}} = \frac{\mu_{1,T}^{0,O_2}}{RT} + \ln fO_2 + 2\lambda_3 = 0. \tag{9.100}$$

The variable n_{O_2} disappears, but we now see why it is necessary to leave it in φ_3 when forming equation (9.98) and calculating the derivative $\partial\varphi_3/\partial n_{O_2}$: it generates the coefficient of the Lagrange multiplier. Finally, the graphite partial derivative is:

$$\frac{\partial\Gamma}{\partial n_{graphite}} = \frac{\mu_{1,T}^{0,graphite} + \int_1^P V_{graphite}\,dP}{RT} + \lambda_1 = 0. \tag{9.101}$$

Equations (9.95), (9.99) and its equivalents, (9.100) and (9.101) constitute a system of 10 non-linear equations in the 10 unknowns: n_{CO_2}, n_{CO}, n_{CH_4}, n_{H_2}, n_{H_2O}, $n_{graphite}$, $\ln(fO_2)$, λ_1, λ_2 and λ_3. The input parameters are P, T and the three bulk gram-atom numbers: N_O, N_H and N_C. As in the chemical equilibrium method, the solution is numerical and iterative. Recalling that the fugacity coefficients are functions of fluid composition, we must initially calculate the standard state fugacity coefficients, use these to obtain an initial solution set, and then use the fluid composition from this solution set to calculate fugacity coefficients in the mixture, iterating until consecutive solutions converge within a desired interval. Implementation in *Maple* is straightforward and is discussed in Software Box 9.4.

Software Box 9.4   Calculation of fluid speciation by Gibbs free energy minimization

The *Maple* worksheet `minG_fluid_species.mw` contains a procedure that calculates fluid speciation in a C–O–H fluid saturated in graphite as a function of bulk O/H ratio, at constant temperature and pressure. The procedure, `COH_graphite_saturation`, implements the solution described in general terms in Section 9.6.2, and in particular solves the problem described in Worked Example 9.8. The procedure is invoked with the following parameters, in this order: (`pressure in bar, temperature in centigrade, name of output file`). It calculates fluid speciation, oxygen fugacity and fluid composition on the graphite saturation boundary, along the O–H join. The content and organization of the output file are described in the *Maple* listing. Thermodynamic properties of the gas species are entered in the spreadsheet `RefStateData`, with a Shomate-type heat capacity equation and volumetric properties for solid phases from Holland and Powell (1998). The properties used in this example can be imported in tab-delimited format from the file `agaspeciesdata`, or they can be copied from a spreadsheet.

Given that we are explicitly studying speciation in graphite-saturated fluids, there is an additional step that we must take, which is to determine the bulk composition of the fluid at which graphite saturation takes place. The graphite saturation boundary defines the boundary of the bulk composition region within which the calculations are valid. For bulk compositions outside of this region the fluid is not saturated in graphite and equation (9.97), and hence (9.98) and (9.101), are no longer valid.

Let us call the ratio N_O/N_H, that we will use as our compositional variable, $N_O/N_H = z$. As long as the system is saturated in graphite, for each value of P, T and z there is one and only one equilibrium fluid composition, because the thermodynamic state of the system is fully determined. This means that species mol fractions, and hence the ratios among them, are fixed and independent of the value of N_C. Along the graphite saturation boundary $n_{graphite}$ vanishes, so that we can write the following two compositional equations valid along the boundary:

$$N_C = n_{CO_2} + n_{CO} + n_{CH_4}$$

$$N_H = 4n_{CH_4} + 2n_{H_2O} + 2n_{H_2}.$$

(9.102)

Using the fact that species mol fractions are fixed at graphite saturation for given P, T and z we can define the following four constant ratios:

$$a_1 = \frac{n_{CO_2}}{n_{CO}}, \quad a_2 = \frac{n_{H_2}}{n_{CH_4}}, \quad a_3 = \frac{n_{CH_4}}{n_{CO}}, \quad a_4 = \frac{n_{H_2O}}{n_{CO}}, \quad (9.103)$$

which, substituting in (9.102), yield the ratio N_C/N_H along the graphite saturation boundary:

$$\frac{N_C}{N_H} = \frac{a_1 + a_3 + 1}{4a_3 + 2a_4 + 2a_2 a_3} \quad (9.104)$$

For any $P-T$ combination we can choose values of $z = N_O/N_H$ from $z \to 0$ to $z \to \infty$, solve for the fluid speciation at each z using any value of N_C large enough to ensure that the system is saturated in graphite (which we can easily verify because the calculation must yield $n_{graphite} > 0$), and then use (9.103) and (9.104) to calculate N_C/N_H along the graphite saturation boundary. The combination of N_O/N_H and N_C/N_H yields a unique point on the graphite saturation boundary. The tedious algebra is easily incorporated in the *Maple* procedure that calculates fluid speciation (Software Box 9.4).

An example of the results obtained by this procedure is shown in Fig. 9.16. The ternary diagram shows three graphite saturation boundaries, calculated at the same temperature, 1200°C, and pressures of 1, 5 and 10 kbar. These conditions may bracket core forming conditions in 100–1000 km planetesimals. Fluids above the boundary are saturated in graphite. Bear in mind that the results depend on the Redlich–Kwong EOS for calculation of the fugacity coefficients, so the 10 kbar values are unlikely to be very accurate, and the same may be true of the 5 kbar results. The qualitative trend is however correct: increasing pressure expands the graphite stability field, as we should expect from the fact that graphite is the low volume phase relative to fluid. The shaded region in the diagram, between the CO_2–H_2O join and the oxygen vertex, represents a range of bulk compositions that cannot be modeled with the equations given above, because in this region oxygen mol number is never negligible (you should explain to yourself why this is the case).

At a given P and T fluids with a constant O/H ratio anywhere within the graphite saturated region (i.e. anywhere along a tie line joining the graphite saturation boundary and the carbon vertex, such as the one shown in the diagram) have the same composition, and therefore also the same oxygen fugacity. Variation of the oxygen fugacity of graphite saturated fluids with O/H ratio is shown in the bottom diagram of Fig. 9.16. Also shown in the plot are oxygen fugacities at the QFM and QFI buffers (the bands correspond to the ranges in $f(O_2)$ between 1 and 10 kbar). We see that $f(O_2)$ of graphite-saturated C–O–H fluids is some 3–4 orders of magnitude lower than at the QFM buffer, which is generally consistent with separation of metallic Fe (Exercise 9.14).

Fluid speciation as a function of O/H ratio is shown in Fig. 9.17, which reveals some interesting trends. At low pressure (1 kbar) the predominant species is CO over a wide range of O/H ratios, and CH_4 for fluids very rich in H. Except over a very narrow interval, H_2O is everywhere subordinate to CO and CH_4. With increasing pressure, however, H_2O becomes the chief species over a progressively wider compositional range. Methane is always the dominant species for H-rich fluids, but carbon monoxide is replaced by carbon dioxide

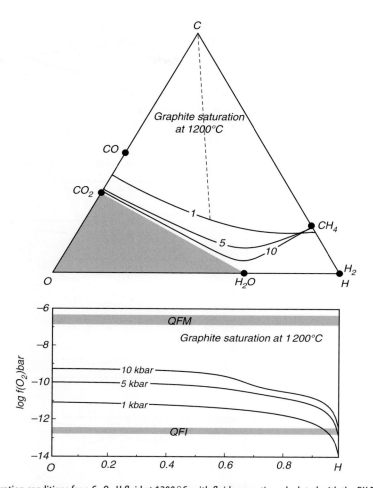

Fig. 9.16 Graphite saturation conditions for a C–O–H fluid at 1200 °C, with fluid properties calculated with the RK EOS. Top: graphite saturation boundaries at 1, 5 and 10 kbar pressure. Increasing pressure expands the graphite stability field, as expected from fundamental thermodynamic relationships. Graphite-saturated fluids anywhere along a join between the graphite saturation boundary and the C vertex, such as the one shown, have the same species distribution, which is shown in Fig. 9.17. The model discussed in the text breaks down in the shaded region, where molecular oxygen concentrations are not negligible. Bottom: oxygen fugacity of graphite-saturated fluids as a function of bulk O/H ratio. Fugacities are in every case several orders of magnitude lower than along the QFM buffer, making crystallization of metallic Fe possible.

as the dominant species in hydrogen-poor fluids at high pressure. These results suggest that retention of volatile components during differentiation of planetesimals smaller than a certain radius may be made very difficult, not only by the low gravitational attraction but also by the fact that H_2O, which could combine with silicates to form hydrous mineral phases, is not an abundant species in the fluid phase (hence its chemical potential will be relatively low). Retention of volatiles in the deep interior of planetesimals by formation of hydrous phases becomes easier with increasing body size, as core formation will take place at higher pressures.

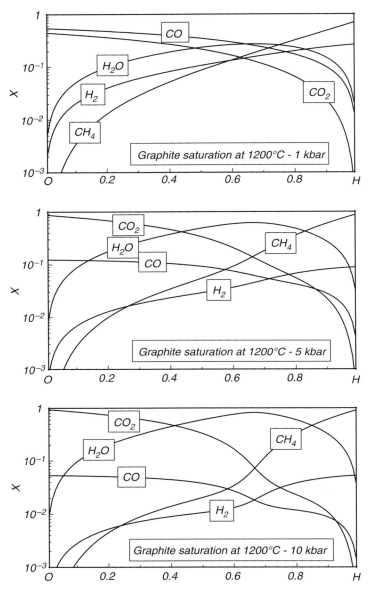

Fig. 9.17 Species distribution in C–O–H fluids saturated in graphite at 1200 °C and 1, 5 and 10 kbar, as a function of bulk O/H ratio. Each horizontal coordinate value corresponds to one tie-line such as the one in the ternary diagram of Fig. 9.16. Carbon monoxide is an important species at 1 kbar, but is replaced by carbon dioxide with increasing pressure. H_2O is not an abundant species at low pressure, which complicates volatile retention during core formation in small planetesimals.

9.7 Fluids at the conditions of giant planet interiors

The equations of state that we discussed to this point are generally not applicable to fluids at the conditions that exist in the interiors of the giant planets (Section 1.15). Fluids at particle densities of $\sim 10^{29}$ m^{-3}, even before they become degenerate (Fig. 1.16), become as incompressible as solids. When electron degeneracy sets in the behavior can only be described in terms of quantum mechanical effects. Explicit EOS often do not exist, and are substituted by tables of thermodynamic properties listed as a function of pressure and temperature. Thermodynamic properties are generally obtained from numerical simulations of interatomic and electronic potentials, and the predictions of the simulations are checked against the results of shock-wave, diamond anvil or other extreme-condition experimental techniques.

Once an EOS for fluids at ultrahigh pressure is available (even if only as a table of thermodynamic values), calculation of chemical potentials, speciation, etc., can be carried out as described in the previous sections, and applied to model processes in the interiors of giant planets, such as unmixing of molecular helium and fluid metallic hydrogen. The EOS for fluid planets is also important to understand their thermal evolution. Let us write the EOS for fluids at ultrahigh pressure as two separate equations, for pressure and internal energy. In each case we consider zero temperature and thermal components (see Section 8.4):

$$P = P_0 + P_{th} = P_0 + \frac{\gamma C_V}{V} T \qquad (9.105)$$

$$E = E_0 + E_{th} = E_0 + C_V T, \qquad (9.106)$$

where γ is the Grüneisen ratio (equations (8.61)) and the variables E, V and C_V are all extensive quantities. We recall that the zero temperature energy is the strain energy of the material when compressed under P_0. We also recall the relationship between pressure and internal energy valid at zero temperature (equation (8.48)) and, using the chain rule, we find:

$$P_0 = -\frac{\partial E_0}{\partial V} = -\frac{\partial E_0}{\partial m}\frac{\partial m}{\partial V} = -\rho \frac{\partial E_0}{\partial m}, \qquad (9.107)$$

where m is mass, and ρ density $= m/V$. We saw in Chapter 2 that in a self-gravitating body at hydrostatic and thermodynamic equilibrium, the gravitational binding energy and the body's thermodynamic state are related to one another by equation (2.42), which we now write as follows:

$$U_B = -3\left[\int_0^M \frac{P_0}{\rho} dm + \int_0^M \frac{P_{th}}{\rho} dm \right]. \qquad (9.108)$$

From (9.107) we have:

$$\int_0^M \frac{P_0}{\rho} dm = -\int_0^M dE_0 = -E_0, \qquad (9.109)$$

i.e. the total strain energy of the planet. We re-write the thermal pressure as follows:

$$P_{th} = \gamma \rho c_V T, \qquad (9.110)$$

where $c_V = C_V/m$ is the specific heat capacity. Substituting in (9.108) and assuming some average values for temperature and heat capacity:

$$\int_0^M \frac{P_{th}}{\rho} dm = \int_0^M \gamma C_V T \, dm = \gamma C_V T = \gamma (E - E_0) = \gamma E_{th}. \tag{9.111}$$

Substituting (9.109) and (9.111) in (9.108) we arrive at:

$$U_B = 3(E_0 - \gamma E_{th}). \tag{9.112}$$

Consider now a planet that contracts by an infinitesimal amount between two states of hydrostatic and thermodynamic equilibrium. The change in gravitational binding energy is given by:

$$dU_B = 3(dE_0 - \gamma dE_{th}) < 0 \tag{9.113}$$

where the negative sign states that the planet is contracting. The thermal energy radiated by the planet is (see equation (2.50)):

$$\begin{aligned} dQ &= dE + dU_B = dE_0 + dE_{th} + 3(dE_0 - \gamma dE_{th}) \\ &= 4dE_0 + (1 - 3\gamma) dE_{th} < 0, \end{aligned} \tag{9.114}$$

where dQ must be a negative quantity because planets do not absorb heat from space. We see that the strain (zero–temperature) and thermal energies of the planet must be related as follows:

$$dE_0 < \frac{3\gamma - 1}{4} dE_{th}. \tag{9.115}$$

This is a general result, that follows from the conditions of thermodynamic and hydrostatic equilibrium. The actual values of dE_0 and dE_{th} are given by the specific equation of state, but they must satisfy (9.115). In the absence of an EOS they remain unknown, but we can analyze some limiting cases.

The Grüneisen ratio is characteristically a number of order 1, so we rewrite (9.115) as follows:

$$dE_0 < \frac{1}{2} dE_{th}. \tag{9.116}$$

Clearly it must be $dE_0 \geq 0$, for the strain energy cannot decrease under compression (equation (9.107)). Suppose first that $dE_0 = 0$. We then get:

$$dE_{th} = -\frac{1}{3} dU_B, \quad dQ = \frac{2}{3} dU_B. \tag{9.117}$$

This case is unphysical, as it implies a material incapable of storing strain energy, or equivalently, a material with no strength. If a planet behaved in this way then upon contraction it would store one third of the gravitational energy dissipated as thermal energy, and radiate the other two thirds to space. Compare this result to the behavior of a monatomic ideal gas ($dQ = -dE$, equation (2.52)), which is as close as we can get to a material with no strength.

Consider now the other limiting case $dE_0 = \frac{1}{2} dE_{th}$. In this case we would have $dQ = 0$, which would imply a planet in thermal equilibrium with its environment and hence not contracting. As long as a planet has a temperature higher than ~ 2.7 K, however, it will radiate thermal energy to space (Chapter 13) and hence contract. We can think of this case, then, as one in which the planet is losing thermal energy stored during an earlier stage of contraction during which $dE_0 < \frac{1}{2}dE_{th}$, whereas at present $dE_0 = \frac{1}{2}dE_{th}$. We then have:

$$dE_{th} = -\frac{2}{3}dU_B, \quad dE_0 = -\frac{1}{3}dU_B, \tag{9.118}$$

i.e. two thirds of the gravitational energy would be stored as thermal energy, and the other third as strain energy. The key point here is that in a planet composed of this type of material all of the gravitational energy released during contraction is stored as internal energy, much of it as heat.

The EOS for hydrogen, helium and H_2O–NH_3–CH_4 mixtures at conditions of the deep interiors of the giant planets suggest that their present day behaviors approach this second limiting case. Today the planets are cooling by releasing either thermal energy stored during a previous stage of contraction during which the properties of the (less dense) fluids were significantly different, or heat generated by radioactive decay, or some combination. Cooling causes contraction, but the gravitational binding energy released in this process is chiefly stored in the planets' deep interiors, as in (9.118).

Exercises for Chapter 9

9.1 Write a *Maple* procedure to calculate $f(H_2O)$ along the serpentinization reaction (9.15). Compare your results to Fig. 9.1.

9.2 The bottom panel of Fig. 9.1 shows that dehydration occurs with increasing pressure. Explain this apparently inconsistent result.

9.3 Write equations for $f(CO_2)$ as a function of $f(O_2)$ at constant P and T along the (hm) and (mt) reactions in Fig. 6.14. Plot the two reactions at 1 bar, 25 °C and 1 kbar, 400 °C and confirm that their intersections coincide with the (sd) reaction (see Fig. 9.3). What is the stable phase assemblage at the Earth's surface? What can you infer about the fluid composition during crystallization of a hydrothermal vein with the equilibrium assemblage siderite + magnetite?

9.4 Prove that for a system at equilibrium the derivative $(\partial P / \partial V)_T$ must be negative. (Hint: start from the definition of Helmholtz free energy.)

9.5 Write a *Maple* procedure to solve the system of simultaneous equations (9.1.1), (9.1.5) and (9.1.6) for the three non-dimensional variables ϕ_{liq}, ϕ_{vap}, and π, as a function of non-dimensional temperature, τ. Verify your results against Fig. 9.7, 9.8 and 9.9.

9.6 Explain why the critical isochore ($\phi = 1$) is the continuation of the univariant phase transition (Fig. 9.9).

9.7 Plot the following three fluids in Figs. 9.7, 9.8 and 9.9: (i) N_2 at 1 bar, 300 K; (ii) N_2 at 1.5 bar, 100 K; (iii) CO_2 at 92 bar, 800 K. Comment on the nature of the atmosphere at the surfaces of Earth, Titan and Venus.

9.8 Compare the behavior of : (i) N_2 at 1 bar, 300 K; (ii) N_2 at 1.5 bar, 100 K; (iii) CO_2 at 92 bar, 800 K, as predicted by the ideal gas EOS and by the Redlich–Kwong EOS. Which EOS would you use to model the atmospheres of Earth, Titan and Venus?

9.9 Use the *Maple* worksheet that you wrote for Exercise 9.1 to justify the choice of $f(H_2O)$ in Worked Example 9.7.

9.10 Recalculate the speciation calculation in Worked Example 9.7 ignoring CO, and compare your results to Fig. 9.16.

9.11 A metamorphic rock contains the equilibrium assemblage: muscovite + sillimanite + quartz + microcline + graphite. During metamorphism this assemblage was at equilibrium with a volatile phase. Construct plots showing the composition of the volatile phase as a function of temperature at 1, 5 and 10 kbar. Begin by finding the upper temperature limit for each plot by determining the location of the muscovite + sillimanite + quartz + microcline + vapor equilibrium in the carbon-free system. Then modify the *Maple* worksheet described in Software Box 9.3 to calculate the diagrams.

9.12 Recalculate the CO–CO_2 speciation calculation (Exercise 5.5) using Gibbs free energy minimization. Plot isobaric graphite saturation boundaries in a diagram of T vs. composition (the binary C–O join). Use your diagram to discuss preservation of elemental carbon during metamorphism of organic sediments.

9.13 Plot log $f(O_2)$ vs. X_{Mg} in olivine at quartz saturation, and log $f(O_2)$ vs. a_{SiO_2} for pure fayalite, along the QFI buffer at 1200°C and constant pressures of 1, 5 and 10 kbar. Discuss the implications for crystallization of metallic Fe during differentiation of planetesimals (assume that the fluids are C–O–H fluids saturated in graphite).

9.14 Plot fugacities of ethane and formaldehyde in the speciation diagrams in Fig. 9.17 and discuss whether including these species in the Gibbs free energy minimization calculations would make any significant difference. Necessary thermodynamic properties can be obtained from NIST's Chemistry WebBook.

The liquid state extends from the melting point to the boiling point. Beyond the critical point fluids with liquid-like densities transition continuously to fluids with gas-like densities. A liquid close to its freezing temperature may differ significantly in such properties as viscosity, microscopic structure and chemical behavior from a liquid of the same composition near its boiling or critical points. For this reason it is convenient to define a melt as a liquid that is at, or very near, its freezing point. A melt is therefore saturated, or nearly so, in a solid phase (or assemblage) of broadly similar bulk composition. The exact meaning of "broadly similar" will remain undefined, but will become clear from the context of this and the following chapter, in which we will discuss electrolyte solutions. There is a parallel between this definition of melt and that of vapor, which is a gas that is at equilibrium with its liquid.

This chapter focuses on the ways in which melts form in planetary interiors. Because several excellent and up-to-date textbooks on igneous petrology are available (see, Winter, 2001; McBirney, 2006; Philpotts & Ague, 2009), and the research literature in the field is vibrant, I will not discuss processes of magma evolution and crystallization. There is no point in repeating here what is explained in much greater detail elsewhere. It is important to recall that a magma is an assemblage of melt, suspended solids and dissolved volatiles. As geologists we are most familiar with the silicate (and minor carbonate) magmatism characteristic of terrestrial planets, but there is no fundamental thermodynamic distinction between that kind of magmatism, ice magmatism in bodies such as Titan, Ganymede or Triton, and equilibrium between molten and solid metals in planetary cores.

10.1 Principles of melting

10.1.1 Melting of simple solids

From the point of view of thermodynamics, melting of a simple substance (e.g. a one-component system) at constant P and T is simply defined as a process that causes the Gibbs free energy of a solid and that of its liquid to become identical. But what is melting at the microscopic scale? Let us begin by looking at elements, which are the simplest possible thermodynamic systems. A crystal of an element consists of an assemblage of identical atoms arranged in a lattice. We can consider the lattice as being made up of *lattice points* that are occupied by atoms, and *interstitial sites* which remain vacant. The latter is the case because, if an atom were to occupy one of these interstitial sites, then it would be closer to some other atoms and the increased repulsive potential (e.g. Section 8.3) would raise the free energy of the crystal. The lattice points are the equilibrium positions of the atoms, that yield the minimum free energy of the crystal. This is a structure with perfect long-range order (Section 7.4). In fact, this description of a crystal of an element is identical to the example that we considered in Section 7.4: there is a total of N sites, of which $N/2$ are occupied lattice points, and $N/2$ are vacant interstitial sites.

As temperature increases the atoms vibrate with increasing amplitude about their equilibrium positions, but remain "anchored" to their corresponding lattice points, so that the configurational entropy of the crystal remains constant. The vibrational, or thermal, entropy of course increases, as additional vibrational energy levels become accessible (Section 4.6.2). At some temperature the amplitude of the vibrations reaches a critical value relative to the lattice spacing, and the atoms become detached from their lattice positions. One particularly fruitful way of modeling this, known as the *Lennard-Jones and Devonshire* theory of melting (Lennard-Jones & Devonshire, 1939a,b), is to postulate that the lattice points and interstitial sites still exist when such vibrational instability sets in, but now the atoms are distributed at random over the two types of sites. This corresponds to the sudden loss of the long range order of the material. There is an entropy discontinuity and hence a first-order phase transition (Section 7.6.3). The loss of long-range order causes the crystal to lose its resistance to shear, because there is no longer an "organized" system of interatomic potentials that generates a restorative force when interatomic bonds are stretched in a given direction. From a mechanical point of view the solid becomes a liquid at the first order phase transition. These are complementary microscopic descriptions of melting or, equivalently, alternative ways of defining the difference between a solid and a liquid, i.e.: (i) as condensed phases that either have long-range order (solids) or not (liquids); (ii) as condensed phases that either have shear strength or not; or (iii) as condensed phases in which the amplitudes and modes of atomic vibrations stay within certain bounds or not. Each of these approaches, and others, have been pursued in order to construct fundamental theories of melting. Excellent discussions can be found in Poirier (1991, most recommended), Ubbelohde (1978), Cotterill (1980), Mulargia (1986) and the remarkably clear and succinct paper by Oriani (1951). Here I will focus only on the point of view of melting as a sudden loss of long range order, as it is the one that is most helpful in understanding the chemical–thermodynamic aspects of melting.

There are several contributions to the increase in entropy during melting. One of them is the configurational entropy that arises from the loss of long-range order, as we discussed in the previous paragraph. In systems more complex than elements there are additional contributions to configurational entropy, arising from chemical mixing. For example, in a silicate crystal there may be ordering between cations occupying different kinds of octahedral sites (say, Ca and Mg in clinopyroxene) that persists to the melting point but not in the melt. There is then an increase in configurational entropy arising from chemical mixing, in addition to the lattice point–interstitial site disorder. For most substances melting is also accompanied by an increase in volume (the low pressure polymorph of H_2O ice is an important exception, that we discuss later). Expansion arises from repulsion among atoms with the same type of charge, that in the ordered crystal are shielded by oppositely charge atoms, and entropy increases in concert with expansion, as additional vibrational energy modes become available. Additional entropy contributions may arise if particles in the melt can acquire rotational degrees of freedom, which do not exist in the solid, or if changes in electron occupation levels become possible (for example, silicon becomes metallic on melting, and this contributes to its entropy of melting).

We can write the entropy of melting of a substance, $\Delta_m S$, as a sum of various contributions:

$$\Delta_m S = \Delta S_{expansion} + \Delta S_{lattice\ disorder} + \Delta S_{chemical\ mixing}$$
$$+ \Delta S_{rotation} + \Delta S_{electronic} + \cdots . \qquad (10.1)$$

For a simple substance only the first two terms are non-zero, and we write them as follows:

$$\Delta_m S = \left(\frac{\partial S}{\partial V}\right)_T \Delta_m V + \Delta S_{lattice\ disorder} \qquad (10.2)$$

where the expansion term equals the rate of increase in entropy (number of vibrational frequencies) with volume at constant temperature, times the increase in volume during melting. Now, from the definition of Helmholtz free energy we have:

$$\left(\frac{\partial S}{\partial V}\right)_T = \left[\frac{\partial}{\partial V}\left(-\frac{\partial F}{\partial T}\right)_V\right]_T = \left[\frac{\partial}{\partial T}\left(-\frac{\partial F}{\partial V}\right)_T\right]_V = \left(\frac{\partial P}{\partial T}\right)_V . \qquad (10.3)$$

Recall that this is one of Maxwell's relations (Section 4.9.1) – you can memorize them, look them up or, better, as this usually brings out their physical meaning, derive them as needed from the cross second derivatives of the appropriate thermodynamic potential. We recall equations (8.59) and (8.60):

$$\left(\frac{\partial P}{\partial T}\right)_V = \alpha K_T = \frac{\gamma C_V}{V} \qquad (10.4)$$

and substituting in (10.2) we get:

$$\Delta_m S = \alpha K_T \Delta_m V + \Delta S_{lattice\ disorder} . \qquad (10.5)$$

In Section 8.4 we saw that above the Debye temperature the product αK_T is nearly constant. This is a good approximation for equation (10.5). If the configurational entropy associated with loss of long-range order is also a constant then the entropy of melting should be a linear function of the volume of melting.

We can test this hypothesis by using the second identity in (10.4) to write (10.5) as follows:

$$\Delta_m S = \gamma C_V \frac{\Delta_m V}{V} + \Delta S_{lattice\ disorder} . \qquad (10.6)$$

It was pointed out by Stishov *et al.* (1973) and Lasocka (1975) that if values of $\Delta_m S$ for many elements are plotted against the corresponding values of $(\Delta_m V/V)$ the points scatter about a straight line with y intercept equal to $R\ln 2$. Figure 10.1a shows this for a number of metals. The volume of melting is divided by the volume of the solid phase. I have also divided equation (10.6) by $R\ln 2$, so that a y intercept equal to 1 corresponds to $\Delta_m S = \Delta S_{lattice\ disorder} = R\ln 2$. The straight line is not a calculated correlation line, but rather an arbitrary line drawn through the point $\{0, 1\}$, intended as a guide for the eye. There is clearly quite a bit of scatter, possibly arising to some extent from experimental uncertainties, but the pattern appears to be robust: for all of these metals the configurational entropy associated with loss of the crystal's long range order is approximately $R\ln 2$. Now, in the previous section we noted that, according to the Lennard-Jones and Devonshire

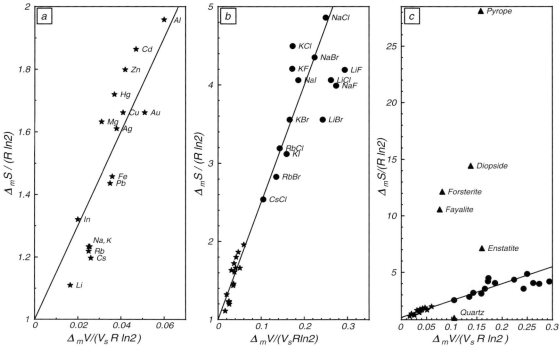

Fig. 10.1 Entropy of melting versus volume of melting for metals (stars), alkali halides (circles) and silicates (triangles). The same group of metals are shown in all three panels, and the same halides in panels (b) and (c). The lines are eyeballed but forced to intersect the vertical axis at $\Delta_m S(R\ln 2) = 1$. Data from Ubbelohde (1978), Bottinga (1985) and Richet and Bottinga (1986).

model of melting, for a simple substance the configurational entropy of the melt corresponds to mixing $N/2$ occupied lattice points and $N/2$ vacant interstitial sites over a total of N sites, *and that this is exactly the same problem that we considered in Section 7.4.* From equation (7.26) we see that, since the solid has no configurational entropy, the entropy increase that we should expect in this case is, indeed, equal to $R\ln 2$. The importance of this result is that it validates the interpretation of the melting phase transition as a sudden loss of long-range order. Note that this is different from a critical phase transition. In a critical phase transition the order parameter approaches zero continuously, so that there are no discontinuities in entropy, or enthalpy, at the phase transition. In the case of melting there is a discontinuous jump in the order parameter (see Fig. 7.10). It has been argued that this discontinuity may vanish at high pressure, leading to a solid–melt critical point analogous to the vapor–liquid critical point. Unequivocal experimental evidence for this is lacking, however, at least for conditions such as those in the interiors of terrestrial planets.

The melting behavior of alkali halides (Fig. 10.1b) appears to follow that of elements fairly closely. This suggests that the alternation of anions and cations is largely preserved in the melt, so that there is no contribution to configurational entropy beyond the redistribution of both types of atoms among lattice points and interstitial sites, as happens in pure elements. Observations such as this one have given rise to the quasicrystalline model of melts. In this model the melt is thought of as a crystal in which the number of defects is of the same order as the total number of crystallographic sites (see Ubbelohde, 1978, Chapter 5).

In contrast to halides, the melting behavior of silicates is clearly different from that of metals (Fig. 10.1c). In this case there is a very large entropy component in excess of the $R \ln 2$ term that arises from lattice point–interstitial site mixing, with the largest contribution likely arising from chemical mixing. Quartz is anomalous, however, in that its configurational entropy of melting is significantly *lower* than that of simple substances. This suggests that SiO_2 melt is strongly polymerized, and preserves much of the long range order of the crystal.

10.1.2 Melting in complex natural systems

Melting in multicomponent systems in which a multi-phase assemblage is stable at the solidus is much more complex. To begin with, if melt–solid equilibrium is multivariant then we must distinguish a solidus and a liquidus (Section 6.6.1). The microscopic picture of melting becomes fuzzy. For example, is the structure of the melt analogous to one of the solid phases in particular, or is it a composite of the various solid phases at equilibrium at the solidus? From a purely thermodynamic point of view the picture may seem clearer, as calculating the composition of a melt at equilibrium with a given solid assemblage as a function of intensive variables is no different from calculating equilibrium among solid phases (Chapters 5 and 8) or between solids and a gas phase (Chapter 9). In fact, in Chapter 6 we calculated some simple melt–solid phase diagrams (Section 6.6). Those calculations, and the resulting phase diagrams, are, however, no more than crude approximations. They are only intended to illustrate the qualitative behavior of the systems that we modeled, for instance, the effect of enthalpy of fusion on the width of a melting loop, or on the magnitude of the melting point depression. Implementing quantitatively rigorous calculations that allow us, for example, to predict basaltic melt compositions as a function of mantle source composition, pressure of melting and oxygen fugacity is *much* more complex than what we did in Chapter 6. Fundamentally, this is so because it is never acceptable to assume that mixing in the melt phase is ideal, and it is not always obvious what the basis for a non-ideal model ought to be. The problem begins by defining what chemical species to use in a multicomponent melt (such as what we did in fluid speciation calculations, Section 9.6). Once the species have been defined, it is not always straightforward to determine their standard state thermodynamic properties and their excess mixing properties. As of this writing, and in the humble opinion of yours truly, only one comprehensive calculation engine *cum* data base exists that generates reliable thermodynamic models of igneous systems over a wide range of bulk compositions and intensive variables. This is MELTS and its extensions, developed by Ghiorso, Sack, and collaborators (see, for example, Ghiorso & Sack, 1995; Ghiorso, 1997; Ghiorso *et al.*, 1983, 2002). Discussing these models is beyond the scope of this book, as well as the available space. I will rather focus on some of the constraints on the conditions of formation of planetary magmas.

10.2 Melting point depression. Eutectics, cotectics and peritectics

Eutectics are a fundamental aspect of planetary magmatism, for two reasons. First, the fact that the solidus temperature of an assemblage of immiscible phases may be hundreds of degrees lower than the melting point of each individual phase (e.g. Figure 6.21) makes igneous activity possible at conditions that otherwise it might not be. Second, the invariant (or nearly so) nature of eutectic melting equilibria restricts the range of common magmatic compositions. Consider the case of silicate planetary bodies. They all have peridotitic mantles. Over a wide range of conditions the initial melting behavior of peridotite is more or less close to eutectic, and the composition of the resulting low-temperature melts is broadly speaking basaltic. Because of this, all silicate planetary bodies have basaltic crusts. Before you get incandescently upset at this sweeping generalization, let me emphasize that this is a book on thermodynamics, not igneous petrology. My purpose is to use simple thermodynamic reasoning to understand many of the first-order characteristics of planetary bodies. Along the way, and because space and time are both finite, we will be forced to ignore important details. Some of these details are very dear to this writer, such as the fact that about one third of the Earth's solid surface is not basaltic, because it is not a simple product of mantle melting.

The fundamental thermodynamic reason for eutectic behavior is the fact that the chemical potential of a component in a stable solution is lower than its standard state chemical potential (e.g. Fig. 5.9). If solid phases do not form a solution, and therefore remain in their standard states, but the components mix stably in the melt, then solid–melt equilibrium is only possible at a temperature lower than the melting temperature of the isolated phases. You can see this graphically by constructing a diagram similar to Fig. 5.3, in which the low entropy curve corresponds to one of the pure solids, and the high-entropy curve to the corresponding component in the melt. If the chemical potential of the melt component decreases equilibrium must shift to a lower temperature. As we saw in Sections 6.5 and 6.6, the magnitude of this effect can be calculated by using the melting point depression equation, (6.49). This equation is important not only for the study of eutectics, but also to study other controls on magma generation. There are reasons to revise equation (6.49) as applied specifically to melting reactions, which we do next.

10.2.1 A revision of the melting point depression equation

Although (6.49) is the traditional way of integrating (6.48), our microscopic understanding of melting, Section 10.1.1, suggests that a better approximation may be to consider the entropy of melting, rather than the enthalpy of melting, to be constant. Using the identity $\Delta_m H = T \cdot \Delta_m S$, we re-write equation (6.48) for the depression of the melting point relative to that of a pure substance as follows:

$$\ln K_T = \int_{T_0}^{T} \frac{\Delta_m S}{RT} dT \tag{10.7}$$

where T_0 is the melting point of a pure one-component substance, and K_T is the ratio of the activity of this component in the melt to its activity in the solid (see equation (6.44)). Equation (10.7) integrates to:

$$K_T = \left(\frac{T}{T_0} \right)^{\frac{\Delta_m S}{R}} \tag{10.8}$$

which is equivalent to (6.49) (but not identical, see Exercise 10.1). We shall use this version of the melting point depression equation throughout this chapter. One advantage of this equation relative to (6.49) is that it is more compact. More importantly, experimental studies of natural rock compositions (see Kojitani & Akaogi, 1997) support the theoretical expectation (Section 10.1.1) that entropy of melting varies less strongly with temperature than enthalpy of melting.

10.2.2 Multicomponent eutectics and cotectics

The conclusion that the eutectic temperature must be lower than the melting point of the isolated phases generalizes immediately to systems of any number of components. Consider a system of c components, in which there are c mutually immiscible solid phases. An assemblage consisting of the c solid phases plus a melt is univariant, or pseudoinvariant at constant pressure. Call the equilibrium temperature T_e. At equilibrium there is one equation like (10.8) for each of the c components. Let i be the component that has the lowest melting temperature, $T_{0,i}$. We then have, because the solid remains pure:

$$a_{i,melt} = \left(\frac{T_e}{T_{0,i}} \right)^{\frac{\Delta_m S}{R}}. \tag{10.9}$$

But if the melt is a stable phase then $a_{i,melt} < 1$ (Fig. 5.9). Because the exponent in (10.9) is always positive for a melting reaction, it follows that $T_e < T_{0,i}$, i.e. the eutectic temperature is lower than the melting point of the least refractory phase.

What happens if one of the phases disappears, but the corresponding component is still present in the melt? We now have two degrees of freedom, so the system becomes pseudounivariant at constant pressure. The $c-1$ solid phases can exist at equilibrium with melt over a temperature range. The melt composition will necessarily change along this temperature range, because the Gibbs free energy of each of the solid phases varies with temperature at a different rate (in order for this not to be the case all the solid phases would have to have the same entropy and heat capacity – why?). The resulting temperature vs. melt composition curve is called a *cotectic*, with which you are almost certainly familiar from igneous petrology. In general, c cotectics radiate away from a eutectic. For our purposes the important point is to recall that temperature along each of the cotectics increases away from the eutectic, and to understand why.

Suppose that the solid phase consisting of component j disappears from the eutectic assemblage. This does not have to be the lowest melting point phase, it can be any of the solid phases. The chemical potential of j in the melt decreases relative to its value when the melt was at equilibrium with the solid phase, because the melt is no longer saturated in this phase. The chemical potentials of the other melt components must increase (equation (5.31)) so that preserving solid–melt equilibrium requires that temperature increase relative to that of the eutectic. Alternatively, you can pretend that temperature does not increase. In that case the solid assemblage has a lower Gibbs free energy than the melt, and the latter will crystallize, unless temperature rises.

The same line of argument can be applied to show that each time that a solid phase disappears, and a degree of freedom is gained, melting temperature must keep increasing (exercise left to the reader). The sequence in which this happens depends on the bulk composition of the system relative to the thermodynamically determined locations of the eutectic and cotectics. It may be different for different bulk compositions, something that is very important in the detailed study of igneous phase relations, but that will not concern us here. The important point for many of our subsequent discussions is that, in general, peridotites consisting of several major components and a comparable number of phases (e.g. lherzolites) are less refractory than simpler ultramafic rocks such as dunites or pyroxenites. Similarly, the mantle of an icy satellite consisting of a mixture of H_2O, NH_3 and CH_4 ices melts at a lower temperature than one consisting of pure H_2O.

10.2.3 Incongruent melting and peritectics

The phase rule requires that an assemblage of c solid phases plus a melt phase in a c-component system be pseudoinvariant at constant pressure, but it is mute with respect to the chemography of the melting reaction. At a eutectic the only phase on the high entropy side of the reaction is melt. Such a reaction is known as a congruent melting reaction. However, as we saw in Worked Example 6.3, there are also incongruent melting reactions in which the high entropy assemblage consists of melt plus one or more solid phases. If the total number of phases at an incongruent melting reaction is $c + 1$ then the variance is still 1. The isobaric pseudoinvariant point is in this case known as *peritectic*. In contrast to eutectics, peritectics are not minimum melting points. It can be shown that requiring a peritectic to be a minimum melting point leads to a violation of Schreinemakers rule (Exercise 10.2).

Our simplified calculation of the temperature and composition of eutectics (e.g. equations (10.9) or (6.70) and (6.71)) assumes that the melt is an ideal mixture of the same chemical species that make up the solid phases. Incongruent melting is impossible with this assumption. In order to see why, consider a binary system. The congruent melting reaction, $A + B \rightarrow$ *liquid*, allows for the liquid to be a combination of species A and B. The incongruent melting reaction, $A \rightarrow$ *liquid* $+ B$, in contrast, makes it impossible for the liquid to be a linear combination of non-negative amounts of species A and B. It follows that calculation of the temperature and composition of a peritectic requires that one specify melt species that are not present in the solid phases, and that may interact non-ideally in the melt phase. How to do this is beyond the scope of this book, but is at the core of thermodynamic models such as MELTS.

10.3 Partitioning of trace components between solids and melts

Virtually all minerals incorporate in their crystalline structures trace amounts of components that are not present in their nominal formulas. For example, end-member forsterite is never Mg_2SiO_4, as a small fraction of the Mg cations are replaced by elements such as Ni and Co. When the mineral melts these components also enter the melt and, under the right circumstances, equilibrium may be established between trace components in the crystal and in the melt, for instance: Ni_2SiO_4 $_{(crystal)} \rightleftarrows Ni_2SiO_4$ $_{(melt)}$. The usual convention is to write the equilibrium of a trace element between crystal and melt with the crystal as product (this convention, which I have always found counterintuitive, probably arises from an excessive preoccupation with magmatic differentiation, as opposed to magma generation):

$$z_{melt} \rightleftharpoons z_{crystal}. \tag{10.10}$$

In this equation z stands for the trace component of interest. It could be an element or some convenient species such as Ni_2SiO_4. If the concentrations of the trace component in the crystal and the melt are low enough that Henry's law is valid (Section 5.9.2), then the thermodynamic equilibrium condition for (10.10) can be written as follows:

$$-\Delta_m G^0_{P,T} + RT \ln \left(\frac{K_H^{z,s} X_{z,s}}{K_H^{z,m} X_{z,m}} \right) = 0, \tag{10.11}$$

where s and m stand for solid and melt, K_H are the respective Henry law constants, X is mol fraction and $\Delta_m G^0_{P,T}$ is the standard state Gibbs free energy of melting of the trace component at the pressure and temperature of interest. We can re-write (10.11) as follows:

$$\frac{X_{z,s}}{X_{z,m}} = \frac{K_H^{z,m}}{K_H^{z,s}} \exp \left(\frac{\Delta_m G^0_{P,T}}{RT} \right). \tag{10.12}$$

As a first approximation it is often assumed that the exponential function is a constant, or at least that it varies very slowly, over a restricted pressure–temperature range. Furthermore, for restricted ranges in the compositions of solid and melt the mol fraction of z can be considered to be related by a constant factor to its mass concentration, expressed, for example, in weight percent or ppm. Using C_z for the mass concentration, we simplify (10.12) to:

$$\frac{C_{z,s}}{C_{z,m}} = D^{s/m}, \tag{10.13}$$

where $D^{s/m}$ is called the solid–melt *partition coefficient*. This equation is sometimes called Nernst's distribution law, and $D^{s/m}$ is Nernst's distribution coefficient (Chapter 11). It incorporates the right-hand side term of equation (10.12), plus the ratio of constant factors that convert mol fraction to mass fraction in the solid and melt. Partition coefficients are determined empirically rather than calculated from thermodynamic properties, so that these different contributions are lumped together and in general cannot be isolated. If enough experimental measurements exist, however, it may be possible to determine the pressure and temperature dependencies of the partition coefficient.

Trace elements are classified as compatible vs. incompatible, depending on whether the value of D is greater or smaller than one, respectively. A compatible trace element prefers the crystal, and an incompatible one the melt. From the point of view of thermodynamics this distinction arises from two independent causes, which can be identified in equation (10.12). The first is the melting point of the trace component relative to that of the crystalline host, or, in other words, whether the trace component is more or less refractory than the host. If it is more refractory then $\Delta_m G^0_{P,T} > 0$, because at the temperature at which the host crystal melts the stable phase for the trace component is the solid. The trace element in this case will tend to behave compatibly. Conversely, it will tend to be incompatible if the trace component is more fusible than the host and $\Delta_m G^0_{P,T} < 0$. The second factor is the relative value of the Henry's law constant in the crystal to that in the melt. Given two trace cations in the same crystalline host, the one closer in size and charge to the essential cation in the crystal will generally have a smaller Henry law constant and be more compatible. It follows that a given trace component can be compatible in some systems and incompatible in others.

Whether a trace element is compatible or incompatible is important. Suppose that we have a crystal that melts progressively, such that melt fraction, φ, varies continuously between 0 and 1. We shall come back to the "melts progressively" statement and analyze it from a rigorous thermodynamic point of view, but for now we note that, if $C_{z,i}$ is the initial concentration of component z in the crystal before melting starts, and melt and crystal remain at equilibrium throughout the melting process, then for any melt fraction we have, by mass balance:

$$C_{z,s}\left(1-\varphi\right)+C_{z,m}\varphi=C_{z,i} \tag{10.14}$$

or:

$$C_{z,m}=\frac{C_{z,i}}{D^{s/m}\left(1-\varphi\right)+\varphi}. \tag{10.15}$$

This equation is known as the *batch melting equation*, meaning that crystal and melt remain at equilibrium from the beginning of melting until (at least) melt fraction φ is attained. Other possibilities exist, and may be more common in nature (Maaløe, 1985; Winter, 2001; Philpotts & Ague, 2009), but (10.15) is sufficient for our purposes. Variation in $C_{z,m}$ with φ for various values of $D^{s/m}$ is shown in Fig. 10.2. We begin by noting that for $\varphi=0$ we get $C_{z,m}=C_{z,i}/D^{s/m}$. This is the concentration in the first infinitesimal fraction of melt; it is very high for a strongly incompatible element (e.g. $D^{s/m}=0.001$, Fig. 10.2) and virtually zero for a strongly compatible one ($D^{s/m}=100$). For $\varphi=1$, $C_{z,m}=C_{z,i}$, as now the composition of the melt is identical with the starting composition.

An incompatible component (Fig. 10.2) is strongly enriched in the first melts and its concentration initially decreases very rapidly. The rate of decrease slows down with increasing melt fraction. For an incompatible trace component: (i) at sufficiently large melt fractions the concentration is more or less independent of the value of D and depends only on melt fraction, and (ii) the same is true for sufficiently small values of D. The reason is that, in both cases, $C_{z,m}\approx C_{z,i}/\varphi$ (see equation (10.11)). A strongly compatible trace element, in contrast, may be virtually absent from the melt until the melt fraction is almost 1. These concepts, and elaborations based on the nature of the melting process (e.g. whether melt and crystal remain at equilibrium or are separated as soon as a new melt increment forms) constitute much of the basis of trace element geochemistry. Application to multicomponent

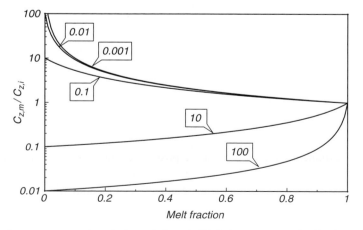

Fig. 10.2 Concentration of trace component z in melt, $C_{z,m}$, relative to bulk initial concentration, $C_{z,i}$. Numbers in boxes are solid–melt partition coefficients, $D^{s/m}$.

systems (i.e. rocks) differs chiefly in the fact that the partition coefficient that is used, called the bulk partition coefficient, is a weighted average of the partition coefficients for each of the minerals in the assemblage. We will not pursue these topics here, but in Section 10.4.2 we will examine the conditions under which trace element partitioning may have important effects on melt generation in planetary mantles.

10.4 The effect of "impurities" on melting temperature

10.4.1 Melting point depression revisited (again!)

Let T_0 be the melting point of a pure one-component crystal, and T the melting point of the crystal in the presence of a component, z, that dissolves in the melt but not in the solid. We shall refer to this component as an "impurity", and to the component that makes up the crystal as the "major" component. The impurity could be an abundant component that is altogether excluded from the solid, or a strongly incompatible trace component – all that matters is that its concentration in the solid is vanishingly small relative to its concentration in the melt and can thus be ignored. The difference with our discussion of eutectics is that now we do not require that this additional component make up a phase of its own. Let the mol fraction of the impurity in the melt be $X_{z,m}$. Because we ignore the concentration of the impurity in the solid the activity of the major component in the solid species is 1, and K_T in the melting point depression equation (10.8) equals the activity of the major component in the melt, just as in the case of a eutectic. Let $\gamma_{r,m}$ be the activity coefficient of the major component in the melt. Then $K_T = \gamma_{r,m} \left(1 - X_{z,m}\right)$ and (10.8) becomes:

$$\gamma_{r,m} \left(1 - X_{z,m}\right) = \left(\frac{T}{T_0}\right)^{\frac{\Delta_m S}{R}} \tag{10.16}$$

or:

$$T = T_0 \left(\gamma_{r,m}\right)^{\frac{R}{\Delta_m S}} \left(1 - X_{z,m}\right)^{\frac{R}{\Delta_m S}}. \tag{10.17}$$

This equation yields the crystal–melt equilibrium temperature as a function of the mol fraction of the impurity in the melt. Calculating an accurate value for T, however, depends on knowing accurate values for (i) $X_{z,m}$ and (ii) $\gamma_{r,m}$. This is the crux of the problem, as it requires that we know the distribution of chemical species in the melt and the possible non-ideal interactions among them, something that is generally far from simple.

We can derive an important generalization, however. If the concentration of the trace component in the melt remains low then $\gamma_{r,m} \sim 1$, and we can ignore it. Because the exponent $R/\Delta_m S$ is always a positive quantity, $T < T_0$. Say that the molecular weight of the impurity is m_z, and that of the major component m_r. If we express concentrations in mass percent (what we normally and incorrectly call weight percent), we have:

$$X_{z,m} = \frac{\frac{C_{z,m}}{m_z}}{\frac{C_{z,m}}{m_z} + \frac{100 - C_{z,m}}{m_r}}. \tag{10.18}$$

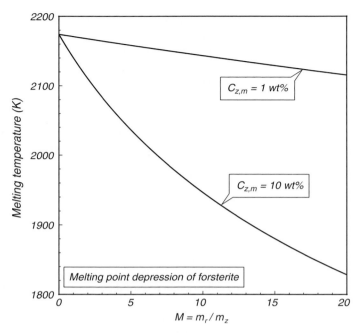

Fig. 10.3 Melting temperature of forsterite as a function of the parameter M = ratio of molecular weight forsterite, m_r, to molecular weight of "impurity" component in the melt, m_z. The concentration of impurity in the melt is constant along each curve, at 1 and 10 wt%.

Defining the ratio of molecular weight of major component to molecular weight of impurity $M = m_r/m_z$, we get:

$$X_{z,m} = \frac{MC_{z,m}}{(M-1)\,C_{z,m} + 100} \tag{10.19}$$

and:

$$T \approx T_0 \left(1 - \frac{MC_{z,m}}{(M-1)\,C_{z,m} + 100}\right)^{\frac{R}{\Delta_m S}}. \tag{10.20}$$

We see that the molecular weight of the impurity relative to that of the major component has a large effect on the magnitude of the melting point depression, simply because a given mass of impurity represents a greater number of mols the lower its molecular weight is relative to that of the host. This is shown by the curves of T vs. M in Fig. 10.3. The curves are calculated with equation (10.20) for values of $T_0 = 2174$ K and $\Delta_m S = 56$ J mol^{-1} K^{-1}, which correspond to melting of forsterite at 1 bar, and constant values of $C_{z,m}$ of 1 and 10 wt%.

10.4.2 Effect of partitioning of incompatible trace components on melt generation

In Section 10.3 I was somewhat careless about the thermodynamics of melting. It is now necessary to clean up the act. Naively, one could think that the "progressive melting" that

I referred to in Section 10.3 corresponds to melting at constant temperature, as a result of slow addition of heat (enthalpy of melting). This is not the case, however. Consider an idealized system in which forsterite is pure Mg_2SiO_4. Melting in this system is univariant. At constant pressure there is a unique temperature at which crystal and melt are at equilibrium, i.e. the solidus and liquidus temperatures are the same. Suppose now that olivine contains a trace component dissolved in its crystalline structure. This could be a cation substituting for Mg, or it could be a molecular species, such as H_2O or some other volatile component, accommodated interstitially in its crystalline structure. In any case, we are now dealing with a two-component system, and since the extra component does not cause a new phase to form, the melting reaction becomes divariant: solidus and liquidus temperatures do not coincide, nor will they in general be the same as the melting temperature of pure forsterite. Whether this thermodynamically inevitable effect of incorporation of trace components is petrologically significant is another question, to which we now seek answers. This topic has been explored in detail by Hirschmann and collaborators in a number of recent and clear contributions (Hirschmann, 2006; Hirschmann *et al.*, 2009; Tenner *et al.*, 2009). Here I will derive a set of very general equations applicable to simple systems of what I shall call "one component plus a trace". By this I mean that the system is binary when the trace component is included, and that we will compare its behavior to that of the unitary trace-free system.

We seek an equation for the solidus temperature of a crystal that contains a trace component dissolved in it. From (10.15) and (10.19) we write:

$$X_{z,m} = \frac{MC_{z,i}}{(M-1)C_{z,i} + 100D^{s/m}(1-\varphi) + \varphi}. \tag{10.21}$$

Substituting this expression in (10.17) and setting $\varphi = 0$ yields the desired equation for the solidus temperature, T_S:

$$T_S = T_0 \left(1 - \frac{MC_{z,i}}{(M-1)C_{z,i} + 100D^{s/m}} \right)^{\frac{R}{\Delta_m S}}. \tag{10.22}$$

This equation allows us to explore the effects of the three parameters, M, $D^{s/m}$ and $C_{z,i}$ on the melting point of a crystal. Figure 10.4 shows curves of T_S vs. $C_{z,i}$ calculated for melting of forsterite at 1 bar ($T_0 = 2174$ K, $\Delta_m S = 56$ J mol^{-1} K^{-1}, same as Fig. 10.3). The horizontal axis extends to a maximum value of 0.1 wt% ($= 1000$ ppm), which makes it clear that even trace amounts of "impurities" can have a dramatic effect on the *solidus* temperature of a crystal (emphasis important, as we shall soon see). The values of the partition coefficient and of the molecular weight of the trace component relative to that of the host both have major effects on the magnitude of the melting point depression. The more incompatible a component is, the more it partitions into the melt and therefore the greater the melting point depression is. We discussed the effect of molecular weight in Section 10.4.1, but let us now consider specific examples.

10.4.3 Volatiles and melting

Volatiles, and in particular H_2O, have a strong effect on the melting point of silicate systems. The key to this behavior is, to a considerable extent, the low molecular weight of volatile species. What determines the maximum magnitude of the melting point depression is the solubility of the volatile in the silicate melt at the conditions of interest, i.e. how large $C_{z,m}$

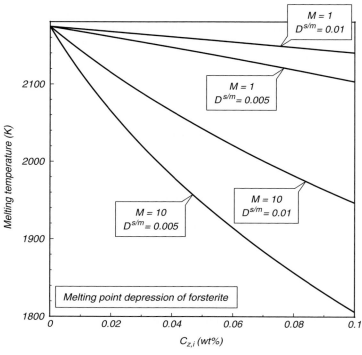

Fig. 10.4 Melting temperature of forsterite as a function of bulk initial concentration of impurity, for different values of the parameter M, and different solid–melt partition coefficients, $D^{s/m}$.

can get. But for a given mass solubility, the lower the molecular weight of the volatile, the stronger its effect on the melting point will be. Volatile solubility in silicate melts is a complicated function of volatile and melt compositions, pressure and temperature. A vast literature on this topic exists, of which a reasonably recent summary can be found in Volume 30 of *Reviews in Mineralogy*. Volatile solubilities in silicate melts increase with pressure at constant temperature. If any generalizations beyond this one are possible, one could say that: (i) H_2O solubility in silicate melts is significant (in the wt% level) beginning at pressures of a few hundred bars, (ii) CO_2 solubility reaches wt% levels only at pressures of order 10 kbar, (iii) fluorine solubility may be comparable to that of H_2O, whereas chlorine may be at least one order of magnitude less soluble, (iv) sulfur solubility is strongly dependent on oxygen fugacity and the oxidation state of iron, and (v) nitrogen solubility may become quite large only at lower mantle pressures.

Another important consideration is the nature of the sub-solidus host for the volatile species. One possibility (see Worked Example 6.3) is that a volatile phase exists at equilibrium with the rock. If the volatile phase is pure (say, pure H_2O), then the melt is saturated in the volatile species. In this case $X_{z,m}$ takes its maximum value at the given P and T, and the solidus temperature takes its minimum value (equation (10.17)). A variation on this theme, that we discuss in Section 10.7, is a situation in which the volatile phase percolates through the rock but the volatile mass flux is low enough that the melt does not become saturated in volatiles, at least initially. Another possibility is that the volatile species is an essential structural component of one or more of the mineral phases in the rock, for example, micas, amphiboles, apatite or carbonates. In this case it does not become available until the host crystalline phase breaks down. This is dehydration melting (discussed in Worked Example 6.3). Chemography requires that in this case the concentration of the volatile in the melt be lower than the saturation concentration. The dehydration-melting solidus temperature must therefore be higher than the vapor-saturated solidus (equation (10.17), Fig. 6.9). Yet another possibility, that may be important in planetary mantles, is that the volatile species is present as a trace component in a nominally anhydrous mineral.

10.4.4 Melting point depression by trace volatiles

Olivine at upper mantle pressures can dissolve a few hundred ppm of H_2O, and measured values of $D^{olivine/melt}$ for H_2O are in the range 0.006 to 0.009; see, for example, Hirschmann (2006); Hirschmann *et al.* (2005); Tenner *et al.* (2009). As pointed out by these authors, a reliable estimate of the melting point depression is complicated by the fact that the speciation of H_2O in silicate melts is far from being completely understood. We can postulate, however, that the molecular weight of H_2O species is likely to be much lower than that of silicate species, so perhaps the curves for $M = 10$ in Fig. 10.4 yield a reasonable estimate of the effect of H_2O on the melting point of forsterite. If this is the case then a few hundred ppm of H_2O dissolved in forsterite will lower its solidus temperature by as much as 150–200 K, which may be significant when considering magma generation in planetary mantles (see Section 10.6). Suppose, on the other hand, that forsterite contains the same mass of a trace incompatible cation in solution, for example potassium. We can reasonably assume that when K enters the melt it does so as a silicate species, so in this case M may be closer to 1. For a partition coefficient similar to that of H_2O the melting point depression would be only $\sim$20 K. These estimates ignore the effect of any excess chemical potential of the major melt component (i.e. $\gamma_{r,m}$ in equation (10.17)). A value of this factor greater than 1 would raise the solidus temperatures relative to those in Fig. 10.4, but the behavior depicted in Figure 10.4 is nonetheless qualitatively correct.

These results correspond to generation of an infinitesimal amount of melt at the solidus. A different question is that of *how much melt* forms as a function of temperature above the solidus. For a strongly incompatible component we expect that, as melt fraction increases above the solidus, the concentration of the trace component in the melt will initially drop very rapidly (Fig. 10.2), giving rise to a negative feedback effect that will limit melt generation. We can examine this by substituting (10.21) in (10.17) and solving for the melt fraction, φ. After some uninspiring algebra:

$$\varphi = \frac{1}{1 - D^{s/m}} \left[\frac{C_{z,i}}{100} \left(\frac{M}{1 - \left(\frac{T}{T_0} \right)^{\frac{\Delta_m S}{R}}} - M + 1 \right) - D^{s/m} \right]. \qquad (10.23)$$

Note that this equation blows up for $T = T_0$, as it must, for the liquidus ($\varphi = 1$) must also be attained at a temperature lower than the melting point of the pure system (T_0). The liquidus temperature can be calculated by substituting (10.21) in (10.17) and setting $\varphi = 1$. It is (obviously) independent of D and for all likely values of $C_{z,i}$ and M it is only a fraction of a degree to a couple of degrees lower than the melting point of the pure system. Equation (10.23) is plotted in Fig. 10.5, for a constant value of $C_{z,i} = 0.05$ ($= 500$ ppm). The plot confirms our suspicion that, even if a strongly incompatible trace component with a low molecular weight can cause a strong depression of the solidus, the increase in melt

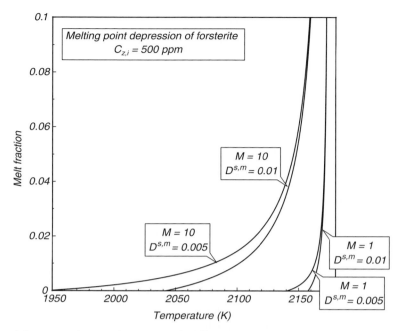

Fig. 10.5 Forsterite melt fraction as a function of temperature, for different values of the parameter M and different solid–melt partition coefficients, $D^{s/m}$. The bulk initial concentration of impurity is 500 ppm in all cases.

fraction above the solidus is painfully slow, requiring perhaps 100 K to generate 1% melt. The rate of melting for a system in which the impurity is closer in molecular weight to the major component is greater, but the melting point depression is relatively minor to begin with. There is no free lunch.

The principles summarized in Figs. 10.4 and 10.5 apply to multicomponent systems, such as rocks, but there is a potentially important difference with the one-component system, which is shown schematically in Fig. 10.6. In this figure I use the labels "dry" solidus and "dry" liquidus to refer to the solidus and liquidus of the system in the absence of a trace component. As we saw, the latter could be a volatile species, but it could also be an incompatible trace element, such as an alkali in a magnesium silicate. Regardless of the number of components in the system, addition of a trace component lowers both the solidus and the liquidus temperatures. Depression of the solidus can be drastic, depending on the nature of the impurity, its abundance, and its partition coefficient. Depression of the liquidus, in contrast, is always vanishingly small, as for $\varphi = 1$ the concentration of the trace in the melt is always small. The initial rate of increase of melt fraction may be very slow (see also Fig. 10.5) but, if the solidus depression is large enough, the melt fraction at the temperature of the dry solidus of a multicomponent system may be substantial. One way of looking at this is that the importance of trace amounts of volatiles in nominally anhydrous mantle phases is not so much that they lower the solidus temperature as the fact that they raise the amount of melt produced at temperatures near that of the dry solidus. Quantifying these effects, however, requires a detailed thermodynamic model of multicomponent silicate melts, which is beyond the scope of this book (see, for example, Hirschmann *et al.*, 1998, 1999a,b).

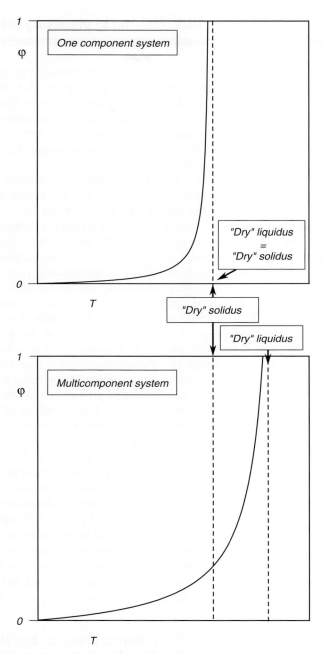

Fig. 10.6 Schematic diagrams comparing variation in melt fraction with temperature in one-component and multicomponent systems. Bulk initial concentration of impurity and all other parameters are the same in both diagrams. Liquidus depression is negligible in both cases.

10.5 Melting in planetary interiors

In Chapter 2 we discussed the sources of thermal energy in planetary interiors. With the exception of catastrophic events, such as planetary-sized impacts, core formation, tidal de-spinning and perhaps decay of abundant short-lived isotopes, all of which may have been common in the early Solar System, the rate of heating of planetary interiors is not sufficient to cause wholesale melting. Formation of magmas is always localized and entails less than complete melting of the parent solid assemblage. We seek to understand what causes localized partial melting in planetary interiors.

The ratio $\Delta H_{fusion}/C_P$ for silicate minerals is of order 10^2–10^3 K. Melting a given amount of rock requires the same amount of energy as raising its temperature by hundreds of degrees. This is a key part of the explanation for why magmas typically do not carry significant superheat, or in other words, that eruption temperatures are not normally above liquidus temperatures. It also tells us that melting must have a considerable effect on a planet's thermal gradient. Let us define our thermodynamic system as a region of the interior of a planetary body in which partial melting takes place. Focusing on its fundamental thermodynamic aspects we can consider three simple end-member situations. First, melting may take place in an open system, in which there is mass and heat transfer across its boundaries. This typically involves influx of a volatile phase that lowers the melting point of the solid assemblage (Section 10.4.4). Melting above terrestrial subduction zones may largely occur in this way. Second, melting may occur in a closed but non-adiabatic system, i.e. one that can exchange heat but not matter with its environment. This may be a magma generation process at major planetary-scale compositional discontinuities, at which jumps in density, melting point and rheological properties may allow the juxtaposition of advective and diffusive heat-transfer regimes. Partial melting of the Earth's continental crust by ponding of mafic magmas near the Moho may be an example. The problem with this mechanism is that it relies on diffusive heat transfer into the region that undergoes melting, and the very long time scales for heat diffusion may render it inoperable (Chapter 12). One way of getting around this difficulty is to decrease the diffusive lengthscale which would be the case, for example, if mafic magmas intrude the crust as a complex of closely spaced dikes. Finally, melting may occur under adiabatic conditions if the solidus of the mineral assemblage intersects the adiabatic thermal gradient. This process, called *decompression melting*, is likely to be the most widespread magma generation process in convective planetary bodies. It is responsible for magmatism at Earth's mid-ocean ridges, hot spots and large igneous provinces, such as continental flood basalts and oceanic plateaus. It is almost certainly responsible for magmatism in Venus and Mars, and perhaps for Io's volcanoes as well. Because of its importance we will discuss this process first.

10.6 Decompression melting

10.6.1 Fundamental relations

In Chapter 3 we derived expressions for the adiabatic temperature gradient, as a function of either depth (equation (3.35)) or pressure (equation (3.32)). In Chapter 4 we saw that these equations correspond to a specific type of adiabatic process, during which entropy is constant (Worked Example 4.6), and that adiabatic but not-isentropic transformations are also possible (Section 4.4). We have calculated and used adiabatic thermal gradients in several discussions without paying much attention to this distinction, which is generally safe to do in systems in which neither inelastic deformation nor phase separation take place. We can ask, however, is decompression melting truly adiabatic (see, for example, Asimow, 2002; Stolper & Asimow, 2007)? Of course, if a transformation is not adiabatic it cannot be isentropic, as heat exchange entails entropy generation, but even if melt generation is an adiabatic process, is it isentropic? We shall address these questions in a later section, but the isentropic approximation is an excellent starting point. This is so, first and foremost, because the mathematics are simple, allowing us to focus on the physics of melt generation during mantle upwelling. Second, in many instances decompression melting is approximately isentropic, at least locally. Third, it is relatively straightforward to start with the isentropic approximation and add to it the effects of entropy gain or loss arising from exchange of heat and matter with the environment, or from energy dissipation. Throughout this discussion, and unless otherwise stated, I will continue to use the terms adiabat, and adiabatic decompression melting, to mean *adiabatic and isentropic,* as this is common throughout the literature. When necessary I will explicitly state whether departures from the constant entropy assumption need to be taken into consideration.

Using equation (3.35) and the thermodynamic properties of forsterite yields a characteristic adiabatic gradient for the Earth's upper mantle of ~ 0.4 K km^{-1}, or, using equation 3.32, ~ 1.5 K kbar^{-1}. The measured volume and entropy of melting of upper mantle minerals (forsterite, diopside, enstatite and spinel) yield Clapeyron slopes for their melting reactions (equation (5.6)) of 50–100 bar K^{-1}, or equivalently 10–20 K kbar^{-1}. Clearly, the adiabat and the melting curve for the mantle can intersect. Whether and where they intersect, however, and what happens next, depend not only on their relative slopes but also on their absolute locations.

We will consider melting under Earth's mid-ocean ridges as our reference model – after all, this is where most of Earth's volcanic activity takes place. Let us assume that the oceanic lithosphere extends to a depth of 150 km, and that the temperature at the base of the lithosphere is 1650 K (= 1377 °C, see Chapter 3). The pressure at that depth is ~ 50 kbar (Chapter 8). From these values and a slope of 1.3 K kbar^{-1} we derive the mantle adiabat shown in Fig. 10.7. The intersection of the mantle adiabat with the Earth's surface ($P = 0$) defines the mantle's *potential temperature*, which in our example is $T_p = 1312$ °C. This is the temperature that the mantle would have if it were allowed to decompress adiabatically (and, remember, isentropically) to the planet's surface, i.e. if the lithosphere did not exist and phase changes did not take place. This may not happen, but knowing the potential temperature is important because, given that an adiabat is fully determined if we specify a single $\{P, T\}$ point on it (Section 3.5), T_p allows us to compare the thermal state of different regions of a convective mantle, as well as the mantles of different planetary bodies. In other words, comparing how much hotter or colder different parts of a convective region are requires that we specify the pressure at which we make the comparison, and we choose zero as the reference pressure.

In Section 3.7.2 we defined the lithosphere as the thermal boundary layer for mantle convection, meaning that heat transfer across the lithosphere is by diffusion. If we ignore radioactive heat production (which, if not quite right, is not altogether unacceptable for the oceanic lithosphere) then the equilibrium lithospheric geotherm is a straight line. In Fig. 10.7 I show this conductive geotherm with a straight line joining the 1650 K temperature at the base of the lithosphere to a surface temperature of 300 K. This translates to a heat flux of

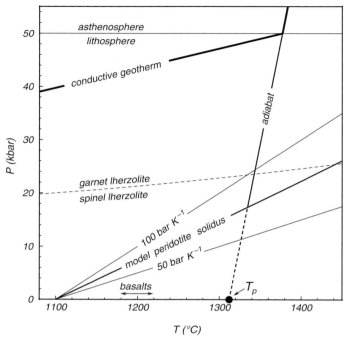

Simplified thermal conditions in the Earth's oceanic upper mantle. The thick "dog-leg" line shows the geotherm far away from a mid-ocean ridge. Under a mid-ocean ridge adiabatic upwelling of the mantle intersects the peridotite solidus. If the adiabat were able to reach the surface unperturbed its temperature would be the potential temperature, T_p. The difference between T_p and basalt eruption temperatures (range shown with an arrow) reflects the enthalpy of melting.

$\sim$50 mW m^{-2}, which is a bit on the low side for old ocean floor. Including radioactive heat production in the lithosphere would probably raise the value to the right ballpark, but this is not important for the present discussion. What is important is that the geotherm in an old sector of ocean floor, effectively infinitely removed from any region of active magma generation, would look like the thick "dog-leg" line in Fig. 10.7, composed of an adiabatic (advective) segment capped by a diffusive segment. There are places, however, where advective flow is able to penetrate to depths that are significantly less than that of the "typical" lithospheric base. This process may be driven either from "above": the lithosphere thins in response to stretching, or from "below": the rate of heat advection is high enough to thermally "erode" the lithosphere. For our purposes it does not matter which is the case. What matters is that under those circumstances the adiabatic thermal gradient will extend to shallower depths, as shown by the line labeled "adiabat" in Fig. 10.7.

The figure also shows the dry model peridotite solidus, after McKenzie and Bickle (1988). The adiabat and the solidus intersect at a pressure of $\sim$17 kbar, indicating where decompression melting begins during mantle upwelling. The intersection occurs in the spinel lherzolite field. This is consistent with the phase equilibrium and geochemistry of MORBs, suggesting that the processes and thermal conditions summarized in Fig. 10.7 are a feasible model for melting under Earth's mid-ocean ridges. The zero-pressure solidus temperature is 1100 °C. Hypothetical solidi going through this temperature and with Clapeyron slopes of 50 and 100 bar K^{-1}, which bracket the slope range for typical mantle minerals at low pressure, are also shown. McKenzie and Bickle's model solidus is almost exactly in the middle, which may be reassuring at some (subjective) level.

If the enthalpy of fusion of rocks were zero and the melts formed by decompression melting rose isentropically then they would reach the surface at the mantle's potential temperature. This is not the case. Basalt eruption temperatures are typically $\sim$1200 °C. The difference between this temperature and the mantle's potential temperature must reflect at least in part the conversion of sensible heat to latent heat during melting, as first-order phase transitions are always accompanied by a non-zero enthalpy change. We will discuss this issue in detail in the next section, but before doing so there are a few more things to learn from Fig. 10.7.

First, the eruption temperature of a magma formed by adiabatic decompression melting must be lower than the potential temperature of its mantle source region. Thus, whereas Fig. 10.7 may be able to account for melting under mid-ocean ridges, it cannot explain high-Mg basaltic lavas with eruption temperatures in excess of 1300 °C, let alone ultramafic komatiite lava flows. These require mantle potential temperatures significantly higher than 1312 °C, which may have been the norm in the Archaean mantle (Worked Example 3.4). Second, mafic lavas exist that have major and trace element characteristics indicative of having formed from garnet-bearing peridotites. Figure 10.7 shows that melting in the garnet lherzolite field can take place through a combination of higher potential temperature, steeper Clapeyron slope and lower solidus temperature at a given pressure. In view of our discussion in Section 10.4, the fact that some of these deep magmas tend to be rich in incompatible elements (e.g. K) and dissolved volatiles may be significant.

10.6.2 Batch decompression melting of a one-component system

We initially consider decompression melting under isentropic conditions. This requires that the melt remain in the system and in equilibrium with the solid, a process that as we saw in Section 10.3 is known as batch melting. We will later relax this constraint and allow the melt to leave the system at the same rate as that at which it is produced. This process, known as fractional melting, is obviously not adiabatic, and hence not isentropic.

Imagine an idealized planetary mantle composed of a single component (Fig. 10.8). At pressures greater than P_i the temperature is below the solidus and the mantle consists of a single solid phase, so that it is a divariant thermodynamic system. Upwelling of solid mantle occurs along an adiabat with potential temperature T_p, from A to B in the figure. At point B the adiabat intersects the solidus at temperature T_i and, as a second phase appears, the system becomes univariant. Adiabatic decompression beyond the intersection point cannot follow the adiabat calculated with equation (3.32). Rather, it must be constrained to the solidus (which in this case is also the liquidus), from B to C in Fig. 10.8. It is important to realize that the two-phase segment of the decompression path, BC, is also adiabatic. However, it is not described by equation (3.32) because this equation does not take into consideration the enthalpy of fusion. At point C, when the upwelling mantle reaches pressure P_f, melting is complete and the system becomes divariant once more. Further ascent, from C to D, occurs along another adiabat calculated with equation (3.32), but with a different potential temperature, $T_{p,m}$, and perhaps a different slope arising from differences in material properties between solid and liquid. The drop in potential temperature from the AB adiabat to the CD adiabat reflects the enthalpy absorbed by the melting phase transition.

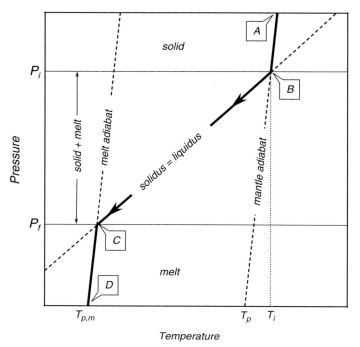

Fig. 10.8 Decompression melting of a one-component system. The solid mantle (divariant in a one-component system) rises adiabatically from A to B. The solidus (which is also the liquidus in a one-component system) is intersected at pressure P_i, temperature T_i (point B). At B the system becomes univariant and adiabatic upwelling is constrained to the melting curve BC. Melt fraction increases from B to C, as sensible heat is converted to latent heat. Melting is complete at C, and melt rises adiabatically from C to D.

We seek an equation that allows us to calculate the extent of melting as a function of pressure along the two-phase segment of the adiabat. We will initially restrict our discussion to the simple one-component system depicted in Fig. 10.8, because this problem has a simple analytical solution that will allow us to examine the fundamental thermodynamic aspects of decompression melting. Later we will see how the results generalize to multicomponent systems in which there is a finite temperature difference between solidus and liquidus.

If decompression melting occurs during isentropic upwelling then we can begin with equation (4.140) and modify it so as to include the thermal effect of the phase transition. We write the new equation on a molar basis, as follows:

$$dS = \left(\frac{\partial S}{\partial T}\right)_P dT + \left(\frac{\partial S}{\partial P}\right)_T dP + \left(\frac{\partial S}{\partial \varphi}\right)_{P,T} d\varphi = 0, \qquad (10.24)$$

where φ is (molar) melt fraction, i.e. $\varphi = 1$ corresponds to melting of one mol of substance. Clearly, $(\partial S/\partial \varphi)_{P,T}$ is the molar entropy of fusion, $\Delta_m S$, so (10.24) becomes (see also (4.141)):

$$\frac{C_P}{T}dT - \alpha V dP + \Delta_m S\, d\varphi = 0. \qquad (10.25)$$

Let the Clapeyron slope of the melting reaction be $y = (dP/dT)_{melting}$. The temperature at any point along the two-phase segment, BC in Fig. 10.8, is then given by:

$$T = T_i + \frac{P - P_i}{y}. \tag{10.26}$$

Taking the derivative of (10.26), substituting in (10.25), rearranging and simplifying we get:

$$\frac{d\varphi}{dP} = -\frac{C_P}{\Delta_m S (y T_i + P - P_i)} + \frac{\alpha V}{\Delta_m S}. \tag{10.27}$$

The material properties C_P, V and α correspond to the two-phase system and must therefore vary between those of the pure solid at B to those of the pure melt at C. Including weighing functions such as $C_p = C_p(\varphi)$, etc., is certainly possible but it complicates the equation tremendously, requires a numerical solution, and obscures the physics contained in 10.27. For simplicity we will use values for C_P, V and α that are intermediate between characteristic solid and melt values for upper mantle phases. Reasonable estimates are: $C_p \approx 200$ J K^{-1} mol^{-1}, $\alpha \approx 6 \times 10^{-5}$ K^{-1} and $V \approx 4.7$ J bar^{-1} mol^{-1}. The other parameters in equation (10.27) are independent of φ. We can choose typical values of 56 J K^{-1} mol^{-1} for the entropy of melting, and 80 bar K^{-1} and 1600 K for the Clapeyron slope and initial melting temperature in the upper mantle, respectively.

Equation (10.27) applies to any point along the two-phase segment of the adiabat: $\{T_i, P_i\}$ could be the point where the solidus is first intersected, or it could be any other point between B and C. Let us consider first a situation in which $\{T_i, P_i\}$ is some arbitrary point in the two-phase segment of the adiabat, and P is infinitesimally close to P_i, so that $P_i - P \approx 0$ and $T \approx T_i$. We can then simplify (10.27) as follows:

$$\frac{d\varphi}{dP} \approx -\frac{C_P}{y T \Delta_m S} + \frac{\alpha V}{\Delta_m S}, \tag{10.28}$$

and in this form it is easier to dissect the physics of decompression melting. Note first that decompression melting means that $d\varphi/dP < 0$. Because all of the variables on the right-hand side of (10.28) are positive quantities, it follows that the first term contributes to melting, whereas the second one opposes melting. The reason is obvious: the first term represents conversion of sensible heat to latent heat (it contains the ratio $C_p/\Delta_m H$) whereas the second term represents the amount of thermal energy that is converted to expansion work. The melting rate is controlled by the *difference* between the Clapeyron slope of the melting reaction and the slope of the isentrope. This can be seen graphically in Fig. 10.8: the energy available for melting is a function of the horizontal distance between the adiabat and the melting curve. The faster the divergence rate between the two curves, the greater the melting rate will be. Using the values for the upper mantle that we chose in the previous paragraph we get:

$$-\frac{C_P}{y T \Delta_m S} \approx -0.028 \text{ kbar}^{-1}, \quad \frac{\alpha V}{\Delta_m S} \approx 0.005 \text{ kbar}^{-1}. \tag{10.29}$$

In this case the energy that goes into expansion work is about half an order of magnitude smaller than the energy used in melting. The resulting melting rate, $d\varphi/dP \approx -0.023$ kbar^{-1} is quite low. Assuming that the melting rate stayed constant (which it doesn't because (10.28) is a linear approximation to (10.27)) complete melting would require $\sim$50 kbar of decompression. Compare this value to our estimate in Fig. 10.7 that melting under Earth's mid-ocean ridges begins at a pressure of $\sim$17 kbar.

Equation (10.28) shows that the rate of decompression melting increases with *decreasing* melting temperature. As we saw in Section 10.1.1, melting is a discontinuous loss of long range order. Because the lattice disorder term is not a strong function of temperature, entropy of melting can be expected to stay more or less constant with temperature (equation (10.6)). Entropy of melting may vary with pressure, but again we can expect such variations to be small, as they would arise from *differences* in the compressibilities of solid and melt (equation (10.6)). Experimental measurements generally confirm these expectations (e.g. Kojitani & Akaogi, 1997). If the entropy of melting stays constant then the enthalpy of melting must increase with temperature, and the melt productivity must concomitantly decrease. This has the somewhat counterintuitive consequence that decompression melting along an adiabat with a higher potential temperature (e.g. at mantle hot spots) requires a greater amount of decompression to produce the same amount of melt that would be produced along a colder adiabat (e.g. at mid-ocean ridges), although melting would begin deeper along the hotter adiabat (Fig. 10.7).

In order to calculate the total amount of melt produced by a given amount of decompression we integrate equation (10.27) between $\varphi = 0$ at $P = P_i$ and any other value of $P \geq P_f$. Ignoring changes in C_P, V and α with φ we get:

$$\varphi = -\frac{C_P}{\Delta_m S} \ln\left(\frac{yT_i + P - P_i}{yT_i}\right) + \frac{\alpha V}{\Delta_m S}(P - P_i). \tag{10.30}$$

Note that φ only depends on the amount of decompression, not on the absolute value of P_i, although of course T_i changes with P_i. Defining the pressure interval over which decompression melting takes place as $\Delta P = P - P_i$, we can re-write (10.30) as follows:

$$\varphi = -\frac{C_P}{\Delta_m S} \ln\left(1 + \frac{\Delta P}{yT_i}\right) + \frac{\alpha V}{\Delta_m S}\Delta P. \tag{10.31}$$

This equation overestimates actual melt production for two reasons. In the first place, (10.31) describes melting of a one-component system, in which the solidus and the liquidus coincide. In reality the mantle is a multicomponent system in which the temperature interval between solidus and liquidus is of the order of several hundred degrees. In order for melting to continue above the solidus temperature must increase relative to the value given by equation (10.26), so that less "sensible heat" is available for conversion to enthalpy of melting. We will estimate the magnitude of this effect in Section 10.6.4. Second, equation (10.31) assumes batch melting, i.e. that melt and solid remain at equilibrium throughout the melting process. If melt is extracted as it is produced then the system is no longer adiabatic. The system in this case loses heat (and thus entropy), and melt productivity must decrease. We prove this formally next.

10.6.3 Fractional decompression melting of a one-component system

Batch melting, in which the melt remains in equilibrium with the solid until a desired melt fraction is attained, or until melting is complete, is one end-member of a spectrum that extends, at the other end, to fractional melting. During fractional melting melt is removed at the same rate as it is produced, so effectively at any given moment there is only an infinitesimal amount of melt in equilibrium with the solid. The process is not adiabatic, but it can be broken down into an infinite number of steps, each of which is adiabatic and isentropic. Equation (10.27) applies to each step: an infinitesimal amount of melt is produced isentropically, then removed, a new melt increment is produced, and so on. The equation needs to be modified, however, to account for the fact that the mass of rock that undergoes partial melting decreases continuously. Each melt increment in equation (10.27) will now be measured relative to the amount of solid that remains. Call this melt infinitesimal $d\varphi_r$. We then re-write equation (10.27) as follows:

$$\frac{d\varphi_r}{dP} = -\frac{C_P}{\Delta_m S\,(yT_i + P - P_i)} + \frac{\alpha V}{\Delta_m S} \tag{10.32}$$

with the understanding that (10.32) applies to each infinitesimal isentropic melting step. If the total amount of melt generated and extracted from one (initial) mol of solid is φ, then a small incremental amount of melt of $\delta\varphi$ mols corresponds to a per mol increment in melt fraction, $\delta\varphi_r$ (i.e. measured relative to the remaining solid) of:

$$\delta\varphi_r = \frac{\delta\varphi}{1 - \varphi} \tag{10.33}$$

so, passing to the limit:

$$\frac{d\varphi}{d\varphi_r} = 1 - \varphi. \tag{10.34}$$

From (10.32) and (10.34), using the chain rule:

$$\frac{d\varphi}{dP} = \frac{d\varphi}{d\varphi_r}\frac{d\varphi_r}{dP} = (1 - \varphi)\left(-\frac{C_P}{\Delta_m S\,(yT_i + P - P_i)} + \frac{\alpha V}{\Delta_m S}\right), \tag{10.35}$$

which we integrate by rearranging as follows:

$$\frac{d\varphi}{(1 - \varphi)} = \left(-\frac{C_P}{\Delta_m S\,(yT_i + P - P_i)} + \frac{\alpha V}{\Delta_m S}\right)dP. \tag{10.36}$$

The integral of the right-hand-side term is identical to (10.30), so, after rearranging:

$$\varphi = 1 - \exp\left[\frac{C_P}{\Delta_m S}\ln\left(1 + \frac{\Delta P}{yT_i}\right) - \frac{\alpha V}{\Delta_m S}\Delta P\right]. \tag{10.37}$$

Melt production by fractional decompression melting of a one-component system (equation (10.37)) is compared to batch melting (equation (10.30)) in Fig. 10.9. Material properties and other physical parameters are the same as those used in equations (10.29). As we expected, fractional melting generates less melt than batch melting, because migratory coconuts err.... melts extract thermal energy from the system. The two curves diverge strongly for large values of ΔP. Since the enthalpy of fusion comes exclusively from cooling, the system can never become completely molten if it loses melt at the same rate as it is produced, so the fractional melting curve approaches $\varphi = 1$ asymptotically. The two curves, for batch and fractional melting, bound a region of possible decompression melting behaviors. Melting in nature is likely to proceed in some intermediate fashion. One possibility is that melt is extracted continuously once some finite melt fraction is attained – this differs from fractional melting in that a finite, as opposed to infinitesimal, amount of

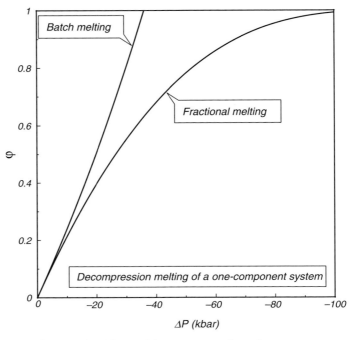

Batch melting

Fractional melting

Decompression melting of a one-component system

φ

ΔP (kbar)

Fig. 10.9 Variation in melt fraction during batch and fractional decompression-melting of a one-component system. Batch melting is assumed to be isentropic. During fractional melting melt leaves the system at the same rate as it is produced, carrying thermal energy, and entropy, with it.

melt is always present. Another possibility is that melt is generated in small batches that are extracted discontinuously. Whatever the actual melting mechanism, however, melt productivity by decompression melting of a one-component system is bound by two curves such as in Fig. 10.9.

10.6.4 Decompression melting in multicomponent systems

Melting of rocks takes place over multivariant intervals, extending from the solidus to the liquidus. A rather simplistic view of this process is presented in Fig. 10.10. As in the one-component case (Fig. 10.8) melting begins at point B, at temperature T_i and pressure P_i, where the adiabat with potential temperature T_p intersects the solidus. The melt + solid assemblage is in this case multivariant, so temperature increases above the solidus as melting progresses. Adiabatic upwelling therefore causes the P–T conditions of the system undergoing melting to describe a path such as BC in the figure. The liquidus is intersected at point C, and further upwelling occurs along the adiabat with potential temperature $T_{p,m}$. Compared to the one-component system (Fig. 10.8), in this case there is less thermal energy available for melting, because some enthalpy must be used to raise the temperature above the solidus. We seek to estimate the magnitude of this effect.

For simplicity we will assume that the solidus and the liquidus are parallel, and that melt fraction varies linearly between 0 at the solidus and 1 at the liquidus. As it turns out, this by itself is not a bad assumption for melting of peridotite, but it masks another effect that is

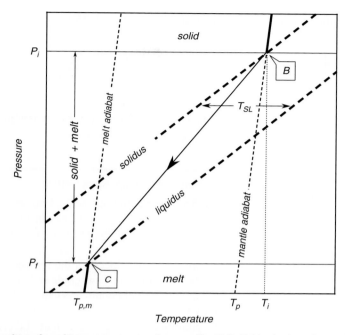

Fig. 10.10   Decompression melting of a multi-component system (compare Fig. 10.8). Melting begins at B, as in the
one-component system, but now in order for additional melting to take place the temperature must increase above
the solidus, as shown by the arrow. The temperature interval between solidus and liquidus, T_{SL}, is assumed to be
independent of pressure. In reality the gap between peridotite solidus and liquidus narrows with increasing pressure.

important, and that we will completely ignore. This is the fact that the nature of the melting
reaction changes from the solidus to the liquidus, and also with pressure. We are ignoring
differences in the enthalpy of melting of the various reactions, as well as the enthalpies
of possible reactions among solid phases. The only way of including these effects is by
means of a sophisticated thermodynamic model for silicate melts such as pMELTS, which
is beyond the scope of this book.

Let T_{SL} be the constant temperature difference between solidus and liquidus. Ignoring all
compositional effects, and assuming a linear increase in melt fraction over the temperature
interval T_{SL}, we write the equation for the BC melting curve in Fig. 10.10 as follows:

$$T = \varphi T_{SL} + T_i + \frac{P - P_i}{y}. \qquad (10.38)$$

The total differential of T is:

$$dT = \left(\frac{\partial T}{\partial \varphi}\right) d\varphi + \left(\frac{\partial T}{\partial P}\right) dP = T_{SL} d\varphi + \frac{dP}{y}. \qquad (10.39)$$

Substituting (10.38) and (10.39) in (10.28), making $\Delta P = P - P_i$, and rearranging we
obtain the following expression for batch melting:

$$\left[\frac{C_P T_{SL}}{\varphi T_{SL} + T_i + \frac{\Delta P}{y}} + \Delta_m S\right] d\varphi + \left[\frac{C_P}{y(\varphi T_{SL} + T_i) + \Delta P} - \alpha V\right] d(\Delta P) = 0 \qquad (10.40)$$

and similarly for fractional melting, using (10.36):

$$\left[\frac{C_P T_{SL}}{\varphi T_{SL} + T_i + \frac{\Delta P}{y}} + \Delta_m S \right] \frac{d\varphi}{1 - \varphi} + \left[\frac{C_P}{y(\varphi T_{SL} + T_i) + \Delta P} - \alpha V \right] d(\Delta P) = 0. \quad (10.41)$$

These equations cannot be integrated analytically, but generating numerical approximations is a simple exercise with *Maple* (Software Box 10.1).

Software Box 10.1   Calculation of decompression melting

The *Maple* worksheet **decompression_melting.mw** contains two procedures, batch_melting and fractional_melting, that use *Maple's* fsolve to solve equations (10.40) and (10.41), respectively. The two procedures are identical except for the core equation.

When invoking the procedure one specifies the temperature and pressure at which the adiabat first intersects the solidus (in Centigrade and kbar) and a filename for *Maple* to send the output to. All other model parameters can be changed by editing the procedure. This includes: the Clapeyron slope of the solidus, the solidus–liquidus temperature interval, heat capacity, entropy of melting, coefficient of thermal expansion, molar volume, and the pressure increment for the solution. The latter must be a negative number (decompression) and must be expressed in bar.

The procedure generates a text file in which each line corresponds to a decompression step, extending from the initial intersection of the solidus with the adiabat to the planet's surface. The first field in each line is the total amount of decompression (in kbar), the second field is melt fraction, the third field is temperature (in Centigrade), and the fourth field is pressure (in kbar).

Worked Example 10.1   Decompression melting in the Earth's mantle

We apply numerical solutions of equations (10.40) and (10.41) to study decompression melting of the Earth's upper mantle under different thermal regimes. We start with a model dry peridotite solidus, simplified after McKenzie and Bickle (1988) for $P < 45$ kbar, and Herzberg *et al.* (2000) at higher pressure, as shown in Figs. 10.11 and 10.12. A characteristic value of T_{SL} for mantle peridotites is $\sim$500 K. We therefore draw the peridotite liquidus parallel to the solidus and displaced 500 K towards higher temperature. We study decompression melting along three possible mantle adiabats, with potential temperatures of 1300, 1500 and 1700 °C. The first two may represent conditions under present-day terrestrial mid-ocean ridges and oceanic islands, respectively. The last one represents possible conditions in the Archaean Earth. The three adiabats intersect the peridotite solidus at temperatures of approximately 1600, 1820 and 2050 K, and pressures of 16, 34 and 57 kbar, respectively. We use the same material properties and other physical parameters as in previous examples (e.g. equations (10.29)), except that we make the Clapeyron slope $y = 80$ bar K^{-1} for melting at mid-ocean ridges and oceanic islands, and 85 bar K^{-1} along the hotter adiabat, to reflect the fact that the peridotite solidus steepens with increasing pressure (we discuss this in more detail in the next example).

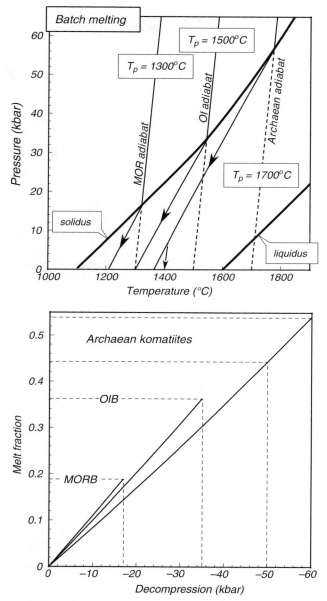

Fig. 10.11   Simplified model for batch melting of the terrestrial mantle. Solidus after McKenzie and Bickle (1988) for
$P < 45$ kbar, and Herzberg *et al.* (2000) at higher pressure. The liquidus drawn parallel to the solidus with
$T_{SL} = 500$ K. Three mantle adiabats are shown, with potential temperatures of 1300, 1500 and 1700°C. Solid lines
with arrows show melting paths, corresponding to the segment BC in Fig. 10.10. The broken arrow forking from the
hottest adiabat shows adiabatic ascent of melt with eruption temperature of 1400°C. The bottom diagram shows
melt fraction as a function of decompression. A range is shown for the hottest adiabat ("Archaean komatiites"), bound
between the decompressive path that generates melt with 1400°C eruption temperature and the maximum possible
decompressive path (to $P = 0$).

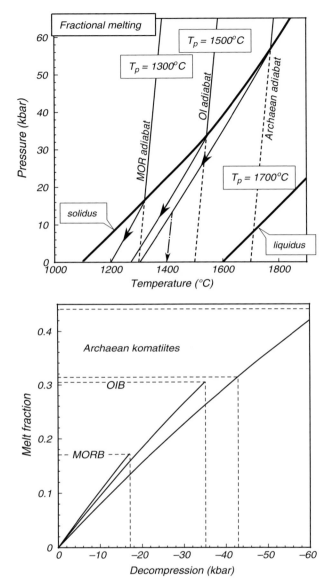

Fig. 10.12   Same as Fig. 10.11, but for fractional melting.

The solid arrows in Figs. 10.11 and 10.12 are isentropic melting paths, corresponding to BC in Fig. 10.10. The $P-T$ paths are calculated by first solving equations (10.40) or (10.41) for φ as a function of P, and then substituting these values of φ in equation (10.38). In every case the decompression melting paths intersect the surface before reaching the liquidus. The batch melting paths reach the surface at temperatures of $\sim$1200 °C and $\sim$1300 °C for mid-ocean ridges and oceanic islands, respectively. The corresponding temperatures assuming fractional melting are $\sim$10–20 °C lower. If mantle upwelling caused decompression melting to continue to the surface erupted magmas would have these temperatures. Eruption temperatures would be higher if melts separated from their source regions at depth. The agreement with observed basalt eruption temperatures is quite good, but this should be hardly surprising, as the potential temperature of the mantle is to a large extent constrained from basalt eruption temperatures.

Calculated melt fractions along the isentropic melting paths are shown in the bottom panels of the figures. Maximum melt production at mid-ocean ridges, assuming that the upwelling mantle melts continuously from 17 kbar (where the solidus is intersected) to zero pressure, ranges from 17%–19%, depending on whether one assumes fractional or batch melting, respectively. This happens to be a reasonably good estimate. The corresponding values for melting under oceanic islands are 30%–36%, for a maximum decompression interval of 35 kbar. These values possibly overestimate actual melt fractions to some extent, although the generally more mafic nature of ocean island basalts compared to MORBS is consistent with them representing a higher melt fraction of peridotite.

Melting of the mantle along our hypothetical Archaean adiabat, from the solidus at 57 kbar to the surface, would generate 44%–54% melt with eruption temperatures of 1300–1350 °C. Komatiite eruption temperatures appear to have been somewhat higher than this, perhaps ~1400 °C. Melts with this eruption temperature could have been extracted from the upwelling mantle region at pressures of 7–14 kbar, for batch and fractional melting, as shown by the broken arrows in the figures. The more restricted decompression melting intervals would generate between 32% and 44% melt, which is not altogether unreasonable for komatiites. The geometry of the figure shows that an adiabat with a potential temperature higher than 1700 °C would intersect the solidus deeper and the resulting melting path would reach the 1400 °C melt adiabat at a shallower depth than the one emanating from the 1700 °C adiabat. Thus, it would generate a higher fraction of melt with the same eruption temperature. No surprise here – it is simply energy conservation.

Worked Example 10.2   Melting in the Martian mantle

We can use equations (10.40) and (10.41) to assess the plausibility of the Martian mantle models that we developed in Chapter 3. In Section 3.9 we saw that two possible models for the Martian interior are a relatively thin lithosphere (~250 km thick) capping a hot mantle, or a thick lithosphere (~700 km) overlying a colder mantle, but still hotter than the Earth's. Let us assume that the mantle potential temperatures for the two models are 1800 °C and 1500 °C, respectively, which are approximately 500 °C and 200 °C hotter than the terrestrial mantle under mid-ocean ridges. Intersection of these adiabats with the model peridotite solidus (shown in Fig. 10.11) occurs at temperatures of 1543 °C and 1935 °C, and pressures of ~34 and 105 kbar, respectively. We assume that this solidus curve is applicable to the Martian mantle, which is not strictly true but there are larger sources of uncertainty. The Clapeyron slope of the solidus steepens sharply with pressure above about 60 kbar. The adiabat also steepens, although to a lesser extent. Equations (10.40) and (10.41) assume that both of these quantities are constant, and although it is possible to modify them to account for their pressure dependency this complicates the equations considerably and is left as an exercise for the reader. As we saw, the degree of melting is ultimately determined by the *difference* between the slope of the isentrope and that of the melting reaction, and this difference varies less strongly with pressure than the Clapeyron slope of the solidus.

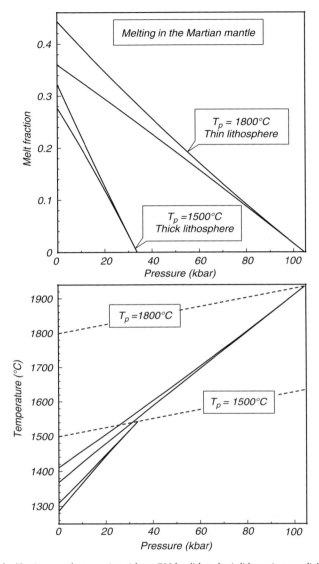

Fig. 10.13 Melt generation in the Martian mantle, assuming either a 700 km lithospheric lid capping an adiabatic mantle with $T_P = 1500\,°C$, or a 250 km thick lithospheric lid over an adiabatic mantle with $T_P = 1800\,°C$. Envelopes spanning batch melting to fractional melting are shown in both cases. Melting in the thin lithosphere model begins within the asthenosphere and is probably inconsistent with the lack of planet-wide volcanism. Melting in the thick lithosphere model would begin well within the lithosphere, and would only be made possible by the existence of large and long-lived mantle plumes. This may be the reason why only two large and isolated young volcanic provinces, Tharsis and Elysium, exist on Mars.

As a first approximation, then, we will assume that these quantities stay constant along each decompression path, but choose different Clapeyron slopes, of 80 and 130 bar K^{-1} for the melting paths beginning at 34 and 105 kbar, respectively.

The calculated decompression melting paths are shown in Fig. 10.13, as a function of pressure in the Martian interior (*not* ΔP, compare Figs. 10.11 and 10.12). Envelopes are shown for each adiabat, spanning the range between batch and fractional melting. The pressure at the bottom of a 700 km thick Martian lithosphere is approximately 86 kbar. If the potential temperature of the underlying convective mantle is 1500 °C melting would begin at a pressure of ∼34 kbar, or more than halfway across the lithosphere. In other words, the asthenosphere must be able to penetrate some 400 km of lithosphere before it starts melting. One could expect that volcanic activity in this model would be sparse, and limited to places where large and very long-lived hot spots are capable of transferring enough heat to "thermally drill" through the lithosphere. If a decompression melting regime can be established melt fractions of up to about 30% would be achieved, and the magmas could have eruption temperatures of at least 1300 °C (higher if magmas are segregated at depth). A hotter Martian mantle, with a potential temperature of 1800 °C, would start melting at about 105 kbar. The pressure at the bottom of the 250 km lithosphere capping that mantle would be about 30 kbar. Melting in this case would be widespread, and melt fractions of 40% or higher would not be uncommon.

Although the results shown in Fig. 10.13 are only rough approximations, the qualitative distinction between the magmatic activity that is to be expected from the two Martian mantle models is probably robust. The thick lithosphere model is consistent with the existence of few, large and long-lived magmatic provinces, which is what is observed in Mars. The thin lithosphere would generate a much more volcanically active planet than is observed.

I wish to emphasize that the equations and numerical examples presented in this section are extremely simplified pictures of decompression melting in nature. They are only intended to convey the thermodynamic underpinnings of the process, and in that sense I believe that their simplicity is hard to argue with. The fundamental physics are contained in the equations, even if they cannot provide any petrological detail. Thorough mathematical and numerical treatments of decompression melting in real muticomponent systems have been presented, among others, by McKenzie (1984), McKenzie and Bickle (1988, and most recommended), Iwamori *et al.* (1995), Asimow and Stolper (1999) and Asimow (2002). The reader is urged to consult these works for what is missing above.

10.6.5 Departures from the constant-entropy assumption

During fractional decompression melting heat loss from the magma source region lessens melt production relative to batch melting. There are other reasons why the results of the previous sections, which assume either global or local isentropic conditions, may have to be revised. For example, we may ask, what must be the size of the region undergoing decompression melting so that diffusive heat loss can be safely ignored, i.e. so that the system is truly adiabatic? If diffusive heat loss is not negligible then melt production will be lower than that predicted by the equations derived above. We may also wonder whether the melting region is isentropic, even if it is adiabatic. If energy dissipation takes place inside an adiabatic system then $dS > 0$, and in this case melt production will be higher than what those equations predict. We can generate order of magnitude estimates for some of these effects.

For the sake of expository synthesis I will call the ascending region of the mantle that undergoes decompression melting a "diapir", without necessarily implying that I know what a diapir is, nor that there is consensus that diapirs as envisioned by Ramberg (1967) exist. We can estimate the effect of diffusive heat exchange between the diapir and its environment by means of the Péclet number (Section 3.6.2), and we can do this in two ways. First, consider vertical heat flow, i.e. in the same direction that the diapir is moving. Figure 10.7 shows that the characteristic depth at which melting begins under mid-ocean ridges is of order 50 km. Using this value for Λ in equation (3.46), assuming $u = 5$ cm yr$^{-1} \approx 1.6 \times 10^{-9}$ m s^{-1} and $\kappa = 10^{-6}$ m^2 s^{-1} we get $Pe \approx 80$. Thus, mantle upwelling is 80 times more efficient than diffusion in transporting heat in the direction in which progressive melting takes place, which says that we are justified in considering the system adiabatic in this direction. But what about horizontal heat diffusion? Equation 3.48 relates the lengthscale of advective heat transfer (Λ) to the lengthscale of diffusive heat transfer in a direction perpendicular to advection (λ). This can be applied to a situation in which hot asthenosphere penetrates, and is surrounded by, colder lithosphere (Fig. 10.7). Substituting the definition of Péclet number (equation (3.46)) in (3.48) we obtain:

$$\lambda \sim 2 \left(\frac{\Lambda \kappa}{u} \right)^{1/2}. \tag{10.42}$$

As before, Λ is the depth at which decompression melting begins ($\sim$50 km), as this is the distance over which the thermal perturbation caused by melting will develop. With the same values of u and κ as before we find $\lambda \approx 50 \, \Lambda^{1/2}$, for λ and Λ in meters. We can expect heat diffusion to affect an approximately 10 km wide rind of a 50 km tall diapir. Thus, if the diapir is at least a few tens of km wide, then its outer part may cool down significantly, but there will always be a core region in which the assumption of adiabatic behavior is warranted. If adiabatic upwelling continues for an extended period of time, moreover, then the environment will heat up, decreasing the horizontal thermal gradient and hence the rate of heat loss. It is significant that λ goes as the square root of Λ. For example, if melting occurred at the base of the thick Martian lithosphere ($\sim$700 km) a diapir would have to be some 3–4 times wider than under Earth's oceans in order to be able to generate eruptable magmas. The size of the Tharsis and Elyseum volcano-tectonic bulges (1000–5000 km) suggests that Martian mantle plumes were (are?) larger than Earth's.

Sources of energy dissipation inside an adiabatically rising diapir include: inelastic rock deformation, dissipation of gravitational potential energy by separation of melt and solid of different densities, viscous flow of melt, radioactive heating and chemical diffusion. We will examine only the first two processes (see Asimow, 2002, for a complete mathematical treatment of all of them). We seek equations for the rate of entropy production with pressure, $(\partial S / \partial P)_{pr}$. Because this entropy is generated inside the system, and the system is still assumed to be adiabatic, there is no entropy exchange with the environment, and we can modify equation (10.25) as follows:

$$\frac{C_P}{T} dT + \left[\left(\frac{\partial S}{\partial P} \right)_{pr} - \alpha V \right] dP + \Delta_m S \, d\varphi = 0. \tag{10.43}$$

We derived equation (1.83) to quantify frictional heating in a fault or shear zone, but it can also be applied to estimate energy dissipation by inelastic deformation inside a diapir. Let us re-write (1.83) as follows:

$$\frac{dT}{dx} = \frac{\tau V}{z C_P}, \tag{10.44}$$

where τ is the magnitude of the shear stress that causes deformation in the diapir, z is the width over which inelastic strain is distributed, V is the molar volume of the material and

C_P its molar heat capacity (equation (1.83) is written in terms of density and specific heat, but in the present discussion I find it simpler to use molar quantities). Because dx is a displacement in the direction in which the shear stress is applied, and shear stress in an ascending diapir is chiefly vertical, we can interpret the ratio dT/dx as the vertical thermal gradient engendered by viscous dissipation. We seek to convert this into a rate of entropy production with pressure, so, using the chain rule:

$$\left(\frac{\partial S}{\partial P}\right)_{pr} = \frac{dS}{dT}\frac{dT}{dx}\frac{dx}{dP}. \tag{10.45}$$

Writing the condition of hydrostatic equilibrium (equation (3.34)) in terms of molar properties and using (1.3.18) we get :

$$\frac{dx}{dP} = \frac{V}{g\,m}, \tag{10.46}$$

where m is the molecular weight of the material. Recalling that $(\partial S/\partial T)_p = C_P/T$, equation (10.45) becomes:

$$\left(\frac{\partial S}{\partial P}\right)_{pr} = \frac{\tau V^2}{z\,T\,g\,m}. \tag{10.47}$$

For olivine we have $V \approx 4.5 \times 10^{-5}$ m^3 mol^{-1} and $m \approx 0.14$ kg mol^{-1}. Characteristic values for the other parameters for Earth's upper mantle are $\tau \approx 10^2$ bar $= 10^7$ Pa, $T \approx 1600$ K and $g = 9.8$ m s^{-2}. These values yield $(\partial S/\partial P)_{pr} \approx 900/z$ J K^{-1} kbar^{-1} mol^{-1}, with z in meters. If strain is distributed evenly over the width of the diapir, then for a 50 km wide diapir $(\partial S/\partial P)_{pr} \approx 0.02$ J K^{-1} kbar^{-1} mol^{-1}. By comparison, the product αV is of order 0.2 J K^{-1} kbar^{-1} mol^{-1}, i.e. one order of magnitude greater. These two terms are combined in equation (10.43), so we can conclude that the contribution of inelastic rock deformation to melt production *averaged over the entire volume of the diapir* is likely to be small. However, if strain is focused on narrow shear zones (small z) then viscous heating may be locally important and greatly enhance melt production in the shear zones.

Dissipation of gravitational potential energy can take place inside a diapir if the melt, which is less dense than the solid, ascends relative to the latter. We can estimate the thermal effect of gravitational melt segregation as follows. Consider an infinitesimal vertical interval of the diapir of length dx. A volume V_m of melt ascends this distance, displacing an equal volume of solid which sinks the same distance. The amount of gravitational energy that is dissipated, dU_g, is given by:

$$dU_g = V_m \Delta\rho g dx, \tag{10.48}$$

where $\Delta\rho$ is the difference in density between melt and solid and g is gravitational acceleration. The gravitational potential energy dissipated per unit volume of rock, $dU_{g,V}$, must clearly attain its maximum value when the rock is 50% molten. We can then write:

$$dU_{g,V} = \Upsilon \Delta\rho g dx, \tag{10.49}$$

where the parameter Υ is defined as follows. Let φ be the volume fraction of melt. Then $\Upsilon = \varphi$ if $\varphi \leq 0.5$, and $\Upsilon = 1 - \varphi$ if $\varphi > 0.5$. Since gravitational energy is dissipated as thermal energy we can also write:

$$dQ_{g,V} = \frac{C_P}{V}dT = \frac{\rho C_P}{m}dT, \tag{10.50}$$

where C_P, V and m are molar properties, and $d\,\boldsymbol{Q}_{g,V}$ is heat generated per unit volume. Equating (10.49) and (10.50) we get:

$$\frac{dT}{dx} = \Upsilon\frac{\Delta\rho}{\rho}\frac{mg}{C_P} \qquad (10.51)$$

and, by using (10.45) and (10.46):

$$\left(\frac{\partial S}{\partial P}\right)_{pr} = \Upsilon\frac{\Delta\rho}{\rho}\frac{V}{T}. \qquad (10.52)$$

For olivine at mantle conditions $V/T \approx 2.8\,\mathrm{J\,K^{-1}\,kbar^{-1}\,mol^{-1}}$. Now, because (10.52) takes its maximum value for $\varphi = 0.5$, we have $(\partial S/\partial P)_{pr,max} \approx 1.4(\Delta\rho/\rho)\,\mathrm{J\,K^{-1}\,kbar^{-1}\,mol^{-1}}$. For this term to be comparable to $\alpha V \approx 0.2\,\mathrm{J\,K^{-1}\,kbar^{-1}\,mol^{-1}}$ it must be $\Delta\rho/\rho \approx 0.14$, or, for a solid density of 3300 kg m^{-3}, $\Delta\rho \approx 500$ kg m^{-3}, which is about the same as the difference in density between basaltic melt and peridotite. Dissipation of gravitational energy by phase separation could thus make a non-negligible contribution to melt production but only when melt fraction in the diapir approaches 50%, and assuming that there is complete separation of the two phases. Even in this case the effect of phase separation would only be enough to cancel out the expansion term, αV (see equation (10.43)), which itself is about one fifth the magnitude of the cooling term (equations (10.29)). One can conclude that under most circumstances the isentropic model of melt production (equation (10.25)) is a reasonably good approximation to decompression melting in planetary mantles.

10.7 Open system melting

At the other end of the spectrum from isentropic decompression melting is open system melting, in which the system is open to influx of mass, typically in the form of a fluid phase, and energy. The simplest example of this is a two-component system with a binary melt phase, in which one of the components makes up a pure solid phase and the other component a pure fluid phase. This system will allow us to look at the fundamental thermodynamic and physical aspects of fluid-fluxed melting. The three phases, solid, fluid and melt, constitute a univariant assemblage or, at constant pressure, a pseudoinvariant assemblage. This assemblage exists at equilibrium at point I in Fig. 10.14, a schematic isobaric diagram in which the coordinates are the chemical potential of the fluid component and temperature. The curve that traces the chemical potential of the pure fluid component ($\mu_{0,fluid}$) bounds a "prohibited region", as the system cannot exist at equilibrium for $\mu_{fluid} > \mu_{0,fluid}$. The two branches of this curve correspond to the pseudounivariant equilibrium assemblages melt + fluid (= solid absent) and solid + fluid (= melt absent), at temperatures higher and lower than that of the pseudoinvariant point, respectively. The vapor-absent melting curve, corresponding to the equilibrium assemblage solid + melt, connects the pseudoinvariant point with the melting point of the pure solid (T_0) in the fluid-free system.

The mol fraction of fluid component in the melt is the variable $X_{z,m}$ in the melting point depression equation (10.16). Assuming ideal mixing in the melt the vapor-absent curve is

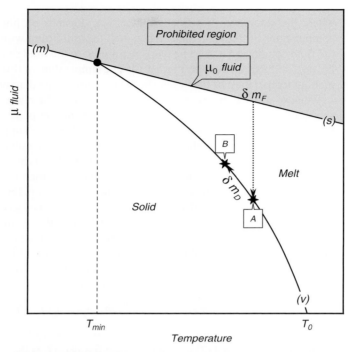

Fig. 10.14 Isobaric $\mu - T$ diagram for fluid-fluxed melting in a two-component system: solid + fluid. Chemical potentials higher than the standard state chemical potential of the pure fluid are impossible at equilibrium. The three phases solid + fluid + fluid-saturated melt exist at equilibrium at point I. The temperature of point I is the minimum melting point of the system. Solid + fluid-undersaturated melt exist along the vapor-absent curve, that ends at the melting temperature of the fluid-absent one-component system. Addition of a small amount of fluid δm_F causes a small amount of solid δm_D to melt. The temperature must drop from A to B to provide the necessary enthalpy of melting.

then given by:

$$\mu_{P,T}^{fluid} = \mu_{P_{(o)},T}^{0,fluid} + RT \ln\left[1 - \left(\frac{T}{T_0}\right)^{\frac{\Delta_m S}{R}}\right]. \tag{10.53}$$

If, as is commonly the case, mixing is not ideal then the term in square brackets would be multiplied by an activity coefficient, but the geometry of the diagram would not be affected. The composition of the melt along the melt–solid coexistence curve (V) is also obtained from the melting point depression equation. Defining m_D and m_F as the number of mols of molten solid and fluid contained in the melt, respectively, we have:

$$T = T_0\left(1 - \frac{m_F}{m_F + m_D}\right)^{\frac{R}{\Delta_m S}} = T_0\left(\frac{m_D}{m_F + m_D}\right)^{\frac{R}{\Delta_m S}}. \tag{10.54}$$

The validity of this equation is limited by the solubility of the fluid at the conditions of melting, $X_{saturation}$, so that we require:

$$\frac{m_F}{m_F + m_D} \leq X_{saturation} \tag{10.55}$$

where the saturation mol fraction $X_{saturation}$ depends on pressure, temperature and the compositions of fluid and melt. When melt in equilibrium with solid becomes saturated in the fluid component it must be $\mu_{fluid} = \mu_{0,fluid}$, so the fluid saturation point corresponds to the pseudoinvariant point in Fig. 10.14. The temperature of this point defines the minimum melting temperature of the system at the pressure of the diagram.

We now add the energy conservation condition to the phase relations. Say that we have a system consisting of melt and solid at equilibrium at point A in the figure. The system is invaded by a small amount of fluid, δm_F, which for simplicity we will assume is at the same temperature as the system (we will address possible differences in temperature shortly). Because the chemical potential of the pure fluid is higher than that of the fluid component in the melt the added fluid must dissolve in the melt in order to restore equilibrium. But the added fluid also lowers the equilibrium temperature for the solid + melt assemblage, so that enthalpy must be released. The enthalpy is absorbed as enthalpy of melting, causing a small amount of dry solid, δm_D, to melt. Addition of δm_F mols of fluid therefore causes equilibrium to shift from point A to point B. The magnitude of this displacement, or equivalently, the amount of solid that melts, δm_D, is determined by two opposing factors: (i) the melting point depression effect of the added fluid (equation (10.54)) and (ii) the energy balance, which is given by the following equation:

$$C_{P,R} m_R \, \delta T + T \, \Delta_m S \, \delta m_D + C_{P,F} \, (T - T_F) \, \delta m_f = 0, \qquad (10.56)$$

where m_R is the total number of mols in the system (solid + melt), T is the temperature after equilibrium is restored (T of point B in the figure), δT is the temperature change required to supply the enthalpy of melting ($T_B - T_A$), $\Delta_m S$ is the entropy of melting, T_F is the initial temperature of the fluid, and $C_{P,R}$ and $C_{P,F}$ are the heat capacities of the system (which for simplicity, and as in the case of decompression melting, we will assume to be constant and independent of melt fraction and melt composition) and of the fluid, respectively.

Equations (10.54) and (10.56) can be solved iteratively for T and m_D as a function of m_F, for both batch and fractional melting. We do that in the following numerical example, but first let us see how the results from the two melting regimes are expected to differ. Differentiating (10.54) we get:

$$\frac{dT}{dm_F} = -T_0 \frac{R}{\Delta_m S} \frac{m_D^{R/\Delta_m S}}{(m_F + m_D)^{R/\Delta_m S + 1}}. \qquad (10.57)$$

The variables m_D and m_F are related by a possibly variable factor k, which, however, is always a finite (as opposed to infinitesimal) quantity. We can write $m_F = k m_D$, and substituting in (10.57) we arrive at:

$$\frac{dT}{dm_F} \sim -T_0 \frac{R}{\Delta_m S} \frac{1}{(k+1)^{R/\Delta_m S + 1}} \frac{1}{m_D^{R/\Delta_m S}}. \qquad (10.58)$$

Recall that m_D is the amount of solid component present in the melt. This is a finite value during batch melting, but tends to an infinitesimally small value during fractional melting. Hence, the rate of change of temperature with added fluid along the vapor-absent curve in Fig. 10.14 *is greater during fractional melting than during batch melting*. This implies a greater enthalpy release and, therefore, for the same total amount of fluid added, greater melt production for fractional melting than for batch melting. This behavior is opposite to that of decompression melting of a dry solid.

Both batch melting and fractional melting can be thought of as being made up of a large number of small increments. The difference between both processes is whether the successive melt increments, δm_D, stay in the system (batch) or leave (fractional). Let T_{i-1} be the equilibrium temperature of the system at the end of an incremental melting step. Substituting (10.54) in (10.56) and simplifying we can write the energy balance for the subsequent melting increment as follows:

$$T_0 \left[C_{P,R} \, m_R + \Delta_m S \, \delta m_D + C_{P,F} \, \delta m_F \right] \left(\frac{m_D}{m_F + m_D} \right)^{\frac{R}{\Delta_m S}} \tag{10.59}$$
$$- C_{P,R} m_R \, T_{i-1} - C_{P,F} \, \delta m_F T_F = 0.$$

The idea now is to solve equations (10.59) and (10.54) iteratively for δm_D and T by adding small fluid increments δm_F. This can be done with a simple *Maple* procedure (Software Box 10.2). It is convenient to define two new variables, the total amount of dry rock that has melted after $(i-1)$ fluid increments, $D_{(i-1)} = \sum_{(i-1)} \delta m_D$, and the total amount of fluid added to that point, $F_{(i-1)} = \sum_{(i-1)} \delta m_F$. The solutions for batch and fractional melting are different from this point on.

Software Box 10.2   Calculation of volatile-fluxed melting

The *Maple* worksheet **volatile_melting.mw** contains the procedures volatile_ batch_melting and volatile_fractional_melting, that calculate melt production by volatile infiltration. The procedures solve equation (10.59) for small melt increments, δm_D, and then calculate temperature with equation (10.54). The algorithms for batch and fractional melting are explained in Worked Example 10.3. The procedure call requires that one specify the initial temperature of the rock, T_0, and the temperature of the fluid, T_F, both in Kelvin, and a filename for the output. Other model parameters can be changed by editing the procedures, following the same conventions discussed in Software Box 10.1. The variable dfluid specifies the added amount of fluid at each step, δm_F (in mols per mol of rock).

The procedures generate text files in which each line gives the state of the system after addition of each fluid increment. There are six output fields, as follows. The first field is the total amount of fluid added, the second is the total amount of solid that has melted, the third is the temperature (in Kelvin), the fourth is the mol fraction of fluid in the melt, the fifth is the size of the system, i.e. initial solid plus added fluid for batch melting, or initial solid minus melt produced for fractional melting, and the sixth field is the melt increment generated at each step, δm_D.

The procedures iterate 2000 times (this can be changed by editing the do loop) and do not check whether the mol fraction of fluid dissolved in melt (the fourth output field) exceeds the likely solubility. One can examine the output and discard whatever portion of it implies unreasonable volatile solubility.

Let us assume that we start with 1 mol of dry solid. During batch melting the melt, including the fluid dissolved in it, stays in the system, so we make $m_R = 1 + F_{(i-1)} + \delta m_F$ (recall that m_R is the total size of the system). Using the same argument we can see that for the batch melting case the melt composition in the melting point depression equations (10.59) and (10.54) is given by $m_D = D_{(i-1)} + \delta m_D$ and $m_F = F_{(i-1)} + \delta m_F$. Choosing some small value for δm_F and making T_{i-1} equal to the value of T calculated in the previous step we solve equation (10.59) for δm_D, and then (10.54) for T. In the case of fractional melting the fraction of solid that melts, and the fluid dissolved in it, leave the system, so we make $m_R = 1 - D_{(i-1)}$. As none of the melt previously formed remains in the system, in this case we make $m_D = \delta m_D$ and $m_F = \delta m_F$. As before, we specify a small value of δm_F and solve (10.59) for δm_D, and then (10.54) for T (Software Box 10.2).

We can study the behavior of these equations by choosing characteristic values for the parameters, as follows: $C_{P,R} = 200$ J K^{-1} mol^{-1} (a decent guess for Mg silicates), $C_{P,F} = 50$ J K^{-1} mol^{-1} (a ballpark figure for H_2O) and $\Delta_m S = 56$ J K^{-1} mol^{-1} (the value for forsterite). Let us also make $T_F = T_0 = 1600$ K (a characteristic upper mantle temperature, which we assume to be also equal to the temperature of the fluid, T_F), and note that, given that the heat capacity of the fluid is only one fourth that of the rock and the total amount of fluid added is well below 1, the solution is not likely to be very sensitive to our choice of T_F. Figure 10.15 shows calculated total melt production (i.e. the fraction of solid that has melted) and melting temperature as a function of the total amount of added fluid, up to 0.1 mol per mol of solid rock.

As expected, our simple model for fluid-fluxed melting produces more melt by fractional melting than by batch melting. The difference is considerable: about 28% of the rock melts during fractional melting, versus some 23% during batch melting. For both batch and fractional melting it is clear that, if the solubility of the fluid in the melt is sufficiently high, then fluid-fluxed melting is an efficient mechanism of magma generation. Between 20 and 30 molar% of dry solid melts by addition of ~0.1 mols of fluid to 1 mol of solid. Solving equation (10.19) for the mass proportion of fluid, C_{fluid}, and assuming that the ratio of molecular weights is $M = 10$ (a possible value for H_2O and Mg silicates) we find that 0.1 mol of H_2O added to 1 mol of silicate rock corresponds approximately to 1 wt% H_2O.

The inset in Fig 10.15 shows the calculated mol fraction of fluid in the melts. The solid curve for fractional melting shows the fluid concentration in each small melt increment, whereas that for batch melting shows fluid concentration in the total amount of melt generated. The equivalent curve for fractional melting is shown with a dashed line, and is of course below that for batch melting, as it must given that the total amount of melt produced is greater (there is an apparent contradiction here that is not such, see Exercise 10.7). The mol fraction of fluid in melts produced with the addition of 0.1 mols of fluid is ~0.32 for batch melting (total melt) and ~0.45 for fractional melting (last melt increment). For $M = 10$ this would correspond to ~4.5 wt% and 7.6 wt%, respectively, which is well below H_2O solubility at mantle pressures. This calculation is admittedly very crude, but even if the ratio M were ~2 (an unlikely low value for H_2O in silicate melts) we would get fluid concentrations of 19 and 29 wt%, respectively, which are comparable to H_2O solubility in silicate melts at upper mantle pressures.

Of course, this calculation is highly simplified and completely ignores the phase equilibrium of melting in complex natural systems. It also ignores effects such as that of excess Gibbs free energy of mixing in the melt, which might cause an even stronger depression of the solidus and potentially supply a greater amount of enthalpy of melting. Recall, finally, that this calculation assumes that the initial temperature of the rock is the dry solidus. Melt production would be less if the rock's initial temperature is lower. Despite these simplifications the calculations expose the fundamental physical aspects of fluid-fluxed melting, and demonstrate the importance of a process without which melting at Earth's convergent plate margins would be virtually impossible.

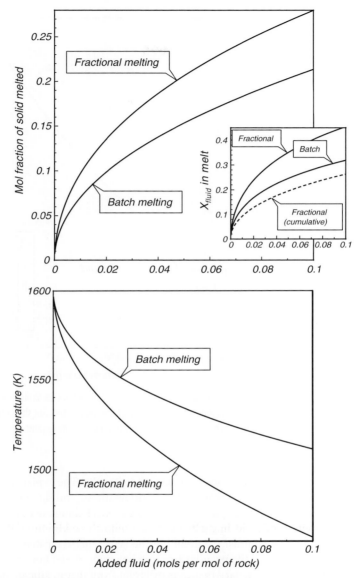

Fig. 10.15 Batch and fractional melting of a two-component system: solid + fluid. The calculation assumes that the solubility of the fluid in the melt is at least as high as the maximum X_{fluid} values shown in the inset diagram.

10.8 The nature of solid–melt equilibrium in icy satellites

In Worked Example 6.10 we discussed how methane–carbon monoxide cryomagmas could form in outer Solar System objects. In less extreme worlds, such as the moons of Jupiter and Saturn, cryomagmatism is more likely to be based on water. A distinct property of water is

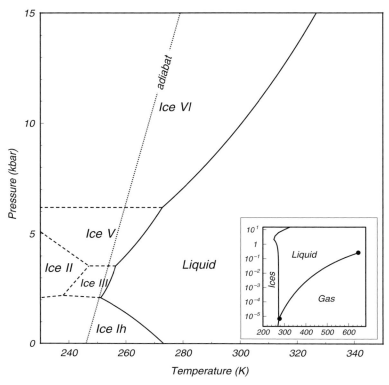

Phase diagram of H_2O, calculated with the thermodynamic model of Choukroun and Grasset (2007). The dotted line shows the hottest ice adiabat along which no melting takes place. The inset shows the ice solidus together with the liquid–vapor coexistence curve, terminating at the critical point (note logarithmic pressure scale, in kbar).

that low-pressure H_2O ice contracts when it melts. The explanation for this resides in the fact that hydrogen bonds have preferred orientations, and this produces a crystalline structure with plenty of open spaces. These open spaces become occupied by H_2O molecules in the disordered melt, causing a decrease in volume upon melting. Higher-pressure polymorphs of ice are, in contrast, "normal" in the sense that their density is greater than that of the melt. Because of the peculiar freezing behavior of H_2O the nature of magmatic activity in H_2O-rich icy planetary bodies is different from that in silicate bodies.

Let us begin by studying the phase diagram of H_2O, shown in Fig. 10.16 (calculated with thermodynamic data from Choukroun & Grasset, 2007). Over the pressure range 0–15 kbar, shown in the figure, four ice polymorphs undergo stable melting reactions: ice Ih (the familiar low-density ice) and three high-pressure ices, denser than liquid: ice III, V and VI. Another polymorph, ice II, is not stable to the solidus. For comparison, the inset in the figure shows the composite solidus curve from the main figure, together with the sublimation and boiling curves of H_2O, the latter terminating at the critical point.

A pressure of 15 kbar in large icy satellites such as Ganymede or Titan corresponds to a depth of $\sim$1000 km, which is about as deep as the outermost H_2O-dominated shell is thought to extend in those worlds. Europa may have a much thinner H_2O veneer, whereas Callisto may be largely undifferentiated, and a mixture of ice and silicates may exist all the way to the satellite's center, at a pressure of $\sim$50 kbar. Maximum pressures in small icy satellites such as Enceladus or Mimas are of order 500 bar, so that high-pressure ices do not form in them.

We now seek answers to a number of questions about the interiors of icy satellites. First, do icy satellites convect? To answer this question, we solve the Rayleigh number (equation 3.62) for the thickness of the convective layer, D:

$$D = \left(\frac{Ra\,\mu\kappa}{g\alpha\rho_0\Delta T} \right)^{1/3}. \tag{10.60}$$

Characteristic values for H_2O ices are: $\mu = 10^{12}$ Pa s, $\kappa = 1.5 \times 10^{-6}$ m^2 s^{-1}, $\alpha = 2 \times 10^{-4}$ K^{-1} and $\rho_0 = 1000$ kg m^{-3}. Recall that the temperature difference ΔT is the temperature drop across the thermal boundary layer. For an icy satellite a conservative value is $\Delta T \sim 10$ K. Assuming that the onset of convection occurs at $Ra \sim 1000$, we get $D \sim 1000\, g^{-1/3}$ m. For a large icy satellite $g \sim 1$ m s^{-2}, so $D \sim 1$ km, whereas for a small icy satellite $g \sim 0.1$ m s^{-2} and $D \sim 2$ km. As these values are orders of magnitude smaller than likely ice thicknesses (100–1000 km), we conclude that icy satellites convect, and in fact do so vigorously (i.e. with a high Nusselt number, equation (3.65)). This conclusion is valid for pure H_2O ice, and probably for mixtures of H_2O–NH_3–CH_4 ices as well.

Because they convect, temperature distribution inside icy satellites must be adiabatic. Using values for high pressure ices, $\alpha = 2 \times 10^{-4}$ K^{-1}, $V = 1.9$ J bar^{-1} mol^{-1}, $C_p = 47$ J K^{-1} mol^{-1} and taking $T = 300$ K we get from equation (3.32) an adiabatic gradient of $\sim$2.5 K kbar^{-1}. This leads to the second question: what is the minimum potential temperature in a convective icy satellite that allows the existence of liquid? For a convecting silicate mantle, the answer is approximately 1400 K ($\approx$ 1100 °C, see Fig. 10.7). An adiabat with this potential temperature would intersect the peridotite solidus at zero pressure. For a satellite composed of pure H_2O we see from Fig. 10.16 that the answer is approximately 247 K. Adiabats with a higher potential temperature will intersect the solidus, but in contrast to silicate systems they will do so at $P > 0$. Moreover, if the potential temperature is between $\sim$247 K and $\sim$273 K then the liquid is only stable at pressures higher than the pressure at the satellites' surface. The conclusion that follows is that, in contrast to a silicate planet, extrusion of cryolavas in a pure H_2O world is impossible.

The surface of our pure H_2O world could be covered by a convective ocean with a potential temperature higher than 273 K. If the ocean was deep enough (equivalent to at least a few kbar) then its bottom would freeze to high-pressure ice. There is no satellite like this in the present-day Solar System. If the potential temperature of the convecting satellite was less than 247 K then it would be solid throughout. This is perhaps the case for most of Saturn's small icy moons, with the exception of Enceladus. Large H_2O satellites with potential temperatures between 247 K and 273 K must contain a layer of liquid H_2O sandwiched between ice Ih at the surface and high pressure ices at depth. In this case there is a maximum possible thickness for the ice Ih shell, which is given by the pressure of the ice Ih–ice III–liquid invariant point, about 2 kbar (see Fig. 10.16). For the Solar System's large icy satellites this corresponds to a depth of $\sim$170 km. This is the greatest possible depth to the top of the liquid layer, if such layer exists. In small icy satellites such as Enceladus liquid H_2O can exist under an ice Ih layer, but the deep high pressure ice layer cannot.

Present day Titan, and quite possibly Ganymede as well, are thought to follow the sandwich model, with a layer of liquid H_2O between low-pressure and high-pressure ice layers. The case of Europa appears to be different, as the H_2O layer is not thick enough to allow crystallization of high-pressure ices, so that liquid H_2O would be sandwiched between the ice Ih shell and the planet's silicate mantle (or crust?). Focusing on Titan and Ganymede, then, we ask our next question: is cryomagmatism possible in these worlds? If the composition of the icy envelope is pure H_2O then, as we saw, the answer is no. Liquid H_2O would freeze when attempting to ascend through the ice Ih shell (this is a convoluted way of saying that ice floats on water). And yet, there is geologic evidence for cryolavas in both of these worlds. The most likely explanation for this is that the material that encases these satellites is a mixture of H_2O with other volatile compounds, chiefly NH_3 and perhaps CH_4 as well. These additional components have two effects that may facilitate cryomagmatism: they lower the melting point of ice, and they lower the density of the liquid.

Melting phase relations for the binary system H_2O–NH_3 at low pressure were summarized by Kargel (1992). Three intermediate crystalline compounds form between Ice Ih and solid ammonia: ammonium dihydrate ($NH_3.2H_2O$), ammonium hydrate ($NH_3.H_2O$) and diammonium hydrate ($2NH_3.H_2O$). These give rise to three eutectics, and a peritectic at which ammonium dihydrate melts incongruently to ice Ih plus a liquid richer in ammonia that $NH_3.2H_2O$. Everywhere between the composition of the peritectic ($\sim$33 wt% NH_3) and the H_2O end of the binary join the H_2O–NH_3 melt is at equilibrium with ice Ih. Therefore, a subsurface liquid layer in a water–ammonia icy satellite containing between 0 wt% and 33 wt% NH_3 can exist at equilibrium with an ice Ih shell.

What is the buoyancy of these H_2O–NH_3 melts relative to the Ice Ih country rock? An equation of state for H_2O–NH_3 liquids has been calibrated by Croft *et al.* (1988). Their equation consists of an isothermal Murnaghan EOS (Section 8.2.1) and a polynomial zero-pressure thermal expansion term. It is somewhat unwieldy because the calibration changes with composition along the H_2O–NH_3 join. We cannot discuss it here owing to space constraints. We can state some of the key results of the Croft *et al.* EOS, though. First, it predicts that the density of H_2O–NH_3 liquids is always lower than that of pure H_2O liquid at the same pressure and temperature, and that density decreases with increasing NH_3 content. Second, it shows that H_2O–NH_3 melts may be more or less dense that ice Ih, depending on pressure and melt composition. Third, it suggests that the melts that form by incongruent melting of ammonium dihydrate are less dense than ice Ih, which is the solid phase that these melts are at equilibrium with. The melt becomes denser than ice Ih with decreasing NH_3 content, but a compositional interval exists within which the Clapeyron slope of the ice Ih melting reaction changes from negative (i.e. as in the pure H_2O system, Fig. 10.16) to positive. Under those circumstances the melt can ascend through the ice Ih crust and eventually be extruded on the surface of the satellite.

And herein lies a significant problem: if a satellite-wide liquid layer of low-density H_2O–NH_3 melt forms, it is gravitationally unstable relative to the ice Ih shell. This would produce a short-lived surge of magmatic activity, at the end of which refractory ice Ih would underlie a less-dense mixture of H_2O–NH_3 ices, making further igneous activity difficult. This is not altogether different from the formation of anorthositic crust from a primordial lunar magma ocean. The problem is that, in contrast to the volcanically dead Moon, there appears to be geologically recent cryovolcanism on Titan, and perhaps on Ganymede as well. One possible way out is that the liquid layer underlying the ice Ih shell is not positively buoyant after all, but merely close to being neutrally buoyant, and that cryomagmas are "squeezed" to the surface by tectonic processes driven by convection of the ice Ih shell (see Mitri & Showman, 2008). Another possibility is that a satellite-girdling liquid layer does not exist at all, because the average internal temperature is lower than the solidus, and that convection causes only local decompression melting of ices, as in silicate planets. If the

NH$_3$ concentration is high enough to preempt the negative Clapeyron slope of the ice Ih melting reaction then the resulting liquids could be extruded on the satellite's surface.

Exercises for Chapter 10

10.1 The two melting point depression equations, (6.49) and (10.8), are approximations. Derive an equation for the difference between both approximations and discuss the conditions under which one or the other, or neither, is a better approximation.

10.2 Show that a peritectic cannot be a minimum melting point because it would violate Schreinemakers' rule.

10.3 Use the *Maple* worksheet described in Software Box 10.1 to study how each of the following variables affects melt production by decompression melting: (i) Clapeyron slope of the solidus; (ii) temperature difference between solidus and liquidus, (iii) ratio of heat capacity to enthalpy of melting.

10.4 Use the *Maple* worksheet described in Software Box 10.1 to study decompression melting in Venus, assuming that the composition of the Venusian mantle is the same as that of the Earth's, and using the thermal structure of the Venusian mantle discussed in Section 3.9.

10.5 Discuss how the "envelope" between batch melting and fractional melting shown in Fig. 10.9 would be affected by the non-isentropic processes discussed in Section 10.6.5. Show this (at least) semi-quantitatively.

10.6 Use the *Maple* worksheet described in Software Box 10.2 to study the effect of fluid temperature on melt production by fluid-fluxed melting. How likely is it that the temperature of the fluid will differ significantly from the temperature of the rock? Use the appropriate equations from Chapter 3 to justify your answer.

10.7 Show that there is no contradiction between the facts that, during fractional volatile-fluxed melting, individual melt increments have higher fluid mol fractions than batch melts, whereas the total melt produced by fractional melting has lower fluid mol fractions than batch melts (see Fig. 10.15).

Dilute solutions

The focus of this chapter is on liquid solutions in which one component is present in much greater abundance, say at least one order of magnitude greater, than all others. Examples that underscore the importance of this type of solutions include seawater, and natural terrestrial waters in general, but one can imagine more exotic possibilities, such as hydrocarbon-based solutions on Titan's surface and ammonia-based solutions in its interior. What all of these examples share is the fact that it is convenient to make a distinction between dilute solutes that may not be liquid in their standard states, and a liquid solvent that is generally close to being in its standard state. Depending on the nature of the solvent and of the solutes the latter may exist as electrically neutral chemical species, as ions, or as a combination of both. Solutions in which solutes dissociate into ions are known as *electrolyte solutions*. Among these, those in which water is the solvent are by far the most important ones, at least in terrestrial environments. The chapter emphasizes aqueous electrolyte solutions, but virtually all of the thermodynamic framework is applicable to any type of dilute solution. We begin with a discussion of dilute solutions in general, and shift the focus to electrolyte solutions in Section 11.3.

11.1 Some properties of dilute solutions

In our discussions so far on the thermodynamic properties of solutions we have made no distinction between the treatment of the different solution components. All our equations have been symmetric, in the sense that we could interchange the components and arrive at the same final result. This was true whether the solution was above or below its critical mixing point (Section 7.1). In the latter case we considered two distinct subcritical phases (for binary systems), and we allowed for the possibility that one or the other, or both, could exist within some region of interest. This is not always the case, however. Consider aqueous solutions as an example. One of the components in such solutions is liquid H_2O. Within the P–T region in which liquid H_2O is stable the solubility of many chemical species of interest in the planetary sciences is small (although there are important exceptions, such as NH_3). One way to look at this is that most aqueous solutions correspond to conditions that are so far below the critical mixing point that their compositional range is restricted to a narrow interval between the solvus and the H_2O end of the composition axis. The other branch of the solvus is generally inaccessible, either because the necessary bulk compositions are not realized in nature, or because the solvus is suppressed by a first-order phase transition and the corresponding liquid component is not stable at the conditions at which liquid H_2O is stable (consider, for example, a solution of NaCl in liquid H_2O). In such cases it is convenient to distinguish between a *solvent* and one or several *solutes,* such that the total solute concentration is much lower than the solvent concentration. "Much lower" is of course not a proper quantitative definition, but a good rule of thumb is "at least one order of magnitude lower". Solvent and solutes are best described by different sets of equations, and the symmetry that characterizes the equations for other types of solutions is lost.

Two points must be clearly understood. First, although the liquid solvent that we most commonly encounter in planetary environments is water this is not necessarily always the case. Second, many aqueous solutions are electrolyte solutions, meaning that the solutes dissociate into ions upon dissolving. Although thermodynamic description of electrolyte solutions requires some specific techniques, that we will study beginning in Section 10.3, there are fundamental aspects of the way in which dilute solutions are handled that are the same regardless of whether or not they are electrolytes.

11.1.1 Concentration scale in dilute solutions

The concentration of the solvent in dilute solutions is expressed in mol fraction, as we have done so far. For the solutes this is inconvenient for two reasons. First, solute mol fractions are small numbers ($\ll 1$). Second and most importantly, many dilute solutions of interest, and in particular aqueous solutions, may contain a large number of solutes. This makes it advisable to describe the concentration of each solute with a function that remains invariant when the amounts of other solutes are varied. Mol fraction does not behave in this way, because if the amount of one component is varied all mol fractions vary. For the solvent in a dilute solution this variation is generally trivial, but the same is not true for the solutes.

Solute concentration in dilute solutions is expressed in *molality*, defined as mols of solute per kg of solvent. We will use the fixed subscript s to refer to the solvent, and the variable subscript i to refer to solutes, where i can take as many different values as there are solutes in the solution. Let the number of mols of solvent be $\boldsymbol{n}_s$, and the number of mols of the ith solute be $\boldsymbol{n}_i$. The mol fraction of the solvent is then given by:

$$X_s = \frac{\boldsymbol{n}_s}{\boldsymbol{n}_s + \sum_i \boldsymbol{n}_i}. \tag{11.1}$$

The number of mols of the ith solute per kg of solvent is given by:

$$\frac{\boldsymbol{n}_i}{\boldsymbol{n}_s M_s}, \tag{11.2}$$

where M_s is the molecular weight of the solvent, in units of kg mol^{-1}. The units of (11.2) are mol kg^{-1}, which agrees with our definition of molality but presents some formal problems. In particular, units of concentration should always be dimensionless (e.g. equation (11.1)) because they end up as arguments of logarithmic functions. The problem is in this case easily solved by defining the dimensionless molecular weight of the solvent, M_s^*, as follows:

$$M_s^* = \frac{M_s}{1\,\text{kg mol}^{-1}}, \tag{11.3}$$

which allows us to define the non-dimensional molality of i, m_i, as follows:

$$m_i \equiv \frac{\boldsymbol{n}_i}{\boldsymbol{n}_s M_s^*}. \tag{11.4}$$

Note that, in contrast to fugacity and partial pressure, which retain their pressure dimension but are divided by the 1 bar standard state pressure when they are the argument of a logarithm (even if we omit the denominators for simplicity), equation (11.4) *defines* a non-dimensional molality. This function has the desired property, that the molality of a solute in a dilute solution depends only on the amount of the solute itself. The amounts of all other solutes $j \neq i$ can be changed and this does not affect m_i.

11.1.2 Standard states in dilute solutions

The standard state for the solvent in a dilute solution is taken as usual, as pure solvent at the temperature and pressure of interest. Commonly the solvent is a liquid at the temperature and pressure of interest, but it could also be a solid or a supercritical fluid. Using the same standard state convention for dilute solutes is not convenient, however. One reason for this is that the pure liquid solute at the temperature and pressure of interest is in many cases not thermodynamically stable. This is the case, for instance, when we consider aqueous solutions of substances that are solids or gases at room temperature. Another reason is that the chemical potential of a very dilute solute may be orders of magnitude lower than that of the pure substance at the same P and T, even if the latter was stable. It is thus more convenient to define the standard state of dilute solutes at the *infinite dilution limit*, rather than at the pure solute state. This requires some arbitrary definitions, and the first one is to postulate that an infinitely dilute solution of a single solute is ideal, and that its molality equals its activity. We write this as follows:

$$\lim_{m_i \to 0} \frac{a^i}{m_i} \equiv 1, \quad m_{j \neq i} = 0, \tag{11.5}$$

where a^i is the activity of solute i. Equation (11.5) carries the assumption that the activity coefficient γ^i of every solute tends to 1 as the total solute molality $\sum m_i$ goes to zero. We now recall the usual relation between μ and μ^0:

$$\mu^i = \mu^{0,i} + RT \ln a^i \tag{11.6}$$

and define the standard state chemical potential of solute i at infinite dilution as follows:

$$\mu^{0,i} \equiv \lim_{m_i \to 0} \left(\mu^i - RT \ln m_i \right), \quad m_{j \neq i} = 0. \tag{11.7}$$

There are two important aspects of this definition. The first one is explicitly stated: the standard state of solute i is defined in an infinitely dilute solution in which i is the only solute. The second one is also contained in (11.7) but it may be less obvious. It is the fact that the standard state of a solute *depends on the identities of both the solute and the solvent*. Recall that in equation (11.5) we *specified* that the infinitely dilute solution is ideal, and that its activity equals its molality. This implies that the molecules of solute and solvent do not interact with one another, which is never the case. The energetic contributions of these interactions, which depend on the nature of the molecules of both solute and solvent, are lumped into the standard state chemical potential by equation (11.7). Consider for example N_2 as a dilute solute in two different solvents, liquid H_2O and methanol (CH_3OH), at the same temperature and pressure. The interaction energies of N_2 molecules with H_2O molecules and CH_3OH molecules are different, and therefore the standard state chemical potential of N_2 in a water solution is different from that in a methanol solution. This is different from the way in which we have dealt with other solutions, where the standard state properties depended only on the identity of the substance itself. Now we need to consider the solute and the solvent but, by requiring that $m_{j \neq i} = 0$, the standard state properties are independent of any other solutes.

You may object that equation (11.7) does not necessarily tell us how to measure the standard state chemical potential at infinite dilution. This is true, but if you think about it it is also true of all the definitions of standard state properties that we have used so far. In this book I take the position that we will not worry about this. It is possible to determine values of standard state chemical potentials, enthalpies, entropies and heat capacities, and extrapolate them to infinite dilution (see, for example, Robinson & Stokes, 1959; Pitzer, 1995; Anderson, 2005). Standard state values for many species in aqueous solution are tabulated at the usual reference conditions: 298.15 K and 1 bar (see Wagman *et al.*, 1982; Robie & Hemingway, 1995). Standard state properties in other solvents are less well known. What matters to us is that equation (11.7) allows us to operate on the chemical potential of dilute solutes, and hence perform chemical equilibrium calculations.

We also note that we can expand (11.6) as follows:

$$\mu^i = \mu^{0,i} + RT \ln \left(\gamma^i m_i \right). \tag{11.8}$$

For a *hypothetical* ideal 1 molal solution we would have $\gamma^i = m_i = 1$, and therefore $\mu^i = \mu^{0,i}$. Because of this relationship the infinite dilution standard state defined by equation (11.7) is also referred to as the "1 molal" standard state. This means that the standard state chemical potential at infinite dilution would also be the chemical potential of the solute in a *hypothetical* 1 molal solution in the same solvent that behaved ideally. The actual 1 molal solution is never ideal, however, even if no other solutes are present. *The chemical potential of a solute in a real 1 molal solution is never equal to its standard state chemical potential*. This may be rather confusing, but it is nothing more than an algebraic manipulation.

11.1.3 Activity coefficient of a dilute solute and excess Gibbs free energy of mixing

From equation (5.129), the excess chemical potential of solute i is given by:

$$\mu^{i,ex} = RT \ln \gamma^i. \tag{11.9}$$

The relationship between excess chemical potential (a partial molar property) and excess Gibbs free energy of mixing, G^{ex} (Section 5.9.1), is given by equation (5.28):

$$\mu^{i,ex} = \left(\frac{\partial G^{ex}}{\partial n_i} \right)_{P,T,n_{j \neq i},n_s} \tag{11.10}$$

where n_s is the number of mols of solvent, and n_j are the number of mols of solutes other than i. From (11.9) and (11.10) we find:

$$RT \ln \gamma^i = \frac{\partial G^{ex}}{\partial n_i} = \frac{\partial G^{ex}}{\partial m_i} \frac{\partial m_i}{\partial n_i} \tag{11.11}$$

and from (11.4):

$$\frac{\partial m_i}{\partial n_i} = \frac{1}{n_s M_s^*} \tag{11.12}$$

from which we get:

$$\ln \gamma^i = \frac{1}{RT n_s M_s^*} \left(\frac{\partial G^{ex}}{\partial m_i} \right)_{P,T,m_{j \neq i}}.$$ (11.13)

Therefore, if one has an explicit formula for the excess Gibbs free energy of the solution (Section 11.6), the activity coefficient of each solute can be calculated with (11.13).

11.1.4 Gases as low-concentration solutes

Consider a molecular species, A, that is present in a gas phase and that dissolves in a coexisting condensed phase. The usual case is that the condensed phase is a liquid, but the following treatment is equally valid for solid–gas equilibrium. At equilibrium we have:

$$\mu^{A(g)} = \mu^{A(s)},$$ (11.14)

where $\mu^{A(g)}$ is the chemical potential of A in the gas phase, and $\mu^{A(s)}$ its chemical potential in the liquid solution. Because at low pressures, say those corresponding to planetary atmospheres and near-surface environments (including planetary oceans), solubilities of most common gas species in water and other possible liquid solvents are rather low, it is convenient to refer the chemical potential of species A dissolved in the liquid phase to the infinite dilution standard state. We therefore expand (11.14) as follows:

$$\mu^{0,A(g)} + RT \ln f_A = \mu^{0,A(s)} + RT \ln a^{A(s)},$$ (11.15)

where $\mu^{0,A(g)}$ is the standard state chemical potential of pure gaseous A at the temperature of interest and 1 bar, $\mu^{0,A(s)}$ is the standard state chemical potential of infinitely dilute A in the solvent of interest and at the temperature of interest (equation (11.7)), f_A is the fugacity of A in the gas phase and $a^{A(s)}$ is the activity of solute A in the liquid solution. Writing activity as the product of molality, m_A, times activity coefficient, γ^A, (equation (11.8)) we rearrange (11.15) as follows:

$$\frac{m_A}{f_A} = \frac{\exp \left(-\frac{\Delta_s G^0}{RT} \right)}{\gamma^A},$$ (11.16)

where $\Delta_s G^0$ is the standard state Gibbs free energy of dissolution at infinite dilution, defined as:

$$\Delta_s G^0 = \mu^{0,A(s)} - \mu^{0,A(g)}.$$ (11.17)

It is an empirical observation that, at constant temperature and for very dilute solutions (i.e. as $m_A \to 0$), the ratio m_A/f_A is approximately constant. This is Henry's law (Section 5.9.2), and the resulting constant, $K_{H,s}^A$, is one way of defining Henry's law constant. We then have:

$$\frac{m_A}{f_A} = K_{H,s}^A, \quad m_A \ll 1,$$ (11.18)

where the constant $K_{H,s}^A$ is equal to the right-hand side of equation (11.16) and therefore includes contributions both from the standard state Gibbs free energy of dissolution and from the excess chemical potential of A in the condensed solution (i.e. the activity coefficient). It follows that the value of Henry's law constant is specific to each solute–solvent pair.

Henry's law (equation (11.18)) describes the equilibrium between a molecular species in a gas and the *same molecular species* in a dilute condensed solution. The quantity m_A refers *only* to the molality of the *molecular species A* in the solution. If the species dissociates, or reacts with the solvent in some other way, then m_A will be less than the total amount of dissolved A, which is a quantity known as the *analytical concentration* of A. This is a key point, that will become clear in several numerical examples.

Henry's law constant as defined by equation (11.18) describes the *solubility* (hence the subscript s) of the gas in the condensed phase. It is also common to use the inverse of equation (11.18) and define Henry's law constant as the ratio f_A/m_A as m_A goes to zero. The resulting constant, $K_{H,v}^A$, describes the *volatility* of the gas and is simply equal to $1/K_{H,s}^A$. This alternate formulation is used in many thermodynamics textbooks (e.g. Pitzer, 1995; Anderson, 2005). I prefer the solubility constant defined by equation (11.18) because it is the convention used in comprehensive data bases for gases in aqueous solutions (see below).

The solubility of most gases in liquids decreases with increasing temperature. Within the temperature range in which aqueous solutions are stable the effect is not negligible. We account for the temperature dependency of Henry's law constant by writing, from equations (11.16) and (11.18):

$$\ln K_{H,s}^A = -\frac{\Delta_s G^0}{RT} - \ln \gamma^A. \tag{11.19}$$

Differentiating with respect to T, and assuming that γ is constant:

$$\frac{\partial \ln K_{H,s}^A}{\partial T} = \frac{1}{R}\left(\frac{\Delta_s S^0}{T} + \frac{\Delta_s G^0}{T^2}\right) \tag{11.20}$$

or:

$$\frac{\partial \ln K_{H,s}^A}{\partial T} = \frac{\Delta_s H^0}{RT^2}, \tag{11.21}$$

where $\Delta_s H^0$ is the molar enthalpy of dissolution at infinite dilution. Using the chain rule we simplify this further to:

$$\frac{\partial \ln K_{H,s}^A}{\partial (1/T)} = \frac{\partial \ln K_{H,s}^A}{\partial T}\frac{\partial T}{\partial (1/T)} = -\frac{\Delta_s H^0}{R}. \tag{11.22}$$

Assuming that the enthalpy of dissolution remains constant we integrate from some reference temperature T_0 (universally taken as 298.15 K) to the temperature of interest:

$$K_{H,s}^A = K_{H,s}^{A,0}\exp\left[-\frac{\Delta_s H^0}{R}\left(\frac{1}{T} - \frac{1}{T_0}\right)\right]. \tag{11.23}$$

Worked Example 11.1   Concentration of atmospheric gases in Earth's oceans

Henry's law constants and their temperature dependency can be calculated from standard state thermodynamic properties and activity coefficients but, more commonly, they are measured directly. Comprehensive tabulations exist for virtually all gases in aqueous solution (Sander, 1999; also NIST Chemistry WebBook). These tables typically list values of $K_{H,s}^{A,0}$ and $(-\Delta_s H^0/R)$, the latter parameter commonly labeled as $(d \ln K /d(1/T))$. A simple application of equations (11.18) and (11.23) is to calculate the concentration of gases dissolved in liquid H_2O in equilibrium with the terrestrial atmosphere. Figure 11.1 shows equilibrium molalities of N_2, O_2, Ar and CO_2 as a function of temperature, calculated with constants and temperature derivatives from Sander (1999). The fugacities are assumed to be constant and equal to 0.8 bar, 0.2 bar, 0.01 bar and 3.8×10^{-4} bar, for N_2, O_2, Ar and CO_2 respectively (the latter figure assumes an atmospheric CO_2 concentration of 380 ppm by volume, and is likely to be a nostalgic value by the time you read this). As expected, all concentrations decrease with increasing temperature, but the temperature dependency of CO_2 concentration is stronger than those of the other gases. This is one of several positive feedback mechanisms in global warming: as the oceans warm they exsolve CO_2, which further raises its atmospheric concentration. Warming oceans also become depleted in oxygen, and the change could be significant for organisms that may have evolved very specific metabolic requirements or oxygen exchange mechanisms.

The molalities shown in the figure correspond to the concentrations of the actual molecular species. For nitrogen, oxygen and argon these are likely to be essentially the same as the total dissolved concentrations ($=$ analytical concentrations) of the corresponding gases. For CO_2 this is not the case, as CO_2 reacts with H_2O to form H_2CO_3, which in turn dissociates into protons and carbonate and bicarbonate anions. The analytical concentration of CO_2 in seawater is thus greater than the concentration of molecular CO_2 calculated from Henry's law (Worked Example 10.4).

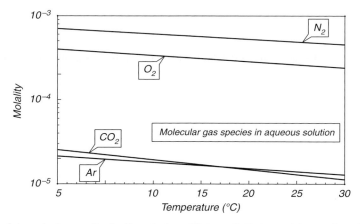

Fig. 11.1   Concentration of atmospheric gases dissolved in water in equilibrium with the terrestrial atmosphere. Calculated with Henry's law constant data from Sander (1999). The plot shows the concentration of the molecular species only.

11.1.5 Nernst's distribution law and trace element partitioning

Although the focus of this chapter is on liquid solutions, and in particular aqueous solutions, it is instructive to take a short detour and revisit the trace element partitioning equations that we discussed in Section 10.3. Consider a solute i in two dilute solutions with different solvents, $s1$ and $s2$. Let the corresponding chemical potentials of the solute be μ^i_{s1} and μ^i_{s2}. At equilibrium we have $\mu^i_{s1} = \mu^i_{s2}$, or:

$$\mu^{0,i}_{s1} + RT \ln\left(\gamma^{i,s1} m_{i,s1}\right) = \mu^{0,i}_{s2} + RT \ln\left(\gamma^{i,s2} m_{i,s2}\right). \tag{11.24}$$

Rearranging:

$$\frac{m_{i,s1}}{m_{i,s2}} = \frac{\gamma^{i,s2}}{\gamma^{i,s1}} \exp\left(\frac{\mu^{0,i}_{s2} - \mu^{0,i}_{s1}}{RT}\right). \tag{11.25}$$

This equation, known as Nernst's distribution law, is the same relationship that we derived in Chapter 10 to describe trace element partitioning between minerals and melt (equation (10.13)). The two standard state chemical potentials are the chemical potentials of the same trace component, at infinite dilution in two different solvents. At constant temperature, therefore, the exponential factor is a constant. If we consider a restricted compositional range then the activity coefficients may also be approximately constant, and we recover (10.13), with the right-hand side of equation (11.25) equal to D, the partition coefficient. Note that we have not specified the aggregation state of the two solvents. Nernst's distribution law is equally valid for crystal–liquid partitioning as it is for partitioning between two different crystals, or two different liquids.

Writing the distribution law in the form of equation (11.25) one can examine some of the complications underlying the modeling of trace element behavior in igneous systems. First, the partition coefficient is a function of temperature. Temperature variability within a given type of magmatic system (e.g. basalts) may be restricted enough that it can be safely ignored, but the same is not true if we compare, for example, basaltic and rhyolitic systems. Second, melts with significantly different major element compositions (consider again basalts vs. rhyolites) effectively behave as different solvents, so that the standard state chemical potential of the same trace element will be different in each solvent, and so will the partition coefficient. Third, the activity coefficients are in general functions of the composition of the solvent and of the concentrations of all the solutes, so if we once again consider the example of basalts vs. rhyolites we see that the activity coefficient for the same solute will almost certainly be different, and so will the partition coefficient.

11.2 Effects of dilute solutes on the properties of the solvent

Dilute solutes can have strong effects on the behavior of the solvent, such as depressing its freezing point, raising its boiling point, or affecting its density and refractive index. We now look at the equations that make it possible to track some of these effects. All of the equations in this section are valid in general, for both molecular and electrolyte solutions, although as we shall see in later sections electrolyte solutions require additional considerations.

11.2.1 The osmotic coefficient

We can write the Gibbs–Duhem equation (equation (6.6)) for a dilute solution at constant temperature and pressure as follows:

$$n_s d\mu^s + \sum_i n_i d\mu^i = 0. \tag{11.26}$$

Dividing by $n_s \, M_s^*$ we get:

$$\frac{1}{M_s^*} d\mu^s + \sum_i m_i d\mu^i = 0. \tag{11.27}$$

If all of the solutes behave ideally (i.e. $\gamma^i = 1$ for all is) then from (11.8) we get:

$$\frac{d\mu^i}{dm_i} = \frac{RT}{m_i} \tag{11.28}$$

so:

$$\frac{1}{M_s^*} d\mu^s + RT \sum_i dm_i = 0. \tag{11.29}$$

Integrating to the composition of the solution:

$$RT M_s^* \sum_i m_i = \mu^{0,s} - \mu^{s,ideal}, \tag{11.30}$$

where $\mu^{0,s}$ is the standard state chemical potential of the solvent (pure solvent at the temperature and pressure of interest) and $\mu^{s,ideal}$ is the chemical potential of the solvent in the ideal dilute solution. The sum $\sum_i m_i$ is the total concentration (molality) of dissolved species.

Equation (11.30) is true only if all of the solutes dissolve ideally, which is in general not the case. In order to account for the aggregate excess mixing behavior of all the solutes one defines a parameter ϕ, called the *osmotic coefficient*, as follows:

$$\phi \equiv \frac{\mu^{0,s} - \mu^s}{RT M_s^* \sum_i m_i} \tag{11.31}$$

from which we get the chemical potential of the solvent (compare equation (11.30)):

$$\mu^s = \mu^{0,s} - RT M_s^* \phi \sum_i m_i. \tag{11.32}$$

The osmotic coefficient ϕ is a function of the molality of each solute, m_i, and of the contribution of each solute to the excess mixing properties of the solution. Calculation of an explicit value is generally far from trivial, but if a value can be measured or calculated then equation (11.32) yields the chemical potential (and other thermodynamic properties) of the solvent.

11.2.2 Osmotic coefficient and excess Gibbs free energy of the solution

The chemical potential of the solvent, $\mu^{s,ideal}$, in equation (11.30) corresponds to an ideal solution, whereas μ^s in equation (11.32) is that of a real solution. The excess chemical potential of the solvent, $\mu^{s,ex} = \mu^s - \mu^{s,ideal}$, is therefore given by:

$$\mu^{s,ex} = (1 - \phi)\, RT M_s^* \sum_i m_i. \tag{11.33}$$

If one has an explicit formula for the excess Gibbs free energy of the solution, $\boldsymbol{G}^{ex}$, then the osmotic coefficient can be calculated as follows. From equation (5.28):

$$\mu^{s,ex} = \left(\frac{\partial \boldsymbol{G}^{ex}}{\partial \boldsymbol{n}_s} \right)_{P,T,\boldsymbol{n}_i} \tag{11.34}$$

where $\boldsymbol{n}_s$ is the number of mols of solvent, and $\boldsymbol{n}_i$ are the number of mols of solutes, which remain constant. Equating (11.33) and (11.34):

$$1 - \phi = \frac{1}{RT M_s^* \sum_i m_i} \left(\frac{\partial \boldsymbol{G}^{ex}}{\partial \boldsymbol{n}_s} \right)_{P,T,\boldsymbol{n}_i}. \tag{11.35}$$

When working with aqueous solutions it is common, however, to measure the mass of the solvent in kg rather than in number of mols. If the mass of $\boldsymbol{n}_s$ mols of solvent is $\boldsymbol{w}_s$, then $\boldsymbol{w}_s = \boldsymbol{n}_s M_s^*$ and (11.35) becomes:

$$1 - \phi = \frac{1}{RT \sum_i m_i} \left(\frac{\partial \boldsymbol{G}^{ex}}{\partial \boldsymbol{w}_s} \right)_{P,T,\boldsymbol{n}_i}. \tag{11.36}$$

Equation (11.36) is the starting point for the calculation of osmotic coefficients in electrolyte solutions (Section 11.6).

11.2.3 Relationship between osmotic coefficient of the solvent and activity coefficients of the solutes

Differentiation of (11.8) for a non-ideal solution yields:

$$\frac{d\mu^i}{dm_i} = \frac{RT}{m_i} + \frac{RT}{\gamma^i} \frac{\partial \gamma^i}{\partial m_i} \tag{11.37}$$

substituting in (11.27) and simplifying:

$$\frac{1}{M_s^*} d\mu^s + RT \sum_i dm_i + RT \sum_i m_i \frac{\partial \gamma^i}{\gamma^i} = 0. \tag{11.38}$$

Differentiating (11.31) we find:

$$d\mu^s = -RT M_s^* \left(\sum_i m_i d\phi + \phi \sum_i dm_i \right). \tag{11.39}$$

Substituting in (11.38) and simplifying:

$$(1-\phi)\sum_i \frac{dm_i}{m_i} - d\phi + \sum_i d\ln\gamma^i = 0, \tag{11.40}$$

which for a solution of a single solute simplifies to:

$$d\ln\gamma = (\phi - 1)\frac{dm}{m} + d\phi. \tag{11.41}$$

Depending on whether one wishes to calculate the activity coefficient of the solute from the osmotic coefficient of the solvent, or vice versa, and recalling that, because the infinitely dilute solution is assumed to be ideal, we have $\gamma = \phi = 1$ as $m \to 0$, we can integrate (11.41) in one of two ways (Exercise 11.1):

$$\ln\gamma = \phi - 1 + \int_0^m \frac{\phi - 1}{m} dm \tag{11.42}$$

or:

$$\phi = 1 + \frac{1}{m}\int_0^m m\, d\ln\gamma. \tag{11.43}$$

Solving the integrals requires either measured values of one of the variables over a concentration range extending to very dilute solutions ($m \to 0$), or an explicit equation for the activity coefficient or osmotic coefficient as a function of composition (Section 11.6).

11.2.4 Freezing and boiling of dilute solutions

One application of osmotic coefficients is to predict how phase transitions of the solvent are affected by the presence of dilute solutes. We shall use the melting point depression equation (6.49) once more. Consider a liquid solution that freezes to pure solid solvent (aqueous solutions behave in this way to an excellent approximation). For this case we write equation (6.49) as follows:

$$\ln a_{liq}^s = -\frac{\Delta_m H_s^0}{R}\left(\frac{1}{T} - \frac{1}{T_{0,f}}\right), \tag{11.44}$$

where a_{liq}^s is the activity of the solvent in the liquid solution, $\Delta_m H_s^0$ is the enthalpy of melting of the solvent and $T_{0,f}$ is the freezing temperature of the pure solvent at the pressure of interest. From equation (11.32) we see that:

$$\ln a_{liq}^s = \frac{\mu^s - \mu^{0,s}}{RT} = -M_s^*\phi\sum_i m_i. \tag{11.45}$$

Substituting (11.45) in (11.44) we find:

$$\frac{1}{T} - \frac{1}{T_{0,f}} = \frac{RM_s^*}{\Delta_m H_s^0}\phi\sum_i m_i. \tag{11.46}$$

If we now assume that T remains close to $T_{0,f}$ we can simplify (11.46) as follows:

$$T = T_{0,f} - \frac{RM_s^* T_{0,f}^2}{\Delta_m H_s^0}\phi\sum_i m_i. \tag{11.47}$$

The first factor in the second term in the right-hand side of equation (11.47) contains properties of the solvent only – its non-dimensional molecular weight, enthalpy of melting and freezing temperature – and the gas constant R. It is therefore a constant for each solvent, which is known as the *cryoscopic constant*, k_{cr}:

$$k_{cr} = \frac{R M_s^* T_{0,f}^2}{\Delta_m H_s^0}.$$ (11.48)

For H_2O at 1 bar the values of the parameters in equation (11.48) are: $T_{0,f} = 273.15$ K, $\Delta_m H_s^0 = 5.94$ kJ mol^{-1}, and $M_s^* = 0.01802$, which yield $k_{cr,H_2O} = 1.882$ K ($P = 1$ bar). Note that the cryoscopic constant is always a positive quantity, as enthalpy of melting is always positive. Therefore, a solution that freezes to pure solvent always does so at lower temperature than the pure liquid solvent. This is of course the same behavior that we found for eutectics in igneous systems (Chapters 6 and 10). We can also re-write (11.47) more simply as:

$$T = T_{0,f} - k_{cr}\phi \sum_i m_i,$$ (11.49)

emphasizing the fact that for a given solvent the magnitude of the freezing point depression depends on the total molality of solutes and on the osmotic coefficient (which is a sort of "aggregate activity coefficient").

If the solutes can be considered to be non-volatile compared to the solvent, i.e. if their vapor pressures are negligible compared to that of the solvent, then when the solution boils the vapor phase is composed to a very good approximation of pure solvent. Aqueous solutions of solid solutes generally behave in this way. The boiling point displacement caused by the presence of non-volatile solutes can also be calculated with equation (6.49). According to the labeling convention adopted at the beginning of Section 6.5.1 regarding the placement of the high-entropy phase we now write equation (6.49) as follows:

$$\ln a_{liq}^s = \frac{\Delta_v H_s^0}{R}\left(\frac{1}{T} - \frac{1}{T_{0,b}}\right).$$ (11.50)

As before, a_{liq}^s is the activity of the solvent in the liquid solution, but now $\Delta_v H_s^0$ is the enthalpy of vaporization of the solvent and $T_{0,b}$ is the boiling temperature of the pure solvent at the pressure of interest. Following the same sequence of steps used in deriving equation (11.47) we arrive at:

$$T = T_{0,b} + \frac{R M_s^* T_{0,b}^2}{\Delta_v H_s^0}\phi \sum_i m_i.$$ (11.51)

The *ebullioscopic constant*, k_{eb}, comprises only properties of the solvent and is defined as:

$$k_{eb} = \frac{R M_s^* T_{0,b}^2}{\Delta_v H_s^0}.$$ (11.52)

The parameters for H_2O at 1 bar are: $T_{0,b} = 373.15$ K, $\Delta_v H_s^0 = 40.6$ kJ mol^{-1}, and $M_s^* = 0.01802$, yielding an ebullioscopic constant $k_{eb,H_2O} = 0.514$ K ($P = 1$ bar). Because the enthalpy of vaporization is always a positive quantity it must be $k_{eb} > 0$. Therefore, it follows from (11.51) that a solution always boils at higher temperature than the pure solvent. Because the osmotic coefficient ϕ is in general a function of temperature its value in equations (11.47) and (11.51) will in general be different, even for solutions with the same total solute molality, $\sum m_i$.

11.2.5 Solubility of a pure solute

Equations (11.47) and (11.51) describe the equilibrium of a dilute solution with pure solid solvent and pure solvent vapor, respectively. It is also of interest to describe the equilibrium between the solution and a pure solute. The composition of the solution along this equilibrium defines the solubility of the solute in the specific solvent. At equilibrium between solution and pure solute we have:

$$\mu^i = \mu^{0,i(s)}, \tag{11.53}$$

where μ^i is the chemical potential of the solute in the solution and $\mu^{0,i(s)}$ is the standard state chemical potential of *pure solute* at the pressure and temperature of interest. Note very carefully that this is the chemical potential of the pure solute, which could be a solid, liquid or gas, *as opposed to the chemical potential of an infinitely dilute solute in the solvent of interest*, which we label $\mu^{0,i}$ (cf. equation (11.7)). Substituting equation (11.8) in (11.53) and rearranging we arrive at:

$$\ln\left(\gamma^i m_i\right) = \frac{\mu^{0,i(s)} - \mu^{0,i}}{RT}. \tag{11.54}$$

Let us now define the molar Gibbs free energy of dissolution at infinite dilution, $\Delta_s G^0$ as follows:

$$\Delta_s G^0 = \mu^{0,i} - \mu^{0,i(s)}. \tag{11.55}$$

Substituting in (11.54) and taking the temperature derivative:

$$\frac{\partial \ln\left(\gamma^i m_i\right)}{\partial T} = \frac{1}{R}\left(\frac{\Delta_s S^0}{T} + \frac{\Delta_s G^0}{T^2}\right), \tag{11.56}$$

which simplifies to:

$$\frac{\partial \ln\left(\gamma^i m_i\right)}{\partial T} = \frac{\Delta_s H^0}{RT^2}, \tag{11.57}$$

where $\Delta_s H^0$ is the *molar enthalpy of dissolution at infinite dilution*. Note that the value of $\Delta_s H^0$ depends on both the solute and the solvent and that, in contrast to the enthalpies of melting and vaporization, it may be a positive or a negative quantity (or, conceivably but very unlikely, zero).

We can integrate (11.57) between some convenient reference temperature T_0 (e.g. 298 K) and any arbitrary temperature T. After some simplification, and assuming that we can make $TT_0 \approx T_0^2$, we arrive at:

$$m_i = m_{i,0}\left(\frac{\gamma^{i,0}}{\gamma^i}\right)\exp\left[\frac{\Delta_s H^0}{RT_0^2}(T - T_0)\right]. \tag{11.58}$$

This equation yields the solubility of the solute in the solvent as a function of temperature (at constant pressure), assuming that we know the solubility at some reference temperature and an equation for the activity coefficient as a function of temperature and composition. If $\Delta_s H^0 > 0$, i.e. if dissolution is endothermic, then solubility varies directly with temperature, and conversely, it varies inversely with temperature if dissolution is exothermic and $\Delta_s H^0 < 0$.

11.2.6 Expansion of the stability field of liquid water by dilute solutes

Water may be the only common liquid solvent in which chemical reactions associated with biological metabolism can take place (Chapter 14). For this reason it is of interest to understand how the presence of dilute solutes expands the stability field of liquid water. In Worked Example 6.5 we used Schreinemakers' rule to derive a schematic phase diagram for the system H_2O–$NaCl$. The qualitative results, summarized in Fig. 6.12, are applicable to aqueous solutions of non-volatile solids in general, regardless of whether or not they are electrolytes. We saw in that example that Schreinemakers' rule causes the solute-absent and solute-present invariant points to be located along the liquid-absent curve in relative positions such that freezing and boiling of brine occur at lower and higher temperatures, respectively, than the corresponding phase transitions for pure H_2O. This is what equations (11.47) and (11.51) require. We can now include some additional information that is missing from Fig. 6.12.

Consider an aqueous solution, which we will label liquid, of a generic non-volatile solid. In most cases of geological interest the solute will be an electrolyte. We study such solutions beginning in the next section, but what we are interested in here is in the behavior of the solvent, and this can be discussed in general terms without specifying the nature of the solute. As far as the solvent is concerned the details of the solute are encapsulated in the osmotic coefficient, which is what determines the behavior of the solvent.

The pressure–temperature phase diagram shown at the bottom of Fig. 11.2 is redrawn from Fig. 6.12. Invariant point O_1 corresponds to the pure H_2O system, whereas O_2 corresponds to the H_2O–salt binary system. The two diagrams at the top of the figure are isobaric temperature–composition sections at two pressures, P_1 and P_2, above and below the pressure of invariant point O_1. We will assume that there are no other volatile components in the system, so that the partial pressure of H_2O in the vapor phase is equal to the total pressure. The diagrams are schematic, however. The composition axes are calibrated in molality and extend from pure H_2O to some unspecified solute molality somewhat higher than the saturation concentrations. For virtually all solutes of interest in the planetary sciences this would correspond to solute mol fractions much smaller than 1.

At pressure P_1 pure H_2O freezes and boils at the temperatures of points b and c, respectively. Addition of a solute shifts the freezing point to lower temperature (equation (11.47)) and the boiling point to higher temperature (equation (11.51)). The graphs of these two functions in the temperature–composition diagram are the freezing and boiling curves, ba and cd, which show the composition of liquid in equilibrium with pure H_2O ice and pure H_2O vapor, respectively. Each of the divariant assemblages (pseudo-univariant at constant pressure), liquid + ice and liquid + vapor, exists inside the field labeled accordingly, with the liquid composition given by the intersection of the temperature coordinate with the corresponding bounding curve. The liquid becomes saturated in solid at points a and d. At each of these two points three phases are stable, so they must be located on the corresponding univariant phase boundaries for the solute-bearing system, as shown in the bottom diagram.

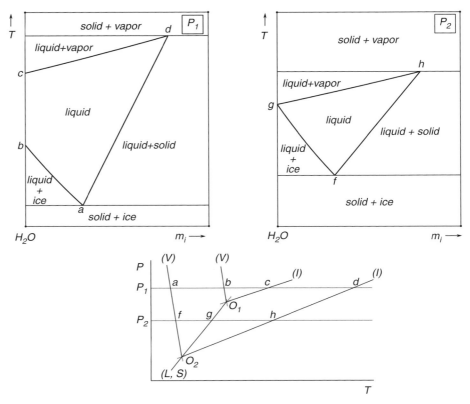

Fig. 11.2 Freezing and boiling phase relations of dilute aqueous solutions of a non-volatile solute. The $P–T$ phase diagram (bottom) is redrawn after Fig. 6.12. The two isobaric temperature–composition phase diagrams (top) correspond to pressures higher and lower than that of the triple point of pure H_2O (O_1). The liquid field shrinks continuously from the pressure of point O_1 to the pressure of point O_2, where it disappears altogether.

They must also appear as pseudo-invariant points on the isobaric sections. Because a and d are saturation points they must both lie on the solubility curve given by equation (11.58). The graph of this function is therefore the curve ad in the figure. The divariant assemblage liquid + solid is stable to the right of the curve, with the composition of the solid-saturated liquid given by the intersection of the temperature coordinate with curve ad. Note that inside the field labeled "liquid" only one phase is stable, so that the assemblage is trivariant, or pseudodivariant at constant pressure. Temperature and liquid composition can be chosen independently as long as no other phase is stable.

Equations (11.47) and (11.51) assure us that, because points a and d are saturated in solid, they must correspond to the minimum freezing temperature and maximum boiling temperature at pressure P_1. Point a is therefore a eutectic (compare with Figs. 6.21 and 6.23). For all temperatures lower than that of a the system consists of the divariant assemblage ice + solid. Similarly, in a neighborhood above the temperature of point d the stable assemblage is solid + vapor, but this divariant field must terminate at some higher temperature, at which the solid melts. The eutectic and maximum boiling temperatures can be calculated by solving simultaneously equations (11.47) and (11.58), or (11.51) and (11.58), respectively.

Consider now the temperature–composition phase relations at pressure P_2, lower than that of the triple point of pure H_2O (invariant point O_1). At pressure P_2 pure liquid H_2O is not stable, so the liquid + vapor and liquid + ice fields must meet at the H_2O composition end-member. This happens at point g, which corresponds to a point along the stable univariant sublimation curve for H_2O (see bottom figure). As in the previous case, the freezing and boiling curves, gf and gh, are the graphs of functions (11.47) and (11.51). They map the liquid compositions in equilibrium with ice and vapor, respectively, and terminate at points f and h on the solubility curve, given by equation (11.58). As pressure decreases from O_1 to O_2 the temperatures of the eutectic point, f, and of the maximum boiling point, h, approach each other, so that the temperature interval within which liquid is stable shrinks. The two temperatures become identical, and equal to the sublimation temperature, at O_2. For pressures lower than that of O_2 liquid is never stable and the only possible univariant equilibrium in the two-component system is the degenerate H_2O sublimation reaction.

There is a subtle point that we must address. In deriving equations (11.47) and (11.51) we assumed that the pure liquid solvent is stable, and therefore that freezing and boiling of the pure solvent take place. For pressures between O_1 and O_2 liquid is stable only if it contains dissolved solutes in it, so that freezing and boiling of the pure solvent do not take place. Is it then "legal" to use equations (11.47) and (11.51) to construct the freezing and boiling curves? Moreover, what temperatures should be chosen for $T_{0,f}$ and $T_{0,b}$, the freezing and boiling point of the pure solvent? The graphical construction in Fig. 11.2 suggests that both temperatures should be the same, and equal to the sublimation temperature of the solvent at the pressure of interest. We now demonstrate formally that this is indeed the case. Equilibrium along the freezing curve can be written as (see equation (11.32)):

$$\mu_{liquid}^{H_2O} = |\mu_{liquid}^{0,H_2O}| - RTM_s^*\phi\sum_i m_i = \mu_{ice}^{0,H_2O}, \qquad (11.59)$$

where the standard state chemical potential of liquid H_2O enclosed inside vertical lines is a fictive value, corresponding to a hypothetical pure liquid H_2O phase at pressure P_2. Similarly, for the boiling curve we have:

$$\mu_{liquid}^{H_2O} = |\mu_{liquid}^{0,H_2O}| - RTM_s^*\phi\sum_i m_i = \mu_{vapor}^{0,H_2O}. \qquad (11.60)$$

The two curves meet at the pure H_2O end-member (point g), at which $\sum m_i = 0$, so:

$$\mu_{ice}^{0,H_2O} = \mu_{vapor}^{0,H_2O} = |\mu_{liquid}^{0,H_2O}|. \qquad (11.61)$$

The sublimation temperature (g in the figure) therefore corresponds to both the (fictive) freezing and boiling temperatures of pure H_2O. Moreover, the fictive standard state chemical potential of liquid H_2O is well defined, and is equal to the standard state chemical potentials of ice and vapor at the sublimation temperature. This makes it possible to define the activity of H_2O in liquid as in equation (11.45), and thus confirms that functions (11.47) and (11.51) represent the freezing and boiling curves of liquid at pressures below the triple point of pure H_2O too.

The chemical experiments aboard the *Phoenix* spacecraft, that reached Mars in May of 2008, detected what appears to be a large concentration of the perchlorate ion, ClO_4^-, in her landing site on the planet's northern plains (Hecht *et al.*, 2009; Kounaves *et al.*, 2009). The importance of this finding is that alkali- and alkali-earth perchlorates are quite soluble in water. The maximum freezing point depression of a solution (i.e. its eutectic temperature) depends in part on the location of the solubility curve *ad* in Fig. 11.2 (given by equation (11.58)) – the more soluble a solid is, the lower the eutectic temperature is likely to be. The question arises, could perchlorates be soluble enough, and abundant enough on Mars, to extend the stability field of liquid water to the conditions of the Martian surface? And why would perchlorates be so abundant on Mars in the first place? Formation of perchlorates requires an oxidant more powerful than molecular oxygen (Exercise 11.2). The most likely explanation for its presence on Mars, and in some desert terrestrial environments, is the oxidation of chloride anions in the atmosphere, either by ozone or by some of the products of ozone photochemistry (Catling *et al.*, 2010; see also Chapter 12).

The calculation of the eutectic point of an aqueous solution entails the simultaneous solution, for temperature and solute molality, of equations (11.49) and (11.58). The melting and solubility curves, *ba* and *ad* in Fig. 11.2, are then constructed by solving each of these equations for one of the variables (e.g. temperature) as a function of the other (molality). The procedure is in principle the same one that we used to calculate eutectic phase relations in Chapter 6, and that we implemented in Software Box 6.1. The phase diagrams for magnesium perchlorate and sodium perchlorate aqueous solutions were calculated in this manner by Chevrier *et al.* (2009). For solutions of electrolytes the calculations are far from trivial, however. This is so because, as we shall see later in this Chapter, the activity and osmotic coefficients of electrolyte solutions are rather complex functions of composition. We will therefore not repeat the calculations of Chevrier *et al.* (2009), but will rather study their phase diagrams, from which I extract the diagrams at the top of Fig. 11.3.

Chevrier *et al.* found that the system $Mg(ClO_4)_2$–H_2O has a 1-bar eutectic temperature of 206 K, and that the concentration of magnesium perchlorate in the eutectic melt is 44 wt%. The corresponding values for the system $NaClO_4$–H_2O are 236 K and 52 wt%. These temperatures must lie on the vapor-absent curves of the respective binary systems, such as *a* or *f* in Fig. 11.2. We can construct the vapor-absent curves by assuming that their slope is the same as that of the freezing curve of pure H_2O. This is not rigorously true, as both the entropy and volume of the solutions will in general be different from those of pure liquid water (recall that $dP/dT = \Delta S/\Delta V$, and that ice is a pure phase in all cases). The error that might be introduced between 1 bar and the near zero pressure of the Martian surface is, however, almost certainly negligible. The bottom diagram in Fig. 11.3 shows the *P–T* locations of the vapor-absent equilibria, which are the Mg-perchlorate and Na-perchlorate eutectics, inferred from the temperatures calculated by Chevrier *et al.* (2009) and the Clapeyron slope of the freezing reaction of pure H_2O. Also shown is the ice sublimation curve, calculated from the H_2O data of Wagner *et al.* (1993). Since this is the liquid-absent curve its location is independent of the composition of the solution. The intersection of each of the freezing curves with the sublimation curve defines the invariant point for each system (the invariant point for pure H_2O is also shown for comparison). A third phase boundary, the boiling curve, emanates from each invariant point. These curves are shown schematically, as their slopes remain unknown.

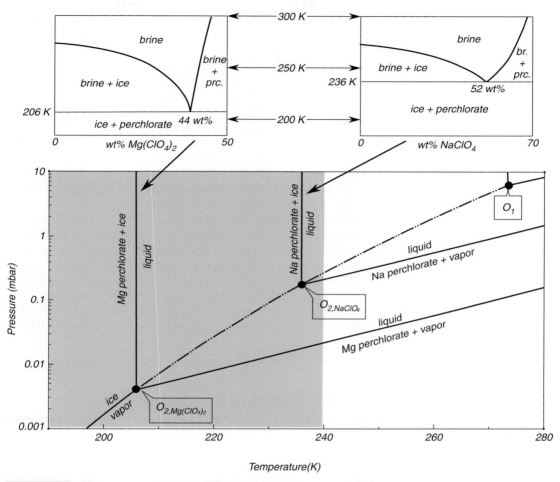

Fig. 11.3 Melting phase relations for the binary systems Mg(ClO$_4$)$_2$–H$_2$O and NaClO$_4$–H$_2$O (top), simplified from Chevrier *et al.*
(2009). The corresponding *P–T* phase relations for the two systems are shown in the bottom diagram. Compare
Fig. 11.2.

The *P–T* diagram in Fig. 11.3 corresponds to a system in which no additional volatile
components are present, so that the partial pressure of H$_2$O in the vapor phase is equal to
the total pressure on the system. The Martian atmosphere, however, consists predominantly
of components other than H$_2$O, so the partial pressure of H$_2$O in the gas phase is much
lower than atmospheric pressure, which is about 7 mbar at the planet's mean surface level.
The diagram can nevertheless be applied to understand conditions on the Martian surface,
if we keep the following in mind. The freezing curve is a vapor-absent equilibrium so it
is not affected by changes in the composition of the vapor phase. Its *P–T* location remains
unchanged whether the pressure axis represents total pressure or partial pressure. The boiling
and sublimation curves represent equilibrium between vapor and condensed phases, and
the condensed phases will not undergo any significant change in volume in response to
pressure differences of order 1 bar or less. Therefore, the location of the phase boundaries

within the restricted range of conditions shown in the figure depends only on the partial pressure of H_2O, and we can think of the pressure coordinate as representing this variable, rather than atmospheric pressure.

The shaded region in the diagram shows a range of possible surface conditions on Mars. Martian surface temperatures are within the range in which perchlorate brines can exist. Whether such brines are thermodynamically stable depends on the H_2O content of the Martian atmosphere. In the Mg-perchlorate system the liquid can be stable down to an H_2O partial pressure of ~ 0.003 mbar, which would correspond to an H_2O mol fraction of $\sim 10^{-4}$ in the Martian atmosphere. The values for the Na-perchlorate system are ~ 0.2 mbar H_2O partial pressure, equivalent to an H_2O mol fraction of about 0.03. For comparison, the saturation vapor pressure of pure H_2O at 298 K is ~ 31 mbar, so that 100% relative humidity in the terrestrial atmosphere at this temperature corresponds to an H_2O mol fraction of about 0.03. The corresponding mol fraction value at 273 K is ~ 0.006.

The minimum atmospheric H_2O concentration required to stabilize Na perchlorate brine is almost certainly higher than anything that is common, or even possible, in the Martian atmosphere, but Mg perchlorate brines might be stable in unusually "humid" Martian environments. We can speculate that stable perchlorate brines probably do not exist over most of the surface of Mars. Chevrier *et al.* (2009) showed, however, that evaporation rates of metastable brine pools could be slow enough that they could persist for a few hours. Even if long-lived pools of stable perchlorate brines did exist on Mars, however, it is not clear what their biological significance might be, given that perchlorates are powerful oxidizers (Chapter 14).

11.3 Electrolyte dissociation

A very important class of dilute solutions are electrolyte solutions. These are solutions in which the solutes dissociate, partially or fully, into electrically charged mobile species, or ions. The evidence that this happens is the higher electrical conductivity of electrolyte solutions compared to that of the pure solvent. The thermodynamic treatment of electrolyte solutions is based on the general formulation for dilute liquid solutions that we developed in Sections 11.1 and 11.2, but includes some significant extensions.

A thermodynamic description of electrolyte solutions should be independent of any microscopic model of the solution, but it will be helpful to have in mind at least a rudimentary microscopic model. In fact, accurate prediction of excess mixing properties in electrolyte solutions is impossible without recourse to their microscopic properties. Let us begin with a simple question: why do substances dissociate into ions? This happens when the thermal energy of the ions, of order $k_B T$, becomes comparable to the work that must be performed to separate them. In an ionic solid the forces holding the atoms together are purely electrostatic. Therefore, an ionic solid will dissociate into ions either if the thermal energy increases or if the electrostatic attractive forces decrease. The former case corresponds to melting and gives rise to what is called a pure liquid electrolyte, i.e. a liquid made up of free anions and cations. Molten alkali halides are the archetypal examples of this (see also Section 10.1). Alternatively, electrostatic attractive forces may decrease as a result of interactions with the molecules of the solvent. Dissociation in this case occurs at temperatures at which the substance would otherwise be a solid. The result is the formation of an electrolyte solution.

11.3.1 The dielectric constant and the structure of polar liquids

We recall Coulomb's law, equation (1.38), that gives the magnitude of the electrostatic force between two charged particles:

$$|\overline{f}_e| = \frac{1}{4\pi\epsilon} \frac{|q_1||q_2|}{x^2}.$$ (11.62)

The constant ϵ is a material property called the permittivity, except if the charges are located in vacuum, in which case it is a universal constant called the permittivity of free space and symbolized by ϵ_0. For all substances it is found that $\epsilon > \epsilon_0$. We define a non-dimensional quantity ε_r, called the *dielectric constant*, as the ratio:

$$\varepsilon_r = \frac{\epsilon}{\epsilon_0} > 1.$$ (11.63)

From (11.62) it follows that the dielectric constant is also a measure of how much the electrostatic force decreases in response to the introduction of a material between charged particles. The higher the dielectric constant, the lower the electrostatic force, everything else being equal.

The physical explanation of this behavior is easy to see. Consider two oppositely charged plates, initially separated by a vacuum (Fig. 11.4). By (11.62), intercalation of an insulating substance between the plates decreases the electrostatic force between them. The decrease in the electrostatic force happens because all substances become polarized to some extent. What this means is that the electric field between the plates causes the charge distribution in the interstitial material to become asymmetric, such that there is some excess of negative charge closer to the positive plate, and an equal excess of positive charge closer to the negative plate. These charge excesses generate an induced electrostatic field that opposes the primary field between the plates, and hence lowers the net attractive force between them. Of course, if the substance is an electrical conductor (i.e. if the potential difference between the plates is higher than its dielectric strength, see Worked Example 1.4) then current will flow, the plates will lose their charge, and this picture is no longer valid.

Non-polar materials consist of molecules with symmetric charge distribution. In the absence of an external field the molecules do not display any preferred charge distribution, or in other words, they are not polarized. When a non-polar substance is immersed in an electric field, however, it acquires induced polarization as described in the last paragraph. The magnitude of the induced field that opposes the primary field depends on the characteristics of the individual molecules and on their density. In diatomic elemental gases at low pressure (low molecular density) the induced field is extremely weak. Hence, their dielectric constant may be indistinguishable from 1 for most practical purposes. For instance, the dielectric constant of dry air at 298 K and 1 bar is ~ 1.0005, so that the force between charged particles immersed in air is virtually indistinguishable from what the force would be in vacuum. In non-polar liquids, such as hydrocarbons or liquid CO_2, the dielectric constant is somewhat higher (typically of order 2), largely as a result of the higher molecular density, which engenders a stronger induced field.

In contrast to non-polar substances, polar materials consist of molecules with an inherent charge asymmetry that exists regardless of the presence of an external electrostatic field. Water is both an excellent example and the most important one in the planetary sciences. The two OH bonds in H_2O subtend an angle of $\sim 105°$. Because the bonds arise from

Non-polar molecule

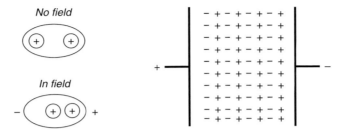

Polar molecule

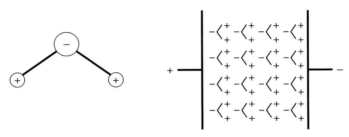

Fig. 11.4 Schematic diagram comparing the behavior of a non-polar substance (top) with that of a polar substance (bottom). In the absence of an external electric field charges in the molecules of non-polar substances are distributed symmetrically, but the symmetry is broken when an external field is applied (right). Polar substances such as water have molecules with permanent charge asymmetry, that acquire a preferred orientation in an external field. In both cases an induced electric field appears that opposes the external field, but this field is much stronger in polar substances. Because thermal agitation tends to disrupt molecular orientation the intensity of the induced field (and the dielectric constant) decreases with increasing temperature.

sharing of the H electrons, each of the two H ends of the molecule acquires a fractional positive charge, and the O vertex the complementary negative charge. When H_2O molecules are immersed in an electric field between charged plates the molecules orient themselves such that the negatively charged vertices face the positive plate and the positively charged ends the negative plate. The result is a strong induced field and therefore a large decrease in the electrostatic force between the plates. For example, the dielectric constant of H_2O at 298 K is $\varepsilon_{r,H_2O} = 78.3$, meaning that the force between charged particles immersed in water is almost two orders of magnitude less than in vacuum. The dielectric constant varies inversely with temperature. Its value for liquid water at 1 bar is 87.7 at the freezing point and 55.7 at the boiling point. It decreases further at higher temperatures, in liquid water kept from boiling by increased pressure, and in supercritical H_2O. This behavior is to be expected from the simple fact that increasing thermal agitation tends to disrupt the molecular orientation responsible for the induced field that opposes the external field. Other polar liquids generally have dielectric constants comparable to or lower than that of water. For instance, liquid ammonia has a dielectric constant of 22 at its 1 bar boiling point (239.7 K).

11.3.2 Free energy of ions in dielectric materials

At an intuitive level we can think of ionic dissociation in polar solvents as taking place because the dielectric constant of the solvent lowers the attractive force between ions in the crystal to the point where thermal agitation is able to break the electrostatic bonds. The problem with this simple model is that it considers the solvent to be a structureless continuum, when in reality it is made up of molecules of comparable size to the ions of the solute. The dielectric constant is a macroscopic property, so that its value cannot be expected to remain unchanged down to molecular lengthscales. In spite of this, the continuum model for the solvent forms the basis of what was historically the first successful physical model for electrolyte solutions, proposed by the great German physicist Max Born in the early 1920s. Born's model seeks to calculate the change in free energy that takes place when an isolated ion is transferred from vacuum to a material. The result provides fundamental physical insight into the nature of electrolytic dissociation, so we will reconstruct the main aspects of Born's reasoning.

First we need to define the electrostatic potential, Φ_e. As for gravitational potential, this is defined as the electrostatic potential energy per unit of electric charge. From equation (1.47) we get

$$\Phi_e = \frac{1}{4\pi\epsilon}\frac{q}{r},\tag{11.64}$$

where q is the electric charge that engenders an electric field in a medium of permittivity ϵ, and Φ_e is the electrostatic potential at a distance r from the charge. Note that the potential can be positive or negative, depending on the sign of the charge. Born's idea is to consider ions as spheres of radius r_i that are initially electrically neutral, and calculate how much energy is required to charge the sphere with the electric charge of the ion, q_i. This energy equals the work, w_e, done against the electrostatic field by the charging process. The energy, which is stored in the field as electrostatic potential energy, is given by the following integral:

$$w_e = \int_0^{q_i} \Phi_e dq = \frac{1}{4\pi\epsilon r_i}\int_0^{q_i} q dq = \frac{1}{4\pi\epsilon r_i}\frac{q_i^2}{2}.\tag{11.65}$$

The procedure is somewhat analogous to calculating the gravitational binding energy of a body (Section 2.4), but there is a crucial difference. This is that, whereas gravitational assembly of a body always extracts potential energy from the gravitational field, charging a body stores energy in the electrostatic field, because work must be performed to overcome the electrostatic repulsion between the charge already present in the body and the newly arriving charge increment. The square in q_i in equation (11.65) ensures that w_e is always positive, whether the charge of the ion is positive or negative.

We recall from Section 4.8.4 that, if energy transformations other than heat exchange, expansion work and transfer of chemical components take place in a system, then their contributions must be included in the thermodynamic potentials. Charging a sphere at constant temperature, pressure and composition therefore changes its Gibbs free energy by an amount $\Delta_{ch}G$ equal to w_e, or:

$$\Delta_{ch}G = \frac{1}{4\pi\epsilon r_i}\frac{q_i^2}{2}.\tag{11.66}$$

As always with Gibbs free energy, what we are interested in is not its absolute value, but rather how its value changes between different states of a system. In this case we ask, how does the Gibbs free energy of an ion differ between a state in which the ion is in vacuum and another state in which the ion is immersed in a material with dielectric constant ε_r? Calling this Gibbs free energy difference $\Delta_{ds}G$, we see that it is given by:

$$\Delta_{ds}G = (\Delta_{ch}G)_{material} - (\Delta_{ch}G)_{vacuum} = \frac{1}{4\pi\epsilon r_i}\frac{q_i^2}{2} - \frac{1}{4\pi\epsilon_0 r_i}\frac{q_i^2}{2} \qquad (11.67)$$

which, simplifying and introducing the definition of the dielectric constant (equation (11.63)), becomes:

$$\Delta_{ds}G = \frac{1}{4\pi\epsilon_0}\frac{q_i^2}{2r_i}\left(\frac{1}{\varepsilon_r} - 1\right) < 0. \qquad (11.68)$$

This is known as Born's equation, and it packs a physics punch in a beautifully compact expression. First, because for all materials the dielectric constant is greater than 1, the Gibbs free energy change is always negative. Thus, an isolated ion is always more stable if it is immersed in a solvent than if it is in vacuum. Moreover, the tendency for an ion to enter solution becomes stronger (i.e. $\Delta_{ds}G$ becomes more negative) the greater the dielectric constant of the solvent is. This says that electrolyte solutions form more readily the more polar a solvent is. For a given solvent, as temperature increases and the dielectric constant decreases electrolytic dissociation becomes less likely. Born's equation also shows that the tendency towards electrolytic dissociation varies directly with the square of ionic charge and inversely with ionic radius.

By differentiating (11.68) one can obtain an expression for the enthalpy change associated with immersing an ion into a solvent. With some additional algebraic manipulations and reference state conventions (see below) this enthalpy becomes an experimentally accessible quantity. Experimental measurements show that, although Born's equation is a brilliant conceptual model, it is not quantitatively successful, as it predicts enthalpies of solution that are generally too high by a factor of ~2. The failure is to a significant extent due to the fact that the macroscopic dielectric constant is not valid at the range typical of molecular interactions. The presence of a charged ion must affect the local arrangement of molecules in a polar solvent, and therefore the local value of its dielectric constant. For example, a cation in liquid water surrounds itself with a sheath of H_2O molecules arranged with their oxygen vertices towards the cation. This structure is known as a solvation sphere (hydration if the solvent is water), and the number of solvent molecules attached to the ion as the solvation (or hydration) number. Solvation causes the effective dielectric constant at molecular lengthscales to be lower than the macroscopically determined value.

11.4 Thermodynamic formulation of electrolyte solutions

It is possible to distinguish two contrasting behaviors among electrolytes. There are electrolytes that undergo complete dissociation, meaning that there is no evidence that undissociated molecules exist in the solution. These are called strong electrolytes. The classic examples of these are alkali halides in aqueous solution at room temperature, and in particular NaCl. On the other hand, there are electrolytes, called weak electrolytes, that undergo only partial dissociation, such that ions coexist in the solution with neutral undissociated molecules. Room temperature aqueous solutions of weak acids and of many transition metal salts behave in this way. The distinction is not invariable. For example, some electrolytes are fully dissociated at low concentrations and only partially dissociated as concentration increases. The nature of the solvent also plays a fundamental role, as described by Born's equation (equation (11.68)). A substance may behave as a strong electrolyte in a solvent with a high dielectric constant, and as a weak one if the solvent's dielectric constant is lower. In particular, substances that behave as strong electrolytes in liquid water at room temperature become weak electrolytes, or fail to dissociate altogether, in high temperature liquid water and supercritical H_2O. The thermodynamic treatment of the two types of electrolytes differs in the definition of the standard states and of quantities that depend on the standard state, such as activity and activity coefficient.

11.4.1 Thermodynamic functions of ionic species

We begin by discussing general thermodynamic relations that are the same for all types of electrolytes. The standard state thermodynamic properties of ionic species are defined as in Section 11.1.2, at the infinite dilution limit (equation (11.7)), which, as we saw, can also be thought of as a hypothetical one-molal ideal solution. In the case of ionic species there is an additional restriction, which is that the solution must be electrically neutral. The condition of electrical neutrality of an electrolyte solution is written as follows:

$$\sum_i m_i z_i = 0, \tag{11.69}$$

where m_i is the molality of ion i and z_i is its charge, positive or negative depending on whether it is a cation or an anion. Even if a solution contains many different ions, it will be convenient to consider anion–cation pairs that correspond to dissociation of specific electrolytes. Take for instance an ionic substance composed of anions $A-$, of charge z_{A-}, and cations $B+$ of charge z_{B+}, and let the formula unit of the substance be $B_\beta A_\alpha$, such that:

$$\beta z_{B+} + \alpha z_{A-} = 0. \tag{11.70}$$

Dissociation of the substance corresponds to the reaction:

$$B_\beta A_\alpha \rightleftharpoons \beta (B+)_{aq} + \alpha (A-)_{aq}. \tag{11.71}$$

This equation is valid regardless of whether the substance is a strong or weak electrolyte. I have specified the state of the ionic species by means of the subscript aq, meaning aqueous solution. All of the equations that we will develop here are valid for any polar solvent, but since liquid H_2O is the most common one in natural environments we will use "aqueous" as a generic designation. If we were to consider solutions in liquid ammonia, for instance, then am could replace aq as the subscript. Note that, in contrast to the ionic species, I have *not* specified the state of the compound $B_\beta A_\alpha$ in equation (11.71). This will take different meanings in strong and weak electrolytes. For now we are only concerned with the ionic species, and equation (11.71) is in this sense best thought of as an algebraic identity, not as a chemical reaction.

We write the chemical potentials of the ionic species as follows:

$$\mu_{aq}^{B+} = \mu_{aq}^{0,B+} + RT \ln a_{aq}^{B+}$$
$$\mu_{aq}^{A-} = \mu_{aq}^{0,A-} + RT \ln a_{aq}^{A-} \tag{11.72}$$

where the standard state chemical potentials correspond to the usual infinite dilution limit. The electrical neutrality constraint (equation (11.69)) introduces the complication that concentrations of individual ions cannot be varied independently of those of opposite charge. As we shall see later this is solved by arbitrarily assigning a value of zero to the standard state properties of the cation H^+.

The ionic activities in equations (11.72) are given by:

$$a_{aq}^{B+} = \gamma_{aq}^{B+} m_{B+}$$
$$a_{aq}^{A-} = \gamma_{aq}^{A-} m_{A-}. \tag{11.73}$$

The $B-A$ ion pair appears in the proportions β to α in every equilibrium equation involving the compound $B_\beta A_\alpha$. This makes it convenient to define the mean ionic molality, $m_\pm$, and mean ionic activity coefficient, $\gamma_{aq}^\pm$, of dissociated electrolyte as follows:

$$m_\pm = \left[(m_{B+})^\beta (m_{A-})^\alpha \right]^{\frac{1}{\beta+\alpha}} \tag{11.74}$$

and:

$$\gamma_{aq}^\pm = \left[\left(\gamma_{aq}^{B+} \right)^\beta \left(\gamma_{aq}^{A-} \right)^\alpha \right]^{\frac{1}{\beta+\alpha}} \tag{11.75}$$

The three following points need to be emphasized. First, the mean ionic molality and mean ionic activity coefficient correspond specifically to electrolyte $B_\beta A_\alpha$. Second, equation (11.74) does not imply that the molalities of B^+ and A^- in the solution are in the proportion β to α. This would be the case *only* if the solution was one of pure electrolyte $B_\beta A_\alpha$. In a mixed electrolyte solution the relative amounts of the ions are only constrained by the bulk charge balance condition (11.69), and will in general not correspond to the stoichiometric coefficients β and α. However, the mean ionic molality and activity coefficient for electrolyte $B_\beta A_\alpha$ are still given by (11.74) and (11.75). Third, for reasons that will become apparent in the following section we *have not* defined a "mean ionic activity" analogous to the mean ionic molality and activity coefficient.

11.4.2 Standard states and activities of strong and weak electrolytes

A solution of a weak electrolyte contains undissociated molecules, which we shall label $(B_\beta A_\alpha)_{aq}$, as well as ions. For equilibrium among the aqueous species:

$$\left(B_\beta A_\alpha \right)_{aq} \rightleftharpoons \beta \left(B+ \right)_{aq} + \alpha \left(A- \right)_{aq} \tag{11.76}$$

it must be:

$$\mu_{aq}^{B_\beta A_\alpha} = \beta \mu_{aq}^{B+} + \alpha \mu_{aq}^{A-}. \tag{11.77}$$

We can also write (11.77) as:

$$\beta \mu_{aq}^{0,B+} + \alpha \mu_{aq}^{0,A-} - \mu_{aq}^{0,B_\beta A_\alpha} + RT \ln K_{ds} = \Delta_r G_{P,T}^0 + RT \ln K_{ds} = 0, \tag{11.78}$$

where $\Delta_r G^0_{P,T}$ is the standard state Gibbs free energy change for the dissociation reaction (11.76) at infinite dilution and the temperature and pressure of interest, so that:

$$K_{ds} = \exp\left(-\frac{\Delta_r G^0_{P,T}}{RT}\right). \tag{11.79}$$

The equilibrium constant K_{ds} is known as the *dissociation constant* and is also equal to:

$$K_{ds} = \frac{\left(a^{B+}_{aq}\right)^{\beta} \left(a^{A-}_{aq}\right)^{\alpha}}{a^{B_\beta A_\alpha}_{aq}}. \tag{11.80}$$

Because in the case of a weak electrolyte the species $(B_\beta A_\alpha)_{aq}$ exists, its standard state properties are defined in the same way as those for any other molecular or ionic dilute solute (equation (11.7)). In this case the molality of the undissociated solute is a physically significant quantity. We can then write the activity of the undissociated electrolyte in terms of its molality and activity coefficient, as follows:

$$a^{B_\beta A_\alpha}_{aq} = \gamma^{B_\beta A_\alpha}_{aq} m_{B_\beta A_\alpha} \tag{11.81}$$

and using the mean ionic molality and mean ionic activity coefficient of the dissociated species (equations (11.74) and (11.75)) we get the following expression for the dissociation constant:

$$K_{ds} = \frac{\left(\gamma^{\pm}_{aq} m_{\pm}\right)^{\beta+\alpha}}{\gamma^{B_\beta A_\alpha}_{aq} m_{B_\beta A_\alpha}}. \tag{11.82}$$

In contrast to weak electrolytes, when a strong electrolyte dissolves in a polar solvent it undergoes complete dissociation. If this is the case then the molecular species $(B_\beta A_\alpha)_{aq}$ is not present in the solution, and one cannot specify its molality. As we shall see, it is nevertheless convenient to define thermodynamic properties for a fictive species that we can call the *bulk aqueous electrolyte*, $(B_\beta A_\alpha)_{bl}$. We can give a precise thermodynamic definition of $(B_\beta A_\alpha)_{bl}$ by stating that, if $B_\beta A_\alpha$ is a strong electrolyte and Y stands for any of the thermodynamic potentials, the entropy, the heat capacity or the volume of the bulk aqueous electrolyte, then Y is given by:

$$Y^{B_\beta A_\alpha}_{bl} \equiv \beta Y^{B+}_{aq} + \alpha Y^{A-}_{aq}. \tag{11.83}$$

Thus, the chemical potential of $(B_\beta A_\alpha)_{bl}$ is:

$$\mu^{B_\beta A_\alpha}_{bl} \equiv \beta \mu^{B+}_{aq} + \alpha \mu^{A-}_{aq}. \tag{11.84}$$

Note that there is a subtle difference with equation (11.77), for a weak electrolyte: whereas in that case the equation is the statement of chemical equilibrium, equation (11.84), for a strong electrolyte, is the *definition* of the chemical potential of the fictive bulk species. Expanding (11.84) in terms of standard state chemical potentials and activities we find:

$$\mu^{0,B_\beta A_\alpha}_{bl} + RT \ln a^{B_\beta A_\alpha}_{bl} = \beta\left(\mu^{0,B+}_{aq} + RT \ln a^{B+}_{aq}\right) + \alpha\left(\mu^{0,A-}_{aq} + RT \ln a^{A-}_{aq}\right). \tag{11.85}$$

Since we required that all thermodynamic potentials of $(B_\beta A_\alpha)_{bl}$ be given by an equation of the form of (11.83), the standard state chemical potential of the bulk aqueous electrolyte is given by:

$$\mu_{bl}^{0,B_\beta A_\alpha} \equiv \beta \mu_{aq}^{0,B+} + \alpha \mu_{aq}^{0,A-}. \tag{11.86}$$

This definition of standard state properties is the crucial difference with the treatment of weak electrolytes. A consequence of equation (11.86) is that we can think of the dissociation constant of a strong electrolyte as being identically equal to 1 at all pressures and temperatures for which the electrolyte undergoes complete dissociation. *Equation (11.86) is, however, never valid for a weak electrolyte in which, as we saw, the standard state is defined in the usual way for a dilute solute.*

From (11.85) and (11.86) we find that the activity of the bulk electrolyte is given by:

$$a_{bl}^{B_\beta A_\alpha} = \left(a_{aq}^{B+}\right)^\beta \left(a_{aq}^{A-}\right)^\alpha. \tag{11.87}$$

Note that this is different from the activity of an undissociated weak electrolyte, given by (11.81). Substituting mean ionic molality and mean ionic activity coefficient, (11.87) becomes:

$$a_{bl}^{B_\beta A_\alpha} = \left(\gamma_{aq}^\pm m_\pm\right)^{\beta+\alpha}, \tag{11.88}$$

which should be compared with equation (11.81) for a weak electrolyte. Exercise 11.3 asks you to apply these relations to the equations that relate activity and osmotic coefficients (Section 11.2.3), and may be helpful in clarifying the meaning of (11.87) and (11.88).

11.4.3 The reference for standard state properties of ionic species

A problem that arises in the treatment of ionic species is that their concentrations cannot be varied independently of those of ions of opposite charge. It is possible to determine the infinite dilution standard state properties of dissolved molecular species for a weak electrolyte, or the standard state properties of (fictive) bulk strong electrolytes, for example by measurements of enthalpies of dissolution. Once these data are available one knows the sum of the standard state properties of the corresponding anion and cation, e.g. via equations (11.78) or (11.86). In all cases we have one equation with two free parameters, so that there is no unique solution unless one specifies the value of one of the parameters. The convention is to make the standard state enthalpy, entropy, Gibbs free energy and heat capacity of the H^+ cation in aqueous solution at infinite dilution equal to zero, i.e.:

$$\Delta_f H_{aq}^{0,H^+} = \Delta_f G_{aq}^{0,H^+} = S_{aq}^{0,H^+} = C_P^{H^+} = 0. \tag{11.89}$$

Because the heat capacity is set equal to zero equation (11.89) is valid at all temperatures. Of course, the entropy of protons at any temperature other than zero is not zero, but since all we care about when performing thermodynamic calculations are entropy (or free energy) differences at constant temperature and pressure the convention does not introduce any difficulties. Also, H^+ ions in aqueous solutions are always hydrated, i.e. attached electrostatically to one or more H_2O molecules, and do not exist as free protons. The symbol H^+ in equation (11.89) and all subsequent discussions must be understood as thermodynamic shorthand for the actual molecular entity that exists in solution.

11.4.4 Dissociation of water and pH

Water is a polar liquid, so it should not be surprising that H_2O molecules dissociate in liquid water to some extent (Section 11.3). We can write this dissociation reaction as follows:

$$(H_2O)_{liq} \rightleftharpoons H_{aq}^+ + OH_{aq}^-. \qquad (11.90)$$

Choosing the standard state of water as the pure liquid at the temperature and pressure of interest, we get the following equation for the dissociation constant of water, K_w (compare equation (11.80) for the dissociation of a weak electrolyte):

$$K_W = a_{aq}^{H^+} \cdot a_{aq}^{OH^-} = \left(m_{H^+} \cdot m_{OH^-} \right) \left(\gamma_{aq}^{H^+} \cdot \gamma_{aq}^{OH^-} \right). \qquad (11.91)$$

At 298.15 K the constant K_W, also known as the ionization product of water, has a value of $K_W = 1.011 \times 10^{-14}$, which is commonly rounded off to $K_W \approx 10^{-14}$. The ionic concentration in pure water is very low. Therefore, as a first approximation that is consistent with the infinite dilution standard state, we can assume that the activity coefficients of the ionic species are unity. The charge balance constraint, equation (11.69), then requires:

$$m_{H^+} = m_{OH^-} \approx 10^{-7}. \qquad (11.92)$$

This was the basis of the original definition of pH as the negative of the decimal logarithm of the hydrogen ion concentration:

$$pH = -\log_{10} m_{H^+} \qquad (11.93)$$

and the definition of a neutral solution, with $pH = 7$, as one in which the hydrogen ion molality is 10^{-7}. Rigorously, however, pH is defined on the basis of hydrogen ion activity:

$$pH \equiv -\log_{10} a_{aq}^{H^+} = -\log_{10} \left(\gamma_{aq}^{H^+} m_{H^+} \right) \qquad (11.94)$$

and a neutral solution is one in which the hydrogen ion *activity* is 10^{-7}. In dilute electrolyte solutions the difference between (11.93) and (11.94) is negligible, but in concentrated solutions it may be necessary to calculate the activity coefficient of H^+ and use the rigorous definitions (11.91) and (11.94).

Worked Example 11.3   Atmospheric CO₂ and the pH of rainwater

Dissolution of atmospheric gases in water is described by equation (11.18). In Worked Example 11.1 we calculated the concentrations of the molecular species N_2, O_2, Ar and CO_2 in water at equilibrium with the terrestrial atmosphere. When CO_2 dissolves in water, however, it reacts with H_2O and produces bicarbonate and carbonate ions, according to:

$$CO_{2\,aq} + H_2O \rightleftharpoons HCO_{3\,aq}^- + H_{aq}^+ \quad (I)$$

$$HCO_{3\,aq}^- \rightleftharpoons CO_{3\,aq}^{2-} + H_{aq}^+ \quad (II). \qquad (11.95)$$

If the solution is dilute and we take the activity coefficients to be unity then we can write the dissociation constants as:

$$K_{ds,1} = \frac{m_{HCO_3^-}\, m_{H^+}}{m_{CO_2}} \qquad (11.96)$$

and:

$$K_{ds,II} = \frac{m_{CO_3^{2-}} m_{H^+}}{m_{HCO_3^-}}. \tag{11.97}$$

At equilibrium the chemical potential of gaseous CO_2 in the atmosphere is the same as the chemical potential of molecular CO_2 dissolved in water. Therefore, by choosing the standard state of CO_2 as the pure gas at the temperature of interest and 1 bar, rather than infinitely dilute CO_2 in aqueous solution, we can re-write equation (11.96) as follows:

$$K_{ds,I} = \frac{m_{HCO_3^-} m_{H^+}}{f_{CO_2}}. \tag{11.98}$$

Equations (11.96) and (11.98) are equivalent, differing only in the choice of standard state for CO_2 and, therefore, in the numerical value of the dissociation constant. Although it is customary to use an equation of the form of (11.96), I believe that (11.98) is preferable, as this gets around the problem of calculating activity coefficients for dissolved molecular gases. If one wishes to calculate the concentration of dissolved gases this is simply accomplished by means of Henry's law (equation (11.18)), i.e. in this case:

$$m_{CO_2} = f_{CO_2} \cdot K_{H,s}^{CO_2}. \tag{11.99}$$

There are two additional equations that must be satisfied by the molalities of dissolved species at equilibrium. One is the water ionization equation (11.91) (taking the activity coefficients to be unity), and the other is the charge balance equation (11.69) which in this case is:

$$m_{H^+} - m_{HCO_3^-} - 2m_{CO_3^{2-}} - m_{OH^-} = 0. \tag{11.100}$$

If we fix the fugacity of CO_2 in equilibrium with water then equations (11.91) (setting the activity coefficients equal to 1), (11.97), (11.98) and (11.100) constitute a system of four equations in which the four aqueous molalities are the unknowns. The equations are non-linear, but a numerical solution with *Maple* is straightforward (Software Box 11.1). The two ionic dissociation constants can be calculated from standard state properties for aqueous species given, for example, by Wagman *et al.* (1982) or Robie and Hemingway (1995). If one wishes to calculate equilibrium at 298.15 K then the dissociation constants are simply obtained from the listed standard state Gibbs free energy of formation of the species, as in equation (11.79). Equilibrium at other temperatures can be calculated from listed standard state enthalpy and entropy values: $\Delta_r G^0 = \Delta_r H^0 - T \Delta_r S^0$, including heat capacity integrals if the C_P values are known. The constants at 298.15 K and 1 bar are: $K_{ds,I} = 1.465 \times 10^{-8}$ (with a standard state of pure CO_2 gas) and $K_{ds,II} = 4.685 \times 10^{-11}$.

Software Box 11.1   Calculation of carbonate speciation in aqueous solution, assuming ideal behavior

Procedure `zeco2` in *Maple* worksheet **`aq_spec_ideal.mw`** calculates molalities of dissolved carbonate, bicarbonate and carbon dioxide, in equilibrium with a 1 bar atmosphere at 298.15 K. The aqueous solution is assumed to be ideal (see Worked Example 11.3). A range of CO_2 atmospheric concentrations (in ppm) is specified in the procedure call, as well as the number of intermediate values to calculate between these boundaries, and a file name to send the output to. Output fields are: CO_2 atmospheric concentration (ppm), pH, m_{CO_2}, $m_{HCO_3^-}$, $m_{CO_3^{2-}}$, total dissolved CO_2 molality.

A second procedure, `zecalcite`, adds calcite saturation (equation (11.106), see Worked Example 11.4) but is otherwise identical to `zeco2`. An additional field at the end of each line of output contains $m_{Ca^{2+}}$.

The results of the calculations as a function of f_{CO_2} for values ranging from 0 to 5×10^{-4} bar (≈ 500 ppm CO_2 by volume in a 1 bar atmosphere) are shown in Fig. 11.5. The top panel shows molalities of dissolved molecular CO_2 (calculated from f_{CO_2} and Henry's law constant given by Sander, 1999, $K_H^{CO_2} = 0.034$) and of carbonate and bicarbonate ions, as well as the sum of all three species, which is the total amount of dissolved CO_2, called the carbonate analytical concentration. At very low CO_2 fugacity the dominant species is the bicarbonate anion, but as CO_2 concentration increases and the solution becomes more acidic (bottom panel) the increased H^+ concentration displaces equilibrium (I) to the left and limits dissociation, so that molecular CO_2 is the most abundant species. The bottom diagram

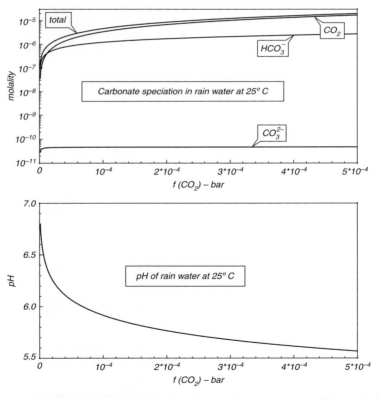

Fig. 11.5 Carbonate species distribution (top) and pH (bottom) in "pure" water ($\sim$ rainwater) in equilibrium with the terrestrial atmosphere at 25 °C. The analytical CO_2 concentration equals the total concentration of CO_2-bearing species $= m_{CO_2} + m_{HCO_3^{(-)}} + m_{CO_3^{(2-)}}$.

shows that the *pH* of water in equilibrium with the present-day terrestrial atmosphere (380 ppm CO_2) should be about 5.6. This is approximately the *pH* of rainwater in regions far removed from sources of pollution (particularly burning of high-sulfur coal and diesel fuel). It is not the *pH* of ocean water, nor of surface waters in general, as these waters contain other solutes in addition to CO_2 and may also be in equilibrium with solid phases.

11.4.5 Equilibrium between electrolyte solutions and solid phases

A question that repeatedly comes up when studying aqueous solutions is whether the solution becomes saturated in one or more crystalline phases. This is the same problem that is at the core of igneous petrology, but aqueous solutions are generally better understood from a theoretical point of view than silicate melts. We can represent equilibrium between an ionic crystalline solid, $(B_\beta A_\alpha)_{xs}$, and its dissociation products by:

$$\left(B_\beta A_\alpha \right)_{xs} \rightleftharpoons \beta \left(B+ \right)_{aq} + \alpha \left(A- \right)_{aq}. \tag{11.101}$$

This equilibrium is equally valid whether the substance is a strong or weak electrolyte. In the latter case we could also, if we wished, write an equilibrium equation with the undissociated aqueous species. As usual, the standard state for the crystalline phase is the pure solid at the temperature and pressure of interest. Note that this is not equal to the standard state chemical potential of the molecular species $(B_\beta A_\alpha)_{aq}$ for a weak electrolyte. The equilibrium condition for (11.101) is:

$$\mu_{xs}^{0, B_\beta A_\alpha} = \beta \mu_{aq}^{B+} + \alpha \mu_{aq}^{A-}$$
$$= \beta \mu_{aq}^{0, B+} + \alpha \mu_{aq}^{0, A-} + RT \ln \left[\left(a_{aq}^{B+} \right)^\beta \left(a_{aq}^{A-} \right)^\alpha \right]. \tag{11.102}$$

Let us call the equilibrium constant for this reaction K_{sp}, so that:

$$K_{sp} = \exp\left(-\frac{\Delta_r G^0}{RT} \right) = \exp\left(-\frac{\beta \mu_{aq}^{0, B+} + \alpha \mu_{aq}^{0, A-} - \mu_{xs}^{0, B_\beta A_\alpha}}{RT} \right) \tag{11.103}$$

and:

$$K_{sp} = \left(a_{aq}^{B+} \right)^\beta \left(a_{aq}^{A-} \right)^\alpha = (m_{B+})^\beta (m_{A-})^\alpha \left(\gamma_{aq}^\pm \right)^{\beta+\alpha}. \tag{11.104}$$

These equations are valid in general, but it is convenient to distinguish between relatively soluble and relatively insoluble substances. For the former the molalities at saturation are large enough that the activity coefficients cannot be ignored. For relatively insoluble electrolytes the molalities are small enough that it may be acceptable to ignore the activity coefficients. In such cases K_{sp} is known as the *solubility product* and it provides a convenient way of checking for saturation of specific phases. At a given temperature and pressure the solubility product is a constant for each phase in equilibrium with an aqueous solution. It is independent of the composition of the solution, as it is a combination of standard state properties only (see equation (11.103), but recall that if the solvent is a liquid other than water then the standard states for the solutes will be different, and so will the solubility product). At saturation the product of the activities (or molalities if the solution is sufficiently dilute) of the ionic species, with the stoichiometric coefficients as exponents, is equal to the solubility product. If the activity (or molality) product is less than the solubility product then the solution is not saturated in that particular phase. If it is greater then the solution is supersaturated with respect to the phase, which is a non-equilibrium condition.

Shallow ocean water is saturated in calcium carbonate, or at least nearly so. We can be reasonably certain of this because, for instance, the shells of marine invertebrates do not dissolve when the animals die. We will study how calcium carbonate saturation affects our calculation of carbonate speciation and *pH* (Worked Example 11.3). Three of the equations that we used in those calculations, (11.97), (11.98) and the water ionization equation (11.91), remain unchanged. Because now we must also consider aqueous Ca^{2+} ions, the electrical neutrality equation (11.100) is modified as follows:

$$m_{H^+} + 2m_{Ca^{2+}} - m_{HCO_3^-} - 2m_{CO_3^{2-}} - m_{OH^-} = 0 \qquad (11.105)$$

and we have an additional equation, which is the solubility product of calcite:

$$K_{sp,cc} = m_{Ca^{2+}} \cdot m_{CO_3^{2-}} \cdot \left(\gamma_{aq}^{\pm,CaCO_3} \right)^2 . \qquad (11.106)$$

There are now five equations in five unknowns: the five ionic molalities listed in equation (11.105). The solubility product of calcite at 298.15 K calculated with data from Wagman *et al.* (1982) is $K_{sp,cc} = 4.965 \times 10^{-9}$ (using aragonite instead of calcite makes a small difference). For now we will assume that we can ignore the activity coefficient in equation (11.106), and will return to this in Section 11.6. Modifying the *Maple* procedure discussed in Software Box 11.1 to include the additional equation (11.106) is straightforward (Exercise 11.4).

For an atmospheric CO_2 concentration of 380 ppm we calculate an oceanic *pH* of $\sim$8.3, which is similar to typical values measured in the Earth's oceans. The agreement is remarkably good, considering that we have ignored all excess mixing properties. Table 11.1 compares the calculated solution compositions with and without calcite saturation (Worked Example 11.3), for a fixed atmospheric CO_2 concentration of 380 ppm. The concentration of dissolved molecular CO_2 is of course the same in both cases, as this is fixed by equilibrium with the gas phase. Total dissolved carbonate, however, is almost two orders of magnitude higher in water saturated with calcite. The calculated Ca^{2+} molality is $\sim$5.4 $\times$ 10^{-4}, which is more than one order of magnitude lower than measured molality in seawater ($\sim 10^{-2}$). The discrepancy arises to a large extent from ignoring the activity coefficient in equation (11.106).

Why is the *pH* of seawater controlled by $CaCO_3$ saturation? The answer is that among the most abundant species in seawater carbonate is the only weak electrolyte, for which it is possible to write a partial dissociation reaction such as (11.95)(II). Table 11.2 (taken from Millero, 2004) shows the concentrations of the most abundant ions in seawater. The four most abundant cations are strong bases, and the halide and sulfate anions are strong acids. None of these ions react with H_2O to consume or produce H^+ ions. The *pH* of seawater is therefore fixed by the solid carbonate for which the solubility product is first exceeded, and for present day seawater this is calcite.

Table 11.1 Calculated ionic speciation in simplified natural waters				
	Rainwater		Calcite-saturated water	
	ideal	Debye–Hückel	ideal	Debye–Hückel
CO_2	1.29×10^{-5}	1.29×10^{-5}	1.29×10^{-5}	1.29×10^{-5}
HCO_3^-	2.36×10^{-6}	2.36×10^{-6}	1.05×10^{-3}	1.15×10^{-3}
CO_3^{2-}	4.68×10^{-11}	4.71×10^{-11}	9.28×10^{-6}	1.22×10^{-5}
Total CO_2	1.53×10^{-5}	1.53×10^{-5}	1.07×10^{-3}	1.18×10^{-3}
Ca^{2+}	–	–	5.35×10^{-4}	5.89×10^{-4}
Ionic strength	2.36×10^{-6}	2.37×10^{-6}	1.61×10^{-3}	1.78×10^{-3}
pH	5.63	5.63	8.28	8.30

Calculations are based on 380 ppm atmospheric CO_2 at 1 bar and 25 °C, assuming either ideal aqueous solutions or Debye–Hückel activity coefficients. Concentrations given in molality.

Table 11.2 Most abundant ionic species in seawater (after Millero, 2004)			
Cation	Molality	Anion	Molality
Na^+	0.4691	Cl^-	0.5459
Mg^{2+}	0.0528	SO_4^{2-}	0.0282
Ca^{2+}	0.0103	HCO_3^-	0.0018
K^+	0.0102	Br^-	0.0008
		F^-	0.0007
		CO_3^{2-}	0.0003

11.5 Speciation in ionic solutions. Iron solubility in ocean water as an example

Iron cations occur in two oxidation states, Fe^{2+} and Fe^{3+}, that have vastly different solubilities in water. Both ferrous and ferric ions react with water to form several ionic species whose relative abundances are a function of pH. These properties make iron aqueous solutions an excellent test subject on which to apply the concepts that we have discussed so far. We will estimate the solubility of iron in ocean water, and study how this may have varied with changes in atmospheric composition. The problem was introduced in Worked Example 6.7, where we discussed banded iron formations and the information that they preserve about conditions in the early Earth.

The solubility of iron in the present day shallow oceans, i.e. in ocean water that is in equilibrium with the atmosphere, is limited by precipitation of Fe^{3+} species. This follows from the fact that the present day atmospheric oxygen fugacity is about 70 orders of magnitude higher than the oxygen fugacity along the hematite–magnetite buffer at room temperature (Fig. 9.3). We can expect the solubility-limiting phase to be either ferric oxide or ferric hydroxide. The Gibbs free energy change for the reaction $2\,Fe(OH)_3 \rightarrow Fe_2O_3 + 3\,H_2O$ at 298 K is ~ -30 kJ mol^{-1}, which means that hematite is the stable phase. The phase that usually precipitates from seawater is not hematite, however, but rather an amorphous solid or a hydrated crystalline ferric oxide with a composition intermediate between those of hematite and ferric hydroxide. Over time the precipitate dehydrates and becomes hematite, and this may have been the mechanism by which Precambrian hematite iron formations formed. This is a kinetic problem, that we will not address here (see Chapter 12). We will use hematite in our calculations, in order to be consistent with Worked Example 6.7. Because the metastable phase that actually forms has higher Gibbs free energy than hematite, the results of our calculations are an absolute minimum of solubility.

In these calculations we will only consider dissolved iron species that exist in an aqueous solution with no other solutes, except carbonate anions. The effect of this simplification is to underestimate the solubility of iron relative to the actual solubility in seawater, in which formation of other Fe-bearing ionic species is possible. We will also assume that the activity of dissolved iron species is ideal (i.e. $a = m$). Our estimated total iron contents in solution will be about two orders of magnitude lower than the analytical concentrations of iron in present day seawater ($\sim 10^{-10}$ molal in oxygen-rich near-surface water). The qualitative behavior with changes in oxygen fugacity and pH that we will uncover are, however, robust, and will illuminate the significance of banded iron formations (see also Worked Example 6.7). A rigorous discussion of iron solubility in seawater is given, for example, by Millero *et al.* (1995) and Liu and Millero (2002).

We can think of dissolution of Fe ionic species in water as occurring in one of two ways. The one that is perhaps more intuitively appealing is to think of iron hydroxides as weak bases. Just as dissolved CO_2, which is a weak acid, undergoes successive ionization reactions (Worked Example 11.3), so a molecular aqueous species such as $Fe(OH)_{3aq}$ can be thought to dissociate in steps to $Fe(OH)_2{}^+$, $FeOH^{2+}$ and finally Fe^{3+}. All of these stoichiometric species are known to exist in ferric iron solutions. Alternatively, we may imagine that Fe^{3+} exists in solution as free ions that associate electrostatically with OH^- groups to form complex ions with the same stoichiometries as the molecular species that would form by dissociation. The difference is that whereas in the former case the intermediate ionic species, as well as neutral $Fe(OH)_3$, are taken to be covalently bonded molecules, in the latter case the ionic species are complexes in which Fe^{3+} and OH^- ions are attached electrostatically. The more accurate physical picture is the latter one, but from a macroscopic thermodynamic point of view the two approaches are equivalent. I will therefore use the "weak base" model, because of its intuitive appeal. According to this model dissolution of hematite in water proceeds as follows:

$$\frac{1}{2}Fe_2O_3 + \frac{3}{2}H_2O \rightleftharpoons Fe(OH)_{3\,aq} \quad (III, hm)$$

$$Fe(OH)_{3\,aq} \rightleftharpoons Fe(OH)_2^+{}_{aq} + OH^-_{aq} \quad (III, 1)$$

$$Fe(OH)_2^+{}_{aq} \rightleftharpoons FeOH^{2+}_{aq} + OH^-_{aq} \quad (III, 2) \qquad (11.107)$$

$$FeOH^{2+}_{aq} \rightleftharpoons Fe^{3+}_{aq} + OH^-_{aq} \quad (III, 3)$$

for which the equilibrium constants are (assuming an ideal electrolyte solution):

$$K_{III,hm} = m_{Fe(OH)_3}$$

$$K_{III,1} = \frac{m_{Fe(OH)_2^+} \cdot m_{OH^-}}{m_{Fe(OH)_3}}$$

$$K_{III,2} = \frac{m_{FeOH^{2+}} \cdot m_{OH^-}}{m_{Fe(OH)_2^+}} \qquad (11.108)$$

$$K_{III,3} = \frac{m_{Fe^{3+}} \cdot m_{OH^-}}{m_{FeOH^{2+}}}.$$

The values of the equilibrium constants at 298.15 K and 1 bar, calculated with data from Wagman *et al.* (1982) are: $K_{III,hm} = 1.499 \times 10^{-12}$, $K_{III,1} = 5.997 \times 10^{-12}$, $K_{III,2} = 1.011 \times 10^{-9}$, $K_{III,3} = 1.515 \times 10^{-12}$. From equations (11.108) it follows that the solubility product for hematite is $K_{sp,hm} = K_{III,hm} \cdot K_{III,1} \cdot K_{III,2} \cdot K_{III,3} = 1.38 \times 10^{-44}$. If we were to calculate the solubility of hematite from this figure we would get a molality of Fe^{3+} of $\sim 10^{-23}$ at $pH = 7$. This is the correct concentration of Fe^{3+} cations in equilibrium with hematite, but the analytical concentration of dissolved ferric iron is many orders of magnitude higher than this, as the solubility product fails to account for the abundances of the other ionic species. This is a point that one must always be careful with when calculating solubilities: failure to account for speciation leads to grossly erroneous results. In fact, at $pH = 7$ the most abundant species is neutral $Fe(OH)_3$, and the second most abundant one, with a concentration over four orders of magnitude lower, is $Fe(OH)_2^+$. Fe^{3+} becomes the dominant aqueous species only in very acidic solutions (Exercise 11.5).

The total concentration (= analytical concentration) of dissolved ferric iron is obtained by solving each equation in (11.108) for one of the species molalities, and adding up the resulting equations. Using the ionization product of water (equation (11.91)) to convert m_{OH^-} to m_{H^+} we get the following equation for total dissolved ferric iron, $m_{Fe^{3+},total}$, in equilibrium with hematite, as a function of pH:

$$\left[m_{Fe^{3+},total} \right]_{hm\,sat} =$$

$$= K_{III,hm} \left(1 + \frac{K_{III,1}}{K_W} 10^{-pH} + \frac{K_{III,1} \cdot K_{III,2}}{K_W^2} 10^{-2pH} + \frac{K_{III,1} \cdot K_{III,2} \cdot K_{III,3}}{K_W^3} 10^{-3pH} \right).$$

$$(11.109)$$

Although in the present day terrestrial oceans the concentration of Fe^{2+} is vanishingly small (at least in those parts of the ocean that are close to equilibrium with the atmosphere), the same may not have been true during the Archaean. To account for this we write equations for equilibrium between hematite and dissolved ferrous iron, which will necessarily involve oxygen. In contrast to $Fe(OH)_3$, there is no experimental evidence for the existence of the neutral species $Fe(OH)_2$ in aqueous solution. We therefore write the reactions for hematite saturation from Fe^{2+} as follows:

$$\frac{1}{2}Fe_2O_3 + H_2O \rightleftharpoons FeOH_{aq}^+ + OH_{aq}^- + \frac{1}{4}O_2 \quad (II, hm)$$

$$FeOH_{aq}^+ \rightleftharpoons Fe_{aq}^{2+} + OH_{aq}^- \quad (II, 2). \qquad (11.110)$$

Taking the standard state of oxygen as the pure gas at the pressure and temperature of interest the equilibrium constants are:

$$K_{II,hm} = \left(f_{O_2}\right)^{1/4} \cdot m_{FeOH^+} \cdot m_{OH^-}$$

$$K_{II,2} = \frac{m_{Fe^{2+}} \cdot m_{OH^-}}{m_{FeOH^+}} \qquad (11.111)$$

and the values of the equilibrium constants (calculated with data from Wagman *et al.*, 1982) are $K_{II,hm} = 7.431 \times 10^{-32}$ and $K_{II,2} = 5.920 \times 10^{-8}$. We can now derive an equation for total dissolved ferrous iron in equilibrium with hematite, analogous to (11.109), which is:

$$\left[m_{Fe^{2+},total}\right]_{hm\,sat} = \frac{K_{II,hm}}{K_W} \frac{10^{-pH}}{\left(f_{O_2}\right)^{1/4}} \left(1 + \frac{K_{II,2}}{K_W} 10^{-pH}\right). \qquad (11.112)$$

Equations (11.109) and (11.112) are plotted in Fig. 11.6 as a function of oxygen fugacity, at a constant $pH = 8$, which is approximately that of present day ocean water. Also shown

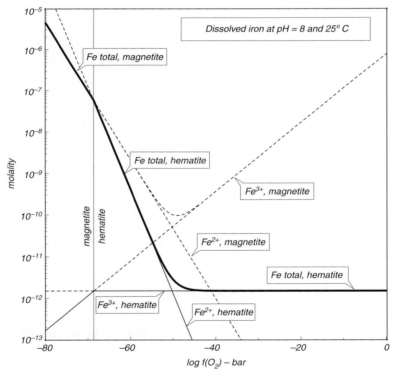

Fig. 11.6 Concentration of dissolved iron species in water at 25 °C as a function of oxygen fugacity at $pH = 8$ ($\sim$ present day seawater). The diagram shows total Fe^{3+} species (Fe^{3+} analytical concentration), total Fe^{2+} species (Fe^{2+} analytical concentration) and total dissolved iron ($= Fe^{3+} + Fe^{2+}$ species analytical concentrations) in equilibrium with either hematite or magnetite. The stable phase is the one which attains saturation at lower concentration, shown with the solid curves. The dashed curves show calculated concentrations in equilibrium with the metastable phase. The thick solid curve shows total dissolved iron in equilibrium with the stable phase, which is hematite at $\log f(O_2) > -68.59$ and magnetite at $\log f(O_2) < -68.59$.

is a curve for total dissolved iron $= m_{Fe^{3+},total} + m_{Fe^{2+},total}$. Recall that the calculations are only appropriate for a very dilute solution of iron in pure water, and that concentrations in ocean water are 2–3 orders of magnitude higher. The qualitative behavior shown in the figure is, however, applicable to seawater. The concentration of total dissolved ferric iron in equilibrium with hematite is of course independent of oxygen fugacity, but that of total dissolved ferrous iron is not. For the present day atmospheric oxygen fugacity the concentration of total dissolved ferrous iron is vanishingly small, and only becomes comparable to that of total ferric iron at oxygen fugacities $\sim 10^{-45}$ bar. Below 10^{-50} bar $f(O_2)$ ferrous species become the dominant forms of dissolved iron, and iron solubility increases rapidly with decreasing oxygen fugacity.

At 298 K magnetite becomes stable at $f(O_2) \sim 10^{-69}$ bar (Fig. 9.3). In order to see the effect of this phase transition on iron solubility (see also Worked Example 6.7) we have to write the corresponding saturation equations. This is accomplished by replacing each of the first equations in (11.107) and (11.110) with the corresponding magnetite equations:

$$\frac{1}{3}Fe_3O_4 + \frac{3}{2}H_2O + \frac{1}{12}O_2 \rightleftharpoons Fe(OH)_{3\,aq} \quad (III, mt) \tag{11.113}$$

and:

$$\frac{1}{3}Fe_3O_4 + H_2O \rightleftharpoons FeOH_{aq}^+ + OH^- + \frac{1}{6}O_2 \quad (II, mt) \tag{11.114}$$

with equilibrium constants:

$$K_{III,mt} = \frac{m_{Fe(OH)_3}}{\left(f_{O_2}\right)^{1/12}} \tag{11.115}$$

and:

$$K_{II,mt} = \left(f_{O_2}\right)^{1/6} \cdot m_{FeOH^+} \cdot m_{OH^-}. \tag{11.116}$$

Numerical values (with standard state data from Wagman *et al.*, 1982) are: $K_{III,mt} = 1.356 \times 10^{-6}$, $K_{II,mt} = 6.723 \times 10^{-26}$. The other equations in (11.107) and (11.108) remain unchanged, so we find the following equations for total dissolved ferric and ferrous iron in equilibrium with magnetite:

$$\left[m_{Fe}^{3+}, total\right]_{mt\,sat} = K_{III,mt} \cdot \left(f_{O_2}\right)^{1/12} \cdot$$

$$\cdot \left(1 + \frac{K_{III,1}}{K_W}10^{-pH} + \frac{K_{III,1} \cdot K_{III,2}}{K_W^2}10^{-2pH} + \frac{K_{III,1} \cdot K_{III,2} \cdot K_{III,3}}{K_W^3}10^{-3pH}\right)$$

$$\tag{11.117}$$

and:

$$\left[m_{Fe^{2+}}, total\right]_{mt\,sat} = \frac{K_{II,mt}}{K_W}\frac{10^{-pH}}{\left(f_{O_2}\right)^{1/6}}\left(1 + \frac{K_{II,2}}{K_W}10^{-pH}\right). \tag{11.118}$$

Equations (11.117) and (11.118), and the sum of both, are also shown in Fig. 11.6. The curves for total dissolved iron in equilibrium with hematite and magnetite intersect at

$\log f(O_2) = -68.59$. This is the oxygen fugacity at which the hematite–magnetite phase transition takes place, but the careful reader may note that there is a slight discrepancy between this value and the one shown in Fig. 9.3. This arises from the use of standard state properties from different data sets. I used data from Holland and Powell (1998) to calculate the hematite-magnetite buffer in Fig. 9.3, and Wagman *et al.* (1982) to calculate the equilibrium constants in this example.

At oxygen fugacity higher than that of the magnetite–hematite equilibrium the calculated concentrations of total dissolved ferrous and ferric species in equilibrium with magnetite are higher than those in equilibrium with hematite. This means that magnetite is not stable, and that iron solubility is controlled by hematite saturation. There are two equivalent ways of seeing this: you can think that with increasing iron content the solution becomes saturated in hematite first, or that the chemical potentials of iron species at equilibrium with hematite are lower than those at equilibrium with magnetite. At $f(O_2)$ lower than that of the phase transition the converse is true.

If we now "clean up" the diagram, leaving only the stable saturation curves, we get the phase diagram shown in Fig. 11.7. The fields labeled "hematite" and "magnetite" correspond to total dissolved iron concentrations above the solubility curves. These are "prohibited regions", in the sense that solution compositions inside these regions are thermodynamically unstable relative to precipitation of the corresponding crystalline phase. Conditions below the solubility curves correspond to solutions undersaturated in iron oxides. For oxygen fugacities lower than $\sim 10^{-50}$ bar, i.e. in the region where iron solubility varies strongly with oxygen fugacity, the solubility is controlled by oxidation of Fe^{2+} to a ferric crystalline phase. This is so because in this region the saturation concentration of total dissolved

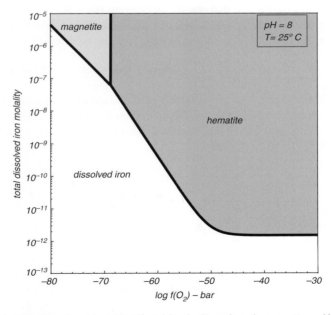

Fig. 11.7 Same as Fig. 11.6, but showing only concentration of total dissolved iron along the magnetite and hematite saturation curves (the thick curve in Fig. 11.6). Conditions inside the shaded regions are metastable relative to precipitation of the corresponding oxide phase.

ferric iron remains very low and constant, at close to 10^{-12} molal, until magnetite starts precipitating, where it decreases further with decreasing oxygen fugacity. Above $\sim 10^{-50}$ bar oxygen fugacity the concentration of total dissolved ferrous iron becomes negligibly small and iron solubility is determined by the solubility of ferric species.

The total concentration of dissolved iron in Archaean seawater in equilibrium with an anoxic atmosphere in which $f(O_2)$ was less than about 10^{-50} bar may have been several orders of magnitude higher than the solubility of iron in today's shallow oceans. Dissolved iron in Archaean oceans would have been overwhelmingly present as ferrous species, that could precipitate to solid ferric phases upon oxidation. Iron solubility remains essentially unchanged over an $f(O_2)$ range of ~ 40 orders of magnitude, from 10^{-40} bar to its present-day value. Deposition of banded iron formations by oxidation of dissolved ferrous iron must have taken place under oxidation conditions ($f(O_2) < \sim 10^{-50}$ bar) that are not even remotely comparable to those that are needed for aerobic metabolism (Chapter 14).

The equations show that iron solubility in water is a function not only of oxygen fugacity but also of pH, and the question arises, could the deposition of banded iron formations primarily reflect a change in oceanic pH, rather than in oxygen fugacity? In Exercise 11.6 you can explore this and see why, although possible, this is a very unlikely explanation – BIF deposition must be the response to oxidation of the Earth's surficial environments.

Figure 11.7 shows a subtle detail that is not evident from the schematic phase diagrams that we constructed in Worked Example 6.7. This is the fact that formation of magnetite iron formations requires more concentrated iron solutions than formation of hematite iron formations. For a total iron concentration lower than that at the magnetite–hematite transition ($\sim 6 \times 10^{-8}$ molar in this simplified model) magnetite iron formations cannot form at any oxygen fugacity, but hematite iron formations would form by oxidation.

In Worked Example 6.7 we saw that magnetite iron formations are sometimes associated with siderite, and that some Precambrian BIFS are in fact composed chiefly of siderite. In order to study the conditions that lead to the formation of siderite iron formations we begin by writing the following magnetite–siderite equilibrium:

$$2Fe_3O_4 + 6CO_2 \rightleftharpoons 6FeCO_3 + O_2. \qquad (11.119)$$

The oxygen fugacity at which siderite replaces magnetite as the iron solubility-limiting phase depends on the fugacity of carbon dioxide. Using the pure gases at the temperature of interest and 1 bar as the standard states for O_2 and CO_2 we write the equilibrium condition for (11.119) as follows:

$$\frac{f_{O_2}}{\left(f_{CO_2}\right)^6} = K_{mt-sd} \qquad (11.120)$$

with $K_{mt-sd} = \exp(-\Delta_r G^0/RT) = 2.887 \times 10^{-70}$ at 298 K (data from Wagman et al., 1982). As an example, let us assume that CO_2 fugacity in the Archaean atmosphere was 0.02 bar. A value considerably higher than today's is suggested by the fact that Archaean glaciations appear to be non-existent, despite the fact that the early Sun was fainter than today's (Sagan & Mullen, 1972; Kasting, 1987; Gilliland, 1989). From this CO_2 fugacity we calculate an oxygen fugacity at the magnetite–siderite transition of $\sim 1.9 \times 10^{-80}$ bar, i.e. well below the hematite phase transition. An important question is whether CO_2 is stable at this low oxygen fugacity, relative to, for example, CO. The answer, which you are asked to prove in Exercise 11.7, is yes.

It would appear that we could now simply add another phase boundary to Fig. 11.7, showing the stability of siderite below this oxygen fugacity. The problem is that this diagram is calculated for $pH = 8$, which is approximately correct for the present day atmospheric CO_2 concentration but not for the higher concentration that we are now assuming. Under 0.02 bar of CO_2 the ocean must have been more acidic than today. The effect of pH on the saturation boundaries is small (Exercise 11.6) but not entirely negligible. We therefore must first calculate the pH of seawater in equilibrium with 0.02 bar of CO_2, and then adjust Fig. 11.7 accordingly.

Suppose that the Archaean ocean was saturated in both calcite and siderite. We can then add two more equations to the system of equations that we solved in Worked Example 11.4. One of them is the solubility product of siderite:

$$FeCO_3 \rightleftharpoons Fe^{2+}_{aq} + CO^{2-}_{3aq} \qquad (11.121)$$

for which $K_{sp,sd} = 3.13 \times 10^{-11}$. The other is the Fe^{2+} hydrolysis reaction:

$$Fe^{2+}_{aq} + H_2O \rightleftharpoons FeOH^+_{aq} + H^+_{aq}, \qquad (11.122)$$

which can be converted to the second dissociation reaction in (11.110) using the ionization product of water (equation (11.91)). We must also modify the charge balance equation (11.105) to include the molalities of Fe^{2+} and $FeOH^+$. Modifying the *Maple* routine to do the calculations is easy, and is left as an exercise for the reader (Exercise 11.8).

For a CO_2 fugacity of 0.02 bar we calculate the pH of our simplified ocean to be 7.13. The solution set of the system of equations for siderite saturation (Exercise 11.8) also yields the total concentration of dissolved ferrous iron, which in this case is 4.11×10^{-5} molal. Figure 11.8 shows the hematite and magnetite saturation boundaries redrawn for $pH = 7.13$ (compare Fig. 11.7). The concentration of ferrous aqueous species in equilibrium with siderite is independent of oxygen fugacity (equation (11.121)), so the siderite saturation boundary is parallel to the $f(O_2)$ axis, at a total ferrous iron molality of 4.11×10^{-5}. This line intersects the magnetite saturation boundary at $f(O_2) = 1.9 \times 10^{-80}$ bar, which is the same oxygen fugacity that we calculated for the magnetite–siderite equilibrium from equation (11.120). Consistency between two sets of calculations is always reassuring.

The resulting diagram shows the stability fields of the three most common phases in banded iron formations. Reaction (11.119) determines the maximum oxygen fugacity that allows siderite precipitation, which varies as the sixth power of CO_2 fugacity. An increase in CO_2 fugacity expands the siderite field both towards the right (at the expense of magnetite) and downwards, by lowering the concentration of Fe^{2+} required for siderite crystallization (solubility product for reaction (11.121)). We can calculate with equation (11.120) the CO_2 fugacity needed to eliminate the magnetite field altogether, i.e. so that the siderite field adjoins the hematite field. Making $\log f(O_2) = -68.59$ (the oxygen fugacity at the magnetite–hematite transition) we get $f(CO_2) = 1.44$ bar. The rarity of hematite–siderite iron formations suggests that this is a reasonable upper bound for CO_2 fugacity in the Archaean atmosphere at the time at which banded iron formations precipitated.

All of these calculations ignore the activity coefficients of aqueous species, which are not negligible in a solution as concentrated as seawater. The rest of this chapter is focused on this problem.

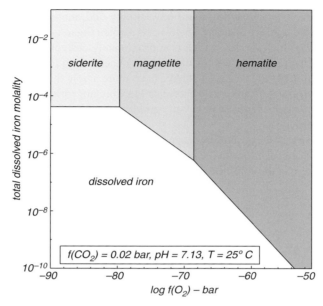

Fig. 11.8 Same as Fig. 11.7, but at $pH = 7.13$, corresponding to a calcite-saturated solution at $f(CO_2) = 0.02$ bar. Increasing CO_2 fugacity expands the siderite field downwards and rightwards, along the magnetite saturation curve. The magnetite field disappears at $f(CO_2) \approx 1.44$ bar. This represents the minimum chemical potential of CO_2 at which siderite and hematite can precipitate at equilibrium. The rarity of this assemblage in terrestrial banded iron formations places an upper bound on Archaean atmospheric CO_2 fugacity.

11.6 Activity coefficients in electrolyte solutions

The key difficulty when calculating chemical equilibrium in electrolyte solutions is in estimating reliable values for the activity coefficients of the many charged and neutral species that exist in solution. The preceding section shows that even in a highly idealized solution of Fe in water one must deal with almost ten species, and the number of species grows rapidly with the number of components in the solution. The problem is not unlike that of calculating speciation in a gas phase (Chapter 9), but it is more complicated because of (i) the greater number of species, (ii) the existence of a solvent whose properties change with temperature and (iii) the nature of the interactions among charged particles, and between ions and solvent molecules.

The fact that electrolyte solutions behave differently from solutions of neutral species was recognized early in the twentieth century. The different behaviors are compared schematically in Fig. 11.9, which shows the logarithm of the activity coefficient as a function of molality for very dilute solutions of a neutral species ($\ln\gamma$) and an electrolyte ($\ln\gamma^{\pm}$). In both cases the activity coefficient becomes unity at $m = 0$. It is important to recall that this is conventional, and that it arises from choosing to define the standard state at infinite dilution. In other words, and as we discussed in Section 11.1.2, any excess Gibbs free energy that may exist at infinite dilution remains unknown and is lumped into the standard state Gibbs free energy. What matters now is that, whereas at very great dilution the activity coefficient

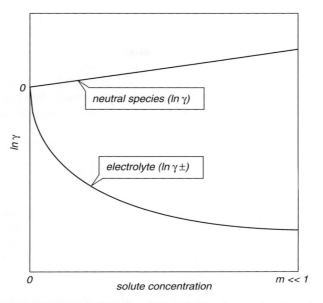

Fig. 11.9   Schematic diagram showing the contrasting behaviors of dilute solutions of a neutral species and an electrolyte. The key difference is that, whereas in a dilute solution of a neutral species $\ln \gamma$ varies linearly with concentration, the variation in a dilute electrolyte solution is of the form $\ln \gamma = -\alpha I^{\beta}$, with $\alpha > 0$ and $0 < \beta < 1$.

of neutral species varies linearly with molality, the same is not true for electrolytes. The activity coefficient of electrolytes in dilute solutions varies in a strongly non-linear fashion with concentration, and approaches the limiting value at infinite dilution with an infinite slope. For neutral species the activity coefficient is generally greater than 1, whereas for electrolytes activity coefficients in dilute solutions are less than one, but may become greater than one at higher concentrations.

The distinct behavior of electrolytes can be thought of as arising from the formation of "ionic atmospheres". These are regions around each ion that, owing to electrostatic attraction, carry an excess of charge of a sign opposite to that of the central ion. As long as the individual ionic atmospheres remain distant, i.e. at low concentration, they shield the central ions and make it less likely that they will interact with other ions. This is expressed macroscopically as a decrease in their chemical potential. As the solution becomes less dilute the ionic atmospheres interfere with one another and shielding becomes less effective. Particles of a neutral solute, in contrast, are not surrounded by ionic atmospheres so one would expect that any energetic effect arising from interactions among them would vary more or less linearly, or at least monotonically, with concentration.

11.6.1 Debye–Hückel theory

The first theory capable of predicting activity coefficients for electrolytes on the basis of a rigorous physical description of the microscopic structure of electrolyte solutions was proposed by Debye and Hückel in 1923. The place of Debye–Hückel theory in the thermodynamics of electrolytes is not unlike that of the van der Waals equation of state in the thermodynamics of fluids. The theory is quantitatively accurate only for dilute solutions, however. For example, it is already grossly inaccurate for seawater, in which the total concentration of dissolved electrolytes is less than 1 mol kg^{-1}. It is, however, the basis for many of the more elaborate formulations that are used to calculate activity coefficients in concentrated electrolyte solutions. In contrast to most of the latter formulations, which are empirical to varying extents, the Debye–Hückel approach has a strong theoretical foundation.

A rigorous derivation of the remarkably simple final equation of Debye and Hückel for the mean ionic activity coefficient of an electrolyte is beyond the scope of this book, but it is instructive to construct a semi-formal *a posteriori* justification of their equation. We begin by defining the ionic strength of a solution, symbolized by I, as:

$$I = \frac{1}{2} \sum_i z_i^2 m_i, \tag{11.123}$$

where z_i is the charge of ion i, m_i its molality, and the sum is over all of the ionic species present in the solution. The ionic strength is a measure of the total concentration of dissolved electrolytes. If we apply (11.123) to a solution of a single electrolyte we see that the behavior of an electrolyte solution as it approaches the infinite dilution limit (Fig. 11.9) can be represented by a function of the form:

$$\ln \gamma = -\alpha I^\beta, \quad \alpha > 0, \quad 0 < \beta < 1. \tag{11.124}$$

The derivative of this function relative to I diverges as I goes to zero, as required by the observed behavior of very dilute electrolyte solutions schematized in Fig. 11.9. Debye and Hückel proved rigorously that the actual equation valid as the solution approaches infinite dilution is:

$$\log_{10} \gamma_i = -A |z_{i+} z_{i-}| I^{1/2}, \tag{11.125}$$

where z_{i+} and z_{i-} are the charges of the cation and anion in electrolyte i, and A is a constant that depends on the solvent only, but varies with temperature and pressure. I is the ionic strength, given by (11.123), and includes the concentration of i as well as of any other ions present in the solution. Equation (11.125) is known as Debye–Hückel's limiting law. It is only accurate for ionic strengths lower than $\sim 10^{-3}$, but at such low concentrations it is accurate enough that it is routinely used to extrapolate measurements of thermodynamic functions done at very low dilutions to the infinite dilution limit. Its main use for our purposes will be to understand the meaning of the parameter A and, with it, some of the physical aspects of Debye–Hückel theory. Their theory shows that A is given by:

$$A = \left(2\pi N \rho s^3\right)^{1/2}, \tag{11.126}$$

where N is Avogadro's number, ρ is the density of the solvent, and s is a length given by:

$$s = \frac{e^2}{4\pi \epsilon_0 \varepsilon_r k_B T} \tag{11.127}$$

with e the elementary unit of charge, ϵ_0 the permittivity of free space, ε_r the dielectric constant of the solvent and k_B Boltzmann's constant. The meaning of the length s becomes clear if we re-write (11.127) as follows:

$$k_B T = \frac{e^2}{4\pi \epsilon_0 \varepsilon_r s} \tag{11.128}$$

and note that the equation now has units of energy. The length s is the distance between two units of charge at which their electrostatic energy (see also equations (11.65)) equals their thermal energy. The higher the temperature, or the dielectric constant, the closer the charges have to be in order for the electrostatic force to overcome thermal agitation. We next note that we can re-write the product $N\rho$ as:

$$N\rho = \frac{M}{V/N} \approx \frac{M}{\lambda^3}, \tag{11.129}$$

where M is the molecular weight of the solvent, V its molar volume and λ is a length of the order of the intermolecular distance in the solvent. The parameter A is a function of the ratio (s/λ), i.e.:

$$A \sim \left[2\pi M \left(\frac{s}{\lambda} \right)^3 \right]^{1/2}. \tag{11.130}$$

If (s/λ) is small then the distance over which electrostatic forces are effective is small compared to the distance between solvent molecules, and formation of an ionic atmosphere would tend to be restricted: the value of the parameter A decreases and the activity coefficient approaches unity (equation (11.125)). On the other hand, a large value of the ratio (s/λ) implies that electrostatic forces are effective over distances greater than intermolecular separation in the solvent, allowing the formation of ionic atmospheres that lower the chemical potential of the ion.

Before we proceed note two formal aspects of equation (11.125). First, it is customary to write it in terms of decimal logarithm, rather than natural logarithm. Second, there is the usual problem with units that arises when expressing concentration as molality. As written, equation (11.126) implies that A has units of $(\text{kg mol}^{-1})^{\frac{1}{2}}$. If we express molality in mol kg^{-1} then the units of $I^{1/2}$ are $(\text{mol kg}^{-1})^{\frac{1}{2}}$ and the logarithm is dimensionless, as it must be. Alternatively, if we choose to use dimensionless molality (Section 11.1.1) then the term in parentheses in equation (11.126) must be multiplied by 1 mol kg^{-1} (see also equation (11.4)) and A becomes a dimensionless parameter. This is the usual convention. Numerically it makes no difference, but mathematical rigor requires that we worry about this.

Debye–Hückel's parameter A is a property of the solvent only (and temperature, both directly, see equation (11.127), and via the effect of T on the dielectric constant and density of the solvent). The limiting law, equation (11.125), does not contain any solute properties besides their charges. In particular, it ignores ionic radius. We can expect this to be a reasonable model at very high dilution, where ions are separated by distances that are large enough compared to their radii that they can be thought of as point charges. As concentration increases the radius of the ions becomes significant relative to their separation and we must expect the activity coefficient to become a function of ionic radius. This leads to the full Debye–Hückel equation:

$$\log_{10}\gamma_i = \frac{-A|z_{i+}z_{i-}|I^{1/2}}{1 + \mathring{a}_i B I^{1/2}}, \tag{11.131}$$

where $\mathring{a}_i$, known as the *distance of maximum approach*, is a function of the effective ionic radii of the ions that constitute electrolyte i, and B is a parameter that depends only on the identity of the solvent and temperature. The product $BI^{1/2}$ has dimension of length^{-1} (necessarily, as $\mathring{a}_i$ has dimension of length) and can be thought of as the inverse of the effective radius of the ionic atmosphere. If the distance of maximum approach is much smaller than this radius then the denominator of (11.131) approaches 1 and we recover the limiting law, equation

(11.125). Note that the radius of the ionic atmosphere is a function both of a solvent property (B) and of the ionic strength of the solution. For historical reasons, the unit that is universally used for $\mathring{a}_i$ and B is the Ångstrom $= 10^{-10}$ m. Values of the $\mathring{a}$ parameter for many common ions were tabulated by Kielland (1937) and this is still the standard reference more than seven decades later. The values of the A and B parameters for water have been calculated over a large temperature range by Helgeson and Kirkham (1974).

Although equation (11.131) has theoretical justification it does not reproduce the observed behavior of electrolytes at ionic strengths greater than ~ 0.1. Figure 11.10 shows measured activity and osmotic coefficients for four strong electrolytes at 298 K (data from Robinson & Stokes, 1959). Also shown are activity coefficients calculated with the Debye–Hückel equation (11.131), and osmotic coefficients calculated by integrating (11.131) according to (11.43) (Exercise 11.9). The discrepancy above $I \sim 0.1$ is very large. Note that, as equation (11.131) does not have a minimum for any value of I, it is incapable of reproducing the behavior of actual electrolytes, which generally show a minimum in γ at moderate ionic strengths.

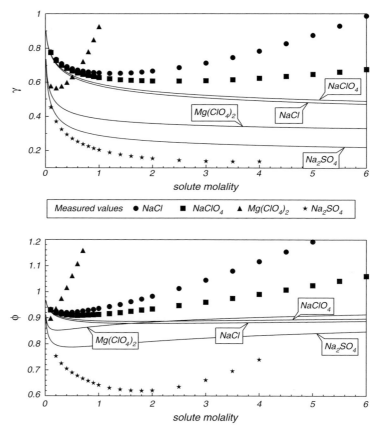

Fig. 11.10 Measured activity and osmotic coefficients (symbols) in aqueous solutions of sodium chloride, sulfate and perchlorate, and magnesium perchlorate (data from Robinson & Stokes, 1959), compared to the values of the activity coefficients calculated with Debye–Hückel equation (11.131), and osmotic coefficients calculated by integrating (11.131) according to (11.43).

We can refine the calculations in Worked Examples 11.3 and 11.4 by incorporating activity coefficients calculated with the full Debye–Hückel equation, (11.131). In order to calculate the *pH* of rainwater we re-write the first dissociation reaction, (11.98) as follows:

$$K_{ds,I} = \frac{m_{HCO_3^-} m_{H^+}}{f_{CO_2}} \cdot \gamma_{aq}^{HCO_3^-} \cdot \gamma_{aq}^{H^+} \tag{11.132}$$

or, using the definition of mean ionic activity coefficient, equation (11.75)

$$K_{ds,I} = \frac{m_{HCO_3^-} m_{H^+}}{f_{CO_2}} \cdot \left(\gamma_{aq}^{HCO_3-H^+}\right)^2 \tag{11.133}$$

where:

$$\log_{10}\left(\gamma_{aq}^{HCO_3-H^+}\right) = \frac{-A|z_{H^+} z_{HCO_3^-}| I^{1/2}}{1 + \frac{1}{2}\left(\mathring{a}_{H^+} + \mathring{a}_{HCO_3^-}\right) B I^{1/2}}. \tag{11.134}$$

The distance of maximum approach between two ions equals the sum of their radii, but Kielland (1937) and all subsequent writers use ionic diameters – hence the division by 2 in the denominator of (11.134). The second dissociation reaction, (11.97), becomes

$$K_{ds,II} = \frac{m_{CO_3^{2-}} m_{H^+}}{m_{HCO_3^-}} \cdot \frac{\gamma_{aq}^{CO_3^{2-}} \cdot \gamma_{aq}^{H^+}}{\gamma_{aq}^{HCO_3^-}}. \tag{11.135}$$

In contrast to (11.132), the individual ion activity coefficients in (11.135) do not work out to a mean ionic activity coefficient. We have two options to deal with this situation. We can choose to calculate single ion activity coefficients by applying (11.131) to each ion, so that the product of charges in the numerator becomes the square of the charge of the ion, and the distance of maximum approach is the ionic diameter. Or we can multiply the numerator and denominator in (11.135) by $\gamma_{aq}^{H^+}$ and re-write this equation as:

$$K_{ds,II} = \frac{m_{CO_3^{2-}} m_{H^+}}{m_{HCO_3^-}} \cdot \frac{\left(\gamma_{aq}^{CO_3^{2-}2H^+}\right)^3}{\left(\gamma_{aq}^{HCO_3-H^+}\right)^2}. \tag{11.136}$$

Mathematically the two approaches are identical. Physically, however, equation (11.136) is preferable because single ion activity coefficients cannot be measured, whereas mean ionic activity coefficients can. Finally, we include activity coefficients in the water ionization reaction, (11.91), which becomes:

$$K_W = m_{H^+} \cdot m_{OH^-} \cdot \left(\gamma_{aq}^{H^+OH^-}\right)^2 \tag{11.137}$$

and we calculate *pH* with equation (11.94).

We are now ready to re-do the calculation in Worked Example 11.3, by using equations (11.133), (11.136) and (11.137), plus the charge balance equation (11.100), which of course remains unchanged. The solution must in this case be found by iteration, because in order to calculate activity coefficients we need to know the ionic strength. A *Maple* procedure that does this is described in Software Box 11.2. The calcite saturation reaction (Worked Example 11.4) is handled in an identical fashion, by adding the solubility product equation (11.106) to the system of equations, calculating the mean ionic activity coefficient of $Ca^{2+}CO_3^{2-}$ with (11.131), and using the charge balance equation (11.105).

Software Box 11.2   Calculation of carbonate speciation in aqueous solution. Non-ideal behavior calculated with Debye–Hückel equation.

Maple worksheet **aq_spec_DH.mw** contains two procedures, `zeco2_DH` and `zecalcite_DH`, that parallel the procedures in **aq_spec_ideal.mw,** but add calculation of aqueous activity coefficients with the Debye–Hückel equation, (11.131) (see Worked Example 11.5). The Debye–Hückel equations are contained in their own procedures, at the top of the worksheet. Input for `zeco2_DH` and `zecalcite_DH` is the same as for the equivalent procedures in **aq_spec_ideal.mw**. Output is also the same, except that an additional field at the end of each line gives ionic strength.

The soda lake calculations in Worked Example 11.5 are contained in the *Maple* worksheet **soda_lakes_DH.mw**. There are two procedures, `soda_lake_Ca` and `soda_lake_CO2`. The first procedure calculates the composition of the soda lake as a function of the parameter d_a (equation (11.139)) at constant CO_2 fugacity (see Fig. 11.11). Input is $f(CO_2)$ in bar, high d_a, low d_a, number of intermediate values and output file name. The output fields are: d_a, pH, $m_{Ca^{(2+)}}$, m_{Na^+}, $m_{HCO_3^-}$, $m_{CO_3^{(2-)}}$, Na bicarbonate ion activity product, Na carbonate ion activity product, ionic strength. The second procedure performs the same calculations as a function of CO_2 fugacity at constant d_a (Fig. 11.12). Input is low $f(CO_2)$ in bar, high $f(CO_2)$ in bar, d_a, number of intermediate values and output file name. The output fields are: $f(CO_2)$ in bar, pH, $m_{Ca^{(2+)}}$, m_{Na^+}, $m_{HCO_3^-}$, $m_{CO_3^{(2-)}}$, Na bicarbonate ion activity product, Na carbonate ion activity product, ionic strength.

The results of the refined calculations are compared in Table 11.1 to the estimates that we obtained assuming that the electrolyte solutions were ideal, for a constant CO_2 fugacity of 3.8×10^{-4} bar. For rainwater in equilibrium with atmospheric CO_2 we calculate an ionic strength of 2.4×10^{-6}, so the Debye–Hückel equation can be expected to yield accurate results – in fact, even the limiting law, equation (11.125), would work in this case. Calculated electrolyte concentrations and pH are virtually indistinguishable from the ideal estimates. The solution saturated in calcite has an ionic strength of $\sim 1.8 \times 10^{-3}$, so equation (11.131) should provide accurate results. In this case pH increases by 0.02 and the concentration of Ca^{2+} increases by about 10% relative to the ideal calculation. These values would be appropriate for "pure" water in equilibrium which calcite, but not for seawater, which has an ionic strength about three orders of magnitude higher than what calcite dissolution by itself generates.

An interesting extension of these calculations is to study the chemical evolution of *soda lakes*. These are bodies of water that generally form in closed basins in arid climates, subject to strong evaporation. The defining characteristic of soda lakes is that the supply of alkali-earth cations, Ca^{2+} and Mg^{2+}, is severely limited (see

Risacher & Fritz, 1991, 2009; Lowenstein & Risacher, 2009; Millero, 2009; Reimer *et al.*, 2009). This is the case, for example, if the drainage basin feeding the lake consists predominantly of felsic volcanic or plutonic rocks. Charge balance of the carbonate and bicarbonate anions in soda lakes is in such case chiefly accomplished by the alkali cations Na^+ and K^+. Evaporation concentrates the solution to the point where carbonates precipitate but, because sodium and potassium carbonates and bicarbonates are quite soluble, the crystallizing carbonate assemblage typically consists of calcite, magnesite and dolomite. Because the system is depleted in Mg^{2+} and Ca^{2+}, however, the *pH* of soda lakes is not buffered as effectively as that of seawater, and can in fact reach fairly extreme values. We can construct a simplified thermodynamic model of a soda lake by considering only the Na^+ and Ca^{2+} cations and carbonate and bicarbonate anions, and writing the following charge balance equation:

$$m_{H^+} + 2m_{Ca^{2+}} + m_{Na^+} - m_{HCO_3^-} - 2m_{CO_3^{2-}} - m_{OH^-} = 0. \qquad (11.138)$$

We seek the values of the six ionic molalities, so we need six equations, one of which is (11.138). The carbonate and bicarbonate dissociation reactions and the water ionization reaction are also applicable, so we include (11.133), (11.136) and (11.137). We assume that the soda lake is saturated in calcite, so we also use equation (11.106). These five equations do not contain information about the degree of Ca^{2+} depletion of the soda lake, yet this is what defines them and what makes their chemistry so radically different from that of seawater. We therefore write a final equation that describes the Ca to carbonate ratio, as follows:

$$\frac{2m_{Ca^{2+}}}{m_{HCO_3^-} + 2m_{CO_3^{2-}}} = d_a. \qquad (11.139)$$

A value of $d_a = 1$ means that carbonate and bicarbonate charges are exactly balanced by Ca^{2+}. In our simplified model this means that there can be no Na^+ in solution, but in nature of course sodium (and potassium) would be present, balanced by other anions such as chloride and sulfate, as in seawater. If $d_a < 1$ then the soda lake must contain Na^+ in solution. We will assume that sodium supply to the lake is unrestricted, so that Na^+ concentration in the solution is limited only by the charge balance constraint, equation (11.138).

The six equations: (11.106), (11.133), (11.136), (11.137), (11.138) and (11.139) contain two free parameters: the fugacity of CO_2 and the calcium to carbonate ratio, d_a. We study the evolution of soda lakes as a function of these parameters. Numerical solution of the set of equations with *Maple* is a straightforward extension of the procedure that we used for the first part of this example, and is described in Software Box 11.2. Let us assume first that CO_2 fugacity is fixed by equilibrium with the atmosphere, so we make $f(CO_2) = 3.8 \times 10^{-4}$ bar. Figure 11.11 shows changes in the chemical composition of the solution as a function of the calcium to carbonate ratio, d_a. The top panel shows *pH* and ionic strength. For $d_a = 1$ we recover the results for seawater, *pH* ∼8.3 and I ∼10^{-3}. Both *pH* and ionic strength increase with decreasing Ca^{2+} content. At $d_a \sim 5 \times 10^{-5}$ the *pH* is about 9.7 and the ionic strength is about 0.1. The results for our simplified soda lake should be fairly accurate up to this point, but become less so as Ca^{2+} content decreases further and ionic strength increases (e.g. Fig. 11.10). Ionic strength in real soda lakes is of course much higher, so the results are not strictly applicable to complex natural systems, but the behavior is qualitatively the correct one.

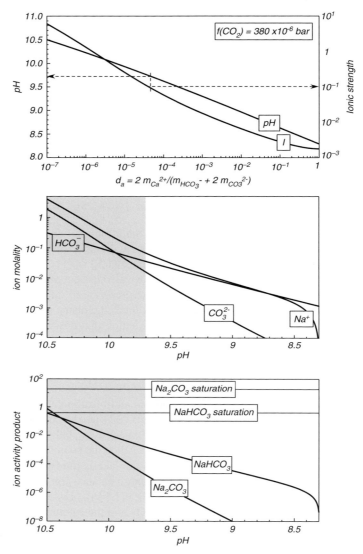

Fig. 11.11 The chemistry of an idealized soda lake in equilibrium with atmospheric CO_2 ($f(CO_2) = 3.8 \times 10^{-4}$ bar) at 25 °C. The lake is assumed to contain only Na^+ and Ca^{2+} cations and HCO_3^- and CO_3^{2-} anions. Changes in pH and species distribution are tracked as a function of the Ca^{2+} deficit, given by the parameter d_a. The pH axes of the center and bottom diagrams are scaled to correspond approximately to the d_a values in the top diagram. The bottom diagram shows that the first sodium species to precipitate is nahcolite (sodium bicarbonate), at $pH \approx 10.5$. The shaded regions correspond to ionic strengths greater than 0.1 (shown with the dashed arrows in the top diagram), for which Debye–Hückel theory is not accurate, so the results are only semi-quantitative.

The middle panel of the figure shows the molalities of Na^+, HCO_3^- and CO_3^{2-} ions as a function of pH, with the pH axis scaled so as to correspond approximately to the values of the d_a parameter in the top panel. The shaded area on the left of the figure corresponds to the region outside the range of validity of Debye–Hückel theory, i.e. $I > 0.1$. The dominant carbonate species changes from HCO_3^- for $pH <\sim 10$ to CO_3^{2-} for $pH > \sim 10$. All three ions become strongly concentrated with increasing pH, so one may wonder, do sodium carbonates precipitate? In order to answer this question we write the saturation reactions:

$$NaHCO_{3(xs)} \rightleftharpoons Na_{aq}^+ + HCO_{3aq}^- \quad (1)$$

$$Na_2CO_{3(xs)} \rightleftharpoons 2Na_{aq}^+ + CO_{3aq}^{2-} \quad (2) \tag{11.140}$$

and the corresponding equilibrium conditions:

$$K_1 = m_{Na^+} \cdot m_{HCO_3^-} \cdot \left(\gamma_{aq}^{HCO_3^- Na^+} \right)^2$$

$$K_2 = m_{Na^+}^2 \cdot m_{CO_3^{2-}} \cdot \left(\gamma_{aq}^{CO_3^{2-} 2Na^+} \right)^3. \tag{11.141}$$

In order to determine whether the soda lake becomes saturated in sodium carbonate or bicarbonate we compare the values of the equilibrium constants calculated from standard state thermodynamic properties (the left-hand side of equations (11.141)) with the ion activity products (the right hand side of the equations) that result from the thermodynamic model of the soda lake. The values of the equilibrium constants (calculated with standard state properties from Wagman *et al.*, 1982) are $K_1 = 3.915 \times 10^{-1}$ (Na bicarbonate saturation) and $K_2 = 18.11$ (Na carbonate saturation). As long as the ion activity products are smaller than these numbers the lake is not saturated in sodium carbonates. The bottom panel in Fig. 11.11 shows the values of the ion activity products as a function of pH, as well as the values of the equilibrium constants. At a constant CO_2 fugacity of 3.8×10^{-4} bar the soda lake trends towards sodium carbonate saturation with *decreasing* acidity. It becomes saturated in sodium bicarbonate (the mineral nahcolite) at $pH \sim 10.5$, but at these conditions it is still far from being saturated in sodium carbonate.

If we keep the calcium to carbonate ratio constant and vary $f(CO_2)$ the behavior of the soda lake is strikingly different. Figure 11.12 shows this, for a constant value of the parameter $d_a = 10^{-3}$. The solution in this case becomes more acidic as it becomes more concentrated, in response to increasing CO_2 fugacity. This is opposite to the effect of starving the lake of calcium at constant CO_2 fugacity. A consequence of this is that the lake is driven towards saturation in sodium carbonates with *increasing* acidity, as can be seen in the middle and bottom panels of Fig. 11.12. For $d_a = 10^{-3}$ nahcolite precipitates at $pH \sim 6.5$, which corresponds to a CO_2 fugacity of about 10 bar. This result is thermodynamically correct, but is a consequence of the constraints imposed on this simplified model, in particular the assumption that an unlimited supply of sodium is available to balance the charge of whatever carbonate concentration one chooses to apply. There may be few natural environments in which conditions that allow sodium bicarbonate to precipitate from a mildly acidic solution are realized. Environments of this kind could exist in lakes formed in craters or calderas of active felsic and alkaline volcanoes, composed of rocks such as trachytes and pantellerites. These rocks are rich in alkalis and relatively poor in calcium and magnesium, and volcanic gases could provide a high CO_2 flux.

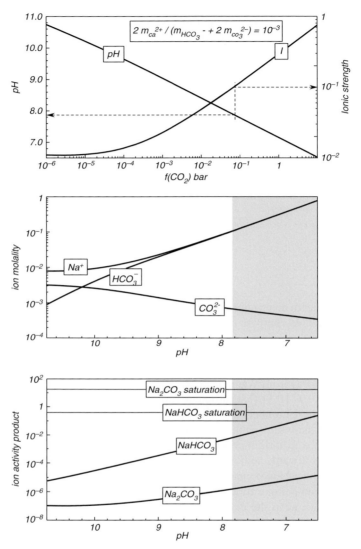

Fig. 11.12 Same as Fig. 11.11, but tracking changes in the soda lake as a function of CO_2 fugacity, for a constant value of $d_a = 0.001$. The behavior of the soda lake is remarkably different. Precipitation of nahcolite takes place in this case in response to increasing acidity. This may occur in response to influx of CO_2-rich volcanic gases.

11.6.2 Activity coefficients in concentrated electrolyte solutions

That the complete Debye–Hückel equation, (11.131), is not accurate for solutions with ionic strengths greater than ~0.1 has been known since the equation was proposed. Debye–Hückel theory, however, performs well for dilute solutions, in which key assumptions, such as treating the solvent as a continuum (i.e. assuming that the macroscopic dielectric constant is valid) and ignoring all characteristics of the ions beyond their charge and radius, can be expected to be better approximations than in concentrated solutions. This suggests that equation (11.131) encapsulates important aspects of the behavior of real electrolytes, and that concentrated solutions may be amenable to be described by extensions of Debye–Hückel's theory. Two such theories are widely used to study natural systems. One of them, due to Pitzer and co-workers, is largely based on calibrating non-ideal empirical interaction parameters among individual ions. The other one, proposed by Helgeson and collaborators, seeks to preserve the Debye–Hückel philosophy as much as possible, by identifying the species that form by ion association, defining their standard state properties as a function of temperature and pressure, and accounting for the effects of temperature and pressure on the dielectric properties of the solvent. Both formulations lead to equations of significant operational complexity (as opposed to mathematical complexity), the full development of which is beyond the space available here. Excellent summaries of the contrasting Pitzer and Helgeson approaches, as well as of a few other simpler alternatives, can be found in the textbooks by Anderson (2005) and Nordstrom and Munoz (1986).

The formulation of Pitzer and co-workers starts with the limiting law of Debye and Hückel and expands the excess Gibbs free energy of mixing, beyond the amount that is accounted for by Debye–Hückel theory, as a virial-like, but largely empirical, series in composition. Philosophically this is the same approach used in the Pitzer and Sterner equation of state (Section 9.4.4). The earliest attempt along these lines is due to Guggenheim (1935; 1967, p. 286), who suggested a characteristically simple and clever approach. His idea was based on the fact that, because the Debye–Hückel equation is chiefly (or only, for the limiting law) a function of solvent properties and ionic strength, different solutes of the same charge type (and approximately the same ionic size) should have the same excess chemical potential in any solution of the same ionic strength. This is not what is observed. Guggenheim proposed that some arbitrarily defined and well-characterized electrolyte, for example NaCl, be used as a reference, and that one then measure the *difference* in the excess Gibbs free energy of mixing of other electrolytes relative to that of the standard. He then suggested that the difference in excess Gibbs free energy of mixing between an arbitrary solution and a solution of pure standard electrolyte be expanded as a sum of terms, each of which has the form $\beta_{ca}m_cm_a$, where m_c and m_a are cation and anion molalities, and β_{ca} is an empirical interaction parameter that characterizes each cation–anion pair. Guggenheim did not include terms for interactions between ions with the same charge. This is known as the *principle of specific ion interactions* and was first stated by Brønsted (1922) who hypothesized that ions of the same charge never come close enough together to have an effect on the free energy of the solution beyond that which arises from their charge (i.e. their identity does not matter). This may be true in dilute solutions, but not in concentrated ones. By differentiating Guggenheim's expression for excess Gibbs free energy of mixing one obtains expressions for the osmotic coefficient (equation (11.36)) and activity coefficients (equation (11.13)), relative to those of the standard. The properties of the standard electrolyte are given by the Debye–Hückel theory. These will not be accurate in absolute terms, but since the activity coefficients for all electrolytes are referred to this same standard the *differences* among them can be expected to reproduce the observed behavior fairly well.

Pitzer's approach is an extension of these ideas, differing in three important ways: (i) the binary interaction terms contain more than one empirical parameter, which are functions of ionic strength (Guggenheim's β_{ca} is a constant, except perhaps for an unknown temperature dependency), (ii) interactions between ions of the same sign are included, and (iii) triple particle interactions, which may be negligible in dilute solutions but not at high concentrations, are also included. Pitzer's general equation for the excess Gibbs free energy of the solution has the form:

$$\frac{\boldsymbol{G}^{ex}}{RT} = \boldsymbol{w}_s f(I) + \frac{1}{\boldsymbol{w}_s} \sum_{ij} \lambda_{ij}(I) \boldsymbol{n}_i \boldsymbol{n}_j + \frac{1}{\boldsymbol{w}_s^2} \sum_{ijk} \mu_{ijk} \boldsymbol{n}_i \boldsymbol{n}_j \boldsymbol{n}_k, \qquad (11.142)$$

where $\boldsymbol{w}_s$ is the mass of solvent (in kg), $f(I)$ is a function of ionic strength and solvent properties that is similar, but not identical, to the Debye–Hückel equation (11.131), and the $\boldsymbol{n}_i, \boldsymbol{n}_j, \boldsymbol{n}_k$ are number of mols of ionic species. The interaction parameters $\lambda_{ij}(I)$ are a function of ionic strength, they include more than one adjustable parameter, and they apply to interactions among same charge as well as opposite charge ions, and neutral species too if they exist. The ternary interaction parameters μ_{ijk}, on the other hand, are not a function of ionic strength and are set to zero if all three ions are of the same charge. Equation (11.142) is manipulated so as to express the interaction parameters as functions of measurable electrolyte properties, and the resulting expression differentiated relative to $\boldsymbol{w}_s$ or m_i in order to obtain the osmotic coefficient of the solvent or the activity coefficient of each solute (equations (11.13) and (11.36)). The details, which are rather ponderous, can be found, for example, in Pitzer (e.g. 1987, 1995), and the final equations are also summarized by Anderson (2005). Important applications of Pitzer's equations include to terrestrial evaporites (see Harvie & Weare, 1980; Harvie et al., 1984) and ocean water (see Millero, 2004, 2009, and references therein), Martian brines and evaporites (see Chevrier & Altheide, 2008; Chevrier et al., 2009; Marion et al., 2003, 2008, 2009), and icy-satellite cryomagmas (Marion et al., 2005).

Pitzer's model is calibrated at 298 K and 1 bar. Extrapolation to other conditions is accomplished by expressing the interaction parameters as polynomial functions of temperature and pressure, although at present these functions are calibrated for only a relatively small set of electrolytes of geological importance (see, for example, Marion et al., 2008). An important aspect of the model, which is expressed in equation (11.142), is that it assumes that interactions occur among individual ions. If a specific ionic association (i.e. species formation) takes place this is not considered explicitly but is rather reflected in the values of the interaction parameters, which generate a lower activity coefficient for the affected ions. The model is therefore purely thermodynamic, in the sense that it focuses only on macroscopic properties, regardless of what the microscopic mechanism for these properties may be.

Helgeson's approach (see Helgeson, 1969; Helgeson et al., 1981; Tanger & Helgeson, 1988) differs significantly from Pitzer's in this last respect. Ion association and complex formation are explicitly accounted for. Infinite dilution standard state properties are generated for all species, together with heat capacity and volumetric data. This allows a more straightforward extrapolation to high temperatures and pressures. In fact, at high temperature Helgeson's model is physically more satisfactory than Pitzer's, because the decrease in the dielectric constant of water with increasing temperature, that limits electrolyte dissociation, is explicitly accounted for (all electrolytes at high temperature tend to behave as weak electrolytes). Speciation calculations in Helgeson's model are accomplished by writing chemical reactions among the species and solving for the species concentrations that make the Gibbs free energy change of all the homogeneous equilibrium reactions vanish (this is equivalent to the chemical equilibrium approach to fluid speciation, see Section 9.6.1). Activity coefficients are based on the Debye–Hückel model, complemented by empirical species-specific terms. The computational details are intricate and are nicely summarized by Anderson (2005).

Exercises for Chapter 11

11.1 Integrate (11.41) in two different ways, so as to obtain (11.42) and (11.43).

11.2 Study various pathways for the formation of ClO_2 (chlorine peroxide gas), e.g. from various combinations of the elements in their molecular and atomic states, as well as from the respective hydrides (HCl and H_2O). What are likely pathways for formation of perchlorates in planetary atmospheres? What makes these reactions possible (i.e. where is the "missing free energy" coming from?) What are likely constraints on the types of natural environments where perchlorates can form? Necessary thermodynamic properties can be obtained from Wagman *et al.* (1982) or NIST's Chemistry WebBook.

11.3 Show that the equations derived in Section 11.2.3, that lead to the relations between activity and osmotic coefficients, equations (11.41) through (11.43), are also valid for electrolyte solutions, using bulk electrolyte properties.

11.4 Modify the *Maple* procedure used to calculate carbonate speciation in rainwater (Fig. 11.5) to include calcite saturation equation (11.106).

11.5 Study the distribution of ferric species in aqueous solution as a function of *pH* at some constant value of oxygen fugacity that is high enough to make concentrations of ferrous species negligible (e.g. the present day terrestrial atmosphere). Find the conditions under which Fe^{3+} becomes the dominant aqueous species. Discuss how analytical Fe^{3+} concentration varies with *pH* at constant oxygen fugacity. How would your conclusions be affected by changes in oxygen fugacity, as long as its value is oxidizing enough to suppress ferrous species? What is the likely significance of primary hematite deposits, i.e. hematite that precipitated from aqueous solution directly, without formation of intermediate metastable ferric hydroxides?

11.6 Discuss why, although possible, it is unlikely that most terrestrial BIF's formed in response to a change in oceanic *pH*, and that a change in oxidation conditions is the simplest and most probable explanation.

11.7 Show that in an atmosphere as reducing as $f(O_2) \sim 10^{-70}$ bar the concentration of carbon monoxide is negligible relative to that of carbon dioxide, so that CO_2 is the dominant oxidized carbon species. If the atmosphere was also saturated in H_2O, could CH_4 concentration be significant at this oxygen fugacity? A review of Chapter 9 can be useful. We will discuss these calculations in greater detail in Chapter 14.

11.8 Modify the *Maple* procedures discussed in Software Box 11.1 to include saturation in both calcite and siderite. The calculation must include the Fe^{2+} hydrolysis reaction, equation (11.122). Assume an ideal aqueous solution. Plot the concentrations of the various aqueous ferrous species, and total Fe^{2+} analytical concentration as a function of *pH*. Assume that oxygen fugacity is low enough to make Fe^{3+} concentration negligible.

11.9 Write a *Maple* procedure that calculates Debye–Hückel osmotic coefficients, by integrating (11.131) according to (11.43).

11.10 Study the phase relations of saturation of calcite, dolomite and magnesite from aqueous solution as a function of dissolved Mg^{2+} and Ca^{2+} concentrations, atmospheric CO_2 concentration and temperature. Write the necessary *Maple* procedures, using Debye–Hückel activity coefficients. Your calculations are not applicable to seawater (why?), but are the results likely to be reasonably accurate for a simple solution like the one you are modeling? Molalities of Mg^{2+} and Ca^{2+} in present-day ocean water are ~ 0.053 and 0.010, respectively (from Millero, 2004). What carbonate should one expect to precipitate according to the simplified model? How does this compare with carbonate precipitation from seawater? Discuss your findings.

12 Non-equilibrium thermodynamics and rates of natural processes

With the exceptions of Chapters 2 and 3, our discussions have largely focused on equilibrium conditions, and chiefly on the equilibrium of systems in which the only kind of mechanical work that takes place is expansion work. Even with these severe restrictions thermodynamics can provide a wealth of information about the nature and evolution of planetary bodies. But these are by no means the limits of thermodynamics. In fact, one could argue that thermodynamics only begins to get interesting when these restrictions are lifted. An in-depth discussion of the possibilities that open up would demand an entire book, at least as long as this one. That is a fight for another day. The goal of this chapter is to lift a corner of the proverbial veil. I will introduce some of the principles of linear non-equilibrium thermodynamics and use them to examine chemical diffusion and chemical reaction mechanisms and rates. These are two classes of processes that are responsible for displacing systems towards equilibrium in a wide range of situations.

12.1 Non-equilibrium thermodynamics

By definition, equilibrium thermodynamics concerns itself with static systems and with "quasi-static" transformations between equilibrium states. These are abstractions, but if our discussions so far are any indication, they are very useful ones. Throughout the previous chapters we have also come across non-equilibrium processes, however. Examples include catastrophic planetesimal collisions (Chapter 2), heat transfer (Chapter 3) and non-isentropic melting (Chapter 10). Processes such as these are the purview of non-equilibrium thermodynamics. I will present here a short introduction to non-equilibrium thermodynamics of systems close to equilibrium (the precise meaning of this will become clear in an instant), along the lines initiated by de Donder, Onsager, Mazur, Prigogine, de Groot and collaborators (see, for example, de Donder, 1936; de Groot, 1959; de Groot & Mazur, 1984; Prigogine, 1961, 1962, 1967; Prigogine & Defay, 1965). This topic is not only intellectually exciting, but it also provides the framework that underlies processes as apparently distinct as heat and mass diffusion, the progress and rate of chemical reactions, viscous deformation and electrical currents.

12.1.1 Fundamental concepts

We begin by revisiting our definition of entropy (Chapter 4). Let us split the entropy change of a system into an external part, $d_e S$ and internal part, $d_i S$:

$$dS = d_e S + d_i S. \tag{12.1}$$

We define the external contribution as the entropy change that arises from exchange of heat with the environment, i.e.:

$$d_e S \equiv \frac{dQ}{T} = \frac{dE + PdV}{T}.$$ (12.2)

This component can be positive, negative or, in the case of adiabatic or isolated systems, zero. The internal contribution to entropy arises from processes that occur inside the system. These could be, for example, heat transfer, chemical diffusion, chemical reactions, viscous dissipation, or dissipation of electric currents. An important part of the study of non-equilibrium thermodynamics consists of finding rigorous mathematical expressions for these and other entropy production processes. At this point we note that $d_i S$ must obey the following relationship:

$$d_i S \geq 0.$$ (12.3)

The equality holds for a system at equilibrium, i.e. static. If a system is not static then $d_i S > 0$. Of course, there is nothing new here: if we apply (12.1) to an isolated system ($d_e S = 0$) then (12.3) recovers the property of entropy given by equation (4.7). The importance of separating external and internal entropy contributions as in (12.1) is that internal entropy production ($d_i S$) is *always* positive in a non-static system, regardless of how the system interacts with its environment. Moreover, it is positive not only for the system as a whole but also for any (non-static) part of the system that we may wish to analyze independently.

If a system is not at internal equilibrium then gradients in intensive variables must exist within it. If entropy gradients exist then the concept of molar entropy for the system as a whole loses meaning. If the system is not far from equilibrium, however, it is possible to define a lengthscale within which local equilibrium holds (i.e. such that gradients in intensive variables can be neglected over distances of this magnitude). It is then convenient to work with entropy per unit volume, so that the entropy of sufficiently small volume elements is well-defined. More precisely, because we are interested in non-static conditions, we consider the *rate of entropy production per unit volume*, σ, defined as:

$$\sigma \equiv \frac{1}{V} \frac{d_i S}{dt}.$$ (12.4)

We note that the units of σ are entropy per unit volume per unit time, for instance: J K^{-1} m^{-3} s^{-1}.

Entropy production is the result of *flows* that occur as long as the system is not at equilibrium, and that cease when the system reaches equilibrium. We shall call these *thermodynamic flows*. They could be, for instance, flow of heat, matter, electric charge or momentum, or the change in the number of molecules of a given species during a chemical reaction (a special case of mass flow). The thermodynamic flows are driven by *potential gradients*, which are also called *thermodynamic forces*. For a general system in which several different entropy production processes operate simultaneously we have:

$$\sigma = \Sigma_i \boldsymbol{J}_i \boldsymbol{F}^i,$$ (12.5)

where the $\boldsymbol{J}_i$ are flows (vectors), the $\boldsymbol{F}^i$ are potential gradients (one-forms) and (12.5) is an inner product that produces the scalar quantity σ (Box 1.1). Consider heat flow as an example. In this case $\boldsymbol{J}_q$ is the heat flux vector (Section 3.1), which has units of J s^{-1} m^{-2}. On dimensional grounds we see that the units of the potential gradient that drives this flow must be $K^{-1}\,m^{-1}$. Restricting our discussion to flow in one spatial dimension, z, we conclude that the thermodynamic gradient that drives heat flow is:

$$F^q = \frac{\partial}{\partial z}\left(\frac{1}{T}\right). \tag{12.6}$$

We shall now make the assumption that, if the potential gradients are small, then the flows are linear functions of the potential gradients. *This linear relationship is the formal definition of a system close to equilibrium.* If the linear relationship between flows and forces does not hold then the system is far from equilibrium and the discussion in this and subsequent sections does not hold. We write the linear relationship as follows:

$$J_i = \Sigma_k L_k^i F^k, \tag{12.7}$$

where the L_k^i are constants called *phenomenological coefficients*. Note that this equation says that a flow J_i may be driven not only by the gradient F^i but also by all other potential gradients that may exist in the system. For instance, if gradients in temperature, F^q (equation (12.6)) and chemical potential, F^d, exist in a system, then according to (12.7), both heat flow, J_q, and mass flow, J_d, each includes two separate contributions, one driven by F^q and the other by F^d. Expanding (12.7) for this case we would have:

$$\begin{aligned} J_q &= L_q^q F^q + L_d^q F^d \\ J_d &= L_q^d F^q + L_d^d F^d. \end{aligned} \tag{12.8}$$

Each phenomenological coefficient is a scalar, and the matrix composed of all phenomenological coefficients arranged as in equations (12.8) is a geometric object called a *tensor*. The two diagonal coefficients are easily interpreted: L_q^q relates temperature gradient to heat flow, so it must be somehow related to the heat conductivity, k (Chapter 3), whereas L_d^d links the gradient in chemical potential to mass flow, so it must be related to chemical diffusivity (Section 12.2.2).

The other two coefficients are more obscure and perhaps unexpected. $L_d^q \neq 0$ implies that a gradient in chemical potential drives heat flow, a phenomenon know as the Dufour effect, whereas $L_q^d \neq 0$ means that a gradient in temperature causes chemical diffusion, which is known as the Soret effect. These and other "cross-flow" phenomena have been known since the nineteenth century. For example, a temperature engenders an electric current (the Seebeck effect) and a gradient in electrical potential gives rise to heat flow at constant temperature (the Peltier effect). Cross chemical diffusion terms arise in systems in which there are gradients in the chemical potentials of more than one component (Section 12.2.3). Lars Onsager (1931a, and b) demonstrated that these effects are not random, but rather a fundamental property of non-equilibrium systems. His work is one of the cornerstones of non-equilibrium thermodynamics. We shall not discuss it here, but we will mention a fundamental result that is due to Onsager. This is the fact that the matrix of phenomenological coefficients is symmetric. In other words, all cross coefficients obey the identity:

$$L_k^i = L_i^k. \tag{12.9}$$

This is known as *Onsager's reciprocal relation.*

12.1.2 Heat diffusion revisited and the principle of minimum entropy production rate

Consider a one-component system, in which compositional gradients are by definition impossible. If we impose a thermal gradient on this system then because $F^d = J_d = 0$ it must be $L_d^q = L_q^d = 0$, and (12.8) collapses to:

$$J_q = L_q^q F^q. \tag{12.10}$$

The thermodynamic flow is in this case the heat flux, given by equation (3.5):

$$J_q = q = -k\frac{dT}{dz}. \tag{12.11}$$

The potential gradient F^q is given by equation (12.6), which we can also write as:

$$F^q = \frac{\partial}{\partial z}\left(\frac{1}{T}\right) = -\frac{1}{T^2}\frac{dT}{dz}. \tag{12.12}$$

We then obtain a relationship between the phenomenological coefficient L_q^q and the thermal conductivity, k:

$$k = \frac{L_q^q}{T^2} \tag{12.13}$$

and, using (12.5), the rate of entropy production per unit volume is:

$$\sigma = \frac{k}{T^2}\left(\frac{dT}{dz}\right)^2. \tag{12.14}$$

As expected, σ is a non-negative quantity, and vanishes only if temperature is uniform and hence there is no heat flow.

We recall from Chapter 3 that the steady state for heat diffusion, i.e. $\partial T/\partial t = 0$, is attained, in a system with no heat generation, when the thermal gradient is uniform, i.e. $\partial T/\partial z = $ constant (e.g. equation (3.15)). It can be proved beginning from equation (12.14) that, for any non-zero temperature gradient, σ is minimized if $T(z)$ is a linear function, so that $\partial T/\partial z = $ constant. The formal demonstration of this will not be presented here (see, for example, Kondepudi & Prigogine, 1998, p. 399). The result is, however, general, and is known as the *theorem of minimum entropy production rate*, originally due to Prigogine. In words, it states that any non-equilibrium system in which at least some of the thermodynamic forces do not vanish, and in which the linear phenomenological law (12.70) and Onsager reciprocal relations are valid, evolves to a non-equilibrium steady state in which the rate of entropy production is minimum. One way to think of this, suggested by Onsager, is that the rate of entropy production behaves as a potential, that is minimized when a non-equilibrium system reaches a dynamic state that remains stationary with time. Although the reality of this result is not in question, different opinions exist on whether minimum entropy production is a principle (i.e. non-demonstrable from simple statements) or a theorem, as envisioned by Prigogine (see, for example, Jaynes, 1980).

12.2 Chemical diffusion

12.2.1 Fundamental relationships

Transport of chemical species down a chemical potential gradient is a non-equilibrium process. It therefore generates entropy. Mass transfer can take different forms. One of them is chemical diffusion without chemical reaction. We seek an equation that relates this type of matter flow to entropy production. We begin with the fundamental equation (4.101), which we re-write as follows:

$$dS = \frac{dE + PdV}{T} - \frac{1}{T}\sum_i \mu^i d\boldsymbol{n}_i. \tag{12.15}$$

Comparing to (12.1) and (12.2) we find that internal entropy production is given by:

$$d_i S = -\frac{1}{T}\sum_i \mu^i d\boldsymbol{n}_i. \tag{12.16}$$

Consider diffusive mass transfer of a single component, which requires that there be a gradient in the chemical potential of only that component. Physically this could be possible, for example, in a one-component system that is not in equilibrium in a gravitational field (Chapter 13), or in a two-component system in which the solute is dilute enough that the concentration of the solvent can be considered to be constant even if the solute concentration varies (Chapter 11). The latter case is known as *tracer diffusion*. For simplicity we will consider diffusion in a single spatial dimension, z, but the equations are easily extended to diffusion in three dimensions (see Kondepudi & Prigogine, 1998; Zhang, 2008; Borg & Dienes, 1988).

Consider two parallel surfaces of cross-sectional area a, separated by a small distance δz, and such that the chemical potentials of the diffusing component at each surface are μ_1 and μ_2, with $\mu_1 > \mu_2$ (Fig. 12.1). Define $\delta\mu = \mu_2 - \mu_1$. Matter flows from 1 to 2, so that if we call $d\boldsymbol{n} = d\boldsymbol{n}_2 = -d\boldsymbol{n}_1 > 0$, we have:

$$\sum_i \mu_i d\boldsymbol{n}_i = \mu_1 d\boldsymbol{n}_1 + \mu_2 d\boldsymbol{n}_2 = \delta\mu d\boldsymbol{n}. \tag{12.17}$$

Substituting in (12.16):

$$d_i S = -\frac{1}{T}\delta\mu d\boldsymbol{n}, \tag{12.18}$$

which is always positive, as $\delta\mu$ and $d\boldsymbol{n}$ always have opposite signs. Passing to the limit and noting that this amount of entropy is produced inside a volume of size $a\delta z$, we get, by using (12.4):

$$\sigma = \frac{1}{a}\frac{d\boldsymbol{n}}{dt}\left(-\frac{1}{T}\frac{d\mu}{dz}\right). \tag{12.19}$$

Comparing (12.19) to (12.5), and allowing for the possibility that temperature gradients may also exist (Section 12.2.4), we identify the thermodynamic flow and the thermodynamic

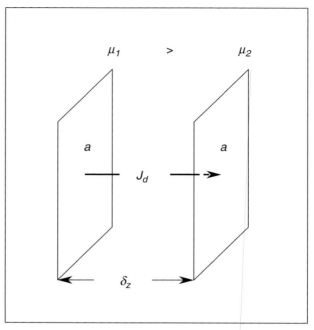

Fig. 12.1 Matter flow, J_d, between two parallel surfaces of area a on which the chemical potentials of the diffusing species are $\mu_1 > \mu_2$.

potential gradient as follows:

$$J_d = \frac{1}{a}\frac{dn}{dt}$$

$$F^d = -\frac{d}{dz}\left(\frac{\mu}{T}\right). \tag{12.20}$$

For isothermal diffusion of a single component we write the phenomenological relationship (12.7) as follows:

$$J_d = -\frac{L}{T}\frac{d\mu}{dz}. \tag{12.21}$$

12.2.2 Fick's laws of chemical diffusion

Equation (12.21) describes isothermal chemical diffusion of a single chemical species, but it is not convenient because chemical potential is not a directly measurable quantity. We seek to recast this equation in terms of concentration of the diffusing species. If c_i is the molar concentration of the diffusing species per unit volume and V is the molar volume of the system, then the mol fraction X_i is $X_i = V c_i$. Taking the standard state as the pure substance at the temperature and pressure of interest:

$$\mu^i = \mu^{0,i} + RT\ln\left(\gamma_i c_i V\right). \tag{12.22}$$

By the chain rule, and assuming that the specis concentration is low enough that it follows Henry's law (γ_i approximately constant):

$$\frac{d\mu^i}{dz} = \frac{d\mu^i}{dc_i}\frac{dc_i}{dz} = \frac{RT}{c_i}\frac{dc_i}{dz}. \tag{12.23}$$

Substituting in (12.21):

$$\boldsymbol{J}_d = -\frac{LR}{c_i}\frac{dc_i}{dz} \tag{12.24}$$

and defining:

$$D \equiv \frac{LR}{c_i} \tag{12.25}$$

we arrive at:

$$\boldsymbol{J}_d = -D\frac{dc_i}{dz}, \tag{12.26}$$

which is known as *Fick's first law of diffusion*. Equation (12.26) is identical to Fourier's law of heat conduction (equation (3.5)) and it is no more of a "law" than the latter. Rather, it is another constitutive equation that is a special case of the general transport relation expressed by equation (3.4).

We now recall that c_i has units of mols per unit volume, and that the matter flux $\boldsymbol{J}_d$ has units of mols per unit area per unit time. It follows that the dimension of D is area per unit time, e.g. $m^2\ s^{-1}$, which are units of diffusivity (Section 3.2.3). The parameter D is called the *chemical diffusivity* or *diffusion coefficient*. In particular, for the dilute one-component case that we are considering here it is called the *tracer diffusion coefficient*. Diffusion coefficients are a strong function of temperature (Section 12.4.1) and equation (12.25) shows that they are also a function of composition. This latter point can cause considerable complications in the mathematics of diffusion, but the compositional dependency can usually be ignored in tracer diffusion problems.

Equation (12.26) still has the disadvantage that it includes a matter flux term that is generally not easily measured, especially if D is small. A better alternative would be to measure the change in concentration with time. Consider a volume element of unit cross-sectional area and width δz (Fig. 12.2), and let the matter fluxes across its two faces be $\boldsymbol{J}_z$ and $\boldsymbol{J}_{z+\delta z}$. The change per unit time in the number of mols of solute contained in the volume is $\boldsymbol{J}_z - \boldsymbol{J}_{z+\delta z}$, so that the rate of change of concentration (mols per unit volume) is:

$$\frac{dc_i}{dt} = \frac{1}{\delta z}\left(\boldsymbol{J}_z - \boldsymbol{J}_{z+\delta z}\right). \tag{12.27}$$

Using (12.26):

$$\frac{\partial c_i}{\partial t} = \frac{D}{\delta z}\left(\left.\frac{\partial c_i}{\partial z}\right|_{z+\delta z} - \left.\frac{\partial c_i}{\partial z}\right|_z\right) = \frac{D}{\delta z}\frac{\partial^2 c_i}{\partial z^2}\delta z \tag{12.28}$$

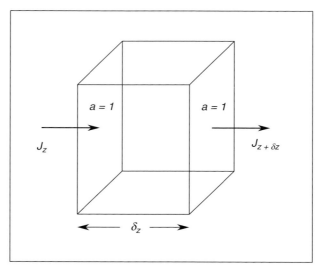

Fig. 12.2   Geometry of the chemical diffusion equation in one dimension. The volume element has unit cross-sectional area perpendicular to the matter flow direction and thickness δz along this direction.

or:

$$\frac{\partial c_i}{\partial t} = D \frac{\partial^2 c_i}{\partial z^2}. \tag{12.29}$$

Equation (12.29), which is identical to the heat diffusion equation (3.15) without a source term, is known as *Fick's second law of diffusion*. You may also recognize in equations (12.27) and (12.28) a compact version of the derivation of equation (3.15) in Section 3.2.2. As I mentioned there, equation (12.29) is a differential equation that shows up in many branches of physics and is simply known as the diffusion equation. The mathematics of diffusion are the same regardless of what is the physical entity that is transported.

Worked Example 12.1   Chemical diffusion on planetary time and lengthscales

We can get a feeling for the relevance of chemical diffusion in planetary processes by focusing on a few examples. In particular, we will look at diffusion in the atmosphere, the oceans, magmatic systems and minerals at metamorphic conditions. As for heat diffusion, we have the relationship:

$$\frac{\lambda^2}{\tau} \sim D, \tag{12.30}$$

where λ is the characteristic diffusive lengthscale, τ the characteristic time scale, and D the diffusion coefficient. Diffusivities of trace gas components in air at 298 K and 1 bar are of the order of 10^{-5} m^2 s^{-1}, whereas for molecular and ionic species in aqueous solution at 298 K typical values are $\sim 10^{-9}$ m^2 s^{-1} (see Kondepudi & Prigogine, 1998; Zhang, 2008). Diffusivities in silicate melts are somewhat more variable; we will use a typical value for

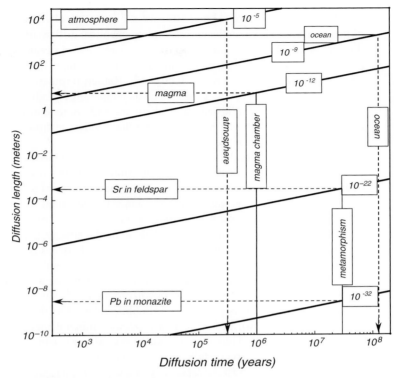

Fig. 12.3 Some examples of chemical diffusion in planetary environments. The lines labeled with diffusion coefficients (in m^2 s^{-1}) are plots of equation (12.30). For the atmosphere and ocean we start from their characteristic lengthscales and infer unrealistically long chemical diffusion time scales, implying that other homogenization mechanisms operate, such as eddy diffusion. For igneous and metamorphic processes we start with estimates of characteristic time scales, e.g. 10^6 years for crystallization of a magma chamber and 20 $\times$ 10^6 years for a metamorphic event, and infer lengthscales for various processes (see text).

H$_2$O in rhyolite melts at 900°C, which is $\sim$10^{-12} m^2 s^{-1} (Zhang *et al.*, 2007). Tracer diffusion coefficients in minerals are much more variable. For example, Pb in monazite at 700°C has a diffusivity of $\sim$10^{-32} m^2 s^{-1} (Cherniak *et al.*, 2004), whereas the value for Sr in feldspars at the same temperature is $\sim$10^{-22} m^2 s^{-1}(Cherniak & Watson, 1994). Figure 12.3 shows plots of equation (12.30) with each of these diffusivity values.

The terrestrial atmosphere and oceans have lengthscales of $\sim$10 km and 2 km, respectively. We see from Fig. 12.3 that diffusive homogenization of the atmosphere would take about 300 000 years. The corresponding value for the ocean would be about 130 million years. Clearly, some other mass transfer mechanism must operate in both of these systems. For instance, changes in CO$_2$ concentration with a periodicity of less than one year are seen in the atmosphere. A diffusive time for the ocean of 130 million years would imply that it should not be possible to detect differences between Cretaceous and present-day ocean chemistry, yet marine sediments record changes on much shorter time scales. If we assume that the atmosphere mixes over times of order 1 year, and the ocean over times of $\sim$100 years (rates are probably faster than these), then we can calculate with equation (12.30) that the effective diffusivities are $\sim$3 and 10^{-3} m^2 s^{-1}, respectively. These values are 5 to 6 orders of magnitude greater than the corresponding chemical diffusivities. The explanation for the discrepancy is that mass transfer in the atmosphere and ocean is chiefly controlled by a process known as *eddy diffusivity*. The idea is that, owing to their relatively low viscosities, neither air nor water are ever perfectly still. Rather, turbulent motions, i.e. eddies, occur at all lengthscales, driven by causes such as temperature gradients and motion of animate or inanimate bodies. Eddies cause local stirring and homogenization, and coupling among eddies effectively diffuses matter at a rate that renders chemical diffusion in oceans and atmospheres inconsequential.

A reasonable time scale for crystallization of a large igneous system may be 1 million years. Over this time diffusive homogenization of compositional gradients would extend over a distance of about 6 m. At least some magmatic systems are known to be homogeneous over greater distances, as suggested for instance by the composition of km-size plutons and of large volcanic eruptions. Again, some other mechanism for chemical mixing appears to be required, but eddy diffusivity is unlikely to be the answer, given the very high viscosity of silicate melts. Convective stirring in magma chambers is a possible explanation.

Let us now assume that a typical high-grade metamorphic recrystallization event lasts 20 million years. Over this time Sr in feldspar would diffuse a distance of perhaps 0.5 mm, whereas Pb in monazite would homogenize over 10^{-9}–10^{-8} m. Radiogenic Pb formed by decay of U and Th is essentially immobile relative to the size of a monazite crystal (say, 10^{-5} to 10^{-4} m). Monazite is a refractory mineral that commonly survives high-grade metamorphism. Therefore, monazite crystal cores can be used to date events that preceded metamorphic recrystallization, even if rims of neoformed monazite grow during metamorphism. In contrast, feldspar may break down and regrow in response to changes in metamorphic conditions. Chemical reactions such as these are mass transfer mechanisms, which may be much faster than simple chemical diffusion. However, even if feldspar growth did not take place, Sr would homogenize over a length not too different from the size of feldspar crystals. Rb–Sr dating may then yield the age of metamorphism.

An important generalization that follows from these examples is that, whereas chemical diffusion is not an important mass transfer mechanism at planetary lengthscales, it provides the physical underpinnings and constraints for many powerful techniques used to study planetary processes, such as radiometric dating and estimation of rates of geological processes. A comprehensive and up to date treatment of these topics is given by Zhang (2008).

12.2.3 Interdiffusion

Consider now the case of a binary system in which two components are present in comparable concentrations. If there is a compositional gradient then the two components will diffuse in opposite directions. This is known as *interdiffusion* or, also, *binary diffusion*. It is the process by which crystals that grow during metamorphism or igneous crystallization tend to homogenize.

Let the matter fluxes be J_1 and J_2. As always we consider diffusion in one dimension. In some systems the following relationship is valid:

$$J_1 + J_2 = 0. \tag{12.31}$$

Recall that we consider flux in units of mols, or particles, per unit of surface per unit of time. Interdiffusion in a mixture of ideal gases at constant temperature and pressure obeys (12.31). The same can be expected to be at least approximately true in a crystal, in which particles exchange places among fixed sites in a crystalline lattice. But there may be instances in which (12.31) is not valid, for example if the partial molar volumes of the two components are significantly different. This requires only a slight modification to the equations, as we shall see.

Calling the mol fractions of the two components X_1 and X_2 we have, from Gibbs–Duhem's equation (6.7) at constant temperature and pressure:

$$X_1 d\mu^1 + X_2 d\mu^2 = 0 \tag{12.32}$$

or:

$$X_1 \frac{d\mu^1}{dz} + X_2 \frac{d\mu^2}{dz} = 0. \tag{12.33}$$

We can combine (12.31) and (12.33) as follows:

$$\frac{\boldsymbol{J}_1 X_2}{d\mu^1/dz} = \frac{\boldsymbol{J}_2 X_1}{d\mu^2/dz} \tag{12.34}$$

and define a phenomenological coefficient, L, as follows:

$$-\frac{L}{T} \equiv \frac{\boldsymbol{J}_1 X_2}{d\mu^1/dz} = \frac{\boldsymbol{J}_2 X_1}{d\mu^2/dz}. \tag{12.35}$$

Calling the molar concentrations per unit volume c_1 and c_2, and using (12.23), we find:

$$\begin{aligned} \boldsymbol{J}_1 &= -\frac{LR}{X_2 c_1} \frac{dc_1}{dz} \\ \boldsymbol{J}_2 &= -\frac{LR}{X_1 c_2} \frac{dc_2}{dz} \end{aligned} \tag{12.36}$$

so that the diffusion coefficients are (see equation (12.26)):

$$\begin{aligned} D_1 &= \frac{LR}{X_2 c_1} \\ D_2 &= \frac{LR}{X_1 c_2}. \end{aligned} \tag{12.37}$$

For $X_2 \to 1$ the coefficient D_1 becomes the tracer diffusion coefficient given by equation (12.25), and as long as $X_1 \neq 0$, D_2 remains bound and the flux $\boldsymbol{J}_2$ vanishes as $dc_2/dz \to 0$. In other words, we recover the tracer diffusion case. We also note that, if the molar volume of the mixture is V, then for all values of X_1 and X_2 we have $X_1 = V c_1$ and $X_2 = V c_2$. Substituting these identities in (12.37) we find:

$$D_1 = D_2 = D_i, \tag{12.38}$$

where D_i is the *interdiffusion coefficient* of species 1 and 2 in a binary mixture. Note that identity (12.38) does not mean that the interdiffusion coefficient is constant: equations (12.37) show that it is a function of the concentration product $c_1 c_2$. Because the phenomenological coefficient L is a constant, however, it is in principle possible to calculate the interdiffusion coefficient for any arbitrary composition from measurements of tracer diffusion coefficients of one of the two species. From this we can infer that interdiffusion coefficients are at most of the same order of magnitude as tracer diffusion coefficients. For example, if we take species 1 as the trace component we have (equation (12.25)):

$$D_{1,tr} = \frac{LR}{c_{1,tr}},\tag{12.39}$$

where $c_{1,tr}$ is a trace concentration of 1, and $D_{1,tr}$ is the corresponding tracer diffusion coefficient. Using 12.37 we calculate the interdiffusion coefficient for $c_1 \gg c_{1,tr}$:

$$D_i = D_{1,tr} \frac{c_{1,tr}}{V c_1 c_2},\tag{12.40}$$

which is at most of the same order as $D_{1,tr}$, and generally smaller. It follows that the conclusions that we reached in Worked Example 12.1 are valid for chemical diffusion in general, not just tracer diffusion.

Consider now a slightly different case, in which rather than balancing the number of particles, as in equation (12.31), the opposing flows of matter keep the volume of the system constant. Equation (12.31) is then modified as follows:

$$v_1 \boldsymbol{J}_1 + v_2 \boldsymbol{J}_2 = 0,\tag{12.41}$$

where v_1, v_2 are the partial molar volumes of the two species. In this case we find that the diffusion coefficients are given by:

$$D_1 = \frac{LR}{X_2 v_1 c_1}$$
$$D_2 = \frac{LR}{X_1 v_2 c_2}\tag{12.42}$$

(note that the dimension of L is not the same as in equations (12.37), which of course follows from the fact that we are now considering volume fluxes rather than particle fluxes). The interdiffusion coefficient is now given by:

$$D_i = v_1 D_1 = v_2 D_2,\tag{12.43}$$

which, if the two species have the same partial molar volume, as is the case for ideal gases, collapses to (12.38).

The fact that binary diffusion can be fully described with a single phenomenological coefficient (equations (12.37) or (12.42)) means that there are no "cross-flow" terms, or, in other words, that the Onsager reciprocal relationship does not apply to binary diffusion. This is so because the chemical potential gradient of each component is unequivocally determined by the concentration gradient of the other. Reciprocal matter flow terms appear in systems of three or more components (Exercise 12.1), in which there is at least one concentration that can vary independently.

12.2.4 Coupling of mass and heat transfer

Let us consider a system in which there are gradients in temperature and composition. For simplicity we will assume that a compositional gradient exists in only one component, i.e. we consider tracer diffusion, and we restrict the treatment to one spatial dimension. Using J_q for heat flux and J_i for matter flux of component i, the rate of entropy production σ is (see equations (12.5), (12.6) and (12.20)):

$$\sigma = J_q \frac{\partial}{\partial z}\left(\frac{1}{T}\right) - J_i \frac{\partial}{\partial z}\left(\frac{\mu^i}{T}\right). \tag{12.44}$$

We have:

$$\frac{\partial}{\partial z}\left(\frac{\mu^i}{T}\right) = \frac{1}{T}\frac{\partial \mu^i}{\partial z} + \mu^i \frac{\partial}{\partial z}\left(\frac{1}{T}\right) \tag{12.45}$$

but chemical potential is a function of both composition and temperature, both of which vary, so:

$$\begin{aligned}
\frac{\partial \mu^i}{\partial z} &= \frac{\partial \mu^i}{\partial c_i}\frac{\partial c_i}{\partial z} + \frac{\partial \mu^i}{\partial T}\frac{\partial T}{\partial z} = \frac{\partial \mu^i}{\partial c_i}\frac{\partial c_i}{\partial z} - s^i \frac{\partial T}{\partial z} \\
&= \frac{\partial \mu^i}{\partial c_i}\frac{\partial c_i}{\partial z} + s^i T^2 \frac{\partial}{\partial z}\left(\frac{1}{T}\right),
\end{aligned} \tag{12.46}$$

where s^i is the partial molar entropy of component i, and c_i its concentration, which we choose to express in mols per unit volume. Substituting in (12.45):

$$\begin{aligned}
\frac{\partial}{\partial z}\left(\frac{\mu^i}{T}\right) &= \frac{1}{T}\frac{\partial \mu^i}{\partial c_i}\frac{\partial c_i}{\partial z} + \left(s^i T + \mu^i\right)\frac{\partial}{\partial z}\left(\frac{1}{T}\right) \\
&= \frac{1}{T}\frac{\partial \mu^i}{\partial c_i}\frac{\partial c_i}{\partial z} + h^i \frac{\partial}{\partial z}\left(\frac{1}{T}\right),
\end{aligned} \tag{12.47}$$

where h^i is the partial molar enthalpy of i. We now substitute (12.47) in (12.44) and group terms as follows:

$$\sigma = \left(J_q - h^i J_i\right)\frac{\partial}{\partial z}\left(\frac{1}{T}\right) - J_i \frac{1}{T}\frac{\partial \mu^i}{\partial c_i}\frac{\partial c_i}{\partial z}. \tag{12.48}$$

The product $h^i J_i$, known as the *heat of transport,* is the thermal energy content of the matter flow. Defining the *reduced heat flux,* $J_{q,r}$ as:

$$J_{q,r} \equiv J_q - h^i J_i \tag{12.49}$$

we write the linear phenomenological relationships, (12.8), as follows:

$$\begin{aligned}
J_{q,r} &= L_q^q \frac{\partial}{\partial z}\left(\frac{1}{T}\right) - L_i^q \frac{1}{T}\frac{\partial \mu^i}{\partial c_i}\frac{\partial c_i}{\partial z} \\
J_i &= L_q^i \frac{\partial}{\partial z}\left(\frac{1}{T}\right) - L_i^i \frac{1}{T}\frac{\partial \mu^i}{\partial c_i}\frac{\partial c_i}{\partial z}
\end{aligned} \tag{12.50}$$

or, using the definitions of heat conductivity k (equation (12.13)) and chemical diffusivity D_i (equation (12.25)), and identity (12.23):

$$\boldsymbol{J}_{q,r} = -k\frac{\partial T}{\partial z} - L_i^q \frac{R}{c_i}\frac{\partial c_i}{\partial z}$$
$$\boldsymbol{J}_i = -L_q^i \frac{1}{T^2}\frac{\partial T}{\partial z} - D_i\frac{\partial c_i}{\partial z}. \tag{12.51}$$

The two cross terms are the heat flux carried by matter flow, which is known as the Dufour effect, and the flow of matter driven by the temperature gradient, which is known as the Soret effect, or also as thermal diffusion. Defining the Dufour coefficient, D_D, and the coefficient of thermal diffusion, D_T, as follows (do not confuse D_T with the thermal diffusivity, κ!):

$$D_{D,i} \equiv \frac{L_i^q R}{c_i^2}$$
$$D_{T,i} \equiv \frac{L_q^i}{c_i T^2} \tag{12.52}$$

equations (12.51) become:

$$\boldsymbol{J}_{q,r} = -k\frac{\partial T}{\partial z} - c_i D_{D,i}\frac{\partial c_i}{\partial z}$$
$$\boldsymbol{J}_i = -c_i D_{T,i}\frac{\partial T}{\partial z} - D_i\frac{\partial c_i}{\partial z} \tag{12.53}$$

in which we have from the Onsager reciprocal relationship, $L_i^q = L_q^i$:

$$\frac{D_{D,i}}{D_{T,i}} = \frac{RT^2}{c_i}. \tag{12.54}$$

Worked Example 12.2 Soret and Dufour diffusion in planetary processes

The derivations in this section make for some rather elegant, if elementary, algebra (even more so in three dimensions). They lead to equation (12.54), which suggests an experimental test of the Onsager reciprocal relation. But how important are these effects in planetary processes? During the 1980s it was thought that Soret diffusion could be an important, perhaps even dominant, mechanism of igneous differentiation (see Walker & DeLong, 1982; Lesher, 1986; Lesher & Walker, 1991). This is now known not to be the case, except perhaps in very specific and localized situations, and then only over minuscule lengthscales. Let us see why.

It is convenient to define the Soret coefficient, s_T, as the ratio of thermal to chemical diffusivity:

$$s_T \equiv \frac{D_{T,i}}{D_i}. \tag{12.55}$$

We note that s_T has dimension of $[T]^{-1}$, i.e. K^{-1}. The matter flux equation becomes:

$$\boldsymbol{J}_i = -s_T c_i D_i \frac{\partial T}{\partial z} - D_i \frac{\partial c_i}{\partial z}. \tag{12.56}$$

The condition of no matter flow, i.e. $\boldsymbol{J}_i = 0$, results if the thermal and chemical gradients have opposite signs and their effects exactly balance each other out. Thus, by setting $\boldsymbol{J}_i = 0$ we get an estimate of the magnitude of the compositional gradient that can be caused by a thermal gradient. It is convenient to do this in terms of mol fraction rather than molar concentration, so using the relationship $X_i = c_i V$ we write:

$$-\frac{s_T X_i D_i}{V} \frac{\partial T}{\partial z} - \frac{D_i}{V} \frac{\partial X_i}{\partial z} = 0 \tag{12.57}$$

which simplifies to:

$$\frac{d X_i}{d T} = -s_T X_i. \tag{12.58}$$

Thus, approximately:

$$\delta X_i \sim -s_T X_i \delta T. \tag{12.59}$$

The key is, of course, the value of s_T. Soret coefficients are poorly understood functions of composition, temperature and temperature gradient (Lesher & Walker, 1991; Huang *et al.*, 2010). For a wide range of materials, including silicate melts, aqueous electrolyte solutions and gases, s_T is of order 10^{-2} to 10^{-3} K^{-1}. This means that we can expect Soret diffusion to be a significant effect only in settings in which there are strong temperature differences, of order 100 K or more. Such temperature differences could exist, for example, across the thermal boundary layer of a convective magma chamber. Soret diffusion could be important in that instance (Sonnenthal, 2004).

This is not the whole story, however, as the rate of mass transfer is still determined by the chemical diffusivity, regardless of whether the driving force is a gradient in composition or temperature. Given that thermal diffusivity, $\kappa \sim 10^{-6}$ m^2 s^{-1} is typically six orders of magnitude greater than chemical diffusivity in silicate melts, $D_i \sim 10^{-12}$ m^2 s^{-1} (see Worked Example 12.1), thermal gradients decay much faster than chemical gradients. The question is, then, how far can Soret diffusion extend in the time that it takes for a thermal gradient to dissipate? Using λ_q for the thermal diffusion lengthscale and λ_i for the chemical diffusion lengthscale we have, from equations 3.16 and 12.30:

$$\frac{\lambda_i}{\lambda_q} \sim \left(\frac{D_i}{\kappa}\right)^{1/2} \sim 10^{-3}. \tag{12.60}$$

Suppose that a thermal gradient of ~ 100 K exists across a 10-m thick thermal boundary layer of a magma chamber. Soret diffusion would cause a significant compositional change over a thickness of only ~ 1 cm. An essentially identical result was obtained by Cygan and Carrigan (1992) by means of a rigorous numerical solution of the mass flow equations (i.e. (12.56)). Soret diffusion is not important in nature, although it could have noticeable effects on isotopic fractionation during high-temperature experiments (Richter *et al.*, 2008, 2009), and also has technical applications.

The Dufour effect is smaller than the Soret effect. Using (12.54) and (12.55) in the heat flow equation (12.53), and setting $J_{q,r} = 0$, we find:

$$\delta T \sim -\frac{s_T D_i R T^2}{kV} \delta X_i.$$

(12.61)

Substituting typical values we find $\delta T \sim 10^{-4} \delta X_i$. We can interpret this result as meaning that the Dufour effect is equivalent to the heat flow driven by a temperature difference of order 10^{-4} K times the mol fraction difference. It is unlikely that there are natural environments where this is a significant effect.

12.3 Rate of chemical reactions

12.3.1 Fundamental concepts

Chemical reactions displace systems towards equilibrium. Therefore, chemical reactions take place while a system is not at equilibrium. The study of the rate of chemical reactions is a huge field and we can only cover some of the basic concepts. In particular, we will focus only on chemical reactions in homogeneous systems, e.g. a gas phase or a liquid solution. We begin by considering *elementary reactions*. These can be defined as reactions that go from reactants to products without any intermediate steps. At a molecular level, elementary reactions take place by collisions between the actual reactant molecules. Most chemical reactions are not elementary. Rather, they are combinations of elementary steps. Here we focus on elementary reactions in homogeneous systems only, whether or not I state this explicitly.

We can expect on physical grounds that there are likely to be some simple constraints on the nature of elementary reactions, in particular with regards to the number of participating molecules. This number is known as the *molecularity* of the reaction. Elementary reactions are known to have molecularities of 1, 2 or 3 only, and are designated unimolecular, bimolecular or termolecular, respectively. Of these, bimolecular reactions are the most straightforward, as they simply require a collision between two reactant molecules, for example A and B:

$$A + B \rightarrow products.$$

In contrast to this simple case, unimolecular and termolecular reactions each present problems of their own, and it is arguable whether they are rigorously elementary. Let us focus on termolecular reactions first, which we can schematize as:

$$A + B + C \rightarrow products.$$

The probability that three different molecules will converge on the exact same spot at exactly the same time is very low, and in fact termolecular elementary reactions are much rarer than bimolecular ones. The simplest picture of how termolecular reactions take place is that they consist of two collisions in rapid succession, such that one of the collisions produces a temporary "agglutination" of two of the reactant molecules, for example:

$$A + B \rightarrow AB,$$

$$AB + C \rightarrow products.$$

An important point is that A and B remain stuck for only a very short period of time. If they collide with C within this time then the products form; if not A and B separate. The fact that termolecular reactions become more common with increasing pressure is consistent with this model, as higher molecular density makes it more likely that C will collide with the AB cluster before it has time to break down. We can expect that reactions with molecularities higher than three must be exceedingly rare, and indeed there is no experimental evidence for such reactions taking place.

Unimolecular reactions correspond to:

$$A \rightarrow products$$

and they present a different problem. If a reaction is truly unimolecular, then what makes a molecule react? Some fraction of the reactant molecules must acquire higher energy, sufficient to cause the energized molecules to break apart into the product molecules (if all reactant molecules were energized then the reactant would disappear instantaneously and we would not have to worry about chemical kinetics). One way in which this can happen is by irradiation with photons of the appropriate wavelength – this is known as *photodissociation* (Section 12.4.2). But unimolecular reactions that are not photo-activated also exist. The path of unimolecular decomposition in this case is known as the Lindemann mechanism and can be schematized as follows:

$$A + M \rightarrow A^* + M$$

$$A^* + M \rightarrow A + M$$

$$A^* \rightarrow products.$$

The first step, known as the activation step, involves a collision with another molecule, M, known as a collision partner. M could be another molecule of A, it could be a product molecule, or it could be a molecule of some other species. The important point is that as a result of the collision some of the energy of M is transferred to A, which becomes the activated molecule A^*. This molecule can lose its energy excess either by colliding with another collision partner in a deactivation step (second step) or by breaking apart into the products (third step). The reaction is not truly unimolecular, but if we add up the first and third steps it looks unimolecular.

One can legitimately argue whether a unimolecular reaction of this kind is an elementary reaction, and the same is true of termolecular reactions which consist of two collisions in rapid succession. Such arguments may not be all that important, however, as what we observe macroscopically is not the molecularity but a related quantity, known as the *order of the reaction*. In the simple examples discussed above the molecules that participate in elementary reactions were labeled A, B and C, but they do not necessarily have to be different. We can therefore write a generalized elementary reaction as follows:

$$\sum_i v_i Y_i \rightarrow products, \tag{12.62}$$

where ν_i is the stoichiometric coefficient of reactant species Y_i. We can expect that the rate of an elementary reaction must be proportional to the number of molecules that are available to react, which in the simplest case means the concentration of each of the reactant molecules. A thermodynamically more rigorous choice would be their activities (see Section 12.3.2), but concentration is universally used in kinetic studies, because it is a directly observable quantity, whereas activity is not. We therefore write the *rate law* of reaction (12.62) as:

$$r = k \prod_i [Y_i]^{\nu_i}, \tag{12.63}$$

where r is the reaction rate expressed in units of number of (reacting) mols per unit of volume per unit of time, and $[Y_i]$ is the concentration of reactant Y_i in mols per unit volume. This notation for concentration is different from what we have used so far, but is standard in the literature of chemical kinetics, and convenient too. The parameter k is called the *rate constant* for the reaction. It is important to understand that the rate constant is not a thermodynamic quantity, and that it in fact encapsulates aspects of chemical kinetics that cannot be addressed by thermodynamics. We shall have more to say about this in later sections, but we note at this point that the rate constant is a measure of the fraction of molecules that are reactive. Owing to the statistical distribution of molecular energies (e.g. Section 1.14) we can expect that there will always be some fraction of the total ensemble of molecules that do not carry enough energy for the molecular bonds to be broken during a collision. Those molecules that are energetic enough are the reactive molecules. The fraction of reactive molecules, and thus the rate constant, increases with temperature in the case of thermally activated reactions (Section 12.4.1), but molecules can become activated by non-thermal processes as well, such as absorption of photons within a specific wavelength. This is the basis of photochemical reactions (Section 12.4.2).

Reaction (12.63) is said to be of order ν_i in species Y_i, and the order of the reaction as a whole is the sum $\sum_i \nu_i$. At first sight the order of an elementary reaction is the same as its molecularity, and indeed elementary reactions can be of first, second or third order only, but the detailed microscopic picture is a bit different. It is clear that bimolecular reactions must be of order two. Regardless of whether an elementary termolecular reaction occurs by a single triple collision or by two collisions in rapid succession, we may expect it to be of order three, because in either case the rate should vary with the product of the concentrations of the three species. The point is that what is actually measured is the order of the reaction, which is a macroscopic variable, and the molecularity is one possible microscopic interpretation of this observation. Finally, it must be noted that the overall order of a compound reaction, composed of an assemblage of elementary steps, is defined by equation (12.63), but in such case there is no connection with any simple microscopic picture, and the order need not be 1, 2, or 3, in fact not even an integer (Logan, 1996; Houston, 2006).

An important part of the study of chemical kinetics focuses on the rate laws for reactions of the various orders, and on how these laws combine in non-elementary reactions. We discuss these topics below, but before doing that it is important to clarify some of the relationships between thermodynamics and kinetics.

12.3.2 Thermodynamics, kinetics and entropy production by chemical reactions

All of the arguments that we made in the preceding section are strictly kinetic. By writing an elementary reaction, for example as $A + B \rightarrow products$, we are eschewing any thermodynamic content, as we are implying that the reaction proceeds in only one direction, and that there is no possibility of equilibrium between reactants and products. Thermodynamics tells us that this is not the case, and that in fact there is always the possibility, at least in principle, that the free energies of products and reactants will be the same, and equilibrium will be attained. The key to understanding the relationship between thermodynamics and kinetics can be found by focusing on equilibrium situations, and more specifically on how a chemical reaction approaches equilibrium.

Let us write a somewhat different version of reaction (12.62) as follows:

$$\sum_i v_i Y_i \rightleftarrows \sum_j v_j Z_j, \tag{12.64}$$

where now v_i is the stoichiometric coefficient of reactant species Y_i, and v_j is the stoichiometric coefficient of product species Z_j. The implication of equation (12.64) is that there are two elementary reactions taking place simultaneously: the forward reaction (Ys going to Zs) and the reverse reaction (Zs going to Ys). At equilibrium the rates of the two reactions are the same, and in fact this is one possible definition of chemical equilibrium, but in general this need not be so. We also define a variable, ξ, called the *extent of reaction* or *progress variable*, as follows:

$$d\xi \equiv \frac{dn_{Z_j}}{v_j} = -\frac{dn_{Y_i}}{v_i}, \ all \ i,j, \tag{12.65}$$

where dn_{Z_j} is the number of mols of species Zj that are produced, dn_{Y_i} is the number of mols of species Yi that are consumed, and v_j, v_i are the corresponding stoichiometric coefficients. By defining the progress variable in this way we avoid any ambiguities about the extent of reaction that might arise from the variable stoichiometric coefficients. Note that the progress variable is always a positive quantity. We also define the *net rate* of reaction (12.64), r_n, as follows:

$$r_n \equiv \frac{1}{V}\frac{d\xi}{dt} \tag{12.66}$$

with dimension of mols per unit volume per unit time. If the reaction rates of the forward and reverse reactions in (12.64) are r_f and r_r, respectively, then the net rate of the reaction, r_n, must be equal to the difference between the forward and reverse rates, i.e.:

$$r_n = r_f - r_r. \tag{12.67}$$

We now seek to relate kinetics to thermodynamics, so we will write the rate laws (equation (12.63)) in terms of activity rather than concentration. One way to think about this is that in (12.63) the activity coefficients are subsumed in the rate constants, whereas now we include them in the activities. We therefore write the forward and reverse reaction rates as follows:

$$r_f = k_f \prod_i \left(a^{Y_i}\right)^{v_i}$$
$$r_r = k_r \prod_j \left(a^{Z_j}\right)^{v_j}, \tag{12.68}$$

where k_f, k_r are the rate constants for the forward and reverse reactions.

The double arrow in equation (12.64) is always true, but we commonly think of a reaction as "proceeding to the right" or "proceeding to the left". What these statement refer to is to the relative magnitudes of the forward and reverse rates. In particular, we can define chemical equilibrium as the condition for which $r_n = 0$. From (12.68) this implies:

$$\frac{k_f}{k_r} = \frac{\prod_j \left(a^{Z_j}\right)_{eq}^{v_j}}{\prod_i \left(a^{Y_i}\right)_{eq}^{v_i}} = \exp\left(-\frac{\Delta_r G^0_{P,T}}{RT}\right), \tag{12.69}$$

where I have added the subscript *eq* to specify that these are the activities at equilibrium. Up until this point we have only dealt with equilibrium situations, so that this notation was not necessary, but this is no longer the case. If the rate laws given by equations (12.68) are valid, then equation (12.69) says that the ratio between the forward and reverse rate constants is determined by equilibrium thermodynamics, and is in fact equal to the equilibrium constant. *What thermodynamics cannot do is provide the values of the individual rate constants.*

We now seek a thermodynamic function that gives the distance of a chemical reaction from equilibrium. This function is called the *affinity*, which I will represent with $\mathscr{E}$. It is defined as the difference between the sum of the chemical potentials of the reactants and those of the products. Using the notation of equation (12.64) we have:

$$\mathscr{E} \equiv \sum_i v_i \mu^{Y_i} - \sum_j v_j \mu^{Z_j}. \tag{12.70}$$

If the affinity is positive then reaction (12.64) proceeds from left to right, whereas $\mathscr{E} = 0$ corresponds to equilibrium. We can also write (12.70) as follows:

$$\mathscr{E} = -\Delta_r G^0_{P,T} + RT \ln\left(\frac{\prod_i \left(a^{Y_i}\right)^{v_i}}{\prod_j \left(a^{Z_j}\right)^{v_j}}\right), \tag{12.71}$$

where in this case the activities are the actual values, i.e. the values as the chemical reaction is taking place, and *not necessarily the equilibrium values.* Using (12.69) we see that the affinity is also given by:

$$\mathscr{E} = RT \ln\left(\frac{k_f \prod_i \left(a^{Y_i}\right)^{v_i}}{k_r \prod_j \left(a_{p_j}\right)^{v_j}}\right) \tag{12.72}$$

or, from (12.68):

$$\frac{r_f}{r_r} = \exp\left(\frac{\mathscr{E}}{RT}\right), \tag{12.73}$$

which should be compared to (12.69): the ratio between the rate constants is a constant (the equilibrium constant), but the ratio between the reaction rates varies with the affinity, or, in other words, with the progress of the reaction. Using (12.73) we can write (12.67) as follows:

$$r_n = r_f\left[1 - \exp\left(-\frac{\mathscr{E}}{RT}\right)\right], \tag{12.74}$$

which shows that the net rate of the reaction approaches a maximum value, equal to the rate of the forward reaction, when $\mathscr{E} \to \infty$, i.e. when the products are infinitely dilute. The net reaction rate decreases exponentially as equilibrium is approached and $\mathscr{E} \to 0$, but *equation (12.74) yields a value for the net reaction rate only if r_f is known.* This in turn requires that we know the value of the rate constant k_f (equation (12.68)), which is something that thermodynamics cannot supply. Equation (12.74) does say that the net reaction rate increases exponentially with distance from the equilibrium state, *but it does not say that the rates of different reactions can be compared on the basis of their affinities*, as the rate constants for different reactions are generally different, and not predictable from thermodynamics. Equations (12.69) and (12.74) encapsulate the relationship between thermodynamics and kinetics.

We can calculate the rate of entropy production by a chemical reaction. Using equation (12.65) we expand (12.16) as follows:

$$d_i S = -\frac{1}{T} \left(\sum_j v_j \mu^{Z_j} - \sum_i v_i \mu^{Y_i} \right) d\xi \tag{12.75}$$

or:

$$d_i S = \frac{\mathscr{E}}{T} d\xi \tag{12.76}$$

from which the rate of entropy production per unit volume (equation (12.4)) is:

$$\sigma = \frac{\mathscr{E}}{T} \frac{1}{V} \frac{d\xi}{dt}. \tag{12.77}$$

Note that this is always positive, i.e. chemical reactions always produce entropy. Comparing with equations (12.5) and (12.66) we identify the thermodynamic flow with the net reaction rate, i.e.:

$$J = \frac{1}{V} \frac{d\xi}{dt} = r_n. \tag{12.78}$$

We can think of a chemical reaction as a mass transfer process in which chemical components migrate from the reactant species to the product species, and in which the net reaction rate is the mass transfer rate. The thermodynamic potential gradient is given by:

$$F = \frac{\mathscr{E}}{T} \tag{12.79}$$

so that the linear phenomenological relationship (12.7) implies:

$$r_n = L \frac{\mathscr{E}}{T}. \tag{12.80}$$

This appears to be at odds with (12.74), in which we found a non-linear relationship between the thermodynamic flow and the potential gradient. The apparent contradiction arises from the fact that (12.74) is valid in general, whereas the linear relation (12.80) is valid only close to equilibrium. This is explored further in Exercise 12.2.

12.3.3 Differential and integral rate laws

We now return to the strictly kinetic question of how the concentrations of chemical species change with time. The simplest case is that of an elementary first-order reaction $A \rightarrow products$. From (12.63) the rate law of a first order reaction is $r = k[A]$, where $[A]$ is molar concentration per unit volume. Setting $r_r = 0$ in (12.68), and using (12.67) and (12.66):

$$k[A] = \frac{1}{V}\frac{d\xi}{dt} \tag{12.81}$$

and from (12.65):

$$\frac{d\xi}{dt} = -V\frac{d[A]}{dt} \tag{12.82}$$

so:

$$\frac{d[A]}{dt} = -k[A], \tag{12.83}$$

which is the first-order rate law.

Let x be the change in concentration of A (mols per unit volume), so that $\xi = xV$. If the initial concentration of A is $[A_0]$, and we make $x = 0$ at $t = 0$, then we have:

$$[A] = [A_0] - x \tag{12.84}$$

which allows us to re-rewrite equation (12.83) as:

$$\frac{dx}{x - [A_0]} = -kdt. \tag{12.85}$$

This integrates to:

$$x = [A_0]\left(1 - e^{-kt}\right) \tag{12.86}$$

or equivalently:

$$[A] = [A_0]e^{-kt}. \tag{12.87}$$

This is the first-order rate law in integral form, and is identical to the radioactive decay law (Section 2.9). Radioactive decay, although not a chemical reaction, is the archetypal example of a process that follows first-order kinetics. As for radioactive decay, we can define the *half life* of a chemical reaction, $\tau_{1/2}$, as the time required for the concentration of the reactant to decay to half of its initial value. Setting $[A] = 1/2[A_0]$ in (12.87) we find:

$$\tau_{1/2} = \frac{\ln 2}{k}. \tag{12.88}$$

An alternative estimate of the characteristic rate of a process, that is commonly used for chemical reactions, is the *reaction time scale*, which is the time required for the concentration of the reactants to become "exponentially close" to the equilibrium concentration. The definition of the reaction time scale τ, is given by:

$$[A_\tau] - [A_{eq}] = \frac{1}{e}\left([A_0] - [A_{eq}]\right), \tag{12.89}$$

where $[A_{eq}]$ is the equilibrium concentration of A and $[A_\tau]$ is its concentration at time τ. Equivalently:

$$x_\tau = x_{eq}\left(1 - \frac{1}{e}\right).\tag{12.90}$$

Substituting in (12.86), and noting that the equilibrium concentration is attained as $t \to \infty$, we find for a first-order elementary reaction:

$$\tau = \frac{1}{k}.\tag{12.91}$$

The definitions of reaction half life and reaction time scale are always the same, but the specific equations (12.88) and (12.91) are valid only for reactions that follow first-order kinetics.

When we consider higher order reactions the number of possible rate laws multiplies. Equation (12.83) is the only possible first-order rate law, but for a second-order reaction we have two possibilities: $A + B \to products$ and $2A \to products$. For third-order reactions there are three possibilities. As an example we will look at the second-order reaction $A + B \to products$, others are left as exercises. We can write the rate law for this reaction as follows:

$$\frac{d[A]}{dt} = -k[A][B]\tag{12.92}$$

or, in terms of the progress variable x and the initial concentrations $[A_0]$ and $[B_0]$:

$$\frac{dx}{(x - [A_0])([B_0] - x)} = -kdt.\tag{12.93}$$

The integral is messier than that for a first-order reaction, but, thanks to *Maple*:

$$\ln\left(\frac{([B_0] - x)}{[B_0]}\right) - \ln\left(\frac{([A_0] - x)}{[A_0]}\right) = -([A_0] - [B_0])kt\tag{12.94}$$

or, with a bit of rearrangement:

$$x = \frac{[A_0][B_0](1 - \vartheta)}{[B_0] - \vartheta[A_0]}\tag{12.95}$$

where:

$$\vartheta = e^{([A_0] - [B_0])kt}.\tag{12.96}$$

The result is more informative if we express it in terms of the concentrations of the reactants:

$$\frac{[A]}{[B]} = \vartheta\frac{[A_0]}{[B_0]}.\tag{12.97}$$

Equations 12.96 and 12.97 show that, as we should expect, the system becomes enriched in the reactant that is present in excess. Consider the limiting case in which $[B_0] \gg [A_0]$, so that even when all of reactant A has reacted it is $[B] \approx [B_0]$. Then equation (12.97) simplifies to:

$$[A] = [A_0]e^{-[B_0]kt}.\tag{12.98}$$

This is known as pseudo-first-order kinetics. It differs from true first-order behavior (equation (12.87)) in that the exponential factor is multiplied by the concentration of the abundant species, B. Calculation of reaction half life and time scale for second- and third-order reactions is left as an exercise.

12.3.4 Some simple composite reactions

We will look at two examples, sequential reactions and parallel reactions, of how the rate laws of elementary reactions combine to yield more complex behaviors. Consider first the case in which a reactant A goes to a product C with formation of an intermediate compound, B. For simplicity let all of the reactions be first-order:

$$A \rightarrow B, \quad \text{rate constant} = k_1$$
$$B \rightarrow C, \quad \text{rate constant} = k_2.$$

The concentrations of the three species vary as:

$$\frac{d[A]}{dt} = -k_1[A]$$
$$\frac{d[B]}{dt} = k_1[A] - k_2[B] \qquad (12.99)$$
$$\frac{d[C]}{dt} = k_2[B].$$

The behavior of this reaction depends on the stability of the intermediate species, B. This must be reflected in the relative values of the two rate constants. If B is relatively unstable then it is likely to be short-lived, which requires that k_2 be much larger than k_1, so that the concentration of B always stays close to zero. Conversely, if k_2 is much smaller than k_1 then B must be relatively long-lived and its concentration may build to the point where it becomes the dominant species in the system. We seek an equation for the change of $[B]$ with time.

Let us assume that the initial concentrations are $[A_0] \neq 0$ and $[B_0] = [C_0] = 0$. The concentration of A as a function of time is given by equation (12.87), with $k = k_1$. Substituting (12.87) in (12.99):

$$\frac{d[B]}{dt} = k_1[A_0]e^{-k_1 t} - k_2[B]. \qquad (12.100)$$

Solution of this differential equation by hand is not immediate, but *Maple*'s differential equation solver does it in a single step. The final result is:

$$[B] = [A_0]\frac{k_1}{k_2 - k_1}\left(e^{-k_1 t} - e^{-k_2 t}\right) \qquad (12.101)$$

from which we can also calculate:

$$\frac{d[B]}{dt} = [A_0]\frac{k_1}{k_2 - k_1}\left(-k_1 e^{-k_1 t} + k_2 e^{-k_2 t}\right). \qquad (12.102)$$

If $k_1 \gg k_2$ then these equations become, approximately:

$$[B] = [A_0] \left(e^{-k_2 t} - e^{-k_1 t} \right)$$

$$\frac{d[B]}{dt} = [A_0] \left(k_1 e^{-k_1 t} - k_2 e^{-k_2 t} \right) \tag{12.103}$$

whereas for $k_2 \gg k_1$ we have:

$$[B] = [A_0] \frac{k_1}{k_2} \left(e^{-k_1 t} - e^{-k_2 t} \right)$$

$$\frac{d[B]}{dt} = [A_0] \frac{k_1}{k_2} \left(k_2 e^{-k_2 t} - k_1 e^{-k_1 t} \right). \tag{12.104}$$

These two distinct behaviors are plotted in Fig. 12.4. If $k_1 \gg k_2$ then A quickly decays to B. The concentration of B builds up to a value comparable to $[A_0]$, and then decays as B reacts to C. The derivative $d[B]/dt$ (bottom panel) varies strongly and approaches zero only after a long time, when most of B has decayed. In this case B is a long-lived intermediate species. In contrast, if $k_2 \gg k_1$ $[B]$ reaches a maximum value that is only a small fraction of $[A_0]$ and then remains approximately constant. This is emphasized by the graph of $d[B]/dt$, which after the initial "build-up" period stays close to zero. B is in this case a short-lived, or relatively unstable, intermediate species, and its concentration after the initial "build-up" period can be considered to be approximately constant. This leads to the *steady-state approximation,* from which it is possible to calculate the behavior of the entire system in a relatively straightforward fashion, by converting the differential equations into algebraic equations. Setting $d[B]/dt = 0$ we get, from (12.99) and (12.87):

$$[B] = \frac{k_1}{k_2} [A_0] e^{-k_1 t} \tag{12.105}$$

and, from mass balance:

$$[C] = [A_0] - [A] - [B] = [A_0] \left(1 - \frac{k_1 e^{-k_1 t} - k_2 e^{-k_2 t}}{k_2} \right). \tag{12.106}$$

This approach yields a tremendously simplified solution. In this particular case, which consists of only two first-order reactions, the exact solution is easy to obtain, but this is not the case in general, and the steady state approximation, if it can be justified from the relative values of the decay constants, is a powerful shortcut.

A different case is that of parallel reactions, in which the same species follow different reaction pathways. These could be either the same reactants giving rise to different product assemblages, or different reactant assemblages generating the same products. Consider the case of the following two competing first-order reactions:

$$A \to B, \quad \text{rate constant} = k_B$$

$$A \to C, \quad \text{rate constant} = k_C.$$

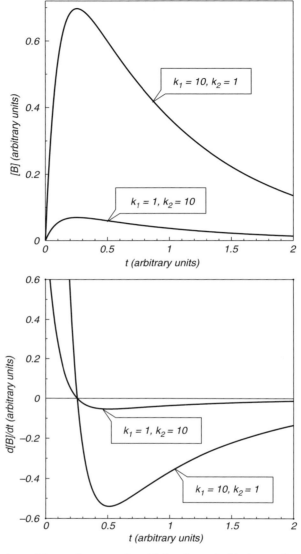

Fig. 12.4   Concentration (top) and rate of change of concentration with time (bottom) of the intermediate species, B, in a sequential reaction. If B is a short-lived species ($k_1 \ll k_2$) $d[B]/dt$ approaches zero after a relatively short initial build-up period and $[B]$ remains approximately constant after this. If B is a long-lived species ($k_1 \gg k_2$) $d[B]/dt$ does not approach zero until the reaction is almost complete and its concentration cannot be considered to remain constant.

We assume that $[A_0] \neq 0$, $[B_0] = [C_0] = 0$. The rate of decay of A is given by:

$$\frac{d[A]}{dt} = -k_B[A] - k_C[A] \qquad (12.107)$$

so that, from equation (12.87):

$$[A] = [A_0]e^{-(k_B + k_C)t} \qquad (12.108)$$

Because A is consumed by both reactions, the total rate constant is the sum of the individual rate constants. For product B we have:

$$\frac{d[B]}{dt} = k_B[A] = k_B[A_0]e^{-(k_B+k_C)t} \tag{12.109}$$

which integrates to:

$$[B] = [A_0]\frac{k_B}{k_B+k_C}\left[1 - e^{-(k_B+k_C)t}\right] \tag{12.110}$$

and similarly for C:

$$[C] = [A_0]\frac{k_C}{k_B+k_C}\left[1 - e^{-(k_B+k_C)t}\right]. \tag{12.111}$$

We see that the ratio between the concentrations of B and C is a constant, known as the *branching ratio*, and given by:

$$\frac{[B]}{[C]} = \frac{k_B}{k_C}. \tag{12.112}$$

As usual, the mathematics become considerably more complicated as soon as we consider anything but first-order kinetics. In many cases closed analytical solutions do not exist, and the problems must be solved numerically.

Worked Example 12.3   Ozone in planetary atmospheres

Ozone is a minor but essential component in the terrestrial stratosphere (Chapters 13 and 14), and is also found in trace amounts in the Martian atmosphere. Its atmospheric concentration is not in equilibrium with oxygen, as can be seen from the homogeneous gas phase reaction:

$$3O_2 \rightleftarrows 2O_3. \tag{12.113}$$

Atmospheric pressure in Earth at an elevation of 30 km is approximately 20 mbar (Chapter 13), so we can estimate a characteristic partial pressure of O_2 in the stratosphere of $\sim$4 mbar. The standard state Gibbs free energy change for reaction (12.113) at 298 K is $\Delta_r G^0_{1,298} = 326.4$ kJ, from which we calculate an equilibrium ozone partial pressure in the stratosphere of $\sim 6 \times 10^{-33}$ bar. Measured values are in the range 10^{-8}–10^{-7} bar, i.e., some 25 orders of magnitude higher. This huge ozone excess is the result of a dynamic process sustained by a constant supply of energy, which was first proposed by Chapman (1930a, b). In its simplest form the *Chapman cycle* begins with photodissociation of molecular oxygen according to:

$$O_2 + h\nu_{uv} \rightarrow O + O,$$

where $h\nu_{uv}$ represents an ultraviolet photon (Chapter 13). Atomic oxygen can be consumed by any of the following three reactions:

$$O + O \rightarrow O_2$$
$$O + O + M \rightarrow O_2 + M$$
$$O + O_2 + M \rightarrow O_3 + M,$$

where M is a collision partner, which in the terrestrial atmosphere is typically N_2 or O_2 (i.e. one of the dominant species). An intuitive argument based on the fact that $[M] \ggg [O]$ and $[O_2] \ggg [O]$ suggests that the dominant reaction is likely to be the third one. A rigorous analysis using the values of the corresponding rate constants confirms this expectation (see, for example, de Pater & Lissauer, 2001, p. 114). Ozone is produced by a sequential reaction with a short-lived intermediate product, atomic oxygen, and is in turn consumed by two reactions, photochemical dissociation:

$$O_3 + h\nu_{uv} \rightarrow O + O_2$$

and chemical recombination with atomic oxygen:

$$O + O_3 + M \rightarrow 2O_2 + M.$$

The simplest version of the Chapman cycle, in an atmosphere in which all components other than oxygen can be considered to be inert, can then be summarized as follows, where the symbols following the reactions are the corresponding rate constants:

$$\begin{aligned}
O_2 + h\nu &\rightarrow O + O, \quad j_1 \\
O + O_2 + M &\rightarrow O_3 + M, \quad k_2 \\
O_3 + h\nu &\rightarrow O + O_2, \quad j_3 \\
O + O_3 &\rightarrow O_2 + O_2, \quad k_4
\end{aligned} \tag{12.114}$$

The rate constants for the two photochemical reactions, symbolized by j_1 and j_3 for reasons that we discuss later, are functions of the solar energy flux and of the *absorption cross sections* of oxygen and ozone molecules at ultraviolet wavelengths (Section 12.4.2 and Chapter 13). They vary in a complex fashion with latitude, season, time of day and elevation. For this example we will take characteristic values for the terrestrial stratosphere, averaged over time, altitude and latitude: $j_1 \sim 10^{-12}\,\text{s}^{-1}$ and $j_3 \sim 10^{-4}\,\text{s}^{-1}$ (de Pater & Lissauer, 2001, p. 112). The other two reactions are thermally activated and follow Arrhenius-like behavior (Section 12.4.1). Characteristic values for the corresponding rate constants at stratospheric conditions are: $k_2 \sim 10^{-33}\,\text{cm}^6\,\text{molecule}^{-2}\,\text{s}^{-1}$ and $k_4 \sim 10^{-15}\,\text{cm}^3\,\text{molecule}^{-1}\,\text{s}^{-1}$. Note that the units of the rate constants are determined by the order of the reaction. For example, because reaction 2 is third order its concentration product has units of $[\text{molecules}]^3 \times [\text{volume}]^{-3}$, so that the rate constant must have units of $[\text{molecules}]^{-2} \times [\text{volume}]^2 \times [\text{time}]^{-1}$, in order to yield a reaction rate in the proper units: $[\text{molecules}] \times [\text{volume}]^{-1} \times [\text{time}]^{-1}$. The use of molecules (rather than mols) and cm (rather than m) is common in chemical kinetics.

In the Chapman cycle there are two species that are likely to have short half lives: atomic oxygen and ozone. We can expect this on thermodynamic grounds. As we saw, ozone is not stable relative to molecular oxygen, and a similar calculation shows the same to be true for atomic oxygen. We can then assume that the concentrations of these two species are in a steady state, i.e.:

$$\frac{d[O]}{dt} = 2j_1[O_2] - k_2[O][O_2][M] + j_3[O_3] - k_4[O][O_3] = 0 \tag{12.115}$$

and

$$\frac{d[O_3]}{dt} = k_2[O][O_2][M] - j_3[O_3] - k_4[O][O_3] = 0. \tag{12.116}$$

Adding the two equations and rearranging:

$$[O] = \frac{j_1}{k_4}\frac{[O_2]}{[O_3]} \tag{12.117}$$

substituting in (12.116) and rearranging:

$$[O_3]^2 + \frac{j_1}{j_3}[O_2][O_3] - \frac{j_1 k_2}{j_3 k_4}[O_2]^2[M] = 0. \tag{12.118}$$

For a pressure of 20 mbar we calculate from the ideal gas EOS that $[M] \approx 5 \times 10^{17}$ molecules cm^{-3}, so that $[O_2] \approx 10^{17}$ molecules cm^{-3}. Using the values for the rate constants listed above we also find:

$$\frac{k_2}{k_4}[M] \sim 0.5. \tag{12.119}$$

We therefore note that it must be:

$$\frac{\frac{j_1}{j_3}[O_2][O_3]}{\frac{j_1 k_2}{j_3 k_4}[O_2]^2[M]} = \frac{[O_3]}{\frac{k_2}{k_4}[O_2][M]} \ll 1 \tag{12.120}$$

so that we can drop the linear term in (12.118) and solve for $[O_3]$ as follows:

$$[O_3] \approx [O_2]\left(\frac{j_1 k_2}{j_3 k_4}[M]\right)^{1/2}. \tag{12.121}$$

Substituting numerical values we find that $[O_3] \approx 7 \times 10^{12}$ molecules cm^{-3}, which corresponds to an ozone partial pressure of $\sim 3 \times 10^{-7}$ bar.

This calculation yields an ozone concentration that is close to the measured value, but not quite right. It overestimates the actual atmospheric concentration by up to one order of magnitude. The reason for this discrepancy is that other species that are present in the terrestrial atmosphere are not inert with respect to ozone destruction, but they do not affect ozone formation, which relies exclusively on the photochemical dissociation of oxygen, i.e. the first reaction in (12.114). Three such species are the radicals OH and NO, and atomic Cl. These species form by photodissociation of H_2O, N_2O and halomethanes such as CCl_2F_2, respectively. The latter are exclusively of anthropogenic origin, whereas nitrogen oxide has both natural and anthropogenic sources (e.g., jet engine exhaust). Each of these species gives rise to an ozone-consuming sequence of reactions, that can be written as follows:

$$OH + O_3 \rightarrow HO_2 + O_2$$
$$HO_2 + O \rightarrow OH + O_2$$

and:

$$NO + O_3 \rightarrow NO_2 + O_2$$
$$NO_2 + O \rightarrow NO + O_2$$

and:

$$Cl + O_3 \rightarrow ClO + O_2$$
$$ClO + O \rightarrow Cl + O_2.$$

The net result of all three reaction sequences is the same, namely:

$$O + O_3 \rightarrow 2O_2$$

with regeneration of the active species. For this last reason these are known as *catalytic cycles*. If we use X to designate a catalyst in general then all of these reactions correspond to:

$$X + O_3 \rightarrow XO + O_2, \quad k_5$$
$$XO + O \rightarrow X + O_2, \quad k_6. \tag{12.122}$$

We can modify (12.115) and (12.116) as follows:

$$\frac{d[O]}{dt} = 2j_1[O_2] - k_2[O][O_2][M] + j_3[O_3] - k_4[O][O_3] - k_6[XO][O] = 0 \tag{12.123}$$

and

$$\frac{d[O_3]}{dt} = k_2[O][O_2][M] - j_3[O_3] - k_4[O][O_3] - k_5[X][O_3] = 0 \tag{12.124}$$

and add a rate law for any one of the two X-bearing species (they don't vary independently), for example:

$$\frac{d[XO]}{dt} = k_5[X][O_3] - k_6[XO][O] = 0. \tag{12.125}$$

Eliminating [O] between (12.123), (12.124) and (12.125) we get a quadratic equation in $[O_3]$:

$$[O_3]^2 + \left(\frac{k_2 k_5}{j_3 k_4}[O_2][M][X] + \frac{j_1}{j_3}[O_2]\right)[O_3] - \frac{j_1 k_2}{j_3 k_4}[O_2]^2[M] = 0. \tag{12.126}$$

Comparing the two contributions to the coefficient of the linear term, and using the same values for the rate constants and concentration of the collision partner as before, we find:

$$\frac{\frac{k_2 k_5}{j_3 k_4}[O_2][M][X]}{\frac{j_1}{j_3}[O_2]} = \frac{j_1 k_4}{k_2[M]} \frac{1}{k_5[X]} \sim \frac{10^{-12}}{k_5[X]} \ll 1, \tag{12.127}$$

where the last inequality is true unless the rate constant k_5 is very small – this will be justified further *a posteriori*. We can then simplify (12.126) as follows:

$$[O_3]^2 + \frac{k_2 k_5}{j_3 k_4}[O_2][M][X][O_3] - \frac{j_1 k_2}{j_3 k_4}[O_2]^2[M] = 0 \tag{12.128}$$

which, using the quadratic formula and noting that the only physical root is the positive one, yields:

$$[O_3] = \left[\left(\frac{k_2 k_5}{2 j_3 k_4}[O_2][M][X]\right)^2 + \frac{j_1 k_2}{j_3 k_4}[O_2]^2[M]\right]^{1/2} - \frac{k_2 k_5}{2 j_3 k_4}[O_2][M][X]. \tag{12.129}$$

Comparing the second term in the square root in (12.129) to (12.121) we see that this is the square of the ozone concentration in the absence of an active species, which we can call $[O_3]_0$, so that we have:

$$[O_3] = \left[\left(\frac{k_2 k_5}{2 j_3 k_4}[O_2][M][X]\right)^2 + ([O_3]_0)^2\right]^{1/2} - \frac{k_2 k_5}{2 j_3 k_4}[O_2][M][X] \tag{12.130}$$

or:

$$\frac{[O_3]}{[O_3]_0} = \left[\left(\frac{k_2 k_5}{2 j_3 k_4}\frac{[O_2][M][X]}{[O_3]_0}\right)^2 + 1\right]^{1/2} - \frac{k_2 k_5}{2 j_3 k_4}\frac{[O_2][M][X]}{[O_3]_0}. \tag{12.131}$$

The ratio is equal to one for $[X] = 0$, as we should expect, and it is less than one for any value of $[X] > 0$. Thus, equation (12.131) confirms a decrease in the ozone concentration when active species are added to the atmosphere. This behavior is general, but in order to study it in more detail it is necessary to substitute numerical values. Using once more the rate constants and gas concentrations in the terrestrial stratosphere discussed earlier we find:

$$\frac{k_2}{2 j_3 k_4}[O_2][M] \sim 10^{20} \tag{12.132}$$

so that, for these particular parameter values, (12.131) becomes:

$$\frac{[O_3]}{[O_3]_0} = \left[\left(k_5 \times 10^{20}\frac{[X]}{[O_3]_0}\right)^2 + 1\right]^{1/2} - k_5 \times 10^{20}\frac{[X]}{[O_3]_0}. \tag{12.133}$$

The behavior of this function is shown in Fig. 12.5, for values of the rate constant k_5 ranging from 10^{-8} to 10^{-14} cm^3 molecule^{-1} s^{-1}, which ensure the validity of (12.127). We see that a sufficiently small concentration of the active catalytic species has a negligible effect on ozone concentration, but that there is a threshold value beyond which ozone concentration becomes quite sensitive to increased concentration of the active species and declines steeply. The value of this concentration threshold depends on the magnitude of the rate constant k_5. For example, k_5 for the Cl-initiated cycle is of the order of 10^{-11} cm^3 molecule^{-1} s^{-1}, which means that significant ozone depletion begins when the concentration of Cl atoms is as low as $\sim 10^{-10}$ times the initial ozone concentration, and that a Cl concentration of only 10^{-8} $[O_3]_0$ is sufficient to virtually deplete stratospheric ozone. The ozone-depleting catalytic

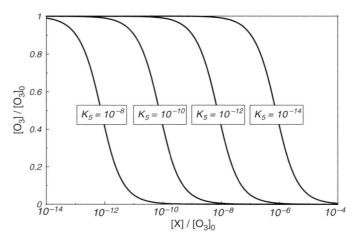

Fig. 12.5   Atmospheric ozone depletion caused by a catalytic cycle initiated by an active species X (X could be, for instance, OH, NO or Cl). Numbers in boxes are values of the rate constant k_5, for the reaction: $X + O_3 \rightarrow XO + O_2$, in cm^3 molecule^{-1} s^{-1}. $[O_3]_0$ is the concentration of ozone in the absence of species X. Note that minuscule concentrations of the active species X can wipe-out atmospheric ozone, and that there is a threshold value of [X] above which ozone destruction increases rapidly.

cycles are in reality considerably more complex that the examples that we considered here but the behavior depicted in Fig. 12.5 is qualitatively correct.

It is important to point out that the laws of thermodynamics are not "violated" by the fact that ozone concentration in the atmosphere is not at equilibrium with that of oxygen. This non-equilibrium ozone concentration is kept at a (more or less) constant level by a continuous supply of energy, in the form of solar photons that split oxygen atoms in the first reaction of the Chapman cycle (equation (12.114)). Production of ozone, and its persistence in a concentration far greater than the equilibrium concentration, are dynamic processes that depend on the continuous operation of this photochemical reaction. If ultraviolet radiation were to be turned off permanently then ozone, and atomic oxygen, would convert to the stable species O_2, at rates dependent on the specific reaction mechanism.

The two first reactions of the Chapman cycle can be combined to yield reaction (12.113). The enthalpy change for this reaction is 285.4 kJ. We can calculate the total energy of the photons consumed by this reaction with the formula:

$$U_v = \frac{hcA}{\lambda}, \tag{12.134}$$

where h is Planck's constant, c is the speed of light, A is Avogadro's number, and λ is the wavelength of the photon. It is found experimentally that photodissociation of O_2 occurs for photons of $\lambda < 240$ nm (see also Chapter 13), from which we calculate $U_v > 498$ kJ mol^{-1}. There is more than enough energy to account for the higher enthalpy of ozone relative to oxygen. Where does the energy excess go? Because of conservation of momentum, the photons' momentum $(= h\nu/c)$ must be transferred to the oxygen atoms, increasing their translational kinetic energy. This energy influx is then distributed throughout the gas by molecular collisions, increasing its temperature and therefore its entropy. We return to this discussion in the next chapter.

12.4 Controls on rate constants

Much of this chapter has focused on dynamic processes, by which we mean that the thermodynamic state of the system changes with time. Rates are quantified by means of "constants", such as chemical diffusivities and reaction rate constants. But what determines the values of these "constants"? We shall briefly discuss two important classes of processes: thermally activated processes and photoactivated processes.

12.4.1 Thermally activated processes. Arrhenius law

The rates of many natural processes, such as chemical diffusion, viscous flow (which we discussed in Chapter 3 and which, as we saw there, corresponds to diffusion of momentum) and thermally activated chemical reactions, vary exponentially with temperature according to:

$$\Psi = A \exp\left(-\frac{E_a}{RT}\right), \tag{12.135}$$

where A and E_a are positive constants and Ψ is a parameter that describes the characteristic rate of the process. For example, Ψ can be the chemical diffusion coefficient D (Section 12.2.2), the rate constant of a chemical reaction, k (Section 12.3.1), or the viscous flow shear rate du/dx (Section 3.7.4). In the latter case the parameter Ψ is the rate of shearing, which at constant shear stress is the inverse of the viscosity μ (equation (3.7)). Because viscosity, rather than its inverse, is the parameter that is used to measure the rate of momentum transfer the negative sign does not appear in the exponential factor of equation (3.3.1). Otherwise that equation is identical to (12.135).

Equation (12.135) is known as *Arrhenius law*, after the Swedish physical chemist Svante Arrhenius, who discovered it experimentally. If we think of Arrhenius law as an empirical macroscopic relationship then it is not necessary to attach any physical meaning to the constants A and E_a. We can nonetheless note that E_a has units of energy, and it is a quantity known as the *activation energy* for the process in question. The units of A, known as the pre-exponential factor, vary depending on what the parameter Ψ represents, but they always include the factor $[T]^{-1}$. From a purely macroscopic point of view we can think of A as the limiting value of Ψ at very high temperature. We can, however, attach more specific physical meanings to the two constants, which relate to the microscopic nature of the processes in question. In this view A is a quantity that is proportional to the rate at which microscopic events take place. The exponential factor $\exp(-E_a/RT)$ is the probability that a given event will produce a specific outcome. For example, if Ψ corresponds to the rate constant of a chemical reaction then the events would be molecular collisions, and the exponential factor would describe the probability that a given collision has enough energy to break atomic bonds and result in chemical recombination. If Ψ describes diffusion of matter or momentum then the events would be excursions of atoms away from their equilibrium positions and the exponential factor would be the probability that a given excursion is energetic enough to cause an atom to jump from its initial equilibrium position to a neighboring one. In every case the probability tends to zero as temperature goes to zero. Then no events have enough energy to accomplish the desired result, and the macroscopic process stops. The probability tends to one at high temperature, so that all events accomplish the desired result and the rate of the macroscopic process approaches some maximum, but finite, value.

Arrhenius law is a reasonably good first approximation to the behavior of thermally activated processes. Chemical diffusivities and reaction rate constants are commonly tabulated in the form of values of the A and E_a parameters in equation (12.135). An example is the large database of reaction rate constants, from which the values used in Worked Example 12.3 were obtained, maintained on-line by the U.S. National Institute of Science and Technology (NIST Chemical Kinetics Database). A compilation of chemical diffusivity parameters of geological interest (used in Worked Example 12.1) can be found in Zhang (2008).

Equation (12.135) is not a complete physical picture, however. For example, it is found that the pre-exponential factor is generally a function of temperature, and pressure dependencies of the pre-exponential factor and the activation energy also exist. These macroscopic effects point to a microscopic mechanism that is more complex than what I describe here. Rigorous discussions of these topics can be found, for example, in the books by Logan (1996); Houston (2006); Borg and Dienes (1988), and with specific focus on geological processes, Poirier (1985) and Zhang (2008).

12.4.2 Photochemical processes

Thermal energy is incapable of initiating a chemical reaction if the resulting kinetic energy of the molecules is not sufficient to break interatomic bonds. Even at low temperature, however, individual molecules may acquire excess energy and become reactive by absorbing photons of specific wavelengths (see also Chapter 13). An example of this is the initiation step of the Chapman cycle. At the temperature of the stratosphere, and even at temperatures considerably higher than this, O_2 molecules do not have sufficient kinetic energy to dissociate as a result of collisions, but a molecule of O_2, regardless of its temperature, dissociates when it absorbs an ultraviolet photon. This is not a thermal process, which means that it does not rely on a statistical distribution of molecular speeds and on the probability that the speeds of some molecules will exceed some characteristic threshold. The rate constants for photochemical reactions are therefore not described by equation (12.135). Partly for this reason it is customary to use a different symbol for the rate constants (j instead of k), even if the rate laws are the same regardless of the nature of the activation process.

Photochemical reactions are activated by radiation within a specific range of wavelengths. For example, photodissociation of O_2 (the first reaction in the Chapman cycle) requires ultraviolet photons with wavelengths shorter than 240 nm. There are reactions, called radiolytic reactions, that require higher energy photons, in the X-ray and gamma part of the spectrum. We can expect that the rate constant will vary directly with the intensity of the radiation of the required wavelength (number of photons per unit of area per unit of time, i.e., photon flux) and with the absorption cross section, which we can think of as the effective target area offered by the molecules to the photon flux (more on this in Chapter 13). We write this relation as follows:

$$j = \int I(\lambda)\sigma(\lambda)d\lambda, \qquad (12.136)$$

where $I(\lambda)$ is the photon flux at wavelength λ, $\sigma(\lambda)$ is the absorption cross section, which also varies with wavelength, and the integral is over the range of wavelengths within which photoactivation takes place.

This equation is appropriate, for example, to photoactivation of a chemical reaction in a laboratory environment. Photochemical reactions in planetary atmospheres are more complicated, because the photon flux varies with elevation. It is maximum, and equal to the incident solar flux, at the top of the atmosphere, and then decreases as photons are absorbed by atmospheric gases. If, for the sake of simplicity, we consider only a hypothetical situation in which solar radiation is always perpendicular to the planet's surface (e.g. noon at the equator on the day of the equinox), then we have, by Beer–Lambert's law of radiation (Section 13.2.4):

$$I(\lambda) = I_0(\lambda)e^{-\tau_\lambda}, \tag{12.137}$$

where $I_0(\lambda)$ is the photon flux at the top of the atmosphere and τ_λ is a non-dimensional absorption length, known as the optical thickness (Section 13.2.4), and given by (see also equation (13.55)):

$$\tau_\lambda = \sum_i \sigma_i(\lambda) \int N_i(z)dz, \tag{12.138}$$

where σ_i is the absorption cross section of chemical species i, N_i is the number density of species i (molecules per unit volume), the integral is from the top of the atmosphere to the elevation of interest, and the summation is over all atmospheric species that absorb at wavelength λ. If the incident solar radiation is not perpendicular to the planet's surface then a geometric correction must be applied to Beer–Lambert's law.

Substitution of (12.138) in (12.137), and then in (12.136), yields the value of the rate constant as a function of elevation:

$$j_a(z) = \int I_0(\lambda)e^{-\sum_i \sigma_i(\lambda)\int N_i(z)dz}\sigma_a(\lambda)d\lambda, \tag{12.139}$$

where σ_a is the absorption cross section of the photoactivated molecule of interest. As we should expect, this equation states that the rate constant decreases with increasing optical thickness, i.e. as the planet's surface is approached. Photochemistry is therefore a more important process in the stratosphere than in the troposphere, for the simple reason that there is a higher photon flux. The rate constants will obviously also vary with time of day and season, as the angle of the incident solar radiation and hence the photon flux will vary. Rate constants for atmospheric photochemistry vary over many orders of magnitude across a planetary atmosphere. The values of j_1 and j_3 used in Worked Example 12.3 are approximate average values for the terrestrial stratosphere at an elevation of $\sim$30 km. Since the rate of production and destruction of ozone is critically dependent on these rate constants, it follows that the numerical results obtained there change with elevation.

12.5 An introduction to kinetics of heterogeneous processes

The kinetic behavior of homogeneous processes, such as chemical reaction in a homogeneous phase or diffusion in a single continuous phase, can be set-up mathematically in a relatively straightforward and unambiguous fashion – which of course does not mean that the resulting equations are necessarily easy to solve. The mathematical description of the kinetics of heterogeneous processes is less straightforward. It commonly involves sets of coupled differential equations that describe rates of chemical reaction, heat transfer and mass transfer. Simplifying assumptions are sometimes possible, though. Consider the following three simple examples.

(i) Crystallization of enstatite from a basaltic melt, which we can schematize as follows:

$$MgO_{melt} + SiO_{2\ melt} \rightleftharpoons MgSiO_{3\ crystal}.$$

(ii) Devolatilization of the assemblage magnesite + quartz:

$$MgCO_{3\ crystal} + SiO_{2\ crystal} \rightleftharpoons MgSiO_{3\ crystal} + CO_{2\ vapor}.$$

(iii) Weathering of enstatite at the Earth's suface:

$$MgSiO_{3\ crystal} + 3H_2O \rightleftharpoons Mg_{aq}^{2+} + H_4SiO_{4\ aq} + 2OH^-.$$

One way of analyzing the kinetic behavior of heterogeneous reactions such as these is to assume that the locus of the chemical reaction is a homogeneous region of infinitesimal extent, which we may call the reaction volume or reaction interface (Zhang, 2008). The rate of the chemical reaction is then a function of its rate constant and of the affinity of the reaction inside the infinitesimal reaction volume (equation (12.74)). The rate constants of all three of these reactions are functions of temperature (equation (12.135)), and the affinity, which is a linear combination of chemical potentials, varies with temperature, pressure and composition. As the reactions proceed enthalpy is liberated or absorbed, and chemical species are consumed and produced. The temperature and affinity in the reaction volume are therefore determined by the relative rates of chemical reaction and heat and mass exchanges between the reaction volume and its environment.

Crystallization of enstatite from a melt (reaction i) liberates enthalpy ($\Delta_r H < 0$) and consumes MgO and SiO$_2$ melt components. If the reaction volume were a closed system then its temperature would increase and the concentrations of the reactants would decrease. From the definition of affinity, equation (12.70), we can see that:

$$\frac{\partial \mathscr{E}}{\partial T} = \Delta_r S = \frac{\Delta_r H}{T} \tag{12.140}$$

and:

$$\frac{\partial \mathscr{E}}{\partial X_i} = \frac{\nu_i RT}{X_i} > 0, \quad i : reactant. \tag{12.141}$$

In a closed reaction volume the affinity would decrease and hence the reaction would slow down, and eventually stop. There would also be an effect on the rate constant that would act in the opposite direction, but this is likely to be minor at magmatic temperatures, at which chemical reaction rates are always fast. In reality, however, crystallization sets up gradients in temperature and chemical potential between the reaction volume and the surrounding melt, so that heat is carried away from the reaction volume and chemical components are transported towards it. If we assume that diffusion is the transport mechanism for both heat and chemical components then we can conclude that mass transfer is the rate-limiting process, as heat diffusivity at magmatic temperatures is ~6 orders of magnitude greater than chemical diffusivity (see Worked Examples 12.1 and 12.2). As a first approximation, then, the kinetics of reaction (i) could be treated as a chemical diffusion problem.

Consider now reaction (ii). This reaction is endothermic ($\Delta_r H > 0$) so that, if it took place in a closed system, the temperature would decrease and the fugacity of CO_2 would increase. If the gas phase is a product then differentiation of (12.70) yields:

$$\frac{\partial \mathscr{E}}{\partial f_j} = \frac{v_j RT}{f_j} < 0, \quad j : product \tag{12.142}$$

so that it follows from (12.140) and (12.142) that the affinity in a closed reaction volume would go down. Let us assume for the sake of argument that the decarbonation reaction (ii) occurs in a permeable setting, in which, for example, a network of cracks allows the gas phase to escape as soon as it forms. If, as in the previous case, the temperature is high enough that the reaction rate is much faster than heat diffusion, then we can consider two end-member situations. If grain size is "infinitesimally fine" and the mixture of minute magnesite and quartz crystals is perfectly random then we might be able to ignore mass transfer rates and assume that the rate-limiting process is heat diffusion. For any "geologically reasonable" grain size, however, it is likely that mass transfer will determine the rate of reaction (ii), but there is an important qualitative difference with the crystallization reaction (i). In that case diffusion of Mg and Si cations in the melt phase does not present a conceptual problem, but in the case of the solid-state reaction (ii) this is not so. How do SiO_2 and $CaCO_3$ components migrate to the reaction volume where enstatite and CO_2 are produced? Does this happen by solid-state diffusion? If so, which are the chemical species that actually diffuse, and what is the underlying atomic lattice through which they diffuse? Or does mass transfer take place by dissolution of the reactants in a fluid phase, for example the CO_2 produced by the reaction itself, and migration of the fluid phase? The rate-determining process may be difficult to pin down, and even more difficult to quantify.

In the case of the weathering reaction (iii) we may assume that circulation of meteoric water, i.e. advection, is the mechanism that accounts for both heat and mass transfer. Given that advection is much faster than diffusion, and that from Arrhenius law (equation (12.135)) we can expect a small value of the rate constant for the chemical reaction at room temperature, we may infer that weathering rates are likely to be controlled by chemical reaction rates. The chief difficulty in quantifying the process in this case would be in the experimental determination of the required rate constants.

Worked Example 12.4   Relationships between affinity and progress variable

The attentive reader should have noticed that the statement that I made regarding reactions (i) and (ii), to the effect that in a closed system the affinity of a reaction decreases as the reaction proceeds, is always true. In fact, it is an alternative way of stating the Second Law of Thermodynamics, which we can summarize as follows:

$$\frac{\partial \mathscr{E}}{\partial \xi} \leq 0. \tag{12.143}$$

The inequality holds true in a closed system – it simply states the fact that in a spontaneous chemical reaction the Gibbs free energy of the products is lower than that of the reactants. In order for the equality to be true the rates of heat and mass transfer between the reaction volume and its environment must exactly match the rates of change of enthalpy and chemical species concentrations caused by the reaction. Assume for simplicity that a chemical reaction occurs in the linear regime defined by equation (12.80) (i.e. close to equilibrium, see also Exercise 12.2). We can re-write this equation as follows:

$$\frac{\partial \xi}{\partial t} = k_1 \mathscr{E}, \tag{12.144}$$

where k_1 is a positive constant. To begin with, we note that (12.144) says that a chemical reaction can only take place beginning from a non-equilibrium condition, as the reaction rate vanishes for $\mathscr{E} = 0$. This means that some degree of *overstepping* is always required in order for reactions to occur in nature. For example, melting can only begin at a temperature higher than the solidus, and crystallization cannot begin until temperature drops below the liquidus. According to equation (12.144), the amount by which the reaction is overstepped (i.e. the value of $\mathscr{E}$) will determine the rate at which melting or crystallization proceeds. We can now look at different ways in which $\mathscr{E}$ may vary with the progress variable ξ.

Assume first that $\mathscr{E}$ is a constant. This could happen, for instance, during crystallization of a well-stirred one-component system (e.g. ice from pure H_2O), or during crystallization from a very large reservoir of low-viscosity liquid (e.g. ocean water). Defining the value of the progress variable at $t = 0$ as $\xi_0 = 0$ we have:

$$\xi = k_1 \mathscr{E} t, \tag{12.145}$$

which states that the reaction will proceed indefinitely. The behavior becomes more interesting if we consider the inequality in (12.143). This means that either heat or mass transfer are not able to keep up with the chemical reaction rate. The exact functional form of the derivative $\partial \mathscr{E} / \partial \xi$ may be difficult to determine, but we can choose the simplest possible function as an example. This is the linear law:

$$\frac{\partial \mathscr{E}}{\partial \xi} = -k_2, \tag{12.146}$$

where k_2 is another positive constant. This constant must contain information about the efficiency of the rate controlling process. For instance, if the rate is controlled by chemical diffusion then k_2 must vary inversely with diffusivity: the lower the diffusivity, the more the affinity will decrease with reaction progress, and thus the larger the magnitude of k_2 must be. Let the affinity at $\xi_0 = 0$ be $\mathscr{E}_0$. Then:

$$\mathscr{E} = \mathscr{E}_0 - k_2 \xi \tag{12.147}$$

so that, substituting in (12.144) and integrating:

$$\xi = \frac{\mathscr{E}_0}{k_2} \left(1 - e^{-k_1 k_2 t} \right). \tag{12.148}$$

We now note that the reaction goes to completion as $t \to \infty$ and that, because k_1 and k_2 are positive constants:

$$\lim_{t \to \infty} \xi = \frac{\mathscr{E}_0}{k_2}. \tag{12.149}$$

Say that the process that we are modeling is crystallization from a silicate melt and that we assume, rather simplistically, that the magnitude of the progress variable ξ represents crystal size. Chemical diffusivity in melts varies inversely with viscosity, so that k_2 varies directly with viscosity. Equation (12.149) then states that crystals that grow from a viscous melt (e.g. a water-poor rhyolite) should be smaller than those that grow from a fluid melt (e.g., a water-saturated pegmatite). Although during magmatic crystallization there are additional complications that we have ignored, most notably the generation and availability of nucleation sites, equation (12.149) provides at least a qualitative kinetic explanation for an observation that we became familiar with during our introductory physical geology course.

Exercises for Chapter 12

12.1 Derive the full set of interdiffusion equations, including the Onsager reciprocal terms, for a ternary system consisting of two dilute solutes and a solvent (Section 12.2.3).

12.2 Find a linear approximation to (12.74) by writing $r_f = r_{f,eq} + \delta r_f$, where $r_{f,eq}$ is the rate of the forward reaction at equilibrium, by lineariting the exponential function for $\frac{\mathscr{E}}{RT} \ll 1$. Use your linearized expression to find a relationship between the phenomenological coefficient L and r_f. Establish a criterion by which a reaction can be considered to be close to equilibrium.

12.3 Write differential and integral rate laws for the elementary reactions:
   (i) $2A \rightarrow products$
   (ii) $3A \rightarrow products$
   (iii) $2A + B \rightarrow products$
   (iv) $A + B + C \rightarrow products$.
   (Hint: use *Maple* wherever possible.)

12.4 Find expressions for the half lifes and time scales of the elementary reactions:
   (i) $2A \rightarrow products$
   (ii) $A + B \rightarrow products$
   (iii) $2A + B \rightarrow products$.
   For (ii) and (iii) consider three cases: $[A_0] = [B_0]$, $[A_0] >> [B_0]$ and $[A_0] << [B_0]$.

12.5 Compare the exact solution for the concentration of a short-lived species $[B]$ (equation (12.104)) with the steady state solution (12.105). Establish a criterion for how much time must be allowed to elapse for the steady state solution to become an acceptable approximation. Also, describe some conditions under which the steady state solution is never an acceptable approximation.

12.6 Estimate what would be the half life of ozone in the terrestrial atmosphere if ultraviolet radiation were to stop suddenly. Comment on likely night vs. day variations in ozone concentration.

13

Topics in atmospheric thermodynamics and radiative energy transfer

This chapter seeks to answer three basic questions about the physical nature of planetary atmospheres. First, why do some planets have atmospheres and others do not? Second, what controls the mass and the thickness of a planetary atmosphere? Third, how is a planet's surface temperature affected by the existence of an atmosphere? Some of the answers that we find here will be the starting point for our discussion, in the last chapter, of the chemical evolution of atmospheres in the terrestrial planets and the possible relationships between atmospheric composition and the origin of life.

13.1 Gravitational binding of planetary atmospheres

Whether or not a planet has an atmosphere depends on the relationship between the gravitational potential at the planet's surface and the translational kinetic energy of gas molecules. The latter is of order $k_B T$ (Section 1.14), whereas the gravitational potential at the planet's surface can be written as $\Phi = 4/3\pi G r^2 \rho$ (from equation (1.8)). Recalling that Φ is gravitational potential energy per unit mass, we take the product Φu, where $u = 1.66054 \times 10^{-27}$ kg is the atomic mass unit, and form the non-dimensional ratio:

$$\Xi = \frac{3k_B T}{4\pi G r^2 \rho u}. \tag{13.1}$$

The parameter Ξ is an estimate of the tendency of a planetary atmosphere to escape. For reasons that we will see later the absolute value of Ξ does not have a rigorous interpretation, but the relative value among different planets is a good indicator of whether or not a planet is likely to have an atmosphere, and, to some extent, of what type of volatile species may be most abundant. The smaller the value of Ξ, the more strongly bound an atmosphere is, and the more likely it is to contain a large proportion of low molecular weight species.

Calculated values of Ξ for planets and major moons are shown in Fig. 13.1, arranged in order of increasing value of Ξ. The value of T used in this graph is the temperature at the planet's surface (or its equilibrium temperature, Worked Example 13.1). As we shall see this is not the best choice, but it is the simplest one. Regardless, we note that, with one exception, the parameter Ξ discriminates correctly between bodies with "substantial" atmospheres (to the left of Triton) and those without one (to the right of Triton). Triton's faint atmosphere marks the boundary. The exception is Pluto, which is thought to have an atmosphere comparable to that of Triton, but given the non-trivial uncertainties in Pluto's physical parameters the significance of this is not clear – it will likely remain unresolved until the *New Horizons* spacecraft arrives in the Pluto–Charon system in 2015. Ignoring this exception, equation (13.1) shows, for instance, that the reason why Titan has an atmosphere but none of the Galilean satellites of Jupiter does is the fact that they lie closer to the

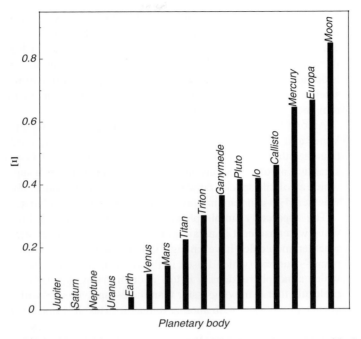

Fig. 13.1 Value of the non-dimensional parameter Ξ (equation (13.1)) for the planets and large moons of the Solar System. Triton marks the boundary between bodies with atmospheres (to the left) and airless bodies (to the right), the one outlier being Pluto.

Sun. Interestingly, the plot suggests that Europa is as unlikely to be able to hold on to an atmosphere as Mercury, and that the Moon is even more hopeless in this respect.

Focusing on the left side of the plot, we note that the four giant planets have very small Ξ values, ranging from 0.0007 to 0.003. This means that their atmospheres are bound very tightly, but it also means something else. In equation (13.1) I used a constant mass, u, to obtain the non-dimensional parameter Ξ, but the masses of different gas species, and hence their gravitational binding energies, differ. For instance, the gravitational potential energy of an H_2 molecule (mass $= 2$) is more than twenty times smaller than that of a CO_2 molecule (mass $= 44$). At a given temperature, however, their kinetic energies are the same, so that the hydrogen molecule is much more likely to escape than the carbon dioxide molecule. We can also think of this in terms of velocities: the gravitational potential determines the planet's escape velocity (Section 2.4.1), whereas molecular speeds are a function of temperature and molecular mass (Section 1.14). Molecular speeds follow a statistical distribution with long "high speed" tails (Fig. 1.12), so that it is generally the case that only some fraction of the total number of gas molecules in a planetary atmosphere have speeds greater than the planet's escape velocity. The rate at which gases can escape a body's gravitational attraction varies directly with the fraction of molecules that have speeds that exceed the escape velocity, and at a given temperature this fraction is greater for light species (e.g. H_2) than for heavy ones (e.g. CO_2). We can then (qualitatively) interpret a very small value of the parameter Ξ as signifying that even light molecules will generally be able to remain gravitationally bound to the planet, because only a very small portion of the high velocity tail represents molecules with enough velocity to exceed the escape velocity. The giant planets can therefore be expected to be rich in the lightest volatiles, hydrogen and helium, which is indeed the case. By comparison, Titan has a much higher Ξ value, so that even though its temperature, given by its solar distance, is similar to that of Saturn, its atmosphere is dominated by the much heavier species N_2. The value of the parameter for the Galilean satellites of Jupiter is higher still, consistent with the fact that not even N_2 molecules have remained gravitationally bound.

Among the three large terrestrial planets Earth has a significantly lower Ξ value than Venus and Mars. This is part of the explanation for why Earth is so much richer in water than the other two planets. However, the fact that Europa, Ganymede, Callisto and even the small icy moons of Saturn are richer in water than Earth suggests that the temperature at which a volatile species condenses and/or freezes is also an important factor in determining whether H_2O can remain gravitationally bound to the planet.

The apparently simple physical picture suggested by Fig. 13.1 becomes complicated when we consider it in detail. For instance, the characteristic temperature that determines whether molecules have enough kinetic energy to escape is not the temperature at the planet's surface (which I used in the figure), but rather the temperature at an elevation at which atmospheric density is low enough that molecular collisions are unlikely, so that a molecule that is moving faster than the escape velocity will indeed escape before it has a chance of colliding with another molecule. The elevation where this becomes true, and the temperature at that elevation, are not as simply defined as the surface temperature (or the equilibrium temperature, Worked Example 13.1). Also, the escape mechanism on which equation (13.1) is based, which is known as thermal escape, is not the only way in which a planet can lose its atmosphere, and perhaps not even the dominant way. Further discussion of these topics is beyond the scope of this book but clear mathematical descriptions, including rigorous treatments of thermal and other escape mechanisms, can be found, for example, in Hunten (1973), Chamberlain and Hunten (1987), Zahnle and Kasting (1986), Zahnle *et al.* (1990) and Bohren and Albrecht (1998).

Should we expect the mass of a planet's atmosphere to be related in some simple fashion to the parameter Ξ? Atmospheric pressure at a planet's surface is given by the weight of the gas column. Calling the atmospheric mass density per unit of area m_a, the mean atmospheric pressure at the planet's surface, P_0, is given by:

$$P_0 = m_a g, \tag{13.2}$$

where g is the gravitational acceleration at the planet's surface, assumed to remain constant throughout the atmospheric thickness. Substituting numerical values of P_0 and g for the five solid planetary bodies with substantial atmospheres we find that atmospheric mass densities vary by six orders of magnitude, as follows:

$$\text{Venus} \sim 1.1 \times 10^6 \text{ kg m}^{-2}$$

$$\text{Titan} \sim 1.1 \times 10^5 \text{ kg m}^{-2}$$

$$\text{Earth} \sim 1.0 \times 10^4 \text{ kg m}^{-2}$$

$$\text{Mars} \sim 1.7 \times 10^2 \text{ kg m}^{-2}$$

$$\text{Triton} \sim 2.1 \times 10^0 \text{ kg m}^{-2}$$

Comparing these values with Fig. 13.1 reveals no correlation between atmospheric mass and Ξ. What determines atmospheric mass, then? We begin by considering planetary accretion, the bulk of which must have taken place while the growing solid bodies were immersed in the solar nebula. Rocky planets are likely to have had primordial atmospheres of solar composition, i.e. dominated by hydrogen and helium. These primordial atmospheres persist in the giant planets (which have a rocky core – the image that comes to mind is that of a dandelion), but they have been lost from the solid planets (think of dandelions after you blow on them). Loss of primordial atmospheres probably took place by a combination of gradual processes, such as thermal and non-thermal escape (see Hunten, 1973; Chamberlain & Hunten, 1987), and catastrophic processes such as atmospheric blowoff caused by large impacts (Pepin, 1997). This physical picture is supported by the observed abundances and isotopic compositions of noble gases in planetary atmospheres (Pepin, 2006). It is unlikely that whatever atmosphere the proto-Earth may have had would have been capable of surviving the Moon-forming event.

Present day atmospheres are therefore "secondary", in the sense that they were acquired after, and perhaps long after, planetary accretion was substantially completed, and the primordial atmospheres had been lost. The present day atmospheric masses and compositions are therefore determined by the relative rates of addition, removal and modification of individual volatile species since loss of the nebular atmosphere. Buildup of the secondary atmospheres must have occurred via a combination of late accretion of volatile-rich materials (e.g. comets) and volcanic outgassing. The rates of these two processes during the formative stages of the present-day atmospheres, almost certainly more than 4 billion years ago, are very difficult to pin down with any degree of certainty. Yet those rates are largely responsible for determining the initial masses and chemical compositions of the secondary atmospheres. Crucially, the compositions of volcanic gases and cometary volatiles are likely to have been different, and are poorly constrained.

Removal of volatile species is perhaps easier to constrain. It takes place by three distinct pathways: escape to space, condensation, and chemical reaction with the planet's surface materials. As we saw above, the effectiveness of the first of these pathways depends at least in part on the planet's mass (i.e. its gravitational attraction) and distance from the Sun (i.e. temperature), and may selectively remove light volatile species. For instance, we can expect that molecular and atomic hydrogen escape planetary atmospheres much more effectively than, say, molecular nitrogen or carbon dioxide, and that hydrogen loss will be more severe from Venus and Mars than from Earth (Fig. 13.1). Hydrogen-bearing molecules such as H_2O and CH_4 are photodissociated in the upper atmosphere (Section 12.4.2). Escape of the resulting hydrogen atoms is equivalent to an irreversible loss of water or methane, and an increase in the oxidation state of the planet's surface. Replenishment of these species in the atmosphere by evaporation of liquid or solid reservoirs provides a continuous pathway for planetary desiccation and oxidation (Chapter 14), which may have gone to near completion in Venus and perhaps somewhat less in Mars.

Condensation of volatile species removes them from the atmosphere and tends to protect them from escape. The process is particularly efficient if volatile species freeze, as in Europa, Ganymede, Callisto and Triton. The surfaces of these bodies can be thought of as collapsed atmospheres, that have been protected from escape by virtue of the very low vapor pressures in equilibrium with solid phases. Since the rate of photodissociation is proportional to concentration (e.g. equation (12.83)), the low vapor pressure of H_2O in equilibrium with a planetary surface composed of ice hinders hydrogen loss.

Volatile species can also be removed from the atmosphere by reaction with surface materials. One example is precipitation of carbonates by reaction of atmospheric CO_2 with aqueous cations leached from silicate minerals. This process is thought to have scavenged 10–60 bars' worth of CO_2 from the terrestrial atmosphere (see for example, Walker, 1985; Kasting & Ackerman, 1986; Kasting, 1987; Tajika & Matsui, 1993), which would otherwise have present-day composition, mass, surface pressure and temperature not all that different from those of Venus (and Gustav Mahler would not have existed, which would have made the universe a much poorer place). Another example is the effect of Fe^{2+} as a sink for atmospheric O_2. Oxidation of Fe^{2+} has sequestered oxygen liberated by photodissociation of H_2O in the Martian atmosphere over billions of years, generating the strongly oxidized Martian surface. At least in the case of Earth, present day atmospheric composition is also the result of modification by biological activity, including photosynthesis and also unnecessary and irresponsible use of pick-up trucks, minivans, jet-skis, snowmobiles, street lighting, insanely cold air conditioning, suffocating heating, and other sources of anthropogenic greenhouse gases.

13.2 Equilibrium thermodynamics in a gravitational field

The equations of equilibrium thermodynamics that we have discussed to this point ignore the existence of a gravitational field. This is generally acceptable locally, for example, in a laboratory setting or at the scale of heterogeneous phase equilibrium in a planetary body. However, if we are interested in the equilibrium distribution of species in a planetary atmosphere, then the effect of the gravitational force must be taken into account. Each of the terms in the fundamental equation for Gibbs free energy:

$$dG = -SdT + VdP + \sum_i \mu_i dn_i \qquad (13.3)$$

is the product of a pair of conjugate variables, one intensive and the other extensive. In every case the intensive variable can be thought of as a field, such that gradients in the field drive displacement of the corresponding extensive quantity. Thus, a temperature gradient generates heat flow, or, equivalently, entropy flow; a pressure gradient causes volume change, and a gradient in chemical potential drives mass transfer. Another way of stating these conditions is that $dT = 0$ implies thermal equilibrium, $dP = 0$ implies mechanical equilibrium relative to expansion work, and $d\mu = 0$ implies chemical equilibrium.

If the system is immersed in a non-uniform gravitational field then there is an additional contribution to its energy, which arises from the work associated with displacement of matter in the gravitational field. This is mechanical work, but it is distinct from the expansion work that is encapsulated in the VdP term in equation (13.3). If Φ is the gravitational potential ($=$ gravitational potential energy per unit mass, equation (1.8)) at a point, and m_i the molecular weight of component i, then the product $m_i \Phi dn_i$ is the work associated with an infinitesimal change in the amount of component i at the point. This contribution must be included in the Gibbs free energy of a system embedded in a gravitational field, which now becomes:

$$dG = -SdT + VdP + \sum_i \left(\mu^i + m_i \Phi \right) dn_i. \qquad (13.4)$$

By analogy with equation (5.24), a system in a gravitational field is in equilibrium relative to transfer of component i if:

$$d\left(\mu^i + m_i \Phi\right) = 0 \tag{13.5}$$

or, as molecular weight is a constant:

$$d\mu^i + m_i d\Phi = 0. \tag{13.6}$$

In the absence of a gravitational field, or if the field can be considered to be uniform, $d\Phi = 0$ and we recover (5.24). *However, if $d\Phi \neq 0$, which is the general case, and is in particular true in the neighborhood of planetary bodies, then equilibrium with respect to mass transfer requires that $d\mu \neq 0$.* Thus, equilibrium distribution of matter in a non-vanishing gravitational field implies a gradient in chemical potentials. Moreover, because chemical potential is in general a function of temperature, pressure and composition, it follows that gradients in at least some of these variables must exist in matter at equilibrium in a gravitational field.

We can write the total change in chemical potential of component i as follows:

$$d\mu^i = \left(\frac{\partial \mu^i}{\partial T}\right)_{P,X_i} dT + \left(\frac{\partial \mu^i}{\partial P}\right)_{T,X_i} dP + \sum_{j \neq i} \left(\frac{\partial \mu^i}{\partial X_j}\right)_{P,T} dX_j \tag{13.7}$$

or, by using (5.37) and (5.38), in terms of partial molar entropy and partial molar volume:

$$d\mu^i = -s^i dT + v^i dP + \sum_{j \neq i} \left(\frac{\partial \mu^i}{\partial X_j}\right)_{P,T} dX_j. \tag{13.8}$$

Substituting in (13.6):

$$-s^i dT + v^i dP + \sum_{j \neq i} \left(\frac{\partial \mu^i}{\partial X_j}\right)_{P,T} dX_j + m_i d\Phi = 0, \tag{13.9}$$

which is the equilibrium condition for each component i of a multicomponent phase immersed in a gravitational field. Summing over all components we get the equilibrium condition for the phase:

$$-SdT + VdP + \sum_i \left(X_i \sum_{j \neq i} \left(\frac{\partial \mu^i}{\partial X_j}\right)_{P,T} dX_j\right) + Md\Phi = 0, \tag{13.10}$$

where S and V are the molar entropy and volume of the phase, and M is the molecular weight of the phase ($=$ weighted average of the molecular weights of the phase components). From the Gibbs–Duhem equation (6.7) we have:

$$\left(\sum_i X_i d\mu^i\right)_{P,T} = \sum_i \left(X_i \sum_{j \neq i} \left(\frac{\partial \mu^i}{\partial X_j}\right)_{P,T} dX_j\right) = 0. \tag{13.11}$$

So (13.10) simplifies to:

$$-SdT + VdP + Md\Phi = 0. \tag{13.12}$$

These differential equations are general, but they are not necessarily simple to solve. In particular, the entropy term always introduces significant conceptual and computational difficulties. Here we restrict ourselves to isothermal processes. At constant temperature, and noting that density, $\rho = M/V$, we re-write (13.12) as follows:

$$\frac{dP}{d\Phi} = -\rho, \tag{13.13}$$

which is the condition of hydrostatic equilibrium. In order to see this, use equation (1.8) to calculate $d\Phi/dr$, apply the chain rule to calculate dP/dr, and compare to equation (2.34).

13.2.1 Pressure in a one-component isothermal atmosphere. Atmospheric scale height

Let us assume that a planetary atmosphere is composed of a single ideal gas species, so that M is constant with elevation, and that the atmosphere is isothermal (see also Exercise 13.1). From the ideal gas EOS we have:

$$\rho = \frac{PM}{RT}. \tag{13.14}$$

Substituting in (13.13):

$$\frac{dP}{P} = -\frac{M}{RT}d\Phi \tag{13.15}$$

and integrating:

$$\frac{P}{P_0} = \exp\left[-\frac{M}{RT}(\Phi - \Phi_0)\right], \tag{13.16}$$

where P_0, Φ_0 could be the pressure and gravitational potential at the planet's surface, for instance. With this convention, as we move up in the planet's atmosphere it is $\Phi > \Phi_0$ (recall equation (1.8)), so that $P < P_0$, as expected. Using the approximate expression for gravitational potential energy that we discussed in Worked Example 1.1, and using $h > 0$ for elevation above the planet's surface we get:

$$P = P_0 \exp\left(-\frac{Mgh}{RT}\right). \tag{13.17}$$

Provided that atmospheric temperature does not vary greatly, so that it makes sense to talk of a "characteristic" temperature for the atmosphere, this equation provides a first approximation to the variation in atmospheric pressure with elevation. In particular, it leads to the definition of the *scale height* of the atmosphere, $\boldsymbol{H}$:

$$\boldsymbol{H} \equiv \frac{RT}{Mg} \tag{13.18}$$

so that:

$$P = P_0 \exp\left(-\frac{h}{\boldsymbol{H}}\right) \tag{13.19}$$

Thus, when elevation changes by H atmospheric pressure varies by a factor e. This gives an indication of how quickly an atmosphere "fades" with elevation: the greater the scale height, the more the atmosphere extends into space. Substituting appropriate values for the five planetary bodies with substantial atmospheres in equation (13.18) we calculate scale heights of: 7.9 km for Earth, 10.7 km for Mars, 13.3 km for Triton, 14.9 km for Venus and 19.8 km for Titan. Of all the bodies with air in the Solar System the Earth is the one that "holds" its atmosphere closer to the solid surface. At the other end are Venus and Titan, that have the most "stretched out" atmospheres, although for different reasons (see equation (13.18)): Venus because of its high temperature, and Titan because of its low gravitational acceleration. In the case of Triton the low gravitational acceleration is offset by the very low temperature. Note that the definition of scale height, equation (13.18), is similar to that of the parameter Ξ, equation (13.1), except that H is a dimensional quantity, and the composition of the atmosphere is accounted for by means of the mean molecular weight, M. The scale height is, however, undefined for an airless body, but Ξ is independent of the existence of an atmosphere.

13.2.2 Compositional stratification in a fluid immersed in a gravitational field

Consider a fluid composed of an arbitrary number of chemical species. The chemical potential of species i is given by:

$$\mu^i = \mu^{0,i} + RT \ln f_i \qquad (13.20)$$

from which, at constant temperature:

$$d\mu^i = RT \, d\ln f_i. \qquad (13.21)$$

Substituting in (13.6):

$$RT \, d\ln f_i + m_i d\Phi = 0 \qquad (13.22)$$

and integrating:

$$f_i = f_{i,0} \exp\left[-\frac{m_i}{RT}(\Phi - \Phi_0)\right] \qquad (13.23)$$

where $f_{i,0}$ is the fugacity at the reference level, *not* the standard state fugacity. This relationship is general, and must be satisfied by every species in a multicomponent gas phase immersed in a gravitational field. Suppose that the atmosphere behaves as an ideal gas. We can then substitute partial pressure for fugacity, which simplifies things considerably. By using, as in the previous example, h for elevation relative to the planet's surface we find:

$$p_i = p_{i,0} \exp\left(-\frac{m_i g h}{RT}\right), \qquad (13.24)$$

where p_i is the partial pressure of species i.

If we now consider two species in the gas phase, call them 1 and 2, with different molecular weights we see that the ratio between the partial pressures of the two species varies as a function of elevation as:

$$\frac{p_1}{p_2} = \frac{p_{1,0}}{p_{2,0}} \exp\left[(m_2 - m_1)\frac{gh}{RT}\right], \qquad (13.25)$$

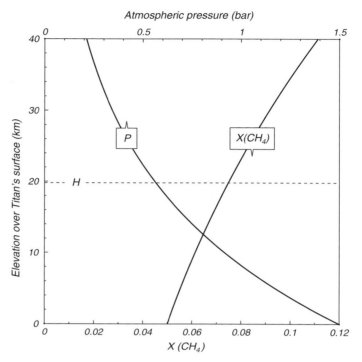

Fig. 13.2 Variation with height of atmospheric pressure and methane mol fraction in Titan's atmosphere, assuming surface pressure = 1.5 bar and constant temperature = 90 K. H is the atmospheric scale height (equation (13.18)).

which shows that the composition of the gas phase will be stratified according to the molecular weights (or densities) of the component species. For instance, if $m_2 > m_1$, then the ratio p_1/p_2 increases with elevation. The gas becomes enriched in the lighter component towards the top – which of course we already knew, but we can now quantify the effect. Equation (13.25) shows that the extent to which a fluid phase fractionates in a gravitational field varies directly with the difference in molecular weights of the component species and with the gravitational acceleration, and inversely with temperature.

Figure 13.2 shows an application to Titan's atmosphere. We assume that the atmosphere consists of a binary mixture of CH_4 and N_2, with $X(CH_4)$ at the surface equal to 0.05, a surface pressure of 1.5 bar and a characteristic atmospheric temperature of 90 K. The partial pressure of each gas as a function of elevation is calculated with (13.24), the mol fraction of methane is $X(CH_4) = p(CH_4)/(p(CH_4) + p(N_2))$, and the total atmospheric pressure is $p(CH_4) + p(N_2)$. The latter value is virtually indistinguishable from the value calculated with (13.19) because, even if the methane/nitrogen ratio changes substantially with elevation, nitrogen is always the dominant species. The mol fraction of methane more than doubles between the surface and the upper atmosphere, which is important given that methane is destroyed by photochemical reactions in the upper atmosphere (Section 12.4.2), leading to irreversible hydrogen loss. Since the photochemical reaction rate is proportional to methane concentration, atmospheric stratification enhances the rate of hydrogen loss.

13.3 Radiative energy transfer

Although advection is an important heat transfer mechanism in some atmospheric layers, solar heating of atmospheres and planetary surfaces, and heat loss of planets to space, take place by radiation. We therefore devote this section to a discussion of radiative heat transfer, and apply it to construct a simple quantitative model of greenhouse warming.

13.3.1 Fundamental concepts and equations of thermal radiation

All materials, at any temperature, emit and absorb electromagnetic radiation. The rates at which a body emits and absorbs electromagnetic energy are not necessarily the same, however. If they are then there is no net conversion between internal and electromagnetic energies, but if, say, the rate of emission is higher than the rate of absorption then internal energy is converted to electromagnetic energy, and conversely if absorption outpaces emission. At the microscopic level, conversion between internal energy and electromagnetic radiation corresponds to exchanges between the vibrational, rotational and/or translational kinetic energy modes of atoms and molecules on one side and energy of photons on the other.

Here we will assume that all radiant surfaces behave as *diffuse emitters*, which means that they emit radiation with the same intensity in all directions. This is not necessarily true in nature, but this simplification makes it possible to avoid a significant amount of terminology, algebra and solid geometry, and concentrate on the fundamental physics of radiative heat transfer. We define the *irradiance*, F, as the total flux (energy per unit area per unit time) of electromagnetic radiation over the entire spectrum and traveling in all directions. The *monochromatic irradiance*, F_λ, is the energy flux for radiation of a single wavelength λ. The two variables are obviously related by:

$$F = \int_0^\infty F_\lambda d\lambda. \tag{13.26}$$

It is often necessary to distinguish between emitted and incident energy flux, which we will do explicitly (special terms exist for these different quantities, but we will not introduce them here).

Electromagnetic radiation travels unimpeded in vacuum, but it interacts with matter. Interactions between radiant energy and matter can be described in terms of the absorptivity, A_λ, the transmissivity, Θ_λ, and the reflectivity, R_λ, of the medium, all of which are functions of wavelength. These macroscopic parameters arise from microscopic interactions between photons and particles of matter (molecules for the range of wavelengths of interest in planetary processes, see Section 13.3.3). Each of these parameters varies between 0 and 1 and represents the fraction of the irradiance of wavelength λ that is absorbed, transmitted and reflected, respectively. Their sum is always equal to 1, i.e.:

$$A_\lambda + \Theta_\lambda + R_\lambda = 1 \tag{13.27}$$

and also:

$$A + \Theta + R = 1, \tag{13.28}$$

where A, Θ and R are the total absorptivity, transmissivity and reflectivity, integrated over all wavelengths. These equations state energy conservation: the total incident flux of radiant energy must be accounted for in terms of a fraction that is reflected, a fraction that is absorbed and a fraction that is transmitted.

The spectrum and intensity of the electromagnetic radiation emitted by a body depends on its temperature and on another macroscopic parameter called the *emissivity*, ϵ ($\epsilon \leq 1$), which is a function of temperature and of wavelength. A black body is defined as a substance that emits and absorbs the maximum possible intensity of radiation at all wavelengths and in all directions. Thus, for a black body $A_\lambda = \epsilon_\lambda = 1$ and $\Theta_\lambda = R_\lambda = 0$, for all wavelengths and at all temperatures. The black body monochromatic emission flux, F_λ^*, also called the spectral emissive power, is given by Planck's radiation law:

$$F_\lambda^* = \frac{2\pi h c^2}{\lambda^5 \left(e^{\frac{hc}{\lambda k_B T}} - 1 \right)}, \tag{13.29}$$

where h is Planck's constant, k_B is Boltzmann's constant and c is the speed of light. This equation was first proposed semi-empirically by Planck in 1901 and became one of the foundational pillars of quantum mechanics. Its derivation is beyond the scope of this book but can be found, for example, in Incropera and DeWitt (1996) or Jones (2000). Equation (13.29) is a function of two variables, temperature and wavelength. It yields the spectrum of the electromagnetic radiation emitted by a black body at a constant temperature T, i.e. the distribution of emitted energy flux as a function of wavelength (Fig. 13.3). The curve is a "skewed bell". From the peak, radiation flux falls off steeply towards shorter wavelengths and more gently towards longer wavelengths. The wavelength at which the maximum flux occurs can be found by taking the derivative $dF_\lambda^*/d\lambda$ and equating it to zero (see Exercise 13.3). The result, known as Wien's displacement law, is:

$$\lambda_{peak} \approx \frac{2898}{T} \tag{13.30}$$

with T in Kelvin and λ in μm. The peak shifts towards shorter wavelengths with increasing temperature.

Figure 13.3 shows the regions into which the electromagnetic spectrum is conventionally subdivided within the interval 10^{-2}–10^4 μm. The Sun radiates (approximately) as a black body at a temperature of $\sim$6000 K. Because radiation emitted at this temperature peaks at $\sim$0.5 μm it was evolutionarily advantageous for life on Earth to develop acute sensory organs that respond to wavelengths in this region. For this reason we call the 0.4–07 μm range visible radiation. It also happens that animal perception senses radiation extending from the near ultraviolet to the near infrared (roughly, 0.1–10 μm) as heat. Conventionally, we extend this range to the far infrared ($\sim$ 100 μm) and call the range 0.1–100 μm thermal radiation. However, any electromagnetic radiation that has a spectrum given by Planck's equation (13.29) is strictly speaking thermal radiation.

The integral of F_λ^* over all wavelengths (i.e. the area under the curves in Fig. 13.3) yields the total black body emission flux, F^*, at temperature T and across the full spectrum. Substituting (13.29) in (13.26) and solving the integral (see Exercise 13.4) we get:

$$F^* = \int_0^\infty F_\lambda^* d\lambda = \sigma T^4. \tag{13.31}$$

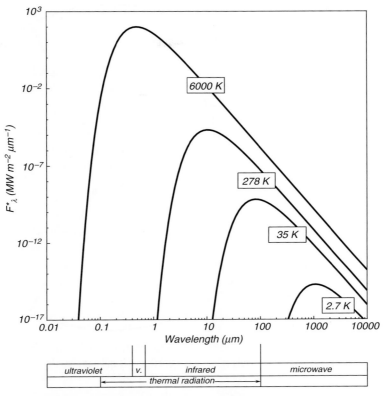

Fig. 13.3 Blackbody emission power spectrum calculated with Planck's radiation law (equation (13.29)) for the temperature of the solar photosphere ($\sim$6000 K), the terrestrial equilibrium temperature (278 K), the temperature at which the Earth's internal energy flux would radiate to space if the Sun disappeared (35 K) and the cosmic microwave background radiation (2.7 K). The narrow wavelength interval labeled "v." is what we call visible radiation, because we evolved around a star whose spectrum peaks in this region, making sensory organs that respond to these wavelengths evolutionarily advantageous.

This is Stefan–Boltzmann's law (equation (2.1), with $\epsilon = 1$). From the integral it also follows that the Stefan–Boltzmann constant, σ, is given by (Exercise 13.4):

$$\sigma = \frac{2}{15}\pi^5 h^{-3} c^{-2} k_B^4. \qquad (13.32)$$

13.3.2 Radiant energy exchange

The geometry sketched in Fig. 13.4 yields some results that are useful in solving problems of thermal radiation that arise in planetary sciences. We consider a spherical body of radius r_b and surface area a_b, at a uniform temperature T_b, concentric with a spherical cavity of radius $r_c \gg r_b$ and surface area a_c, at temperature T_c. We assume that the intervening space has transmissivity $\Theta = 1$ and we will initially also assume that both the sphere and the cavity are black bodies, i.e. $A_\lambda = \epsilon_\lambda = 1$ for all λ. From equation (13.31) and the surface areas

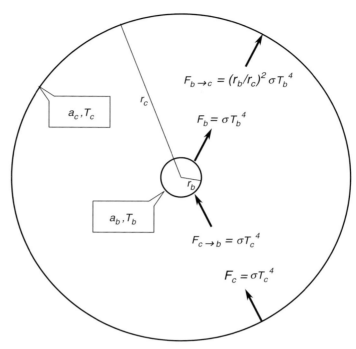

Fig. 13.4 Radiant energy exchange between a cavity at temperature T_c and a body at temperature T_b. Both are assumed to behave as black bodies, and the intervening medium has unit transmittivity.

of body and cavity, the total rates of energy radiation are $a_b \sigma T_b^4$ and $a_c \sigma T_c^4$, respectively. There is an important difference between the two, though: the cavity radiates onto itself, but the body does not. Heat transfer by radiation between the two is thus not symmetrical. In particular, the total amount of energy radiated by the body is absorbed by the cavity, but the converse is not true.

The radiation emitted by the body gives rise to an incident energy flux $F_{b \to c}$ on the surface of the cavity given by:

$$F_{b \to c} = \frac{a_b \sigma T_b^4}{a_c} = \left(\frac{r_b}{r_c} \right)^2 \sigma T_b^4 \tag{13.33}$$

which is the inverse square law of radiation, a.k.a. the equation of energy conservation. The flux of electromagnetic radiation emitted by the cavity, σT_c^4, bathes its interior uniformly. This can be demonstrated formally (Winterton, 1997) or you can accept it intuitively on the basis of symmetry. The body interposes a surface area a_b to this energy flux, so that the total amount of energy emitted by the cavity that is absorbed by the body is $a_b \sigma T_c^4$, and the flux of radiation incident on the body $F_{c \to b}$, is, therefore:

$$F_{c \to b} = \frac{a_b \sigma T_c^4}{a_b} = \sigma T_c^4. \tag{13.34}$$

Because the body is a black body it absorbs this energy, so that the net flux of electromagnetic energy that leaves the body, $F_{b,net}$, equals the energy emitted minus the energy absorbed, i.e.:

$$F_{b,net} = F_b - F_{c \to b} = \sigma \left(T_b{}^4 - T_c{}^4 \right). \tag{13.35}$$

If $T_b > T_c$ then the body is losing internal energy, and must either be cooling down, if this energy is not being replenished, or there is an internal source that supplies energy at this rate. Conversely, if $T_c > T_b$ radiant energy is being transformed to internal energy in the body. Think of this conversion in terms of photons colliding with particles of matter, whereupon the kinetic energy of the photons is added to the translational, vibrational or rotational energies of the particles. As an aside, photons carry not only kinetic energy but also momentum, and momentum conservation must be obeyed too. Transfer of momentum from photons to particles of matter gives rise to *radiation pressure*.

Equation (13.35) shows that radiative heat flux varies as the difference between the fourth powers of temperature, in contrast to diffusive and convective heat flux, which are linear functions of temperature difference, or nearly so (e.g. equations (3.5) and (3.89)). It justifies equation (2.16): the planet at temperature T is immersed in the solar nebula, which we can think of as the "cavity", at temperature T_0.

From (13.35) we can define the equilibrium temperature for the body, $T_{b,eq}$, as the temperature at which there is no net flux of electromagnetic radiation from the body and hence no change in its internal energy content with time. This is simply:

$$T_{b,\, eq} = T_c. \tag{13.36}$$

A body at the same temperature as its environment does not exchange electromagnetic energy with it.

Worked Example 13.1   Radiative energy balance at a planet's surface

The Sun can be approximated as a spherical black body with an emission temperature of ~ 6000 K, and planetary orbits can be thought of as circumferences on spherical cavities centered on the Sun. Setting $r_b = r_s =$ solar radius, $r_c = r_o =$ orbital radius and $T_b = T_s =$ Sun's emission temperature, equation (13.33) gives the flux of solar radiation across a planet's orbit, which is called the *solar constant* (Fig. 13.5). The solar constant for the Earth is ~ 1368 W m^{-2}. This is the absolute maximum rate at which energy can be extracted from sunlight at the Earth's surface. Even if all of this radiant energy could be converted to mechanical energy, it represents somewhat less than 2 horse-power per square meter. The energy density of solar power is quite low, which is a reality that economic development of this energy source must cope with. In practice, moreover, only a fraction of the solar energy flux can be converted to usable energy at the Earth's surface, partly because for the atmosphere $A > 0$ (see below), and also because conversion to electrical or mechanical energy is never 100% efficient (Chapter 4).

We can calculate the equilibrium temperature of the planet, T_{eq}, that we referred to in Section 2.1. This is a temperature such that the planet emits electromagnetic radiation at the same rate as it absorbs it from the Sun. The total amount of solar radiation that reaches a planet is equal to the solar constant multiplied by the cross section of the planet, πr_p^2, where r_p is the planet's radius. Note that this is not the surface area of the hemisphere facing the sun, but the cross section that intersects the solar energy flux, i.e. the stream of solar photons (Fig. 13.5). Think of the planet as the "body" in Fig. 13.4, inside a cavity of infinite extent.

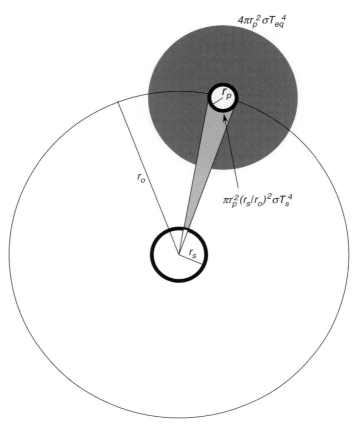

$$4\pi r_p^2 \sigma T_{eq}^4$$

$$\pi r_p^2 (r_s/r_o)^2 \sigma T_s^4$$

Radiant energy exchange between the Sun (radius r_s) and a planet of radius r_P in a circular orbit of radius r_o. The planet absorbs the solar flux contained in the light grey solid angle, defined by the cross section area πr_p^2 at the planet's orbital distance. Thermalized radiation is radiated back to space (a cavity of infinite extent symbolized by the dark grey circle) over the entire surface area of the planet, $4\pi r_p^2$.

We see from equation (13.35) that, because during the day $T_b < T_c$, the planet absorbs sunlight, whereas at night $T_b > T_c$ and the planet emits radiation. This radiation is what we called thermalized radiation, or thermalized sunlight, in Chapter 2 . If the planet rotates fast enough we can assume that it radiates at the same equilibrium temperature, T_{eq}, over its entire surface area, $4\pi r_p^2$. We will ignore the planet's internal energy flux. The Earth's internal energy flux is $\sim 10^{-5}$ times the solar constant, so it cannot have any noticeable effect on its surface equilibrium temperature. The same is true of the other terrestrial planets but, as we saw in Chapter 2, it is not the case for the fluid planets nor for Io. Assuming that the planet behaves like a black body, T_{eq} must satisfy the following energy conservation condition:

$$4\pi r_p^2 \sigma \left(T_{eq}\right)^4 = \pi r_p^2 \left(\frac{r_s}{r_o}\right)^2 \sigma T_s^4 \qquad (13.37)$$

or:

$$T_{eq} = \left(\frac{1}{4}\right)^{1/4} \left(\frac{r_s}{r_o}\right)^{1/2} T_s. \tag{13.38}$$

At this temperature the planet radiates energy at the same (average) rate as it receives it from the Sun. For the Earth we calculate an equilibrium temperature of $\sim$278 K. We can use Planck's law (equation (13.29)) to compare the spectrum of sunlight with that of thermalized sunlight (Fig. 13.3). The emission peak shifts from 0.5 μm, in the visible range, to 10 μm, in the mid infrared. Because of the very strong dependency of emitted flux on temperature there is a difference of seven orders of magnitude between the height of the peaks at 6000 K and 278 K.

The Earth's equilibrium temperature is about 10 K lower than the actual mean terrestrial surface temperature ($\sim$288 K). Why the discrepancy? In short, because the Earth has an atmosphere, oceans and a climate system. For example, some trace components in the troposphere such as H_2O, CO_2 and CH_4 absorb in the infrared part of the spectrum, causing warming of the lower atmosphere (the greenhouse effect, see below). Additional complications arise from variations in solar radiation with latitude and seasons (that drive convective heat transport) and with changes in albedo related to ground cover and cloud cover.

I stated above that at night $T_b > T_c$, but what is the "cavity temperature" at night? Disregarding for now the planet's atmosphere (and heterogeneities that may arise from interplanetary and galactic energy sources) this is the temperature of the microwave background radiation, $\sim$2.7 K (Fig. 13.3). In terms of heat transfer, you can think of this radiation as being emitted by an infinitely distant cavity at this uniform temperature, although this is of course not its true physical nature (it is strongly red-shifted radiation from the early Universe). If the Earth were to be ejected from the Solar System into the depths of intergalactic space, would it still radiate electromagnetic energy to space? Yes, at its current average internal energy output ($\sim$87 mW m^{-2}) it would radiate as a black body at a temperature of $\sim$35K (equation (13.31)) with a peak at a wavelength of $\sim$82 μm (equation (13.30)), corresponding to the far infrared part of the spectrum (Fig. 13.3). Emission flux originating from the Earth's internal heat (at 35 K) is some five orders of magnitude lower than the flux of thermalized solar radiation at 278 K (Fig. 13.3). This explains why it is generally not possible to use remote sensing to determine heat flux from terrestrial planets. It is interesting (and pointless) to speculate that the low surface temperature that would result from the Earth being ejected to intergalactic space would increase (slightly) the mantle's Rayleigh number and hence the rate of plate tectonics.

To conclude this section we derive a fundamental relationship between emissivity and absorptivity. Assume that the body in Fig. 13.4 is not a black body. Then, for each wavelength λ it has an emissivity $\epsilon_\lambda \neq 1$ and an absorptivity $A_\lambda \neq 1$. The cavity is a black body that emits an energy flux $F^*_{c,\lambda}$, given by equation (13.29). From equation (13.34) and the definition of absorptivity, the energy flux absorbed by the body at a given wavelength is:

$$F_{c \to b, \lambda} = A_\lambda F^*_{c,\lambda}. \tag{13.39}$$

From the definition of emissivity, the emission from the body is given by:

$$F_{b,\lambda} = \epsilon_\lambda F^*_{b,\lambda}, \tag{13.40}$$

where $F^*_{b,\lambda}$ is the black body emission at the body's temperature. The net radiant flux leaving the body is then given by:

$$F_{b,net,\lambda} = F_{b,\lambda} - F_{c \to b,\lambda} = \epsilon_\lambda F^*_{b,\lambda} - A_\lambda F^*_{c,\lambda}. \qquad (13.41)$$

If the temperatures of the body and the cavity are the same then $F^*_{c,\lambda} = F^*_{b,\lambda}$ and also, from equation 13.35, $F_{b,net,\lambda} = 0$. The following result, known as Kirchoff's law, follows:

$$A_\lambda = \epsilon_\lambda. \qquad (13.42)$$

In other words, absorptivity and emissivity *at a given wavelength* are equal. There are some important caveats. First, if a body absorbs radiation that was emitted at a temperature different from that of the body itself then the integrated absorptivity and emissivity are not equal. For example (see Fig. 13.3), because sunlight peaks at $\sim 0.5\ \mu$m whereas thermalized sunlight on Earth peaks at $\sim 10.4\ \mu$m, the absorptivity and emissivity of the Earth's surface integrated over all wavelengths are not equal. Second, it is in general the case that $A \neq \epsilon$ even if emitted and absorbed radiation correspond to the same black body temperature. The reasons for this have to do with the fact that, in contrast with a black body, radiation from real bodies is in general not isotropic and may also be polarized differently from absorbed radiation. Discussion of these effects is beyond the scope of this book.

13.3.3 Molecular absorption and emission of electromagnetic radiation

Electromagnetic radiation transports heat because photons can interact and exchange energy with particles of matter. Heat transport by radiation is relatively minor in liquids and generally negligible in solids, but *absorption and emission* of radiant energy by planetary liquids and solids is important. In contrast, gases are important in terms of radiant energy transported as well as absorbed and emitted. We saw in Chapter 1 that molecules in gases carry energy as translational, rotational and vibrational kinetic energy, and that, as temperature increases, all three modes come to participate in the heat capacity of gases. Molecules also store energy in their electronic configurations. This mode does not show up in the heat capacity of gases at "normal" planetary conditions because at these conditions there is not enough energy to bring about changes in the electronic configuration of a molecule or atom, or, in the terminology of quantum mechanics, to excite the electronic energy modes.

In order to discuss emission, absorption and transport of radiant energy in gases we must take a closer look at the various modes in which energy is stored in molecules. The rotational, vibrational and electronic modes are all quantized, which means that they can carry energy only in specific levels that are separated by discrete energy differences. Molecules can absorb or emit energy only in discrete packages that correspond to the differences between quantum energy levels. Photons carry discrete amounts of energy too, given by $(hc)/\lambda$. A photon can interact with a molecule only if its energy corresponds to some of the possible energy transitions in the molecule. Only photons of the appropriate wavelength ($=$ energy) can be absorbed and when they are, one of the molecular energy modes is excited to a higher energy level. Conversely, when the molecule "drops" from an excited state to the ground state it emits a photon of a wavelength that corresponds to this energy difference.

It is perhaps intuitively apparent that it takes relatively little energy to change the state of rotation of a molecule, more energy to change the vibrations of the atomic bonds, and even more energy to alter the electronic configuration of the molecule. This is formally proved in quantum mechanics and is important for our purposes. The energies required to excite electronic modes are carried by photons with wavelengths of order 0.1–0.5 μm, within the ultraviolet and visible part of the spectrum (see Fig. 13.3). Electronic excitation corresponds to breaking or forming of chemical bonds, i.e. chemical reactions. Vibrational energy transitions are associated with lower energies, corresponding to infrared photons with wavelengths of 1–10 μm. Rotational transitions occur in response to even lower energy photons, in the microwave region of the spectrum ($10^2 - 10^4$ μm). Excitations of rotational modes are generally unimportant in planetary processes, as they correspond to exceedingly low temperatures, but are important in astrophysics, where non-thermal mechanisms for the emission of microwave electromagnetic radiation exist (they are also what makes microwave ovens work).

Molecular gases absorb and emit radiation of wavelengths extending from the ultraviolet to the infrared, by exciting electronic and vibrational energy modes. The mechanisms are different for the two types of energy transitions. Absorption of ultraviolet radiation by electronic transitions is associated to photodissociation reactions. Oxygen–ozone reactions in the Earth's stratosphere are a good example (Worked Example 12.3). The first and third reactions in the Chapman cycle (equation (12.144)) absorb photons with wavelengths in the 0.2–0.25 μm range. In both cases absorption of a photon excites an electronic transition which results in breakage of an atomic bond. Absorption of ultraviolet radiation by these reactions in the Earth's stratosphere has two important effects. First, because complex organic molecules such as proteins and DNA can also break up by absorbing photons in this energy range, photoactivated reactions that produce and destroy ozone allow us to be here discussing these things. Second, absorption of ultraviolet radiation heats the stratosphere and inverts the temperature gradient that drives convection in the troposphere.

Infrared photons are not energetic enough to break atomic bonds and facilitate chemical reactions. They are absorbed by exciting vibrational modes. There is an additional restriction in this case, that arises from quantum mechanics *selection rules*. The selection rule for vibrational excitations is that they can only happen in molecules in which the electrostatic dipole moment (i.e. the distribution of electric charge across the molecule) is asymmetric. What this means is that homonuclear diatomic molecules such as O_2 and N_2 cannot absorb infrared radiation, because the electric charge of both atoms is identical. In contrast, diatomic molecules made up of different atoms (e.g. CO or HCl) and polyatomic molecules have dipole moments that are not symmetric relative to the molecular structure. By the selection rule they can absorb infrared photons as long as they have the correct energy to excite one of the possible vibrational transitions. A well-known example of this is the capability of molecules such as CO_2, H_2O and CH_4 to absorb infrared radiation at wavelengths that are close to the emission peak of thermalized solar radiation (see below).

Gases emit and absorb photons of specific wavelengths only, that correspond to allowed energy transitions in the molecules. Emission and absorption of electromagnetic radiation by molecular gases gives rise to line spectra. The molecules and atoms in solids and liquids, in contrast, are close enough that the quantum states of individual atoms are not independent of one another and the discrete energy transitions become smoothed out. The result is that absorption and emission of electromagnetic radiation in solids and liquids extend over continuous regions of the spectrum. This is also true of gases in which electrons are free, because in such case electrons do not have set energy levels, and it is the reason why the outer envelopes of stars, made up of ionized gas, radiate as black bodies.

13.3.4 Absorption and emission of electromagnetic radiation: the macroscopic description

Consider a layer of material of infinitesimal thickness, dx (Fig. 13.6) and a beam of electromagnetic radiation of wavelength λ incident on one of its sides. Recall that the irradiance, F_λ, is the total flux of radiation traveling in all directions. We will consider only the simplest case of changes in radiation intensity in a single direction. The intensity of the radiation, I_λ, is defined as the flux of radiation traveling in a single direction, which in this case is perpendicular to the layer of material. As in all radiation problems, a complete analysis requires that we consider the full geometry of the problem and how radiation intensity changes with direction, but the physical principles involved are easier to see if we ignore these complications.

In general, the material may both absorb and emit radiation of a given wavelength, so that the total change in the intensity of the beam over the thickness dx is given by:

$$dI_\lambda = dI_{\lambda,absorbed} + dI_{\lambda,emitted}. \tag{13.43}$$

From the definition of absorptivity we see that:

$$dI_{\lambda,absorbed} = -A_\lambda I_\lambda. \tag{13.44}$$

It is reasonable to assume that the absorptivity is a function of some intrinsic material property, on how much matter there is, and of the thickness of the layer, so we write:

$$A_\lambda = k_\lambda \rho \, dx, \tag{13.45}$$

where ρ is density and k_λ is called the mass absorption coefficient for radiation of wavelength λ. Because absorptivity is a non-dimensional number, the mass absorption coefficient must have dimensions $[L]^2 \, [M]^{-1}$, e.g. $m^2 \, kg^{-1}$. In this derivation we will consider only absorption and emission by molecular mechanisms such as those that we discussed in the previous section. The intensity of electromagnetic radiation can also change as a result of scattering, as when photons interact with solid or liquid particles suspended in the atmosphere. This process will not be considered here, but it is straightforward to add the effects

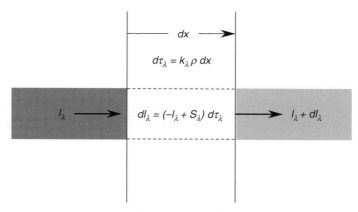

Fig. 13.6 Interaction between a monochromatic beam of electromagnetic radiation of intensity I_λ and a slab of material of thickness dx.

of scattering to the absorption coefficient, in which case the name changes to *extinction coefficient* to account for the combined effects of absorption and scattering.

We define a non-dimensional variable, called the *optical thickness*, τ_λ, as follows:

$$\tau_\lambda = k_\lambda \rho x, \tag{13.46}$$

and from equations (13.44), (13.45) and (13.46), we write the change in beam intensity due to absorption as follows:

$$dI_{\lambda, absorbed} = -k_\lambda \rho I_\lambda dx = -I_\lambda d\tau_\lambda. \tag{13.47}$$

We can also write an expression for $dI_{\lambda, emitted}$ similar to (13.47) by defining a variable S_λ, called the source term, which, as I_λ, has units of energy flux. Thus:

$$dI_{\lambda, emitted} = S_\lambda d\tau_\lambda. \tag{13.48}$$

The source term is a measure of how much radiation the layer emits per unit of non-dimensional optical thickness. Using equations (13.47) and (13.48), (13.43) becomes:

$$dI_\lambda = -I_\lambda d\tau_\lambda + S_\lambda d\tau_\lambda \tag{13.49}$$

The solution of this differential equation is straightforward if S_λ is a constant (see Exercise Problem 13.12, or differentiate (13.50) and substitute in (13.49) to verify that it is a solution):

$$I_\lambda = \left(I_{\lambda(0)} - S_\lambda\right) e^{-\tau_\lambda} + S_\lambda, \tag{13.50}$$

where $I_{\lambda(0)}$ is the intensity of the incident beam and I_λ is the intensity at optical thickness τ_λ.

The meaning of the optical thickness becomes apparent if we consider the case in which $S_\lambda = 0$. Equation (13.50) then simplifies to the following, which is known as the Beer–Lambert law (see also equation (12.137)):

$$I_\lambda = I_{\lambda(0)} e^{-\tau_\lambda}. \tag{13.51}$$

The optical thickness is the non-dimensional absorption length. About 2/3 of the incident radiation is absorbed at $\tau_\lambda = 1$, and at five optical thicknesses more than 99% of the incident radiation has been absorbed. As an example, consider transmission of solar radiation through the terrestrial atmosphere. Solar radiation peaks at visible and ultraviolet wavelengths. The temperature of the atmosphere is of the order of 250 K. We can see from Planck's and Wien's laws that the atmosphere does not emit any significant amount of radiation in this region of the spectrum, so that $S_\lambda \approx 0$. From our discussion in Section 13.3.3 it follows that atmospheric oxygen has a large value of k_λ for $\lambda \sim 0.2\,\mu\text{m}$ (ultraviolet photons). For the Earth's atmosphere $\tau_\lambda \gg 1$ for ultraviolet radiation. We say that the atmosphere is optically thick in the ultraviolet, or, equivalently that the atmosphere is nearly opaque to ultraviolet radiation.

Returning to the full absorption–emission equation, (13.50), we consider what happens if the temperature of the medium is such that emission at the wavelength of the incident radiation (the term S_λ) cannot be ignored (this can be the case for infrared radiation in CO_2-rich planetary atmospheres). In this case, for small optical thicknesses the source term cancels out and the radiation flux is dominated by the incident flux, whereas for large optical thicknesses radiation flux is dominated by thermal emission in the layer.

13.3.5 Absorption cross section and mean free path

The mass absorption coefficient, k_λ, is a macroscopic parameter that describes the interaction between electromagnetic radiation and matter. We can give it a microscopic interpretation, as follows. From the definition of optical thickness, equation 13.46, we see that the product $k_\lambda \rho$ has dimension of length^{-1}, so that $(k_\lambda \rho)^{-1}$ is the natural lengthscale for absorption of radiation of wavelength λ. We now search for a product of two microscopic variables with the same dimension. We define the *absorption cross section* at wavelength λ, σ_λ, as the effective target area that each molecule offers to photons of this wavelength. Thus, if a substance is inert to a particular wavelength, such as molecular oxygen to infrared, σ_λ vanishes. In contrast, if a substance absorbs radiation of a given wavelength (e.g. CO_2 in the infrared) its absorption cross section for that wavelength is a finite and potentially large number. The dimension of σ_λ is length^2 molecule^{-1}. If we multiply the absorption cross section by the number density of molecules, N, defined as the number of molecules per unit volume, we get a product with dimension of length^{-1}. This product is the total absorption cross section per unit volume, so we can suggest the following relationship between the macroscopic absorption coefficient and the microscopic absorption cross section:

$$k_\lambda \rho = \sigma_\lambda N. \tag{13.52}$$

Imagine now that we have a slab of absorbing substance of cross section a^2 and thickness ι_λ, such that ι_λ is the characteristic distance that a photon of wavelength λ can penetrate into the substance before it becomes certain that it will be absorbed. This distance is called the photon's *mean free path*. The mean free path must be of the same order as the distance into the layer at which the sum of the absorption cross sections of all of the molecules equals the actual cross section of the slab. Since $\sigma_\lambda N$ is the absorption cross section per unit volume, multiplying this by the volume of the slab we get the total absorption cross section of the slab, which we require to be equal to the physical cross section of the slab, i.e.:

$$\sigma_\lambda N a^2 \iota_\lambda = a^2, \tag{13.53}$$

which shows that the mean free path is given by:

$$\iota_\lambda = \frac{1}{\sigma_\lambda N} = \frac{1}{k_\lambda \rho}. \tag{13.54}$$

The mean free path thus equals the lengthscale for absorption, $(k_\lambda \rho)^{-1}$. From equations (13.46) and (13.54) we have:

$$\tau_\lambda = \frac{x}{\iota_\lambda}. \tag{13.55}$$

An optical thickness of 1 is reached, and most of the incident radiation is absorbed, when the distance traveled by the electromagnetic radiation is of the order of the mean free path. Other things being equal, the mean free path varies inversely with the number density, N (equation (13.54)), explaining in part why solids tend to be more opaque (absorptivity $= 1$) than gases.

In a mixture of ideal gases, the number density N of a gas species at constant temperature is proportional to the partial pressure of the species. For dilute gases at low pressure, therefore, mean free path varies inversely with the partial pressure of the absorbing species (equation (13.54)) and optical thickness can be approximated with a linear function of partial pressure (equations (13.54) and (13.55)).

13.3.6 A radiative toy model of greenhouse warming

We close our discussion of radiative heat transfer by constructing a "toy model" of greenhouse warming. The goal of this section is not to develop a complete and quantitatively accurate model of the process, which is far beyond the scope of this book. Rather, I simply want to highlight the aspects of radiative heat transport that underlie the planetary greenhouse effect. With this caveat, we proceed as follows (see Fig. 13.7).

Consider a planet such that the solar constant at its orbit is F_s. The planet has an atmospheric layer in which radiative heat exchanges take place. The elevation and thickness of this active layer are unspecified, except for the fact that it is close enough to the planet's solid surface, and thin enough, that the radius of the planet, r_p, and the mean radius of the active atmospheric layer can be considered to be equal. We will simplify the mathematical treatment by assuming that we can describe the interactions between electromagnetic radiation, the atmosphere and the surface on the basis of only two sets of A, Θ, R and ϵ. As we saw, all of these parameters depend on wavelength but here we will assume that we can

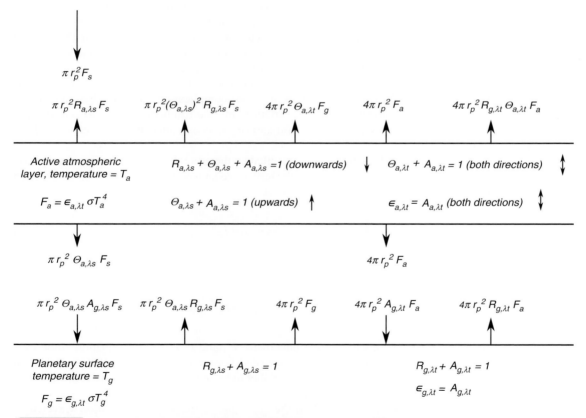

Fig. 13.7   Energy balance for a simple radiative model of greenhouse warming of a planet of radius r_P and solar constant F_s. The planet has an atmospheric layer that interacts with solar and thermal radiation, and attains an equilibrium temperature T_a. The equilibrium temperature of the planet's surface is T_g. The arrows illustrate the direction of the various energy fluxes, but not their relative intensities.

define two "average" values for each parameter, one for the ultraviolet and visible wavelengths that make up most of solar radiation, and the other one for infrared wavelengths radiated by the planet's surface and atmosphere.

The active atmospheric layer is at a constant and uniform temperature T_a. It reflects a fraction $R_{a,\lambda s}$ of the incoming short wavelength solar radiation, with characteristic wavelength λs, and absorbs a fraction $A_{a,\lambda s}$ of this radiation. This absorption could take place, for example, as a result of ozone-forming and ozone-destroying reactions. The transmitted fraction of the solar energy flux, $\Theta_{a,\lambda s} = 1 - (A_{a,\lambda s} + R_{a,\lambda s})$, reaches the planet's surface, which is opaque to all electromagnetic radiation. We will assume that the atmosphere does not reflect sunlight downwards, i.e., for sunlight reflected from the planet's surface, we will set $R_{a,\lambda s} = 0$ and $\Theta_{a,\lambda s} = 1 - A_{a,\lambda s}$. The reason for this is to avoid dealing with second and higher order internal reflections, which obscure the calculations without providing any important insight, nor affecting the results in any significant way. The average absorptivity of the surface for solar radiation is $A_{g,\lambda s}$ and its reflectivity is $R_{g,\lambda s} = 1 - A_{g,\lambda s}$. Solar radiation is thermalized and re-emitted at the average temperature of the planet's surface, T_g. The emissivity of the ground for long-wavelength thermalized radiation, characterized by wavelength λt, is $\epsilon_{g,\lambda t}$. Abusing somewhat equation (13.42), we will assume that the absorptivity of the ground for thermalized radiation is equal to the emissivity, i.e. $A_{g,\lambda t} = \epsilon_{g,\lambda t}$ and since the ground is opaque at all wavelengths, $R_{g,\lambda t} = 1 - A_{g,\lambda t}$. The absorptivity and emissivity of the atmospheric layer for thermalized radiation are also considered to be equal to each other, i.e. $A_{a,\lambda t} = \epsilon_{a,\lambda t}$, and we assume that the atmosphere does not reflect long wavelength radiation either, i.e. $A_{a,\lambda t} + \Theta_{a,\lambda t} = 1$. We assume that convection does not take place in the atmosphere, so heat transport is by radiation only, and that there are no other complications such as phase transitions (e.g. cloud formation) and latitudinal variations in solar irradiation. Our goal is to find the equilibrium temperatures of the ground and of the active atmospheric layer, T_g and T_a, respectively.

We will solve for the energy fluxes emitted by the ground and by the active atmospheric layer, F_g and F_a, which are related to the respective temperatures by the Stefan–Boltzmann law:

$$F_a = \epsilon_{a,\lambda t} \sigma T_a{}^4 \qquad\qquad (13.56)$$

and:

$$F_g = \epsilon_{g,\lambda t} \sigma T_g{}^4. \qquad\qquad (13.57)$$

We need two equations in the two unknowns, F_a and F_g. Our first equation is the bulk planetary energy balance. This is an expanded version of the problem that we solved in Worked Example 13.1. In order for the planet to be in thermal equilibrium it must radiate energy to space at the same average rate as it receives it from the Sun, but now there is radiation from both the surface and the atmosphere, and some of the short wavelength sunlight is reflected without being thermalized. Recall from Worked Example 13.1 that incident and reflected solar energy fluxes act over the cross section of the planet that intersects the solar photon flux, $\pi r_p{}^2$, whereas thermalized fluxes are radiated over the entire surface area of the planet and atmosphere, $4\pi r_p{}^2$. The energy flows that we need to consider, shown in Fig. 13.7, are as follows.

Solar radiation reflected to space by atmosphere: $\pi r_p^2 R_{a,\lambda s} F_s$.

Solar radiation reflected by planet surface: $\pi r_p^2 (\Theta_{a,\lambda s})^2 R_{g,\lambda s} F_s$
  (recall that we assume that the atmosphere does not reflect sunlight downwards).

Thermalized radiation from planet's surface: $4\pi r_p^2 \Theta_{a,\lambda t} F_g$.

Thermalized radiation from top of atmosphere: $4\pi r_p^2 F_a$.

Thermalized radiation from atmosphere reflected from planet's surface:
  $4\pi r_p^2 R_{g,\lambda t} \Theta_{a,\lambda t} F_a$.

The sum of all of these terms is the energy radiated by the planet, so that at thermal equilibrium this sum must equal the total energy received from the sun, $\pi r_p^2 F_s$. With some algebra we find that the planetary energy balance simplifies to:

$$k_1 F_s = 4 k_2 F_a + 4 \Theta_{a,\lambda t} F_g, \tag{13.58}$$

where the two parameters k_1 and k_2 combine terms as follows:

$$k_1 = 1 - R_{a,\lambda s} - \Theta_{a,\lambda s}{}^2 R_{g,\lambda s}$$
$$k_2 = 1 + \Theta_{a,\lambda t} R_{g,\lambda t}. \tag{13.59}$$

The parameter k_1 is the fraction of solar energy that is effectively absorbed by the planet, and k_2 is the total thermal energy radiated by the atmosphere that escapes to space, including the direct upwards flux and the fraction of the downwards flux that is reflected by the planet's surface.

For our second equation we can choose to balance energy either at the planet's surface or in the atmosphere. We can choose either one, but regardless of which one we choose we must get the same results. I will choose energy balance at the surface, as the equations are simpler. As an exercise, you should redo the calculations using energy balance in the atmosphere (Exercise 13.13). In order for the temperature of the planet's surface to be constant with time the flow of thermalized energy emitted by the planet's surface, $4\pi r_p^2 F_g$, must equal the total radiant energy absorbed by the surface, which comprises the following two terms (see Fig. 13.7).

Thermalized radiation from bottom of atmosphere: $4\pi r_p^2 A_{g,\lambda t} F_a$

Absorbed solar radiation: $\pi r_p^2 \Theta_{a,\lambda s} A_{g,\lambda s} F_s$.

Equating and simplifying, we find:

$$\Theta_{a,\lambda s} A_{g,\lambda s} F_s = 4 \left(F_g - A_{g,\lambda t} F_a \right). \tag{13.60}$$

The solutions that we seek for the fluxes of thermal energy emitted by the surface and the atmosphere are:

$$F_g = \frac{F_s}{4} \frac{k_1 A_{g,\lambda t} + k_2 \Theta_{a,\lambda s} A_{g,\lambda s}}{k_2 + \Theta_{a,\lambda t} A_{g,\lambda t}} \tag{13.61}$$

$$F_a = \frac{F_s}{4} \frac{k_1 - \Theta_{a,\lambda t} \Theta_{a,\lambda s} A_{g,\lambda s}}{k_2 + \Theta_{a,\lambda t} A_{g,\lambda t}}, \tag{13.62}$$

which we convert to ground and atmospheric temperatures with equations (13.56) and (13.57).

We now use these equations to analyze the effects of some of the radiative energy transfer parameters on average global temperature. Among these, the following are the ones that are likely to be most variable.

(i) $A_{a,\lambda t}$, the long-wavelength absorptivity of the atmosphere, which is sensitive to the atmospheric concentration of infrared active molecules such as CO_2, H_2O and CH_4.

(ii) $R_{g,\lambda s}$, the reflectivity of ground for solar radiation (i.e. the albedo of the planet's surface), which is strongly affected by ice and snow cover, as well as by vegetation.

(iii) $R_{a,\lambda s}$, the fraction of solar radiation reflected by the upper atmosphere, which can be affected by fine dust, soot and sulfur dioxide crystals such as can be produced by volcanic eruptions, meteorite impacts and missile-launch-button-happy individuals.

We study the effects of these three parameters, and assume that the absorptivity of the atmosphere for solar radiation, which is controlled to a significant extent by oxygen concentration, and the absorptivity of the ground for infrared radiation have the constant values $A_{a,\lambda s} = 0.2$ and $A_{g,\lambda t} = 0.95$ (hence, $R_{g,\lambda t} = 0.05$). We will also assume that the solar constant equals its present-day value of 1368 Wm^{-2}.

Figure 13.8 shows the effects of $A_{a,\lambda t}$ and $R_{g,\lambda s}$. The solid circle on the leftmost panel shows the equilibrium temperature for a black body Earth ($\sim$ 278 K, see Worked Example 12.3). We then add $A_{a,\lambda s} = 0.2$ and $A_{g,\lambda t} = 0.95$ and reasonable values (taken from de Pater & Lissauer, 2001, and Chamberlain & Hunten, 1987) for the present day surface albedo ($R_{g,\lambda s} = 0.3$) and atmospheric reflectivity ($R_{a,\lambda s} = 0.1$). Without greenhouse warming ($A_{a,\lambda t} = 0$) the equilibrium ground temperature drops to $\sim$ 249 K, shown by the open circle. The diagram shows that a temperature comparable to the present-day average surface temperature of the Earth ($\sim$288 K, shown by the diamond) is attained with $A_{a,\lambda t} = 0.9$. This is our starting point for the other panels in the figure. The one on top shows the effect of increasing $A_{a,\lambda t}$ beyond this value while holding the albedo, $R_{g,\lambda s}$, constant. An increase of $\sim$10% in the infrared absorptivity of the atmosphere raises ground temperature by about 8 K. To put this number in perspective, the increase in global temperatures since the Pleistocene is about 6 K, of which 1–2 K are the result of anthropogenic causes since the Industrial Revolution. Rising global temperatures cause glaciers and sea ice to melt, lowering the planetary albedo. The effect of decreasing $R_{g,\lambda s}$ at constant $A_{a,\lambda t} = 0.9$ is shown in the bottom center panel. In reality both parameters are coupled: as $A_{a,\lambda t}$ increases so does global temperature, causing $R_{g,\lambda s}$ to drop. The functional relationship between the two parameters is complex and is in fact one of several sources of uncertainty in global warming models. The coupling does not depend on radiative heat transport, but rather on physical, chemical and biological interactions between the atmosphere, the oceans and the solid surface. The figure suggests that a strong coupling between the two parameters can plausibly raise average global temperatures by 10–15 K, which may result in ice-free conditions comparable to those that existed during much of the Mesozoic.

The effect of changes in the solar radiation reflectivity of the atmosphere, $R_{a,\lambda s}$, is shown in Fig. 13.9. Here we keep $A_{a,\lambda t}$ constant at 0.9. In the top panel we vary $R_{a,\lambda s}$ from 0.1 to 0.2, while keeping $R_{g,\lambda s}$ constant. As global temperatures drop, one expects that ice and snow cover will increase, raising the planetary albedo and thus further lowering surface temperatures (bottom right panel). As in the previous case, coupling between the two effects is very complex and I will not attempt to include it. Strong coupling in this case could lead

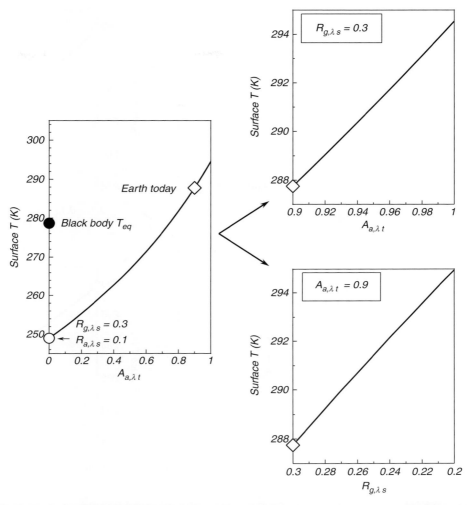

Fig. 13.8 Left: the Earth's equilibrium black body temperature (278 K) decreases to 249 K when the non-zero reflectivities of the atmosphere ($R_{a,\lambda s}$) and planetary surface ($R_{g,\lambda s}$) are taken into account. Present-day average temperature (diamond) requires long-wavelength absorption coefficient in the atmosphere $A_{a,\lambda t} = 0.9$. Right top: effect of increasing atmospheric long-wavelength absorption $A_{a,\lambda t}$ while leaving other parameters constant. Right bottom: effect of decreasing surface short-wavelength reflectivity $R_{g,\lambda s}$ (i.e. decreasing the albedo of the planet's surface) while leaving other parameters constant. One should expect positive feedback between $A_{a,\lambda t}$ and $R_{g,\lambda s}$: the increase in temperature caused by greater long-wavelength atmospheric absorption melts ice, which lowers short-wavelength reflectivity of the planet's surface. This effect is not included in the figure.

to *average* global temperatures that are near freezing, perhaps as happened on Earth during widespread late Proterozoic glacial periods.

Atmospheres are much more complicated systems than this simple toy model. Nonetheless, because radiative heat transfer is the only way in which a planet exchanges energy with its space environment, the model does capture some important qualitative aspects. First, without greenhouse warming the Earth might not be a habitable planet. Second, planetary

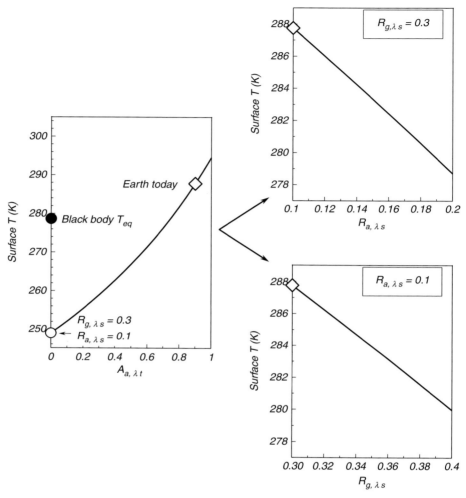

Fig. 13.9 Left panel is the same as in Fig. 13.8. Right top: effect of increasing atmospheric short-wavelength reflectivity $R_{a,\lambda s}$ while leaving other parameters constant. Right bottom: effect of increasing surface short-wavelength reflectivity $R_{g,\lambda s}$ while leaving other parameters constant. There is coupling between both parameters, as the temperature drop caused by increased reflection of sunlight by the atmosphere increases ice cover and therefore surface albedo. This effect is not included in the figure.

surface temperatures are determined by several radiative heat transfer parameters, of which atmospheric absorptivity in the infrared is one. Other parameters that have comparably strong effects are the short-wavelength reflectivity of the planet's surface and of the atmosphere. Relatively small changes in the values of any of these parameters can cause shifts in surface planetary temperatures of the order of 1–10 K. Although this may not sound like much, temperature changes of this magnitude can have potentially catastrophic effects on biological systems that have evolved to match very specific environmental conditions, especially if the rate of temperature change is significantly faster than the rate at which evolution operates. If significant areas of a planet have average surface temperatures that straddle the freezing point of abundant planetary volatiles (e.g. H_2O on Earth, CH_4 or C_2H_6 on Titan) temperature variations triggered by changes in one heat transfer parameter may have profound effects on planetary albedo that amplify the temperature perturbation, both towards cooling and warming.

Exercises for Chapter 13

13.1 Modify the equations in Section 13.2.1 to include adiabatic temperature change with elevation. Discuss how this affects the estimate of atmospheric scale height.

13.2 Using the equations in Section 13.2.2 as a guide, derive an approximate expression for Fe/Mg fractionation in an isothermal planetary magma ocean of composition $(Mg,Fe)_2SiO_4$. Estimate the extent of Fe/Mg fractionation in a 1000-km deep magma ocean in Earth, Mars and the Moon, assuming $T = 1800$ K.

13.3 Derive Wien's law, equation (13.30).

13.4 Derive Stefan–Boltzmann's law, equation (13.31), and find the value of the Stefan–Boltzmann constant as a function of Planck's constant, Boltzmann's constant and the speed of light (equation (13.32)).

13.5 Can you suggest a reason why humans do not perceive radiation with wavelengths longer than about 10 μm as heat?

13.6 Calculate the orbital radius that a planet must have in order to receive the same energy flux as the Earth from its star (i.e. the solar constant, ~ 1368 W m^{-2}) if the planet orbits:
  (i) a red dwarf with $T = 2600$ K and $r = 0.15\ r_{Sun}$
  (ii) a blue giant with $T = 10\,000$ K and $r = 1.7\ r_{Sun}$
  (iii) a blue supergiant with $T = 25\,000$ K and $r = 37\ r_{Sun}$,
where the stellar radii, r, are given in terms of the solar radius, r_{Sun}. Express your answer in astronomical units and km.

13.7 What is the equilibrium temperature of each of the three planets in **13.6**?

13.8 In what section of the spectrum does the electromagnetic energy received by each of the planets in **13.6** peak? For each of the three planets, discuss the possible nature of their atmospheres (e.g. are molecules likely to be stable?), the likelihood of life, and the nature of any possible life.

13.9 Comment on how much a star can conceivably differ from the Sun and still be able to sustain life as we know it.

13.10 Plot thermalized solar radiation emission flux as a function of heliocentric distance. Discuss the implications for remote sensing of internal energy flux.

13.11 A Dyson sphere is a stellar engineering project, conceived by the physicist Freeman Dyson, consisting of an artificial spherical cavity centered on the Sun. Design a Dyson sphere with an equilibrium temperature of 288 K, assuming that the medium between the Sun and the internal surface of the sphere does not interact with electromagnetic radiation of any wavelength.

13.12 Solve differential equation (13.49).

13.13 Solve for the energy fluxes emitted by the ground and by the active atmospheric layer, F_g and F_a (Section 13.3.6) using energy balance in the atmosphere rather than the surface. Compare to equations (13.61) and (13.62).

13.14 Estimate a range of possible surface temperatures for Venus that arise from equations (13.61) and (13.62). How do these temperatures compare with the observed temperature of the surface of Venus ($\sim$800 K)? Modify the toy model in Section 13.3.6 so as to include additional terms that might improve the agreement. Comment on your results.

14

Thermodynamics of life

In this final chapter we examine life, and in particular how life may have originated, from a strictly thermodynamic point of view. I will not get anywhere close to biochemistry, biophysics or genetics, nor will I offer a definition of life. Rather, I begin from a concept that everybody must agree upon. This is the fact that a necessary (but not sufficient!) component of the definition of life is that it is a process that never reaches thermodynamic equilibrium, for if thermodynamic equilibrium is reached then the process stops, and life is no more. Life must therefore be powered by a gradient in free energy, which for the only type of life that we know takes the form of a chemical potential gradient, i.e. a non-zero affinity. Catabolic metabolism (henceforth simply metabolism, as I will not discuss anabolic metabolism in detail) is a chemical reaction (or rather a set of coupled chemical reactions) that transfers chemical energy from reactants in an organism's inorganic environment, known as the substrate, to complex organic molecules inside the organism, such as ATP (adenosine triphosphate), that are capable of delivering this energy to structures where the chemical energy is transformed to mechanical energy (e.g. motion), electrical energy (e.g. conscience), electromagnetic energy (e.g. fireflies), etc.

Atmospheric composition, and in particular the oxidation state of the atmosphere, is one of the factors in understanding the origin of life. Strongly reducing atmospheres, rich in species such as CH_4, NH_3 and H_2S, are thought to be best suited for the synthesis of complex organic molecules. As we shall see, however, an atmosphere with these characteristics may present near insurmountable obstacles to the establishment of metabolic pathways that supply energy to living organisms. A somewhat more oxidized atmosphere, in which CO_2 is an abundant species, is better suited to the inception of catabolic processes. In the first section of this chapter we use the machinery of fluid equilibrium that we developed in Chapter 9 to place some constraints on the atmospheric evolution of the terrestrial planets.

14.1 Chemical evolution of post-nebular atmospheres

We seek to understand the chemical evolution of post-nebular atmospheres, i.e. of those atmospheres that accumulated on rocky planets after loss of the original hydrogen-dominated nebular atmospheres. The greatest problem that we face in trying to reconstruct these early environments is that we know very little about the nature of volatile sources (impactors and/or volcanoes?), their compositions, and the rate of supply of material. We are also uncertain about the volatile sinks at that time. Thermodynamics provides a way of analyzing the problem. By constructing a model of speciation in a gas phase we can examine how the nature of the atmosphere responds to changes in the bulk composition of the phase, temperature, density and gravitational acceleration.

We will use the Gibbs free energy minimization procedure described in Section 9.6.2 and Worked Example 9.8. The equations and discussion in this section follow closely what we did there, with a few modifications. We assume a simple atmosphere composed of C, H, O and N (four system components), and we consider a two-phase system, consisting of a gas phase in equilibrium with liquid water. The justification for assuming H_2O saturation is that oxygen isotopic compositions of detrital zircons suggest that liquid water existed on the Earth's surface as early as 4.2 billion years ago, and perhaps even earlier (Valley *et al.*, 2005; Valley, 2006). We wish to know what type of atmosphere could have been in equilibrium with planetary oceans. This is a two-phase four-component system, so it has four degrees of freedom. We choose to specify temperature and three compositional variables (concentration of the fourth component follows by difference). Pressure is not specified *a priori*, but is rather one of the variables that is solved for, via equation (13.2). The model ignores dissolution of gas species in liquid water, which may have non-negligible effects on the calculated gas phase concentrations of fairly soluble species, such as CO_2 and NH_3, but the qualitative trends are almost certainly correct despite this simplification.

Let the total number of mols per unit of planetary surface area of each of the system components be N_C, N_H, N_O and N_N (in atomic proportions, not molecules). *Note that these variables have units of number of mols per unit area, so their values reflect the total amount of volatile material added to the planet's surface.* We consider the following chemical species: $H_2O_{(vapor)}$, $H_2O_{(liquid)}$, CH_4, H_2, CO_2, CO, N_2, NH_3 and O_2, and use n_i for the number of *mols per unit area* of species i. Other species that could be present in small concentrations include formaldehyde (CH_2O) and hydrogen cyanide (HCN). We will show that their concentrations can be reliably estimated *a posteriori*. In fact, the same is true of carbon monoxide, but we choose to leave it in the Gibbs free energy minimization calculation anyway.

The total number of mols of gas species per unit area is given by:

$$n_t = n_{H_2O(vapor)} + n_{H_2} + n_{CH_4} + n_{CO_2} + n_{CO} + n_{N_2} + n_{NH_3}. \tag{14.1}$$

As in Worked Example 9.8, oxygen is not included in (14.1) because its concentration is vanishingly small, so that it is simpler to write its Gibbs free energy in terms of oxygen fugacity (equation (9.96), see also below). The pressure at the planet's surface is given by equation (13.2), which we expand as follows using (14.1) and molecular weights in kg atoms:

$$P = \left[0.018 n_{H_2O(vapor)} + 0.002 n_{H_2} + 0.0016 n_{CH_4} + 0.044 n_{CO_2} \right.$$
$$\left. + 0.028 n_{CO} + 0.028 n_{N_2} + 0.017 n_{NH_3} \right] g. \tag{14.2}$$

We have four mass balance equations of the type of (9.92), as follows:

$$\varphi_1 = RT \left[2 n_{H_2O(vapor)} + 2 n_{H_2O(liquid)} + 2 n_{H_2} + 4 n_{CH_4} + 3 n_{NH_3} - N_H \right] = 0$$
$$\varphi_2 = RT \left[n_{H_2O(vapor)} + n_{H_2O(liquid)} + 2 n_{CO_2} + n_{CO} - N_O \right] = 0$$
$$\varphi_3 = RT \left[n_{CO_2} + n_{CO} + n_{CH_4} - N_C \right] = 0 \tag{14.3}$$
$$\varphi_4 = RT \left[2 n_{N_2} + n_{NH_3} - N_N \right] = 0.$$

Atmospheric pressures are typically low enough that the ideal gas approximation is perfectly acceptable. Saturation of the atmosphere in liquid water is therefore given by:

$$\frac{\mu_{1,T}^{0,H_2O(liquid)}}{RT} - \frac{\mu_{1,T}^{0,H_2O(vapor)}}{RT} - \ln P - \ln\left(\frac{n_{H_2O(vapor)}}{n_t}\right) = 0. \qquad (14.4)$$

We write six partial derivatives of the function Γ (equation (9.93)), one each for H_2, CH_4, CO_2, CO, N_2 and NH_3. Because we assume ideal gas behavior we omit the fugacity coefficients from the partial derivatives (compare equation (9.94)). This leads to a considerable simplification in the calculation procedure (Software Box 14.1). The six derivatives $\partial\Gamma/\partial n_i$ are:

$$\frac{\mu_{1,T}^{0,H_2}}{RT} + 1 + \ln P + \ln\left(\frac{n_{H_2}}{n_t}\right) - \frac{n_{H_2}}{n_t} + 2\lambda_1 = 0$$

$$\frac{\mu_{1,T}^{0,CH_4}}{RT} + 1 + \ln P + \ln\left(\frac{n_{CH_4}}{n_t}\right) - \frac{n_{CH_4}}{n_t} + 4\lambda_1 + \lambda_3 = 0$$

$$\frac{\mu_{1,T}^{0,CO_2}}{RT} + 1 + \ln P + \ln\left(\frac{n_{CO_2}}{n_t}\right) - \frac{n_{CO_2}}{n_t} + 2\lambda_2 + \lambda_3 = 0$$

$$\frac{\mu_{1,T}^{0,CO}}{RT} + 1 + \ln P + \ln\left(\frac{n_{CO}}{n_t}\right) - \frac{n_{CO}}{n_t} + \lambda_2 + \lambda_3 = 0 \qquad (14.5)$$

$$\frac{\mu_{1,T}^{0,N_2}}{RT} + 1 + \ln P + \ln\left(\frac{n_{N_2}}{n_t}\right) - \frac{n_{N_2}}{n_t} + 2\lambda_4 = 0$$

$$\frac{\mu_{1,T}^{0,NH_3}}{RT} + 1 + \ln P + \ln\left(\frac{n_{NH_3}}{n_t}\right) - \frac{n_{NH_3}}{n_t} + 3\lambda_1 + \lambda_4 = 0.$$

Software Box 14.1   Speciation calculation in C–H–O–N atmospheres

The *Maple* worksheet **highway_to_hell.mw** contains a procedure, `atmos`, that calculates speciation in a C–H–O–N gas phase in equilibrium with liquid H_2O at low pressure (ideal gas behavior is assumed) by Gibbs free energy minimization. It solves the system of equations (14.2) to (14.7). The procedure call is as follows.

```
atmos (T in C, mols H, mols O, mols C, mols N, g in m s⁻²)
```

Output is sent to the terminal and is self-explanatory. The procedure can be included in a `do` loop that varies the input parameters, in order to calculate diagrams such as those in Figures 14.1 to 14.5.

Thermodynamic data are contained in tab-delimited format in the file **aatmosdata**, and also in the spreadsheet sh_data. The order in which the data are to be loaded (row id numbers) is listed in the heading of procedure `atmos`.

The error message "Water boiled off" means that a solution of the system of equations saturated in liquid water was not found. The error message "No convergence" means that the procedure failed to find a solution within the maximum number of iterations specified by the variable `hmax`.

The partial derivative of the Gibbs free energy of oxygen follows equations (9.96) and (9.100):

$$\frac{\mu_{1,T}^{0,O_2}}{RT} + \ln fO_2 + 2\lambda_2 = 0 \tag{14.6}$$

and the partial derivative for liquid water, ignoring the effect on pressure on the chemical potential of liquid H_2O, is (compare equation (9.101) for graphite):

$$\frac{\mu_{1,T}^{0,H_2O(liquid)}}{RT} + 2\lambda_1 + \lambda_2 = 0. \tag{14.7}$$

Equations (14.2)–(14.7) constitute a system of 14 equations in the 14 unknowns: $n_{H_2O(liquid)}$, $n_{H_2O(vapor)}$, n_{H_2}, n_{CH_4}, n_{CO_2}, n_{CO}, n_{N_2}, n_{NH_3}, $\ln(fO_2)$, P, λ_1, λ_2, λ_3 and λ_4 (n_t is given by (14.1)). Numerical solution with *Maple* is straightforward and is briefly discussed in Software Box 14.1. Recall that the four Lagrange multipliers are necessary in order to solve the constrained minimum problem (Section 9.6.2), but we have no use for their numerical values.

Let us first examine solution sets at 25 °C for a planet with Earth's gravitational acceleration and with total volatile content adjusted so as to yield an atmospheric mass of order 10^4 kg m^{-2}, comparable to the present day Earth. We start with the following values.

$$N_H = 1650 \times 10^3 \text{ mols m}^{-2}$$

$$N_O = 650 \times 10^3 \text{ mols m}^{-2}$$

$$N_C = 350 \times 10^3 \text{ mols m}^{-2}$$

$$N_H = 100 \times 10^3 \text{ mols m}^{-2}$$

The total volatile mass that results is equal to 1.765×10^4 kg m^{-2}, which is almost twice the present-day terrestrial atmospheric mass, but as we shall see some of this mass condenses as liquid water. The relative proportions of the four components are similar to cometary material (see Lodders & Fegley, 1998, Tables 15.3 and 15.4), but I emphasize that there is no special significance to this starting composition, other than being plausible. The point of this exercise is to examine how atmospheric speciation responds to *changes* in the model parameters.

Figure 14.1 shows the effect of changing the bulk contents of H, C and O independently, i.e. the bulk content of one each of these elements changes, while all the others remain constant and equal to the values listed above. The independent variable in each graph is the bulk atomic fraction of the component that varies. Temperature is kept constant at 25 °C, and $g = 9.8$ m s^{-2}. The top panels show the oxygen fugacity and mol fraction of each of the gas species except CO, which never rises above $\sim 10^{-10}$ and which we discuss later. The bottom panels display atmospheric pressure at the planet's surface and the amount of H_2O that condenses as liquid, converted to thickness of the water column (55.56×10^3 mols m^{-2} = 1 m of water depth).

The calculations show a steep jump in species distribution, between a H_2–NH_3–CH_4 atmosphere with virtually no CO_2, and a CH_4–CO_2 atmosphere with small but non-zero contents of NH_3 and H_2. The atmosphere becomes less reduced in response to decreasing bulk H content or increasing bulk O or C contents, but in every case there is a sudden jump in species abundances over very narrow bulk composition intervals. The behavior is in some

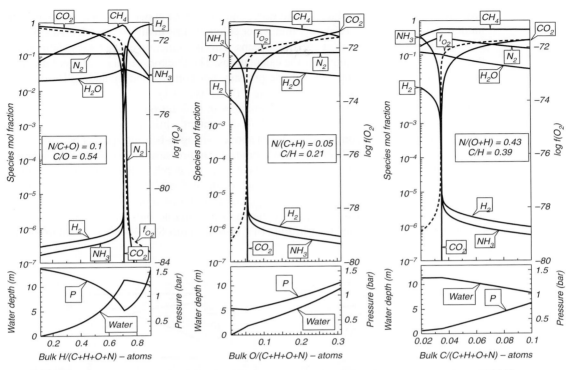

Fig. 14.1   Species distribution in a C–H–O–N atmosphere calculated by Gibbs free energy minimization at 25 °C. The top diagrams show mol fractions in the gas phase (solid curves, port axes) and oxygen fugacity (dashed curves, starboard axes). Bottom diagrams show thickness of the liquid water column averaged over the planet's surface, and atmospheric pressure at the planet's surface. Bulk content of H, O or C is varied in each plot, while keeping bulk contents of all other components constant. Note the sharp transition between H_2–NH_3 atmospheres and CH_4–CO_2 atmospheres.

ways analogous to a phase transition (Chapter 7), in the sense that there are two possible "phases": a CO_2-bearing atmosphere with very little ammonia and molecular hydrogen, or a CO_2-free atmosphere dominated by methane, ammonia and molecular hydrogen. Atmospheres in which CO_2, NH_3 and H_2 are all significantly abundant are not thermodynamically stable. Over a very narrow compositional interval spanning the transition the equilibrium species distribution consists chiefly of $CH_4 + H_2O \pm N_2$. Atmospheres with this composition are unlikely to be common, however, as they require very fine-tuned conditions (Fig. 14.1).

The CO_2/CH_4 ratio increases with the oxidation state. CO_2 becomes the dominant atmospheric species for bulk compositions sufficiently rich in O or poor in H. Oxygen fugacity, however, remains $< \sim 10^{-70}$ bar even in atmospheres with $\sim$90 mol% CO_2. This oxygen fugacity is below the stability limit of hematite (e.g. Fig. 11.7 and 11.8). Thus, Fe^{2+}-rich Archaean oceans are consistent with a CO_2-dominated atmosphere (Section 11.5). Increasing oxidation of a CH_4–CO_2 atmosphere always raises atmospheric pressure, as the light carbon species, CH_4, is replaced by the much heavier CO_2. The thickness of the water column, however, responds differently depending on whether oxidation is driven by hydrogen loss, which produces CO_2 at the expense of liquid water and CH_4, or by an increase in bulk oxygen content, which produces liquid water and CO_2 at the expense of CH_4. Note that the models in Fig. 14.1 yield relatively thin water columns, less than $\sim$10 m, which are orders of magnitude thinner than the terrestrial oceans. This is a function of the total volatile mass that I arbitrarily chose as input for the calculations. The thickness of the water column can be varied without changing atmospheric pressure nor composition by adding H and O to the bulk composition in a proportion of 2 to 1 (Exercise 14.1).

These trends are largely unaffected by changes in bulk nitrogen content. The left panel in Fig. 14.2 shows the effect of varying bulk N content while keeping bulk H, O and C contents fixed at the starting values chosen above. Nitrogen-bearing species become more abundant with increasing N content, and atmospheric pressure increases, but the relative proportions of C–H–O species, oxygen fugacity and the depth of the liquid water column remain constant. An ammonia-dominated atmosphere forms if the bulk composition is sufficiently rich in both nitrogen and hydrogen. This is shown in the right panel of Fig. 14.2, in which gas speciation is tracked as a function of variable bulk H content, for a bulk N content ten times greater than in Fig. 14.1 (atomic ratio $N/(C+O) = 1$, compared to

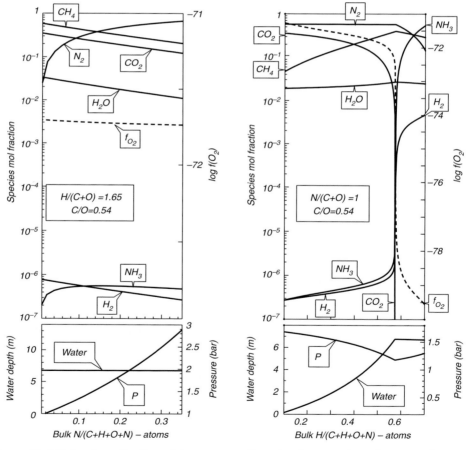

Fig. 14.2   Same as Fig. 14.1, focusing on the effects of nitrogen content. Formation of an ammonia–methane atmosphere requires a bulk composition very rich in N and H (right panel).

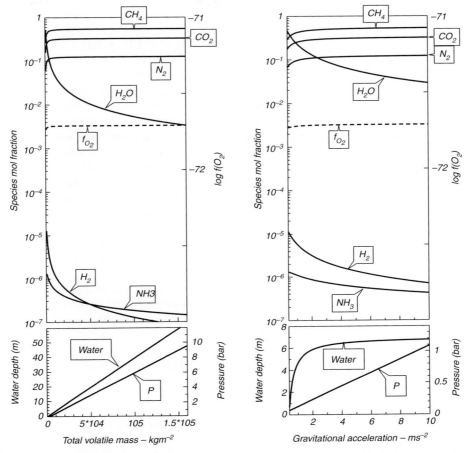

Fig. 14.3 Effects of total volatile mass and gravitational acceleration on atmospheric speciation. Atmospheric pressure varies directly and almost linearly with both variables. Decrease in atmospheric pressure is accompanied by evaporation of liquid water, and increase in X(H$_2$O) in the gas phase. The effect of gravitational acceleration is strongly non-linear for small planets, accounting at least in part for Mars's predicament.

N/(C + O) = 0.1 in the equivalent plot in Fig. 14.1). In this case too there is a rapid transition from a strongly reducing NH$_3$–CH$_4$ atmosphere to a N$_2$–CO$_2$ atmosphere, in response to hydrogen loss.

The effects of changing total volatile mass and gravitational acceleration are shown in Fig. 14.3. Increasing both of these variables raises atmospheric pressure almost linearly but has negligible effects on atmospheric composition, except at very low pressure, or for H$_2$O vapor content, at all pressures. The effect on X(H$_2$O) is a consequence of saturation in liquid water: the chemical potential, and hence partial pressure, of H$_2$O vapor is fixed, so its concentration must decrease as pressure increases. Decreasing pressure is accompanied by an increase in the concentration of molecular hydrogen because the methane oxidation reactions:

$$CH_4 + H_2O \rightleftarrows CO + 3H_2$$

and

$$CH_4 + 2H_2O \rightleftharpoons CO_2 + 4H_2$$

both have large and positive $\Delta_r V$, so they are favored by low pressure. As the concentration of H_2 increases with decreasing pressure so does the ammonia concentration. Note in passing that although all the calculations are performed at constant pressure, assumed to be the pressure at the planet's surface given by the atmospheric mass, the results in Fig. 14.3 can also be used to infer that the *equilibrium* species distribution does not change significantly with elevation. The emphasis is on equilibrium because, as we saw in Chapter 12, photoactivated processes affect non-equilibrium species distribution with elevation.

The depth of the water column varies linearly with total volatile mass, but it responds in a non-linear fashion to changes in gravitational acceleration at constant volatile mass. In the latter case evaporation of water (required to preserve the chemical potential of H_2O in the atmosphere) depletes a liquid reservoir of definite size. Of course gravitational acceleration does not change with time once a planet has accreted. Rather, this result should be seen as a demonstration of the difficulty that a small planet such as Mars may have had in holding on to its oceans and lakes.

The concentrations of minor species can be calculated *a posteriori*, because they have a negligible effect on the Gibbs free energy of the system and on the mass balance constraints (equations (14.3)). For example, we can calculate the concentrations of hydrogen cyanide and formaldehyde from the equilibria:

$$NH_3 + CH_4 \rightleftharpoons HCN + 3H_2 \tag{i}$$

and:

$$H_2O + CH_4 \rightleftharpoons CH_2O + 2H_2 \tag{ii}$$

for which we have:

$$
\begin{aligned}
X_{HCN} &= \frac{X_{NH_3} \cdot X_{CH_4}}{\left(X_{H_2}\right)^3 \cdot P^2} \exp\left(-\frac{\Delta_r G^{0,(i)}}{RT}\right) \\
X_{CH_2O} &= \frac{X_{H_2O} \cdot X_{CH_4}}{\left(X_{H_2}\right)^2 \cdot P} \exp\left(-\frac{\Delta_r G^{0,(ii)}}{RT}\right).
\end{aligned}
\tag{14.8}
$$

Representative results are shown in Fig. 14.4 (left panel), together with CO concentrations, which were calculated as part of the Gibbs free energy minimization procedure, but could also have been calculated as in equations (14.8). Formaldehyde and other organic molecules in which carbon is present in the same oxidation state (CH_2O) are essential biological building blocks (Section 14.2.1). These results suggest that their equilibrium atmospheric concentrations are unlikely to ever have been significant, even in reducing atmospheres. The synthesis of these molecules must have been the outcome of other processes that took place at the inception of life on Earth.

For simplicity sulfur is not included in the model calculations, because it is likely to have always been less abundant than C, H, O and N. Even if we ignore the absolute abundances of sulfur species, it is important to constrain their oxidation state, partly for biological reasons.

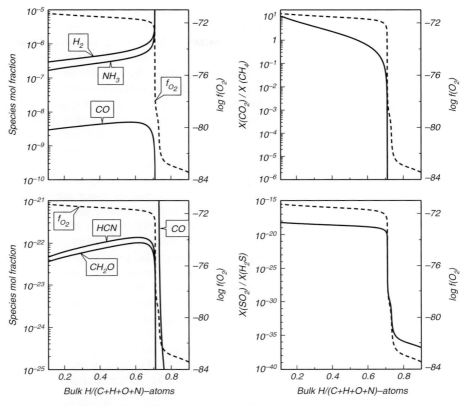

Fig. 14.4 Left: atmospheric concentrations of the minor species CO, CH_2O and HCN (note scale break between both diagrams). Ammonia and molecular hydrogen shown for comparison. Right: SO_2/H_2S concentration ratio, compared to CO_2/CH_4 ratio. Sulfur is present as a reduced species even in atmospheres that have ten times more CO_2 than CH_4.

We do this by means of the equilibrium:

$$H_2S + O_2 \rightleftarrows SO_2 + H_2 \tag{iii}$$

which yields:

$$\frac{X_{SO_2}}{X_{H_2S}} = \frac{f_{O_2}}{X_{H_2} \cdot P} \exp\left(-\frac{\Delta_r G^{0,(iii)}}{RT}\right). \tag{14.9}$$

Representative results are shown in the right panel of Fig. 14.4, where they are compared to the CO_2/CH_4 ratios for the same range of atmospheric compositions. An important conclusion is that carbon oxidizes more readily than sulfur, so that a CO_2-dominated atmosphere can still be reducing enough to contain H_2S and virtually no SO_2. We will return to this later in this chapter.

Photodissociation of CH_4, NH_3 and H_2O molecules produces free hydrogen which, as we saw in Chapter 13, can be lost by thermal escape. This is an irreversible process, for the same relationship between kinetic energy and gravitational binding energy (equation (13.1)) that makes hydrogen loss possible makes it impossible for a planet to capture hydrogen atoms.

If surface conditions are such that the atmosphere contains significant H_2O vapor, then hydrogen loss leads inevitably to a dry planet with an atmosphere dominated by carbon dioxide and nitrogen, regardless of how reduced the initial atmospheric composition might have been (Fig. 14.1 left panel, and Fig. 14.2 right panel). *In the absence of other processes a CO_2-rich atmosphere is a terminal state, as carbon–oxygen gas species are not readily lost by atmospheric escape processes.* Protracted hydrogen loss is the most likely explanation for the nature of the present-day atmospheres of Venus and Mars. Neither the Earth nor Titan fit this picture, however. Titan's atmosphere consists chiefly of nitrogen, with minor amounts of CH_4, H_2 and other reduced carbon species. Thus, it must be located on the reduced side of the speciation transition (Fig. 14.1 and 14.2). Yet according to equation (13.1) and Fig. 13.1 hydrogen loss from Titan's atmosphere must be at least as efficient as in Venus and Mars. Oxidation of Titan's atmosphere is prevented by its very low surface temperature, which keeps the partial pressure of H_2O ($\sim$ the saturation vapor pressure over ice at very low temperature) virtually equal to 0. The example of Titan emphasizes the importance of H_2O vapor as the oxygen source for atmospheric oxidation. It is the only abundant molecule that contains both H and O, so that photodissociation followed by hydrogen loss makes oxygen available. If there is no H_2O in the atmosphere, for instance because it is sequestered in low-temperature ice, then there is simply no source of oxygen.

The Gibbs free energy minimization model also allows us to examine the effect of temperature on the equilibrium species distribution. Representative results are shown in Fig. 14.5, in which I have adjusted the bulk composition so that there are $\sim$100 m of liquid water in equilibrium with a CO_2-rich atmosphere at 25 °C and $\sim$1.5 bar pressure. As temperature increases water must evaporate in order to preserve the equilibrium saturation vapor pressure. This raises the concentration of H_2O vapor in the atmosphere, and also atmospheric mass and hence atmospheric pressure. At temperatures approaching 100 °C, H_2O becomes the dominant atmospheric component, even if water may be kept from boiling by the high atmospheric pressure. This condition, known as a steam atmosphere, greatly accelerates the rate of hydrogen loss because it increases both the concentration of H_2O in the atmosphere, and hence the rate of photodissociation (equation (12.83)), and its temperature (equation (13.1)). Temperature increase also raises the equilibrium H_2 and NH_3 concentrations by several orders of magnitude.

Buildup of a steam atmosphere is thought to be a self-reinforcing process, by virtue of the strong infrared absorption of H_2O molecules (Section 13.3.6). It is possible that once the concentration of H_2O vapor exceeds certain threshold it causes a runaway temperature increase that results in complete dessication of the planet's surface and CO_2 accumulation in the atmosphere. As long as there is liquid water CO_2 may be scavenged by carbonate precipitation at a rate that is largely controlled by the rate at which silicate weathering supplies cations such as Ca^{2+}, Mg^{2+} and Fe^{2+} in aqueous solution. This process is known as the *Urey reaction* (Urey, 1952). Once liquid water disappears this scrubbing mechanism is no longer possible, and CO_2 atmospheric concentration cannot decrease. Hydrogen escape rates are likely to have always been lower in Earth than in Venus and Mars (e.g. Fig. 13.1). Whether or not Venus ever had a steam atmosphere, it may have lost its water early on, and with it the capability of controlling its atmospheric CO_2 concentration. Mars probably underwent slower dessication and oxidation, and eventual freezing of its remaining surface water. Hydrogen loss from Earth has been slow enough to allow much of its surface water to persist over the age of the solar system.

If atmospheric methane concentration decreases to trace levels then the atmosphere loses its ability to buffer oxygen fugacity. As long as there is water vapor available that can

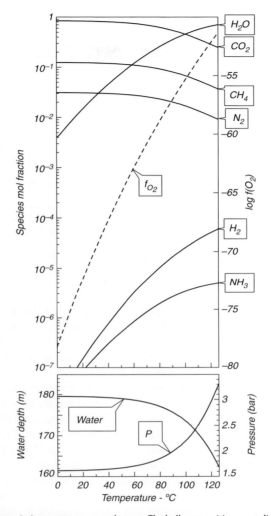

Fig. 14.5 Response of atmospheric speciation to temperature changes. The bulk composition was adjusted so as to generate ~180 m of liquid water in equilibrium with a CO$_2$-dominated atmosphere at 25 °C. Increasing temperature raises the mol fraction of H$_2$O vapor in the atmosphere and atmospheric pressure. For this bulk composition H$_2$O becomes the dominant atmospheric component at ~100 °C, giving rise to a steam atmosphere.

undergo photodissociation oxygen concentration could in principle increase freely, but this is generally not the case. Under equilibrium conditions atmospheric oxidation is limited by the availability of reduced species at the planet's surface, chiefly ferrous iron, sulfides and reduced carbon. Figure 11.8 shows that oxidation of iron to its terminal ferric state buffers equilibrium oxygen fugacity to $\log(f\mathrm{O}_2) \sim -68.6$ (at 25 °C). Sulfide oxidation can be modeled by means of the following reaction, that produces anhydrite and hematite at the expense of pyrite and calcite:

$$4\mathrm{FeS}_2 + 8\mathrm{CaCO}_3 + 15\mathrm{O}_2 \rightleftarrows 8\mathrm{CaSO}_4 + 2\mathrm{Fe}_2\mathrm{O}_3 + 8\mathrm{CO}_2,$$

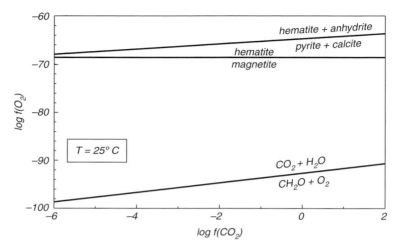

Fig. 14.6 Equilibrium oxygen fugacities buffered by organic matter, ferrous–ferric equilibrium, and sulfide–sulfate equilibrium.

and that is at equilibrium for: $\log(f O_2) = (8/15)\log(f CO_2) - 64.7$. These two oxidation boundaries are shown in Fig. 14.6. If a planet has internal activity and is able to resupply its surface with reduced species such as pyrite, olivine, pyroxene and other iron silicates via volcanism and tectonism then its equilibrium atmospheric oxygen fugacity cannot exceed these boundaries. Venus is still active today, so that the oxygen liberated by photodissociation of H_2O and hydrogen escape in the geologic past must be present in the lithosphere, combined with iron, sulfur and other elements of variable oxidation state. There is no H_2O left to supply oxygen, and there is almost certainly an excess of reduced species of internal origin. The atmosphere of Venus is a thermodynamic dead end. The case of Mars is somewhat different. First, it still has frozen H_2O on its surface, and perhaps liquid or frozen H_2O in its shallow subsurface. Second, Mars may have largely lost its capability to transport reduced species from the planet's interior to its surface, and the surface is already thoroughly oxidized. The possibility therefore exists that oxygen fugacity in the Martian atmosphere exceeds the iron and sulfur oxidation boundaries (Fig. 14.6), as attested by the existence of perchlorates on its surface (Chapter 11).

The Earth is of course different, because its atmosphere is very far from thermodynamic equilibrium with its surface, and in particular with the biosphere. Carbon in organic matter typically has the same oxidation state as in formaldehyde, CH_2O, so that equilibrium between organic matter and atmospheric oxygen can be modeled by the reaction:

$$CH_2O + O_2 \rightleftarrows CO_2 + H_2O.$$

which results in: $\log(f O_2) = \log(f CO_2) - 92.67$. This reaction, also shown in Fig. 14.6, is located more than 90 orders of magnitude (!) below the actual oxygen fugacity in the terrestrial atmosphere. In a planet covered with organic matter it would not be even possible for ferrous iron or sulfide to oxidize, if the atmosphere was in thermodynamic equilibrium with the biosphere. Upkeep of the oxygen content of the terrestrial atmosphere is a non-equilibrium process driven by supply of solar energy.

14.2 Thermodynamics of metabolic processes

14.2.1 Carbon fixing and respiration of reduced organic carbon

Chemical interactions between living organisms and their substrate are of two principal types. One of these consists of reactions that reduce oxidized carbon absorbed from the environment, typically CO_2, in which C has an oxidation state of $+4$. The products of these reactions are organic molecules in which the oxidation state of C is characteristically 0. The simplest such compound is formaldehyde: CH_2O. Reactions of this type are called *carbon-fixing reactions*. The reduced carbon species generated by these reactions fulfill two distinct biological roles. Firstly, they are used by living matter to synthesize all of the complex organic molecules needed for life – we will not discuss this role any further but we will keep in mind that this process, which is the essence of anabolic metabolism, is the reason why carbon fixation is essential for life as we know it. Secondly, reduced carbon species can serve as reactants for metabolic reactions that transfer energy from the substrate to the living organism (i.e. catabolic metabolism). These reactions are called *respiration reactions* and are the other type of chemical interactions that take place between living organisms and their substrate. Some respiration reactions use reduced carbon as a reactant. This is most commonly organic carbon that was fixed by reactions of the first type, although reduced carbon of inorganic origin (e.g. CH_4 in volcanic gases) can also be used. However, many respiration reactions exist in which no carbon species are involved (more on this in the next section).

Because their role is to transfer energy from the substrate to energy-carrier molecules within living cells, a necessary condition of respiration reactions is that they must be thermodynamically spontaneous. Any combination of chemical species with $\Delta_r G < 0$ (or $\mathscr{E} > 0$) could, in principle, constitute the basis for respiration. In contrast, with one important exception that we will discuss in detail below, carbon-fixing reactions have $\Delta_r G > 0$. They must therefore be sustained by a constant energy flux from the environment. Perhaps the most familiar carbon-fixing reaction is oxygenic photosynthesis, which we can model as follows:

$$CO_{2(g)} + H_2O_{(l)} + h\nu \rightarrow CH_2O_{(g)} + O_{2(g)}, \tag{14.10}$$

where $h\nu$ represents electromagnetic radiation in the visible part of the spectrum and $CH_2O_{(g)}$ is formaldehyde gas. Actual photosynthesis produces more complex molecules with the same oxidation state as formaldehyde, such as glucose ($C_6H_{12}O_6$), but the thermodynamic relations are qualitatively the same as those that can be derived from equation (14.10). The standard state Gibbs free energy change for this reaction is 528.96 kJ per mol of formaldehyde. At equilibrium we thus have:

$$\frac{f_{CH_2O} \cdot f_{O_2}}{f_{CO_2}} \approx 10^{-93} \tag{14.11}$$

which, as we saw in the previous section, means that the terrestrial atmosphere ($f_{O_2} \approx$ 0.2 bar) is very far from being in thermodynamic equilibrium with reduced organic matter. Reaction (14.10) is a photochemical reaction that fixes carbon and liberates molecular oxygen against a huge chemical potential gradient by converting an electromagnetic energy flux to chemical bonding energy (see also Worked Example 12.3). Photosynthesis based on other elements that can change oxidation state also exists. Two important examples are based on sulfur and iron. The former is carried out by green sulfur bacteria, that ingest H_2S and excrete elemental sulfur and which we may model as:

$$CO_{2(g)} + 2H_2S_{(g)} + h\nu \rightarrow CH_2O_{(g)} + H_2O_{(l)} + 2S_{(xtal)} \qquad (14.12)$$

while iron-based photosynthesis occurs in purple bacteria and we model it as follows:

$$CO_{2(g)} + 4Fe_{(aq)}^{2+} + 4H_{(aq)}^{+} + h\nu \rightarrow CH_2O_{(g)} + H_2O_{(l)} + 4Fe_{(aq)}^{3+}. \qquad (14.13)$$

With the standard states defined as in (14.12) and (14.13), the values of $\Delta_r G^0$ for these reactions are 121.8 and 351.5 kJ per mol of formaldehyde, respectively. As for oxygenic photosynthesis, these reactions are not spontaneous for any reasonable value of the chemical potential of formaldehyde, and rely on a flux of electromagnetic energy in order to fix carbon.

Reduced organic molecules are oxidized by respiration reactions. A simple example is the inverse of (13.1):

$$CH_2O + O_2 \rightarrow CO_2 + H_2O. \qquad (14.14)$$

This reaction proceeds spontaneously with a large affinity value (e.g. equation (14.11)). Note, however, that even though respiration reactions are thermodynamically spontaneous their rate constants are very low (or equivalently, their activation energies are very high, Section 12.4.1). If this were not the case then organic matter would spontaneously (and rather unpleasantly) combust in the terrestrial atmosphere. Respiration reactions are catalyzed in living cells by organic molecules known as enzymes, that effectively lower their activation energies. Reaction (14.14) is the respiration process of choice of, for instance, self-aware eukaryotes, but many prokaryotes are able to oxidize reduced carbon with other oxidants. Thus, iron-reducing bacteria carry out the inverse of (14.13), in bogs and other euxinic environments:

$$CH_2O + H_2O + 4Fe^{3+} \rightarrow CO_2 + 4Fe^{2+} + 4H^{+}. \qquad (14.15)$$

Denitrifying bacteria use nitrate in order to oxidize organic matter, in a process that is crucial in maintaining a constant atmospheric concentration of N_2 (which is in turn oxidized to nitrate inorganically during electrical storms, and also by organic processes):

$$5CH_2O + 4NO_3^{-} + 4H^{+} \rightarrow 5CO_2 + 2N_2 + 7H_2O. \qquad (14.16)$$

Sulfate-reducing bacteria can make a living by spoiling organic matter, including eggs:

$$2CH_2O + SO_4^{2-} + 2H^{+} \rightarrow 2CO_2 + H_2S + 2H_2O. \qquad (14.17)$$

Reactions (14.14)–(14.17) have two common characteristics: they are all thermodynamically spontaneous, and they all oxidize carbon to its terminal oxidation state, corresponding to CO_2. The ultimate source of the energy that is liberated in these reactions is solar electromagnetic radiation. The reduced carbon species that are produced by photochemical carbon-fixing reactions act as energy carriers. There are other respiration reactions, however, that do not rely on the availability of reduced carbon of organic origin. As we shall see, a strong case can be made that a reaction of this kind must have been the primordial metabolic process that powered the first living organisms.

14.2.2 Respiration without reduced organic carbon

Respiration is a continuous chemical reaction. This is only possible in an open system, in which there is an uninterrupted supply of reactants and evacuation of products, for in a closed system equilibrium would unavoidably be reached. Any spontaneous chemical reaction ($\Delta_r G < 0$, or $\mathscr{E} > 0$) could, in principle, constitute the basis for respiration, but we can pare down the list. Life is based on fundamental units, the cells, and metabolism takes place inside cells, where the energy liberated by the respiration reaction is used to produce energy-carrier molecules such as ATP. The supply of reactants to, and evacuation of products from, the open system must take place by diffusion across the cell walls. This means that respiration must be based on substances that are either (a) soluble in the substance that makes up most of the cell (H_2O for the only type of life that we know) or (b) able to exist as colloidal suspensions in that substance or (c) gases (which generally dissolve in the cell's constitutive substance anyway). We shall return to this in a moment.

A requirement for life is that metabolic reactions be able to proceed continuously. This necessitates the maintenance of a continuous chemical potential gradient inside the cell, which relies on an uninterrupted supply of reactants. By this I mean that some process, or processes, external to the system in which metabolic reactions take place, is able to supply the system with a steady flow of chemical species that are out of equilibrium and that can therefore react to a different set of chemical species with lower Gibbs free energy – this is the essence of respiration. Terrestrial life exploits two sources of out-of-equilibrium chemical species to power respiration. One of them, that we discussed in the previous section, is based on production of reduced carbon species by photosynthetic reactions. This source is kept active by electromagnetic energy from the Sun. The other source, that is kept active by the planet's internal energy, is the difference in thermodynamic states between the surface of the planet and its interior.

Internal processes such as volcanism and orogeny supply the surface of the Earth with mineral assemblages and fluids that are stable at temperatures higher than, and oxygen fugacities lower than, those at the surface of the Earth. The bulk of these minerals are silicates, which at the surface of the Earth react (albeit slowly) to yield assemblages stable at lower temperature – this is the process that we call weathering. The following reaction is a simple model for weathering of forsterite on the ocean floor:

$$Mg_2SiO_4 + 2H_2O \rightleftarrows 2Mg^{2+} + SiO_2 + 4OH^-. \tag{14.18}$$

Formation of ocean floor at mid-ocean ridges continuously replenishes the reaction with forsterite, so that in principle weathering of Mg silicates could be the basis for a metabolic process. Yet, as far as I know, there is no respiration based on silicate weathering. The chief reason for this is, I think, the fact that high-temperature crystalline silicates cannot dissolve in water while preserving their crystalline structure, and hence their "high-temperature" thermodynamic properties, nor do they form colloidal suspensions, so there is no pathway for these reactants (i.e. the high-temperature phases) to enter cells. This essentially rules out the possibility of using the difference between the equilibration temperature of igneous and metamorphic assemblages and the temperature at the planet's surface as a source of reactants for respiration, i.e. of chemical species out of equilibrium.

The difference in oxidation state between the Earth's interior and its surface is, on the other hand, the basis for a large number of respiration processes in many groups of bacteria and archaea. In the present-day Earth this difference corresponds to many orders of magnitude in oxygen fugacity. The fugacities of reduced species such as H_2, CH_4 and H_2S in mantle and crustal fluids may be high enough that chemical reactions with large affinities result when these fluids enter surficial environments in which there are non-negligible chemical potentials of species such as CO_2, SO_4^{2-}, Fe^{3+}, NO_3^- or O_2. Also important, and used in respiration by some microorganisms, is the fact that transition elements can exist in multiple oxidation states, and that their oxidation states in igneous and metamorphic assemblages are characteristically low. Ferrous iron is the most common example, but respiration processes based on reduced forms of Mn, Se, As, Cu and other elements are also known. Respiration reactions in these cases depend on the ability of the reduced cation to dissolve in water and preserve its low oxidation state until the solution enters a cell and the oxidation reaction can take place inside the cell.

Respiration based on the difference in thermodynamic states between the Earth's surface and its interior is always based on redox reactions. The continuous supply of metabolic reactants, which we call nutrients, depends on the ability of the planet to transport reduced chemical species from its interior to its surface, and this ability depends on the planet's internal heat.

14.2.3 Speculations about the origin and evolution of metabolism on Earth

There is strong evidence that all life on Earth has a common ancestor (see, for example, Theobald, 2010). This does not necessarily mean that life originated only once, but rather that only our lineage managed to survive the conditions of the early Earth. If there was more than one independent biogenesis then the product of one of these processes was able to outcompete the rest. It is interesting to speculate on what may be some of the reasons for the common ancestry of all extant terrestrial life.

Whether replication and inheritance mechanisms originated before or after metabolism (and this is intensely debated by origin of life researchers), the first living organisms had to metabolize. What is the primordial respiration process likely to have been? We can begin with an inventory of available nutrients in the present-day Earth, and see how it extrapolates to the early Earth. At present most reduced species originate in one or more of the following four ways.

(i) As relatively minor components in volcanic gases, chiefly H_2, H_2S and CH_4, but one could also include NH_3.

(ii) As reduced cations with variable oxidation state in volcanic rocks (chiefly ocean floor). The most important by far is Fe^{2+}, but Mn^{2+} and other less abundant trace elements are also used by metabolic processes.

(iii) As products of photosynthesis. As we saw, these are organic molecules in which the oxidation state of carbon is lower than that in CO_2.

(iv) As products of decay of organic matter under low oxygen fugacity. These processes produce species such as CH_4, H_2S, NH^{4+}, Fe^{2+}, etc., and are typically carried out by microorganisms.

If we wish to discuss primordial metabolism then we can immediately discard (iv). Although the argument has been made that sulfide-oxidizing photosynthesis (reaction (14.12)) or iron-oxidizing photosynthesis (reaction (14.13)) may represent primordial processes (see Hartman, 1998; Mulkidjanian & Galperin, 2009), photosynthesis is a complex process, that requires specialized enzymes and pigments that absorb electromagnetic radiation of the required wavelengths. It also must take place in an environment bathed in solar radiation, that includes a high flux of ultraviolet photons capable of photodissociating complex organic molecules. Pre-biotic and early biotic organic synthesis may have been difficult, if not impossible, in such high energy environment (present-day organisms have evolved defensive mechanisms to cope with this problem and ozone, which is ultimately of organic origin, serves as a UV shield). It thus appears unlikely that photosynthesis is a primordial process (see Fenchel, 2002; Schulze–Makuch & Irwin, 2004), and we can with some confidence discard (iii) as a source of reduced species for primordial metabolic processes. Reduced chemical species corresponding to categories (i) and (ii), on the other hand, were certainly present in the early Earth, and perhaps more abundant than today.

The most common oxidizers at the Earth's surface today are O_2, Fe^{3+}, SO_4^{2-}, NO_3^- and CO_2. If the only source of molecular oxygen is photosynthesis then its chemical potential in the early Earth, before photosynthetic organisms evolved, must have been vanishingly small (Section 14.1). Molecular oxygen can be produced by photodissociation of water vapor in the stratosphere. The temporal and spatial distributions of Archaean banded iron formations (BIF), however, place an upper bound on likely paleoArchaean atmospheric oxygen fugacity of the order of 10^{-70} bar (e.g. Figs. 10.7 and 10.8), which implies that molecular oxygen was not present in the atmosphere. Fe^{3+} content in pre-BIF oceans must have been virtually zero (Figs. 11.7 and 11.8), and SO_4^{2-} concentration must also have been very small (Fig. 14.6). Thus, these oxidizers are also unlikely to have been present at the inception of primordial metabolism. Production of nitrate requires much higher oxygen fugacity. By simple elimination one is left with CO_2 as the most likely oxidizer at the beginning of life. This inference is consistent with the results of Section 14.1, that suggest that the Archaean atmosphere must have been rich in CO_2. We are led to the tentative conclusion that the best candidate for primordial respiration is a reaction that used CO_2 to oxidize H_2, H_2S, CH_4, Fe^{2+}, NH_3, or some combination of these species.

Primordial life must also have been able to produce CH_2O-type molecules, that were then used as the building blocks for more complex organic compounds. One could infer that the following reaction, that is one of the tentative respiration reactions that we identified in the previous paragraph, might have been able to accomplish this:

$$CH_4 + CO_2 \rightarrow 2CH_2O \qquad (14.19)$$

but the problem is that this reaction has a large and positive $\Delta_r G^0 \approx 240\,\text{kJ}$. Reaction (14.19) is not spontaneous – but what exactly does this mean? Using equation (12.71), and recalling that a reaction is spontaneous if its affinity is positive, we find that reaction (14.19) will proceed to the right only as long as:

$$\frac{\left(f_{CH_2O}\right)^2}{f_{CH_4} \cdot f_{CO_2}} < 10^{-42}. \qquad (14.20)$$

From a purely thermodynamic point of view we interpret this as meaning that, in a closed system, the reaction reaches equilibrium, and therefore stops, once a vanishingly small amount of formaldehyde is produced. From a biological point of view we can say that, in order for the metabolic reaction to be spontaneous, inequality (14.20) must apply to the fugacities inside the cell. If the ratio of fugacities becomes equal to, or greater than, 10^{-42}, then the cell "chokes" on its own metabolic products and metabolism stops. The point is that reaction 14.19 cannot be a respiration reaction. It could act as a carbon-fixing reaction if a constant supply of energy were available (for instance, via photosynthetic reactions such as (14.10), (14.12) and (14.13)), but for the reasons that we discussed above photosynthesis is out of the question for primordial metabolism.

Could some of the other reactions between CO_2 and reduced endogenous species serve as an energy source? Consider hydrogen oxidation:

$$4H_2 + CO_2 \rightarrow CH_4 + 2H_2O. \tag{14.21}$$

The Gibbs free energy change for this reaction is $\Delta_r G^0 \approx -130\,\text{kJ}$ per mol of CH_4 produced, from which we infer that the reaction is spontaneous if:

$$\frac{f_{CH_4}}{f_{CO_2} \cdot \left(f_{H_2}\right)^4} < 10^{23}. \tag{14.22}$$

The reaction proceeds to the right even if the fugacity of the metabolic product (methane) inside the cell is orders of magnitude higher than those of the nutrients (hydrogen and carbon dioxide). This means that (14.21) is a possible, and in fact thermodynamically very favorable, respiration reaction. The reaction liberates energy, but it does not produce the CH_2O-type molecules that are needed by biological processes.

Take now a linear combination of (14.19) and (14.21) of the form:

$$CH_4 + CO_2 \rightarrow 2CH_2O$$

$$+n(4H_2 + CO_2 \rightarrow CH_4 + 2H_2O)$$

$$\overline{\rule{0pt}{0pt}\hspace{2cm}}$$

$$4nH_2 + (n+1)CO_2 \rightarrow (n-1)CH_4 + 2CH_2O + 2nH_2O. \tag{14.23}$$

From the standard state Gibbs free energy changes of (14.19) and (14.21) we see that for any value of n greater than ~ 2, $\Delta_r G^0$ of reaction (14.23) is negative, or:

$$\frac{\left(f_{CH_4}\right)^{n-1} \cdot \left(f_{CH_2O}\right)^2}{\left(f_{CO_2}\right)^{n+1} \cdot \left(f_{H_2}\right)^{4n}} < 10^Q \tag{14.24}$$

with $Q > 0$. If there is a constant supply of hydrogen and carbon dioxide, and the cell is able to dispose of the metabolic products, for instance, by excreting methane and converting formaldehyde into complex organic molecules, then (14.23) acts as a respiration reaction that at the same time fixes carbon. Note that, given the large energy requirement of the carbon fixing reaction (14.19), it *must* be $n > 2$, so that the complete reaction (14.23) always produces an excess of methane, that must be excreted.

Discussing the enzymatic pathways that may make a reaction such as (14.23) possible, and how such pathways may have developed, is beyond the scope of this book (see, for example, Nitschke & Russell, 2009). The crucial fact is that such reactions exist. They are known as methanogenesis, and are used as the energy source by a group of present-day Archaea known as methanogens. Similar reactions, that excrete acetate (CH_3COO^-) rather than methane, are the metabolic processes that sustain a group of bacteria known as acetogens. *Methanogenesis and acetogenesis are unique, in that they are the only metabolic processes in extant organisms that fix carbon while simultaneously liberating energy* (see Lane et al., 2010; Nitschke & Russell, 2009). They are also the only known respiration processes that reduce carbon rather than oxidizing it. All other known carbon-fixing pathways absorb energy (e.g. reactions (14.10), (14.12) and (14.13)), so that organisms that depend on them must rely on distinct processes in order to fix carbon (photosynthesis) and liberate energy (by respiration processes that oxidize carbon). Autotrophic organisms are able to carry out both types of processes, whereas heterotrophic organisms can only respire, and depend on carbon fixed by autotrophs.

The sheer simplicity and efficiency of a reaction such as (14.23) makes it an excellent candidate for a primordial metabolic process, but there are other lines of evidence that support this hypothesis. Methanogenesis depends on a constant supply of hydrogen, for example derived from volcanic gases (perhaps via serpentinization reactions, see Worked Example 9.7), and carbon dioxide, which is likely to have dominated the Hadean and early Archaean atmosphere (Section 14.1), and must therefore have been abundant in solution in the primeval terrestrial oceans. Many extant methanogens (though not all) are thermophilic, which is what one would expect if they are descended from primordial life that originated near submarine hydrothermal vents, where H_2 was abundant. There is also geochemical evidence for methanogenesis from fluid inclusions in paleo-Archaean rocks (Ueno et al., 2006). More directly, the genomes of methanogenic organisms place them close to the root of the tree of life (see Di Giulio, 2009).

Reaction (14.23) is a possible pathway to methanogenesis, but is it the only one that was available at the beginning of life? Recall that other reduced species are also present in volcanic products. Could they reduce CO_2 to CH_4, so as to yield a reaction that can be combined with the CH_2O-forming reaction (14.19)? We can immediately discard the following two reactions:

$$4H_2S + CO_2 \rightarrow CH_4 + 4S + 2H_2O \tag{14.25}$$

$$CO_2 + 8Fe^{2+} + 8H^+ \rightarrow CH_4 + 8Fe^{3+} + 2H_2O \tag{14.26}$$

as they both have positive standard state Gibbs free energy changes: 3.6 and 463 kJ per mol of CH_4. Ammonia oxidation is feasible, as the reaction:

$$3CO_2 + 8NH_3 \rightarrow 3CH_4 + 4N_2 + 6H_2O \tag{14.27}$$

has $\Delta_r G^0 = -86.8$ kJ per mol of CH_4. It could thus be combined with (14.19) to yield a self-sustaining carbon-fixing reaction. If this metabolic pathway ever existed on Earth it appears to have left no descendants. Perhaps it never arose, or perhaps it did but it was out-competed by methanogenesis. The reason in either case is likely to have been that terrestrial volcanic gases never contained large enough ammonia concentrations to make it competitive with hydrogen oxidation (reaction (14.23)). Ammonia oxidation might be the starting point for biology in other worlds.

Another respiration reaction that is a feasible energy source for primordial organisms is oxidation of sulfur according to:

$$8H_2S + 4Fe^{2+} + CO_2 \rightarrow CH_4 + 4FeS_2 + 2H_2O + 8H^+. \tag{14.28}$$

This reaction has a negative standard state Gibbs free energy change (-186.9 kJ per mol of CH_4), so it can be combined with (14.19) in a manner similar to (14.23), to yield the following methanogenesis reaction that also precipitates pyrite:

$$8nH_2S + 4nFe^{2+} + (n+1)CO_2 \rightarrow$$

$$\rightarrow (n-1)CH_4 + 2CH_2O + 4nFeS_2 + 2nH_2O + 8nH^+, \tag{14.29}$$

where $n > \sim 1.5$ (because of the relative values of $\Delta_r G^0$ of reactions (14.19) and (14.28)). A reaction of this kind has been proposed as a metabolic pathway for primordial metabolism (e.g. by Wächtershäuser, 1988, and many subsequent publications). It oxidizes S^{2-} to S^- by reducing carbon. The oxidation state of iron does not change, but the reaction is driven strongly to the right by the very low solubility of pyrite. It relies on the availability of Fe^{2+}, for example extracted from basaltic ocean floor and dissolved in anoxic ocean water, and H_2S, which may be abundant in reduced volcanic gases. There is thus no *a priori* reason why (14.29) could not have been the primordial metabolic pathway, rather than (14.23), except that modern methanogens metabolize by oxidation of hydrogen according to (14.23), without precipitating pyrite. One could argue that the vanishingly small concentration of Fe^{2+} in the present-day ocean is an impediment that did not exist during the Archaean, but there are other environments, such as euxinic lakes, where the required nutrients (hydrogen sulfide and ferrous iron) are abundant and yet microorganisms in such environments use other metabolic pathways. Lane *et al.* (2010) and Nitschke and Russell (2009), among others, review a number of biochemical constraints that indicate that all modern life forms derive from a common ancestor that oxidized hydrogen by the simpler methanogenic reaction (14.23). If life based on pyrite-producing methanogenesis (reaction (14.29)) did arise on Earth then for some reason it was not as successful as life based on direct hydrogen oxidation (reaction (14.23)), and it became extinct.

Thermodynamic arguments do not constitute proof that life on Earth originated by methanogenesis, but this may be as close as we may be able to get to a reasonably plausible answer. In my view this is a simple process of elimination. There is a limited list of possible reduced nutrients of endogenous origin that are abundant enough, and an even more restricted list of possible exogenous oxidizers, essentially limited to CO_2. Thermodynamics then forces us to discard all but two of the possible metabolic pathways: methanogenesis by hydrogen oxidation (14.23) and methanogenesis by sulfur oxidation and pyrite precipitation (14.29). On the reasonable postulate that primordial metabolism is likely to have been as simple as possible, we may choose reaction (14.23). This thermodynamic choice is supported by genetics and molecular biology.

Even if we settle on methanogenesis as the most likely primordial metabolism, there are a number of other tantalizing questions. Among them: (1) what was the source, or sources, of molecular hydrogen; (2) would the origin of life have been possible if the terrestrial atmosphere had been dominated by reduced species such as hydrogen and ammonia? (Section 14.1); and (3) could the origin of life have been dependent on delivery of extraterrestrial reduced carbon species?

Molecular hydrogen is present in reduced volcanic gases and it can also be produced by the serpentinization process discussed in Worked Example 9.7 (see in particular Fig. 9.15). This process is known to occur at present in hydrothermal vents located a few km from active mid-ocean ridge volcanoes (see Kelley *et al.*, 2001). Fluids in these "secondary" hydrothermal vents are considerably colder than those emanating from black smokers near sites of volcanic activity, and this would have been an advantage for the synthesis and preservation of complex organic molecules.

With regard to question (2) it is hard to see what type of metabolic reactions could be powered by the difference in thermodynamic states between the Earth's interior and a methane–ammonia–hydrogen atmosphere, as in that case there would have been no significant difference in oxidation conditions between the Earth's interior and its surface. A possible pathway exists, however, for a nitrogen-dominated atmosphere such as would form if the temperature is low enough to lock H_2O as ice (Section 14.1). We return to this possibility in Section 14.3.2.

Question (3) is prompted by the suggestion that cometary impacts in the early Earth could have supplied the young planet with abundant complex organic molecules (see Chyba *et al.*, 1990). Such compounds are indeed known to occur in comets and chondritic meteorites, so we could add extraterrestrial reduced carbon to the list of possible nutrients that we discussed at the beginning of this section. The problem with this hypothesis is the lack of oxidizing species in the early Earth. The strongest abundant oxidizer is likely to have been CO_2, and oxidation of organic matter with CO_2 is thermodynamically very unfavorable, as shown by the large and positive standard state Gibbs free energy change of reaction (14.19). Methanogenesis is the only present-day metabolic process that uses CO_2 as the oxidizer, and it works only because H_2 is a powerful reducing agent, capable of producing reduced carbon as a metabolic product. All other known metabolic processes, including those in which the reduced nutrient is organic matter, rely on species more oxidized than CO_2, such as SO_4^{2-}, Fe^{3+}, NO_3^- or O_2. These reactions *generate* CO_2 as a metabolic product of organic nutrients. Methanogenesis is distinctly different in this respect, and is the only metabolic reaction for which there was ample supply of nutrients in the early Earth. From a thermodynamic point of view the largest stumbling block for the origin of life by any means other than methanogenesis is lack of oxidizers, so cometary supply of reduced carbon does not appear to be of much help.

14.3 Speculations about extraterrestrial life

As a long-time fan of "hard" science fiction, I could not resist this one. But I will keep the discussion firmly ground on thermodynamics. Before we get to thermodynamics, though, it is important that I state the following: there are at least two good reasons why it is very unlikely that life can be based on any element other than carbon. The only other element that is able to generate a large variety of complex molecular structures is, or course, silicon. Yet what makes organic chemistry, and biochemistry, unique is the capability of carbon atoms to bond with other carbon atoms *without intervening atoms of other elements*. Polymerization in silicate structures – minerals and melts – requires bridging anions (typically oxygen), making a silicon-based analog of organic chemistry impossible. This is the first reason. The second one is more obvious, yet rarely invoked: Earth is a silicate planet. If silicon-based life were possible, then why did carbon-based life originate instead? The answer, I think, is because there is no other possibility.

14.3.1 Extraterrestrial life based on free energy gradients other than chemical potentials

Having established why, in my opinion, life must be based on carbon, we can now focus on thermodynamics. The *sine qua non* requirement for life is that the substrate be able to sustain an uninterrupted free energy gradient. For all known life forms this corresponds to a chemical potential gradient. What other possibilities are there? Good reviews are offered by Schulze–Makuch and Irwin (2004) and Lunine (2005). Here I will focus on only a few of the possibilities, which are the ones that I consider to be less unlikely.

The first possibility that usually comes to mind is a thermal gradient – and "gradient" is the key word here. A living organism that feeds on thermal energy (let us call this process thermofagy) is a heat engine, that absorbs heat from a high temperature reservoir and excretes heat to a lower temperature reservoir (Chapter 4). Thus, internal planetary heat by itself cannot sustain life, just as a reduced chemical species by itself (e.g. hydrogen) can't. Internal planetary heat can be used as a biological nutrient only if the organism is able to transfer heat to a colder environment faster than the rate at which heat would flow by inorganic means (e.g. by diffusion or advection). This is equivalent to the concept that aerobic life can use reduced carbon and oxygen in respiration only because enzymatic pathways alter the kinetics of the oxidation reaction and make it faster than inorganic oxidation at room temperature. A thermofagous organism could conceivably accomplish this in two distinct ways. First, by swimming rather quickly between, say, a hydrothermal vent on the sea floor and the surface of the sea. Second, by being a very large sessile organism that is able to sustain internal convection, while extending across a substrate in which heat moves only by diffusion. I don't think that there is much hope for Jabba the Hutt, but thermal swimmers may stand a chance. I will not worry about the biochemical pathways that could make it possible to harvest energy from a thermal gradient (but see Schulze–Makuch & Irwin, 2004, pp. 54–56). Thermal swimmers, if they are at all possible, may be the best chance for life in Europa, Ganymede, Enceladus, Titan and other icy moons with subsurface oceans, where they could exploit the temperature difference between hydrothermal vents or other warm areas of the ocean floor and the bottom of the ice lid. Could they have arisen on Earth and then been out-competed by the type of life that we know? We will probably never find out.

An idea that is closely allied to thermofagy is the exploitation of a chemical concentration gradient. A swimming organism might be able to move between regions in an ocean in which the concentration of a given solute is different. The organism may be enveloped by an osmotic membrane (like all organisms are) and its internal fluid may have an intermediate concentration of the solute, so that solvent (for instance, water) will be ingested from the low-concentration region and excreted into the high concentration region. These we could call osmotic swimmers, and they may also stand a chance in icy worlds, though exactly how solute diffusion across an osmotic membrane could be converted to biological molecules is hard to say (again, see Schulze–Makuch & Irwin, 2004, for some ideas).

A somewhat different possibility is to harvest kinetic energy. This could be accomplished, for instance, by ciliate organisms that could have somehow evolved the capability of using the movement of the cilia to power metabolic processes. The organisms would feed off convective flow of the surrounding fluid, which ultimately would be driven by temperature or chemical potential gradients. I think that thermal swimmers or osmotic swimmers would be more efficient, but who knows?

More outlandish possibilities have been considered (see the references cited above). These include: feeding off magnetic fields, electromagnetic induction, gravitational potential gradients, tectonic stress, radioactive decay, etc. I do not think that any of these putative nutrients warrants serious consideration. In fact, I don't think that any of the other possibilities considered in this section are very likely either. Chemical potential gradients are able to generate a much higher energy flow, so the question is, what other chemical pathways to metabolism that do not exist on Earth are possible?

14.3.2 Extraterrestrial life based on chemical potential gradients uncommon or non-existent in Earth

The fact that all extant life on Earth can trace its ancestry to a methanogenic common ancestor probably reflects the fact that methanogenesis is a relatively simple and efficient metabolic process. As such, one could expect that methanogenic life will arise anywhere that the inorganic substrate is able to provide the required nutrients, i.e. hydrogen and carbon dioxide. If active volcanoes still exist on Mars then the Red Planet (there – I said it) may very well support methanogenic life, perhaps in subterranean environments in which water is kept liquid by internal heat. The presence of methane in the Martian atmosphere has been interpreted as the signature of Martian life (Krasnopolsky *et al.*, 2004), but the problem is that methane can also be produced by inorganic serpentinization reactions (Worked Example 9.7), that would also be fueled by active internal processes.

If an endogenous supply of reduced species such as hydrogen does exist in Mars, however, the strongly oxidized nature of the Martian surface makes other metabolic pathways possible, such as:

$$H_2 + 2Fe^{3+} \rightarrow 2H^+ + 2Fe^{2+}. \tag{14.30}$$

This is a respiration reaction with $\Delta_r G^0 = -148.4$ kJ per mol of H_2. It is used by anaerobic terrestrial microorganisms, but of course it requires a separate metabolic pathway to fix carbon. It may not be an ideal candidate for the origin of life, but it may have evolved from a methanogenic ancestor on Mars, as it did on Earth. Other respiration reactions that use Mn^{4+}, SO_4^{2-}, ClO_4^-, etc., to oxidize hydrogen are also possible on Mars.

There is no lack of likely metabolic pathways, most of them tested by terrestrial life, that could sustain life in present-day Mars, but there is no proof of Martian life. In this respect it is instructive to examine the supposed presence of microfossils in Martian meteorite ALH84001 (McKay *et al.*, 1996). This claim is now discredited, and the following simple argument (after Fenchel, 2002) shows why it should never have been taken seriously in the first place. The supposed Martian nanofossils have a diameter of $\sim$100 nm, which yields a characteristic volume of $\sim 4 \times 10^{-21}$ m^3. If the organism consisted only of H_2O then it would contain $\sim 2 \times 10^{-16}$ mols of H_2O. Assuming that the organisms were adapted to neutral pH conditions, they would contain $\sim 2 \times 10^{-23}$ mols of protons, or about one proton! Such an organism would not be able to maintain a constant pH, which would make metabolism impossible. One could improve the viability of these putative organisms by assuming that they are adapted to acid conditions, but the fact remains that the smallest terrestrial organisms have a volume of $\sim 5 \times 10^{-19}$ m^{-3}, i.e. two orders of magnitude greater than the artifacts in the Martian meteorite.

Consider now a planet with a nitrogen atmosphere and active volcanism that supplies reduced fluids containing hydrogen. There is no CO_2 available to act as an oxidizer. The following is a possible respiration reaction in such a planet, with $\Delta_r G^0 = -16$ kJ per mol of NH_3:

$$3H_2 + N_2 \rightarrow 2NH_3. \tag{14.31}$$

The atomic bond in the N_2 molecule is very strong, so the kinetics of this reaction are very unfavorable, but perhaps enzymatic pathways might evolve that would make it possible, as happens on Earth with bacteria responsible for oxidizing atmospheric nitrogen. Production of organic molecules in our hypothetical planet could begin with the reaction:

$$CH_4 + H_2O \rightarrow CH_2O + 2H_2 \tag{14.32}$$

for which $\Delta_r G^0 = 185$ kJ per mol of formaldehyde. By analogy with terrestrial methanogenesis (reaction (14.23)), reactions (14.31) and (14.32) could be coupled to yield the following respiration reaction that also produces formaldehyde:

$$n\,N_2 + (3n - 2)H_2 + CH_4 + H_2O \rightarrow CH_2O + 2n\,NH_3 \qquad (14.33)$$

with $n > 6$. This reaction requires a supply of the reduced species H_2 and CH_4 (e.g. volcanic gases), liquid H_2O, and a dense nitrogen atmosphere, which can generate a significant concentration of N_2 dissolved in water. Earth is a candidate planet, but this metabolic pathway does not exist. If it arose in the primeval Earth it must have been driven to extinction by the much more efficient methanogenesis process. Present day Titan may have the necessary ingredients, but the liquid water requirement would restrict life based on reaction (14.33) to deep environments, where the supply of atmospheric nitrogen may be problematic. Future warming of Titan (as solar energy output increases) may make it more feasible, as long as tidal and/or radioactive heating continues to support some sort of endogenous activity. Reaction (14.33) may be a pathway to life in extra-solar worlds with less oxygen than the early Earth, but whether the rich biology that we are familiar with could spring from these beginnings is another question.

The possibility of life in the ice-mantled oceans of Europa, Ganymede and Enceladus is a popular discussion topic, but the fundamental thermodynamic constraints should temper our enthusiasm for this possibility. A source of internal heat, that undoubtedly exists in all these three worlds, *is not sufficient* – what is required is a sustained free energy gradient. Leaving aside the fanciful non-chemical possibilities discussed in the previous section, how could a chemical potential gradient arise in an ocean covered by ice? If the ice layer is thin enough to allow some high-energy electromagnetic radiation to get through then photodissociation of H_2O molecules could generate molecular oxygen, which could in turn oxidize species such as ferrous iron or sulfide. Convection could transport these oxidized components to the deep ocean, where a chemical potential gradient relative to reduced species would arise. In principle some of the same respiration pathways that exist on Earth could be sustained in this way, but the low efficiency and unpredictability of the process would make life for putative Europan microbes rather miserable.

What I conclude, and this is opinion rather than a logically rigorous conclusion, is that when we find extraterrestrial life we will immediately recognize it, as it will be chemically very familiar. Yes, there are other ways to make a living, and it is possible, perhaps even likely, that some of these other biological pathways arose in the early Earth. But there is a good reason why we all share a common methanogenic ancestor, which is natural selection of the most efficient metabolic pathway.

14.4 Entropy and life

Metabolism, and therefore life, is only possible if the inorganic reactants in the substrate can be related by a chemical reaction with positive affinity. Metabolism therefore generates entropy, just as any chemical reaction that proceeds from a state of non-equilibrium towards equilibrium does (equation (12.77)). I am always mystified by statements to the effect that life and biological evolution "violate" the Second Law because, in producing "ordered" structures, they "lower the entropy of the universe". Such statements denote a profound ignorance of thermodynamics, as they fail to appreciate the fact that entropy is a macroscopic function, whereas the probabilistic foundation of entropy (the "order–disorder" idea) is an exclusively microscopic concept (Chapter 4). Evolution, assembly and growth of complex and organized organisms (be it a bacterium or one of my tabbies) is a macroscopic process, and has no bearing on the state of microscopic order of the system, which is what determines its entropy. The correct way to see any biological process is that both catabolic metabolism (responsible for energy transformations) and anabolic metabolism (responsible for synthesizing cellular structures from material gathered from the substrate) are chemical reactions, and *all chemical reactions generate entropy*, as equation (12.77) shows.

The same is true of the origin and evolution of life, although in this case we can make an additional argument. Say that the origin of life from inanimate matter is an ordering process, in the same sense that growth of a crystal from a solution or melt is an ordering process – the random distribution of atoms and simple molecules of nutrients in the "primordial soup" becomes a more ordered arrangement of atoms in complex organic compounds, including RNA, DNA, etc. We can use Landau theory (Section 7.6) to examine this process. We set the order parameter $\phi = 0$ for the inanimate precursor of life (the disordered phase), and $\phi > 0$ for the first living organism (the ordered phase). *Note that the order–disorder concept applies to the microscopic nature of the system, as it must*. Using (7.48) we write (7.49) as follows:

$$G\left(\phi\right) = 0, \quad (\textit{disordered phase})$$
$$G\left(\phi\right) = -\frac{1}{4}g_4\phi^4, \quad g_4 > 0, \quad (\textit{ordered phase}) \tag{14.34}$$

and (7.50) as:

$$S\left(\phi\right) = 0, \quad (\textit{disordered phase})$$
$$S\left(\phi\right) = -\frac{\alpha}{2}\phi^2, \quad (\textit{ordered phase}). \tag{14.35}$$

The enthalpy of ordering is given by:

$$\Delta H_{ordering} = G\left(\phi\right)_{ordered} + TS\left(\phi\right)_{ordered} - G\left(\phi\right)_{disordered} - TS\left(\phi\right)_{disordered}$$
$$= -\frac{1}{4}g_4\phi^4 - T\frac{\alpha}{2}\phi^2 < 0. \tag{14.36}$$

Ordering is always an exothermic process. The system that becomes ordered (in this case, a biological molecule) releases heat, that is absorbed by the environment. Because the environment must be at a lower temperature in order for heat to flow, its entropy increase is greater than the entropy decrease associated with ordering of the primordial cell. Heat transfer generates entropy (equation (12.14)), and the "entropy of the Universe" increases.

We may not yet have a full understanding of how life on Earth originated, but we can be certain that it did not involve intervention of some ad-hoc and imaginary entity capable of violating the Second Law of Thermodynamics. The origin of life was a chemical reaction, however complex, and as such it generated entropy.

Exercise for Chapter 14

14.1 Consider a planet with $g = 9.8$ m s^{-2} and a water column of thickness 2 km averaged over the surface area of the planet. The planet has an atmosphere in which the partial pressure of N_2 is 0.8 bar. The planet's early atmosphere was rich in CO_2 (later in its history most of that CO_2 was scrubbed by formation of carbonates and organic matter, but we will ignore this). Use the *Maple* code described in Software Box 14.1 to estimate a possible range of bulk C–H–O–N contents if the partial pressure of CO_2 in the early atmosphere was (i) 1 bar, (ii) 10 bar, (iii) 50 bar. Compare your results to chondritic and cometary abundances, and discuss the implications for the origin and early evolution of the terrestrial atmosphere.

Appendix 1 Physical constants and other useful numbers and conversion factors

Constant	Symbol	Value
Gas constant	R	$8.3145 \text{ J K}^{-1} \text{ mol}^{-1}$
Avogadro's number	N	$6.0221 \times 10^2 \text{ mol}^{-1}$
Boltzmann constant	k_B	$1.3807 \times 10^{-23} \text{ J K}^{-1}$
Atomic mass unit	u	$1.6605 \times 10^{-27} \text{ kg}$
Planck constant	h	$6.6261 \times 10^{-34} \text{ J s}$
Speed of light in vacuum	c	$2.9979 \times 10^8 \text{ m s}^{-1}$
Gravitation constant	G	$6.6726 \times 10^{-11} \text{ m}^3 \text{ kg}^{-1} \text{ s}^{-2}$
Stefan–Boltzmann constant	σ	$5.6705 \times 10^{-8} \text{ J s}^{-1} \text{ m}^{-2} \text{ K}^{-4}$
Permittivity of free space	ϵ_0	$8.8542 \times 10^{-12} \text{ C}^2 \text{ N}^{-1} \text{ m}^{-2}$
Electron charge	e	$1.6022 \times 10^{-19} \text{ C}$
Electron mass	m_e	$9.1094 \times 10^{-31} \text{ kg}$
Million years	my	$3.1558 \times 10^{13} \text{ s}$
bar	bar	$10^5 \text{ N m}^{-2} = 10^5 \text{ Pa}$
kilobar	kbar	$10^8 \text{ N m}^{-2} = 100 \text{ MPa} = 0.1 \text{ GPa}$

Solving thermodynamic equations often entails simplifications and combinations of partial derivative relationships. We often need to write equations in terms of variables that are either easily measured or that we want to solve for, such as T, P and V, and of the three material properties, K_T (or β_T), α and C_P. From the material properties we extract the following three partial derivatives, which are the ones that we wish to appear in our equations, to the exclusion of other partial derivatives:

$$\left(\frac{\partial V}{\partial T}\right)_P = \alpha V \tag{A2.1}$$

$$\left(\frac{\partial V}{\partial P}\right)_T = -\beta_T V = -\frac{1}{K_T} V \tag{A2.2}$$

$$\left(\frac{\partial H}{\partial T}\right)_P = C_P. \tag{A2.3}$$

There is a straightforward method that allows one to obtain any partial derivative as a combination of these material properties and the variables P, T and V. A "black-box" recipe to do this was given almost a century ago by the American physicist P. W. Bridgman (1914), who made pioneering contributions to high-pressure research. The recipe relies on using tabulated "half identities" (see, for example, Glasstone, 1946). Bridgman's relations can be derived in an elegant fashion using Jacobian matrices and determinants, as discussed by Shaw (1935) and Guggenheim (1967, p. 41). A detailed explanation of how Bridgman obtained his equations can also be found in Tunell (1985).

Whenever possible, I find it more satisfying to derive the relations one may need from first principles. This is the case with all thermodynamic identities used in this book. Many of the most commonly used relations follow directly from the identity of the mixed second derivatives of the thermodynamic potentials (Maxwell's relations, see Section 4.9.1). Other relations at constant T or P can be found as follows (this derivation largely follows that of Tunell, 1985.

The idea is that thermodynamic state functions can be written as functions of the variables P, T, V, E and S. The first task is to find expressions for the first derivatives of E and S relative to T and P, as a function of some combination of the variables P, V, T and the three material properties, α, β_T and C_p. The second derivatives of the Gibbs free energy function (equations (4.135) and (4.139)) give us the first two equations:

$$\left(\frac{\partial S}{\partial T}\right)_P = \frac{C_P}{T} \tag{A2.4}$$

$$\left(\frac{\partial S}{\partial P}\right)_T = -V\alpha. \tag{A2.5}$$

Recall that (A2.5) arises from one of Maxwell's relations (Section 4.9.1). Using (1.3.3) we write the total differentials of E, Q and V relative to T and P:

$$dE = \left(\frac{\partial E}{\partial T}\right)_P dT + \left(\frac{\partial E}{\partial P}\right)_T dP \tag{A2.6}$$

$$dQ = \left(\frac{\partial Q}{\partial T}\right)_P dT + \left(\frac{\partial Q}{\partial P}\right)_T dP = C_P dT + \left(\frac{\partial Q}{\partial P}\right)_T dP \tag{A2.7}$$

$$dV = \left(\frac{\partial V}{\partial T}\right)_P dT + \left(\frac{\partial V}{\partial P}\right)_T dP = \alpha V dT - \beta_T V dP. \tag{A2.8}$$

We next substitute these expressions into the First Law: $dE = dQ - P dV$ (equation (1.56)) and collect terms as follows:

$$\left(\frac{\partial E}{\partial T}\right)_P dT + \left(\frac{\partial E}{\partial P}\right)_T dP = (C_P - P\alpha V) dT + \left[\left(\frac{\partial Q}{\partial P}\right)_T + P\beta_T V\right] dP. \tag{A2.9}$$

Because dP and dT are linearly independent variables we have the following identities:

$$\left(\frac{\partial E}{\partial T}\right)_P = C_P - P\alpha V \tag{A2.10}$$

$$\left(\frac{\partial E}{\partial P}\right)_T = \left(\frac{\partial Q}{\partial P}\right)_T + P\beta_T V. \tag{A2.11}$$

Equation (A2.10) is the third equation that we seek. To reduce (A2.11) to material properties we use the definition of entropy, equation (4.6), to write:

$$\left(\frac{\partial Q}{\partial P}\right)_T = T\left(\frac{\partial S}{\partial P}\right)_T = -\alpha V T \tag{A2.12}$$

so:

$$\left(\frac{\partial E}{\partial P}\right)_T = (P\beta_T - T\alpha) V. \tag{A2.13}$$

Summarizing the four equations:

$$\left(\frac{\partial E}{\partial T}\right)_P = C_P - P\alpha V \qquad \left(\frac{\partial E}{\partial P}\right)_T = (P\beta_T - T\alpha) V$$
$$\left(\frac{\partial S}{\partial T}\right)_P = \frac{C_P}{T} \qquad \left(\frac{\partial S}{\partial P}\right)_T = -V\alpha. \tag{A2.14}$$

As an example, suppose that we need an expression for $(\partial H/\partial P)_T$, see Section 3.5. From equation (4.122):

$$\left(\frac{\partial H}{\partial P}\right)_T = \left(\frac{\partial E}{\partial P}\right)_T + P\left(\frac{\partial V}{\partial P}\right)_T + V \tag{A2.15}$$

using (A2.2) and (A2.13):

$$\left(\frac{\partial H}{\partial P}\right)_T = (P\beta_T - T\alpha) V - P\beta_T V + V = V(1 - \alpha T). \tag{A2.16}$$

For a different example, suppose we need the pressure and temperature derivatives of the Helmholtz free energy. From (4.125), (A2.13) and (A2.5) we have:

$$\left(\frac{\partial F}{\partial P}\right)_T = \left(\frac{\partial E}{\partial P}\right)_T - T\left(\frac{\partial S}{\partial P}\right)_T = PV\beta_T \qquad (A2.17)$$

and from (4.125), (A2.10) and (A2.4):

$$\left(\frac{\partial F}{\partial T}\right)_P = \left(\frac{\partial E}{\partial T}\right)_P - T\left(\frac{\partial S}{\partial T}\right)_P = -PV\alpha. \qquad (A2.18)$$

The equations summarized in (A2.14) work only if we seek derivatives of other thermodynamic functions at constant temperature or constant pressure. If different conditions are required, such as constant entropy, then other equations must be found, in which case it may be convenient to use a different set of material properties, such as the adiabatic bulk modulus or the Grüneisen parameter. Some examples are presented in the main text, see especially Chapters 7, 8 and 9. If all else fails, then one can always use Bridgman's formulas or, better, Shaw's method.

References

Acton, G., Yin, Q-Z, Verosub, K. L. *et al.* (2007). Micromagnetic coercitivity distributions and interactions in chondrules with implications for paleointensities of the early solar system. *J. Geophys. Res.*, **112**, doi:10.1029/2006JB004655.

Acuña, M. H. *et al.* (1999). Global distribution of crustal magnetization discovered by the Mars Global Surveyor MAG/ER experiment. *Science*, **284**, 790–793.

Anderson, D. L. (1989). *Theory of the Earth.* Blackwell, Boston, 366 pp.

Anderson, G. M. (2005). *Thermodynamics of Natural Systems.* Second Edition. Cambridge University Press, Cambridge, 648 pp.

Anderson, O. L. (1995). *Equations of State of Solids for Geophysics and Ceramic Science.* Oxford University Press, New York, 405 pp.

Anderson, O. L., Isaak, D. & Oda, H. (1992). High-temperature elastic constant data on minerals relevant to geophysics. *Rev. Geophys.*, **30**, 57–90.

Asimow, P. D. (2002). Steady-state mantle–melt interacctions in one dimension: II. Thermal interactions and irreversible terms. *J. Petrol.*, **43**, 1707–1724.

Asimow, P. D., Hirschmann, M. M., Ghiorso, M. S., O'Hara, M. J. & Stolper, E. M. (1995). The effect of pressure induced solid–solid phase transitions on decompression melting in the mantle. *Geochim. Cosmichim. Acta*, **59**, 4489–4506.

Asimow, P. D. & Stolper, E. M. (1999). Steady-state mantle–melt interactions in one dimension: I. Equilibrium transport and melt focusing. *J. Petrol.*, **40**, 475–494.

Barron, T. H. K., Berg, W. T., & Morrison, J. A. (1959). On the heat capacity of crystalline magnesium oxide. *Proc. R. Soc. London*, **A250**, 70–83.

Baylin, M. (1994). *A Survey of Thermodynamics.* American Institute of Physics, New York, 657 pp.

Berg, W. T. & Morrison, J. A. (1957). The thermal properties of alkali halide crystals. I. The heat capacity of potassium chloride, potassium bromide, potassium iodide and sodium iodide between 2.8 and 270°K. *Proc. R. Soc. Lond.*, **A242**, 467–477.

Berkeley Labs Commons. *https://commons.lbl.gov/display/ALSBL6/Dielectric+strength+of+air*

Berman, R. G. (1988). Internally consistent thermodynamic data for minerals in the system $Na_2O–K_2O–CaO–MgO–FeO–Fe_2O_3–Al_2O_3–SiO_2–TiO_2–H_2O–CO_2$. *J. Petrol.*, **29**, 445–522.

Beyer, W. H. (1987) *CRC Standard Mathematical Tables.* 28th Edition. CRC Press, Boca Raton, 674 pp.

Birch, F. (1952). Elasticity and constitution of the Earth's interior. *J. Geophys. Res.*, **57**, 227–286.

Bizzarro, M., Baker, J. A. & Haack, H. (2004). Mg isotope evidence for contemporaneous formation of chondrules and refractory inclusions. *Nature*, **431**, 275–578.

Bohren, C. F. & Albrecht, B. A. (1998). *Atmospheric Thermodynamics.* Oxford University Press, New York, 402 pp.

Borg, R. J. & Dienes, G. J. (1988). *An Introduction to Solid State Diffusion*. Academic Press, San Diego, 360 pp.

Bottinga, Y. (1985). On the isothermal compressibility of silicate liquids at high pressure. *Earth Pl. Sci. Letters*, **74**, 350–360.

Bridgman, P. W. (1914). A complete collection of thermodynamic formulas. *Phys. Rev.*, **3**, 273–281.

Brodholt, J. & Wood, B. (1993). Simulations of the structure and thermodynamic properties of water at high pressures and temperatures. *J. Geophys. Res.*, **98**, 519–536.

Bronsted, J. N. (1922). Studies on solubility, IV. The principle of specific interaction of ions. *J. Am. Chem. Soc.*, **44**, 877–898.

Buffett, B. A. & Bloxham, J. (2002). Energetics of numerical geodynamo models. *Geophys. J. Int.*, **149**, 211–224.

Bullard, E. C. & Gellman, H. (1954). Homogeneous dynamos and terrestrial magnetism. *Phil. Trans. R. Soc. Lond.*, **A 247**, 213–278.

Burke, W. L. (1985). *Applied Differential Geometry*. Cambridge University Press, Cambridge, 414 pp.

Burnham, C. W., Holloway, J. R. & Davis, N. F. (1969). Thermodynamic properties of water to 1000°C and 10,000 bars. *Geol. Soc. Am. Spec. Paper* **132**, 96 pp.

Callen, H. B. (1985). *Thermodynamics and an Introduction to Thermostatistics*. Wiley, New York, 493 pp.

Cameron, A. G. W. & Benz, W. (1991). The origin of the Moon and the single impact hypothesis. *Icarus*, **92**, 204–216.

Carr, M. H. (1999). Mars: surface and interior. In: *Encyclopedia of the Solar System*. Eds. P. R. Weissman, L. McFadden & T. V. Johnson. Academic Press, New York, pp. 291–308.

Carslaw, H. S. & Jaeger, J. C. (1959). *Conduction of Heat in Solids*. Oxford University Press, Oxford, 510 pp.

Catling, D. C., Claire, M. W., Zahnle, K. J. *et al.* (2010). Atmospheric origins of perchlorate on Mars and in the Atacama. *J. Geophys. Res.*, **115**, doi:10.1029/2009JE003425.

Cemič, L. & Dachs, E. (2006). Heat capacity of ferrosilite, $Fe_2Si_2O_6$. *Phys. Chem. Minerals*, **33**, 457–464.

Chabrier, G., Saumon, D., Hubbard, W. B. & Lunine, J. I. (1992). The molecular–metallic transition of hydrogen and the structure of Jupiter and Saturn. *Astrophys. J.*, **391**, 817–826.

Chamberlain, J. W. & Hunten, D. M. (1987). *Theory of Planetary Atmospheres. An Introduction to their Physics and Chemistry*. Second Edition. Academic Press, Orlando, 481 pp.

Chambers, J. E. (2005). Planet formation. In: *Meteorites, Comets and Planets. Treatise on Geochemistry, v.1*. Ed A. M. Davis. Elsevier, Amsterdam, pp. 461–475.

Chandrasekhar, S. (1958). *An Introduction to the Study of Stellar Structure*. Dover, New York, 501 pp.

Chapman, S. (1930a). A theory of upper-atmospheric ozone. *Mem. R. Meteorol. Soc.*, **3**, 103–125.

Chapman, S. (1930b). On ozone and atomic oxygen in the upper atmosphere. *Phil. Mag.*, **10**, 369–383.

Chatterjee, N. D., Kruger, R., Haller, G. & Olbricht, W. (1998). The Bayesian approach to an internally consistent thermodynamic database: theory, database and generation of phase diagrams. *Contrib. Mineral. Petrol.*, **133**, 149–168.

Cherniak, D. J. & Watson, E. B. (1994). A study of strontium diffusion in plagioclase using Rutherford backscattering spectroscopy. *Geochim. Cosmochim. Acta*, **58**, 5179–5190.

Cherniak, D. J., Watson, E. B., Grove, M. & Harrison, T. M. (2004). Pb diffusion in monazite: a combined RBS/SIMS study. *Geochim. Cosmochim. Acta*, **68**, 829–840.

Chevrier, V. F. & Altheide, T. S. (2008). Low temperature aqueous ferric sulfate solutions on the surface of Mars. *Geophys. Res. Letters*, **35**, doi:10.1029/2008GL035489.

Chevrier, V. F., Hanley, J. & Altheide, T. S. (2009). Stability of perchlorate hydrates and their liquid solutions at the Phoenix landing site, Mars. *Geophys. Res. Letters*, **36**, doi:10.1029/2009GL037497.

Choukroun, M. & Grasset, O. (2007). Thermodynamic model for water and high-pressure ices up to 2.2 GPa and down to the metastable domain. *J. Chem. Phys.*, **127**, DOI: 10.1063/1.2768957.

Chyba, C. F., Thomas, P. J., Brookshaw, L. & Sagan, C. (1990). Cometary delivery of organic molecules to the early Earth. *Science*, **249**, 366–373.

Colburn, D. S. (1980). Electromagnetic heating of Io. *J. Geophys. Res.*, **85**, 7257–7261.

Cotterill, R. M. J. (1980). The physics of melting. *J. Crystal Growth*, **48**, 582–588.

Croft, S. K., Lunine, J. I. & Kargel, J. (1988). Equation of state of ammonia–water liquid: derivation and planetological applications. *Icarus*, **73**, 279–293.

Cygan, R. T. & Carrigan, C. R. (1992). Time-dependent Soret transport: applications to brine and magma. *Chem. Geol.*, **95**, 201–212.

Darken, L. S. & Gurry, R. W. (1953). *Physical Chemistry of Metals*. McGraw-Hill, New York, 525 pp.

Davies, A. G. (2001). Volcanism on Io: the view from Galileo. *Astronomy & Geophysics*, **42**, 10–15.

Davies, G. F. (1999). *Dynamic Earth. Plates, Plumes and Mantle Convection*. Cambridge University Press, Cambridge, 458 pp.

de Donder, Th. (1936). *L'affinite'*. Gauthier-Villars, Paris, 142 pp.

de Groot, S. R. (1959). *Thermodynamics of Irreversible Processes*. North-Holland, Amsterdam, 242 pp.

de Groot, S. R. & Mazur, P. (1984). *Non-equilibrium Thermodynamics*. Dover, New York, 510 pp.

de Pater, I. & Lissauer, J. J. (2001). *Planetary Sciences*. Cambridge University Press, Cambridge, 528 pp.

Di Giulio, M. (2009). A methanogen hosted the origin of the genetic code. *J. Theor. Biol.*, **260**, 77–82.

Donati, J.-F., Paletou, F., Bouvier, J. & Ferreira, J. (2005). Direct detection of a magnetic field in the innermost regions of an accretion disk. *Nature*, **438**, 466–469.

Drake, R. P. (2010). High-energy-density physics. *Physics Today,* **63**(6), 28–33.

Duan, Z., Moller, N. & Weare, J. H. (1992). Molecular dynamics simulation of *PVT* properties of geological fluids and a general equation of state of nonpolar and weakly polar gases up to 2000 K and 20,000 bar. *Geochim. Cosmichim. Acta*, **56**, 3839–3845.

Duan, Z., Moller, N. & Weare, J. H. (1996). A general equation of state for supercritical fluid mixtures and molecular dynamics simulation of mixture *PVTX* properties. *Geochim. Cosmichim. Acta*, **60**, 1209–1216.

Duan, Z., Moller, N. & Weare, J. H. (1992). Accurate prediction of the thermodynamic properties of fluids in the system $H_2O–CO_2–CH_4–N_2$ up to 2000 K and 100 kbar from a corresponding states/one fluid equation of state. *Geochim. Cosmichim. Acta*, **64**, 1069–1075.

Dziewonski, A. M. & Anderson, D. L. (1981). Preliminary reference Earth model. *Phys. Earth. Pl. Interiors*, **25**, 297–356.

Emanuel, K. (1986). An air–sea interaction theory for tropical cyclones. Part I: steady-state maintenance. *J. Atm. Sci.*, **43**, 585–605.

Emanuel, K. (2006). Hurricanes: tempests in a greenhouse. *Physics Today,* **59**(8), 74–75.

England, P., Molnar, P. & Richter, F. (2007). John Perry's neglected critique of Kelvin's age of the Earth: a missed opportunity in geodynamics. *GSA Today,* **17**, 4–9.

Fenchel, T. (2002). *The Origin and Early Evolution of Life*. Oxford University Press, Oxford, 171 pp.

Flasar, F. M. (1973). Gravitational energy sources in Jupiter. *Astrophys. J.*, **186**, 1097–1106.

Fortney, J. J. & Hubbard, W. B. (2004). Effects of helium phase separation on the evolution of extrasolar giant planets. *Astrophys. J.*, **608**, 1039–1049.

Frost, D. J., Langenhorst, F. & van Aken, P. A. (2001). Fe–Mg partitioning between ring-woodite and magnesiowüstite and the effect of pressure, temperature and oxygen fugacity. *Phys. Chem. Minerals*, **28**, 455–470.

Ganguly, J. & Saxena, S. K. (1987). *Mixtures and Mineral Reactions: Minerals and Rocks*. Springer-Verlag, New York, 291 pp.

Giauque, W. F. & Powell, T. M. (1939). Chlorine. The heat capacity, vapor pressure, heats of fusion and vaporization, and entropy. *J. Am. Chem. Soc.,* **61**, 1970–1974.

Ghiorso, M. S. (1997). Thermodynamic models of igneous processes. *Ann. Rev. Earth Planet. Sci.*, **25**, 221–241.

Ghiorso, M. S., Carmichael, I. S. E., Rivers, M. L. & Sack, R. O. (1983). The Gibbs free energy of mixing of natural silicate liquids: An expanded regular solution approximation for the calculation of magmatic intensive variables. *Contrib. Mineral. Petrol.*, **84**, 107–145.

Ghiorso, M. S., Hirschmann, M. M., Reiners, P. W. & Kress, V. C. (2002). The pMELTS: A revision of MELTS for improved calculation of phase relations and major element partitioning related to partial melting of the mantle to 3 GPa. *Geochem. Geophys. Geosyst.*, **3**, 1–36.

Ghiorso, M. S. & Sack, R. O. (1995). Chemical mass transfer in magmatic processes, IV. A revised and internally consistent thermodynamic model for the interpolation and extrapolation of liquid−solid equilibria in magmatic systems at elevated temperatures and pressures. *Contrib. Mineral. Petrol.*, **119**, 197–212.

Gilliland, R. L. (1989). Solar evolution. *Palaeogeog. Palaeoclimatol. Palaeoecol.*, **75**, 35–55.

Glasser, L. & Brooke Jenkins, H. D. (2000). Lattice energies and unit cell volumes of complex ionic solids. *J. Am. Chem. Soc.,* **122**, 632–638.

Glasstone, S. (1946). *Textbook of Physical Chemistry*. Second Edition. Van Nostrand, New York, 1320 pp.

Glazer, M. & Wark, J. (2001). *Statistical Mechanics. A Survival Guide.* Oxford University Press, Oxford, 142 pp.

Graboske, H. C., Pollack, J. B., Grossman, A. S. & Olness, R. J. (1975). The structure and evolution of Jupiter: the fluid contraction stage. *Astrophys J.*, **199**, 265–281.

Gradie, J. & Tedesco, E. (1982). Compositional structures of the asteroid belt. *Nature*, **216**, 1405–1407.

Green, D. H. & Ringwood, A. E. (1967). The stability field of aluminous pyroxene peridotite and garnet peridotite and their relevance in upper mantle structure. *Earth. Pl. Sci. Letters*, **3**, 151–160.

Griffiths, D. J. (1999). *Introduction to Electrodynamics*. Third Edition. Prentice Hall, Upper Saddle River, N. J., 576 pp.

Griffiths, R. B. (1973). Proposal for notation of tricritical points. *Phys. Rev. B*, **7**, 545–551.

Griffiths, R. B. & Wheeler, J. C. (1970). Critical points in multicomponent systems. *Phys. Rev. A*, **2**, 1047–1064.

Grimm, R. E. & Mcsween, Jr, H. Y. (1989). Water and the thermal evolution of carbonaceous chondrite parent bodies. *Icarus*, **82**, 244–280.

Guillot, T. (1999). A comparison of the interiors of Jupiter and Saturn. *Planet. Space Sci.*, **47**, 1183–1200.

Guillot, T. (2005). The interiors of the giant planets: models and outstanding questions. *Annu. Rev. Earth Planet. Sci.*, **33**, 493–530.

Guggenheim, E. A. (1935). The specific thermodynamic properties of aqueous solutions of strong electrolytes. *Phil. Mag.*, **19**, 588–643.

Guggenheim, E. A. (1952). *Mixtures; the Theory of the Equilibrium Properties of Some Simple Classes of Mixtures, Solutions and Alloys*. Clarendon Press, Oxford, 270 pp.

Guggenheim, E. A. (1967). *Thermodynamics. An Advanced Treatment for Chemists and Physicists*. Fifth Edition. North Holland, Amsterdam, 390 pp.

Hanks, T. C. & Anderson, D. L. (1969). The early thermal history of the Earth. *Earth Planet. Interiors*, **2**, 19–29.

Hansen, C. J. & nine others (2008). Water vapour jets inside the plume of gas leaving Enceladus. *Nature*, **456**, 477–479.

Hartman, H. (1998). Photosynthesis and the origin of life. *Orig. Life Evol. Biosph.*, **28**, 515–521.

Hartmann, W. K. & Davis, D. R. (1975). Satellite-sized planetesimals and lunar origin. *Icarus*, **24**, 504–512.

Harvie, C. E., Moller, N. & Weare, J. H. (1984). The prediction of mineral solubilities in natural waters: the Na–K–Mg–Ca–H–Cl–SO$_4$–OH–HCO$_3$–H$_2$O system to high ionic strengths at 25°C. *Geochim. Cosmochim. Acta.*, **48**, 723–751.

Harvie, C. E. & Weare, J. H. (1980). The prediction of mineral solubilities in natural waters: the Na–K–Mg–Ca–Cl–SO$_4$–H$_2$O system from zero to high concentration at 25°C. *Geochim. Cosmochim. Acta.*, **44**, 981–997.

Head, J. W. & Basilevsky, A. T. (1999). Venus: surface and interior. In: *Encyclopedia of the Solar System*. Eds. P. R. Weissman, L. McFadden & T. V. Johnson. Academic Press, New York, pp. 161–189.

Hecht, M. H. *et al.* (2009). Detection of perchlorate and the soluble chemistry of Martian soil at the Phoenix lander site. *Science*, **325**, 64–67.

Helgeson, H. C. (1969). Thermodynamics of hydrothermal systems at elevated temperatures and pressures. *Am. J. Sci.*, **267**, 729–804.

Helgeson, H. C. & Kirkham, D. H. (1974). Theoretical prediction of the thermodynamic behavior of aqueous electrolytes at high pressures and temperatures: II. Debye–Hückel parameters for activity coefficients and relative partial molal properties. *Am. J. Sci.*, **274**, 1199–1261.

Helgeson, H. C., Kirkham, D. H. & Flowers, G. C. (1981). Theoretical prediction of the thermodynamic behavior of aqueous electrolytes at high pressures and temperatures: IV.

Calculation of activity coefficients, osmotic coefficients, and apparent molal and standard and relative partial molal properties to 600°C and 5 KB. *Am. J. Sci.*, **281**, 1249–1516.

Herbert, F. (1989). Primordial electrical induction heating of asteroids. *Icarus*, **78**, 402–410.

Herbert, F. & Sonett, C. P. (1979). Electromagnetic heating of minor planets in the early solar system. *Icarus*, **40**, 484–496.

Herzberg, C., Raterron, P. & Zhang, J. (2000). New experimental observations on the anhydrous solidus for peridotite KLB−1. *Geochem. Geophys. Geosyst.*, **1**. doi:10.1029/2000GC000089.

Hill, T. L. (1986). *An Introduction to Statistical Thermodynamics*. Dover, New York, 508 pp.

Hirschmann, M. M., Asimow, P. D., Ghiorso, M. S. & Stolper, E. M. (1999a). Calculation of peridotite partial melting from thermodynamic models of minerals and melts. III. Controls on isobaric melt production and the effect of water on melt production. *J. Petrol.*, **40**, 831–851.

Hirschmann, M. M., Ghiorso, M. S. & Stolper, E. M. (1999b). Calculation of peridotite partial melting from thermodynamic models of minerals and melts. II. Isobaric variations in melts near the solidus and owing to variable source composition. *J. Petrol.*, **40**, 297–313.

Hirschmann, M. M., Ghiorso, M. S., Wasylenski, L. E., Asimow, P. D. & Stolper, E. M. (1998). Calculation of peridotite partial melting from thermodynamic models of minerals and melts. I. Review of methods and comparison with experiments. *J. Petrol.*, **39**, 1091–1115.

Hirschmann, M. M. (2006). Water, melting and the deep H_2O cycle. *Ann. Rev. Earth Planet. Sci.*, **34**, 629–653.

Hirschmann, M. M., Aubaud, C. & Withers, A. C. (2005). Storage capacity of H_2O in nominally anhydrous minerals in the upper mantle. *Earth Pl. Sci. Letters*, **236**, 167–181.

Hirschmann, M. M., Tenner, T., Aubaud, C. & Withers, A. C. (2009). Dehydration melting of nominally anhydrous mantle: the primacy of partitioning. *Phys. Earth Planet. Int.*, **176**, 54–68.

Holbrook, J. B., Sabry-Grant, R., Smith, B. C. & Tandel, T. V. (1990). Lattice enthalpies of ionic halides, hydrides, oxides, and sulfides: second-electron affinities of atomic oxygen and sulfur. *J. Chem. Educ.*, **67**, 304–307.

Holland, T. J. B., & Powell, R. (1998). An internally consistent thermodynamic data set for phases of petrological interest. *J. Met. Geol.*, **16**, 309–343.

Holloway, J. R. (1987). Igneous fluids. In: *Thermodynamic Modeling of Geological Materials: Minerals, Fluids and Melts. Reviews in Mineralogy*, **17**, 211–233.

Holloway, J. R. & Wood, B. (1988). *Simulating the Earth: Experimental Geochemistry.* Unwin-Hyman, Boston, 196 pp.

Houston, P. L. (2006). *Chemical Kinetics and Reaction Dynamics*. Dover, New York, 330 pp.

Huang, F., Chakraborty, P., Lundstrom, C. C., *et al.* (2010). Isotope fractionation in silicate melts by thermal diffusion. *Nature*, **464**, 396–400.

Hubbard, W. B. (1970). Structure of Jupiter: chemical composition, contraction and rotation. *Astrophys. J.*, **162**, 687–697.

Hubbard, W. B. (1980). Intrinsic luminosities of the Jovian planets. *Rev. Geophys. Space Phys.*, **18**, 1–9.

Hubbard, W. B. (1984). *Planetary Interiors.* Van Nostrand Reinhold, New York, 334 pp.

Hubbard, W. B., Guillot, T., Marley, M. S. *et al.* (1999). Comparative evolution of Jupiter and Saturn. *Planet. Space Sci.*, **47**, 1175–1185.

Hunten, D. M. (1973). The escape of light gases from planetary atmospheres. *J. Atm. Sci.*, **30**, 1481–1494.

Incropera, F. P. & DeWitt, D. P. (1996). *Introduction to Heat Transfer*. Wiley, New York, 801 pp.

Ito, E. & Takahashi, E. (1989). Postspinel transformations in the system Mg_2SiO_4–Fe_2SiO_4 and some geophysical implications. *J. Geophys. Res.*, **94**, 10 637–10 646.

Iwamori, H., McKenzie, D. & Takahashi, E. (1995). Melt generation by isentropic mantle upwelling. *Earth. Pl. Sci. Letters*, **134**, 253–266.

Jackson, I. & Rigden, S. M. (1996). Analysis of P–V–T data: constraints on the thermoelastic properties of high-pressure minerals. *Phys. Earth. Planet. Inter.*, **96**, 85–112.

Jacobs, G. K. & Kerrick, D. H. (1981). Methane: an equation of state with application to the ternary system CO_2–H_2O–CH_4. *Geochim. Cosmochim. Acta*, **45**, 607–614.

Jaynes, E. T. (1980). The minimum entropy production principle. *Ann. Rev. Phys. Chem.*, **31**, 579–601.

Jones, C. A., Longbottom, A. W. & Hollerbach, R. (1995). A self-consistent convection driven geodynamo model using a mean-field approximation. *Phys. Earth Planet. Inter.*, **92**, 119–141.

Jones, H. R. N. (2000). *Radiation Heat Transfer*. Oxford University Press, Oxford, 86 pp.

Kapustinskii, A. F. (1956). Lattice energy of ionic crystals. *Quart. Rev. Chem. Soc.*, **10**, 283–294.

Kargel, J. S. (1992). Ammonia–water volcanism on icy satellites: phase relations at 1 atmosphere. *Icarus*, **100**, 556–574.

Kargel, J. S. & Lewis, J. S. (1993). The composition and early evolution of Earth. *Icarus*, **105**, 1–25.

Kasting, J. F. (1987). Theoretical constraints on oxygen and carbon dioxide concentrations in the Precambrian atmosphere. *Prec. Res.*, **34**, 205–229.

Kasting, J. F. & Ackerman, T. P. (1986). Climatic consequences of very high CO_2 levels in the Earth's early atmosphere. *Science*, **234**, 1383–1385.

Kaula, W. M. (1964). Tidal dissipation by solid friction and the resulting orbital evolution. *Rev. Geophys.*, **2**, 661–685.

Kelley, D. S., Karson, J. A., Blackman, D. K. *et al.* (2001). An off-axis hydrothermal vent field near the Mid-Atlantic Ridge at 30 degrees N. *Nature*, **412**, 145–149.

Kerrick, D. H. & Darken, L. S. (1975). Statistical thermodynamic models for ideal oxide and silicate solid solutions, with applications to high temperature plagioclase. *Geochim. Cosmochim. Acta*, **39**, 1431–1442.

Kerrick, D. H. & Jacobs, G. K. (1981). A modified Redlich–Kwong equation for H_2O, CO_2 and H_2O–CO_2 mixtures at elevated pressures and temperatures. *Am. J. Sci.*, **281**, 735–767.

Kielland, J. (1937). Individual activity coefficients of ions in aqueous solutions. *J. Am. Chem. Soc.*, **59**, 1675–1678.

Klemme, S. (2004). The influence of Cr on the garnet–spinel transition in the Earth's mantle: experiments in the system MgO–Cr_2O_3–SiO_2 and thermodynamic modelling. *Lithos*, **77**, 639–646.

Klemme, S. & O'Neill, H. S. (2000a). The near-solidus transition from garnet lherzolite to spinel lherzolite. *Contrib. Mineral. Petrol.*, **138**, 237–248.

Klemme, S. & O'Neill, H. S. (2000b). The effect of Cr on the solubility of Al in orthopyroxene: experiments and thermodynamic modelling. *Contrib. Mineral. Petrol.*, **140**, 84–98.

Kojitani, H. & Akaogi, M. (1997). Melting enthalpies on mantle peridotite: calorimetric determinations in the system CaO–MgO–Al$_2$O$_3$–SiO$_2$ and application to magma generation. *Earth Pl. Sci. Letters*, **153**, 209–222.

Kondepudi, D. K. & Prigogine, I. (1998). *Modern Thermodynamics: from Heat Engines to Dissipative Structures*. John Wiley, Chichester, 486 pp.

Korzhinskii, D. S. (1959). *Physicochemical Basis of the Analysis of the Paragenesis of Minerals*. Consultants Bureau, New York, 142 pp.

Kounaves, S. P. *et al.* (2009). The wet chemistry experiments on the 2007 Phoenix: data analysis and results. *J. Geophys. Res.*, **115**, doi:10.1029/2009JE003424.

Krasnopolsky, V. A., Maillard, J. P. & Owen, T. C. (2004). Detection of methane in the Martian atmosphere: evidence for life? *Icarus*, **172**, 537–547.

Kress, V. C. (2003). On the mathematics of associated solutions. *Am. J. Sci.*, **303**, 708–722.

Kreyszig, E. (1991). *Differential Geometry*. Dover, New York, 352 pp.

Krupka, K. M., Robie, R. A., Hemingway, B. S., Kerrick, D. M. & Ito, J. (1985). Low-temperature heat capacities and derived thermodynamic properties of anthophyllite, diopside, enstatite, bronzite, and wollastonite. *Am. Mineral.*, **70**, 249–260.

Kuang, W. & Bloxham, J. (1997). An Earth-like numerical dynamo model. *Nature*, **389**, 371–374.

Kuiper, G. P. (1952). Planetary atmospheres and their origin. In: *The Atmospheres of the Earth and Planets*. Ed. G. P. Kuiper. University of Chicago Press, Chicago.

Lane, N., Allen, J. F. & Martin, W. (2010). How did LUCA make a living? Chemiosmosis in the origin of life. *BioEssays*, **32**, 271–280.

Lasocka, M. (1975). On the entropy of melting. *Phys. Letters A*, **51**, 137–138.

Lennard-Jones, J. E. & Devonshire, A. F. (1939a). Critical and cooperative phenomena. III. A theory of melting and the structure of liquids. *Proc. R. Soc. Lond.*, **A 169**, 317–338.

Lennard-Jones, J. E. & Devonshire, A. F. (1939a). Critical and cooperative phenomena. IV. A theory of order–disorder in solids and liquids and the process of melting. *Proc. R. Soc. Lond.*, **A 170**, 464–484.

Lesher, C. E. (1986). Effects of silicate liquid composition on mineral–liquid element partitioning from Soret diffusion studies. *J. Geophys. Res*, **91**, 6123–6141.

Lesher, C. E. & Walker, D. (1991). Thermal diffusion in petrology. In: *Diffusion, Atomic Ordering, and Mass Transport. Advances in Physical Geochemistry, vol. 8*. Ed. J. Ganguly. Springer, New York, pp. 396–451.

Lewis, G. N. & Randall, M. R. (1961). *Thermodynamics*. Second Edition, rev. by K. S. Pitzer and L. Brewer. McGraw-Hill, New York, 723 pp.

Liu, X. & Millero, F. J. (2002). The solubility of iron in seawater. *Marine Chem.*, **77**, 43–54.

Lodders, K. & Fegley Jr., B. (1997). An oxygen isotope model for the composition of Mars. *Icarus*, **126**, 373–394.

Lodders, K. & Fegley Jr., B. (1998). *The Planetary Scientist's companion*. Oxford University Press, New York, 371 pp.

Logan, S. R. (1996). *Fundamental of Chemical Kinetics*. Longman, Essex, 264 pp.

Lopes, R. M. C. & 25 others (2010). Distribution and interplay of geologic processes on Titan from Cassini radar data. *Icarus*, **205**, 540–558.

Lorenzen, W., Holst, B. & Redmer, R. (2009). Demixing of hydrogen and helium at megabar pressures. *Phys. Rev. Letters,* **102.** doi: 10.113/PhysRevLett.102.115701.

Lowenstein, T. K. & Risacher, F. (2009). Closed basin brine evolution and the influence of Ca–Cl inflow waters: Death Valley and Bristol Dry Lake California, Qaidam Basin, China, and Salar de Atacama, Chile. *Aquat. Geochem.*, **15**, 71–94.

Lunine, J. I. (2005). *Astrobiology. A Multidisciplinary Approach*. Addison Wesley, San Francisco, 586 pp.

Maaløe, S. (1985). *Principles of Igneous Petrology*. Springer-Verlag, Berlin, 374 pp.

Machida, M. N., Inutsuska, S.-I. & Matsumoto, T. (2007). Magnetic fields and rotations of protostars. *Astrophys J.,* **670**, 1198–1213.

Malvern, L. R. (1969). *Introduction to the Mechanics of a Continuous Medium*. Prentice Hall, Englewood Cliffs, N. J., 713 pp.

Marion, G. M., Catling, D. C. & Kargel, J. S. (2003). Modeling aqueous ferrous iron chemistry at low temperatures with application to Mars. *Geochim. Cosmochim. Acta*, **67**, 4251–4266.

Marion, G. M., Crowley, J. K., Thomson, B. J., *et al.* (2009). Modeling aluminum–silicon chemistries and application to Australian acidic playa lakes as analogues to Mars. *Geochim. Cosmochim. Acta*, **73**, 3493–3511.

Marion, G. M., Kargel, J. S. & Catling, D. C. (2008). Modeling ferrous–ferric iron chemistry with application to Martian surface geochemistry. *Geochim. Cosmochim. Acta*, **72**, 242–266.

Marion, G. M., Kargel, J. S., Catling, D. C. & Jakubowski, S. D. (2005). Effects of pressure on aqueous chemical equilibria at subzero temperature applications to Europa. *Geochim. Cosmochim. Acta*, **69**, 259–274.

Marliacy, P., Solimando, R., Bouroukba, M. & Schuffenecker, L. (2000). Thermodynamics of crystallization of sodium sulfate decahydrate in H_2O–$NaCl$–Na_2SO_4: application to $Na_2SO_4.10H_2O$-based latent heat storage materials. *Thermochimica Acta*, **344**, 85–94.

Mason, E. A. & Spurling, T. H. (1969). *The Virial Equations of State*. Pergamon Press, Oxford, 297 pp.

Mathews, J. F. (1972). The critical constants of inorganic substances. *Chem. Rev.*, **72**, 71–100.

Matsuzaka, K., Akaogi, M., Suzuki, T. & Suda, T. (2000). Mg–Fe partitioning between silicate spinel and magnesiowüstite at high pressure: experimental determination and calculation of phase relations in the system Mg_2SiO_4–Fe_2SiO_4. *Phys. Chem. Minerals*, **27**, 310–319.

Mattern, E., Matas, J., Ricard, Y. & Bass, J. (2005). Lower mantle composition and temperature from mineral physics and thermodynamic modelling. *Geophys. J. Int.*, **160**, 973–990.

McBirney, A. R. (2006). *Igneous Petrology*. Edition 3. Jones & Bartlett Learning, 545 pp.

McKay, D. S., Everett, K. G., Thomas-Keprta, K. L. *et al.* (1996). Search for past life on Mars: possible relic of biogenic activity in Martian meteorite ALH84001. *Science*, **273**, 924–930.

McKenzie, D. (1984). The generation and compaction of partially molten rock. *J. Petrol.*, **25**, 713–765.

McKenzie, D. & Bickle, M. J. (1988). The volume and composition of melt generated by extension of the lithosphere. *J. Petrol.*, **29**, 625–679.

McSween, H. Y. (2005). Mars. In: *Encyclopedia of the Solar System*. Eds. P. R. Weissman, L. McFadden & T. V. Johnson. Academic Press, New York, pp. 601–621.

Meyer, J. & Wisdom, J. (2007). Tidal heating in Enceladus. *Icarus*, **188**, 535–539.

Meyer, J. & Wisdom, J. (2008). Tidal evolution of Mimas, Enceladus, and Dione. *Icarus*, **193**, 213–223.

Millero, F. J. (2004). Physicochemical controls on seawater. In: *The Oceans and Marine Geochemistry. Treatise on Geochemistry, v.6.* Ed. H. Elderfield. Elsevier, Amsterdam, pp. 1–21.

Millero, F. J. (2009). Thermodynamic and kinetic properties of natural brines. *Aquat. Geochem.*, **15**, 7.41.

Millero, F. J., Yao, W. & Aicher, J. (1995). The speciation of Fe(II) and Fe(III) in natural waters. *Marine Chem.*, **50**, 21–39.

Misner, C. W., Thorne, K. S. & Wheeler, J. A. (1973). *Gravitation.* W. H. Freeman, San Francisco, 1279 pp.

Mitri, G. & Showman, A. P. (2008). Thermal convection of ice-I shells of Titan and Enceladus. *Icarus*, **193**, 387–396.

Miyashiro, A. (1994). *Metamorphic Petrology.* Oxford University Press, New York, 404 pp.

Moody, G. J. & Thomas, J. D. R. (1965). Lattice energy and chemical prediction. Use of the Kapustinskii equations and the Born–Haber cycle. *J. Chem. Educ.*, **42**, 204–210.

Morales, M. A., Schwegler, E., Ceperley, D. *et al.* (2009). Phase separation in hydrogen–helium mixtures at Mbar pressure. www.pnas.org /cgi/doi/10.1073/pnas.0812581106.

Morey, G. W. & Williamson, E. D. (1918). Pressure-temperature curves in univariant systems. *J. Am. Chem. Soc.*, **40**, 59–84.

Mulargia, F. (1986). The physics of melting and temperatures in the Earth's outer core. *Q. J. R. Astr. Soc.*, **27**, 383–402.

Mulkidjanian, A. Y. & Galperin, M. Y. (2009). On the origin of life in the zinc world. 2. Validation of the hypothesis on the photosynthesizing zinc sulfide edifices as cradles of life on Earth. *Biology Direct*, **4**, doi:10.1186/1745−6150−4−27.

Murnaghan, F. D. (1937). Finite deformation of an elastic solid. *Am. J. Math.*, **59**, 235–239.

Navrotsky, A. (1986). Cation distribution energetics and heats of mixing in $MgFe_2O_4$–$MgAl_2O_4$, $ZnFe_2O_4$–$ZnAl_2O_4$ and $NiFe_2O_4$–$NiAl_2O_4$ spinels: study by high temperature calorimetry. *Am. Mineral.*, **71**, 1160–1169.

Navrotsky, A. (1986). Models of crystalline solutions. In: *Thermodynamic Modeling of Geological Materials: Minerals, Fluids and Melts. Reviews in Mineralogy*, **17**, 35–69.

Nimmo, F., Spencer, J. R., Pappalardo, R. T. & Mullen, M. E. (2007). Shear heating as the origin of the plumes and heat flux on Enceladus. *Nature*, **447**, 289–291.

NIST Chemical Kinetics Database. http://kinetics.nist.gov/kinetics/welcome.jsp

NIST Chemistry WebBook. http://webbook.nist.gov/chemistry/

Nitschke, W. & Russell, M. J. (2009). Hydrothermal focusing of chemical and chemiosmotic energy, supported by delivery of catalytic Fe, Ni, Mo/W, Co, S and Se, forced life to emerge. *J. Mol. Evol.*, **69**, 481–496.

Nordstrom, D. K. & Munoz, J. L. (1986). *Geochemical Thermodynamics.* Blackwell, Palo Alto, 477 pp.

Olson, P., Christensen, U. & Glatzmaier, G. A. (1999). Numerical modeling of the geodynamo: Mechanisms of field generation and equilibration. *J. Geophys. Res.*, **104**, 10 383–10 404.

Onsager, L. (1931a). Reciprocal relations in irreversible processes. I. *Phys. Rev.*, **37**, 405–426.

Onsager, L. (1931b). Reciprocal relations in irreversible processes. II. *Phys. Rev.*, **38**, 2265–2279.

Oriani, R. A. (1951). The entropies of melting of metals. *J. Chem. Phys.*, **19**, 93–97.

Palme, H. and Beer, H. (1993). Abundances of the elements in the solar system. In: *Landolt–Börnstein Group VI: Astronomy and Astrophysics: Instruments; Methods; Solar System.* Ed. H. H. Voigt. Springer, Berlin, pp. 196–221.

Patiño Douce, A. E. (2005). Vapor—absent melting of tonalite at 15–32 kbar. *J. Petrol.*, **46**, 275–290.

Patiño Douce, A. E., Humphreys, E. D. & Johnston, A. D. (1990). Anatexis and metamorphism in tectonically thickened continental crust exemplified by the Sevier hinterland, western North America. *Earth Pl. Sci. Letters*, **97**, 290–315.

Peale, S. J. & Cassen, P. (1978). Contribution of tidal dissipation to lunar thermal history. *Icarus*, **36**, 245–269.

Peale, S. J., Cassen, P., & Reynolds, R. T. (1979). Melting of Io by tidal dissipation. *Science*, **203**, 892–894.

Pepin, R. O. (1997). Evolution of Earth's noble gases: consequences of assuming hydrodynamic loss driven by giant impacts. *Icarus*, **126**, 148–156.

Pepin, R. O. (2006). Atmospheres of the terrestrial planets: clues to origin and evolution. *Earth. Pl. Sci. Letters*, **252**, 1–14.

Philpotts, A. R. & Ague, J. J. (2009). *Principles of Igneous and Metamorphic Petrology.* Second Edition. Cambridge University Press, Cambridge, 667 pp.

Pitzer, K. S. (1987). A thermodynamic model for aqueous solutions of liquid-like density. In: *Thermodynamic Modeling of Geological Materials: Minerals, Fluids and Melts. Reviews in Mineralogy*, **17**, 97–142.

Pitzer, K. S. (1995). *Thermodynamics.* Third Edition. McGraw-Hill, New York, 626 pp.

Pitzer, K. S. & Sterner, S. M. (1994). Equations of state valid continuously from zero to extreme pressures for H_2O and CO_2. *J. Chem. Phys.*, **101**, 3111–3116.

Poirier, J. P. (1985). *Creep of Crystals: High Temperature Deformation Processes in Metals, Ceramics and Minerals.* Cambridge University Press, Cambridge, 260 pp.

Poirier, J. P. (1991). *Introduction to the Physics of the Earth's Interior.* Cambridge University Press, Cambridge, 264 pp.

Prigogine, I. (1961). *Introduction to Thermodynamics of Irreversible Processes.* Second Edition. Interscience Publishers, New York, 119 pp.

Prigogine, I. (1962). *Non-equilibrium Statistical Mechanics.* Interscience Publishers, New York, 319 pp.

Prigogine, I. (1967). *Introduction to Thermodynamics of Irreversible Processes.* Third Edition. Interscience Publishers, New York, 147 pp.

Prigogine, I. & Defay, R. (1965). *Chemical Thermodynamics.* Longmans, London, 543 pp.

Purucker, M. E., Sabaka, T. J., Solomon, S. C. *et al.* (2009). Mercury's internal magnetic field: Constraints on large- and small-scale fields of crustal origin. *Earth Pl. Sci. Letters*, **285**, 340–346.

Putnis, A. (1992). *Introduction to Mineral Sciences.* Cambridge University Press, Cambridge, 457 pp.

Ramberg, H. (1967). *Gravity and Deformation of the Earth's Crust, as Studied by Centrifuged Models.* Academic Press, London, 214 pp.

Randall, L. (2007). The case for extra dimensions. *Physics Today*, **60**, 80–81.

Redlich, O. & Kwong, J. N. S. (1949). An equation of state. Fugacities of gaseous solutions. *Chem. Rev.*, **44**, 233–244.

Reimer, A., Landmann, G. & Kempe, S. (2009). Lake Van, Eastern Anatolia, hydrochemistry and history. *Aquat. Geochem.*, **15**, 195–222.

Ricci, J. E. (1966). *The Phase Rule and Heterogeneous Equilibrium*. Dover, New York, 505 pp.

Richet, P. & Bottinga, Y. (1986). Thermochemical properties of silicate glasses and liquids. *Rev. Geophys.*, **24**, 1–25.

Richter, F. M., Watson, E. B., Mendybaev, R. A., Teng, F.-Z. & Janney, P. E. (2008). Magnesium isotope fractionation in silicate melts by chemical and thermal diffusion. *Geochim. Cosmochim. Acta*, **72**, 206–220.

Richter, F. M., Watson, E. B., Mendybaev, R. *et al.* (2009). Isotopic fractionation of the major elements of molten basalt by chemical and thermal diffusion. *Geochim. Cosmochim. Acta*, **73**, 4250–4263.

Risacher, F. & Fritz, B. (1991). Geochemistry of Bolivian salars, Lipez, southern Altiplano: origin of solutes and brine evolution. *Geochim. Cosmochim. Acta*, **55**, 687–705.

Risacher, F. & Fritz, B. (2009). Origin of salts and brine evolution of Bolivian and Chilean salars. *Aquat. Geochem.*, **15**, 123–157.

Robie, R. A. & Hemingway, B. S. (1995). *Thermodynamic properties of minerals and related substances at 298.15 K and 1 bar (10^5 pascals) pressure and at higher temperatures*. U. S. Geological Survey Bulletin, **2131**.

Robinson, J. A. C. & Wood, B. J. (1998). The depth of the spinel to garnet transition and the peridotite solidus. *Earth Pl. Sci. Letters*, **164**, 277–284.

Robinson, R. A. & Stokes, R. H. (1959). *Electrolyte Solutions, the Measurement and Interpretation of Conductance, Chemical Potential, and Diffusion in Solutions of Simple Electrolytes*. Second Edition. Butterworths, London, 559 pp.

Rodríguez-Navarro, C., Doehne, E. & Sebastian, E. (2000). How does sodium sulfate crystallize? Implications for the decay and testing of building materials. *Cem. Concrete. Res.*, **10**, 1527–1534.

Ross, M. N. & Schubert, G. (1988). Viscoelastic models of tidal heating in Enceladus. *Icarus*, **78**, 90–101.

Ruiz, J. (2005). The heat flow of Europa. *Icarus*, **177**, 438–446.

Sagan, C. & Mullen, G. (1972). Earth and Mars: evolution of atmospheres and surface temperatures. *Science*, **177**, 52–56.

Sander, R. (1999). *Compilation of Henry's Law constants for inorganic and organic species of potential importance in environmental chemistry*. http://www.mpch–mainz.mpg.de/~sander/res/henry.html

Sanloup, C., Jambon, A. & Gillet, P. (1999). A simple chondritic model for Mars. *Earth Planet. Sci. Letters*, **112**, 43–54.

Saumon, D., Chabrier, G. & van Horn, H. M. (1995). An equation of state for low-mass stars and giant planets. *Astrophys. J. Suppl. Series*, **99**, 713–741.

Saumon, D., Hubbard, W. B., Chabrier, G. & van Horn, H. M. (1992). The role of the molecular–metallic transition of hydrogen in the evolution of Jupiter, Saturn, and brown dwarfs. *Astrophys. J.*, **391**, 827–831.

Saxena, S. K., Chatterjee, N., Fei, Y. & Shen, G. (1993). *Thermodynamic Data on Oxides and Silicates: an Assessed Data Set Based on Thermochemistry and High Pressure Phase Equilibrium*. Springer-Verlag, Berlin, 428 pp.

Schreinemakers, F. A. H. (1965). *In-, Mono-, and Di-variant Equilibria*. The Pennsylvania State University, University Park, Pennsylvania.

Schubert, G., Anderson, J. D., Travis, B. J. & Palguta, J. (2007). Enceladus: present internal structure and differentiation by early and long-term radiogenic heating. *Icarus*, **188**, 345–355.

Schubert, G., Turcotte, D. L. & Olson, P. (2001). *Mantle Convection in the Earth and Planets*. Cambridge University Press, Cambridge, 940 pp.

Schultz, P. H., Burns, J. A. & Greeley, R. (1976). Ancient lunar tides and the emplacement of the maria. *Lunar Science*, **VII**, 785–787.

Schulze-Makuch, D. & Irwin, L. N. (2004). *Life in the Universe. Expectations and Constraints*. Springer, Berlin, 172 pp.

Scott, E. R. D. (2007). Chondrites and the protoplanetary disk. *Annu. Rev. Earth Planet. Sci.*, **35**, 577–620.

Segatz, M., Spohn, T., Ross, M. N. & Schubert, G. (1988). Tidal dissipation, surface heat flow and figure of viscoelastic models of Io. *Icarus*, **75**, 187–206.

Shaw, A. N. (1935). The derivation of thermodynamic relations for a simple system. *Phil. Trans. R. Soc. Lond.*, **A 234**, 299–328.

Shalom, E., Ghatak, A. & Hora, H. (2002). *Fundamentals of Equations of State*. World Scientific, River Edge, N. J., 366 pp.

Shomate, C. H. (1954). A method for evaluating and correlating thermodynamic data. *J. Physical Chem.*, **58**, 368–372.

Shomate, C. H. & Cohen, A. J. (1955). High temperature heat content end entropy of lithium oxide and lithium hydroxide. *J. Am. Chem. Soc*, **77**, 285–286.

Sokolnikoff, I. S. & Redheffer, R. M. (1966). *Mathematics of Physics and Modern Engineering*. Second Edition. McGraw-Hill, New York, 752 pp.

Sonnenthal, E. L. (2004). A numerical model of chemical and Soret diffusion at crystallizing boundaries. *Abs. Am. Geophys. Union*, doi:10.1029/2004AGUFM.V53A0620S

Stacey, F. D. (1976). Paleomagnetism of meteorites. *Annu. Rev. Earth Planet. Sci.*, **4**, 147–157.

Stacey, F. D. (1992). *Physics of the Earth*. Third Edition. Brookfield Press, Brisbane, 513 pp.

Stevenson, D. J. (1982a). Saturn's luminosity and magnetism. *Science*, **208**, 746–748.

Stevenson, D. J. (1982b). Interiors of the giant planets. *Annu. Rev. Earth Planet. Sci.*, **10**, 257–295.

Stevenson, D. J. & Salpeter, E. E. (1977). The dynamics and helium distribution in hydrogen–helium fluid planets. *Astrophys. J. Suppl. Series*, **35**, 239–261.

Stishov, S. M., Makarenko, I. N., Ivanov, V. A. & Nikolaenko, A. M. (1973). On the entropy of melting. *Phys. Letters A.*, **45**, 18.

Stolper, E. & Asimow, P. (2007). Insights into mantle melting from graphical analysis of one-component systems. *Am. J. Sci.*, **307**, 1051–1139.

Strom, R. G. (1999). Mercury. In: *Encyclopedia of the Solar System*. Eds. P. R. Weissman, L. McFadden & T. V. Johnson. Academic Press, New York, pp. 123–145.

Tajika, E. & Matsui, T. (1993). Degassing history and carbon cycle of the Earth: from an impact-induced steam atmosphere to the present atmosphere. *Lithos*, **30**, 267–280.

Tanger, J. C. & Helgeson, H. C. (1988). Calculation of the thermodynamic and transport properties of aqueous species at high pressures and temperatures: revised equations of state for the standard partial molal properties of ions and electrolytes. *Am. J. Sci.*, **288**, 19–98.

Taylor, G. J. & Scott, E. R. D. (2005). Mercury. In: *Meteorites, Comets and Planets. Treatise on Geochemistry, vol.1*. Ed. A. M. Davis. Elsevier, Amsterdam, pp. 477–485.

Tenner, T., Hirschmann, M. M., Withers, A. C. & Hervig, R. L. (2009). Hydrogen partitioning between nominally anhydrous upper mantle minerals and melt between 3 and 5 GPa and applications to hydrous peridotite partial melting. *Chem. Geol.*, **262**, 42–56.

Theobald, D. L. (2010). A formal test of the theory of universal common ancestry. *Nature*, **465**, 219–222.

Thommes, E. W., Duncan, M. J. & Levison, H. F. (2003). Oligarchic growth of giant planets. *Icarus*, **161**, 431–455.

Topor, L., Navrotsky, A., Zhao, Y. & Weidner, D. J. (1997). Thermochemistry of fluoride perovskites: heat capacity, enthalpy of formation, and phase transition of NaMgF. *J. Solid State Chem.*, **132**, 131–138.

Truesdell, C. (1984). *Rational Thermodynamics*. Second Edition. Springer, New York, 578 pp.

Tunell, G. (1985). *Condensed Collections of Thermodynamic Formulas for One-component and Binary Systems of Unit and Variable Mass.* Carnegie Institution of Washington, publication 408 B, Washington D. C., 294 pp.

Turcotte, D. L. (1995). How does Venus lose heat? *J. Geophys. Res.*, **100**, 16 931–16 940.

Turcotte, D. L., Cisne, J. L. & Nordmann, J. C. (1977). On the evolution of the lunar orbit. *Icarus*, **30**, 254–266.

Turcotte, D. L. & Schubert, G. (2002). *Geodynamics*. Second Edition. Cambridge University Press, Cambridge, 456 pp.

Ubbelohde, A. R. (1978). *The Molten State of Matter.* Wiley, Chichester, 454 pp.

Ueno, Y., Yamada, K., Yoshida, S., Maruyama, S. & Isozaki, Y. (2006). Evidence from fluid inclusions for microbial methanogenesis in the early Archaean era. *Nature,* **440**, 516–519.

Urey, H. C. (1952). *The Planets, their Origin and Development.* Yale University Press, New Haven, 245 pp.

Valley, J. W. (2006). Early Earth. *Elements*, **2**, 201–204.

Valley, J. W., Lackey, J. S., Cavosie, A. J. *et al.* (2005). 4.4 billion years of crustal maturation: oxygen isotope ratios in magmatic zircons. *Contrib. Mineral. Petrol.*, **150**, 561–580.

Wächtershäuser, G. (1988). Before enzymes and templates: theory of surface metabolism. *Microbiol. Rev.*, **52**, 452–484.

Wagman, D. D., Evans, W. H., Parker, V. B. *et al.* (1982). The NBS tables of chemical thermodynamic properties. Selected values for inorganic and C_1 and C_2 organic substances in SI units. *J. Phys. Chem. Ref. Data*, **11**, **Suppl. 2**, 1–392.

Wagner, W., Saul, A. & Pruß, A. (1993). International equations for the pressure along the melting and along the sublimation curve of ordinary water substances. *J. Phys. Chem. Ref. Data*, **23**, 515–525.

Walker, D. & DeLong, S. E. (1982). Soret separation of mid-ocean ridge basalt magma. *Contrib. Mineral. Petrol.*, **79**, 231–240.

Walker, J. C. G. (1985). Carbon dioxide on the early Earth. *Origins of Life*, **16**, 117–127.

Wanke, H. & Dreibus, G. (1988). Chemical composition and accretion history of terrestrial planets. *Phil. Trans. R. Soc. London*, **A325**, 545–557.

Weidenschilling, S. J. (1974). A model for accretion of the terrestrial planets. II. *Icarus*, **22**, 426–435.

Weidenschilling, S. J. (1976). A model for accretion of the terrestrial planets. *Icarus,* **27**, 161–170.

Weidenschilling, S. J. (1980). Dust to planetesimals: settling and coagulation in the solar nebula. *Icarus*, **44**, 172–189.

Weidenschilling, S. J. (2000). Formation of planetesimals and accretion of the terrestrial planets. *Space Sci. Rev.*, **92**, 295–310.

Weisstein, E. W. (2003). *CRC Concise Encyclopedia of Mathematics*. Second Edition. Chapman & Hall/CRC Press, Boca Raton, 3242 pp.

Wetherill, G. W. (1985). Occurrence of giant impacts during growth of the terrestrial planets. *Science*, **228**, 877–879.

Wetherill, G. W. (1990). Formation of the Earth. *Annu. Rev. Earth Planet. Sci.*, **18**, 205–256.

Wetherill, G. W. (1994). Provenance of the terrestrial planets. *Geochim. Cosmochim. Acta*, **58**, 4513–4520.

Wetherill, G. W. & Inaba, S. (2000). Planetary accumulation with a continuous supply of planetesimals. *Space Sci. Rev.*, **92**, 311–320.

Williamson, E. D. & Morey, G. W. (1918). The laws of chemical equilibrium. *J. Am. Chem. Soc.*, **40**, 49–59.

Winter, J. D. (2001). *An Introduction to Igneous and Metamorphic Petrology*. Prentice Hall, Upper Saddle River, 697 pp.

Winterton, R. H. S. (1997). *Heat Transfer*. Oxford University Press, Oxford, 85 pp.

Wood, B. J. (1987). Thermodynamics of multicomponent systems containing several solid solutions. In: *Thermodynamic Modeling of Geological Materials: Minerals, Fluids and Melts. Reviews in Mineralogy*, **17**, 71–95.

Wood, B. J. & Nicholls, J. (1978). The thermodynamic properties of reciprocal solid solutions. *Contrib. Mineral. Petrol.*, **66**, 389–400.

Yamamura, Y., Tsuji, T., Saito, K. & Sorai, M. (2004). Heat capacity and order–disorder phase transition in negative thermal expansion compound ZrW_2O_8. *J. Chem. Thermodynamics*, **36**, 525–531.

Yoder, C. H. & Flora, N. J. (2005). Geochemical applications of the simple salt approximation to the lattice energies of complex materials. *Am. Mineral.*, **90**, 488–496.

Zahnle, K. J. & Kasting, J. F. (1986). Mass fractionation during transonic hydrodynamic escape and implications for loss of water from Venus and Mars. *Icarus,* **68**, 462–480.

Zahnle, K. J., Kasting, J. F. & Pollack, J. B. (1990). Mass fractionation of noble gases in diffusion-limited hydrodynamic hydrogen escape. *Icarus,* **84**, 502–527.

Zen, E-an (1984). *Construction of Pressure–Temperature Diagrams for Multicomponent Systems After the Method of Schreinemakers: a Geometric Approach*. U. S. Geological Survey Bulletin, **1225**.

Zhang, C. & Duan, Z. (2009). A model for C–O–H fluid in the Earth's mantle. *Geochim. Cosmochim. Acta*, **73**, 2089–2102.

Zhang, Y. (2008). *Geochemical Kinetics*. Princeton University Press, Princeton, 631 pp.

Zhang, Y., Xu, Z., Zhu, M. & Wang, H. (2007). Silicate melt properties and volcanic eruptions. *Rev. Geophys.*, **45**, doi:10.1029/20069RG000216.

Zhao, Y., Weidner, D. J., Parise, J. B. & Cox, D. E. (1993). Thermal expansion and structural distortion of perovskite – data for $NaMgF_3$ perovskite. Part I. *Phys. Earth Pl. Interiors*, **76**, 1–16.

Index